A Tool Kit for Groupoid C^*-Algebras

Mathematical
Surveys
and
Monographs

Volume 241

A Tool Kit for Groupoid C^*-Algebras

Dana P. Williams

AMERICAN MATHEMATICAL SOCIETY
Providence, Rhode Island

2010 *Mathematics Subject Classification*. Primary 46L55, 46L05;
Secondary 46L35, 46L08, 22D25.

Partially supported by a Simons Foundation Collaboration grant #507798.

For additional information and updates on this book, visit
www.ams.org/bookpages/surv-241

Library of Congress Cataloging-in-Publication Data

Names: Williams, Dana P., 1952- author.
Title: A tool kit for groupoid C*-algebras / Dana Williams.
Description: Providence, Rhode Island : American Mathematical Society, [2019] | Series: Mathematical surveys and monographs ; volume 241 — Includes bibliographical references and index.
Identifiers: LCCN 2019016171 | ISBN 9781470451332 (alk. paper)
Subjects: LCSH: Operator algebras. | Groupoids. | Group theory. | C*-algebras. | AMS: Functional analysis – Selfadjoint operator algebras (C^*-algebras, von Neumann (W^*-) algebras, etc.) – Noncommutative dynamical systems. msc | Functional analysis – Selfadjoint operator algebras (C^*-algebras, von Neumann (W^*-) algebras, etc.) – General theory of C^*-algebras. msc | Functional analysis – Selfadjoint operator algebras (C^*-algebras, von Neumann (W^*-) algebras, etc.) – Classifications of C^*-algebras. msc | Functional analysis – Selfadjoint operator algebras (C^*-algebras, von Neumann (W^*-) algebras, etc.) – C^*-modules. msc | Topological groups, Lie groups – Locally compact groups and their algebras – C^*-algebras and W^*-algebras in relation to group representations. msc
Classification: LCC QA36 .W55 2019 | DDC 512/.556–dc23
LC record available at https://lccn.loc.gov/2019016171

∞ The paper used in this book is acid-free and falls within the guidelines
established to ensure permanence and durability.
Visit the AMS home page at `https://www.ams.org/`

10 9 8 7 6 5 4 3 2 1   24 23 22 21 20 19

To my lovely wife Letha

Contents

Introduction

The algebraic notion of a groupoid was distilled over time as it was realized that varied situations could be usefully studied by taking advantage of the underlying algebraic structure. Ronald Brown's survey article [**Bro87**] is an excellent overview of both the history and the myriad areas in which groupoids play a role. There Brown credits Brandt [**Bra27**] for starting the party in 1927. Good bibliographies for topological groupoids appear in [**BDNH76**] and [**BH76**]. For the smooth setting—involving Lie groupoids, Lie Algebroids, and their connection to foliation theory—see [**MM03**] and [**Mac05**].

However the focus in this book is strictly on applications to operator algebras and their representations. Hence a reasonable starting point for our story would be Mackey's virtual groups [**Mac68**, **Mac66**] in the mid 1960s. Full fledged groupoids appear in Westman's work [**Wes67**, **Wes68**], Peter Hahn's work [**Hah78**] as well as Arlan Ramsay's [**Ram82**]. However, the tale told here has its origins in Jean Renault's [**Ren80**] which initiated the idea of associating a C^*-algebra to a locally compact groupoid in a manner exactly paralleling the situation for a locally compact group. One of Renault's primary motivations was to find a C^*-algebra analogue of the work of Feldman and Moore in [**FM75**, **FM77a**, **FM77b**] associating a von-Neumann algebra to a twisted equivalence relation. Renault's work also inspired Kumjian's [**Kum86**] which is also a foundational work in the subject.

As the name suggests, a groupoid is a generalization of a group. The primary difference is that multiplication is not everywhere defined. As a result, there will be multiple identities or units. The elements of the groupoid can be thought of as "arrows" between units and multiplication as composition of the arrows. Since groups and transformation groups (arising from a group action on a space) are fundamental examples, we expect to construct a C^*-algebra from continuous, compactly supported functions on the underlying topological space of the groupoid. To ensure that there are sufficiently many such functions, we need the groupoid to have a locally compact topology such that the groupoid operations are continuous. Then we can give $C_c(G)$ a $*$-algebra structure and we realize $C^*(G)$ as an appropriate completion exactly as in the case of a group C^*-algebra or a transformation group C^*-algebra. Renault laid all this out very carefully in Chapter II of [**Ren80**].

The business of associating interesting C^*-algebras to locally compact groupoids has been very active since the publication of Renault's thesis in 1980. Naturally, there have been a number of important developments and refinements since that time such as groupoid equivalence, the disintegration theorem, and great advancements in our understanding of amenability of groupoids. Some of these are covered in Paterson's book [**Pat99**], Muhly's unpublished CBMS lectures notes [**Muh99**], and of course Anantharaman-Delaroche and Renault's [**ADR00**]. The purpose of this work is to both highlight and summarize these advancements, as well as exhibit some sharpening of the original ideas and techniques from the literature. In addition to reviewing the basic constructions, the new work we want to include is as follows.

We need to see that bounded $*$-representations of $C_c(G)$ are actually the integrated forms of a suitably defined unitary representations of G in analogy with the case for groups—this is often referred to as the "Disintegration Theorem". Renault had a basic version of this result in [**Ren80**] and provided a very general version later in [**Ren87**] with additional details appearing in the unpublished notes by Muhly [**Muh99**]. The techniques demand a heavy dose of measure theory and involve direct integrals. As a result, separability is required. One of our goals here is to describe this theory in detail.

Another important development was the notion of equivalence for groupoids and Renault's Equivalence Theorem implying that equivalent groupoids have Morita equivalent C^*-algebras. The proof in [**MRW87**] relied heavily on the Disintegration Theorem and required considerable overhead to construct the requisite bimodules. Recently, a more direct proof using linking algebras and linking groupoids appeared in [**SW12**], and we use that approach here.

The Morita equivalence provided by the Equivalence Theorem also allows a sharpening of the notion of induced representations for groupoids and their C^*-algebras using the Rieffel theory from [**RW98**, §2.4]. This was done on an *ad hoc* basis for some years and was formalized in [**Goe09**, §6.1] and [**IW09b**, §2].

Because amenability plays such an important role for groups, it is natural to try to formalize a notion of amenability for groupoids. Although Renault stressed in [**Ren80**] that his first steps in defining such a notion there were only meant as first approximations, the fact is that his initial definitions have stood up well. In fact, groupoid amenability has become important in many applications. Nevertheless, the situation is still in flux and far from satisfactory. There are three distinct flavors—if not more—to deal with. The primary notion is topological amenability—called simply amenability here—and it is based on the existence of well-behaved functions of positive type on G. There is a second notion called measurewise amenability which is based on a generalized notion of an invariant mean on $L^\infty(G)$ for suitable measures on G. Then there is simply the question of whether the universal norm and reduced norm agree on $C_c(G)$. I call this property metric amenability although the term is not in general use. We know that amenability implies measurewise amenability and that measurewise amenability implies metric amenability. For groups, these three notions coincide. Until recently, it seemed possible that they were the same for all locally compact groupoids. But recent work starting with Willett's example in [**Wil15**] has revealed that metric amenability is not equivalent to measurewise amenability. It is still possible that measurewise amenability and amenability are equivalent, but we currently only know this in special cases. The general theory took a huge jump forward with the *tour de force* [**ADR00**] published in 2000 by Anantharaman-Delaroche and Renault. However, the focus in [**ADR00**] was on Borel groupoids rather than locally compact ones. Furthermore, Renault's recent work in [**Ren15**] has allowed some dramatic improvements and new applications. Hence it seemed wise to include an extensive summary of their theory here specialized to locally compact groupoids.

One of the important results for transformation group C^*-algebras is the Gootman-Rosenberg-Sauvageot result verifying the Effros-Hahn conjecture that asserts every primitive ideal of a transformation group C^*-algebra is induced from an isotropy group provided that the group acting is amenable [**Sau77**, **Sau79**, **GR79**]. In [**Ren91**], Renault proved some closely related results for groupoids which lead to some powerful characterizations of simplicity for certain groupoid C^*-algebras. So it also seemed worthwhile to revisit some of these simplicity results from [**Ren91**].

Reader's Guide. I have tried to keep the required prerequisites for reading this book to a minimum with the goal that a reasonably well-prepared graduate student working in functional analysis can work through the material here without spending significant time with other sources. Here "reasonably well-prepared" has to be defined to include familiarity with Morita theory as exposed in Chapters 2 and 3 of [**RW98**]. Some familiarity with with the group C^*-algebra construction and a basic knowledge of crossed products would be especially helpful, and I have not hesitated to make frequent use of [**Wil07**] for background material. At the very least, it will be necessary to learn a bit about Borel Hilbert bundles and direct integrals as in [**Wil07**, Appendix F].

In Chapter 1, we begin with the necessary definitions and constructions to build a C^*-algebra, $C^*(G)$, out of a locally compact Hausdorff groupoid G. To do this we need an analogue of Haar measure on a group and define a Haar system for this purpose. Then we define a $*$-algebra structure on $C_c(G)$ and complete with respect to the universal norm arising from suitably bounded $*$-representations of $C_c(G)$ as bounded operators on a Hilbert space. Although producing interesting examples is difficult without more technology in place, we at least describe how to show that our C^*-algebra constructions generalize both the group C^*-algebra construction as well as the transformation group C^*-algebra construction.

In Chapter 2 we define groupoid equivalence and state the Equivalence Theorem. For this we need to take an in depth look at groupoid actions on spaces which is a interesting and important topic in its own right. Since the concept will be crucial to the proof of the Equivalence Theorem, we also introduce linking groupoids. Then we state the various parts of the Equivalence Theorem and explore some immediate consequences in order to provide some interesting examples.

Chapter 3 is reserved for some measure theoretic results needed in the sequel. It is just a fact of life that it is hard to avoid measure theory when studying the C^*-algebras of locally compact groupoids. It is certainly not an unreasonable option to skip this chapter and come back to the appropriate section only as needed.

Chapter 4 is devoted to the proofs of the various parts of the Equivalence Theorem.

In Chapter 5, we consider the representation theory of $C^*(G)$. This chapter avoids mention of unitary representations and hence the associated measure theory and direct integral theory. This still allows us to develop a full Rieffel theory of induced representations. We also examine the basic structure of $C^*(G)$, especially in the case that the action of G on its unit space is well-behaved.

In Chapter 6, we consider the question of when a given locally compact groupoid has a Haar system and if it does, to what extent is the Haar system unique. Unlike the case of a locally compact groups and Haar measure, the situation appears very subtle and a number of interesting open questions persist.

In Chapter 7 we start working seriously with unitary representations of a groupoid and the associated measure theory and direct integrals. We need this background to finally tackle the Disintegration Theory in Chapter 8. While unitary representations themselves do not play ubiquitous role in the theory, they are crucial to establishing certain continuity criteria on representations that are in turn fundamental. In fact, we need this material to fully justify assertions made in the proof of the Equivalence Theorem in Chapter 4.

Chapters 9 and 10 are devoted to distilling out a thorough summary of Anantharaman-Delaroche and Renault's theory of amenability from [**ADR00**] and [**Ren15**] specialized to the setting of locally compact groupoids.

Finally, in Chapter 11, we look at the Renault's Effros-Hahn results and the symplicity of $C^*(G)$ following Renault's [**Ren91**].

Further Reading. There is vast and ever-growing literature on groupoids and their C^*-algebras, and I am not going to attempt a thorough review here. As a "Took kit", this book is meant to be a start-up tool and a point of access to the literature. Nevertheless, there are some omissions in this book that should be mentioned. An important application of groupoid theory is to associate a C^*-algebra to a foliation. However the groupoids associated to foliations, while locally compact, are not necessarily Hausdorff—only locally Hausdorff.[1] If G is locally compact, locally Hausdorff, then there may be very few functions in $C_c(G)$. Connes was able to find a suitable replacement $\mathscr{C}(G)$ for $C_c(G)$ and show how to build a suitable C^*-algebra, still denoted $C^*(G)$, out of $\mathscr{C}(G)$ which reduces to our constructions when G is Hausdorff [**Con79, Con82**]. To reduce the overhead and clarify the main issues, I have opted not to include the locally compact, locally Hausdorff case here. A summary of the general case with further references can be found in [**MW08b**, §2]. As shown in [**MW08b**, Theorem 7.8], there is a full version of the Disintegration Theorem in this generality. Furthermore, the Equivalence Theorems (via linking groupoids) can also be proved for locally compact, locally Hausdorff groupoids (see [**SW12**, §§4–5]). Since Renault and Anantharaman-Delaroche's [**ADR00**] is written for Borel groupoids, much of their work is valid for locally compact, locally Hausdorff groupoids as well and can be extracted from [**ADR00**] by the dedicated reader. In particular, the results in [**Ren15**] are valid in the locally Hausdorff setting as are those in [**Ren91**] which spawned our simplicity results in Chapter 11.

C^*-dyanamical systems have a straightforward generalization to a groupoid acting on a C^*-algebra which fibres suitably over the unit space of G—that is, a $C_0(G^{(0)})$-algebra. The details can be found in [**MW08b**] where we also generalize the Equivalence Theorem to groupoid C^*-dynamical systems with G locally compact, locally Hausdorff. For the sake of exposition, I have not included a continuous 2-cocycle $\sigma : G^{(2)} \to \mathbf{T}$ in the construction of $C^*(G,\sigma)$ as in [**Ren80**]. In fact, $C^*(G,\sigma)$ is a special case of the C^*-algebra associated to a Twist as defined by

[1] For those bound by the Bourbaki tradition, a topological space is called "compact" only when it satisfies the Heine-Borel property *and* is Hausdorff. Spaces that satisfy the Heine-Borel property and are not necessarily Hausdorff are called "quasi-compact". In the Bourbaki world, a locally compact space is necessarily locally Hausdorff. While this has the advantage of never having to write "locally compact, locally Hausdorff", it does obscure the rather dramatic fact that the underlying spaces need not be Hausdorff so I prefer the messier term—and never having to say quasi-compact.

Kumjian (see [**Kum86**, **Kum85**]). Such constructions and many more can be subsumed under the umbrella of C^*-algebras built from a Fell bundle over a groupoid G. As described in [**MW08a**, §2], the Fell bundle C^*-algebra construction subsumes not only the construction of C^*-algebras associated to Kumjian's twists but also groupoid crossed product variations paralleling such as Green twisted systems as defined in [**Ren87**] as well as Busby-Smith twisted systems for group C^*-dynamical systems as in [**PR89**]. The theory and details (for a locally compact second countable Hausdorff groupoid G) can be found in [**MW08a**] where we again extend the Equivalence Theorem and the Disintegration Theorem to this general setting.

Using Renault's work in [**Ren91**], we were able to prove in [**IW09a**], in analogy with the Gootman-Rosenberg-Sauvageot Theorem for transformation group C^*-algebras, that every primitive ideal of $C^*(G)$ is induced from an isotropy group provided G is amenable. However, the proof requires a dose of groupoid crossed products, so it was not possible to include it here.

Even though this book is meant to be a start-up tool, space considerations made it impossible to include many important recent examples and applications of groupoid theory. Some samples include the following and the references therein. There is a Baum-Connes conjecture for groupoids [**Tu00**] which has generated important work [**Tu12**, **STY02**]. Other examples and applications in noncommutative geometry can be found in [**vEY19**, **vEY17**, **LGTX07**, **TXLG04**]. Very important examples of groupoids arise in the study of graph C^*-algebras [**KPRR97**, **Pat02**] and higher-rank graph C^*-algebras and further generalizations [**Dea95**, **RW17**, **KL17**, **Yee07**, **FMY05**]. The classification program means that questions about purely infinite and simple groupoid C^*-algebras generate considerable interest [**BCSS16**, **BCS15**, **BCFS14**].

Errata. As I become aware of mistakes and typos in this book, I will add them to a list available at `http://math.dartmouth.edu/groupoids/`. If you find a mistake, typo, or obscurity that is not listed there, I would very much appreciate your emailing me via the link provided there so that I can add your contribution to the list.

Assumptions and Conventions. Since some of basic underlying results in the theory—such as the Disintegration Theorem and the Equivalence Theorem—require separability, almost all the proofs in this book assume that the topological spaces in play are second countable. In particular, "groupoid" almost always means a second countable, locally compact Hausdorff groupoid. I use the usual convention that homomorphisms between C^*-algebras are $*$-preserving. Representations of C^*-algebras are presumed to be nondegenerate unless stated otherwise. Ideals in C^*-algebras are always closed and two-sided.

Acknowledgments. This subject, let alone this book, would not exist without the pioneering work of Jean Renault. He along with Paul Muhly introduced me to the subject early on and have remained mentors throughout my career. I want to also thank Michael Firrisa, Scott Smedinghoff, Daniël van Wyk, and Jean Renault for examining some early drafts and providing valuable feedback.

CHAPTER 1

From Groupoid to Algebra

In this chapter, we want to see how to produce a C^*-algebra $C^*(G)$ from a locally compact Hausdorff groupoid G and a Haar system λ. In particular, we need to define precisely what a locally compact groupoid is, and explain what a Haar system is. First however, we need to establish some basic terminology and background.

1.1. Preliminaries

It will be helpful to review some preliminary material and set out some basic tools to be used in the sequel. We can't get anywhere without locally compact spaces. A good summary of what we need is given in [**Wil07**, §1.2]. Here we just recall that a space X is *locally compact* if every point has a neighborhood basis consisting of compact sets. If X is Hausdorff, this is equivalent to every point having a compact neighborhood [**Wil07**, Lemma 1.17]. For reasons to be made clear shortly, we will almost always assume that our spaces are not only locally compact and Hausdorff, but second countable as well. Even so, we often work with nets in place of sequences. A good reference for nets is Pedersen's [**Ped89**, §1.3]. As usual, we'll write $C(X)$ for the vector space of all continuous complex-valued functions on X and

$$C_b(X) \supset C_0(X) \supset C_c(X)$$

for the subspaces, respectively, of bounded functions, of functions that vanish at infinity, and of functions with compact support.

In the theory, open maps play a key role—recall that $f : X \to Y$ is open if $f(V)$ is open in Y whenever V is open in X. The use of *Fell's Criterion* for open maps is ubiquitous. The formal statement of Fell's criterion below comes from [**FD88**, Proposition II.13.2] (see also [**Wil07**, Proposition 1.15]).

PROPOSITION 1.1 (Fell's Criterion). *Suppose that $f : X \to Y$ is a surjection. Then f is open if and only if given a net $\{y_i\}_{i\in I}$ converging to $f(x)$ in Y, then is a subnet $\{y_{i_j}\}_{j\in J}$ and a net $\{x_j\}_{j\in J}$ (indexed by the same set) such that $x_j \to x$ in X and $f(x_j) = y_{i_j}$.*

We often use Fell's Criterion to "lift" convergent nets; suppose that $f : X \to Y$ is an open surjection and $y_i \to f(x)$. Then we can pass to subnet $\{y_{i_j}\}$ and find $x_j \to x$ such that $f(x_j) = y_{i_j}$. If we can do so conveniently, we often simply replace the original net $\{y_i\}$ with the appropriate subnet $\{y_{i_j}\}$ and dispense with the distracting subnet notation. This is what is meant by the phrase "passing to a subnet, relabeling, and assuming there exists $x_i \to x$ with $f(x_i) = y_i$".

If X is a locally compact Hausdorff space, then by a measure on X we always mean a *Radon measure*. By definition (see [**Wil07**, §B.1]), a Radon measure is the regular measure on the Borel sets $\mathscr{B}(X)$ associated to a positive linear functional

$\mu : C_c(X) \to \mathbf{C}$ by the Riesz Representation Theorem [**Rud87**, Theorem 2.14]. We will usually not distinguish between the linear functional and its corresponding measure μ. If X is second countable, then *any measure* μ on the Borel sets of X which is finite on compact sets is a Radon measure [**Rud87**, Theorem 2.18].

One of the reasons locally compact spaces play a key role in analysis is that the vector space $C_c(X)$ of compactly supported $\mathbf{C}$-valued continuous functions on X is rich as a result of Urysohn's Lemma [**Wil07**, Lemma 1.41] and the Tietze Extension Theorem [**Wil07**, Lemma 1.42]. A standard technique in this subject is to construct "interesting" completions of $C_c(X)$ as follows. Suppose that ρ is a semi-norm on $C_c(X)$. Let N be the subspace of vectors of zero length. Then we get a well-defined norm on $C_c(X)/N$ in the expected way: $\|f + N\| = \rho(f)$. Then we complete $(C_c(X)/N, \|\cdot\|)$ to get a Banach space $\mathcal{B}$.[1] To more easily work with $\mathcal{B}$, we want to equip $C_c(X)$ with a topology that is more natural with respect to the given topology on X, but still respects the image of $C_c(X)$ in $\mathcal{B}$. For example, we could let μ be a Radon measure on X and

$$\rho(f) := \int_X |f(x)|\, d\mu(x).$$

Then we can identify $\mathcal{B}$ with $L^1(X, \mu)$. Note that if $f_i \to f$ uniformly on X and there is a compact set K such that we eventually have $\operatorname{supp} f_i \subset K$, then f_i converges to f in $L^1(X, \mu)$. (I really should say the image of f_i converges to the image of f in $L^1(X, \mu)$, but I don't think this pedantry is helpful here.) In general, it will always be the case that our seminorm ρ will have the property that if $\{f_i\} \subset C_c(X)$ is net converging uniformly to f with supports eventually contained in a fixed compact set, then $\rho(f_i - f) \to 0$. We'll say that such a net *converges to f properly in the inductive limit topology* on $C_c(X)$. The obvious question is whether there is a topology on $C_c(X)$ for which these are all the convergent nets? The answer, unfortunately, is "almost". The best we can do is to use the inductive limit topology as defined in [**RW98**, §D.2]. The upshot is that if $\{f_i\}$ converges to f uniformly with supports eventually contained in a fixed compact set, then $f_i \to f$ in the inductive limit topology. But the converse can fail. Nevertheless, as proved in [**RW98**, Lemma D.10], if $\rho(f_i - f) \to 0$ for any net converging uniformly with supports eventually contained in a fixed compact set, then $\rho(f_i \to f) \to 0$ for any net $\{f_i\}$ converging to f in the inductive limit topology. So even though it is not technically correct, there is usually no harm in thinking that convergence in the inductive limit topology is tantamount to uniform convergence with supports all in a fixed compact set. I'll add the adjective "properly" when I want to stress that the convergence can really be taken to be uniform with supports eventually in a fixed compact set. Some additional details about the inductive limit topology are provided in Appendix C.

1.2. Getting Started

Rather than start with a formal algebraic definition of a groupoid, it is helpful to get an idea of what is intended. I like to think of a groupoid as a group-like object in which multiplication is not everywhere defined. If this seems like an odd idea, it shouldn't. Anyone who has had a basic course in algebraic topology has

[1] If Y is a normed vector space, let $\iota : Y \to Y^{**}$ be the isometric embedding of Y into its double dual: $\iota(a)(\varphi) = \varphi(y)$. Then the completion of Y is the closure of $\iota(Y)$ in Y^{**}.

already worked with the concept even if the term "groupoid" was never mentioned. If X is a topological space, then we can consider the set Γ of equivalence classes of paths where two paths are considered equivalent if there is a end-point fixing homotopy between them. Note that each equivalence class $\gamma \in \Gamma$ has a well-defined initial point or "source" as well as a well-defined final point or "range". Then Γ has a partially defined multiplication given by concatenation: if the source of η is the same as the range of γ, then $\eta \cdot \gamma$ is the obvious class of paths from the source of γ to the range of η. (The details behind these assertions are usually verified in any elementary text on the fundamental group of a space. For example, consult [**Mas91**, §§II.2–3]. But we're meant to proceed intuitively for now.) The classes of the constant paths ξ_x at points $x \in X$ act like identities: if γ is a path from x to y, then $\xi_y \cdot \gamma = \gamma$ and $\gamma \cdot \xi_x = \gamma$. Furthermore, each path η has an inverse path η^{-1} obtained by reversing the orientation. Then if γ is as above, $\gamma \cdot \gamma^{-1} = \xi_y$ and $\gamma^{-1} \cdot \gamma = \xi_x$. If we restrict to paths with the same range and source x_0, then we recover the fundamental group $\pi(X, x_0)$ of X at x_0. While the fundamental group is ubiquitous in topology, Γ—the Fundamental Group*oid*—seems to me at least to be the more natural object. For example, we don't have to choose a point x_0 and worry about what happens if we choose a different base point.

If we keep the concept of the fundamental groupoid Γ of a space X in mind, then the formal definition of a groupoid is quite natural. The particular form of the definition below comes from [**Muh99**] via [**Goe09**].

DEFINITION 1.2. A *groupoid* is a set G and a subset $G^{(2)} \subset G \times G$ together with maps $(\gamma, \eta) \mapsto \gamma\eta$ from $G^{(2)}$ to G and $\gamma \mapsto \gamma^{-1}$ from G to G such that

(a) (associativity) if (γ, η) and (η, ζ) are in $G^{(2)}$, then $(\gamma\eta, \zeta)$ and $(\gamma, \eta\zeta)$ are in $G^{(2)}$ with $(\gamma\eta)\zeta = \gamma(\eta\zeta)$,
(b) (involution) for all $\gamma \in G$, $(\gamma^{-1})^{-1} = \gamma$ and
(c) (identity) for all $\gamma \in G$, $(\gamma, \gamma^{-1}) \in G^{(2)}$ and $(\gamma, \eta) \in G^{(2)}$ implies $\gamma^{-1}(\gamma\eta) = \eta$ and $(\gamma\eta)\eta^{-1} = \gamma$.

We say that γ and η are *composable* if $(\gamma, \eta) \in G^{(2)}$ and call $G^{(2)}$ the set of *composible pairs*. We call γ^{-1} the *inverse* of γ.

We say that a groupoid G is a *topological groupoid* if the underlying set G comes with a topology such that inversion is continuous from G to G and multiplication is continuous from $G^{(2)}$ (with the relative product topology) to G.

Although every element is assigned an inverse, there are generally lots of elements such as $\gamma\gamma^{-1}$ and $\gamma^{-1}\gamma$ that act like identity elements. Hence we make the following definition.

DEFINITION 1.3. We let $G^{(0)}$ be the set

$$\{ u \in G : u = u^{-1} = u^2 \}.$$

We define the *range map* $r : G \to G^{(0)}$ by $r(\gamma) = \gamma\gamma^{-1}$ and the *source map* $s : G \to G^{(0)}$ by $s(\gamma) = \gamma^{-1}\gamma$.

Then we have the following.

LEMMA 1.4. *Let G be a groupoid.*

(a) *We have $(\gamma, \eta) \in G^{(2)}$ if and only if $s(\gamma) = r(\eta)$.*
(b) *If $(\gamma, \eta) \in G^{(2)}$, then $(\eta^{-1}, \gamma^{-1}) \in G^{(2)}$ and $(\gamma\eta)^{-1} = \eta^{-1}\gamma^{-1}$.*

PROOF. See Exercise Ex 1.2.1. □

It is not hard to see that the images of r and s are exactly $G^{(0)}$. We call $G^{(0)}$ the *unit space* and the elements of $G^{(0)}$ are called units. It is clear that units act as identities: we have $u\gamma = \gamma$ if $\gamma \in r^{-1}(u)$.

We call $H \subset G$ a *subgroupoid* of G if H is a groupoid with respect to the inherited operations (Ex 1.2.9). Note that $H^{(0)} \subset G^{(0)}$ (Ex 1.2.10).

If $A \subset G^{(0)}$, then we let

$$G^A = \{\, \gamma \in G : r(\gamma) \in A \,\} \quad \text{and} \quad G_A = \{\, \gamma \in G : s(\gamma) \in A \,\}.$$

We normally write G^u in place of $G^{\{u\}}$ and similarly with G_u. We also let $G^A_B := G^A \cap G_B$.

DEFINITION 1.5. If G is a groupoid and $A \subset G^{(0)}$, then

$$G^A_A := G|_A := G(A)$$

is called the *reduction of G to A*. The term *restriction* is also used.

It is not hard to see that the reduction $G|_A$ is a subgroupoid of G with unit space A.

By convention, if U and V are subsets of G then,

$$UV := \{\, \gamma\eta : \gamma \in U,\ \eta \in V \text{ and } (\gamma, \eta) \in G^{(2)} \,\}.$$

In particular, $G^A = AG$, $G_B = GB$, and $G^A_B = AGB$. If $u \in G^{(0)}$, then an important special case is $G(u) = G^{\{u\}}_{\{u\}}$, usually written G^u_u. Note that $G(u)$ is a group, and we'll call $G(u)$ the *isotropy group* at u.

LEMMA 1.6. *If G is a topological groupoid, then r and s are continuous. If G is Hausdorff, then $G^{(0)}$ is closed in G.*

PROOF. The first assertion is immediate. If $\{u_i\} \subset G^{(0)}$ and $u_i \to u$, then $\{u_i^{-1}\}$ must converge to both u and u^{-1}. If G is Hausdorff, then we conclude $u = u^{-1}$. Similarly for u^2. □

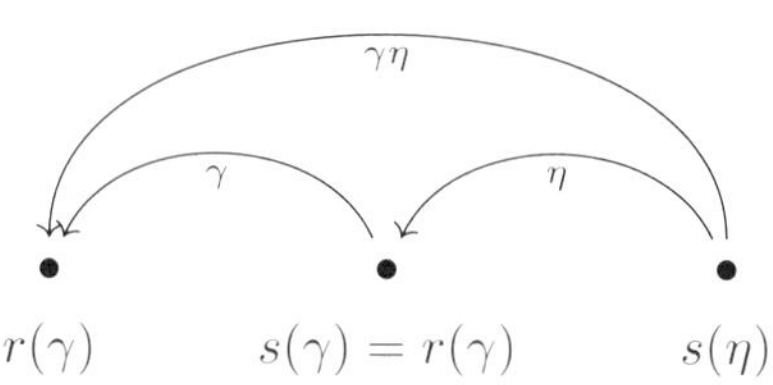

FIGURE 1.1. Groupoid Composition

REMARK 1.7. Rather than thinking of elements of a groupoid as paths in the fundamental groupoid of a topological space, many experts prefer thinking of the elements of a groupoid as *arrows* or morphisms from their source to their range. Thus if we think of γ as a function from $s(\gamma)$ to $r(\gamma)$, then if $r(\beta) = s(\gamma)$ we can compose γ with β to get a function from $s(\beta)$ to $r(\gamma)$ as in Figure 1.1. (This is justified by Lemma 1.4 on the previous page.) More formally, we can think of the units as objects in a category and the elements of the groupoid as morphisms in the category from the source to the range. This leads to abstract nonsense such as described in Ex 1.2.2 where we make the observation that "a groupoid is just a small category with inverses". After looking at Ex 1.2.2, you can join the debate as to whether that definition should be paraded in public or not.

DEFINITION 1.8. If G and H are groupoids, then a map $\varphi : G \to H$ is a *groupoid homomorphism* if $(\gamma, \eta) \in G^{(2)}$ implies $(\varphi(\gamma), \varphi(\eta)) \in H^{(2)}$ and $\varphi(\gamma)\varphi(\eta) = \varphi(\gamma\eta)$. We call φ a *groupoid isomorphism* if φ is bijective and φ^{-1} is a groupoid homomorphism. If G and H are topological groupoids, we would also insist that φ, and if it exists, φ^{-1} are continuous. See Ex 1.2.4.

Of course, every group is a groupoid with a single unit, and conversely, a groupoid with a single unit is a group. But these are hardly informative examples. In fact, there are, in some sense, too many varieties of groupoids in that general statements about groupoids are rare and uninteresting.[2] Consider Figure 1.2 on the following page. We start with some fundamental examples which are very helpful in getting a hold on the subject and its terminology.

EXAMPLE 1.9 (Equivalence Relations). Let $R \subset X \times X$ be an equivalence relation in a set X. We declare (x, y) and (w, z) to be composible if $y = w$ and then $(x, y)(y, z) = (x, z)$ (which is well-defined precisely because R is an equivalence relation). Clearly we want $(x, y)^{-1} = (y, x)$. Then $s(x, y) = (y, y)$ and $r(x, y) = (x, x)$. Therefore the unit space $R^{(0)}$ is the diagonal $\Delta \subset X \times X$. For better or worse, it is common practice to identify $R^{(0)}$ with X. (Despite this practice, it should always be remembered that $G^{(0)}$ is a *subset* of G—see Remark 1.18 on page 7.) Of course, if X is a topological space, then we can give R the relative product topology and then R is a topological groupoid. Surprisingly, we will discover that this topology is rarely the "right" one—or at least, it is rarely the one we need to work with in examples.

A couple of special cases of Example 1.9 deserve to be singled out.

DEFINITION 1.10. The equivalence relation $X \times X$ is called the *trivial equivalence relation* on X or the *pair groupoid* associated to X.

One could argue that the title "trivial equivalence relation" belongs to the diagonal equivalence relation $\Delta \subset X \times X$. In [**Muh99**], Muhly calls Δ the *co-trivial equivalence relation* on X. It is worth noting here that via co-trivial equivalence relations, the class of groupoids includes the class of all sets as a sub-class.

A slight upgrade to Example 1.9 is the following.

EXAMPLE 1.11 (Generalized Equivalence Relations). Let $\mathcal{G}$ be a group and X a set. Let $G = X \times \mathcal{G} \times X$. Then we make G into a groupoid analogously to the equivalence relation example: $(x, g, y)(y, h, z) = (x, gh, z)$ and $(x, g, y)^{-1} = (y, g^{-1}, x)$. Once again, we can identify the unit space $G^{(0)} = \{(x, e, x)\}$ with X so that the range of $\gamma = (x, g, y)$ is x and the source is y. If X is a topological space and $\mathcal{G}$ is a topological group, then G is a topological groupoid with respect to the product topology.

[2]The pleasantly paradoxical quote, "any statement about all countable groups is either trivial or false," is often attributed to Gromov. Whether or not Gromov said it, it is clear that interesting general statements about increasingly more general algebraic objects get increasingly more rare.

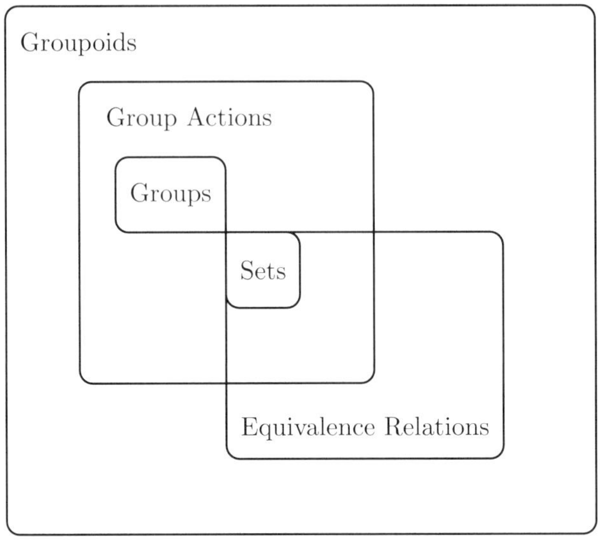

FIGURE 1.2. Groupoids

Recall that a group $\mathcal{G}$ acts on a space X if there is a map $(g, x) \mapsto g \cdot x$ from $\mathcal{G} \times X$ to X such that $e \cdot x = x$ for all $x \in X$ and $(gh) \cdot x = g \cdot (h \cdot x)$ for all $g, h \in \mathcal{G}$ and $x \in X$. If $\mathcal{G}$ is a topological group and X is a topological space, then we say X is a $\mathcal{G}$-space if $(g, x) \mapsto g \cdot x$ is continuous. The pair $(\mathcal{G}, X)$ is called a transformation group.

EXAMPLE 1.12 (Group Actions). Let X be a $\mathcal{G}$-space. Let $G = G(\mathcal{G}, X) = \{ (x, g, y) : x = g \cdot y \}$. Then G is, in an obvious sense, a subgroupoid of the groupoid exhibited in Example 1.11 on the preceding page: $(x, g, y)(y, h, z) = (x, gh, z)$, etc. Naturally, we equip G with the relative product topology. Notice that $(g, x) \mapsto (x, g, g^{-1} \cdot x)$ is a homeomorphism of $\mathcal{G} \times X$ with G. We call G either the *action groupoid* or the *transformation groupoid* associated to $(\mathcal{G}, X)$. Note that we can still identify $G^{(0)}$ with X as in Example 1.11.

One can't go too far wrong using transformation groupoids as a model for groupoids. In fact, the study of algebras associated to groupoids originated from transformation group algebras. Unfortunately, this leads to some peculiar notational choices in the early literature.

REMARK 1.13 (Notation Woes). As pointed out in Example 1.12, $G(\mathcal{G}, X)$ is naturally identified with $\mathcal{G} \times X$. There are lots of ways to do this. The map I chose in Example 1.12 is, oddly enough, the one that returns the standard notation for transformation groups. Therefore one can port the groupoid structure on $G(\mathcal{G}, X)$ to $\mathcal{G} \times X$. Then, continuing to identify the unit space with X, we have $s(g, x) = g^{-1} \cdot x$ and $r(g, x) = x$. The multiplication becomes $(g, x)(h, g^{-1} \cdot x) = (gh, x)$ and $(g, x)^{-1} = (g^{-1}, g^{-1} \cdot x)$. Perhaps the only good thing about this peculiar choice of notation is that it will make connecting up with crossed products easier when the time comes. In the literature, it is common to write $\mathcal{G} \ltimes X$ in place of $G(\mathcal{G}, X)$.

REMARK 1.14. Let $(\mathcal{G}, X)$ be a locally compact transformation group.[3] Suppose that $\mathcal{G}$ acts freely in that $g \cdot x = x$ implies that $g = e$. Then we get an equivalence relation on X given by $R = \{ (x, y) : \mathcal{G} \cdot x = \mathcal{G} \cdot y \}$. Note that R will be locally compact in the relative product topology if and only if R is locally closed [**Wil07**, Lemma 1.25]. Thus R is a locally compact groupoid only if R is the intersection of an open set and a closed set in $X \times X$. However, $G = G(\mathcal{G}, X)$ is always locally compact and the map π given by $(y, g, x) \mapsto (y, x)$ is a (algebraic) groupoid isomorphism of G onto R. The map is clearly continuous, but can easily fail to be a homeomorphism. But we could use this map to give R a locally compact topology which is potentially finer than the relative product topology. As a dramatic example, consider $\mathbf{Z}$ action on $\mathbf{T}$ by an irrational rotation: $n \cdot z = e^{2\pi i n \theta} z$ where $\theta \in [0, 1] \setminus \mathbf{Q}$. Then every orbit is dense in $\mathbf{T}$ (cf., e.g., [**Wil07**, Lemma 3.29]), and the image of π is pathological. If $z \in \mathbf{T}$, then we can find a sequence $\{n_k\}$ such that $n_k \nearrow \infty$ and $n_i \cdot z \to z$. Hence $\{(n_i \cdot z, n_i, z)\}$ diverges to infinity in G, while its image under π converges to (z, z) in R. Hence π is certainly not bi-continuous.

EXAMPLE 1.15. Let $G = G(\mathbf{Z}, \mathbf{T})$ be the action groupoid for an irrational rotation as in Remark 1.14 above. Let $A = \{ z \in \mathbf{T} : \operatorname{Im}(z) \geq 0 \}$. For the sake of this example, realize G as the product $\mathbf{Z} \times \mathbf{T}$ as in Remark 1.13 on the preceding page. Then the reduction $G|_A$ of G to A, $\{ (n, z) \in \mathbf{Z} \times A : \operatorname{Im}(n \cdot z) \geq 0 \}$, is a groupoid with a rich and very subtle structure that is far from an action groupoid.

DEFINITION 1.16. We say that a locally compact groupoid G is a *groupoid bundle* over a locally compact space T if there is a continuous open surjection $p : G^{(0)} \to T$ which is invariant in that $p \circ r = p \circ s$.

To see where the terminology comes from, observe that if $t \in T$, then we can form the reduction $G(t) := G|_{p^{-1}(t)}$ of G to the closed subspace $p^{-1}(t)$ of $G^{(0)}$. The map $p \circ r$ (or equivalently $p \circ s$) can be viewed as a bundle map $p \circ r : G \to T$ and $p^{-1}(t) = G(t)$. So a groupoid bundle is indeed a bundle of groupoids.

EXAMPLE 1.17 (Group Bundle Groupoids). As a special case of a groupoid bundle, consider a locally compact groupoid such that $r = s$. Then G is a groupoid bundle over $G^{(0)}$ and each fibre $G(u)$ is a group. We single this case out as *group bundle groupoid*.

REMARK 1.18. The term "group bundle groupoid" does not roll off the tongue. But it is useful to keep in mind that the groupoid context comes with restrictions. In topology, a group bundle $p : B \to T$ need not admit any continuous sections $c : T \to B$. But in the case of group bundle groupoid $p : G \to G^{(0)}$, the inclusion map $\iota : G^{(0)} \hookrightarrow G$ is always a continuous section.

EXERCISES

Ex 1.2.1. Prove Lemma 1.4.

—Answer on page 353

Ex 1.2.2. A Category $\mathscr{C}$ consists of objects $\mathscr{C}^{(0)}$ and morphisms $\mathscr{C}^{(1)}$ satisfying certain axioms [**Lan02**, §1.11]. We say that $\mathscr{C}$ is a *small category* if the collection of objects $\mathscr{C}^{(0)}$ is a set. In that case, the collection $\mathscr{C}^{(1)}$ of all morphisms is also

[3]This just means that $\mathcal{G}$ is a locally compact group and that X is a locally compact Hausdorff space.

a set. If we identify each object with the corresponding identity morphism, then we can view $\mathscr{C}^{(0)}$ as a subset of $\mathscr{C}^{(1)}$. Show that if $\mathscr{C}$ is a small category in which each morphism is invertible, then $\mathscr{C}^{(1)}$ is a groupoid. Conversely, we can view any groupoid G as the morphisms in a small category with inverses. (This leads to the definition of a groupoid as a small category with inverses. This is probably a useful definition only if you already know what a groupoid is.) Compare with Remark 1.7 on page 4 and recall that we often refer to morphisms in a category as "arrows".

Ex 1.2.3. Suppose that G and H are groupoids and that $\varphi : G \to H$ is a homomorphism. Then $\varphi(G^{(0)}) \subset H^{(0)}$ and $\varphi(\gamma)^{-1} = \varphi(\gamma^{-1})$.

—Answer on page 353

Ex 1.2.4. Suppose that G and H are groupoids and that $\varphi : G \to H$ is a bijective groupoid homomorphism. Prove that φ^{-1} is a homomorphism and hence that φ is a groupoid isomorphism.

—Answer on page 353

Ex 1.2.5. Suppose that A and B are compact subsets of G. Then so is AB.

—Answer on page 353

Ex 1.2.6. Let G be a groupoid with open range and source maps. Suppose that U and V are open subsets of G. Then UV is open.

—Answer on page 353

Ex 1.2.7. In a group, if *either* U or V is open, then UV is open. Show that this fails for groupoids.

—Answer on page 353

Ex 1.2.8. Let G be a groupoid with open range and source maps. If A and B are closed, must AB be closed? What if in addition A or B is compact?

—Answer on page 353

Ex 1.2.9. Show that a nonempty subset H of a groupoid G is a subgroupoid if and only if

(a) $\gamma \in H$ implies $\gamma^{-1} \in H$, and

(b) $(\gamma, \eta) \in G^{(2)} \cap (H \times H)$ implies $\gamma\eta \in H$.

Ex 1.2.10. Show that if H is a subgroupoid of G, then $H^{(0)} \subset G^{(0)}$.

—Answer on page 353

Ex 1.2.11. Suppose that G is a locally compact Hausdorff groupoid and that $F \in C_c(G^u \times G^u)$. Given $\epsilon > 0$, show that there is an open neighborhood V of u in G such that

$$|F(\gamma, \eta) - F(\gamma', \eta')| < \epsilon$$

provided $(\gamma^{-1}\gamma', \eta^{-1}\eta') \in V \times V$. (Compare with [**Wil07**, Lemma 1.62].)

—Answer on page 353

1.3. Haar Systems

In this book, groupoids are used as a means to build interesting C^*-algebras. This requires that we have a $*$-algebra structure on $C_c(G)$. Hence, at the very least, we need $C_c(G)$ to be a nontrivial vector space. Therefore, from now on, we will assume that G is locally compact and Hausdorff. To form a convolution product that generalizes that used in the group C^*-algebra construction, we need something to play the role of a Haar measure. However, rather than fixing a single measure on all of G, the correct idea is to take a family of measures indexed by the unit space.

DEFINITION 1.19. Let G be a locally compact Hausdorff groupoid. Then a *Haar system* on G is a family of Radon measures $\{\lambda^u\}_{u\in G^{(0)}}$ such that

(HS1) for all u, $\operatorname{supp}\lambda^u = G^u$,

(HS2) for all $f \in C_c(G)$, the map

$$u \mapsto \lambda^u(f) = \int_G f(\gamma)\, d\lambda^u(\gamma)$$

is continuous, and

(HS3) for all $\eta \in G$, $\eta \cdot \lambda^{s(\eta)} = \lambda^{r(\eta)}$ where $\eta \cdot \lambda^{s(\eta)}(E) := \lambda^{s(\eta)}(\eta^{-1}E)$.

Property (HS3) is not so useful as written. Instead, we usually invoke it in its "integrated form":

(HS3)$'$ For all $f \in C_c(G)$,

$$\int_G f(\eta\gamma)\, d\lambda^{s(\eta)}(\gamma) = \int_G f(\gamma)\, d\lambda^{r(\eta)}(\gamma).$$

The trouble is that (HS3)$'$ takes a moment to digest. The function $\gamma \mapsto f(\eta\gamma)$ is only defined on $G^{s(\eta)}$. But this is the support of $\lambda^{s(\eta)}$ so (HS3)$'$ is meant to be interpreted in this way.

REMARK 1.20 (Measures on G). Until we get to the definition of the convolution product on $C_c(G)$ in the next section, it might seem troubling not to be defining a measure on all of G. But a Haar system λ gives rise to a measure on G if we also prescribe a measure μ on $G^{(0)}$. Then we can form a measure $\nu = \mu \circ \lambda$ as follows. Since λ is a Haar system,

$$\lambda(f)(u) = \int_G f(\gamma)\, d\lambda^u(\gamma)$$

defines an element of $C_c(G^{(0)})$ by (HS2). As a result, we can define a positive linear functional ν on $C_c(G)$ by $\nu(f) = \mu(\lambda(f))$. As usual, we also denote the corresponding Radon measure on G by ν. We normally summarize all this by writing

$$\nu(f) = \int_{G^{(0)}} \int_G f(\gamma)\, d\lambda^u(\gamma)\, d\mu(u) \quad \text{for all } f \in C_c(G). \tag{1.1}$$

The relationship between the measure ν and the linear functional in (1.1) is explored in more detail in Section 3.1. See, in particular, Corollary 3.9 on page 72 which implies that (1.1) can be interpreted in the natural way if f is a suitable Borel function.

In the sequel, when we do require a measure on G, it will be necessary to allow a variety of choices for μ. Furthermore, Haar systems do not play at all with right-translation. Hence we have to accept the reality that if G is not a group, then there is no global modular function such as there is for groups (see [**Wil07**, Lemma 1.61]). However, if we choose the right sort of Radon measure on $G^{(0)}$, then we do get a modular function for inversion that depends on the choice of μ—see Definition 7.1 on page 157.

In our two basic examples, the choice of a Haar system seems clear.

EXAMPLE 1.21 (Generalized Equivalence Relations). Let ν be any measure on X with full support and let μ be a left Haar measure on $\mathcal{G}$. Let $G = X \times \mathcal{G} \times X$ as in Example 1.11 on page 5. Identify $G^{(0)}$ with X as before. Let $\lambda^x = \delta_x \times \mu \times \nu$. That is,

$$\lambda^x(f) = \int_{\mathcal{G}} \int_X f(x, g, y)\, d\mu(g)\, d\nu(y)$$

for $f \in C_c(X \times \mathcal{G} \times X)$. Then it is not hard to see that $\{\lambda^x\}_{x \in X}$ is a Haar system (Ex 1.3.5).

EXAMPLE 1.22. [Action Groupoids] Suppose that X is a $\mathcal{G}$-space. Then we can form the action groupoid $G = G(\mathcal{G}, X)$ and identify $G^{(0)}$ with X as usual. Let μ be a left Haar measure on $\mathcal{G}$. Then define λ^x by

$$\lambda^x(f) = \int_{\mathcal{G}} f(x, g, g^{-1} \cdot x)\, d\mu(g). \tag{1.2}$$

It is a straightforward exercise to see that $\{\lambda^x\}_{x \in X}$ is a Haar system for G (Ex 1.3.6). Since it is so common to identify $G(\mathcal{G}, X)$ with $\mathcal{G} \times X$, it is important to observe that under this identification (1.2) becomes

$$\lambda^x(F) = \int_{\mathcal{G}} F(g, x)\, d\mu(g)$$

for $F \in C_c(\mathcal{G} \times X)$.

It is natural to ask whether all groupoids have Haar systems. The above examples suggest that there is little uniqueness involved. We'll explore these issues in more detail in Chapter 6. However, here is one necessary condition to keep in mind.

PROPOSITION 1.23. *If G is a locally compact groupoid admitting a Haar system, then the range map, and hence the source map, is open.*

PROOF. Let $\{\lambda^u\}_{u \in G^{(0)}}$ be a Haar system on G. Let $V \subset G$ be a nonempty open set. Fix $u \in r(V)$ and $\gamma \in V$ such that $r(\gamma) = u$. Let U be an open set in G such that $\gamma \in U \subset \overline{U} \subset V$ with $\overline{U}$ compact. Let $f \in C_c^+(G)$ be such that $f(\gamma) = 1$ and $\operatorname{supp} f \subset U$. Then since λ^u has full support, $\lambda(f)(u) > 0$. Furthermore, $\operatorname{supp} \lambda(f) \subset r(\overline{U}) \subset r(V)$. Since $\lambda(f)$ is continuous, u is an interior point of $r(V)$. Since u is arbitrary, $r(V)$ is open. □

There are groupoids that don't have open range and source maps (see Example 6.1 on page 141), so openness is a real obstruction. However, the question of which groupoids with open range maps have Haar systems is very subtle. We'll have more to say about this in Chapter 6. But there is an important class of groupoids—étale groupoids—that always have a very easily described Haar system. First some definitions and observations.

DEFINITION 1.24. If G is a locally compact Hausdorff groupoid, then we call G *r-discrete* if $G^{(0)}$ is open in G.

A justification for the terminology is the following.

LEMMA 1.25. *If G is r-discrete, then for each $u \in G^{(0)}$, G^u and G_u are discrete spaces.*

PROOF. See Ex 1.3.2. □

Using Lemma 1.25, it is not hard to show that *if* an r-discrete groupoid has a Haar system, then the family of counting measures on the G^u form a Haar system (Ex 1.3.4). We know from Proposition 1.23 on the facing page, that to have a Haar system, we need r to be open. To connect up with the exposition in [**Ren80**, §I.2], we need the following.

LEMMA 1.26. *Suppose that G is an r-discrete groupoid with r an open map. Then r is a local homeomorphism.*

PROOF. It suffices to see that every $\gamma \in G$ has an open neighborhood V such that $r|_V$ is injective. Suppose to the contrary there is a $\gamma \in G$ such that r fails to be injective on every open neighborhood V of γ. Then there are $\gamma_V, \eta_V \in V$ such that $r(\gamma_V) = r(\eta_V)$ and $\gamma_V \neq \eta_V$. But then

$$\gamma_V^{-1}\beta_V \to \gamma^{-1}\gamma = s(\gamma) \in G^{(0)}.$$

Since $G^{(0)}$ is open, we must eventually have $\gamma_V^{-1}\beta_V \in G^{(0)}$. But that means we eventually have $\gamma_V = \beta_V$ which is a contradiction. □

DEFINITION 1.27. We call $B \subset G$ a *bisection* or a *G-set* if both $r|_B$ and $s|_B$ are homeomorphisms onto their ranges.

EXAMPLE 1.28. Let X be a locally compact space. Let h be a homeomorphism from $U \subset X$ onto $V \subset X$. Then $B = \{\, (x, h(x)) : x \in U \,\}$ is a bisection in the trivial equivalence relation $R = X \times X$ and every bisection is of this form.

PROPOSITION 1.29 ([**Ren80**, Proposition I.2.8]). *Let G be an r-discrete groupoid. The following are equivalent.*

(a) *Counting measures form a Haar system for G.*
(b) *r is an open map.*
(c) *r is a local homeomorphism.*
(d) *G has a basis for its topology consisting of open bisections.*

Moreover, if G is any locally compact Hausdorff groupoid for which r is a local homeomorphism, then G is r-discrete.

PROOF. We get (a)$\Longrightarrow$(b) from Proposition 1.23 on the preceding page. We get (b)$\Longrightarrow$(c) from Lemma 1.26.

(c)$\Longrightarrow$(d): Let U be an open neighborhood of $\gamma \in G$. By assumption, we can find an open neighborhood U' of γ such that $r|_{U'}$ is a homeomorphism. Since s is also a local homeomorphism, we can shrink U' if necessary so that $s|_{U'}$ is also a homeomorphism. Then U' is a bisection and we've proved (d).

(d)$\Longrightarrow$(a): Let λ^u be counting measure on G^u. To see that $\lambda = \{\lambda^u\}_{u\in G^{(0)}}$ is a Haar system, it will suffice to verify the continuity condition (HS3). Let $f \in C_c(G)$. Then we can cover $\operatorname{supp} f$ with finitely many open bisections. Using a partition of unity, we can assume that f is supported in a single bisection B. Then we have a continuous section $s : r(B) \to B$ such that $r(s(u)) = u$. In particular, $\lambda^u(f) = f(s(u))$ is continuous.

Now suppose that G is any locally compact Hausdorff groupoid and r is a local homeomorphism. Let $u \in G^{(0)}$ and let V be an open neighborhood of u in G such that $r|_V$ is injective. Then $U := V \cap r(V)$ is an open neighborhood of u in $G^{(0)}$ and $U = r^{-1}(U)$ is open in G. Hence $G^{(0)}$ is open in G. $\square$

DEFINITION 1.30. We call a locally compact Hausdorff groupoid *étale* if the range map (or equivalently the source map) is a local homeomorphism.

Note that an étale groupoid G is necessarily r-discrete and satisfies (a), (b), (c), and (d) of Proposition 1.29 on the previous page. In particular, an étale groupoid always has a Haar system given by the appropriate collection of counting measures. The class of étale groupoids is extremely well studied.

EXAMPLE 1.31. Let $G = G(\mathcal{G}, X)$ with $\mathcal{G}$ a discrete group. Then G is an étale groupoid.

REMARK 1.32 (The Literature). The term "étale" did not enter the literature until some time after the publication of Renault's thesis [**Ren80**]. Because of the general preoccupation with Haar systems in our C^*-world, the term "r-discrete groupoid" was often used in contexts where the existence of a Haar system was assumed by default. Part of the point of Proposition 1.29 on the preceding page is that such groupoids are necessarily étale. Conversely, now that the use of étale is common, it is not always made clear that results about étale groupoids apply to r-discrete groupoids provided only that the range and source maps are open.

EXERCISES

Ex 1.3.1. Show that the range map on G is open if and only if the source map is open.

—Answer on page 354

Ex 1.3.2. Prove Lemma 1.25. (Consider the map $\gamma \mapsto \eta\gamma$ from G^u to G^v for $\eta \in G^v_u$.)

—Answer on page 354

Ex 1.3.3. We say that $f \in C(G)$ has r-compact support if $r^{-1}(K) \cap \operatorname{supp} f$ is compact for all $K \subset G^{(0)}$ compact. Show that if $f \in C(G)$ has r-compact support then

$$u \mapsto \lambda(f)(u) := \int_G f(\gamma)\, d\lambda^u(\gamma)$$

is continuous on $G^{(0)}$.

—Answer on page 354

Ex 1.3.4. Suppose that G is an r-discrete groupoid with a Haar system $\{\lambda^u\}_{u\in G^{(0)}}$.

(a) Show that there is a strictly positive continuous function $\alpha \in C(G^{(0)})$ such that $\lambda^u(\{\gamma\}) = \alpha(s(\gamma))$ for all $\gamma \in G^u$.

(b) Show that the family of counting measures is also a Haar system on G. (Notice the characteristic function of $G^{(0)}$ is continuous on G and employ Ex 1.3.3.)

—Answer on page 354

Ex 1.3.5. Verify that $\{\lambda^x\}_{x\in X}$ is a Haar system for G in Example 1.21.

—Answer on page 354

Ex 1.3.6. Verify that $\{\lambda^x\}_{x\in X}$ is a Haar system for G in Example 1.22.

—Answer on page 354

Ex 1.3.7. Let $\{\lambda^u\}_{u\in G^{(0)}}$ be a Haar system on a locally compact Hausdorff groupoid G. If $K \subset G$ is compact, show that there is a $M > 0$ such that $\lambda^u(K) < M$ for all $u \in G^{(0)}$.

—Answer on page 355

Ex 1.3.8. Let $G = X \times X$ be the full equivalence relation on a locally compact Hausdorff space X. Let $\lambda = \{\lambda^x\}_{x\in X}$ be a Haar system on X. Show that there is a full Radon measure μ on X such that $\lambda^x = \delta_x \times \mu$ for all x.

—Answer on page 355

1.4. Building C^*-Algebras

To get a C^*-algebra from a groupoid G and a Haar system λ, the first step is to give $C_c(G)$ a $*$-algebra structure. Proceeding on the philosophy that a groupoid is a group-like object, we can use our Haar system λ to define a convolution product

$$f * g(\gamma) = \int_G f(\eta)g(\eta^{-1}\gamma)\,d\lambda^{r(\gamma)}(\eta) \tag{1.3}$$

which at the very least has the advantage that it is the usual thing when G is a group. Our candidate for the involution is

$$f^*(\gamma) = \overline{f(\gamma^{-1})}. \tag{1.4}$$

It is immediate that if $f \in C_c(G)$, then so is f^*. It is not so clear that $f * g$ is back in $C_c(G)$ if f and g are elements of $C_c(G)$. To see that (1.3) is a binary operation on $C_c(G)$, we will need a technical result.

PROPOSITION 1.33. *Let G be a locally compact Hausdorff groupoid with a Haar system $\{\lambda^u\}_{u\in G^{(0)}}$. Let $G *_r G$ be the closed set $\{\,(\eta,\gamma) \in G\times G : r(\eta) = r(\gamma)\,\}$. If $F \in C_c(G *_r G)$, then $\Lambda(F) \in C_c(G)$ where*

$$\Lambda(F)(\gamma) = \int_G F(\eta,\gamma)\,d\lambda^{r(\gamma)}(\eta).$$

PROOF. Suppose that G is second countable so that $G \times G$ is normal. Then the Tietze Extension Theorem implies that there is a function $g \in C_b(G\times G)$ such that $g(\gamma,\eta) = F(\gamma,\eta)$ for all $(\gamma,\eta) \in G *_r G$. Then we multiply g by a function in $C_c(G\times G)$ which is identically 1 on $\operatorname{supp} F$. This gives us a function g' in $C_c(G\times G)$ such that $g'(\gamma,\eta) = F(\gamma,\eta)$ for all $(\gamma,\eta) \in G *_r G$. In particular, $\Lambda(F) = \Lambda(g')$. So we may as well assume that $F \in C_c(G\times G)$ to begin with.

Let $A \times B$ be a compact neighborhood of $\operatorname{supp} F$ with A and B compact in G. Then $\Lambda(F) \leq M\|F\|_\infty$ where M is such that $\lambda^u(A) \leq M$ for all u (Ex 1.3.7). It follows that if $F_i \to F$ in the inductive limit topology on $C_c(G\times G)$, then

$\Lambda(F_i) \to \lambda(F)$ uniformly. Thus the set of F in $C_c(G \times G)$ for which $\Lambda(F)$ is continuous is closed in the inductive limit topology. But the collection of such functions includes finite sums of functions of the form $(\gamma, \eta) \mapsto f(\gamma)g(\eta)$ for $f, g \in C_c(G)$. But such sums are dense in $C_c(G \times G)$ in the inductive limit topology.

If $G \times G$ is not normal, then a bit more care is required.[4] Let $N = \overline{U}$ be a compact neighborhood of $\operatorname{supp} F$ in $G \times G$. Then $C = N \cap \operatorname{supp} f$ is a compact neighborhood of $\operatorname{supp} F$ in $G *_r G$. By [**Wil07**], there is a $g' \in C_c(G \times G)$ such that $g'(\gamma, \eta) = F(\gamma, \eta)$ c for all $(\gamma, \eta) \in C$. Let $u \in C_c^+(G \times G)$ which is one on $\operatorname{supp} F$ and vanishes off U. Let $g = ug'$. Now if $(\gamma, \eta) \in G *_r G$ and $g(\gamma, \eta) \neq 0$, then $(\gamma, \eta) \in C$. Then $g'(\gamma, \eta) = F(\gamma, \eta)$ and it follows that $(\gamma, \eta) \in \operatorname{supp} F$. But then $g(\gamma, \eta) = F(\gamma, \eta)$. This means $\Lambda(F) = \Lambda(g)$ and we can proceed as above. □

PROPOSITION 1.34. *Let G be a locally compact Hausdorff groupoid with a Haar system $\{\lambda^u\}_{u \in G^{(0)}}$. Then the vector space $C_c(G)$ is a $*$-algebra with respect to convolution product given by* (1.3) *and involution given by* (1.4).

PROOF. The continuity of $\gamma \mapsto f * g(\gamma)$ is guaranteed by Proposition 1.33 (with $F(\eta, \gamma) = f(\eta)g(\eta^{-1}\gamma))$. Clearly the support of $f * g$ is contained in $(\operatorname{supp} f)(\operatorname{supp} g)$ which is compact since $\operatorname{supp} f$ and $\operatorname{supp} g$ are (Ex 1.2.5). Hence $f * g \in C_c(G)$ as is f^*.

We still have to check associativity—$f * (g * k) = (f * g) * k$—as well as the properties required of the involution—$(f^*)^* = f$ and $(f * g)^* = g^* * f^*$.

For example,

$$\begin{aligned}(f * g) * k(\gamma) &= \int_G f * g(\eta) k(\eta^{-1}\gamma)\, d\lambda^{r(\gamma)}(\eta) \\ &= \int_G \int_G f(\zeta) g(\zeta^{-1}\eta) k(\eta^{-1}\gamma)\, d\lambda^{r(\eta)}(\zeta)\, d\lambda^{r(\gamma)}(\eta)\end{aligned}$$

which, since $r(\eta) = r(\gamma)$ on the support of $\lambda^{r(\gamma)}$, is

$$= \int_G \int_G f(\zeta) g(\zeta^{-1}\eta) k(\eta^{-1}\gamma)\, d\lambda^{r(\gamma)}(\zeta)\, d\lambda^{r(\gamma)}(\eta)$$

which, by Fubini, is

$$= \int_G \int_G f(\zeta) g(\zeta^{-1}\eta) k(\eta^{-1}\gamma)\, d\lambda^{r(\gamma)}(\eta)\, d\lambda^{r(\gamma)}(\zeta)$$

which, since $r(\gamma) = r(\zeta)$, on the support of $\lambda^{r(\gamma)}$ is

$$= \int_G \int_G f(\zeta) g(\zeta^{-1}\eta) k(\eta^{-1}\gamma)\, d\lambda^{r(\zeta)}(\eta)\, d\lambda^{r(\gamma)}(\zeta)$$

---

[4]I hate to clutter this document with technical details related to non-second countable spaces as we are almost exclusively interested in second countable groupoids. But almost is not always. It was Geoff Goehle who pointed out to me that extra care was needed in the general case, and it seemed ok to give this short digression in case someone is interested. Most folks can skip this bit.

The problem is that if we use [**Wil07**, Lemma 1.42] to produce $g \in C_c(G \times G)$ such that $g(\gamma, \eta) = F(\gamma, \eta)$ for $(\gamma, \eta) \in \operatorname{supp} F$, we do not know that $g(\gamma, \eta) = 0$ for all $(\gamma, \eta) \in G *_r G \setminus \operatorname{supp} f$. Hence we don't know $\Lambda(g) = \Lambda(F)$, and the proof breaks down.

which, using left-invariance to replace η with $\zeta\eta$, is

$$= \int_G \int_G f(\zeta)g(\eta)k(\eta^{-1}\zeta^{-1}\gamma)\,d\lambda^{s(\zeta)}(\eta)\,d\lambda^{r(\gamma)}(\zeta)$$
$$= \int_G f(\zeta)g * k(\zeta^{-1}\gamma)\,d\lambda^{r(\gamma)}(\zeta)$$
$$= f * (g * k)(\gamma).$$

I'll leave the other computations as exercises (Ex 1.4.1). □

EXAMPLE 1.35 (Étale Groupoids). If G is étale, then assuming as we always shall, that we're using the Haar system consisting of counting measures, then we have

$$f * g(\gamma) = \sum_{r(\eta)=r(\gamma)} f(\eta)g(\eta^{-1}\gamma) = \sum_{\alpha\beta=\gamma} f(\alpha)g(\beta).$$

EXAMPLE 1.36 (Action Groupoids). Let $G = G(\mathcal{G}, X)$ be the action groupoid for the $\mathcal{G}$-space X. In order to connect with the transformation group literature, identify G with $\mathcal{G} \times X$. Then using the natural Haar system from Example 1.22 on page 10, we have

$$f * g(s, x) = \int_{\mathcal{G}} f(t, x)g(t^{-1}s, t^{-1} \cdot x)\,dt \quad \text{and} \quad f^*(s, x) = \overline{f(s^{-1}, s^{-1} \cdot x)}.$$

It should be noticed that this is the pay-off for using our peculiar identification of $G(\mathcal{G}, X)$ with $\mathcal{G} \times X$: we have recovered the standard $*$-algebra structure on $C_c(\mathcal{G} \times X)$ viewed as a dense algebra of the transformation group C^*-algebra $C_0(X) \rtimes_{\mathrm{lt}} \mathcal{G}$ as in [**Wil07**, §2.3].

REMARK 1.37 (Modular Worries). It is a fact of life that a (left) Haar measure μ on a locally compact group $\mathcal{G}$ may not be bi-invariant. For example, this is the case for the "$ax + b$" group ([**Wil07**, Example 1.60]). But if $r \in \mathcal{G}$, then $\mu \cdot r$ given by $\mu \cdot r(E) = \mu(Er)$ is also a Haar measure on $\mathcal{G}$ and must be a multiple $\Delta(r)$ of μ. Then $\Delta : G \to (0, \infty)$ is continuous and we get nice formulas like

$$\Delta(r) \int_{\mathcal{G}} f(gr)\,d\mu(g) = \int_{\mathcal{G}} f(g)\,d\mu(g).$$

(For example, see [**Wil07**, Lemma 1.61].) The function Δ is called the *modular function* on G. Since group C^*-algebras were originally thought of as completions of $L^1(G)$, it is standard to incorporate the modular function in the definition of the involution:

$$f^*(g) = \Delta(g^{-1})\overline{f(g^{-1})} \quad \text{for } f \in C_c(G) \text{ or } f \in L^1(G). \tag{1.5}$$

The main reason for this is to ensure that the involution is isometric on $L^1(G)$—see [**Wil07**, Lemma 1.67]. Hence if you are familiar with group algebra or crossed product constructions, you are probably getting sweaty palms over the lack of a modular function in the definition of f^* in the groupoid case; that is, $C_c(\mathcal{G})$ has a different $*$-algebra structure depending on whether we think of $\mathcal{G}$ as a group and define the involution by (1.5), or we think of $\mathcal{G}$ as a groupoid with a single unit and use (1.4). Not to worry. First of all, on most groups Haar measure is bi-inviariant and the modular function is identically equal to 1 so there is no issue in the unimodular case. But even if the group is not unimodular, we will see later that the C^*-algebras are naturally isomorphic (see Section 1.4.1 as well as Proposition 8.6 on page 171).

The process for building a C^*-algebra out of a $*$-algebra is straightforward: we just need to equip $C_c(G)$ with a norm $\|\cdot\|_\alpha$ such that

$$\|f^* * f\|_\alpha = \|f\|_\alpha^2 \quad \text{for all } f \in C_c(G). \tag{1.6}$$

Then the completion and $\overline{(C_c(G), \|\cdot\|_\alpha)}$ will be a C^*-algebra $C^*_\alpha(G)$.[5]

One way to construct such a norm is to consider a $*$-homomorphisms $L : C_c(G) \to B(\mathcal{H}_L)$. Then we get a seminorm satisfying the C^*-identity (1.6) via

$$\|f\|_L := \|L(f)\|.$$

Then we can get a suitable norm by taking the supremum over a class of L that separates points—provided the supremum is finite. To ensure finiteness of the supremum, we need to restrict to a suitably well-behaved class of $*$-homomorphisms L. For $f \in C_c(G)$, we define

$$\|f\|_{I,r} = \sup_{u\in G^{(0)}} \int_G |f(\gamma)|\,d\lambda^u(\gamma) \quad \text{and} \quad \|f\|_{I,s} = \sup_{u\in G^{(0)}} \int_G |f(\gamma^{-1})|\,d\lambda^u(\gamma).$$

REMARK 1.38. We'll often want to work with the push forward of λ^u under inversion. We'll write λ_u for this measure with support G_u. That is

$$\int_G f(\gamma)\,d\lambda_u(\gamma) = \int_G f(\gamma^{-1})\,d\lambda^u(\gamma).$$

If nothing else, this notation makes the definition of $\|\cdot\|_{I,s}$ more elegant looking:

$$\|f\|_{I,s} = \sup_{u\in G^{(0)}} \int_G |f(\gamma)|\,d\lambda_u(\gamma).$$

Then we define the *I-norm* by

$$\|f\|_I = \max\{\, \|f\|_{I,r}, \|f\|_{I,s} \,\}.$$

One advantage of the I-norm is that $\|f^*\|_I = \|f\|_I$. A less obvious fact is that the I-norm is sub-multiplicative and hence an algebra norm—see Ex 1.4.4.

DEFINITION 1.39. A $*$-homomorphism $L : C_c(G) \to B(\mathcal{H}_L)$ is called I-norm bounded if

$$\|L(f)\| \le \|f\|_I \quad \text{for all } f \in C_c(G).$$

We will refer to such maps as *representations* of $C_c(G)$. Note that a representation of $C_c(G)$ is not required to be nondegenerate as in Definition 1.45 on page 18.

The first question[6] that arises is "are there any nontrivial I-norm bounded $*$-homomorphisms"? To build a large and useful class of examples, let μ be any Radon measure on $G^{(0)}$. Then define $\nu = \mu \circ \lambda$ as in Remark 1.20 on page 9 so that

$$\nu(f) = \int_{G^{(0)}} \int_G f(\gamma)\,d\lambda^u(\gamma)\,d\mu(u) \quad \text{for } f \in C_c(G).$$

[5]There are lots of ways to define a completion of a metric space. But for a normed vector space X, it is probably easiest to embed X isometrically into its double dual X^{**} via the map $\iota : X \to X^{**}$ where $\iota(a)(\varphi) = \varphi(a)$. Then we let $\overline{X} := \overline{\iota(X)}$ in X^{**}.

[6]Well, maybe the first question should be "why consider this particular brand of boundedness for $*$-homomorphisms?", but the answer to that is subtle. For the moment, just trust me or skip ahead to Remark 1.46 on page 19.

We let ν^{-1} be the push-forward of ν under inversion. Hence

$$\nu^{-1}(f) = \int_{G^{(0)}} \int_G f(\gamma^{-1}))\, d\lambda^u(\gamma)\, d\mu(u) = \int_{G^{(0)}} \int_G f(\gamma)\, d\lambda_u(\gamma)\, d\mu(u).$$

Now we can form $\mathcal{H} = L^2(G, \nu^{-1})$ and define $\operatorname{Ind}\mu : C_c(G) \to B(\mathcal{H})$ by convolution: for $f \in C_c(G)$ and $h \in L^2(G, \nu^{-1})$ let[7]

$$\operatorname{Ind}\mu(f)(h)(\gamma) = \int_G f(\eta)h(\eta^{-1}\gamma)\, d\lambda^{r(\gamma)}(\gamma). \tag{1.7}$$

REMARK 1.40 (Measure Theory). Sadly, there is lots of measure theory out there in groupoid land, so we should come to grips with formulas like (1.7); that is, (1.7) defines $\operatorname{Ind}\mu(f)(h)$ only ν^{-1}-almost everywhere. Then the computations below will guarantee that it also defines a class in $L^2(\nu^{-1})$. Alternatively, we can simply assume that $h \in C_c(G)$. Since $C_c(G)$ is dense in $L^2(\nu^{-1})$, we can extend to $L^2(G, \nu^{-1})$ once we establish boundedness.

To see the $\operatorname{Ind}\mu$ is I-norm bounded, it will suffice to show that

$$\big|\big(\operatorname{Ind}\mu(f)h \mid k\big)\big| \le \|f\|_I \|h\|_2 \|k\|_2$$

for all $f \in C_c(G)$ and $h, k \in C_c(G)$. To that end,

$$\begin{aligned}\big(\operatorname{Ind}\mu(f)h \mid k\big) &= \int_{G^{(0)}} \int_G \operatorname{Ind}\mu(f)h(\gamma)\overline{k(\gamma)}\, d\lambda_u(\gamma)\, d\mu(u) \\ &= \int_{G^{(0)}} \int_G \int_G f(\eta)h(\eta^{-1}\gamma)\overline{k(\gamma)}\, d\lambda^{r(\gamma)}(\eta)\, d\lambda_u(\gamma)\, d\mu(u)\end{aligned}$$

which, by left invariance, is

$$\begin{aligned}&= \int_{G^{(0)}} \int_G \int_G f(\gamma\eta)h(\eta^{-1})\overline{k(\gamma)}\, d\lambda^u(\eta)\, d\lambda_u(\gamma)\, d\mu(u) \\ &= \int_{G^{(0)}} \int_G \int_G f(\gamma\eta^{-1})h(\eta)\overline{k(\gamma)}\, d\lambda_u(\eta)\, d\lambda_u(\gamma)\, d\mu(u) \qquad (1.8)\end{aligned}$$

Now we introduce a Radon measure ρ on $G *_r G$ given by

$$\rho(F) = \int_{G^{(0)}} \int_G \int_G F(\eta, \gamma)\, d\lambda_u(\eta)\, d\lambda_u(\gamma)\, d\mu(u).$$

Then we can apply the Cauchy-Schwarz inequality in $L^2(\rho)$ to (1.8) to get

$$\begin{aligned}&\big|\big(\operatorname{Ind}\mu(f)h \mid k\big)\big|^2 \\ &\quad \le \Big(\iint_{G *_r G} |f(\gamma\eta^{-1})||h(\eta)|^2\, d\rho(\eta, \gamma)\Big)\Big(\iint_{G *_r G} |f(\gamma\eta^{-1})||k(\gamma)|^2\, d\rho(\eta, \gamma)\Big) \\ &\quad = (A)(B),\end{aligned}$$

where

$$A = \int_{G^{(0)}} \int_G \int_G |f(\gamma\eta^{-1}||h(\eta)|^2\, d\lambda_u(\eta)\, d\lambda_u(\gamma)\, d\mu(u)$$

which by Fubini is

$$= \int_{G^{(0)}} \int_G \Big(\int_G |f(\gamma\eta^{-1})|\, d\lambda_u(\gamma)\Big)|h(\eta)|^2\, d\lambda_u(\eta)\, d\mu(u)$$

---

[7] The notation $\operatorname{Ind}\mu$ will be explained when we introduce induced representations in Section 5.3 on page 113.

which, by right invariance of the λ_u, is

$$= \int_{G^{(0)}} \int_G \Bigl(\int_G |f(\gamma)| \, d\lambda_{r(\eta)}(\gamma) \Bigr) |h(\eta)|^2 \, d\lambda_u(\eta) \, d\mu(u)$$
$$\leq \|f\|_{I,s} \|h\|_2^2$$

A similar computation gives

$$B \leq \|f\|_{I,r} \|k\|_2^2.$$

This completes the proof of the following.

PROPOSITION 1.41. *Let G be a locally compact Hausdorff groupoid with a Haar system. Let μ be any Radon measure on $G^{(0)}$. Then the $*$-homomorphism $\operatorname{Ind}\mu : C_c(G) \to B(L^2(G, \nu^{-1}))$ defined by (1.7) is I-norm bounded and hence a representation of $C_c(G)$.*

The representations of $C_c(G)$ of the form $\operatorname{Ind}\mu$ are called *regular representations*.[8] A critically important subclass of regular representations are those arising from point masses, δ_u, on $G^{(0)}$. Note that $\operatorname{Ind}\delta_u$ can be viewed a representation on $L^2(G_u, \lambda_u)$ and is given by

$$\operatorname{Ind}\delta_u(f)h(\gamma) = \int_G f(\eta)h(\eta^{-1}\gamma) \, d\lambda^{r(\gamma)}(\eta) \quad \text{for } f \in C_c(G) \text{ and } h \in C_c(G_u). \tag{1.9}$$

DEFINITION 1.42. We define the *universal norm* on $C_c(G)$ by

$$\|f\| = \sup\{ \, \|L(f)\| : L \text{ is a representation of } C_c(G) \, \}. \tag{1.10}$$

The supremum in (1.10) is finite since representations are $\|\cdot\|_I$-bounded by definition. It is not hard to see that $\|\cdot\|$ satisfies the C^*-norm identity. And the existence of regular representations guarantee that it is a norm. The completion of $C_c(G)$ with respect to the universal norm is called the *groupoid C^*-algebra $C^*(G)$*. We also sometimes call $C^*(G)$ the full or universal groupoid C^*-algebra.

REMARK 1.43. Definition 1.42 is slightly problematical since the class of all representations of $C_c(G)$ is not a set—for example, it does not make good sense to speak of the direct sum of all representations. There are a number of ways people finesse this problem. Here we'll rely on the observation from set theory that the class of values $\|L(f)\|$ in $\mathbf{R}$ is a set since $\mathbf{R}$ is a set [**Jec78**, §1.1] so that our definition makes sense. See [**Wil07**, Remark 2.28] for further discussion.

REMARK 1.44 (Notation Issues). The construction of $C^*(G)$ certainly depends on the choice of Haar system $\lambda = \{\lambda^u\}_{u \in G^{(0)}}$. So far, I've taken the attitude that the Haar system is part of the groupoid and not cluttered the notation with its presence; e.g., $C^*(G, \lambda)$. I'll reserve this more descriptive notation only for situations where multiple Haar systems are in play. The question of just how much $C^*(G, \lambda)$ depends on the choice of λ is an interesting one which we'll have a bit more to say about in Chapter 6.

DEFINITION 1.45. Let $L : C_c(G) \to B(\mathcal{H}_L)$ be a $*$-homomorphism.

(a) We say that L is *nondegenerate* if

$$\operatorname{span}\{ \, L(f)h : f \in C_c(G) \text{ and } h \in \mathcal{H}_L \, \}$$

is dense in $\mathcal{H}_L$.

[8] See Remark 5.24 on page 119.

(b) We say that L is *continuous in the inductive limit topology* if whenever $f_i \to f$ in the inductive limit topology on $C_c(G)$ and $h, k \in \mathcal{H}_L$, then we have

$$\big(L(f_i)h \mid k\big) \to \big(L(f)h \mid k\big).$$

REMARK 1.46 (Looking Forward). It is not hard to see that any representation $L : C_c(G) \to B(\mathcal{H}_L)$ is continuous in the inductive limit topology (Ex 1.4.3). However, one of the deeper theorems in the subject, , will allow us to deduce that any $*$-homomorphism $L : C_c(G) \to B(\mathcal{H}_L)$ which is continuous in the inductive limit topology is necessarily a representation; that is, such a representation is necessarily I-norm bounded. Hence representations of $C_c(G)$ turn out to be exactly those $*$-homomorphisms which are continuous in the inductive limit topology. This is one strong reason to assert that I-bounded representations are the "right" beasts to be looking at. We'll prove the Disintegration Theorem in Section 8.1.

Even with the apparently somewhat artificial restriction that our representations are I-bounded, the universal norm is going to turn out to have nice functorial properties justifying the "universal" adjective. However, as we'll see shortly, it can be mysterious to get one's hands on. An alternative is the *reduced norm* defined in analogy with the group case.

DEFINITION 1.47. If $f \in C_c(G)$, then its *reduced norm* is

$$\|f\|_r = \sup\{\, \|\operatorname{Ind} \delta_u(f)\| : u \in G^{(0)}\,\}. \tag{1.11}$$

The completion of $C_c(G)$ with respect to the reduced norm is called the *reduced groupoid C^*-algebra $C_r^*(G)$.*

Since the universal norm dominates the reduced norm, there is a natural homomorphism $\iota_r : C^*(G) \to C_r^*(G)$ with kernel $I_{C_r^*(G)} = \{\, a \in C^*(G) : \|a\|_r = 0\,\}$. (Of course, $\|\cdot\|_r$ extends to a seminorm on $C^*(G)$ satisfying a formula analogous to (1.11).)

The next result implies that $C_0(G^{(0)})$ essentially sits inside the multiplier algebra of $C^*(G)$.

LEMMA 1.48. *Let G be a locally compact groupoid with Haar system $\{\lambda^u\}_{u\in G^{(0)}}$. Then there is a homomorphism $V : C_b(G^{(0)}) \to M(C^*(G))$ such that*

$$V(\varphi)f(\gamma) = \varphi(r(\gamma))f(\gamma) \quad \textit{and} \quad fV(\varphi)(\gamma) = f(\gamma)\varphi(s(\gamma))$$

for all $\varphi \in C_b(G^{(0)})$ and $f \in C_c(G)$.

PROOF. Recall that we can view $M(C^*(G))$ as the bounded adjointable operators on $\mathsf{X} = C^*(G)$ viewed as a right Hilbert module over itself so that $\langle f\,,\, g\rangle_{C^*(G)} = f^* * g$ for $f, g \in C_c(G)$.

We can define $V(\varphi)$ on $f \in C_c(G)$ as above. Fix $\varphi \in C_b(G^{(0)})$. Let $\psi = (\|\varphi\|_\infty \cdot 1 - |\varphi|^2)^{\frac{1}{2}}$. Notice that

$$\big\langle V(\varphi)f\,,\, g\big\rangle_{C^*(G)} = \big\langle f\,,\, V(\overline{\varphi})g\big\rangle_{C^*(G)}.$$

In particular,

$$\|\varphi\|^2\langle f\,,\, f\rangle_{C^*(G)} - \big\langle V(\varphi)f\,,\, V(\varphi)f\big\rangle_{C^*(G)} = \big\langle V(\psi)f\,,\, V(\psi)f\big\rangle_{C^*(G)} \geq 0$$

provided $f \in C_c(G)$. It follows that $\|V(\varphi)f\| \leq \|\varphi\|_\infty\|f\|$. Hence $V(\varphi)$ extends to a bounded operator on $C^*(G)$ with adjoint $V(\overline{\varphi})$. This suffices. $\square$

We call a net $\{e_\alpha\} \subset C_c(G)$ a two-sided *approximate identity for the inductive limit topology* on $C_c(G)$ if for all $f \in C_c(G)$ we have both $\lim_\alpha e_\alpha * f = f$ and $\lim_\alpha f * e_\alpha = f$ properly in the inductive limit topology. That is, the convergence is uniform with supports eventually contained in a fixed compact set. Unlike the situation for groups, it is not so easy to see that $C_c(G)$ always has such an approximate identity and that we can arrange both that $e_\alpha^* = e_\alpha$ and that $\limsup \|e_\alpha\|_I \leq 1$ for all α.

PROPOSITION 1.49. *If G is a locally compact Hausdorff groupoid with a Haar system and paracompact unit space, then there is a two-sided self-adjoint approximate identity $\{e_\alpha\}$ for $C_c(G)$ in the inductive limit topology satisfying* $\limsup \|e_\alpha\|_I \leq 1$ *for all α.*

In order to prove Proposition 1.49, it is easiest to appeal to some complicated constructions best left until we need them in the proof of the Equivalence Theorem in Chapter 4. The result is a straightforward consequence of Proposition 4.10 on page 101; the proof of Proposition 1.49 can be found on page 104.

1.4.1. The Classic Examples. It is far from easy to give concrete examples of groupoid C^*-algebras. This is simply a corollary of the fact that is hard to even give concrete examples of either group C^*-algebras or transformation group C^*-algebras, and groupoid C^*-algebras contain these constructions as special cases as we verify below.

EXAMPLE 1.50 (Group C^*-algebras). We want to verify that if our groupoid is a locally compact group $\mathcal{G}$, then our groupoid C^*-algebra $C^*(G)$ can be naturally identified with the classic group C^*-algebra $C^*_{\mathrm{Group}}(\mathcal{G})$. (Then we will once and for all dispense with the silly notation $C^*_{\mathrm{Group}}(\mathcal{G})$.) This is pretty straightforward if $\mathcal{G}$ is unimodular. However if $\mathcal{G}$ is not unimodular, then as we've already observed, $C_c(\mathcal{G})$ has slightly different $*$-algebra structure depending on whether we consider $\mathcal{G}$ to be a group or a groupoid. Let $C_c^{\mathrm{Group}}(\mathcal{G})$ be the $*$-algebra we get when $\mathcal{G}$ is a group, and continue to write $C_c(\mathcal{G})$ for the $*$-algebra defined here. Then in $C_c^{\mathrm{Group}}(\mathcal{G})$, we have

$$f^*(s) = \Delta(s^{-1})\overline{f(s^{-1})}$$

where Δ is the modular function on $\mathcal{G}$ (see [**Wil07**, §2.3]). Define $\varphi(f)(s) = \Delta(s)^{\frac{1}{2}} f(s)$, then $\varphi : C_c^{\mathrm{Group}}(\mathcal{G}) \to C_c(\mathcal{G})$ is a $*$-isomorphism. If $\mathcal{G}$ is unimodular, then φ is the identity. Moreover the L^1-norm on $C_c^{\mathrm{Group}}(G)$, $\|f\|_1$, and the I-norm, $\|f\|_I$, on $C_c(G)$ clearly coincide. Hence the two $*$-algebras have the same bounded representations. Therefore φ is isometric for the C^*-norms and the two C^*-algebras coincide. We want to verify that φ is still isometric for the C^*-norms even if $\mathcal{G}$ is not unimodular. Let L be any nondegenerate representation of $C_c(\mathcal{G})$. (Recall that this means L is $\|\cdot\|_I$-bounded.) Then L is continuous with respect to the inductive limit topology. Then $L \circ \varphi$ is continuous with respect to the inductive limit topology on $C_c^{\mathrm{Group}}(\mathcal{G})$ (as defined in [**Wil07**, Definition 2.43]). Hence $L \circ \varphi$ is a C^*-norm reducing nondegenerate representation of $C_c^{\mathrm{Group}}(\mathcal{G})$ by [**Wil07**, Lemma 2.45]. Since $\|\varphi(f)\|_{C^*(\mathcal{G})}$ is the supremum over $\|L(\varphi(f))\|$ for such L,

$$\|\varphi(f)\|_{C^*(\mathcal{G})} \leq \|f\|_{C^*_{\mathrm{Group}}(\mathcal{G})}$$

Conversely, every nondegenerate bounded representation of $C_c^{\mathrm{Group}}(\mathcal{G})$ is the integrated form of a unitary representation representation of $\pi : \mathcal{G} \to U(\mathcal{H}_L)$ ([**Wil07**,

Proposition 2.39]). Let R^π be the integrated form of π on $C^{\text{Group}}(\mathcal{G})$,[9] Then $L^\pi = R^\pi \circ \varphi^{-1}$ is a $*$-homomorphism on $C_c(\mathcal{G})$ such that

$$(L^\pi(f)h \mid k) = \int_{\mathcal{G}} f(s)\Delta(s)^{-\frac{1}{2}}\big(\pi(s)h|k\big)\,d\lambda(s) \tag{1.12}$$

where λ is a left-Haar measure on $\mathcal{G}$. Using Cauchy-Schwarz, we have

$$\big|\big(L^\pi(f)h \mid k\big)\big| \le \Big(\int_{\mathcal{G}} |f(s)|\Delta(s)^{-1}\,d\lambda(s)\Big)^{\frac{1}{2}}\Big(\int_{\mathcal{G}} |f(s)|\,d\lambda(s)\Big)^{\frac{1}{2}} \le \|f\|_I.$$

Thus L^π is $\|\cdot\|_I$-bounded and extends to a representation of $C^*(G)$. Since $\|f\|_{C^*_{\text{Group}}(\mathcal{G})}$ is the supremum over $\|R^\pi(f)\|$ for all unitary representations π, we have

$$\|f\|_{C^*_{\text{Group}}(\mathcal{G})} \le \|\varphi(f)\|_{C^*(\mathcal{G})}.$$

Hence φ is isometric and implements an isomorphism between the two C^*-completions $C^*_{\text{Group}}(\mathcal{G})$ and $C^*(\mathcal{G})$.

EXAMPLE 1.51 (Reduced Group C^*-algebras). Since a group $\mathcal{G}$ has just a single unit, the reduced norm on $C_c(\mathcal{G})$ is given by the convolution representation, $\operatorname{Ind}\delta_e$, of $C_c(\mathcal{G})$ on $L^2(\mathcal{G},\rho)$ where

$$\rho(f) = \int_{\mathcal{G}} f(s^{-1})\,d\lambda(s) = \int_{\mathcal{G}} f(s)\Delta(s)^{-1}\,d\lambda(s).$$

That is, $\rho = \Delta^{-1}\lambda$ which is right Haar measure on $\mathcal{G}$. Our map φ also implements a unitary transformation U from $L^2(\mathcal{G},\lambda)$ to $L^2(\mathcal{G},\rho)$. In particular, $\operatorname{Ind}\delta_e(\varphi(f)) = U\pi(f)U^*$ where π is the left-regular representation of $\mathcal{G}$ on $L^2(\mathcal{G},\lambda)$. This implies φ is also isometric for the reduced norms and implements an isomorphism of $(C^*_{\text{Group}})_r(\mathcal{G})$ with our $C^*_r(\mathcal{G})$.

REMARK 1.52 (Unitary Representations). The discussion in Example 1.50 on the facing page shows that if $\mathcal{G}$ is a group, then representations of $C_c(\mathcal{G})$ are parametrized by unitary representations of $\mathcal{G}$. The isomorphism $\varphi : C^*_{\text{Group}}(\mathcal{G}) \to C^*(G)$ is natural in the sense that it preserves the parametrization; that is, in the notation of the example, $L^\pi \circ \varphi = R^\pi$. Moreover,

$$L^\pi(f) = \int_{\mathcal{G}} f(s)\pi(s)\Delta(s)^{-\frac{1}{2}}\,d\lambda(s) \quad \text{for } f \in C_c(G). \tag{1.13}$$

(Equation (1.13) is just the vector-valued form of (1.12).) Thus removing the modular function from the formula for the involution requires that it appear in the integrated form of the representations.

Even the easiest example, requires C^*-trickery in the form of the fact that (algebraic) $*$-homomorphisms are automatically norm reducting.

EXAMPLE 1.53. Let $X_n = \{1, 2, \dots, n\}$ be a set with n-elements and R_n the trivial equivalence relation on X_n. We can use counting measures for a Haar system. We can identify $C_c(R_n) = C(R_n)$ easily with the set M_n of $n \times n$ matrices. Furthermore, groupoid multiplication is just matrix multiplication and the involution is the conjugate transpose. Therefore the natural representation id of $C(R_n)$ on $\mathcal{H} = \mathbf{C}^n$ is a $*$-isomorphism of $C(R_n)$ onto $B(\mathcal{H}) \cong M_n$. Hence the universal norm on $C(R_n)$ satisfies $\|f\| \ge \|\operatorname{id}(f)\|$. But if L is any representation of $C(R_n)$,

[9]Usually, the same letter is used for both the unitary representation and its integrated form. I've introduced R^π here for the sake of exposition.

then $L \circ \mathrm{id}^{-1}$ is a representation of M_n and must be norm decreasing (since M_n is a C^*-algebra in its operator norm). Then

$$\|L(f)\| = \|(L \circ \mathrm{id}^{-1}) \circ \mathrm{id}(f)\| \leq \|\mathrm{id}(f)\|.$$

Thus the universal norm on $C(R_n)$ is just the operator norm and $C^*(R_n) \cong M_n$. Of course, M_n is simple and $C^*_r(R_n) = C^*(R_n) \cong M_n$ as well.

Now a surprising subtlety.

EXAMPLE 1.54 (Trivial Equivalence Relation). Let $R = X \times X$ be the trivial equivalence relation on a locally compact Hausdorff space X. Fix a Radon measure μ on X with full support so that the measures $\{\lambda^x\}_{x\in X}$ given by

$$\lambda^x(f) = \int_X f(x, y)\, d\mu(y)$$

are a Haar system. Let $L^x := \mathrm{Ind}\, \delta_x$. Since all the L^x are equivalent (Ex 1.4.8), the reduced norm is given by $\|f\|_r = \|L^x(f)\|$. But after identifying $L^2(\{x\} \times X, \delta_x \times \mu$ with $L^2(X, \mu)$, we have

$$L^x(f)h(y) = \int_X f(x, y)h(y)\, d\mu(y) \quad \text{for } h \in L^2(X, \mu).$$

That is, $L^x(f)$ is the Hilbert-Schmidt operator with kernel $f \in C_c(X \times X)$. Therefore L^x extends to a isomorphism of $C^*_r(R)$ onto the compact operators $\mathcal{K}(L^2(X, \mu))$. But $C_c(R)$ is only dense in $\mathcal{K}(L^2(X, \mu))$, so the C^*-algebra magic of the previous example fails here (Ex 1.4.9). Nevertheless, we still have $C^*(R) = C^*_r(R) \cong \mathcal{K}(L^2(X, \mu))$. There seems to be no simple way to see this, other than, pardon the pun, to show that $C^*(R)$ itself is simple.[10] We'll do this via the *Equivalence Theorem* which is one of the major results in the subject. (We will see that G is equivalent to the trivial group. By the Equivalence Theorem, this will force $C^*(R)$ to be Morita equivalent to the complex numbers. Hence $C^*(R)$ is simple.) We will get to the Equivalence Theorem in Section 2.6.

Now we come to perhaps the most important examples of all.

EXAMPLE 1.55 (Transformation Group C^*-Algebras). Let $G = \mathcal{G} \times X$ be an action groupoid. We'll write $C_c^{\mathrm{Group}}(\mathcal{G} \times X)$ in place of $C_c(\mathcal{G} \times X)$ when we think of it as a $*$-subalgebra of the transformation group C^*-algebra $C_0(X) \rtimes_{\mathrm{lt}} \mathcal{G}$ as in [**Wil07**, §2.3]. As with groups (which are a special case), we want to see that $C^*(G)$ is naturally isomorphic to $C_0(X) \rtimes_{\mathrm{lt}} \mathcal{G}$. As before, we need to address the involution on $C_0(X) \rtimes_{\mathrm{lt}} \mathcal{G}$:

$$f^*(s, x) = \Delta(s)^{-1}\overline{f(s^{-1}, s^{-1} \cdot x)}$$

which includes the modular function on $\mathcal{G}$. We attempt to proceed as in the group case. But now, even if $\mathcal{G}$ is unimodular, the L^1-norm and the I-norm do not play nicely together. Nevertheless, as in Example 1.50 on page 20, we get a $*$-isomorphism $\varphi : C_c^{\mathrm{Group}}(\mathcal{G} \times X) \to C_c(G \times X)$ defined by $\varphi(f)(s, x) = \Delta(s)^{\frac{1}{2}} f(s, x)$. Just as before, if L is a nondegenerate representation of $C_c(G \times X)$, then $L \circ \varphi$ is a representation of $C_c^{\mathrm{Group}}(\mathcal{G} \times X)$ which is continuous in the inductive limit topology. Hence $L \circ \varphi$ is C^*-norm bounded and φ is continuous for the C^*-norms

[10] An alternative to invoking the Equivalence Theorem would be to mimic the proof of the main result in [**Rie72**] and show that every nondegenerate representation of $C^*(R)$ is equivalent to a direct sum of copies of L which clearly forces L to be faithful. But this takes us too far afield.

[**Wil07**, Lemma 2.45]. Conversely, every bounded representation of $C_c^{\mathrm{Group}}(\mathcal{G}\times X)$ in the integrated form of covariant pair (M,π). Hence $L=(M\rtimes\pi)\circ\varphi^{-1}$ is a $*$-homomorphism on $C_c(G)$ such that

$$(L(f)h|k)=\int_{\mathcal{G}}\big(M\big(f(s,\cdot)\big)\pi(s)h\mid k\big)\Delta(s)^{-\frac{1}{2}}\,d\lambda(s). \tag{1.14}$$

Unfortunately, there does seem to be an elementary way to see that L is I-bounded. (Part of the problem is that the I-norm is not so well-behaved and was only introduced to ensure that the universal norm was well-defined.) Rather than get diverted and/or distracted from the importance of this example, we'll invoke the Disintegration Theorem in the form of Corollary 8.5 on page 170 which asserts that $*$-homomorphisms $L:C_c(G)\to B(\mathcal{H})$ which are continuous in the inductive limit topology are necessarily bounded for the C^*-norm. Since this applies to $L=(M\rtimes\pi)\circ\varphi$, L is bounded. Thus φ extends to the desired isomorphism just as in the group case. (See Proposition 8.6 on page 171.)

EXAMPLE 1.56 (Reduced Transformation Group C^*-Algebras). Checking that φ is isometric for the reduced norms is more straightforward. It is not hard to check that the reduced norm on $f\in C_c^{\mathrm{Group}}(\mathcal{G}\times X)$ is given by $\sup_{x\in X}\|R^x(f)\|$ where $R^x=\mathrm{Ind}_e^{\mathcal{G}}$ is the regular representation on $L^2(\mathcal{G},\mu_{\mathcal{G}})$ associated to $\mathrm{ev}_x:C_0(X)\to\mathbf{C}$ where $\mathrm{ev}_x(\psi)=\psi(x)$ (see [**Wil07**, Example 2.14 and Definition 7.7]). That is,

$$R^x(f)h(r)=\int_{\mathcal{G}}f(s,r\cdot x)h(s^{-1})\,d\mu_{\mathcal{G}}(s)$$
$$\text{for } f\in C_c^{\mathrm{Group}}(\mathcal{G}\times X)\text{ and } h\in L^2(G).$$

Using this, it is not hard to check that φ is also isometric for the reduced norms (Ex 1.4.11). Hence $C_r^*(G)\cong C_0(X)\rtimes_{\mathrm{lt},r}\mathcal{G}$.

REMARK 1.57 (Unitary Representations). Part of the reason for the detail in Example 1.55 on the facing page—rather than just appealing to Proposition 8.6 on page 171—is to emphasize that the isomorphism of $C^*(G)$ with $C_0(X)\rtimes_{\mathrm{lt}}\mathcal{G}$ is natural in the sense that the representations of both $C^*(G)$ and $C_0(X)\rtimes_{\mathrm{lt}}\mathcal{G}$ are parametrized by covariant pairs (M,π) and that the isomorphism preserves the correspondence between covariant pairs and their integrated forms. Just as in the group case, we pay for omitting the modular function from the involution in the groupoid world by incorporating it in the integrated form of the representation: if $f\in C_c(\mathcal{G}\times X)$, then in our groupoid world

$$M\rtimes\pi(f)=\int_{\mathcal{G}}M(f(s,\cdot))\pi(s)\Delta(s)^{-\frac{1}{2}}(s)\,d\lambda(s) \tag{1.15}$$

just as in (1.14).

1.4.2. Transformation Group Groupoids. It is not a terrible stretch to assert that the foundational example for groupoid models in C^*-algebra theory is the transformation group C^*-algebra construction. Hence I find needing to appeal to the Disintegration Theorem in Example 1.55 on the facing page very unsatisfying. It "should not" be so hard to see that the C^*-algebra of the fundamental example in the subject coincides with the traditional transformation group C^*-algebra. I want to outline a seemingly different argument that (1.14)—or equivalently (1.15)—defines a $\|\cdot\|_I$-bounded representation $L:C_c(G(\mathcal{G},X))\to B(\mathcal{H}_L)$ (see Proposition 1.60 on page 25). But while it might seem more "hands on", it is in fact just an outline of the proof of the Disintegration Theorem in the case of a transformation group groupoid. Since this section is primarily meant for motivation—and the arguments given in Example 1.55 are perfectly fine—I will go very light on the details. (The full details are given in Chapter 8 when we prove the general version of the Disintegration Theorem.) Even our "light version" is going to require some of the theory of Borel Hilbert Modules we'll review in Section 3.5. If you are not as troubled as I am about Example 1.55, it might make sense to skip this section entirely or postpone it until we get to unitary representations of groupoids in Chapter 7.

We let $G = \mathcal{G} \times X$ be a transformation groupoid and let (M, π) be a covariant representation on $\mathcal{H}$:

$$\pi(s)M(\varphi)\pi(s^{-1}) = M(\mathrm{lt}_s(f)) \quad \text{for all } s \in G \text{ and } \varphi \in C_0(X) \tag{1.16}$$

where $\mathrm{lt}_s(f)(x) = f(s^{-1} \cdot x)$.

As in [**Wil07**, Example F.25], by passing to an equivalent representation, we can assume that $\mathcal{H}$ is the direct integral $L^2(X * \mathscr{H}, \mu)$ for a Borel Hilbert bundle $X * \mathscr{H}$ and a finite Radon measure μ, and that M is just the usual multiplication operator on $\mathcal{H} = L^2(X * \mathscr{H}, \mu)$:

$$M(\varphi)h(x) = \varphi(x)h(x) \quad \text{for } \varphi \in C_0(X) \text{ and } h \in \mathcal{L}^2(X * \mathscr{H}, \mu). \tag{1.17}$$

Note that we can extend M to all of $\mathcal{B}^b(X)$ in the obvious way. Using the Dominated Convergence Theorem, it is not hard to see that $M(C_0(X))$ is dense in $M(\mathcal{B}^b(X))$ in the strong operator topology (Ex 1.4.13).

Let $s \cdot \mu$ be the push-forward by translation by s so that $s \cdot \mu(E) = \mu(s^{-1} \cdot E)$. Then as in [**Wil07**, Lemma H.13]

$$\int_X f(x)\, ds \cdot \mu(x) = \int_X f(s \cdot x)\, d\mu(x).$$

A key point is that the covariance condition forces μ and $s \cdot \mu$ to be equivalent measures. For the proof of this we need to consider the pull-back Borel Hilbert bundle $X * \mathscr{H}_s$ corresponding to translation by s as in [**Wil07**, Example F.18]. Thus $X * \mathscr{H}_s = \{ (x, h) : h \in H(s^{-1} \cdot x) \}$ and the Borel sections of $X * \mathscr{H}_s$ are exactly those of the form $\tilde{h}(x) = h(s^{-1} \cdot x)$ for a Borel section h of $X * \mathscr{H}$.

LEMMA 1.58. *Let G and μ be as above. Then for all $s \in \mathcal{G}$, μ and $s \cdot \mu$ are equivalent measures. In particular, there is a Borel function $d : \mathcal{G} \times X \to (0, \infty)$ such that $d(s, \cdot)$ is representative of the Raydon-Nikodym derivative $\frac{ds \cdot \mu}{d\mu}$ for all s, and*

$$\int_{\mathcal{G}} \int_X f(s, s \cdot x)\, d\mu(x)\, d\mu_{\mathcal{G}}(s) = \int_{\mathcal{G}} \int_X f(s, x) d(s, x)\, d\mu(x)\, d\mu_{\mathcal{G}}(s)$$

for all $f \in C_c(G)$. Moreover, we can also insist that for all $r, s \in G$ we have $d(sr, x) = d(s, x)d(r, s^{-1} \cdot x)$ for μ-almost all x.

PROOF. Fix $s \in \mathcal{G}$. We can define a unitary $V(s) : L^2(X * \mathscr{H}, \mu) \to L^2(X * \mathscr{H}_s, s \cdot \mu)$ by $V(h)(x) = \tilde{h}(x) = h(s^{-1} \cdot x)$. Let N^s be the multiplication representation of $C_0(X)$ on $L^2(X * \mathscr{H}_s, s \cdot \mu)$: $N^s(\varphi)(\tilde{h})(x) = \varphi(x)\tilde{h}(x)$. It is not hard to check that $V\pi(s^{-1})$ intertwines M and N^s. Using Ex 1.4.13, it follows that $\|M(\varphi)\| = \|N^s(\varphi)\|$ for any $\varphi \in \mathcal{B}^b(X)$. Let $\varphi = \mathbb{1}_E$, we see that $\mu(E) = 0$ if and only if $s \cdot \mu(E) = 0$. This proves the first assertion. The rest follows from [**Wil07**, Corollary D.34]. □

We record a simple Corollary for motivation when we get to Section 7.1.

COROLLARY 1.59. *Let $G = G(\mathcal{G}, X)$ and μ be as above. Let $\nu = \mu_G \times \mu$ be the product measure on $G = \mathcal{G} \times X$ and ν^{-1} the push forward by inversion on G. Then ν and ν^{-1} are equivalent and the Radon-Nikodym derivative $\frac{d\nu}{d\nu^{-1}}$ is given by $(s, x) \mapsto \Delta(s)d(s, x)^{-1}$ where Δ is the usual modular function on the group $\mathcal{G}$.*

PROOF. See Ex 1.4.14. □

We can now define a unitary $W(s) : L^2(X * \mathscr{H}_s, \mu) \to L^2(X * \mathscr{H}, \mu)$ by

$$W(s)\tilde{h}(x) = d(s, s \cdot x)^{-\frac{1}{2}}\tilde{h}(s \cdot x).$$

Let $T(s) = \pi(s)W(s)$, and let M^s be the usual multiplication representation of $C_0(X)$ on $L^2(X * \mathscr{H}_s, \mu)$. Using (1.16), it is not hard to see that $T(s)$ intertwines M with M^s. It is a standard result in the theory (see [**Dix81**, II.2.5 Theorem 1]) that operators that commute with multiplication in this way "decompose" over the fibres.[11] Thus there must be a family of operators $u(s, x) : \mathcal{H}(s^{-1} \cdot x) \to \mathcal{H}(x)$ such that

$$T(s)(\tilde{h})(x) = u(s, x)\bigl(\tilde{h}(x)\bigr)$$

with the $u(s, x)$ being determined μ-almost everywhere for a given s. Since $T(s)$ is unitary and $T(s)^* = T(s)^{-1}$ is also decomposable, it follows from the almost everywhere uniqueness of the $u(s, x)$ that the $u(s, x)$ are μ-almost everywhere unitaries. Since $h(x) = W(s)\tilde{k}(x)$ where $\tilde{k}(x) = d(s, s \cdot x)^{\frac{1}{2}}h(x)$, we get

$$\pi(s)h(x) = T(s)\tilde{k}(x) = u(s, x)\tilde{k}(x) = d(s, x)^{\frac{1}{2}}u(s, x)h(s^{-1} \cdot x). \tag{1.18}$$

For the result we want, we also need to know that $(s, x) \mapsto \bigl(u(s, x)h(s^{-1} \cdot x) \mid k(x)\bigr)$ is Borel for all sections h and k. This would be distracting to establish here. Instead we'll assume it and rely on the details of the full Disintegration Theorem in Chapter 8.[12]

PROPOSITION 1.60. *Let $G = G(\mathcal{G}, X)$, (M, π), and μ be as above. Let L be the $*$-homomorphism of $C_c(\mathcal{G} \times X)$ into $B(\mathcal{H})$ given by (1.14). Then L is $\|\cdot\|_I$-bounded and hence a representation of $C_c(G)$.*

PROOF. Starting from (1.14) and (1.17),

$$\begin{aligned}\bigl(L(f)h|k\bigr) &= \int_{\mathcal{G}} \bigl(M\bigl(f(s, \cdot)\bigr)\pi(s)h \mid k\bigr)\Delta(s)^{-\frac{1}{2}}\,d\mu_{\mathcal{G}}(s)\\ &= \int_{\mathcal{G}}\int_X f(s, x)\bigl(\pi(s)h(x) \mid k(x)\bigr)\Delta(s)^{-\frac{1}{2}}\mu(x)\,d\mu_{\mathcal{G}}(s)\end{aligned}$$

which, in view of (1.18), is

$$= \int_{\mathcal{G}}\int_X f(s, x)\bigl(u(s, x)h(s^{-1} \cdot x) \mid k(x)\bigr)d(s, x)^{\frac{1}{2}}\Delta(s)^{-\frac{1}{2}}\,d\mu(x)\,d\mu_{\mathcal{G}}(s).$$

Since $u(s, x)$ is almost everywhere a unitary, we can apply the Cauchy-Schwarz inequality in $L^2(G \times X, \mu \times \mu_{\mathcal{G}})$ to estimate that

$$\bigl|\bigl(L(f)h \mid k\bigr)\bigr| \le A^{\frac{1}{2}} \cdot B^{\frac{1}{2}}, \tag{1.19}$$

[11] This is a generalization of Theorem 3.39 on page 88 and [**Wil07**, Theorem F.21].

[12] Alternatively, the specific derivation here is a special case of [**Wil07**, Proposition G.24] and the details can be extracted from the proof given there.

where

$$A = \int_{\mathcal{G}} \int_X |f(s,x)| \|h(s^{-1} \cdot x)\|^2 d(s,x) \Delta(s)^{-1} \, d\mu(x) \, d\mu_{\mathcal{G}}(s),$$

and

$$B = \int_{\mathcal{G}} \int_X |f(s,x)| \|k(x)\|^2 \, d\mu(x) \, d\mu_{\mathcal{G}}(s).$$

Clearly,

$$B \leq \|k\|^2 \sup_x \int_{\mathcal{G}} |f(s,x)| \, d\mu_{\mathcal{G}}(s) \leq \|k\|^2 \|f\|_{I,r}.$$

On the other hand,

$$\begin{aligned} A &= \int_{\mathcal{G}} \int_X |f(s, s \cdot x)| \|h(x)\|^2 \, d\mu(x) \Delta(s)^{-1} \, d\mu_{\mathcal{G}}(s) \\ &= \int_{\mathcal{G}} \int_X |f(s^{-1}, s^{-1} \cdot x)| \|h(x)\|^2 \, d\mu(x) \, d\mu_{\mathcal{G}}(s) \\ &\leq \|h\|^2 \sup_x \int_{\mathcal{G}} |f^*(s,x)| \, d\mu_{\mathcal{G}}(s) \\ &\leq \|h\|^2 \|f\|_{I,s}. \end{aligned}$$

Hence the left-hand side of (1.19) is bounded by $\|f\|_I \|h\| \|k\|$ as required. □

Exercises

Ex 1.4.1. Complete the proof of Proposition 1.34.

—Answer on page 355

Ex 1.4.2. Show that if $f \in C_c(G)$, then $\|f\|_I < \infty$. If μ is any Radon measure on $G^{(0)}$ and $\nu = \mu \circ \lambda$, then show $|\nu(f)| < \infty$.

—Answer on page 355

Ex 1.4.3. Show that if L is I-norm bounded (that is, if L is a representation of $C_c(G)$), then L is continuous in the inductive limit topology.

—Answer on page 355

Ex 1.4.4. Show that the $\|\cdot\|_I$-norm is a Banach algebra norm. That is, $\|f * g\|_I \leq \|f\|_I \|g\|_I$.

—Answer on page 355

Ex 1.4.5. Suppose that $f, g \in C_c(G)$ and that $\lambda = \{\lambda^u\}_{u \in G^{(0)}}$ is a Haar system on G. Suppose that $f_i \to f$ and $g_i \to g$ properly in the inductive limit topology. Show that $f_i * g_i \to f * g$ properly in the inductive limit topology.

—Answer on page 356

Ex 1.4.6. Suppose that $L : C_c(G) \to B(\mathcal{H}_L)$ is a $*$-homomorphism. Let $\mathcal{H}_{\text{ess}} = \overline{\text{span}\{\, L(f)h : f \in C_c(G) \text{ and } h \in \mathcal{H}_L \,\}}$. Show that $\mathcal{H}_{\text{ess}}$ is an invariant subspace for L and let $\operatorname{ess}(L)$ be the restriction of L to $\mathcal{H}_{\text{ess}}$. Show that L is unitarily equaivalent to $\operatorname{ess}(L) \oplus 0$.

—Answer on page 356

Ex 1.4.7. Show that

$$\|f\| = \sup\{ \|L(f)\| : L \text{ is a nondegenerate representation of } C_c(G) \}.$$

(I suggest using Ex 1.4.6.)

—Answer on page 356

Ex 1.4.8. Suppose that G is a locally compact Hausdorff groupoid with a Haar system $\{\lambda^u\}_{u\in G^{(0)}}$. Show that the regular representations $\operatorname{Ind}\delta_{s(\gamma)}$ and $\operatorname{Ind}\delta_{r(\gamma)}$ are equivalent for any $\gamma \in G$.

—Answer on page 356

Ex 1.4.9. Let G be a compact Hausdorff groupoid admitting a $*$-isomorphism of $C(G)$ onto a C^*-algebra A. Show that $C^*(G) \cong A$.

—Answer on page 357

Ex 1.4.10. In Example 1.53, we showed $C^*(R_n) \cong M_n$ using counting measures for a Haar system. Show that the same holds for any choice of Haar system on R_n. More generally, show that if λ is any Haar system on an étale groupoid G and σ is the usual system of counting measures, then $C^*(G, \lambda)$ is isomorphic to $C^*(G, \sigma)$. (Consider Ex 1.3.4.)

—Answer on page 357

Ex 1.4.11. Verify, as claimed in Example 1.55, that $\varphi : C_c^{\text{Group}}(\mathcal{G}\times X) \to C_c(G\times X)$ defined by $\varphi(f)(s, x) = \Delta(s)^{\frac{1}{2}} f(s, x)$ is isometric for the reduced norms and extends to an isomorphism of $C_r^*(G)$ onto $C_0(X) \rtimes_{\text{lt},r} \mathcal{G}$.

—Answer on page 357

Ex 1.4.12. Let G be a locally compact groupoid with a Haar system. Let $\{e_i\}$ be an approximate identity for $C_c(G)$ in the inductive limit topology for which is $\|\cdot\|_I$-bounded as in Proposition 1.49. Then given $f \in C_c(G)$ and $\epsilon > 0$ there is an i_0 such that $i, j \geq i_0$ implies that $\|e_i * f * e_j - f\|_\infty < \epsilon$.

—Answer on page 357

Ex 1.4.13. Let M be the multiplication representation of $\mathcal{B}^b(X)$ on $L^2(X * \mathscr{H}, \mu)$ as above. The $M(C_0(X))$ is dense in $M(\mathcal{B}^b(X))$ is the strong operator topology.

—Answer on page 357

Ex 1.4.14. Prove Lemma 1.59.

—Answer on page 357

CHAPTER 2

Groupoid Actions and Equivalence

An important goal in this chapter is to carefully state Renault's Equivalence Theorem and related results. We will get to that in Section 2.6. First we need to develop some basic tools and machinery along the way which are of considerable interest and utility on their own.

2.1. Groupoid Actions

Just as for groups, the notion of a groupoid acting on a space is fundamental. Since the multiplication in a groupoid requires that we fibre the groupoid over its unit space, it is natural to ask that the space on which the groupoid acts also be fibred. In these notes, we always require that the space on which the groupoid acts be fibred over the unit space of the acting groupoid. No doubt other variations are possible, but as I don't have decent applications, we'll stick to what is most common in the literature.

DEFINITION 2.1. Suppose that G is a groupoid and that X is a set together with a map $r_X : X \to G^{(0)}$ called the *moment map*. Then a left action of G on X is a map $(\gamma, x) \mapsto \gamma \cdot x$ from $G * X = \{ (\gamma, x) \in G \times X : s(\gamma) = r_X(x) \}$ to X such that

(A1) $r_X(x) \cdot x = x$ for all $x \in X$ and

(A2) if $(\gamma, \eta) \in G^{(2)}$ and $(\eta, x) \in G*X$, then $(\gamma, \eta{\cdot}x) \in G*X$ and $\gamma\eta{\cdot}x = \gamma{\cdot}(\eta{\cdot}x)$.

Note that axiom (A2) also asserts that the moment map is "equivariant" in that $r_X(\eta \cdot x) = r(\eta)$. The action is called *transitive* if given $x, y \in X$, there is a $\gamma \in G$ such that $\gamma \cdot x = y$. The action is called *free* if $\gamma \cdot x = x$ implies that $\gamma = r_X(x)$.

If G is a topological groupoid and X is a topological space, then we say that X is a G-space or that G acts continuously on X if the moment map r_X is continuous and $(\gamma, x) \mapsto \gamma \cdot x$ is continuous from $G * X$, with the relative product topology, to X.

Right actions are defined similarly, except the moment map is usually denoted $s_X : X \to G^{(0)}$ and $x \cdot \gamma$ is only defined if $s_X(x) = r(\gamma)$. Just as with group actions, there is really no difference between left- and right-actions other than convenience and notation: given a left action of G on X, we get a right action by letting $s_X(x) = r_X(x)$ and $x \cdot \gamma := \gamma^{-1} \cdot x$.

The subscripts '$_X$' decorating the moment maps for left- and right-actions usually just clutter up the formulas. Hence it is standard to omit them—assuming the meaning is clear from context—unless they truly add clarity in the situation at hand.

REMARK 2.2 (Moment Maps). The term "moment map" is fairly new and not universally agreed to. I've also seen, and used, "structure map" and no name at all. Renault has also used "anchor map". While it is now reasonably common practice to use the letters 'r' and 's' to stress the analogy with the range and source maps on the groupoids, some authors prefer a clearer distinction and use ρ and σ, or something else entirely. It is a matter of style.

To add to the confusion, in the early literature, it was a standing assumption that moment maps were open maps. In these notes, we will not be assuming this unless specifically stated. But it is a nuisance when referencing results from the literature. Often, but not always, the referenced result remains true without the assumption of openness. An example where openness will be crucial is in the assertion that an imprimitivity groupoid is equivalent to the acting groupoid as in Proposition 2.47 on page 48.

EXAMPLE 2.3. An important and instructive example is the action of a groupoid G on itself either on the left or the right.

EXAMPLE 2.4. Let $\mathcal{G}$ be a group acting on the left of X. Then the action groupoid $G = G(\mathcal{G}, X)$ acts on the left of X: $(y, s, x) \cdot x = y$. Here $r_X = \mathrm{id}_X$. Note that G also acts on the right: $x \cdot (x, s, y) = y$.

DEFINITION 2.5 (Action Groupoids for Groupoid Actions). If G is a groupoid acting a set X, then we can form the *action groupoid* or *transformation groupoid* $A = \{ (x, \gamma, y) \in X \times G \times X : x = \gamma \cdot y \}$ in analogy with group actions: we define $(x, \gamma, y)(y, \eta, z) = (x, \gamma\eta, z)$ provided $s(\gamma) = r(\eta)$. Clearly, $(x, \gamma, y)^{-1} = (y, \gamma^{-1}, x)$. Then $s_A(x, \gamma, y) = (y, s(\gamma), y)$ while $r_A(x, \gamma, y) = (x, r(\gamma), x)$. We often identify $A^{(0)} = \{ (x, r_X(x), x) : x \in X \}$ with X. The notation $G \rtimes X$ is also used for A in the case of left actions, and $X \ltimes G$ in the case of right actions—see Ex 2.1.8.

One of the more important actions around is that of a groupoid on its unit space.

EXAMPLE 2.6. Let G be a groupoid. Then G acts on the left of its unit space $G^{(0)}$ by $\gamma \cdot s(\gamma) = r(\gamma)$. Similarly, G acts on the right of $G^{(0)}$ by $r(\gamma) \cdot \gamma = s(\gamma)$.

DEFINITION 2.7. Suppose that G is a groupoid acting on the left of X. Then the *orbit* of x in X is the set $G \cdot x := \{ \gamma \cdot x : s(\gamma) = r(x) \}$. We let $G\backslash X$ be the set of orbits in X. If X is a G-space, we give $G\backslash X$ the largest topology making the *orbit map* $q : X \to G\backslash X$ continuous (where $q(x) = G \cdot x$). This is also called the *quotient topology*. Then $G\backslash X$ together with its topology is called the *orbit space*. In the case of a right action, we write X/G.

REMARK 2.8 (The Orbit Space Topology). Note that a set $U \subset G\backslash X$ is open if and only if $q^{-1}(U)$ is open in X (Ex 2.1.3). Furthermore $f : G\backslash X \to Y$ is continuous if and only if it "lifts" to a continuous function $\tilde{f} : X \to Y$ where $\tilde{f}(x) = f(q(x))$ (Ex 2.1.4).

DEFINITION 2.9. A groupoid is called *principal* if the action of G on $G^{(0)}$ is free. A groupoid is called *transitive* if the action of G on $G^{(0)}$ is transitive. If $u \in G^{(0)}$, then $G(u) := G_u^u = \{ \gamma \in G : r(\gamma) = u = s(\gamma) \}$ is a subgroup of G called the *isotropy group* at u.[1]

[1]Unfortunately, the terminology for $G(u)$ is not consistent in the literature (and possibly in this book). The term *stability group* is also common.

REMARK 2.10. Principal and transitive groupoids can alternatively be characterized in terms of the natural map $\pi : G \to G^{(0)} \times G^{(0)}$ given by $\pi(\gamma) = \big(r(\gamma), s(\gamma)\big)$ as in Ex 2.1.1.

In the motivating example of the fundamental groupoid Γ of a topological space X, Γ is transitive exactly when X is pathwise connected. Viewing the elements of a groupoid G as arrows, G is transitive if there is at least one arrow connecting any two units: given $u, v \in G^{(0)}$ there is a $\gamma \in G$ such that $s(\gamma) = u$ and $r(\gamma) = v$. A groupoid G is principal if there is at most one arrow between any two units. Equivalently, G is principal exactly when the isotropy groups $G(u)$ are trivial for all $u \in G^{(0)}$ (Ex 2.1.6). In particular, the fundamental groupoid of a path connected space X is principal exactly when the fundamental groups $\pi(X, x)$ are trivial for all $x \in X$; that is, when X is simply connected.

EXAMPLE 2.11 (Transformation Groups). Our choice of terminology becomes more clear if we look at the basic example of a transformation group. Let $G = G(\mathcal{G}, X)$ be an action or transformation groupoid. Then G is transitive exactly when $\mathcal{G}$ acts transitively on X, and G is principal exactly when the $\mathcal{G}$-action is free. The isotropy group $G(x, x) = \{\, (g, x) \in \mathcal{G} \times X : g \cdot x = x \,\} = \mathcal{G}_x \times \{x\}$ where $\mathcal{G}_x$ is the isotropy or stability group for the $\mathcal{G}$-action on X. The orbit of x under the G-action is $\mathcal{G} \cdot x$—the orbit of x under $\mathcal{G}$.

The analogy of groupoids and transformation groups is very strong and a number of basic properties pass from transformation groups to general groupoids. An important one is that the orbit map should be open as well as continuous. But it is important to notice that to get this result, we require that the range and source maps on the groupoid be open. Because of our interest in Haar systems, it is virtually a standing assumption in the rest of these notes. Unfortunately, it is also a standing assumption in a good deal of the literature where the next result is often used without much comment. But the result can fail without the openness assumptions on r and s.[2] Note that *we do not require the moment map to be open.*

PROPOSITION 2.12. *Let G be a topological groupoid with open range and source maps. Let X be a G-space. Then the orbit map $q : X \to G\backslash X$ is an open map.*

PROOF. In view of Ex 2.1.3, it will suffice to see that if V is open in G, then $G \cdot V = q^{-1}(q(V)) = \{\, \gamma \cdot x : x \in V \,\}$ is open in X. Suppose $\gamma \cdot x \in G \cdot V$ with $x \in V$. Suppose that $x_i \to \gamma \cdot x$ in X. It will suffice to see that there is a subnet which is eventually in $G \cdot V$.

But $r_X(x_i) \to r(\gamma)$. Since r is open, we can pass to a subnet, relabel, and assume that there are $\gamma_i \to \gamma$ in G with $r(\gamma_i) = r_X(x_i)$ (by Fell's Criterion). Then $\gamma_i^{-1} \cdot x_i \to x$. Hence $\gamma_i^{-1} \cdot x$ is eventually in the open set V. Hence $x_i = \gamma_i \cdot (\gamma_i^{-1} \cdot x_i)$ is eventually $G \cdot V$. □

COROLLARY 2.13. *Suppose that G and X are locally compact Hausdorff, that X is a G-space, and that G has open range and source maps. Then $G\backslash X$ is locally compact (but not necessarily Hausdorff). If X is second countable, then so is $G\backslash X$.*

PROOF. We need to show that each point in $G\backslash X$ has a neighborhood basis of compact sets. Let W be an open neighborhood of $G \cdot x$. Then $V = q^{-1}(W)$ is an open neighborhood of x in G. Hence there is a pre-compact open set $U \subset X$ such

[2] Consider the action of the groupoid R on its unit space in Example 6.2 on page 141.

that $x \in U \subset \overline{U} \subset V$. Then $q(\overline{U})$ is a compact neighborhood of $G \cdot x$ in W. Thus $G\backslash X$ is locally compact as claimed.

If X is second countable and $\{U_i\}$ is a countable base for the topology, it is not hard to see that $\{q(U_i)\}$ is a countable base for the topology on $G\backslash X$. $\square$

REMARK 2.14. At first glance, it might seem that the G-action on $G^{(0)}$ is rather special. But if G acts on the left of X, then we can form the action groupoid $A = \{ (\gamma \cdot x, \gamma, x) : s(\gamma) = r_X(x) \}$ as in Definition 2.5 on page 30. Let $\varphi : X \to A^{(0)}$ be the identification $x \mapsto (x, r_X(x), x)$. Then we have a well-defined homeomorphism $\bar{\varphi} : G\backslash X \to A\backslash A^{(0)}$, given by $G \cdot x \mapsto A \cdot (x, r_X(x), x)$, such that the diagram

$$\begin{array}{ccc} X & \xrightarrow{\quad\varphi\quad} & A^{(0)} \\ {\scriptstyle q_X}\downarrow & & \downarrow{\scriptstyle q_A} \\ G\backslash X & \xrightarrow{\quad\bar{\varphi}\quad} & A\backslash A^{(0)} \end{array}$$

commutes (where q_G and q_A are the orbit maps). This observation effectively allows us to translate results about groupoids acting on their unit spaces to general groupoid actions. As an added bonus, notice that if X is a G-space and G has an open range map, then $A = G \rtimes X$ has an open range map as well (Ex 2.1.12).

As we shall see in Section 2.2, just as for group actions, there is a Dichotomy dividing pathological actions from those with a host of reasonable properties.[3] But there is one class of especially well-behaved action that we want to single out here as they play a fundamental role in the theory.

DEFINITION 2.15. Let G be a locally compact Hausdorff groupoid acting on a locally compact Hausdorff space X. We say that the action of G on X is *proper*, or that X is a *proper G-space*, if the map $\Theta : G * X \to X \times X$ given by $(\gamma, x) \mapsto (\gamma \cdot x, x)$ is a proper map in the sense that $\Theta^{-1}(W)$ is compact whenever $W \subset X \times X$ is compact.

EXAMPLE 2.16. The left or right action of G on itself is proper. For example, for the left action, we have $G * G = G^{(2)}$. Then Θ is a homeomorphism of $G^{(2)}$ onto $G *_s G = \{ (\gamma, \eta) : s(\gamma) = s(\eta) \}$ with inverse Φ given by $\Phi(\beta, \alpha) = (\beta\alpha^{-1}, \alpha)$. Since Φ is continuous, $\Phi(W) = \Theta^{-1}(W)$ is compact whenever W is.

PROPOSITION 2.17. *Let G be a locally compact Hausdorff groupoid acting on a locally compact Hausdorff space X. Then the following are equivalent.*

(PA1) *G acts properly on X.*

(PA2) *For every pair of compact sets K and L in X,*
$$P(K, L) = \{ \gamma \in G : K \cap \gamma \cdot L \neq \emptyset \}$$
is compact in G.

(PA3) *If $x_i \to x$ and $\gamma_i \cdot x_i \to y$ in X, then $\{\gamma_i\}$ has a convergent subnet.*

PROOF. (PA1)$\Longrightarrow$(PA2): By definition, $\Theta^{-1}(K \times L)$ is compact and $P(K, L)$ is just the projection onto the first factor.

(PA2)$\Longrightarrow$(PA3): Let L be a compact neighborhood of x and K a compact neighborhood of y. Then we eventually have $\{\gamma_i\}$ in $P(K, L)$.

[3]The use of "pathological" is a matter of taste here. There is a sense in which it is exactly these "pathological" actions that are the "interesting" ones.

(PA3)$\Longrightarrow$(PA1): Let $W \subset X \times X$ be compact. Let $\{(\gamma_i, x_i)\}$ be a net in $\Theta^{-1}(W)$. Since $\Theta^{-1}(W)$ is closed, it will suffice to see that $\{(\gamma_i, x_i)\}$ has a convergent subnet. But $\{(\gamma_i \cdot x_i, x_i)\}$ is in the compact set W. After passing to a subnet and relabeling, we can assume that $(\gamma_i \cdot x_i, x_i) \to (y, x) \in W$. By (PA3), we can pass to a subnet, relabel, as assume that $\gamma_i \to \gamma$ in G. Hence $(\gamma_i, x_i) \to (\gamma, x)$. $\square$

PROPOSITION 2.18. *Let G be a locally compact Hausdorff groupoid whose range (and source) map is open. Suppose that X is a locally compact Hausdorff proper G-space. Then $G\backslash X$ is locally compact and Hausdorff.*

PROOF. Suppose that $\{G \cdot x_i\}$ is a net $G\backslash X$ converging to both $G \cdot x$ and $G \cdot y$. Since the orbit map is open, we may as well assume that $x_i \to x$ and that there are γ_i such that $\gamma_i \cdot x_i \to y$. By part (PA3) of Proposition 2.17 on the facing page, we can pass to a subnet, relabel, and assume that $\gamma_i \to \gamma$ in G. Since X is Hausdorff, $y = \gamma \cdot x$ and $G \cdot x = G \cdot y$. This, together with Corollary 2.13 on page 31, suffices. $\square$

DEFINITION 2.19. A locally compact groupoid is called *proper* if and the action of G on $G^{(0)}$ is proper.

LEMMA 2.20. *Let $\pi : Y \to X$ be a continuous equivariant map between G-spaces. If X is a proper G-space, then so is Y.*

PROOF. See Ex 2.1.15. $\square$

LEMMA 2.21. *Suppose that Z is a locally compact Hausdorff space and G a locally compact Hausdorff groupoid with a Haar system $\{\lambda^u\}_{u\in G^{(0)}}$. Suppose also that Z is a free and proper left G-space.*

(a) *If $F \in C_c(Z \times G)$, then*

$$\Phi_F(z, u) = \int_G F(z, \gamma)\, d\lambda^u(\gamma)$$

defines an element of $C_c(Z \times G^{(0)})$.

(b) *If $\varphi \in C_c(Z)$, then*

$$\lambda(\varphi)(G \cdot z) = \int_G \varphi(\gamma^{-1} \cdot z)\, d\lambda^{r(z)}(\gamma)$$

defines a surjection of $C_c(Z)$ onto $C_c(G\backslash Z)$.

PROOF. (a) The assertion follows from the axioms of a Haar system if F is of the form $F(z, \gamma) = f(z)g(\gamma)$ with $f \in C_c(Z)$ and $g \in C_c(G)$. Note that if $\operatorname{supp} F \subset C \times K$ with K compact in G, then $\|\Phi_F\| \le \|F\|_\infty M$ where $M = \sup_{u\in G^{(0)}} \lambda^u(K)$. (We have M finite in view of Ex 1.3.7.) But we can approximate F properly in the inductive limit topology with sums of functions of the form $F(z, \gamma) = f(z)g(\gamma)$ as above. The result follows.

(b) Let $q(z) = G \cdot z$ be the quotient map and let $D = \operatorname{supp} \varphi$. Clearly $\lambda(\varphi)$ vanishes off the compact set $q(D)$. If $\lambda(\varphi)$ is not continuous, then there is a net $G \cdot z_i \to G \cdot z$ such that

$$\bigl|\lambda(\varphi)(G \cdot z_i) - \lambda(\varphi)(G \cdot z)\bigr| \ge \epsilon$$

for some $\epsilon > 0$. Since q is open (Proposition 2.12 on page 31), we may as well assume that $z_i \to z$ and that the z_i are all in a compact neighborhood C of z. Then, since the action is proper,

$$\{ (w, \gamma) \in Z \times G : w \in C \text{ and } \gamma^{-1} \cdot w \in D \}$$

is compact. Hence there is a $F \in C_c(C \times G)$ such that

$$\varphi(\gamma^{-1} \cdot w) = F(w, \gamma) \quad \text{if } w \in C.$$

Then

$$\lambda(\varphi)(G \cdot z_i) = \int_G F(z_i, \gamma) \, d\lambda^{r(z_i)}(\gamma).$$

Hence $z_i \mapsto \lambda(\varphi)(G \cdot z_i)$ is continuous. This leads to a contradiction; hence, $\lambda(\varphi) \in C_c(G\backslash Z)$.

On the other hand, if $d \in C_c(G\backslash Z)$, then there is a compact set $K \subset Z$ such that d vanishes off $q(K)$. Let $g \in C_c^+(Z)$ be strictly positive on K. Then $\lambda(g\lambda(g)^{-1}d) = d$ with a suitable interpretation of $g\lambda(g)^{-1}d \in C_c(Z)$ (Ex 2.1.16). □

For group actions, a slight weakening of the notion of a proper action—called a Cartan action as in [**Pal61**, Definition 1.1.2]—has prove useful. Consequently, we make the following definition.

DEFINITION 2.22. Let G be a locally compact Hausdorff groupoid acting on a locally compact Hausdorff space X. We say that the action is *Cartan*, or that X is a *Cartan G-space*, if every point in X has a compact neighborhood K such that $\Theta^{-1}(K \times K)$ is compact where Θ is as in Definition 2.15 on page 32.

Using (PA2) of Proposition 2.17 on page 32, it follows that Z is a proper G-space if and only if $P(K, K)$ is compact for all compact sets $K \subset Z$. For Cartan actions, we have the following.

LEMMA 2.23. *Let G be a locally compact groupoid acting on a locally compact Hausdorff space X. Then X is a Cartan G-space if and only if every point in X has a compact neighborhood K such that*

$$P(K, K) = \{ \gamma \in G : \gamma \cdot K \cap K \neq \emptyset \}$$

is compact in G.

PROOF. See Ex 2.1.17 □

Clearly every proper action is Cartan, but the converse fails even for group actions—see [**Pal61**, p. 298]. An important observation about Cartan actions is the following. If X is a G-space, let

$$R(G, X) = \{ (x, y) \in X \times X : G \cdot x = G \cdot y \}. \tag{2.1}$$

If the action of G on X is free, then there is a function $\tau : R(G, X) \to G$ such that $\tau(z, w) \cdot w = z$. We call τ the *translation map* for the given free action. The next result is a groupoid version of [**Pal61**, Theorem 1.1.3].

PROPOSITION 2.24. *Let G be a locally compact groupoid acting freely on a locally compact Hausdorff space X with translation map $\tau : R(G, X) \to G$.*

(a) *If the moment map $r_X : X \to G^{(0)}$ is open, then τ is open.*
(b) *If τ is continuous, then X is a Cartan G-space.*

(c) *If either X is a proper G-space, or if X is a Cartan G-space such that G has an open range map, then τ is continuous.*

In particular, if G has an open range map, then τ is continuous if and only if X is a Cartan G-space.

PROOF. (a) We use Fell's criterion. Suppose that $\gamma_i \to \tau(x,y)$. Then $s(\gamma_i) \to r_X(y)$. Since r_X is open, we can pass to subnet, relabel, and find $y_i \to y$ with $r_X(y_i) = s(\gamma_i)$. Then $(\gamma_i \cdot y_i, y_i) \to (\tau(x,y)\cdot y, y) = (x,y)$. Since $\tau(\gamma_i \cdot y_i, y_i) = \gamma_i$, this suffices.

(b) Suppose that τ is continuous and $x \in X$. Let T be a compact neighborhood of $r(z)$ in G. Since τ is continuous, there is a compact neighborhood K of x in X such that $\tau\big((K\times K)\cap R(G,X)\big) \subset T$. Then $\Theta^{-1}(K\times K)$ is a closed subset of $G*X$ contained in $T\times K$. Hence $\Theta^{-1}(K\times K)$ is compact and the action is Cartan.

(c) To see that τ is continuous, it will suffice to see that if $(\gamma_i\cdot x_i, x_i) \to (\gamma\cdot x, x)$ in $R(G,X)$, then $\gamma_i \to \gamma$. For this, it will suffice to see that every subnet of $\{\gamma_i\}$ has a subnet converging to γ. Assume we have already passed to subnet and relabeled. If X is a proper G-space, then the conclusion follows from (PA3) of Proposition 2.17 on page 32 (since X is Hausdorff). Suppose instead that X is a Cartan G-space and $r : G \to G^{(0)}$ is open. Since $r(\gamma_i) \to r(\gamma)$ and the range map is open, we can pass to a subnet, relabel, and assume there are $\eta_i \in G$ such that $\eta_i \to \gamma$ and $r(\eta_i) = r(\gamma_i)$. Then $(\eta_i^{-1}\gamma_i \cdot x_i, x_i) \to (x,x)$. Let K be a compact neighborhood of x such that $\Theta^{-1}(K\times K)$ is compact. Since $\{(\eta_i^{-1}\gamma_i\cdot x_i, x_i)\}$ is eventually in $K\times K$, we can pass to a subnet, relabel, and assume that $\eta_i^{-1}\gamma_i \to r(z)$. Hence $\gamma_i \to \gamma$ as required. □

While Cartan actions are especially nice actions, unlike proper actions, the resulting orbit space need not be Hausdorff. In fact, we know that Palais's example [**Pal61**, p. 298] mentioned above is not proper precisely because the orbit space is not Hausdorff. Nevertheless, we have the following.

PROPOSITION 2.25. *Suppose that G is a locally compact groupoid with open range and source maps and that X is a Cartan G-space. Then the following are equivalent.*

(a) *X is a proper G-space.*
(b) *The orbit space $G\backslash X$ is Hausdorff.*
(c) *$R(G,X)$ is closed in $X\times X$.*

PROOF. See Ex 2.1.18. □

As a corollary, we get the following strengthening of (PA3) of Proposition 2.17 on page 32.

COROLLARY 2.26. *Let G be a locally compact Hausdorff groupoid acting on a locally compact Hausdorff space X. Then G acts properly on X if and only if whenever $x_i \to x$ and $\gamma_i \cdot x_i \to y$, then $\{\gamma_i\}$ is convergent in G.*

PROOF. Suppose X is a proper G-space and $\{x_i\}$ and $\{\gamma_i\}$ are as in the statement of the corollary. Then X is Cartan and the translation map is continuous. Also $R(G,X)$ is closed and $(\gamma_i\cdot x_i, x_i)$ converges in $R(G,X)$. Hence $\{\tau(\gamma_i\cdot x_i, x_i)\} = \{\gamma_i\}$ converges.

The converse follows immediately from Proposition 2.17 on page 32. □

Exercises

Ex 2.1.1. Let G be a topological groupoid and define $\pi : G \to G^{(0)} \times G^{(0)}$ by $\pi(\gamma) = \big(r(\gamma), s(\gamma)\big)$.

(a) Show that the range of π is an equivalence relation R and that π is a continuous groupoid homomorphism when R is given the relative product topology.

(b) Show that G is principal if and only if π is injective.

(c) Show that G is transitive if and only if π is surjective.

—Answer on page 357

Ex 2.1.2. Let G be a transitive groupoid. Show that the isotropy groups $G(u)$ and $G(v)$ are isomorphic for any $u, v \in G^{(0)}$.

—Answer on page 358

Ex 2.1.3. Let X be a G-space with orbit map $q : X \to G\backslash X$. Show that $U \subset G\backslash X$ is open if and only if $q^{-1}(U)$ is open in X.

—Answer on page 358

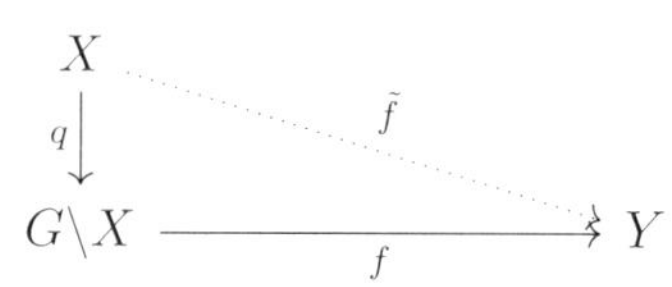

Figure 2.1. Lifting Functions from Quotients

Ex 2.1.4. Let X be a G space and $q : X \to G\backslash X$ the orbit map. Let Y be a topological space and $f : G\backslash X \to Y$ a function. Let $\tilde{f} : X \to Y$ be the lift: $\tilde{f}(x) = f(q(x))$. Show that f is continuous if and only if $\tilde{f}$ is continuous. (See Figure 2.1.)

—Answer on page 358

Ex 2.1.5. Suppose that G has open range and source maps. Let G act on itself by left translation. Show $s : G \to G^{(0)}$ factors through a homeomorphism of $G\backslash G$ with $G^{(0)}$.

—Answer on page 358

Ex 2.1.6. Show that a groupoid G is principal if and only if $\gamma, \eta \in G^u_v$ implies $\gamma = \eta$. Alternatively, show that G is principal if and only if $G(u) = \{u\}$ for all $u \in G^{(0)}$.

—Answer on page 358

Ex 2.1.7. Let G be a locally compact Hausdorff groupoid with a Haar system $\{\lambda^u\}_{u\in G^{(0)}}$. Show that the action group A from Definition 2.5 is a locally compact groupoid with Haar system $\{\beta^x\}_{x\in X}$ where

$$\beta^x(f) = \int_G f(x, \gamma, \gamma^{-1} \cdot x)\, d\lambda^{r(x)}(\gamma).$$

—Answer on page 358

Ex 2.1.8. Define an action groupoid for a right groupoid action.

—Answer on page 358

Ex 2.1.9. Let X be a proper G-space and fix $x \in X$. Then $H = G(x) = \{\, \gamma \in G : \gamma \cdot x = x \,\}$ be the *isotropy group* at x. Note that H acts on the right of $G_{r(x)}$. Show that $\gamma \mapsto \gamma \cdot x$ induces a homeomorphism of $G_{r(x)}/H$ with $G \cdot x$.

—Answer on page 358

Ex 2.1.10. Let G be a locally compact Hausdorff groupoid and let $\pi : G \to G^{(0)} \times G^{(0)}$ be given by $\pi(\gamma) = \big(r(\gamma), s(\gamma)\big)$. Show that π has closed range if and only if the orbit space $G\backslash G^{(0)}$ is Hausdorff.

—Answer on page 359

Ex 2.1.11. Let G be a groupoid with open range map and X a G-space. Show that if $U \subset X$ and $V \subset G$ are open, then $V \cdot U = \{\, \gamma \cdot x : \gamma \in V \text{ and } x \in U \,\}$ is open in X.

—Answer on page 359

Ex 2.1.12. Show that if X is a G-space and G has an open range map, then the range map for $A = G \rtimes X$ is open.

—Answer on page 359

Ex 2.1.13. Show that a locally compact groupoid is proper if and only if the map $\pi : G \to G^{(0)} \times G^{(0)}$ is proper where $\pi(\gamma) = (r(\gamma), s(\gamma))$

—Answer on page 359

Ex 2.1.14. Let G be a locally compact groupoid and Z a locally compact G-space. Show that G acts properly on Z if and only if the action groupoid $G \rtimes Z$ is a proper groupoid.

—Answer on page 359

Ex 2.1.15. Prove Lemma 2.20.

—Answer on page 359

Ex 2.1.16. Provide the details showing that $\lambda(g\lambda(g)^{-1}d) = d$ at the end of the proof of Lemma 2.21.

—Answer on page 359

Ex 2.1.17. Prove Lemma 2.23. (Hint: compare with Proposition 2.17.)

—Answer on page 360

Ex 2.1.18. Prove Proposition 2.25.

—Answer on page 360

Ex 2.1.19. Suppose that G is a locally compact Hausdorff groupoid and that X is a Hausdorff Cartan G-space. Show that every G-orbit is closed in X. (Hint: let $y \in \overline{G \cdot x}$ and pick a compact neighborhood of y such that $P(K, K)$ is compact. Let $\{\gamma_i \cdot x\}$ be a net in K converging to y.)

—Answer on page 360

2.2. The Mackey-Glimm-Ramsay Dichotomy

Let G be a second countable locally compact Hausdorff groupoid with open range and source maps. Suppose that G acts on the left of X. Then $\gamma \mapsto \gamma \cdot x$ defines a continuous map φ_x from $G_{r_X(x)} \to G \cdot x$. If we let $H = \{ \gamma \in G_{r_X(x)} : \gamma \cdot x = x \}$, then we get a continuous bijection $\bar{\varphi}_x$ of the quotient space $G_{r_X(x)}/H$ (with the quotient topology) onto the orbit $G \cdot x$. It is not hard to see that that natural map of $G_{r_X(x)}$ onto $G_{r_X(x)}/H$ is an open map (Ex 2.2.1). Hence $\bar{\varphi}_x$ is a homeomorphism

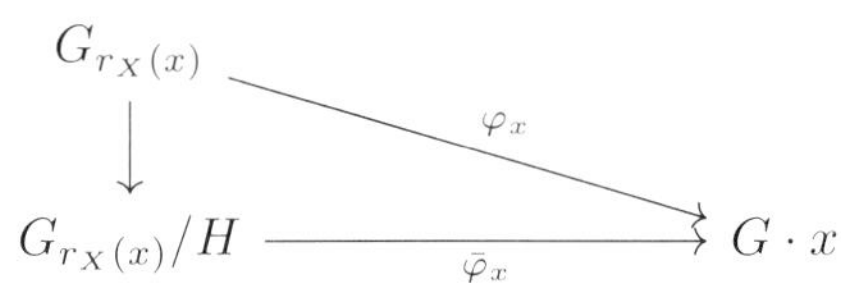

FIGURE 2.2. Map onto $G \cdot x$

if and only if φ_x is an open map (see Figure 2.2).

Simple examples from group actions show that we can't always expect φ_x to be open. Just as for group actions, as long as we stick to second countable spaces, it turns out that the openness of φ_x for all x is equivalent to a number of desirable conditions on the action and the topological structure of the orbit space $G\backslash X$. Hence there is a sharp dichotomy between reasonably well-behaved actions and pathological actions. We record some of the basic equivalences here. For the full story, see [**Ram90**].[4]

Recall that a topological space X is T_0—or more formally, satisfies the T_0 *axiom of separability*—if given distinct points x and y in X, then there is an open set U that either contains x and not y, or which contains y and not x. Alternatively, X is T_0 if and only if distinct points have distinct closures in X (Ex 2.2.3). A subset A of a space X is *locally closed* if A is relatively open in its closure $\overline{A}$ (see [**Wil07**, Lemma 1.25] for other equivalent conditions). We say $A \subset X$ is a G_δ *subset* if it is the countable intersection of open sets in X. We say A is a F_σ *subset* if it is the countable union of closed sets. If X is metrizable, then every closed set is a G_δ. If in addition, X is separable, then every open set is a F_σ (Ex 2.2.2). A not necessarily Hausdorff locally compact space X is said to be *almost Hausdorff* is every locally compact subspace V contains a nonempty relatively open Hausdorff subspace (see [**Wil07**, Lemma 6.3] for a host of equivalent conditions for an almost Hausdorff space). We say that a space is a *Polish space* if it is separable and completely metrizable. A second countable, locally compact Hausdorff space is always Polish [**Wil07**, Lemma 6.5].

THEOREM 2.27 (Mackey-Glimm-Ramsay Dichotomy). *Let G be a second countable locally compact Hausdorff groupoid with open range and source maps acting on the left of a second countable locally compact Hausdorff space X. We endow $G\backslash X$ with the quotient topology. Then the following are equivalent.*

(a) *$G\backslash X$ is a T_0 topological space.*

[4] Ramsay works only with the action of G on $G^{(0)}$. But, as we saw in Remark 2.14 on page 32, we can replace G with the action groupoid A for the action of G on X to see that Theorem 2.27 is a special case of Ramsay's results.

(b) *$G\backslash X$ is almost Hausdorff.*

(c) *Each orbit $G \cdot x$ is locally closed in X.*

(d) *Each orbit $G \cdot x$ is a G_δ-subset of X.*

(e) *For all $x \in X$, $\varphi_x : G_{r_X(x)} \to G \cdot x$ is an open map.*

For the proof we'll need to consider the range R of the continuous map $\pi : G * X = \{ (\gamma, x) \in G \times X : s(\gamma) = r_X(x) \} \to X \times X$ given by $\pi(\gamma, x) = (\gamma \cdot x, x)$. Note that R is an equivalence relation in X. If $q : X \to G\backslash X$ is the orbit map, we can also think of R as the fibred product $X \times_q X = \{ (x, y) : q(x) = q(y) \}$. Since π is continuous and since both G and X are locally compact and second countable (and hence σ-compact), it follows that R is a F_σ-subset of $X \times X$. Since X, and therefore $X \times X$, is a Polish space, R will be Polish (in the relative topology) exactly when it is a G_δ-subset [**Wil70**, Theorem 24.12].

We will also need to recall that a subset A of a topological space X is said to be *meager* or of *first category in* X if it can be written a countable union of nowhere dense sets F_i. (We say F is *nowhere dense* in X if the interior of $\overline{F}$ is empty.) A set that is not of first category in X is said to be of *second category* in X. The Baire Category Theorem implies that any Polish space is of second category in itself [**Wil70**, §25]. Note that X is of second category in itself if and only if it is a *Baire space* in that the countable intersection of open dense sets in X is itself dense [**Wil70**, Theorem 25.2].

Whether a subset is meager or not depends very much on where it lives. The set of integers is of second category as a subset of itself, but is first category as a subset of the real line. We will take advantage of the fact that if $A \subset X \subset Y$ is a second category subset of X, then it remains of second category in $\overline{X}$ (Ex 2.2.6).

If $A \subset X$, then the *quasi-interior* of A in X is the open set

$$A^* = \bigcup \{ U \subset X : U \text{ is open and } U \setminus A \text{ is of first category in } X \}.$$

If X is second countable, then A^* will always have a countable open cover of sets U such that $U \setminus A$ is of first category. Hence $A^* \setminus A$ is of first category.

In general, there is no reason to believe that A^* is nonempty. We say $A \subset X$ has the *Baire property* if there is an open set $U \subset X$ such that

$$A \,\Delta\, U := (A \setminus U) \cup (U \setminus A)$$

is of first category in X. Since open sets clearly have the Baire property and since the sets with the Baire property form a σ-algebra (see [**Wil07**, Lemma D.15]), every Borel set has the Baire property. Hence every Borel set of second category in X must have $A^* \neq \emptyset$.

Recall that the orbit map $q : X \to G\backslash X$ is open by Proposition 2.12 on page 31. The proof of Theorem 2.27 on the facing page is a bit involved and it will be convenient to introduce some additional conditions equivalent to those stated in the theorem:

(1) The equivalence relation R is a G_δ subset of $X \times X$.

(2) Each orbit $G \cdot x$ is a Baire space in its relative topology.

(3) Each orbit $G \cdot x$ is *locally of second category* in $G \cdot x$ in that $U \cap G \cdot x$ is of second category in $G \cdot x$ for every open set $U \subset X$ such that $U \cap G \cdot x \neq \emptyset$.

Then we will prove the following implications:

(2.2)

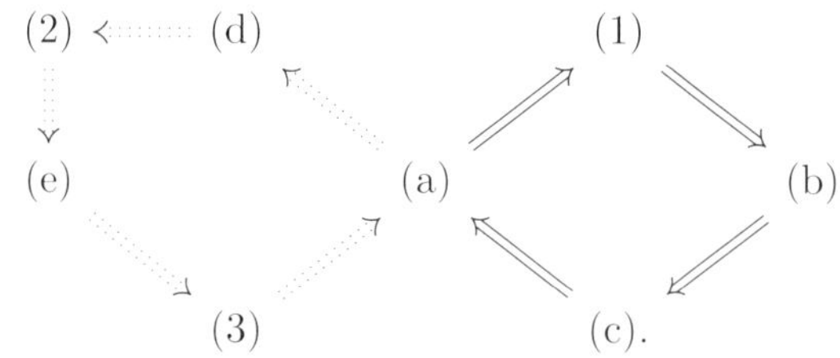

We'll start with the solid arrows, and then consider the dotted implications.

Proof of Theorem 2.27 on page 38. (a)$\Longrightarrow$(1): Since q is open, $G\backslash X$ is a second countable topological space by Corollary 2.13 on page 31. Let $\{ B_n \}$ be a countable collection consisting of a countable base for the topology on $G\backslash X$ together with their complements. Let $A_n = q^{-1}(B_n)$. Note that each A_n is either open or closed in X. If $(x,y) \notin R$, then $q(x) \neq q(y)$. Since $G\backslash X$ is T_0 by assumption, there is a $n(x,y)$ such that $x \in A_{n(x,y)}$ and $y \in A^c_{n(x,y)}$. Thus $(x,y) \in A_{n(x,y)} \times A^c_{n(x,y)}$ and $A_{n(x,y)} \times A^c_{n(x,y)}$ is in the complement of R in $X \times X$. The set

$$D = \{ n(x,y) : (x,y) \notin R \}$$

is countable and

$$(X \times X) \setminus R = \bigcup_{n \in D} A_n \times A^c_n.$$

But each A_n and A^c_n is either closed or an F_σ-set in X. In particular, the complement of R is a F_σ-set. It follows that R is a G_δ-set.

(1)$\Longrightarrow$(b): If view of [**Wil07**, Lemma 6.3], to see that $G\backslash X$ is almost Hausdorff, it will suffice to see that every nonempty closed subset $F \subset G\backslash X$ has a nonempty relatively open Hausdorff subset. Replacing X by $q^{-1}(F)$, it will suffice to show that $G\backslash X$ has a nonempty open Hausdorff subset.

Since R is Polish, it is of second category in itself. Since it is also a F_σ-subset of $X \times X$, there is a closed subset S of R with nonempty interior. Thus there are open sets U and V in X and $(a,b) \in R$ such that

$$(a,b) \in (U \times V) \cap R \subset S \subset R.$$

Since the orbit map $q : X \to G\backslash X$ is open, it will suffice to see that $q(U)$ is Hausdorff. To this end, suppose that $\{q(x_i)\}$ is a sequence converging to both $q(x)$ and $q(y)$ with $x, y \in U$. Since q is open, we can pass to a subsequence, relabel, and assume that there are γ_i and η_i in G such that

$$\gamma_i \cdot x_i \to x \quad \text{and} \quad \eta_i \cdot x_i \to y.$$

Thus we eventually have

$$(\gamma_i \cdot x_i, x_i) \in (U \times U) \cap R \quad \text{and} \quad (\eta_i \cdot x_i, x_i) \in (V \times U) \cap R.$$

Since R is an equivalence relation, we eventually have

$$(\gamma_i \cdot x_i, \eta_i \cdot x_i) \in (U \times V) \cap R \subset S.$$

Since S is closed in $X \times X$, the limit $(x,y) = \lim(\gamma_i \cdot x_i, \eta_i \cdot x_i)$ is in $S \subset R$. This means $q(x) = q(y)$. Hence $q(U)$ is Hausdorff.

(b)$\Longrightarrow$(c): In view of Ex 2.2.7, it will suffice to see that for all x, $q(x)$ is open in $\overline{q(x)}$. But $G\backslash X$ almost Hausdorff implies that $\overline{q(x)}$ has a dense Hausdorff open subset C ([**Wil07**, Lemma 6.3]). Since every neighborhood of every point in $\overline{q(x)}$ must contain $q(x)$, we have $q(x) \in C$. If $q(y) \in C$ and $q(y) \neq q(x)$, then C Hausdorff implies there is an open set U in $\overline{q(x)}$ such that $q(y) \in U$ and $q(x) \notin U$. But then, $q(y) \notin \overline{q(x)}$. This is a contradiction. Hence $C = \{q(x)\}$ is open, and the implication follows.

(c)$\Longrightarrow$(a): Let $q(x)$ and $q(y)$ be distinct points in $G\backslash X$. Suppose that

$$\overline{G \cdot x} \cap G \cdot y = \emptyset. \tag{2.3}$$

Then $U := X \setminus \overline{G \cdot x}$ is an invariant open set disjoint from $G \cdot x$. Hence $q(U)$ is an open neighborhood of $q(y)$ in $G\backslash X$ disjoint from $q(x)$.

Now suppose that (2.3) fails so that $\overline{G \cdot x} \cap G \cdot y \neq \emptyset$. Since both sets are invariant,

$$G \cdot y \subset \overline{G \cdot x}.$$

But if each orbit is locally closed, then there is an open set $U \subset X$ such that

$$G \cdot x = U \cap \overline{G \cdot x}.$$

Then we must have

$$U \cap G \cdot y = \emptyset.$$

Then $V = G \cdot U$ is an open invariant set in X (Ex 2.1.11) and $V \cap G \cdot y = \emptyset$. Thus $q(V)$ is a neighborhood of $q(x)$ disjoint from $q(y)$.

This completes the circle of solid implications in (2.2).

(a)$\Longrightarrow$(d): Let $x \in X$. Let $\{U_n\}$ be a countable collection consisting of a basis for the topology on $G\backslash X$ together with the complement of each basis element. Hence each U_n is either open or closed. Since $G\backslash X$ is T_0, we have

$$(G\backslash X) \setminus q(x) = \bigcup_{q(x) \notin U_n} U_n.$$

Therefore

$$X \setminus G \cdot x = \bigcup_{q(x) \notin U_n} q^{-1}(U_n). \tag{2.4}$$

Note that each $q^{-1}(U_n)$ is either open in X or closed in X. Since open sets in X are F_σ's and since the union in (2.4) is countable, the complement of $G \cdot x$ is a F_σ in X. Hence $G \cdot x$ is a G_δ as required.

(d)$\Longrightarrow$(2): Since any G_δ subset of a locally compact space is Baire [**Wil07**, Lemma 6.4], this implication is immediate.

(2)$\Longrightarrow$(e): We need to see that $V \cdot x$ is open in $G \cdot x$ if V is open in $G_{r_X(x)}$. Clearly we can assume V is open in G. I claim it suffices to see that $W \cdot x$ is a neighborhood of x in $G \cdot x$ for any neighborhood W of $r_X(x)$ in G.

To see that the claim suffices, let V be open in G and let $\sigma \in V$. It will suffice to see that any net $\gamma_i \cdot x \to \sigma \cdot x$ has a subnet eventually in $V \cdot x$. Since $\sigma r_X(x) = \sigma$, and since multiplication is continuous in G, we can find neighborhoods P of σ and N of $r_X(x)$ in G such that $PN \subset V$. Note that $r(\gamma_i) \to r(\sigma)$. Since r is open, we can pass to a subnet, relabel, and assume that there are $\sigma_i \in G$ with $\sigma_i \to \sigma$ and $r(\sigma_i) = r(\gamma_i)$. Then

$$\sigma_i^{-1}\gamma_i \cdot x \to x. \tag{2.5}$$

Since $N \cdot x$ is, by assumption, a neighborhood of x in $G \cdot x$, the left-hand side of (2.5) is eventually in $N \cdot x$. But then

$$\gamma_i \cdot x = \sigma_i(\sigma_i^{-1}\gamma_i) \cdot x$$

is eventually in $PN \cdot x \subset V \cdot x$. This establishes the sufficiency of the claim.

To prove the claim, we need an observation.

LEMMA 2.28. *Suppose that W is open in G and $W \cdot x$ has nonempty interior in $G \cdot x$. Then $W^{-1}W \cdot x$ is a neighborhood of x in $G \cdot x$.*

PROOF. It will suffice to show that any sequence $\gamma_i \cdot x \to x$ has a subsequence which is eventually in $W^{-1}W \cdot x$. Let $\sigma \in W$ be such that $\sigma \cdot x$ is in the interior of $W \cdot x$. As above, we can pass to subsequence, relabel, and find $\eta_i \in G$ with $\eta_i \to \sigma^{-1}$ and $r(\eta_i) = r(\gamma_i)$. Then η_i^{-1} is eventually in W and $\eta_i^{-1}\gamma_i \cdot x \to \sigma \cdot x$ is eventually in $W \cdot x$. Hence

$$\gamma_i \cdot x = \eta_i(\eta_i^{-1}\gamma_i) \cdot x$$

is eventually in $W^{-1}W \cdot x$. This proves the lemma. □

To finish the proof that φ_x is open, we just have to establish the claim; that is, if W is an open neighborhood of $r_X(x)$ in G, then $W \cdot x$ is a neighborhood of x in $G \cdot x$. However, if $\gamma \in G_{r_X(x)}$, then $\gamma^{-1}\gamma = r_X(x)$. Since multiplication is continuous, we can let V_γ be an open neighborhood of γ in G such that

$$V_\gamma^{-1}V_\gamma \subset W.$$

Let N_γ be a compact neighborhood of γ such that $N_\gamma \subset V_\gamma$. Since G, and hence $G_{r_X(x)}$, is second countable, there is a sequence $\{\gamma_n\}$ such that

$$G_{r_X(x)} \subset \bigcup N_{\gamma_n}.$$

Hence

$$G \cdot x = \bigcup N_{\gamma_n} \cdot x.$$

Since $G \cdot x$ is a Baire space, for some n, $N_{\gamma_n} \cdot x$ has interior in $G \cdot x$. Hence $N_{\gamma_n}^{-1}N_{\gamma_n} \cdot x$ is a neighborhood of x in $G \cdot x$ by Lemma 2.28. But

$$N_{\gamma_n}^{-1}N_{\gamma_n} \cdot x \subset V_{\gamma_n}^{-1}V_{\gamma_n} \cdot x \subset W \cdot x.$$

This establishes the claim and hence the proof that φ_x is open.

(e)$\Longrightarrow$(3): Since $G_{r_X(x)}$ is closed in G, it is a second countable, locally compact Hausdorff space. Hence it is a Baire space. Since φ_x is continuous and open, $G \cdot x$ is also a Baire space. More generally, if U is open in X and $U \cap G \cdot x \neq \emptyset$, then $U \cap G \cdot x$ is open and nonempty in $G \cdot x$. Hence it also is the continuous open image of the open nonempty subset $\varphi_x^{-1}(U \cap G \cdot x)$ in $G_{r_X(x)}$. Hence $U \cap G \cdot x$ is a Baire space, and hence of second category in itself. As an open subset of $G \cdot x$, it follows that $U \cap G \cdot x$ is of second category in $G \cdot x$ for all x (Ex 2.2.4).

(3)$\Longrightarrow$(a): Suppose that $G\backslash X$ is not T_0. Then there exist $x, y \in X$ such that $q(x) \neq q(y)$ but $\overline{q(x)} = \overline{q(y)}$. Thus $\overline{G \cdot x} = \overline{G \cdot y}$ (Ex 2.2.7). Let $X = \overline{G \cdot x} = \overline{G \cdot y}$. Since $G \cdot x$ is second category in itself, it is of second category in X (Ex 2.2.6). Since $G \cdot x$ is a F_σ subset of X, it is Borel. Hence the quasi-interior, $(G \cdot x)^*$ in X is open and nonempty such that $(G \cdot x)^* \setminus G \cdot x$ is of first category. But $(G \cdot x)^* \cap G \cdot y$ is a nonempty open subset of $G \cdot y$. Hence it is of second category in $G \cdot y$, and hence in X as well. This is a contradiction and completes the proof that $G\backslash X$ must be T_0. This also completes the proof of the theorem. □

Exercises

Ex 2.2.1. Suppose that X is a left G-space. Let $x \in X$ and $H = \{ \gamma \in G : \gamma \cdot x = x \}$. Show that the natural map of $G_{r_X(x)}$ onto the quotient $G_{r_X(x)}/H$ is open.

—Answer on page 360

Ex 2.2.2. Show that if X is metrizable, then every closed set is a G_δ subset. If in addition X is separable, then every open set is a F_σ subset.

—Answer on page 360

Ex 2.2.3. Show that a topological space X is T_0 if and only if distinct points in X have distinct closures.

—Answer on page 360

Ex 2.2.4. Suppose that U is open in X. Show that if U is of second category in itself, then it is also of second category in X.

—Answer on page 361

Ex 2.2.5. Suppose that D is dense in X. If $A \subset D$ has interior in D, then $\overline{A}$ has interior in X.

—Answer on page 361

Ex 2.2.6. Suppose that $A \subset X \subset Y$ and that A is of second category in X. Then A is also of second category in $\overline{X}$.

—Answer on page 361

Ex 2.2.7. Let X be a G-space and let $q : X \to G\backslash X$ be the orbit map. Suppose that G has an open range map. Show that

$$q^{-1}\big(\overline{q(x)}\big) = \overline{G \cdot x}.$$

(See [**Wil07**, Lemma 6.8].)

—Answer on page 361

2.3. Equivalence

Groupoid equivalence was introduced by Renault in [**Ren82**, §3] as means to show the corresponding groupoid C^*-algebras are Morita equivalent. The current formulation of this theorem and full details came in [**MRW87**]. Currently, the most straightforward proof comes via linking groupoids as introduced in [**SW12**]. We will look at linking groupoids in Section 2.5 and give full statements of the various parts of the Equivalence Theorem(s) in Section 2.6. The proofs are given in Chapter 4 only after exploring some basic consequences of the results and building up a bit more technology.

Since we want to show that equivalent groupoids have Morita equivalent C^*-algebras, we require that both groupoids have Haar systems. Hence in Sections 2.3, 2.5, and 2.6, it will be a standing assumption that all of our groupoids have open range and source maps. Since almost all our applications require separability, we will soon specialize to the case that all topological spaces in sight are second countable. But because I can't help myself, and for future utility, we'll introduce hypotheses as we need them. In keeping with this spirit, we'll take a brief detour in Section 2.4 to explore some notions related to equivalence.

We'll start here with the definition of equivalence, explore some consequences and list some crucial examples.

DEFINITION 2.29. Let G and H be locally compact Hausdorff groupoids with open range maps. We say that a locally compact Hausdorff space Z is a (G, H)-*equivalence* if the following five conditions hold.

(E1) Z is a free and proper left G-space.
(E2) Z is a free and proper right H-space.
(E3) The G- and H-actions commute.
(E4) The moment map $r : Z \to G^{(0)}$ is open and induces a homeomorphism of Z/H onto $G^{(0)}$.
(E5) The moment map $s : Z \to H^{(0)}$ is open and induces a homeomorphism of $G\backslash Z$ onto $H^{(0)}$.

If there is a (G, H)-equivalence, then we say that G and H are *equivalent groupoids*.

REMARK 2.30 (Right-Left Detente). We use both right- and left-actions in the definition of an equivalence just to ease the notational burden.

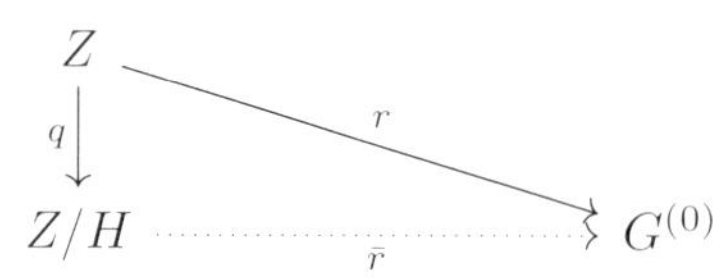

FIGURE 2.3. The Map $\bar{r}$

REMARK 2.31. Let Z be a (G, H)-equivalence as above. Since the actions commute, we must have $r(z) = r(z \cdot \eta)$ for any $\eta \in H^{s(z)}$. In particular, r must factor through a map $\bar{r}$ on Z/H such that the diagram in Figure 2.3 commutes. Since the orbit map is an open surjection, it is easy to see that r is an open map if and only if $\bar{r}$ is an open map. Therefore the hypothesis that the moment maps be open in (E4) and (E5) of Definition 2.29 are technically redundant. But I have included the openness condition to emphasize the necessity (consider Remark 2.2 on page 30).

To see that equivalences arise naturally, we list some of the more important examples. Our first example is a fundamental one—it is used frequently. At the same time it demonstrates that equivalence can be subtle and is very much a topological construct due to the homeomorphism requirement in (E4) and (E5).

EXAMPLE 2.32 (Transitive Groupoids). Let G be a transitive groupoid whose range map is open. Fix $u \in G^{(0)}$ and let $H = G^u_u = G(u)$ be the *isotropy group* at u. Let $Z = G_u$. Then it is not hard to check that Z is a free and proper G space with respect to the moment map $r_Z = r_G|_{G_u}$. Clearly there is a free and proper commuting action of the group H on the right where the moment map is just the constant map onto $\{u\} = H^{(0)}$. It is not hard to see that $\bar{s}$ and $\bar{r}$ are bijections. Hence Z is an equivalence if and only if r_Z is open. However, even though $r : G \to G^{(0)}$ is open (by assumption), its restriction to the closed set G_u need not be—see Ex 2.3.1. However, if G is locally compact Hausdorff *and second countable*, then the Mackey-Glimm-Ramsay Dichotomy (Theorem 2.27 on page 38) comes to the rescue: in this case $G\backslash Z = G\backslash G_u$ is a single point—hence a T_0-space—and $r_z = \varphi_u : G_u \to G^{(0)}$ is open by part (e) of the theorem. Hence we conclude that any *second countable* locally compact Hausdorff transitive groupoid is equivalent to any of its isotropy groups (provided the range map is open).

EXAMPLE 2.33 (Bi-Transformation Groups). Let $\mathcal{G}$ and $\mathcal{H}$ be locally compact groups acting freely and properly on the left and right, respectively, of a locally compact Hausdorff space P such that the actions commute. Then $\mathcal{G}\backslash P$ and $P/\mathcal{H}$ are locally compact Hausdorff spaces. We let $G = G(\mathcal{G}\backslash P, \mathcal{H})$ be the action groupoid for the right $\mathcal{H}$-action on $\mathcal{G}\backslash P$ and let $H = G(\mathcal{G}, P/\mathcal{H})$ be the action groupoid for the left $\mathcal{G}$ action on $P/\mathcal{H}$. It is straightforward to check that P is a (G, H)-equivalence. (The wide applicability of this situation is clearly illustrated in [**Rie82**].)

EXAMPLE 2.34 (Bi-Transformation Groupoids). We can generalize Example 2.33 to the case of commuting free and proper actions of a groupoid G on the left of P and a groupoid H on right. Suppose that G and H have open range maps. Then the orbit spaces $G\backslash P$ and P/H are locally compact Hausdorff by Proposition 2.18 on page 33 and the orbit maps open by Proposition 2.12 on page 31. Let $A(G) = G \ltimes P/H$ be the action groupoid for the left-action of G on P/H, and let $A(H) = G\backslash P \rtimes H$ be the action groupoid for the right-action of H on $G\backslash P$. Then we get a left $A(G)$-action by $(\gamma\cdot p\cdot H, \gamma, p\cdot H)\cdot p = \gamma\cdot p$ and a right $A(H)$-action via $p\cdot(G\cdot p, \eta, G\cdot p\cdot\eta) = p\cdot\eta$. (Note that the moment maps are just the orbit maps.) Then it is straightforward to verify that P is a $(A(G), A(H))$-equivalence.

EXAMPLE 2.35 (Proper Actions). An important special case of both Example 2.33 and Example 2.34 occurs when one of the groupoids is trivial. Then if G acts freely and properly on the right of P, then the action groupoid $G \ltimes P$ is equivalent to the space $G\backslash P$.

EXAMPLE 2.36 (Equivalence Relations). Let X be a locally compact Hausdorff space and $R \subset X \times X$ an equivalence relation such that R is closed in $X \times X$ and hence locally compact. Then is a natural left-action of R on X such that $R\backslash X$ can be identified with the set of orbits. We call R an open equivalence relation if its range map $r : R \to X$ is an open map. Then X implements an equivalence between R and the space $R\backslash X$. (Note that $R\backslash X$ is Hausdorff by Ex 2.1.10.)

EXAMPLE 2.37 (Blowing up the Unit Space). Let G be a locally compact Hausdorff groupoid with open range map. Suppose that Z is locally compact Hausdorff and that $f : Z \to G^{(0)}$ is a continuous *open* map. Then we can form the groupoid

$$G[Z] = \{\, (z, \gamma, w) \in Z \times G \times Z : f(z) = r(\gamma) \text{ and } s(\gamma) = f(w) \,\}.$$

The groupoid operations are the natural ones: $(z, \gamma, w)(w, \eta, x) = (z, \gamma\eta, x)$ and $(z, \gamma, w)^{-1} = (w, \gamma^{-1}, z)$. The unit space $G[Z]^{(0)} = \{\, (z, f(z), z) \,\}$ is naturally identified with Z—hence the terminology; we use f to "blow up" the unit space $G^{(0)}$ up to Z. We get a $(G[Z], Z)$-equivalence

$$W = \{\, (z, \gamma) \in Z \times G : f(z) = r(\gamma) \,\}$$

as follows. The left action of $G[Z]$ is given by $(z, \gamma, w)\cdot(w, \gamma') = (z, \gamma\gamma')$ and the right action of G is given by $(z, \gamma)\cdot\eta = (z, \gamma\eta)$. The openness of the range map of G is required to see that the structure map $r_W : W \to Z$ is open and the openness of f is required to see that $s_W : W \to G^{(0)}$ is open. The remaining details are left as an exercise (Ex 2.3.2).

REMARK 2.38. It turns out that blow-ups as described in Example 2.37 are a more fundamental form of equivalence than one might immediately guess. See Theorem 2.52 on page 49 and Proposition 2.64 on page 55.

A subset $A \subset G^{(0)}$ is called *full* if $G \cdot A = G^{(0)}$ for the usual action of G on $G^{(0)}$ (Example 2.6 on page 30). Alternatively, A is full if and only if it meets every G-orbit in $G^{(0)}$.

EXAMPLE 2.39 (Linking Groupoids). Suppose that G is a locally compact Hausdorff groupoid with open range and source maps, and let A and B be *open* full subsets of $G^{(0)}$. Then $Z = G_B^A$ is an equivalence between the reductions $G(A) = G|_A$ and $G(B) = G|_B$. Two special cases are of interest. The first is when $B = G^{(0)}$. A more important one for us is the case where $G^{(0)}$ is the disjoint union of A and B (so that A and B are clopen), then we call G a *linking groupoid* for $G(A)$ and $G(B)$. (See Ex 2.3.5.)

EXAMPLE 2.40 (Transversals). Let A be a locally compact full subset of $G^{(0)}$. (In the previous example, we assumed A was open.) For example, A could be a *transversal* that meets each G-orbit in $G^{(0)}$ in precisely one point. Then, as in Example 2.39, G_A will be an equivalence between G and $G(A)$ when the restrictions of r and s to G_A remain open. This last requirement is automatic if A is open.

A cursory glance at the above examples does not suggest much of a relationship between the unit spaces of equivalent groupoids. Nevertheless, there is an important restriction on the orbit spaces. If Z is a (G, H)-equivalence, then the homeomorphism $\bar{r} : Z/H \to G^{(0)}$ is G-equivariant. Since we're assuming G has an open range map, the orbit map is open and we get a homeomorphism $\underline{r} : G\backslash(Z/H) \to G\backslash G^{(0)}$ given by $\underline{r}(G \cdot (z \cdot H)) = G \cdot r(z)$. Of course, we also get a homeomorphism $\underline{s} : (G\backslash Z)/H \to H^{(0)}/H$ given by $\underline{s}((G \cdot z) \cdot H) = s(z) \cdot H$. Since we can identify $G\backslash(Z/H)$ with $(G\backslash Z)/H$ via $G \cdot (z \cdot H) \mapsto (G \cdot z) \cdot H$, we have proved the following.

LEMMA 2.41. *Let Z be a (G, H)-equivalence, then the map $G \cdot r(z) \mapsto s(z) \cdot H$ is well-defined and defines a homeomorphism of the orbit spaces $G\backslash G^{(0)}$ and $H^{(0)}/H$.*

Our next result is a critical technical lemma. First, notice that if Z is a (G, H)-equivalence, then $Z *_s Z = \{ (x, y) : s(x) = s(y) \}$ is a closed, hence locally compact, subspace of the product. Moreover, it is a free and proper right H-space for the diagonal action. Similarly, $Z *_r Z = \{ (x, y) : r(x) = r(y) \}$ is a free and proper left G-space.

LEMMA 2.42. *Let Z be a (G, H)-equivalence.*

(a) *Given $(x, y) \in Z *_s Z$ there is a unique $\tau_G(x, y) \in G$ such that $\tau_G(x, y) \cdot y = x$. The map $(x, y) \mapsto \tau_G(x, y)$ is a continuous open map of $Z *_s Z$ onto G which factors through a homeomorphism of the orbit space $(Z *_s Z)/H$ onto G.*

(b) *Given $(x, y) \in Z *_r Z$ there is a unique $\tau_H(x, y) \in H$ such that $x \cdot \tau_H(x, y) = y$. The map $(x, y) \mapsto \tau_H(x, y)$ is a continuous open map of $Z *_r Z$ onto H which factors through a homeomorphism of $G\backslash(Z *_r Z)$ onto H.*

PROOF. (a) Clearly, Z is a free G-space by (E1). In view of (E5), $Z *_s Z = R(G,Z) = \{ (z,w) : G \cdot z = G \cdot w \}$ as in (2.1). Hence τ_G is just the translation map for (G,Z). In particular, τ_G is continuous and open by Proposition 2.24 on page 34.

Since G acts freely, τ_G is H-invariant. If $\tau_G(x,y) = \tau_G(x',y')$, then $s_Z(y) = s_Z(y')$ and $y' = y \cdot \beta$ for some $\beta \in H$ in view of (E5). Hence $x' = \tau_G(x',y') \cdot y' = \tau_G(x,y) \cdot (y \cdot \beta) = x \cdot \beta$. Hence τ_G induces a bijection of the orbit space with H which is a homeomorphism since τ_G is open.

(b) The proof is similar and left as an exercise. □

REMARK 2.43. Note that the first part of Lemma 2.42 imposes a locally compact Hausdorff groupoid structure on the orbit space $H_Z := (Z *_s Z)/H$ simply by pulling back the one on G. Let $[z,w]$ be the orbit of (z,w) in H_Z. Since $r(\tau_G(z,w)) = r(z)$ and $s(\tau_G(z,w)) = r(w)$, we have $[x,y]$ and $[z,w]$ composable when $r(y) = r(z)$. Since Z is an equivalence, this means $w = y \cdot \eta$ for some $\eta \in H$. Then, since $\tau_G(x,y)\tau_G(y, w\cdot\eta^{-1}) = \tau_G(x, w\cdot\eta^{-1})$, we have $[x,y][z,w] = [x, w\cdot\eta^{-1}]$. Alternatively, $[x,y][z,w] = [x,y][y, w \cdot \eta^{-1}] = [x, w \cdot \eta^{-1}]$. Clearly, $[x,y]^{-1} = [y,x]$. We can identify the unit space of H_Z with Z/H. Then $r([z,w]) = z \cdot H$ while $s([z,w]) = w \cdot H$.

Similarly, the second part of Lemma 2.42 on the facing page imposes a similar groupoid structure on $G^Z := G\backslash(Z *_r Z)$. Hence, more or less by construction, we have established the following result which will be important in the sequel.

LEMMA 2.44. *Suppose that Z is a (G,H) equivalence. Let H_Z and G^Z be the groupoids introduced above. Then τ_G implements a topological groupoid isomorphism of H_Z with G, and τ_H implements a topological groupoid isomorphism of G^Z with H.*

It is very useful to be able to "reverse engineer" the construction of H_Z or G^Z. Specifically, suppose that X is a free and proper right H-space. We need to add the hypothesis that H has open range and source maps so that H-orbit maps are open. Since $X *_s X$ is also a free and proper right H-space for the diagonal action, the orbit space $H_X = (X *_s X)/H$ is a locally compact Hausdorff space. If we let $[x,y]$ denote the orbit of (x,y) in H_X, then we get well-defined maps $r : H_X \to X/H$ and $s : H_X \to X/H$ given by $r([x,y]) = x \cdot H$ and $s([x,y]) = y \cdot H$. (Note that X/H is also locally compact Hausdorff.) We declare $[x,y]$ and $[w,z]$ to be composible if $y \cdot H = w \cdot H$. Then if $w = h \cdot \beta$, we define $[x,y][y \cdot \beta, z] = [x, z \cdot \beta^{-1}]$. It is common practice to summarize this composition as simply $[x,y][y,z] = [x,z]$. Similarly, we let $[x,y]^{-1} = [y,x]$. It is straightforward to see that H_X is a groupoid and that we can identify the unit space of H_X with the orbit space X/H.

LEMMA 2.45. *Let X be a free and proper right H-space. Assume the range map on H is open. Then the orbit space $H_X = (X *_s X)/H$ is a locally compact Hausdorff groupoid with the operations $[x,y][y,z] = [x,z]$ and $[x,y]^{-1} = [y,x]$ as defined above. The unit space of H_X can be identified with X/H. We call H_X the* imprimitivity groupoid *of X.*

PROOF. We've already observed that H_X locally compact Hausdorff, and checking that it is a groupoid is not hard. We need to see that the groupoid operations are continuous. Suppose that $\big([x_i,y_i],[y_i',z_i']\big) \to \big([x,y],[y',z']\big)$ in $H_X^{(2)}$. We can choose representatives z_i and z so that $[y_i',z_i'] = [y_i,z_i]$ and $[y',z'] = [y,z]$. To see that multiplication is continuous, it will suffice to see that every subnet of $\{[x_i, z_i]\}$ has a subnet converging to $[x, z]$. Assume that the have already replaced $\{[x_i, z_i]\}$ with a subnet. Since the orbit map is open, we can pass to another subnet and relabel so that there are β_i such that $(x_i \cdot \beta_i, y_i \cdot \beta_i) \to (x, y)$. There is no harm in replacing $(x \cdot \beta_i, y_i \cdot \beta_i)$ with (x_i, y_i). Similarly, we can pass to another subnet, relabel, so that there are η_i so that $(y_i \cdot \eta_i, z_i \cdot \eta_i) \to (y, z)$. Since the H-action is free and proper and since we have both $y_i \to y$ and $y_i \cdot \eta_i \to y$, we can assume that $\eta_i \to s(y) = s(z)$. This means $z_i \to z$. Then $(x_i, z_i) \to (x, z)$.

Showing that inversion is continuous is similar (but easier). □

REMARK 2.46. There is nothing sacred about right actions. If X is a free and proper left G-space and G has open range and source maps, then we can make the orbit space $G\backslash(X *_r X)$ into a groupoid in a manner similar to the above. The resulting groupoid G^X is also called an *imprimitivity groupoid*.

PROPOSITION 2.47. *Let X be a free and proper right H-space. We assume that H has an open range map and* that moment map s_X is open. *Then X becomes a (H_X, H)-equivalence for action of H_X on X: $[x, y] \cdot y = x$.*

REMARK 2.48 (Hidden Hypotheses). Take care to notice that the openness of the moment map s_X was slipped into the hypotheses of Proposition 2.47. This is an excellent example where hypotheses can become buried. This result was first described algebraically but not proved in [**MRW87**]. We gave a proof in [**MW95**, Theorem 3.5]. But in both those papers and in many subsequent papers, groupoids were assumed to have open range maps and *actions were assumed to have open moment maps*. Thus if one wants to work in greater generality, then care must be taken in quoting older results.

REMARK 2.49 (The Left Action). Just as in the statement of Lemma 2.45, the expression $[x, y] \cdot y = x$ is the "colloquial version". Technically, the moment map on X is given by $r_X(x) = x \cdot H$. Then if $z \cdot H = y \cdot H$, we have $z = y \cdot \beta$ and

$$[x, y] \cdot z = [x, y] \cdot y \cdot \beta = [x \cdot \beta, y \cdot \beta] \cdot y \cdot \beta = x \cdot \beta.$$

PROOF OF PROPOSITION 2.47. If $[x, y] \cdot y = y$, then $[x, y] = [y, y]$ which is $s_{H_X}([x, y])$; hence the H_X-action is free. To see that the action is proper, suppose that $y_i \to y$ in X and that $[x_i, z_i] \cdot y_i \to x$ in X. Then $z_i = y_i \cdot \beta_i$ and $[x_i, z_i] \cdot y_i = x_i \cdot \beta_i^{-1}$. Hence $x_i \cdot \beta_i^{-1} \to x$. But the $[x_i, z_i] = [x_i \cdot \beta_i^{-1}, y_i] \to [x, y]$. Therefore the action is proper.

If $r_X(x) = r_X(y)$, then $x \cdot H = y \cdot H$ and $x = y \cdot \beta$ for some $\beta \in H$. Hence r_X induces a bijection of X with X/H and r_X is open as it is an orbit map and H is assumed to have an open range map.

If $s_X(x) = s_X(y)$, then $[x, y] \in H_X$ and $[x, y] \cdot y = x$ so s_X induces a bijection on $H_X\backslash X$ onto $H^{(0)}$. We had to assume that s_X was open, so X is an equivalence as claimed. □

COROLLARY 2.50. *Suppose that H is closed subgroupoid of G such that $H^{(0)} = G^{(0)}$. (We call H a* wide subgroupoid.*) If the range map on H is open, then H is equivalent to the action groupoid $G \rtimes G/H$.*

PROOF. The previous proposition implies that H is equivalent to the imprimitivity groupoid $H_X = (G *_s G)/H$. But $[\gamma, \eta] \mapsto (\gamma \cdot H, \gamma\eta^{-1}, \eta \cdot H)$ is a groupoid isomorphism of H_X onto $G \rtimes G/H$. □

PROPOSITION 2.51. *Groupoid equivalence is an equivalence relation.*[5]

PROOF. Since G is naturally a (G,G)-equivalence, equivalence is reflexive. If Z is a (G,H) equivalence, then we can form the bimodule $Z^{\mathrm{op}} = \{\, \bar{z} : z \in Z \,\}$ where we declare $z \mapsto \bar{z}$ to be a homeomorphism, $r_{Z^{\mathrm{op}}}(\bar{z}) = s_Z(z)$, $s_{Z^{\mathrm{op}}}(\bar{z}) = r_Z(z)$, $\beta \cdot \bar{z} = \overline{z \cdot \beta^{-1}}$ and $\bar{z} \cdot \gamma = \overline{\gamma^{-1} \cdot z}$. Then Z^{op} is a (H,G)-equivalence. Hence equivalence is symmetric. To see that it is transitive, let Z be a (G,H)-equivalence and Y and (H,P)-equivalence. Let $Z * Y = \{\, (z,y) : s_Z(z) = r_Y(y) \,\}$. Then H acts freely and properly on $Z * Y$: $(z,y) \cdot \beta = (z \cdot \beta, \beta^{-1} \cdot y)$. Then you can check that $(Z*Y)/H$ is a (G,P)-equivalence (Ex 2.3.6). □

The following characterization of equivalence was pointed out to me by Jean Renault and Paul Muhly. No doubt it appears explicitly in the literature, but I am unable to find a reference. We say that $G[Z]$ is a blow-up of G if there is an open, continuous map $f : Z \to G^{(0)}$ and $G[Z]$ is the groupoid equivalent to G constructed in Example 2.37 on page 45. Of course, we say that two locally compact groupoids G and H are isomorphic if there is a groupoid isomorphism $\varphi : G \to H$ which is also a homeomorphism. Hence isomorphic groupoids are certainly equivalent.

THEOREM 2.52. *Two groupoids G and H with open range and source maps are equivalent if and only if they have isomorphic blow-ups.*

PROOF. If G and H have isomorphic blow-ups, then they are equivalent by Proposition 2.51.

If G and H are equivalent, let Z be a (G,H)-equivalence. Since the moment maps $r_Z : Z \to G^{(0)}$ and $s_Z : Z \to H^{(0)}$ are open, we can form the blow-ups

$$G[Z] = \{\, (z,\gamma,w) \in Z \times G \times Z : r_Z(z) = r(\gamma) \text{ and } r_Z(w) = s(\gamma) \,\}$$

and

$$H[Z] = \{\, (z,\eta,w) \in Z \times H \times Z : s_Z(z) = r(\eta) \text{ and } s_Z(w) = s(\eta) \,\}.$$

By Lemma 2.42 on page 46, there is a continuous open map $\tau_H : Z *_r Z \to H$ such that $z \cdot \tau_H(z,w) = w$. Then we get a continuous map

$$\varphi : G[Z] \to H[Z]$$

given by

$$\varphi((z,\gamma,w)) = (z, \tau_H(z, \gamma \cdot w), w).$$

Since

$$z \cdot \tau_H(z,\gamma \cdot w)\tau_H(w,\gamma' \cdot y) = \gamma \cdot w \cdot \tau_H(w, \gamma' \cdot y) = \gamma\gamma' \cdot y,$$

it follows that $\tau_H(z,\gamma \cdot w)\tau_H(w,\gamma' \cdot y) = \tau_H(z, \gamma\gamma' \cdot y)$, and that φ is a continuous groupoid homomorphism. Moreover,

$$\varphi^{-1}((z,\eta,w)) = (z, \tau_G(z \cdot \eta, w), w).$$

Hence φ is the required isomorphism. □

The following proposition is in response to a seemingly innocent question of Scott LaLonde's. It would be nice to find a simplier proof than the one supplied here, but I haven't found it.

[5]Of course the collection of all groupoids is not a set, so it doesn't quite make sense to have an equivalence relation on the class of groupoids. So the Proposition should be interpreted as asserting that equivalence is reflexive, symmetric and transitive in the obvious senses.

PROPOSITION 2.53. *Suppose that G and H are second countable locally compact Hausdorff groupoids and that Z is a (G,H)-equivalence. Then Z is second countable.*

We obtain Proposition 2.53 as a corollary of the following.

THEOREM 2.54. *Suppose that G is a second countable, locally compact Hausdorff groupoid acting freely and properly on a locally compact space P such that the orbit space $G\backslash P$ is second countable. Then P is second countable.*

We need some preliminaries. We'll write $q : P \to G\backslash P$ for the orbit map.

LEMMA 2.55. *If P is a free and proper G-space with G and $G\backslash P$ second countable, then P is σ-compact. If fact, if $U \subset G^{(0)}$ is open, then $r_P^{-1}(U)$ is σ-compact.*

PROOF. It suffices to prove the second statement. Since $G\backslash P$ is second countable and locally compact, it is σ-compact. Thus $G\backslash P$ is the countable union $\bigcup C_n$ with each C_n compact. Since the orbit map q is continuous and open, there is a compact set $K_n \subset P$ such that $q(K_n) = C_n$. Since G is σ-compact, so is $r_G^{-1}(U) \subset G$. Hence,

$$r_P^{-1}(U) = \bigcup r_G^{-1}(U) \cdot K_n$$

is σ-compact as claimed. □

The next lemma is key. It strongly uses the properness and the freeness of of the action. Recall that if $K \subset P$, then we let

$$P(K,K) := \{\, \gamma \in G : \gamma \cdot K \cap K \neq \emptyset \,\}.$$

LEMMA 2.56. *Given $p \in P$ and a neighborhood V of $r(p)$ in G, there is precomapct open neighborhood W of p such that*

$$P(\overline{W}, \overline{W}) \subset V.$$

PROOF. If no such neighborhood exists, then for every pre-compact open neighborhood W of p there is $\gamma_W \notin V$ and $p_W \in \overline{W}$ such that $\gamma_W \cdot p_W \in \overline{W}$. Clearly both the nets $\{p_W\}$ and $\{\gamma_W \cdot p_W\}$ converge to p. If K is a compact neighborhood of p, then eventually $\{\gamma_W\} \subset P(K,K)$. Since the action is proper, $P(K,K)$ is compact. Thus $\{\gamma_W\}$ has a subnet converging to γ. But $\gamma \cdot p = p$ forces $\gamma = r(p)$. Thus every subnet of $\{\gamma_W\}$ has a subnet converging to $r(p)$. Hence $\gamma_W \to r(p)$. Of course, that leads to a contradiction as any limit of $\{\gamma_W\}$ must lie in the (closed) complement of V. □

If X is a second countable, locally compact Hausdorff space, then we'll call a countable neighborhood basis $\{B_n\}_{n=1}^{\infty}$ at x a *regular neighborhood basis* if each B_n is open and precompact with $\overline{B_{n+1}} \subset B_n$. Of course, if β is a basis for the topology on X, then β contains a regular neighborhood basis of each point in x.

PROOF OF THEOREM 2.54. Let $\mathcal{E}$ be a countable collection of pre-compact open sets in G consisting of those elements of a fixed basis for the topology on G that have nontrivial intersection with $G^{(0)}$. Then $\mathcal{E}$ contains a regular neighborhood basis in G of every $u \in G^{(0)}$. If $V \in \mathcal{E}$ and $p \in r_P^{-1}(V \cap G^{(0)})$, then Lemma 2.56 implies there is a pre-compact open neighborhood W of p such that

$P(\overline{W},\overline{W}) \subset V$. Lemma 2.55 implies that $r_P^{-1}(V \cap G^{(0)})$ is σ-compact (hence Lindelöf).[6] Thus there is a countable collection $\mathcal{D}_V$ of pre-compact open sets W that cover $r_P^{-1}(V \cap G^{(0)})$ with the property that $P(\overline{W},\overline{W}) \subset V$. Since $\mathcal{E}$ is countable, there is a countable collection $\mathcal{D}$ of pre-compact open sets W in P such that given $V \in \mathcal{E}$ and $p \in r_P^{-1}(V \cap G^{(0)})$ there is a $W \in \mathcal{D}$ such that $p \in W$ and $(\overline{W},\overline{W}) \subset V$. We can even assume that $\mathcal{D}$ is closed under finite intersections.

Let β be a countable basis for $G\backslash P$. We aim to show that the countable collection of open sets

$$U = q^{-1}(B) \cap W$$

with $B \in \beta$ and $W \in \mathcal{D}$ is a basis for the topology on P.

To this end, let O be an open neighborhood of p in P. Let $\{V_n\} \subset \mathcal{E}$ be a regular neighborhood basis of $r(p)$ in G, and let $\{B_n\}$ be a regular neighborhood basis of $q(p)$ in $G\backslash P$. Let $W_n \in \mathcal{D}$ such that $p \in W_n$ and $P(W_n, W_n) \subset V_n$. Using finite intersections, we can assume that $W_{n+1} \subset W_n$.

Now let

$$U_n := q^{-1}(B_n) \cap W_n.$$

Note that $p \in U_{n+1} \subset U_n$. It will suffice to see that there is a $n \in \mathbf{N}$ such that $p \in U_n \subset O$. If not, then for each $n \in \mathbf{N}$ there is

$$p_n \in U_n \setminus O.$$

Then $\{p_n\}$ is in the compact set $\overline{W_1}$. Then there must be a convergent sub*net*,[7] say $\{p_{n_i}\}_{i \in I}$ converging to $p' \in P \setminus O$. Note that given $n \in \mathbf{N}$, there is an $i_0 \in I$ such that $i \geq i_0$ implies $n_i \geq n$. Hence $q(p') \in \overline{B_n}$ for all n. That forces $q(p') = q(p)$. Therefore $p' = \gamma \cdot p$ for some $\gamma \neq r(p)$ (since $p' \neq p$). But then there is a m such that $\gamma \notin V_m$. But as above, then p_{n_i} are eventually in W_m. Hence $p' \in \overline{W_m}$. But then $\gamma \in (\overline{W_m}, \overline{W_m}) \subset V_m$ which is a contradiction. □

Exercises

Ex 2.3.1. Let $G = G(\mathbf{R}_d, \mathbf{R})$ be the action groupoid for the action of the reals with the discrete topology, $\mathbf{R}_d$, acting on the reals with the usual topology, $\mathbf{R}$, by translation. Then G is a transitive locally compact Hausdorff groupoid with trivial isotropy. Show that the argument given in Example 2.32 showing that G is equivalent to the trivial group breaks down.

—Answer on page 361

Ex 2.3.2. Complete the proof that $G[Z]$ is equivalent to G via W as in Example 2.37.

—Answer on page 361

Ex 2.3.3. Provide the details to verify that P is a (G, H)-equivalence in Example 2.33.

—Answer on page 361

---

[6] Recall that a space is called Lindelöf if every open cover has a countable subcover. Clearly, σ-countable spaces are Lindelöf spaces.

[7] At this point, we haven't even establishd that P is first countable—hence we need to allow nets here.

Ex 2.3.4. Provide the details to verify that X is a $(R, R\backslash X)$-equivalence in Example 2.36.

—Answer on page 362

Ex 2.3.5. Provide the details to verify that $Z = G_B^A$ is a $(G(A), G(B))$-equivalence in Example 2.39.

—Answer on page 362

Ex 2.3.6. Show that $(Z * Y)/H$ is a (G, P)-equivalence in Proposition 2.51.

—Answer on page 362

Ex 2.3.7. Suppose that Z is a (G, H)-equivalence and that $z \in Z$. Then the isotropy groups $G(r(z))$ and $H(s(z))$ are isomorphic topological groups.

—Answer on page 363

Ex 2.3.8. Suppose that G and H are equivalent groupoids. Show that if H has open range and source maps, then so does G.

—Answer on page 363

2.4. Generalized Morphisms

This section is an aside to illustrate some interesting properties that arise when we try to find a notion of morphisms for groupoids so that groupoid equivalence becomes an isomorphism. The exposition here is based on conversations with Erik van Erp as well as [**LGTX07**, Lemma 2.10].

Definition 2.57. Suppose that G and H are locally compact Hausdorff groupoids. Then a *generalized morphism* from G to H is a locally compact Hausdorff space Z such that

(M1) Z is a free and proper right H-space.

(M2) Z is a left G-space.

(M3) The G- and H-actions commute.

(M4) The momment map $r : Z \to G^{(0)}$ is open and factors through a homeomorphism of Z/H with $G^{(0)}$.

Remark 2.58. In Definition 2.57, we do not assume that the moment map $s : Z \to H^{(0)}$ for the right H-action is open.

Example 2.59 (Strict Homomorphisms). Let $\varphi : G \to H$ be a continuous groupoid homomorphism. Then the *graph* of φ is the space

$$\mathrm{Gr}(\varphi) = \{ (u, \eta) \in G^{(0)} \times H : \varphi(u) = s(\eta) \}$$

together with the left action of G given by $\gamma \cdot (s(\gamma), \eta) = \big(r(\gamma), \varphi(\gamma)\eta\big)$ and the right H-action given by $(u, \eta) \cdot \eta' = (u, \eta\eta')$. Then if H has open range and source maps, $\mathrm{Gr}(\varphi)$ is a generalized morphism from G to H. A generalized morphism coming from a bona fide homomorphism will be called a *strict homomorphism*.

Example 2.60 (Equivalences). Clearly, a (G, H)-equivalence Z is a generalized morphism from G to H. Note that the opposite module Z^{op}, as in the proof of Proposition 2.51 on page 49, is a generalized morphism from H to G.

We compose generalized morphisms as follows.[8] Let Z be a generalized morphism from G to H and W be a generalized morphism from H to K. Let $Z *_H W$ be the orbit space $(Z*W)/H$ where H acts on $Z*W = \{ (z,w) \in Z \times W : s(z) = r(w) \}$ by $(z,w)\cdot h = (z\cdot h, h^{-1}\cdot w)$. Since H acts freely and properly on the right of Z, H acts freely and properly on the right of $Z * W$. Then, assuming H has open range and source maps, $Z *_H W$ is a locally compact Hausdorff space by Proposition 2.18 on page 33. Let $[z,w]$ denote the orbit of (z,w) in $Z *_H W$. It is not hard to see that we get well-defined commuting G- and K-actions by

$$\gamma \cdot [z,w] = [\gamma \cdot z, w] \quad \text{and} \quad [z,w]\cdot \zeta = [z, w\cdot \zeta].$$

LEMMA 2.61. *Suppose that Z is a generalized morphism from G to H and that W is a generalized morphism from H to K. If H has open range and source maps, then $Z *_H W$ as described above is a generalized morphism from G to K.*

PROOF. To see that the K-action is free, assume $[z, w\cdot\zeta] = [z,w]$. Then there is an η such that $z = z\cdot \eta$ and $\eta^{-1}\cdot w \cdot \zeta = w$. Since the H-action on Z is free, the first equation implies $\eta = s(z)$. But then since the K-action on W is free, we must have $\zeta = s(w)$ as required.

To see that the K-action is proper, we invoke (PA3) of Proposition 2.17 on page 32. Thus suppose that $[z_i, w_i] \to [z,w]$ and $[z_i, w_i\cdot\zeta_i] \to [z', w']$. Since the orbit map from $Z*W$ to $Z *_H W$ is open, we can pass to a subnet and assume that $(z_i, w_i) \to (z,w)$ and that there are $\eta_i \in H$ such that $(z_i \cdot \eta_i, \eta_i^{-1}\cdot w_i \cdot \zeta_i) \to (z', w')$. Since the right H-action on Z is proper, we can pass to a subnet, relabel, and assume that $\eta_i \to \eta$ in H. But then $w_i \to w$ while $w_i \cdot \zeta_i \to \eta^{-1}w'$. Since the right K-action on W is proper, this implies $\{\zeta_i\}$ has a convergent subnet. Thus that K-action on $Z *_H W$ is proper. Since (M2) and (M3) are immediate, it only mains to establish (M4).

Note that the moment map for the left G-action on $Z *_H W$ is given by $r([z,w]) = r(z)$. Thus if $r([z,w]) = r([z',w'])$, then $r(z) = r(z')$. Since Z is a generalized morphism, there is a $\eta \in H$ such that $z' = z\cdot\eta$. Then $s(\eta) = r(w')$ and $r(\eta\cdot w') = r(\eta) = s(z) = r(w)$. Since W is a generalized morphism, $\eta \cdot w' = w\cdot\zeta$ for some $\zeta \in K$. Then

$$[z', w'] = [z\cdot\eta, w'] = [z, \eta\cdot w'] = [z, w\cdot\zeta] = [z,w]\cdot\zeta.$$

Thus $r : Z *_H W \to G^{(0)}$ factors through $(Z *_H W)/K$.

We still need to see that r is open. We will use Fell's criterion. Suppose that $u_i \to r(z) = r([z,w])$. Since $r : Z \to G^{(0)}$ is open, we can pass to a subnet, relabel, and assume that there are $z_i \to z$ such that $r(z_i) = u_i$. Then $s(z_i) \to s(z) = r(w)$. Since $r : W \to H^{(0)}$ is open, we can pass to a subnet, relabel, and assume there are $w_i \to w$ such that $r(w_i) = s(z_i)$. But then $[z_i, w_i] \to [z,w]$. This completes the proof. □

We say that two generalized morphisms Z and W from G to H are *isomorphic* if there is a homeomorphism $\varphi : Z \to W$ intertwining the G- and H-actions.

EXAMPLE 2.62 (The Identity Morphism). Suppose that G has open range and source maps. Then we can view G as a generalized morphism from G to itself. If Z is a generalized morphism from G to H, then $[\gamma, z] \mapsto g\cdot z$ is an isomorphism of $G *_G Z$ and Z. Similarly, $Z *_H H$ and Z are isomorphic.

[8]Compare this construction with the proof of Proposition 2.51 on page 49.

Although we are not going to pursue the details here, we can think isomorphism classes, $[Z]$, of generalized morphisms as morphisms in a category in which the objects are locally compact groupoids with open range and source maps, and the identity maps are given by viewing groupoids and generalized morphisms mapping themselves to themselves as in Example 2.62 on the preceding page. Then a generalized morphism Z from G to H would be invertible in this category exactly when there is a generalized morphism W from H to G such that $[Z *_H W] = [G]$ and $[W *_G Z] = [H]$. For example, suppose that Z is a (G, H)-equivalence—and hence a generalized morphism from G to H. Then Z^{op} is a generalized equivalence from H to G. The map $[z, \overline{w}] \mapsto \tau_G(z, w)$ implements an isomorphism of $Z *_H Z^{\mathrm{op}}$ onto G. (See the proof of Lemma 2.44 on page 47.) Similarly, $Z^{\mathrm{op}} *_G Z$ is isomorphic to H. Hence Z is invertible in this sense. In fact, the converse is true.

PROPOSITION 2.63. *Suppose that Z is a generalized morphism between G and H and that both G and H have open range and source maps. Then Z is invertible—in the sense described above—if and only if Z is a (G, H)-equivalence.*

PROOF. At this point, we just need to see that an invertible generalized morphism Z from G to H is a (G, H)-equivalence. Comparing the definitions, it will suffice to see that the G-action is free and proper and that $s : Z \to H^{(0)}$ is open and factors through a homeomorphism of $G\backslash Z$ with $H^{(0)}$ (see Remark 2.31 on page 44). Let W be a generalized morphism from H to G such that $Z *_H W$ is isomorphic to G and $W *_G Z$ is isomorphic to H. The first assertion implies that the G-action is free. To see that the action is proper, we apply (PA3) of Proposition 2.17 on page 32. Suppose that $z_i \to z$ in Z while $\gamma_i \cdot z_i \to z'$. Let $w \in W$ be such that $r(w) = s(z)$. Since $r : W \to H^{(0)}$ is open, we can pass to a subnet, relabel, and assume that there $w_i \to w$ in W such that $r(w_i) = s(z_i)$. Then $[z_i, w_i] \to [z, w]$ while $\gamma_i \cdot [z_i, w_i] \to [z', w]$ Since $Z *_H W$ is isomorphic to G and the G-action on itself is proper, $\{\gamma_i\}$ must have a convergent subnet. Thus G acts freely and properly on Z.

To see that $r : Z \to G^{(0)}$ is open, we use Fell's criterion. Suppose that $u_i \to r(z)$. Since the range map on G is open, we can pass to a subnet, relabel, and assume there are $\gamma_i \to \gamma \in G^{r(z)}$ such that $r(\gamma_i) = u_i$. Since $Z *_H W$ is isomorphic to G, and $[z, \gamma^{-1} \cdot z]$ is mapped to γ, there are $[z_i, w_i]$ in $Z *_H W$ converging to $[z, \gamma^{-1} \cdot z]$ with $r(z_i) = r(\gamma_i) = u_i$. Since H has open range and source maps, so that that the orbit map from $Z * W$ to $Z *_H W$ is open, we can pass to a subnet, relabel, and assume there are $\eta_i \in H$ such that $(z_i \cdot \eta_i, \eta_i^{-1} \cdot w_i) \to (z, \gamma^{-1} \cdot z)$. Then $r(z_i \cdot \eta_i) = r(z_i) = u_i$ and $z_i \cdot \eta_i \to z$. Hence r is open.

It only remains to check that s induces a bijection of $G\backslash Z$ onto $H^{(0)}$. Hence it will suffice to check that $s(z) = s(z')$ forces z and z' to be in the same G-orbit. Let $\theta : W *_G Z \to H$ be an equivariant isomorphism. Let $w, w' \in W$ be such that $s(w) = r(z)$ and $s(w') = r(z')$. Let $\theta([w, z]) = \eta$ and $\theta([w', z']) = \eta'$. Note that equivariance requires that $s(\eta) = s(z)$, $r(\eta) = r(w)$, $s(\eta') = s(z')$, and $r(\eta') = r(w')$. Then

$$\theta([\eta^{-1} \cdot w, z]) = \eta^{-1}\eta = s(\eta) = s(z) = s(z') = s(\eta') = \theta([(\eta')^{-1} \cdot w', z']).$$

Since θ is injective,

$$[\eta^{-1} \cdot w, z] = [(\eta')^{-1} \cdot w', z'],$$

and there is a $\gamma \in G$ such that $z' = \gamma \cdot z$. This completes the proof. □

We want to finish this section by observing that every generalized morphism is isomorphic to the composition of a blow-up with a strict homomorphism. To that end, let Z be a generalized morphism from G to H. We assume that G and H have open range and source maps. Since $r : Z \to G^{(0)}$ is open, we can form the blow-up

$$G[Z] = \{ (z, \gamma, w) \in Z \times G \times Z : r(z) = r(\gamma) \text{ and } r(w) = s(\gamma) \}.$$

Recall that $G[Z]$ is equivalent to G via the bimodule

$$W = \{ (z, \gamma) \in Z \times G : r(z) = r(\gamma) \}.$$

Note that W is just the graph $\mathrm{Gr}(\varphi)$ where $\varphi : G[Z] \to G$ is the obvious map: $\varphi((z, \gamma, w)) = \gamma$. As in the proof of Lemma 2.42 on page 46, we can use Proposition 2.24 on page 34 to see that we have a continuous map $\tau_H : Z *_s Z \to H$ such that $z \cdot \tau_H(z, w) = w$.

Now almost exactly as in the proof of Theorem 2.52 on page 49, we get a continuous groupoid homomorphism $\psi : G[Z] \to H$ given by

$$\psi((z, \gamma, w)) = \tau_H(z, \gamma \cdot w).$$

Since $\tau_H(z, z) = s(z)$, we can write

$$\mathrm{Gr}(\psi) = \{ (z, \eta) \in Z \times H : s(z) = r(\eta) \}.$$

PROPOSITION 2.64. *Suppose that Z is a generalized morphism from G to H. Let $G[Z]$, φ and ψ be as above. Then Z is isomorphic to*

$$\mathrm{Gr}(\varphi)^{\mathrm{op}} *_{G[Z]} \mathrm{Gr}(\psi).$$

Thus, in the language of this section, Z factors as the composition of a blow-up and a strict homomorphism.

PROOF. Since $\mathrm{Gr}(\varphi)$ is an equivalence, and hence invertible with inverse $\mathrm{Gr}(\varphi)^{\mathrm{op}}$, it will suffice to see that $\mathrm{Gr}(\varphi) *_G Z$ and $\mathrm{Gr}(\psi)$ are isomorphic as generalized morphisms from $G[Z]$ to H.

Note that we can write

$$\mathrm{Gr}(\varphi) * Z = \{ (z, \gamma, w) \in Z \times G \times Z : r(z) = r(\gamma) \text{ and } s(\gamma) = r(w) \}.$$

Thus we get a well-defined continuous map from $\mathrm{Gr}(\varphi) * Z$ onto $\mathrm{Gr}(\psi)$ via

$$(z, \gamma, w) \mapsto (z, \tau_H(z, \gamma \cdot w)).$$

It is clearly equivariant for the right H-action. If $(y, \gamma', z) \in G[Z]$, then

$$(y, \gamma', z) \cdot (z, \gamma, w) = (y, \gamma'\gamma, w) \mapsto (y, \tau_H(y, \gamma'\gamma \cdot w)).$$

But on the other hand,

$$(y, \gamma', w) \cdot (z, \tau_H(z, \gamma \cdot w)) = (y, \tau_H(y, \gamma' \cdot z)\tau_H(z, \gamma \cdot w)) = (y, \tau_H(y, \gamma'\gamma \cdot w)).$$

Hence the map is also $G[Z]$-equivariant. It is not hard to see that it factors through a continuous surjection

$$\Phi : \mathrm{Gr}(\varphi) *_G Z \to \mathrm{Gr}(\psi).$$

To see that Φ is injective, suppose that

$$\tau_H(z, \gamma \cdot w) = \tau_H(z, \gamma' \cdot w').$$

Since the right H-action is free, $\gamma \cdot w = \gamma' \cdot w'$. Therefore (z, γ, w) and (z, γ', w') define the same class in $\mathrm{Gr}(\varphi) *_G Z$.

Since $(z, \eta) \mapsto (z, r(z), z \cdot \eta)$ induces a continuous inverse for Φ, Φ is a homeomorphism and hence and isomorphism of generalized morphisms. □

Exercises

Ex 2.4.1. Suppose that $\mathrm{Gr}(\varphi)$ is the graph of $\varphi : G \to H$ as in Example 2.59. Show that $\mathrm{Gr}(\varphi)$ is a generalized morphism from G to H provided H has open range and source maps.

—Answer on page 363

2.5. Linking Groupoids

In the study of Morita equivalence, the linking algebra has proved to be an useful tool. (See [**RW98**, Theorem 3.19 and its proof].) The fact that the linking algebra has an explicit realization as a C^*-algebra has deep consequences due to the rigid properties of C^*-algebras. The notion of a linking groupoid will allow us to use linking algebra technology to prove the equivalence theorems. The idea that linking groupoids and linking algebras could greatly simplify and enhance the original proofs of the equivalence theorems comes from Alex Kumjian and Aidan Sims. Much of this section is taken from [**SW12**].

As motivation, recall that in Example 2.39 on page 46, we observed that if $G^{(0)}$ is the disjoint union of clopen full sets A and B, then $Z = G_B^A$ is an equivalence between the reductions $G(A) := G|_A$ and $G(B) := G|_B$. We formalize this as follows.

Definition 2.65. If G and H are groupoids, then we call a groupoid L a *linking groupoid* for G and H if $L^{(0)}$ is the disjoint union of full clopen sets A and B such that $L(A)$ is isomorphic to G and $L(B)$ is isomorphic to H.

In this section we want to see that the the set up in Example 2.39 on page 46 is generic: given any two equivalent groupoids H and G, we can find a linking groupoid L for G and H. In keeping with our mission of studying groupoid C^*-algebras, we also want to see that, assuming G and H have Haar systems, then the linking groupoid L has a Haar system which "restricts" to the given Haar systems on G and H, respectively.

In this section our groupoids are always second countable as well as locally compact Hausdorff with open range and source maps.

The Linking Groupoid and its Haar System. Building the linking groupoid is comparatively straightforward. Recall that if Z is a (G, H)-equivalence, then Z^{op} is the (H, G)-equivalence obtained by switching the actions—see the proof of Proposition 2.51 on page 49.

Lemma 2.66. *Suppose that G and H are locally compact Hausdorff groupoids and that Z is a (G, H)-equivalence. Let $L = L(Z)$ be the topological disjoint union*

$$L = G \sqcup Z \sqcup Z^{\mathrm{op}} \sqcup H,$$

and let $L^0 := G^0 \sqcup H^0 \subset L$. Define $r, s : L \to L^0$ to be the maps inherited from the range and source maps on G, Z, Z^{op} and H. Let $L^{(2)} := \{(k, l) \in L \times L : s(k) = r(l)\}$, *and let $(k,l) \mapsto kl$ be the map from $L^{(2)}$ to L which restricts to multiplication on G and H and to the actions of G and H on Z and Z^{op}, and satisfies*

$$z\overline{y} := \tau_G(z,y) \quad \textit{for } (z,y) \in Z *_s Z \quad \textit{and} \quad \overline{y}z := \tau_H(y,z) \quad \textit{for } (y,z) \in Z *_r Z.$$

Define $l \mapsto l^{-1}$ to be the map from L to L which restricts to inversion on G and H and satisfies $z^{-1} = \overline{z}$ and $\overline{z}^{\,-1} = z$ for $z \in Z$. Under these operations, L is a locally compact Hausdorff groupoid called the linking groupoid of Z.

Proof. The inverse map is clearly an involution. Since $\tau_H(z,z) = s(z)$ and $\tau_G(z,z) = r(z)$, it is easy to see that the formulas for r and s are consistent with composition.

The continuity of the inverse map follows from the continuity of the inverse maps on G and H together with the definition of the topology on Z^{op}. The continuity of multiplication follows from continuity of multiplication in G and H, the continuity of the actions of G and H on Z and Z^{op}, and the continuity of $(y,z) \mapsto \tau_G(y,z)$ and $(y,z) \mapsto \tau_H(y,z)$.

The associativity of multiplication follows from routine calculations using the associativity of the groupoid operations and actions, and property (E3) of the definition of groupoid equivalence. For example, if $x,y,z \in Z$ with $s(x) = s(y)$ and $r(y) = r(z)$, then

$$\begin{aligned}(x\overline{y})z &= \tau_G(x,y) \cdot z = \tau_G(x,y) \cdot (y \cdot \tau_H(y,z)) \\ &= (\tau_G(x,y) \cdot y) \cdot \tau_H(y,z) = x \cdot \tau_H(y,z) \\ &= x(\overline{y}z).\end{aligned}$$

□

Remark 2.67. It is not hard to see that $G^{(0)}$ and $H^{(0)}$ are full in $L^{(0)}$. For example, let $v \in H^{(0)}$. Then there is a $z \in Z$ such that $r_{Z^{\mathrm{op}}}(\overline{z}) = s_Z(z) = v$. Let $u = r_Z(z)$. Then $\overline{z} \cdot u = v$. Thus $L \cdot G^{(0)} = L^{(0)}$ as required.

To be useful to us, we are going to need $L = L(Z)$ to have a Haar system if G and H do. Since L is equivalent to both H and G, we could do this abstractly using Theorem 6.20 on page 151.[9] However, it will be convenient to produce one intimately related to the given Haar systems λ and β on G and H, respectively.

Given a (G,H)-equivalence Z, the range map on Z induces a homeomorphism from the orbit space Z/H to $G^{(0)}$. Thus if $u \in G^{(0)}$ and $z \in Z$ with $r(z) = u$, there is a Radon measure σ_Z^u on Z, supported on the orbit $z \cdot H$, determined by

$$\eta)\, d\beta^{s(z)}(\eta) \quad \text{for } \varphi \in C_c(Z).$$

As the notation suggests, σ_Z^u does not depend on the choice of $z \in r^{-1}(u)$: if $y \in Z$ with $r(y) = u$ also, then $y = z \cdot \eta'$ for some $\eta' \in H$ with $r(\eta') = s(z)$, so left-invariance of β gives

$$\int_H \varphi(z \cdot \eta)\, d\beta^{s(z)}(\eta) = \int_H \varphi(z \cdot \eta'\eta)\, d\beta^{s(\eta')}(\eta) = \int_H \varphi(y \cdot \eta)\, d\beta^{s(y)}(\eta).$$

Fix $\varphi \in C_c(Z)$. By Lemma 2.21 on page 33, the map $z \cdot H \mapsto \int_H \varphi(z \cdot \eta)\, d\beta^{s(z)}(\eta)$ is continuous on Z/H. Since r induces a homeomorphism of Z/H onto $G^{(0)}$, it

---

[9] Note that if any *one* of G, H, and L admit a Haar system, then all three do by Theorem 6.20. Once L has a Haar system, we can equip H and G with compatible Haar systems and arrive at the state assumed here.

follows that

$$u \mapsto \int_Z \varphi(z)\, d\sigma_Z^u(z).$$

is a continuous function on $C_c(G^{(0)})$.

By symmetry, we can also define a family of measures $\sigma_{Z^{\mathrm{op}}}^v$ on Z^{op} with $\operatorname{supp} \sigma_{Z^{\mathrm{op}}}^v = r_{Z^{\mathrm{op}}}^{-1}(v)$.

LEMMA 2.68. *For each $w \in L^{(0)}$, let κ^w be the Radon measure on L given on $F \in C_c(L)$ by*

$$\kappa^w(F) = \begin{cases} \lambda^w(F|_G) + \sigma_Z^w(F|_Z) & \text{if } w \in G^{(0)}, \text{ and} \\ \sigma_{Z^{\mathrm{op}}}^w(F|_{Z^{\mathrm{op}}}) + \beta^w(F|_H) & \text{if } w \in H^{(0)}. \end{cases}$$

Then $\{\kappa^w\}_{w\in L^{(0)}}$ is a Haar system for L.

PROOF. It is clear that $\operatorname{supp} \kappa^w$ is $r^{-1}(w) = L^w$. Continuity follows from continuity of σ_Z and $\sigma_{Z^{\mathrm{op}}}$ and of the Haar systems λ and β. It only remains to check left invariance.

Thus, we need to establish that for $k \in L$,

$$\int_L F(l)\, d\kappa^{r(k)}(l) = \int_L F(kl)\, d\kappa^{s(k)}(l).$$

For convenience, assume that $r(k) \in G^{(0)}$. (The case where $r(k) \in H^{(0)}$ is similar.) There are two possibilities: $k \in G$, or $k \in Z$. First suppose $k \in G$. Then for any z satisfying $r(z) = s(k)$,

$$\begin{aligned} \int_L F(kl)\, d\kappa^{s(k)}(l) &= \int_G F(k\gamma)\, d\lambda^{s(k)}(\gamma) + \int_H F(k \cdot z \cdot \eta)\, d\beta^{s(z)}(\eta) \\ &= \int_G F(\gamma)\, d\lambda^{r(k)}(\gamma) + \int_H F\big((k \cdot z) \cdot \eta\big)\, d\beta^{s(k\cdot z)}(\eta) \\ &= \int_G F(\gamma)\, d\lambda^{r(k)}(\gamma) + \int_Z F(w)\, d\sigma^{r(k)}(w) \\ &= \int_L F(l)\, d\kappa^{r(k)}(l). \end{aligned}$$

Now suppose that $k \in Z$. Then

$$\int_L F(kl)\, d\kappa^{s(k)}(l) = \int_{Z^{\mathrm{op}}} F(k\overline{z})\, d\sigma_{Z^{\mathrm{op}}}^{s(k)}(\overline{z}) + \int_H F(k \cdot \eta)\, d\beta^{s(k)}(\eta).$$

Since we can evaluate $\sigma_{Z^{\mathrm{op}}}^{s(k)}$ with any $\overline{w}$ such that $r(\overline{w}) = s(k)$, we may in particular take $\overline{w} = \overline{k}$, giving

$$\int_L F(kl)\, d\kappa^{s(k)}(l) = \int_G F\big(\tau_G(k, \gamma^{-1} \cdot k)\big)\, d\lambda^{r(k)}(\gamma) + \int_Z F(z)\, d\sigma_Z^{r(k)}(z).$$

Since $\tau_G(k, \gamma^{-1} \cdot k) = \gamma$ for all γ, we conclude that

$$\int_L F(kl)\, d\kappa^{s(k)}(l) = \int_L F(l)\, d\kappa^{r(k)}(l).$$

□

THEOREM 2.69. *Two groupoids G and H are equivalent if and only if there is linking groupoid L for G and H. If G and H have Haar systems λ and β, respectively, then there is a linking groupoid L with a Haar system κ such that $\kappa|_{G^{(0)}} = \lambda$ and $\kappa|_{H^{(0)}} = \beta$. Conversely, if L is a linking groupoid for G and H and* *if L has a Haar system κ, then the restrictions $\kappa|_{G^{(0)}}$ and $\kappa|_{H^{(0)}}$ are Haar systems for G and H, respectively.*

PROOF. For the first assertion, it suffices to see that $G^{(0)}$ (and hence $H^{(0)}$) is saturated in $L^{(0)} = L(Z)^{(0)}$ for a (G,H)-equivalence Z. But if $v \in H^{(0)}$, let $z \in Z$ be such that $s_Z(z) = v$. Then $\overline{z} \in Z^{\mathrm{op}} \subset L$ is an element in L with $s_L(\overline{z}) \in G^{(0)}$ and $r_L(\overline{z}) = v$. In other words, $v \in L \cdot G^{(0)}$. It follows that $G^{(0)}$ is saturated.

That assertions about Haar systems follow from Lemma 2.68 in the first case and is straightforward in the second case. □

2.6. The Equivalence Theorem

Informally, the Equivalence Theorem asserts that equivalent groupoids have Morita equivalent C^*-algebras. The basic outline is simple. Let Z be a (G,H)-equivalence where G and H are second countable locally compact Hausdorff groupoids with Haar systems $\lambda = \{\lambda^u\}_{u\in G^{(0)}}$ and $\beta = \{\beta^v\}_{v\in H^{(0)}}$, respectively. If φ and ψ are in $C_c(Z)$, then for any $\eta \in H$, the right-hand side of

$$\langle \varphi \,,\, \psi\rangle_{\star}(\eta) = \int_G \overline{\varphi(\gamma^{-1}\cdot z)}\psi(\gamma^{-1}\cdot z\cdot\eta)\, d\lambda^{r(z)}(\gamma) \tag{2.6}$$

is independent of our choice of z provided $s_Z(z) = r(\eta)$. We can define a right action of $C_c(H)$ on $C_c(Z)$ by

$$\varphi\cdot b(z) = \int_H \varphi(z\cdot\eta)b(\eta^{-1})\, d\beta^{s(z)}(\eta). \tag{2.7}$$

Similarly, for $\gamma \in G$, we can define

$${}_{\star}\langle \varphi \,,\, \psi\rangle(\gamma) = \int_H \varphi(\gamma\cdot w\cdot\eta)\overline{\psi(w\cdot\eta)}\, d\beta^{s(z)}(\eta) \tag{2.8}$$

and a left $C_c(G)$-action on $C_c(Z)$ by

$$f\cdot\varphi(z) = \int_G f(\gamma)\varphi(\gamma^{-1}\cdot z)\, d\lambda^{r(z)}(\gamma). \tag{2.9}$$

To see that these formulae form the actions and inner-products of a pre-imprimitivity bimodule requires some care. We need to see that $\langle \cdot\,,\,\cdot\rangle_{\star}$ and ${}_{\star}\langle \cdot\,,\,\cdot\rangle$ are $C_c(H)$- and $C_c(G)$-valued, respectively, and that our "actions" are really actions. But the serious work is to see that if we view $C_c(H)$ and $C_c(G)$ as dense subalgebras of $C^*(H,\beta)$ and $C^*(G,\lambda)$, respectively, then $C_c(Z)$ becomes a $C_c(G)$ – $C_c(H)$-pre-imprimitivity bimodule as defined in [**RW98**, Definition 3.9] (see Remark 4.4 on page 97). If we can do that, then $C_c(Z)$ completes to a $C^*(G,\lambda)$ – $C^*(H,\beta)$-imprimitivity bimodule by [**RW98**, Proposition 3.12].

However, we could just as easily view $C_c(H)$ and $C_c(G)$ as dense subalgebras of the reduced C^*-algebras $C^*_r(H,\beta)$ and $C^*_r(G,\lambda)$, respectively. If the requirements of [**RW98**, Definition 3.9] are met for the reduced norms, then $C_c(Z)$ could also be completed to a $C^*_r(G,\lambda)$ – $C^*_r(H,\beta)$-imprimitivity bimodule.

Fortunately, both sets of requirements are met by $C_c(Z)$ with the given pre-inner products (2.6) and (2.8) and actions (2.7) and (2.9). Furthermore, these two Morita equivalences are naturally compatible. We will give precise statements of these results here, and postpone the details of the proof to Chapter 4. This will give us the opportunity to explore some of the valuable consequences of these results.

The Statements. As alluded to above, there are really three components to the Equivalence Theorem: the Morita equivalence of the universal algebras, the Morita equivalence of the reduced algebras, and the compatibility between these two results. To emphasize this, and for easy reference, I have broken up the result into three separate statements.

THEOREM 2.70 (Renault's Equivalence Theorem). *Let G and H be second countable locally compact Hausdorff groupoids with Haar systems $\{\lambda^u\}_{u\in G^{(0)}}$ and $\{\beta^v\}_{v\in H^{(0)}}$, respectively. If Z is a second countable locally compact Hausdorff space which is a (G,H)-equivalence, then with respect to the actions and inner products defined in* (2.6), (2.7), (2.7), *and* (2.8), *$C_c(Z)$ has a completion $\mathsf{X} = \mathsf{X}_H^G$ as a $C^*(G,\lambda)$ – $C^*(H,\beta)$-imprimitivity bimodule.*

PROOF. This result is part of Theorem 4.9 on page 99. □

THEOREM 2.71 (Renault's Equivalence Theorem for Reduced Algebras). *We let G and H be second countable locally compact Hausdorff groupoids with Haar systems $\{\lambda^u\}_{u\in G^{(0)}}$ and $\{\beta^v\}_{v\in H^{(0)}}$, respectively. If Z is a second countable locally compact Hausdorff space which is a (G,H)-equivalence, then with respect to the actions and inner products defined in* (2.6), (2.7), (2.7), *and* (2.8), *$C_c(Z)$ has a completion $\mathsf{X}_r = (\mathsf{X}_H^G)_r$ as a $C_r^*(G,\lambda)$ – $C_r^*(H,\beta)$-imprimitivity bimodule.*

PROOF. This result is subsumed by Theorem 4.5 on page 97. □

Since X and X_r are both completions of the same space, we expect a nice relationship between the two. We let $I_{C_r^*(G)}$ be the common kernel of any faithful regular representation of G so that $C_r^*(G) \cong C^*(G)/I_{C_r^*(G)}$.

THEOREM 2.72 (Morita Compatibility). *Let G and H be second countable locally compact Hausdorff groupoids with Haar systems $\{\lambda^u\}_{u\in G^{(0)}}$ and $\{\beta^v\}_{v\in H^{(0)}}$, respectively. If Z is a second countable locally compact Hausdorff space which is a (G,H)-equivalence, then the identity map on $C_c(Z)$ extends to a surjective bimodule map of X onto X_r with kernel Y where Y is a $I_{C_r^*(G)}$ – $I_{C_r^*(H)}$-imprimitivity bimodule. Then X_r is isomorphic to the quotient imprimitivity bimodule X/Y.*

PROOF. This result is proved on page 101. □

2.7. Some Important Examples

Induced representations of C^*-algebras are discussed in [**RW98**, §2.4 and §3.3]. In particular, if X is an (A,B)-imprimitivity bimodule, then we write X–$\operatorname{Ind} L$, or X–$\operatorname{Ind}_B^A L$ if there is ambiguity about the algebras in play, for the representation of A induced from a given representation L of B. We also write X–Ind for the corresponding map on ideals [**RW98**, Proposition 3.24].

PROPOSITION 2.73. *Suppose that G and H are second countable locally compact Hausdorff groupoids with Haar systems $\{\lambda^u\}_{u\in G^{(0)}}$ and $\{\beta^v\}_{v\in H^{(0)}}$, respectively. Let Z be a (G,H)-equivalence and let X be the $C^*(G)$ – $C^*(H)$-imprimitivity bimodule from Theorem 2.70. Then if L is a representation of $C^*(H)$ which factors through $C_r^*(H)$, then X–$\operatorname{Ind} L$ factors through $C_r^*(G)$.*

PROOF. A representation L factors through $C_r^*(H)$ if and only if $I_{C_r^*(H)} \subset \ker L$ where $I_{C_r^*(H)}$ is the kernel of the natural map of $C^*(H)$ onto $C_r^*(H)$. If X is an A – B-imprimitivity bimodule, then X–Ind is a lattice isomorphism of of the ideal lattice $\mathcal{I}(B)$ onto the ideal lattice $\mathcal{I}(A)$ [**RW98**, Theorem 3.22]. By Theorem 2.72 on the facing page and its proof, $\mathsf{X}\text{–}\mathrm{Ind}(I_{C_r^*(H)}) = I_{C_r^*(G)}$. Thus

$$I_{C_r^*(G)} = \mathsf{X}\text{–}\mathrm{Ind}(I_{C_r^*(H)}) \subset \mathsf{X}\text{–}\mathrm{Ind}(\ker L) = \ker(\mathsf{X}\text{–}\mathrm{Ind}\, L)$$

by [**RW98**, Proposition 3.24]. □

We know from simple examples that locally compact groupoids can have very different looking Haar systems. We'll explore questions of uniqueness in detail in Chapter 6. A straightforward application of the Equivalence Theorem guarantees that the C^*-algebras corresponding to different Haar systems are Morita equivalent.

PROPOSITION 2.74. *Suppose that G is a second countable locally compact groupoid with Haar systems $\{\lambda^u\}_{u\in G^{(0)}}$ and $\{\beta^u\}_{u\in G^{(0)}}$. Then $C^*(G,\lambda)$ and $C^*(G,\beta)$ are Morita equivalent as are $C_r^*(G,\lambda)$ and $C_r^*(G,\beta)$.*

PROOF. This is an easy consequence of Theorems 2.70 and 2.71 together with the observation that G is always a (G,G)-equivalence. □

A nice consequence of Proposition 2.74 is that I can be partially exonerated for the sloppy habit of dropping the Haar system from our notation for the groupoid C^*-algebra: that is, writing just $C^*(G)$ rather than $C^*(G,\lambda)$—unless it makes a difference of course.

PROPOSITION 2.75. *Let G be a second countable locally compact Hausdorff groupoid with a Haar system. If G is transitive and if $H = G(u) = \{\gamma \in G : s(\gamma) = u = r(\gamma)\}$ is any isotropy group, then $C^*(G)$ is Morita equivalent to $C^*(H)$ and $C_r^*(G)$ is Morita equivalent to $C_r^*(H)$. Furthermore the reduced norm on $C_c(G)$ is equal to the universal norm if and only if H is amenable.*

PROOF. The statements about Morita equivalence are immediate corollaries of Theorems 2.70 and 2.71. Since $I_{C_r^*(H)}$ is the zero ideal if and only if H is amenable, the second statement follows from the fact that $\mathsf{X}\text{–}\mathrm{Ind}$ is a bijection of the two ideal lattices [**RW98**, Theorem 3.22]. □

REMARK 2.76. Not surprisingly, the equality of the reduced and universal norm on $C_c(G)$ is related to the amenability of G—once we define what that means—see Corollary 10.54 on page 277. Furthermore, this result can be sharpened to an isomorphism result. We'll examine that result in Section 2.8.

Another amusing application of the Morita compatibility of the Equivalence Theorems is the following.

PROPOSITION 2.77 (Bi-Transformation Groups). *Suppose that $\mathcal{G}$ and $\mathcal{H}$ are second countable locally compact groups acting freely and properly on the left and right, respectively, of a locally compact Hausdorff space P such that the actions commute. Then $C_0(\mathcal{G}\backslash P) \rtimes_{\mathrm{rt}} \mathcal{H}$ and $C_0(P/\mathcal{H}) \rtimes_{\mathrm{lt}} \mathcal{G}$ are Morita equivalent as are the reduced crossed products. Furthermore, if $\mathcal{G}$ is amenable, then the reduced norm on $C_c(\mathcal{G}\backslash P \times \mathcal{H})$ equals the universal norm.*

PROOF. Recall that $C^*(\mathcal{G} \rtimes P/\mathcal{H})$ and $C_r^*(\mathcal{G} \times P/\mathcal{H})$ are naturally isomorphic to their crossed product counterparts $C_0(P/\mathcal{H}) \rtimes_{\mathrm{lt}} \mathcal{G}$ and $C_0(P/\mathcal{H}) \rtimes_{\mathrm{lt},r} \mathcal{G})$, respectively (see Example 1.55 on page 22). Hence the assertions about Morita equivalence follow from the Equivalence Theorems and the discussion in Example 1.55 on page 22. On the other hand, if $\mathcal{G}$ is amenable, then it is well-known that the reduced cross product norm on $C_c(\mathcal{G} \times P/\mathcal{H})$ is the same as the universal norm [**Wil07**, Theorem 7.13]. The final assertion follows from Theorem 2.72 on page 60 and the fact that the Rieffel bijection X–Ind is a lattice isomorphism. □

2.8. Transitive Groupoid C^*-Algebras

In this section, we let G be a second countable locally compact Hausdorff *transitive* groupoid with a Haar system $\{\lambda_u\}_{u\in G^{(0)}}$. By Proposition 2.75 on the preceding page, we know that $C^*(G)$ is Morita equivalent to $C^*(G(u))$ for any $u \in G^{(0)}$. Since everything in sight is separable, the Brown-Green-Rieffel Theorem [**RW98**, Theorem 5.55] implies that $C^*(G)$ and $C^*(G(u))$ are stably isomorphic. In this section, we want to provide a significant sharpening of that result generalizing the corresponding result for transitive transformation groups [**Wil07**, Theorem 4.30].

THEOREM 2.78. *Let G be a second countable locally compact Hausdorff* transitive *groupoid with a Haar system. Then $C^*(G)$ is isomorphic to*

$$C^*(G(u)) \otimes \mathcal{K}(L^2(G^{(0)}, \mu))$$

for any unit $u \in G^{(0)}$ and any Radon measure μ on $G^{(0)}$ with full support.

REMARK 2.79. Note that if $u, v \in G^{(0)}$ and $\gamma \in G_u^v$, then $t \mapsto \gamma t \gamma^{-1}$ is a group isomorphism of $G(v)$ onto $G(u)$. Furthermore, unless $G^{(0)}$ is finite, $L^2(G^{(0)}, \mu)$ is an infinite-dimensional separable Hilbert space for any full Radon measure μ. Otherwise $L^2(G^{(0)}, \mu) \cong \mathbf{C}^{|X|}$. In any event, the isomorphism class of $\mathcal{K}(L^2(G^{(0)}, \mu))$ as well as that of $C^*(G(u))$ is determined solely by G (Ex 2.8.1).

The idea for the proof given here goes back to Green [**Gre80**]. The proof for the groupoid version comes from [**MRW87**, Theorem 3.1]. These proofs are based on the following observation concerning Hilbert modules (Ex 2.8.2).

LEMMA 2.80. *Let X and Y be right Hilbert A-modules. Let $W : \mathsf{X} \to \mathsf{Y}$ be an inner-product preserving surjection.*

(a) *Then W is an invertible operator in $\mathcal{L}(\mathsf{X}, \mathsf{Y})$ with adjoint $W^* = W^{-1}$.*
(b) *The map $T \mapsto WTW^{-1}$ is an isomorphism of $\mathcal{L}(\mathsf{X})$ onto $\mathcal{L}(\mathsf{Y})$ taking $\mathcal{K}(\mathsf{X})$ onto $\mathcal{K}(\mathsf{Y})$.*

Recall from Example 2.32 on page 44 that if $u \in G^{(0)}$, then G_u implements an equivalence between G and $G(u)$. Then the Equivalence Theorem 2.70 on page 60 implies that $C_c(G_u)$ can be completed to a right Hilbert $C^*(G(u))$-module $\mathsf{X}_{G(u)}^G$ such that $\mathcal{K}(\mathsf{X}_{G(u)}^G)$ is isomorphic to $C^*(G)$ [**RW98**, Proposition 3.8]. The idea of the proof of Theorem 2.78 is to introduce another Hilbert $C^*(G(u))$-module Y which is isomorphic to X as Hilbert $C^*(G(u))$-modules, and is such that $\mathcal{K}(\mathsf{Y})$ is isomorphic to $C^*(G(u)) \otimes \mathcal{K}(L^2(G^{(0)}, \mu))$ for a full Radon measure μ. This will complete the proof in view of Lemma 2.80.

As is usual, we will work at the level of pre-inner product $C_c(G(u))$-modules (as defined in [**RW98**, Lemma 2.16]). Thus we view $\mathsf{X}_{G(u)}^G$ as the completion of $C_c(G_u)$ with respect to the $C_c(G(u))$-valued inner product[10]

$$\langle f\,,\,g\rangle_\star(t) = \int_G \overline{f(\gamma)}g(\gamma t)\,d\lambda_u(\gamma), \tag{2.10}$$

[10]Note that we are viewing $G(u)$ as a groupoid here. Hence there is no modular function in the involution.

and right $C_c(G(u))$-action

$$f \cdot b(\gamma) = \int_{G(u)} f(\gamma \cdot t) b(t^{-1})\, d\beta^u(t). \tag{2.11}$$

An unusual facet of our method of proof is that we work with bounded Borel functions $\mathcal{B}_c^b(G_u)$ which vanish off a compact set. We will also write $\mathcal{B}^+(G_u)$ for the non-negative Borel functions on G_u and $\mathcal{B}^{+,e}(G_u)$ for the Borel functions on G_u taking values in the extended reals $[0, \infty]$.

Working with Borel functions is problematical since it is generally *not the case* that $\mathcal{B}_c^b(G)$ maps into $C^*(G)$ as in the group case. Just consider $G = X$ where X is a compact space. Then $C_c(G) = C^*(G) = C(X)$ and there is certainly no natural map of $\mathcal{B}^b(X)$ into $C(X)$.[11]

Since G is transitive, the restriction $r : G_u \to G^{(0)}$ is surjective. A *cross section* for this map is a map $\rho : G^{(0)} \to G_u$ such that $r(\rho(v)) = v$ for all $v \in G^{(0)}$. The need for Borel functions arises as there will, in general, be no continuous cross section for the surjection $r : G_u \to G^{(0)}$. But there are Borel ones. We will need a cross section that is not only Borel, but *regular* in that $\rho(K)$ is pre-compact for all $K \subset G^{(0)}$ compact.

LEMMA 2.81. *Let G be a second countable locally compact transitive groupoid. There is a regular Borel cross section $\rho : G^{(0)} \to G_u$ for the map $r : G_u \to G^{(0)}$.*

PROOF. Let $K \subset G_u$ be compact. Since G, and hence K, is second countable, any open set in K is σ-compact. Hence its image under r is σ-compact and therefore Borel. It now follows easily from [**Arv76**, Theorem 3.4.1], that $r|_K$ has a Borel cross section $r_K : r(K) \to K \subset G_u$. But G_u is the increasing union of compact sets K_n such that K_n is contained in the interior of K_{n+1}. (Hence every compact subset of $G^{(0)}$ is contained in some $r(K_n)$.) Now we can piece together the r_{K_n} to obtain the desired regular cross section (Ex 2.8.3). □

Fix $u \in G^{(0)}$ and a regular section as in Lemma 2.81. Let β^u be a left-Haar measure on the isotropy group $G(u) = \{\, t \in G : r(t) = u = s(t) \,\}$. Note that if $f \in \mathcal{B}^+(G_u)$ then Tonelli's Theorem (e.g., [**Fol99**, Theorem 2.37]) implies that

$$\beta(f)(v) := \int_{G(u)} f(\rho(v)s^{-1})\, d\beta^u(s) \tag{2.12}$$

is in $\mathcal{B}^{+,e}(G^{(0)})$.

To get the measure μ in Theorem 2.78 on the facing page, we need the following proposition. It is similar in spirit to the constructions in Section 11.1 which are taken from [**Ren91**, §1].

[11] However, in the transitive case, M. Buneci has shown that $\mathcal{B}_c^b(G)$ does map into $C^*(G)$ [**Bun01**, Propositions 4 and 5] after extending the convolution and involution in the expected way. As in Ex 2.8.5, G is Borel isomorphic to the groupoid $G^{(0)} \times G(u) \times G^{(0)}$. This induces a $*$-homomorphism of $C_c(G^{(0)} \times G(u) \times G^{(0)})$ with respect to the Haar system $\{\delta_u \times \beta^u \times \nu\}_{u \in G^{(0)}}$ (as in Example 1.21 on page 10) onto a $*$-subalgebra of $\mathcal{B}_c^b(G)$ which can be shown to be dense in $C_c(G)$. She gets Theorem 2.78 on the facing page after observing that $C^*(G^{(0)} \times G(u) \times G^{(0)})$ is isomorphic to $C^*(G(u)) \otimes \mathcal{K}(L^2(G^{(0)}, \mu))$ (see [**Bun06**, Proposition 12]). I have chosen to avoid the overhead of [**Bun01**] and use Green's method here.

PROPOSITION 2.82. *Let G be a second countable locally compact Hausdorff groupoid with a Haar system $\{\lambda_v\}_{v\in G^{(0)}}$, and fix $u \in G^{(0)}$. Then there is a Radon measure μ on $G^{(0)}$ with full support such that for all $f \in \mathcal{B}^+(G^u)$ we have*

$$\int_G f(\gamma)\,d\lambda_u(\gamma) = \int_{G^{(0)}} \int_{G(u)} f(\rho(v)s^{-1})\,d\beta^u(s)\,d\mu(v). \tag{2.13}$$

If μ is any Radon measure, then the right-hand side of (2.13) should be interpreted as

$$\int_{G^{(0)}} \beta(f)(v)\,d\mu(v),$$

where $\beta(f)$ is as in (2.12). Since $\beta(f)$ is Borel, the right-hand side of (2.13) is well-defined, and the issue is to properly define μ with the requisite properties.

For the proof, we need the following technical result. Recall that the modular function $\Delta_{G(u)}$ on $G(u)$ satisfies

$$\int_{G(u)} f(t)\,d\beta^u(t) = \Delta_{G(u)}(s) \int_{G(u)} f(ts)\,d\beta^u(t).$$

LEMMA 2.83. *For all $f, g \in \mathcal{B}^+(G_u)$ we have*

$$\lambda_u(g \cdot \beta(f) \circ r) = \lambda_u(\beta(g) \circ r \cdot f). \tag{2.14}$$

PROOF. Starting with the left-hand side of (2.14), we have

$$\lambda_u(g \cdot \beta(f) \circ r) = \int_{G^{(0)}} \int_{G(u)} g(\gamma) f(\rho(r(\gamma))s^{-1})\,d\beta^u(s)\,d\lambda_u(\gamma)$$

which, after sending $s \mapsto s\gamma^{-1}\rho(r(\gamma))$, is

$$= \int_{G(u)} \int_{G^{(0)}} g(\gamma) f(\gamma s^{-1}) \Delta_{G(u)}(\gamma^{-1}\rho(s(\gamma)))\,d\lambda_u(\gamma)\,d\beta^u(s)$$

which, after sending $\gamma \mapsto \gamma s$, is

$$\begin{aligned}
&= \int_{G^{(0)}} \int_{G(u)} g(\gamma s) f(\gamma) \Delta_{G(u)}(s^{-1}) \Delta_{G(u)}(\gamma^{-1}\rho(s(\gamma)))\,d\beta^u(s)\,d\lambda_u(\gamma) \\
&= \int_{G^{(0)}} \int_{G(u)} g(\gamma s^{-1}) f(\gamma) \Delta_{G(u)}(\gamma^{-1}\rho(s(\gamma)))\,d\beta^u(s)\,d\lambda_u(\gamma)
\end{aligned}$$

which, after another right-translation of s, is

$$\begin{aligned}
&= \int_{G^{(0)}} \int_{G(u)} g(\rho(r(\gamma))s^{-1}) f(\gamma)\,d\beta^u(s)\,d\lambda_u(\gamma) \\
&= \lambda_u(\beta(g) \circ r \cdot f).
\end{aligned}$$

□

PROOF OF PROPOSITION 2.82. Since G_u is second countable, it is not hard to see that there is a $d \in \mathcal{B}^+(G_u)$ such that

$$\beta(d)(v) = \int_{G(u)} d(\rho(v)s^{-1})\,d\beta^u(s) = 1 \quad \text{for all } v \in G^{(0)}$$

(Ex 2.8.4). Thus if $h \in \mathcal{B}^+(G^{(0)})$ and if $f(\gamma) = h(r(\gamma))d(\gamma)$, then $\beta(f) = h$. That is, every $h \in \mathcal{B}^+(G^{(0)})$ is of the form $\beta(f)$ for some $f \in \mathcal{B}^+(G_u)$.

In view of Lemma 2.83, for any $f \in \mathcal{B}^+(G_u)$,

$$\lambda_u(f) = \lambda_u\big(\beta(d) \circ r \cdot f\big) = \lambda_u(d \cdot \beta(f) \circ r). \tag{2.15}$$

Hence $\lambda_u(f)$ depends only on $\beta(f)$ and we can define an additive map $\mu : \mathcal{B}^+(G^{(0)}) \to [0,\infty]$ by

$$\mu(\beta(f)) = \lambda_u(f).$$

A couple of applications of the Monotone Convergence Theorem gives

$$\begin{aligned}\mu\Bigl(\sum \beta(f_i)\Bigr) &= \mu\Bigl(\beta\Bigl(\sum f_i\Bigr)\Bigr)\\ &= \lambda_u\Bigl(\sum f_i\Bigr)\\ &= \sum \mu(\beta(f_i)).\end{aligned}$$

It follows that μ determines a measure, also called μ, on $G^{(0)}$. Using (2.15), we can verify that for $b \in \mathcal{B}^+(G^{(0)})$, we have

$$\mu(b) = \int_{G^{(0)}} b(u)\,d\mu \tag{2.16}$$

(Ex 2.8.6). That is, the functional μ is given by integration against the measure μ.

If $K \subset G^{(0)}$ is compact, then $\overline{\rho(K)}$ is compact and we can find $f \in C_c^+(G_u)$ such that $f(\gamma) > 0$ for all $\gamma \in \overline{\rho(K)}$. Since

$$\gamma \mapsto \int_G f(\gamma s^{-1})\,d\beta^u(s)$$

is continuous and nonzero on $\overline{\rho(K)}$ we can multiply f by a constant so that

$$\beta(f)(v) \geq 1 \quad \text{for all } v \in K.$$

Then

$$\mu(K) \leq \int_{G^{(0)}} \beta(f)(v)\,d\mu = \lambda_u(f) < \infty$$

Thus μ is a Radon measure by [**Rud87**, Theorem 2.18].

Let V be a nonempty open set in $G^{(0)}$. Let $K \subset V$ be a nonempty compact set and $h \in C_c^+(G^{(0)})$ be such that $\operatorname{supp} h \subset V$ and $h(v) = 1$ for all $v \in K$. Let $g \in C_c^+(G_u)$ be such that $g(\gamma) > 0$ for all $\gamma \in \rho(K)$. Then $f(\gamma) = g(\gamma)h(r(\gamma))$ is a nonzero function in $C_c^+(G_u)$ and

$$\int_{G^{(0)}} \beta(f)(v)\,d\beta^u(v) = \lambda_u(f) > 0.$$

Since $\beta(f)$ is supported in V, $\mu(V) \neq 0$. Thus μ has full support. □

Observe that the right-hand sides of our pre-inner product (2.10) and our action (2.11) makes good sense even if $f, g \in \mathcal{B}_c^b(G_u)$. In fact, we have the following result.

PROPOSITION 2.84. *With respect to the pairing given by* (2.10) *and right* $C_c(G(u))$*-action given by* (2.11), $\mathcal{B}_c^b(G_u)$ *is a pre-inner product* $C_c(G(u))$*-module as in* [**RW98**, Lemma 2.16]. *The image of* $C_c(G_u)$ *in* $\mathcal{B}_c^b(G_u)$ *under inclusion is dense in the* $\|\cdot\|_{C^*(G(u))}$*-seminorm. Hence we can identify the Hilbert module completion of* $\mathcal{B}_c^b(G_u)$ *with* $\mathsf{X}_{G(u)}^G$.

PROOF. Observe that with respect to the inner product on $L^2(G_u, \lambda_u)$,

$$\langle f\,,g\rangle_{\star}(t) = (g_t \mid f),$$

where $g_t(\gamma) = g(\gamma t)$. Since the usual arguments imply that $t \mapsto g_t$ is continuous from $G(u)$ into $L^2(G_u)$, it follows that (2.10) is $C_c(G(u))$-valued on $\mathcal{B}_c^b(G_u)$.

Moreover,

$$\begin{aligned}\left|\langle f \,,\, f\rangle_{\star}(t) - \langle g \,,\, g\rangle_{\star}(t)\right| &= \left|\langle f \,,\, f-g\rangle_{\star}(t) - \langle f-g \,,\, g\rangle_{\star}(t)\right| \\ &\le \|(f-g)_t\|_2\|f\|_2 + \|g_t\|_2\|f-g\|_2\end{aligned}$$

which, since $\|g_t\|_2 = \|g\|_2$, is

$$= \bigl(\|f\|_2 + \|g\|_2\bigr)\|f-g\|_2.$$

Thus if $f \in \mathcal{B}_c^b(G_u)$ vanishes off K and we let $\{g_i\} \subset C_c(G_u)$ all have support in K and be such that $g_i \to f$ in L^2, then $\langle g_i \,,\, g_i\rangle_{\star} \to \langle f \,,\, f\rangle_{\star}$ in the inductive limit topology, and hence in $C^*(G(u))$. Since each $\langle g_i \,,\, g_i\rangle_{\star}$ is positive in $C^*(G(u))$, so is $\langle f \,,\, f\rangle_{\star}$.

As the remaining axioms are straightforward to check, $\mathcal{B}_c^b(G_u)$ is a pre-inner product module as claimed.

Let $f \in \mathcal{B}_c^b(G_u)$ and $\{g_i\} \subset C_c(G_u)$ be as above. Then

$$\|f - g_i\|^2_{C^*(G(u))} \le \|h\|_I,$$

where

$$h_i(t) = \bigl((f-g_i)_t \mid f - g_i\bigr).$$

Then $\operatorname{supp} h \subset K^{-1}K \cap G(u)$, and

$$\|h_i\|_{I,r} = \int_{G(u)} |h(t)|\,d\beta^u(t) \le \|f-g_i\|_2^2\beta^u(K^{-1}K \cap G(u)).$$

Similarly,

$$\|h_i\|_{I,s} = \int_{G(u)} |h(t^{-1})|\,d\beta^u(t) \le \|f-g_i\|_2^2\beta^u(K^{-1}K \cap G(u)).$$

It follows that $g_i \to f$ in the Hilbert module norm as required. □

If we like, we can view Proposition 2.84 on the preceding page as a generalization of the observation that we can view $\mathcal{B}_c^b(G(u))$ as a pre-Hilbert $C_c(G(u))$-module sitting inside $C^*(G(u))$ viewed as a right Hilbert module over itself. Then we can view $\mathcal{B}_c^b(G(u)) \odot \mathcal{B}_c^b(G^{(0)})$ as a pre-inner product $C_c(G(u))$-module sitting densely inside the external tensor product $C^*(G(u)) \otimes L^2(G^{(0)}, \mu)$. (External tensor products are discussed in [**RW98**, §3.4].) The $C_c(G(u))$-valued inner product on $\mathcal{B}_c^b(G(u)) \odot \mathcal{B}_c^b(G^{(0)})$ is given by

$$\langle h \otimes \varphi \,,\, k \otimes \psi\rangle_{\star}(t) = (\psi \mid \varphi)\langle h \,,\, k\rangle_{\star}(t), \tag{2.17}$$

and the right $C_c(G(u))$-action by

$$(h \otimes \varphi) \cdot b = h * b \otimes \varphi. \tag{2.18}$$

As above, we want to enlarge $\mathcal{B}_c^b(G(u)) \odot \mathcal{B}_c^b(G^{(0)})$ slightly.

PROPOSITION 2.85. *With respect to*

$$\langle \xi \,,\, \zeta\rangle_{\star}(t) = \int_{G^{(0)}} \int_{G(u)} \overline{\xi(s^{-1}, v)}\zeta(s^{-1}t, v)\,d\beta^u(s)\,d\mu(u), \tag{2.19}$$

and

$$\xi \cdot b(t,v) = \int_{G(u)} \xi(s,v)b(s^{-1}t,v)\,d\beta^u(s), \tag{2.20}$$

$\mathcal{B}_c^b(G(u)\times G_u)$ is a pre-inner product $C_c(G(u))$-module. The natural map of $\mathcal{B}_c^b(G(u)) \odot \mathcal{B}_c^b(G^{(0)})$ into $\mathcal{B}_c^b(G(u)\times G^{(0)})$ is isometric with dense range (with respect the Hilbert module norm on $\mathcal{B}_c^b(G(u)\times G^{(0)})$). In particular, the Hilbert module completion of $\mathcal{B}_c^b(G(u)\times G^{(0)})$ is isomorphic to the external tensor product $C^*(G(u))\otimes L^2(G^{(0)},\mu)$.

PROOF. Observe that (2.19) and (2.20) preserve (2.17) and (2.18), respectively, on the image of $\mathcal{B}_c^b(G(u)) \odot \mathcal{B}_c^b(G^{(0)})$. Also, with respect to the inner product on $L^2(G(u)\times G^{(0)},\beta_u\times\mu)$,

$$\langle \xi \, , \zeta\rangle_{\star}(t) = (\zeta_t \mid \xi),$$

where $\zeta_t(s,v) = \zeta(st,v)$

Furthermore, $C_c(G(u)) \odot C_c(G^{(0)})$ is dense in the inductive limit topology on $C_c(G(u)\times G^{(0)})$. Hence we can repeat the argument of Proposition 2.84 on page 65 *mutatis mutandis.* □

If $\gamma \in G_u$, then $\rho(r(\gamma))^{-1}$ and γ are composible. Then $\rho(r(\gamma))^{-1}\gamma \in G(u)$. Thus we get a Borel map

$$\varphi(\gamma) = \big(\rho(r(\gamma))^{-1}\gamma, r(\gamma)\big)$$

from G_u to $G(u)\times G^{(0)}$ with Borel inverse

$$\psi(t,v) = \rho(v)t.$$

Since both φ and ψ take compact sets to pre-compact sets, we get an bijection $W : \mathcal{B}_c^b(G(u)\times G^{(0)}) \to \mathcal{B}_c^b(G_u)$ defined by

$$W(\xi)(\gamma) = \xi(\varphi(\gamma)).$$

Observe that

$$\begin{aligned}\langle W(\xi)\, , W(\zeta)\rangle_{\star}(t) &= \int_G \overline{W(\xi)(\gamma)}W(\zeta)(\gamma t)\,d\lambda_u(\gamma)\\ &= \int_G \overline{\xi(\rho(r(\gamma))^{-1}\gamma, r(\gamma))}\zeta(\rho(r(\gamma))^{-1}\gamma t, r(\gamma))\,d\lambda_u(\gamma)\end{aligned}$$

which, in view of the definition of the measure μ, is

$$\begin{aligned}&= \int_{G^{(0)}}\int_{G(u)} \overline{\xi(s^{-1},v)}\zeta(s^{-1}t,v)\,d\beta^u(s)\,d\mu(v)\\ &= \langle \xi\, , \zeta\rangle_{\star}(t).\end{aligned}$$

Hence W extends to an $C^*(G(u))$-inner product preserving surjection of $\mathsf{X}^G_{G(u)}$ onto $\mathsf{Y} := C^*(G(u))\otimes L^2(G^{(0)},\mu)$. It follows from Lemma 2.80 on page 62 that W is a Hilbert $C^*(G(u))$-module isomorphism of $\mathsf{X}^G_{G(u)}$ onto $C^*(G(u))\otimes L^2(G^{(0)},\mu)$, and that W induces an isomorphism of $C^*(G) \cong \mathcal{K}(\mathsf{X}^G_{G(u)})$ onto $\mathcal{K}(C^*(G(u))\otimes L^2(G^{(0)},\mu)) \cong C^*(G(u))\otimes\mathcal{K}(L^2(G^{(0)},\mu))$. (The second isomorphism follows from [**RW98**, Corollary 3.38].)

This finishes the proof of Theorem 2.78 on page 62.

Exercises

Ex 2.8.1. Let X be a second countable locally compact Hausdorff space and μ a Radon measure on X with full support.

(a) Show that $L^2(X,\mu)$ is a separable Hilbert space.
(b) Show that if X is infinite, then $L^2(X,\mu)$ is infinite dimensional.
(c) Show that if X is finite, then $L^2(X,\mu) \cong \mathbf{C}^{|X|}$.

—Answer on page 363

Ex 2.8.2. Prove Lemma 2.80.

—Answer on page 363

Ex 2.8.3. Complete the proof of Lemma 2.81.

—Answer on page 364

Ex 2.8.4. Show that there is a $d \in \mathcal{B}^+(G_u)$ such that

$$\int_{G(u)} d(c(v)s^{-1})\,d\beta^u(s) = 1 \quad \text{for all } v \in G^{(0)}.$$

(Compare with Corollary 3.31.)

—Answer on page 364

Ex 2.8.5. Let G be a second countable locally compact *transitive* groupoid with a Haar system $\{\lambda_u\}_{u\in G^{(0)}}$. Let $\rho : G^{(0)} \to G_u$ be a regular Borel cross section for $r : G_u \to G^{(0)}$.

(a) Show that

$$\upsilon(\gamma) = \big(r(\gamma), \rho(r(\gamma))^{-1}\gamma\rho(s(\gamma)), s(\gamma)\big)$$

is a Borel isomorphism of G onto the groupoid $H = G^{(0)} \times G(u) \times H$.

(b) Let $\Upsilon : \mathcal{B}(H) \to \mathcal{B}(G)$ be the induced map: $\Upsilon(\xi)(\gamma) = \xi(\upsilon(\gamma))$. Show that Υ maps $C_c(H)$ into $\mathcal{B}^b_c(G)$.
(c) Show that the push forward $\upsilon_*(\lambda^u)$ is the measure $\delta_u \times \beta^u \times \mu$ where μ is the measure constructed in Proposition 2.82.

—Answer on page 364

Ex 2.8.6. Establish (2.16).

—Answer on page 364

CHAPTER 3

Measure Theory

It is a simple truth that one can't study groupoids in depth and still avoid some of the niceties (and complexities) of measure theory and related matters. In this chapter, we'll record some of the technicalities we will need focusing on those that lie outside the content of the standard courses. On the other hand, most of what we need is reasonable enough that it can be assumed at a first reading. So it might be a good idea to skip this chapter and come back to the necessary bits when, or if, you want to fill in the details.

In this chapter, groupoids will always be assumed to have open range and source maps.

3.1. Radon Families

A *Radon measure* on a locally compact space X is the measure associated to a positive linear functional $\mu : C_c(X) \to \mathbf{C}$ by the Riesz Representation Theorem (see [**Fol99**, Theorem 7.2] or [**Rud87**, Theorem 2.14]). Radon measures are such that $\mu(K) < \infty$ for all compact sets $K \subset X$, and are regular in the sense that for every open set $V \subset X$,

$$\mu(V) = \sup\{\, \mu(K) : K \subset V \text{ is compact} \,\} \tag{3.1}$$

and such that for every Borel set $E \subset X$,

$$\mu(E) = \inf\{\, \mu(V) : E \subset V \text{ is open} \,\}.$$

If—as will almost always be the case here—X is second countable, then μ satisfies (3.1) for all Borel sets V, and any Borel measure on X is a Radon measure provided $\mu(K) < \infty$ for any compact set $K \subset X$ [**Rud87**, Theorem 2.18].

If μ is a Radon measure on $G^{(0)}$, then we get a Radon measure ν on G as follows. By definition of a Haar system

$$\Phi(u) = \int_G f(\gamma)\, d\lambda^u(\gamma)$$

defines a function $\Phi \in C_c(G^{(0)})$. Hence we get a positive linear functional on $C_c(G)$ via $f \mapsto \mu(\Phi)$. Naturally, we ignore the intermediate step of introducing the function Φ and just write

$$\nu(f) := \int_{G^{(0)}} \int_G f(\gamma)\, d\lambda^u(\gamma)\, d\mu(u) \quad \text{for } f \in C_c(G). \tag{3.2}$$

Of course, since ν is a Radon measure, the left-hand side of (3.2) extends to all manner of functions including non-negative Borel functions $f \in \mathcal{B}^+(G)$, integrable Borel functions $f \in \mathcal{L}^1(G, \nu)$ and $f \in \mathcal{B}^b_c(G)$, where the latter denotes the set of Bounded Borel functions $f \in \mathcal{B}^b(G)$ that vanish off a compact set. It is less clear what to make of the right-hand side in these cases. In this section we want to see that the right-hand side does make sense for these functions, and that we get equality in (3.2) whenever $f \in \mathcal{B}^+(G)$ or $f \in \mathcal{B}^b_c(G)$ provided the right-hand side is suitably interpreted.

I found it easier to establish this "obvious" formula by taking a more general approach. *I am sticking to the second countable case* for simplicity, and also since that is where our applications to groupoids are in this book. So we let X be a second countable locally compact Hausdorff space, and then we define $\mathcal{B}^b_c(X)$ and $\mathcal{B}^+(X)$ as above. For convenience, we will also use $\mathcal{B}^{+,e}(X)$ to be the Borel functions on X taking values in the extended non-negative reals $[0, \infty]$.

DEFINITION 3.1. Suppose that X and Y are locally compact Hausdorff spaces. A family $\{ \rho^y \}_{y\in Y}$ of Radon measures on X is called a *Radon family* of measures on X if

$$y \mapsto \int_X f(x)\, d\rho^y(x)$$

is in $\mathcal{B}^b_c(Y)$ for all $f \in C_c(X)$.

EXAMPLE 3.2 (Haar Systems). *The* motivating example of a Radon family is a Haar system $\{ \lambda^u \}_{u\in G^{(0)}}$ on a locally compact groupoid G.

One of the standard applications of Fubini type theorems is the measurability of sections such as

$$x \mapsto \int_Y f(x, y)\, d\mu(y)$$

for suitable f. We need the analogue for Radon families.

PROPOSITION 3.3. *Suppose that X, Y and Z are second countable locally compact spaces and let $\{\rho^z\}_{z\in Z}$ be a Radon family on X. If $F \in \mathcal{B}^b_c(X \times Y)$ then*

$$\rho(F)(y, z) := \int_X F(x, y)\, d\rho^z(x)$$

defines an element of $\mathcal{B}^b_c(Y \times Z)$.

Similarly, if $F \in \mathcal{B}^+(X \times Y)$, then $\rho(F) \in \mathcal{B}^{+,e}(Y \times Z)$.

PROOF. Let $\mathscr{A}$ be the vector space of $F \in \mathcal{B}^b_c(X \times Y)$ with $\rho(F) \in \mathcal{B}^b(Y \times Z)$. Then $\mathscr{A}$ clearly contains sums of functions of the form $F(x, y) = \varphi(x)\psi(y)$ with $\varphi \in C_c(X)$ and $\psi \in C_c(Y)$. But if $\operatorname{supp} F \subset C \times K$ with $C \subset X$ and $K \subset Y$ compact, then there are $\varphi \in C^+_c(X)$ and $\psi \in C^+_c(Y)$ such that $\|\varphi\|_\infty \leq 1$, $\|\psi\|_\infty \leq 1$ and $|F(x, y)| \leq \|F\|_\infty \varphi(x)\psi(y)$. Then $\|\rho(F)\|_\infty \leq \|F\|_\infty \|h\|_\infty$, where

$$h(z) = \int_X \varphi(x)\, d\rho^z(x),$$

and $\operatorname{supp} \rho(F) \subset \operatorname{supp} \psi \times \operatorname{supp} h$. Hence, for the first assertion, it will suffice to see that $\mathcal{B}^b_c(X \times Y) \subset \mathscr{A}$.

However, if $F_i \to F$ properly in the inductive limit topology on $C_c(X \times Y)$ with each $F_i \in \mathscr{A}$, then the above discussion implies that $\rho(F_i) \to \rho(F)$ (uniformly). Thus $F \in \mathscr{A}$. As a consequence, we have $C_c(X \times Y) \subset \mathscr{A}$.

If $U \subset X \times Y$ is a precompact open set and $\mathscr{A}(U) = \{ F \in \mathscr{A} : \operatorname{supp} F \subset U \}$, then $\mathscr{A}(U)$ is closed under monotone sequential limits by the Dominated Convergence Theorem. Hence, $\mathcal{B}^b(U) \subset \mathscr{A}(U) \subset \mathscr{A}$ by [**Ped89**, Proposition 6.2.9]. Since this holds for all such U, we have $\mathcal{B}^b_c(X \times Y) \subset \mathscr{A}$ as required.

If $F \in \mathcal{B}^+(X \times Y)$, then there are $F_i \in \mathcal{B}^b_c(X \times Y) \cap \mathcal{B}^+(X \times Y)$ with $F_i \nearrow F$. Then $\rho(F) \in \mathcal{B}^{+,e}(Y \times Z)$ by the Monotone Convergence Theorem. □

REMARK 3.4. If $\{ \rho^y \}_{y \in Y}$ is a Radon family of measures on X and $f \in C_c(X)$, then by assumption

$$y \mapsto \int_X f(x)\, d\rho^y(x) \tag{3.3}$$

defines a function $\rho(f) \in \mathcal{B}^b_c(Y)$. Hence if μ in any Radon measure on Y, then

$$\nu(f) := \int_Y \rho(f)(y)\, d\mu(y) \tag{3.4}$$

defines a positive linear functional on $C_c(X)$. Thus (3.4) defines a Radon measure ν on X. It is standard practice to use the same letter for the functional and the measure. Of course, we usually simply write

$$\nu(f) = \int_Y \int_X f(x)\, d\rho^y(x)\, d\mu(y) \tag{3.5}$$

in place of (3.4).

REMARK 3.5. If $f \in \mathcal{B}^b_c(Y)$, then we can still define $\rho(f)$ as in (3.3). Since $\rho(f) \in \mathcal{B}^b_c(Y)$ by Proposition 3.3, we can make sense of the right-hand side of (3.5) (by interpreting it to be the right-hand side of (3.4)). This raises the question of whether this is the same as the integral of f with respect to ν.

PROPOSITION 3.6. *Suppose that X and Y are second countable locally compact Hausdorff spaces and that $\{ \rho^y \}_{y \in Y}$ is a Radon family of measures on X. Let ν be the Radon measure defined by* (3.4). *If $f \in \mathcal{B}^b_c(X)$ or $\mathcal{B}^+(X)$, then*

$$\nu(f) := \int_X f(x)\, d\nu(x) = \int_Y \int_X f(x)\, d\rho^y(x)\, d\mu(y). \tag{3.6}$$

(Of course, the right-hand side of (3.6) *is to be interpreted* a la *Remark* 3.5.*)*

We'll prove Proposition 3.6 via a series of lemmas.

LEMMA 3.7. *We get a measure $\bar{\nu}$ on X via*

$$\bar{\nu}(E) := \int_Y \int_X \mathbb{1}_E(x)\, d\rho^y(x)\, d\mu(y)$$

for any Borel set $E \subset X$.

PROOF. Since $\rho(\mathbb{1}_E)$ is in $\mathcal{B}^{+,e}(Y)$, $\bar{\nu}(E)$ is defined on all Borel sets E. Clearly $\bar{\nu}(\emptyset) = 0$. To see that $\bar{\nu}$ is countably additive requires only a couple of applications of the Monotone Convergence Theorem. □

LEMMA 3.8. *For all $f \in \mathcal{B}^+(X)$,*

$$\bar{\nu}(f) := \int_X f(x)\, d\bar{\nu}(x) = \int_Y \int_X f(x)\, d\rho^y(x)\, d\mu(y). \tag{3.7}$$

In particular, $\bar{\nu}$ is a Radon measure on X.

Proof. By linearity, (3.7) holds for all nonnegative simple functions. But if $f \in \mathcal{B}^+(X)$, then there are nonnegative simple functions $f_n \nearrow f$. Thus (3.7) holds for all $f \in \mathcal{B}^+(X)$ by the Monotone Convergence Theorem.

Since X is second countable, to see that $\bar{\nu}$ is a Radon measure just requires that we see that $\bar{\nu}(K) < \infty$ for all compact sets $K \subset X$ [**Rud87**, Theorem 2.18]. But we can find $g \in C_c^+(X)$ such that $g(x) = 1$ for all $x \in K$. Then $\bar{\nu}(K) \leq \bar{\nu}(g) = \nu(g) < \infty$. □

Proof of Proposition 3.6. Note that ν and $\bar{\nu}$ are Radon measures that agree on $C_c(X)$. Therefore $\nu = \bar{\nu}$; that is, (3.6) holds as required. □

Corollary 3.9. *Suppose that $\{\lambda^u\}_{u\in G^{(0)}}$ is a continuous Haar system on a second countable locally compact (Hausdorff) groupoid G. Let μ be any Radon measure on $G^{(0)}$ and let ν the Radon measure $\nu = \mu \circ \lambda = \int_{G^{(0)}} \lambda^u \, d\mu$. Then for all $f \in \mathcal{B}_c^b(G)$ or $f \in \mathcal{B}^+(G)$, we have*

$$\nu(f) = \int_{G^{(0)}} \int_G f(\gamma) \, d\lambda^u(\gamma) \, d\mu(u),$$

where we have invoked the conventions in Remark 3.5.

Corollary 3.10. *Let $\{\lambda^u\}_{u\in G^{(0)}}$ be a Haar system on a second countable locally compact groupoid G. Suppose that $F \in \mathcal{B}_c^b(G *_r G)$ where $G *_r G = \{(\gamma, \eta) : r(\gamma) = r(\eta)\}$. Then*

$$g(\eta) = \int_G F(\gamma, \eta) \, d^{r(\eta)}(\gamma)$$

defines an element of $\mathcal{B}_c^b(G)$.

Proof. There is no harm in assuming that we can extend F to an element of $\mathcal{B}_c^b(G \times G)$. Then, using Proposition 3.3 on page 70, we get $f \in \mathcal{B}_c^b(G \times G^{(0)})$ via

$$f(\eta, u) = \int_G F(\gamma, \eta) \, d\lambda^u(\gamma).$$

Then $g(\eta) = f(\eta, r(\eta))$ and we're done. □

3.2. π-systems

Most of the Radon families of measures we need in the sequel are commonly called *π-systems* (see [**Ren87**, p. 69] or [**ADR00**, Definition 1.1.1]). If $\pi : Y \to X$ is a continuous map between locally compact spaces Y and X, then a *π-system* is a family of (positive) Radon measures $\{\beta^x : x \in X\}$ on Y such that $\operatorname{supp} \beta^x \subset \pi^{-1}(x)$ and such that for every $f \in C_c(Y)$, the function

$$x \mapsto \beta(f)(x) := \int_Y f(y) \, d\beta^x(y)$$

is continuous. We say that β is *full* if $\operatorname{supp} \beta^x = \pi^{-1}(x)$ for all $x \in X$. Unless mentioned to the contrary, we will always assume our π-systems are *proper* in the sense that $\beta^x \neq 0$ for all x.[1]

[1]In the literature, it is common to also assume the map π is open as well as continuous. This is automatic if π admits a *full* π-system by Lemma 3.20 on page 77 below. It is also a necessary hypothesis in many of the results to follow.

If Y and X are both (left) G-spaces and π is equivariant, then we say a π-system β is *equivariant* if $\gamma \cdot \beta^x = \beta^{\gamma \cdot x}$ where $\gamma \cdot \beta^x(E) = \beta^x(\gamma^{-1} \cdot E)$ for all $(\gamma, x) \in G * X = \{ (\gamma, x) : s(\gamma) = r(x) \}$. Alternatively,

$$\int_Y f(\gamma \cdot y)\, d\beta^x(y) = \int_Y f(y)\, d\beta^{\gamma \cdot x}(y)$$

for all $f \in C_c(Y)$ and $(\gamma, x) \in G * X$.

EXAMPLE 3.11 (Haar Systems). Note that a Haar system for G is a full equivariant r-system for $r : G \to G^{(0)}$ and the natural left G-actions on itself and $G^{(0)}$.

REMARK 3.12 (Equivariant vs. Invariant). In some of the literature, *equivariant* π-systems are called *invariant* π-systems. To me, "equivariant" seems the proper term, although honesty requires me to admit that I have been inconsistent in the past and even perhaps in this book. In particular, for reasons described in Remark 9.12 on page 198, we are going to find ourselves using the terms interchangeably in (at least) Chapter 9.

LEMMA 3.13. *Let $\pi : Y \to X$ be a continuous open map and let $\{\beta^x\}_{x \in X}$ be a full π-system. Let $U \subset Y$ be open.*

(a) *If $b \in C_c^+(X)$ with $\operatorname{supp} b \subset \pi(U)$, there is a $g \in C_c^+(Y)$ with $\operatorname{supp} g \subset U$ and $\beta(g) = b$.*
(b) *If $g \in C_c^+(Y)$ with $\operatorname{supp} g \subset U$ and if $\epsilon > 0$, then there is a $f \in C_c^+(Y)$ with $\operatorname{supp} f \subset U$ and*

$$|g(y) - f(y)\beta(f)(\pi(y))| < \epsilon \quad \textit{for all } y \in Y.$$

PROOF. (a) Since π is continuous and open, there is a open set $V \subset Y$ with compact closure such that $V \subset \overline{V} \subset U$ and $\operatorname{supp} b \subset \pi(V) \subset \pi(\overline{V}) \subset \pi(U)$. Let $f \in C_c^+(Y)$ be such that $f(y) = 1$ for all $y \in \overline{V}$ and $f(y) = 0$ if $x \notin U$. Since β is full, $\beta(f)(x) > 0$ for all x in the compact set $\pi(\overline{V})$. Hence there is a $m > 0$ such that $\beta(f)(x) \geq m$ for all $x \in \pi(\overline{V})$.

It follows that there is continuous function $c \in C_c^+(X)$ such that c vanishes off $\pi(U)$ and $c(x) = \beta(f)(x)^{-1}$ for all $x \in \pi(\overline{V})$. Now $g(y) = f(y)c(\pi(y))b(\pi(y))$ is what we're looking for.

(b) Let

$$C = \{ y \in Y : g(y) \geq \epsilon \}.$$

Then C is compact, and since π is full, $\beta(g)(x) > 0$ if $x \in \pi(C)$. Since $\pi(C)$ is compact, there is a $m > 0$ such that $\beta(g)(x) \geq m$ if $x \in \pi(C)$. Let

$$V = \{ x \in X : \beta(g)(x) \geq \frac{m}{2} \}.$$

Let $Q \in C_c^+(X)$ be such that $Q(x) = 1$ for $x \in \pi(C)$, $Q(x) = 0$ if $x \notin V$ and $Q(x) \leq 1$ otherwise. Then

$$H(x) = Q(x)\beta(g)(x)^{-\frac{1}{2}}$$

is in $C_c^+(X)$. Let $f(y) = g(y)H(\pi(y)) = g(y)Q(\pi(y))\beta(g)(\pi(y))^{-\frac{1}{2}}$. Observe that

$$f(y)\beta(f)(\pi(y)) = \begin{cases} g(y) & \text{if } \pi(y) \in \pi(C) \text{ and} \\ g(y)Q(\pi(y))^2 & \text{otherwise.} \end{cases}$$

If $\pi(y) \notin \pi(C)$, then $|g(y)| < \epsilon$ and
$$|g(y) - f(y)\beta(f)(\pi(y))| = |g(y) - g(y)Q(\pi(y))^2| = |g(y)||1 - Q(\pi(y))| \leq |g(y)| < \epsilon.$$
Thus f suffices. □

Suppose that the G-actions on X and Y are free and proper. Then the orbit spaces $G\backslash X$ and $G\backslash Y$ are locally compact Hausdorff spaces by Proposition 2.18 on page 33. If $\pi : Y \to X$ is an equivariant, continuous, and open surjection, then the induced map $\dot{\pi} : G\backslash Y \to G\backslash X$ (given by $\dot{\pi}(G \cdot y) = G \cdot \pi(y)$) is also continuous and open. To see this note that that $\bar{\pi} = q_X \circ \pi$ is continuous and open since the orbit

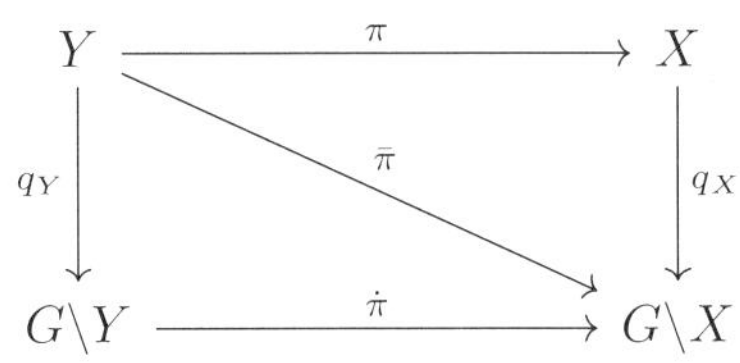

FIGURE 3.1. The map $\dot{\pi}$

map q_X is (see Figure 3.1). Hence $\dot{\pi}$ is continuous and open since q_Y is.

It is not hard to see that for each $x \in X$ that the orbit map $y \mapsto G \cdot y$ induces a homeomorphism of $\pi^{-1}(x)$ onto $\dot{\pi}^{-1}(G \cdot x)$ (Ex 3.2.2). If $\beta = \{\, \beta^x \,\}_{x \in X}$ is an equivariant π-system, then we get a Radon measure α^x on $G\backslash Y$ with support $\dot{\pi}^{-1}(G \cdot x)$ given by
$$\varphi \mapsto \int_Y \varphi(G \cdot y) \, d\beta^x(y).$$
Since β is equivariant, this measure depends only on $G \cdot x$ and we can denote it by $\dot{\beta}^{G \cdot x}$.

This raises the obvious question of whether $\dot{\beta}$ is a $\dot{\pi}$-system. If Y and X are both free and proper G-spaces, then it is not hard to see that $y \mapsto (G \cdot y, \pi(y))$ is a homeomorphism Φ of Y onto the fibred product $G\backslash Y \times_{G\backslash X} X$ (Ex 3.2.3). Thus if $F \in C_c(G\backslash Y)$ and $h \in C_c(X)$, then $f(y) := F(G \cdot y)h(\pi(y))$ is in $C_c(Y)$. Furthermore, since $\operatorname{supp} \beta^x = \pi^{-1}(x)$,
$$\begin{aligned} \beta(f)(x) &= \int_Y F(G \cdot y)h(\pi(y)) \, d\beta^x(y) \\ &= \dot{\beta}(F)(G \cdot x)h(x). \end{aligned} \tag{3.8}$$
Since β is a π-system, the left-hand side of (3.8) is continuous for any $F \in C_c(G\backslash Y)$ and any $h \in C_c(X)$. It follows that $G \cdot x \mapsto \dot{\beta}(F)(G \cdot x)$ is continuous. Hence $\dot{\beta}$ is a $\dot{\pi}$-system.

On the other hand, if we are given a $\dot{\pi}$-system $\dot{\beta}$, we can define a measure $\beta^x = \Phi_*^{-1}(\dot{\beta}^{G \cdot x} \times \delta_{\{x\}})$ on Y with support $\pi^{-1}(x)$; that is,
$$\beta^x(f) = \int_{G\backslash Y} f \circ \Phi^{-1}(G \cdot y, x) \, d\dot{\beta}^{G \cdot x}(G \cdot y).$$

Note that if $K \subset Y$ is compact, then there is a M_K such that $\dot{\beta}^{G \cdot x}(G \cdot K) \leq M_K$ for all $G \cdot x \in G\backslash X$ (Ex 3.2.1). Hence if $\operatorname{supp} f \subset K$, then $\|\beta(f)\|_\infty \leq M_K \|f\|_\infty$. The Stone-Weierstrass Theorem implies that the subspace $\mathcal{F}$ of sums of functions of the form $(G \cdot y, x) \mapsto F(G \cdot y)g(x)$ with $F \in C_c(G\backslash Y)$ and $g \in C_c(X)$ are inductive limit dense in $G\backslash Y \times X$, and hence in the fibred product as well. But if $f \circ \Phi^{-1} \in \mathcal{F}$, then (3.8) implies that $\beta(f)$ is continuous. In general, if $f \in C_c(Y)$, then there are $f_i \to f$ uniformly on Y such that the supports of the f_i are all contained in a fixed compact set K and such that each $f_i \circ \Phi^{-1}$ is in $\mathcal{F}$. Since $\|\beta(f_i) - \beta(f)\|_\infty \leq M_K \|f_i - f\|_\infty$, it follows that $\beta(f_i) \to \beta(f)$ uniformly. Hence $\beta(f)$ is continuous.

We have proved the following from [**Ren87**, Lemme 1.3].

PROPOSITION 3.14. *Let G be a locally compact groupoid with open range and source maps. Suppose that Y and X are free and proper left G-spaces and that $\pi : Y \to X$ be a continuous, open, equivariant surjection. Let $\dot{\pi} : G\backslash Y \to G\backslash X$ be the induced map.*

(a) *Suppose that $\beta = \{\beta^x\}_{x \in X}$ is an invariant π-system. Then there is a $\dot{\pi}$-system $\{\dot{\beta}^{G\cdot x}\}_{G\cdot x \in G\backslash X}$ such that*

$$\dot{\beta}(\varphi) = \int_Y \varphi(G \cdot y)\, d\beta^x(y) \quad \textit{for all } \varphi \in C_c(G\backslash Y).$$

(b) *Conversely, if $\alpha = \{\alpha^{G\cdot x}\}$ is a $\dot{\pi}$-system, then there is a π-system $\beta = \{\beta^x\}$ such that $\dot{\beta} = \alpha$.*

REMARK 3.15. Part (a) can be established without the assumption that the actions are free. See the proof of Proposition 9.23 on page 200.

As an application, since imprimitivity groupoids play an important role in the theory, we make a start on the question of when an imprimitivity groupoid has a Haar system. We'll have considerably more to say about the existence of Haar systems in Chapter 6, but it seems reasonable to record a first step here now that we have Proposition 3.14 in hand.

LEMMA 3.16 ([**KMRW98**, Proposition 5.2]). *Let G and H be second countable locally compact groupoids with open range and source maps. Suppose that Z is a free and proper left G-space and a free and proper right H-space.*

(a) *The imprimitivity groupoid $G^Z = G\backslash(Z *_r Z)$ has a Haar system $\{\lambda^{G\cdot z}\}_{G\cdot z \in G\backslash Z}$ if and only if Z admits a full equivariant r-system $\alpha = \{\alpha^z\}_{z\in Z}$ for the moment map $r : Z \to G^{(0)}$.*

(b) *The imprimitivity groupoid $H_Z = (Z *_s Z)/H$ has a Haar system $\{\underline{\lambda}^{z\cdot H}\}_{z\cdot H \in Z/H}$ if and only if Z admits a full equivariant s-system $\underline{\alpha} = \{\underline{\alpha}^z\}_{z \in Z}$ for the moment map $s : Z \to G^{(0)}$.*

REMARK 3.17. Although it is not needed for the proof, it is instructive in view of Remark 2.48 on page 48 to note that having a full π-systems for the moment maps implies that the moment maps are open by the easy part of Lemma 3.20 on page 77.

PROOF. (a) Suppose that $\alpha = \{\alpha^z\}_{z\in Z}$ is a full equivariant r-system. Let $\pi : Z *_r Z \to Z$ be the projection on the first factor. Note that π is open since r is. If we let $\beta^z = \delta_z \times \alpha^{r(z)}$, then β is full π-system. Since α is equivariant,

$$\int_{Z *_r Z} f(x, y)\, d\beta^{\gamma\cdot z}(x, y) = \int_Z f(\gamma \cdot z, w)\, d\alpha^{r(\gamma\cdot z)}(w)$$

which, since $r(\gamma \cdot z) = \gamma \cdot r(z)$, is

$$= \int_Z f(\gamma \cdot z, \gamma \cdot w)\, d\alpha^{r(z)}(w)$$
$$= \int_{Z *_r Z} f(x, y)\, d\gamma \cdot \beta^z(x, y).$$

Thus, $\gamma \cdot \beta^z = \beta^{\gamma \cdot z}$, and β is equivariant. By Proposition 3.14 on the preceding page, we get a $\dot{\pi} : G_Z \to G\backslash Z$ system $\dot{\beta}$ so that if $f \in C_c(G_Z)$ then

$$\int_{G_Z} f([x, y])\, d\dot{\beta}^{G \cdot z}([x, y]) = \int_Z f([z, w])\, d\alpha^{r(z)}(w).$$

We get a Haar system $\lambda = \{\lambda^{G \cdot z}\}_{G \cdot z \in G \backslash Z}$ by letting $\lambda^{G \cdot z} = \dot{\beta}^{G \cdot z}$. For example, left-invariance is given by

$$\int_{G_Z} f([x, y][y, v])\, d\lambda^{G \cdot y}([y, v]) = \int_Z f([x, v])\, d\alpha^{r(y)}(v)$$

which, since $r(x) = r(y)$, is

$$= \int_Z f([x, v])\, d\alpha^{r(x)}(v)$$
$$= \int_{G_Z} f([x, v])\, d\lambda^{G \cdot x}([x, v]).$$

For the converse, let $\lambda = \{\lambda^{G \cdot z}\}_{G \cdot z \in G \backslash Z}$ be a Haar system on G^Z. Then $\dot{\beta} = \lambda$, is a $\dot{\pi}$-system for the map $r_{G^Z} = \dot{\pi} : G\backslash(Z *_r Z) \to G\backslash Z$. The second part of Proposition 3.14 on the previous page gives us a compatible π system β. But β^z must be of the form $\delta_z \times \alpha^z$ with support $\alpha^z = r^{-1}(z)$. Moreover, if $f \in C_c(Z *_r Z)$ is of the form $f(z, w) = F([z, w])g(w)$ for $F \in C_c(G^Z)$ and $g \in C_c(Z)$, then

$$\beta^z(f) = g(z)\lambda^{G \cdot z}(F). \tag{3.9}$$

Since $(z, w) \mapsto ([z, w], z)$ is a homeomorphism of $\{x\} \times r^{-1}(z)$ onto $(G^Z)^{G \cdot z} \times \{z\}$, and since functions such used in (3.9) span a dense subspace, we must have

$$\lambda^{G \cdot z}(F) = \int_{G^Z} F([z, w])\, d\alpha^z(w).$$

Since $[z, w] = [\gamma \cdot z, \gamma \cdot w]$, replacing z by $\gamma \cdot z$ shows that $\alpha^{\gamma \cdot z} = \gamma \cdot \alpha^z$ and α is equivariant.

(b) The situation is completely symmetric, but we give some details for future reference. Let Z^{op} be as in Proposition 2.51 on page 49 so that Z^{op} is a free and proper left H-space such that $\eta \cdot \bar{z} = \overline{z \cdot \eta^{-1}}$. Since Z^{op} and Z are the same space, if $\{\underline{\alpha}_z\}_{z \in Z}$ is full equivariant s-system of measures on Z, then we get a full equivariant r-system $\{\alpha^{\bar{z}}\}_{\bar{z} \in Z^{\mathrm{op}}}$ on Z^{op} via

$$\alpha^{\bar{z}}(f) = \underline{\alpha}_z(f).$$

By part (a), we get a Haar system on $H^{Z^{\mathrm{op}}}$. Since $[\bar{z}, \bar{w}] \mapsto [z, w]$ is an isomorphism of $H^{Z^{\mathrm{op}}}$ onto H_Z, we get a Haar system on H_Z given by

$$\underline{\lambda}^{z \cdot H}(f) = \int_Z f([z, w])\, d\underline{\alpha}_{s(z)}(w).$$

Conversely, if H_Z has a Haar system, then so does $H^{Z^{\mathrm{op}}}$. Then Z^{op} must have a full equivariant r-system by part (a). Hence Z has a full equivariant s-system. □

Let $\pi : Y \to X$ be a continuous surjection. A subset $A \subset Y$ is called *π-compact* if $A \cap \pi^{-1}(K)$ is compact for every compact set $K \subset X$. We write $C_{c,\pi}(Y)$ for the set of continuous functions on Y whose support is π-compact.

PROPOSITION 3.18 ([**ADR00**, Lemma 1.1.2]). *Let $\pi : Y \to X$ be a continuous surjection and $\beta = \{\beta^x\}_{x\in X}$ a proper π-system. If X is paracompact, then there is a $h \in C^+_{c,\pi}(Y)$ such that $\beta(h)(x) - 1$ for all $x \in X$.*

PROOF. Let $\mathcal{V} = \{ V_i \}_{i\in I}$ be a locally finite cover of X by open sets V_i with compact closure. Let $\{ \alpha_i \}$ be a partition of unity on X subordinate to $\mathcal{V}$ (see [**RW98**, Lemma 4.34]). For each $x \in V_i$ we can find a function $f_x \in C^+_c(Y)$ such that $\beta(f_x)(x) > 0$. Since $\overline{V_i}$ is compact, we can combine finitely many such f_x to produce $f_i \in C^+_c(Y)$ such that $\beta(f_i)(x) > 0$ for all $x \in V_i$. Define

$$g(y) = \sum_i f_i(y)\alpha_i(\pi(y)). \tag{3.10}$$

Since $\mathcal{V}$ is locally finite, for each $y_0 \in Y$ there is a finite set $F \subset I$ such that $g(y)$ is given near y_0 by the sum over $i \in F$ in (3.10). In particular, g is well-defined and continuous on Y. Local finiteness also insures that each compact subset of X meets at most finitely many V_i. Since $\{\alpha_i\}$ is subordinate to $\mathcal{V}$, this ensures g has π-compact support. We have chosen the f_i so that $\beta(g)(x) > 0$ for all x. Thus we can define $h(y) = \beta(g)(\pi(y))^{-1}g(y)$. □

LEMMA 3.19. *Let $\pi : Y \to X$ be a continuous surjection and $\beta = \{\beta^x\}_{x\in X}$ a proper π-system. If Y is second countable then there is a $g \in C(Y)$ such that $g(y) > 0$ for all $y \in Y$ and $\beta(g)(x) = 1$ for all $x \in X$.*

PROOF. Since Y is second countable, $Y = \bigcup_{n=1}^\infty K_n$ with each K_n compact and $K_n \subset K_{n+1}$. Then we can find $g_n \in C^+_c(Y)$ such that $g_n(y) > 0$ for all $y \in K_n$. Multiplying by an appropriate constant allows us to assume both g_n and $\beta(g_n)$ are bounded by 1. Let $g' = \sum_n 2^{-n}g_n$. Then $\beta(g')(x) > 0$ for all x. Let

$$g(y) = \beta(g')(\pi(y))^{-1}g'(y).$$ □

LEMMA 3.20 (Blanchard). *Suppose that $\pi : Y \to X$ is a continuous surjection.*

(a) *If $\beta = \{\beta^x\}_{x\in X}$ is a full π-system, then π is open.*

(b) *If X and Y are second countable and π is open, then there is a full π-system $\alpha = \{\alpha^x\}_{x\in X}$ consisting of probability measures.*

PROOF. Suppose that β is a full π system. Let $V \subset Y$ be a nonempty open subset. Fix $x_0 \in \pi(V)$ and $y_0 \in V$ such that $\pi(y_0) = x_0$. Choose U open such that $y_0 \in U \subset \overline{U} \subset V$ and $\overline{U}$ is compact. Let $f \in C^+_c(Y)$ such that $f(y_0) = 1$ and $\operatorname{supp} f \subset U$. Since β is full, $\beta(f)(x_0) > 0$. But $\beta(f)$ is continuous and $\operatorname{supp} \beta(f) \subset \pi(\overline{U}) \subset \pi(V)$. Hence x_0 is an interior point of $\pi(V)$. Since x_0 was arbitrary, $\pi(V)$ is open.

The proof (b) is very subtle and is due to Blanchard [**Bla96**, Theorem 3.3] who employs a selection result of Michael's [**Mic56**, Theorem 3.2]. A complete proof is given in Appendix B—see Corollary B.18 on page 340.[2] □

REMARK 3.21. The first part of the proof of Lemma 3.20 on the preceding page is routine and does not require separability. The proof of the second part, which requires Blanchard's machinery from Appendix B on page 331, definitely requires separability.

EXERCISES

Ex 3.2.1. Suppose that $\pi : Y \to X$ is a continuous open map and that $\beta = \{\beta^x\}$ is a π-system. Show that for any compact set $K \subset Y$, there is a constant K such that $\beta^x(K) \leq M$ for all $x \in X$.

—Answer on page 365

Ex 3.2.2. Suppose that $\pi : Y \to X$ is a continuous equivariant open surjection between free and proper G-spaces, and that $\dot{\pi} : G\backslash Y \to G\backslash X$ is the induced map. Verify that the orbit map $q_Y : Y \to G\backslash Y$ induces a homeomorphism of $\pi^{-1}(x)$ onto $\dot{\pi}^{-1}(G \cdot x)$. (Recall that G is assumed to have open range and source maps in this section.)

—Answer on page 365

Ex 3.2.3. Suppose that $\pi : Y \to X$ is a continuous equivariant open surjection between free and proper G-spaces. Let

$$G\backslash Y \times_{G\backslash X} X = \{ (G \cdot y, x) : G \cdot \pi(y) = G \cdot x \}$$

be the fibred product. Show that $y \mapsto (G \cdot y, \pi(y))$ is a homeomorphism of Y onto $G\backslash Y \times_{G\backslash X} X$.

—Answer on page 365

Ex 3.2.4. Suppose that $\pi : Y \to X$ is a continuous surjection and that β is a π-system. Show that if $f \in C_{c,\pi}(Y)$, then $\beta(f)$ is continuous.

—Answer on page 365

[2]There is a slight gap in the proof of [**Bla96**, Lemme 3.8]: it is asserted there that if A is a separable C^*-algebra, then its dual A^*, equipped with the topology of uniform convergence on compact subsets of A, is a Fréchet space. However, if A is infinite dimensional, then A^* is not first countable in this topology and so not even a metric space let alone Fréchet. Unfortunately, this invalidates the application of [**Bla96**, Lemme 3.6] in the proof of [**Bla96**, Proposition 3.9]. Fortunately, [**Bla96**, Lemme 3.6] can be modified (using the notation of [**Bla96**, Lemme 3.6]) to case where the topological vector space E contains *complete* convex subset B containing the domain T. Thus the main results of [**Bla96**, §3] remain true as stated. I've added Appendix B to show how this can be done.

3.3. Complex Radon Measures

In the proof of the Disintegration Theorem in Chapter 8 and in our discussion of amenability in Chapters 9 and 10, we will need some facts about *complex Radon measures* that are usually not covered in the standard measure theory courses we all teach. Although much of what we need can be found in authorities like [**Edw65**, Chap. 4], we'll discuss the basics here. (In particular, complex Radon measures are defined in [**Edw65**, Definition 4.3.1], and the Radon-Nikodym Theorem we need can be sorted out from [**Edw65**, §§4.15.7–9].)

Let X be a locally compact *Hausdorff* space. For simplicity, we will assume that X is second countable. By definition, a (positive) *Radon measure* on X is a regular Borel measure associated to a positive linear functional $\mu : C_c(X) \to \mathbf{C}$ via the Riesz Representation Theorem. It is standard practice amongst the cognoscenti to identify the measure and the linear functional, and we will do so here—cognoscente or not. Additionally, we don't add the adjective "positive" unless we're trying to be pedantic. Notice that if μ is a Radon measure on X, then $\mu : C_c(X) \to \mathbf{C}$ is continuous in the inductive limit topology. Motivated by this, we define a *complex Radon measure* on X to be a linear functional $\nu : C_c(X) \to \mathbf{C}$ which is continuous in the inductive limit topology.[3] If ν is actually bounded with respect to the supremum norm on $C_c(X)$, so that ν extends to a bounded linear functional on $C_0(X)$, then ν is naturally associated to a bona fide complex measure on X whose total variation norm coincides with the norm of ν as a linear functional [**Rud87**, Theorem 6.19]. However, in general, a complex Radon measure need not be bounded.[4] While we can't associate an actual set function to ν, we can, with a little work, associate an integral.

If ν is a complex Radon measure, then let $\varphi = \operatorname{Re}\nu$. Then φ is a real linear functional on $C_c(X)$ (viewed as a real vector space). Fix $f \in \mathbf{C}_c^+(X)$ and consider

$$\{\, \varphi(g) \in \mathbf{R} : |g| \leq f \,\}. \tag{3.11}$$

If (3.11) were not bounded, then we could find g_n such that $|g_n| \leq \frac{1}{n} f$ and such that $|\varphi(g_n)| \geq n$. This gives us a contradiction since $g_n \to 0$ in the inductive limit topology. Consequently, φ is *relatively bounded* as defined in [**HR63**, Definition B.31]. Consequently [**HR63**, Theorem B.36] implies that $\varphi = \mu_1 - \mu_2$ where each μ_i is a positive linear functional on $C_c(X)$; that is, each μ_i is a Radon measure. Applying the same analysis to the complex part of ν, we find that there are Radon measures μ_i such that $\nu = \mu_1 - \mu_2 + i(\mu_3 - \mu_4)$, and for each $f \in C_c(X)$, we have

$$\nu(f) = \int_X f(x)\,d\mu_1(x) - \int_X f(x)\,d\mu_2(x) + i\int_X f(x)\,d\mu_3(x) - i\int_X f(x)\,d\mu_4(x).$$

In general, the μ_i will not be finite measures. Hence it makes no sense to talk about $\mu_1 - \mu_2 + i(\mu_3 - \mu_4)$ as a complex measure on X. We nevertheless want a "measure theory" associated to ν. To this end, we can define $\mu_0 := \mu_1 + \mu_2 + \mu_3 + \mu_4$. Since we are assuming that X is second countable, the Radon measures μ_i are

[3] As we shall see in the next paragraph, a complex Radon measure must be relatively bounded. Hence, if X is compact, then ν is always bounded as a linear functional on $C(X)$, and we're back in the standard textbooks.

[4] The distinction between a complex Radon measure and a complex measure is subtle and not always observed in the literature. Sometimes it is necessary to see the context to determine which definition is intended. I have attempted to be careful here, but it is still possible that the term "complex measure" might be used where "complex Radon measure" was meant.

necessarily σ-finite. Then $\mu_i \ll \mu$ for all i and we can invoke the Radon-Nikodym Theorem to assert that there are Borel functions $h_i : X \to [0, \infty)$ such that $\mu_i = h_i\mu$. Since the h_i are finite-valued, we can define a $\mathbf{C}$-valued Borel function by $h = h_1 - h_2 + ih_3 - ih_4$. For each $f \in C_c(X)$, we have

$$\nu(f) = \int_X f(x)h(x)\,d\mu_0(x).$$

We can write $h(x) = \rho(x)p(x)$ for a nonnegative Borel function p and a unimodular Borel function ρ. Replacing $p\mu_0$ by μ, we then have

$$\nu(f) = \int_X f(x)\rho(x)\,d\mu(x) \quad \text{for all } f \in C_c(X). \tag{3.12}$$

If, for example, X is compact, then it is well-known that the measure μ appearing in (3.12) is unique, and that ρ is determined μ-almost everywhere. If X is second countable, and therefore σ-compact, then we see that μ and ρ satisfy the same uniqueness conditions. As in the compact case, we will write $|\nu|$ for μ and call $|\nu|$ the *total variation* of ν. Thus we get a bona fide integral

$$\nu(f) = \int_X f(x)\rho(x)\,d|\nu|(x)$$

with ρ unimodular exactly as in the classical setting.

Since Radon measures are finite on compact subsets, we can certainly make perfectly good sense out of $\nu(f)$ for any $f \in \mathcal{B}^b_c(X)$—that is, for any bounded Borel function f which vanishes outside a compact set—simply by using (3.12). (In fact, we can make sense out of $\nu(f)$ whenever $f \in \mathcal{L}^1(|\nu|)$.) In particular, if B is a *pre-compact*[5] Borel set in X, then we will happily write $\nu(B)$ for $\nu(\mathbb{1}_B)$. We say that a Borel set (possibly not pre-compact) is *locally ν-null* if $\nu(B \cap K) = 0$ for all compact sets $K \subset X$.

We will also need a version of the Radon-Nikodym Theorem for our complex Radon measures. Specifically, we suppose that μ is a Radon measure and that ν is a complex Radon measure such that $\nu \ll \mu$—that is, $\mu(B) = 0$ implies B is locally ν-null. If $\nu \ll \mu$ and if $\mu(E) = 0$, then for each Borel set $F \subset E$, we have

$$\int_F \rho(x)\,d|\nu|(x) = 0.$$

It follows that $\rho(x) = 0$ for $|\nu|$-almost all $x \in E$. Since $|\rho(x)| = 1 \neq 0$ for all x, we must have $|\nu|(E) = 0$. That is $\nu \ll \mu$ if and only if $|\nu| \ll \mu$. Therefore there is a Borel function $\varphi : X \to [0, \infty)$ such that

$$\nu(f) = \int_X f(x)\rho(x)\,d|\nu|(x) = \int_X f(x)\rho(x)\varphi(x)\,d\mu(x),$$

and we call $\frac{d\nu}{d\mu} := \varphi\rho$ the Radon-Nikodym derivative of ν with respect to μ. Of course, $\frac{d\nu}{d\mu}$ is determined μ-almost everywhere.

---

[5]We say that a set is pre-compact if it is contained in a compact subset. Alternatively, B is pre-compact if its closure is compact.

EXERCISES

Ex 3.3.1. Suppose that μ is a (positive) Radon measure on X. Show that the associated linear functional $\mu : C_c(X) \to \mathbf{C}$ is continuous in the inductive limit topology.

—Answer on page 365

3.4. The Fell Topology on the Space of Subgroups

In Chapter 5 and elsewhere, we are going to see that sub*groups* of a groupoid G play a critical role in the representation theory and structure theory for $C^*(G)$. The set of all subgroups can be endowed with a topology that has proven very useful in applications. As described in [**Wil07**, §H.1], the set $\mathscr{C}(G)$ of closed subsets of G has a compact Hausdorff topology—called the *Fell topology* [**Wil07**, Proposition H.3]. A basis for this topology is indexed by compact sets $K \subset G$ and finite families $\mathcal{F} = \{ U_1, \dots, U_n \}$ of open sets $U_k \subset G$ where

$$\mathcal{U}(K;\mathcal{U}) = \{ F \in \mathscr{C}(G) : F \cap K = \emptyset \text{ and } F \cap U_k \neq \emptyset \text{ for all } k \}.$$

Since we are assuming G is second countable, it follows that $\mathscr{C}(G)$ is second countable as well [**Wil07**, Lemma H.7]. The following is a useful criteria for convergence in $\mathscr{C}(G)$ and is taken from [**Wil07**, Lemma H.2].[6]

LEMMA 3.22. *Let $\{F_i\}_{i \in I}$ be a net in $\mathscr{C}(G)$. If $F_i \to F$ in $\mathscr{C}(G)$, then*

(a) *if $\gamma_i \in F_i$ and $\gamma_i \to \gamma$ in G, then $\gamma \in F$, and*

(b) *if $\gamma \in F$ then there is a subnet $\{F_{i_j}\}$ and $t_j \in F_{i_j}$ such that $t_j \to t$.*

On the other hand, if every subnet of $\{F_i\}_{i \in I}$ has properties (a) *and* (b)*, then $F_i \to F$ in $\mathscr{C}(G)$.*

We let Σ_0 be the space of closed subgroups of G viewed as a subset of $\mathscr{C}(G)$. Note that if $H \in \Sigma_0$, then there is a $u \in G^{(0)}$ such that H is a subgroup of the stability group $G(u)$. Using Lemma 3.22, it is not hard to see that $\Sigma_0 \cup \emptyset$ is closed in $\mathscr{C}(G)$ (Ex 3.4.2). Hence Σ_0 is a second countable locally compact Hausdorff space in the Fell topology. Furthermore the map $p_0 : \Sigma_0 \to G^{(0)}$ given by $p(H) = u$ if $H \subset G(u)$ is continuous (Ex 3.4.1). Notice also that if $K \subset G^{(0)}$ is compact, then $p_0^{-1}(K) \subset \Sigma_0$ is compact (Ex 3.4.3). That is, Σ_0 is conditionally compact over $G^{(0)}$ [**Ren91**, §1]. We let $p : \Sigma \to \Sigma_0$ be the associated groupoid group bundle over Σ_0:

$$\Sigma = \{ (u, H, t) \in G^{(0)} \times H \times G : u = p(H) \text{ and } t \in H \}$$

where $p(u, H, t) = H$. (The elements of Σ have been written redundantly to make some of the subsequent computations easier to follow.) Notice that if $(u, H, \sigma) \in \Sigma$, then $H \subset G(u)$. Since Σ is closed in $G^{(0)} \times \Sigma_0 \times G$, it is locally compact Hausdorff.

In the sequel, we will want to make a "continuous choice of Haar measures" on the $H \in \Sigma_0$ so that given any $f \in C_c(G)$ the map

$$H \mapsto \int_H f(h)\, d\beta^H(h) \tag{3.13}$$

is continuous. The existence of a continuous choice of Haar measures is implied by the next proposition.

---

[6]Part of the statement in [**Wil07**] is incorrect as it omits the requirement to pass to a subnet in the converse.

PROPOSITION 3.23. *There is a Haar system* $\{ \beta^H \}_{H \in \Sigma_0}$ *for the groupoid group bundle* Σ.

Since the bundle map from Σ to Σ_0 is clearly open (Ex 3.4.4), the proposition is a special case of Theorem 6.9 on page 143. We will assume it here and postpone the proof to Chapter 6.

To see that $\{\beta^H\}$ satisfies (3.13), note that if $f \in C_c(G)$, then $(u, H, t) \mapsto f(t)$ is in $C_c(\Sigma)$. Hence

$$H \mapsto \int_H f(t)\, d\beta^H(t)$$

is continuous. It will be helpful later to make a more general observation. Note that if $N \subset \Sigma$ is compact, then

$$\{ t \in G : \text{there is an } H \in \Sigma_0 \text{ such that } (p(H), H, t) \in N \}$$

is compact.

LEMMA 3.24. *Let* X *be a locally compact space. Suppose that* $F \in C_c(X \times \Sigma)$. *Then*

$$(x, H) \mapsto \int_H F(x, p(H), H, t)\, d\beta^H(t) \tag{3.14}$$

is continuous on $X \times \Sigma_0$.

PROOF. By taking $f \in C_c^+(G)$ identically one on a compact set K, it follows from (3.13) that $\{\beta^H(K)\}_{U \in \Sigma_0}$ is bounded.

Suppose that $\operatorname{supp} F \subset M \times (C * \Sigma_0 \times K)$ where M is compact in X, C is compact in $G^{(0)}$ and K is compact in G. Let $\Lambda(F)(x, u, H)$ be given by the right-hand side of (3.14). Then $|\Lambda(F)| \leq \|F\|_\infty \sup_{H \in \Sigma_0} \beta^H(K)$ and $\Lambda(F)$ is supported in $M \times C * \Sigma_0$. It follows that the set of $F \in C_c(X \times \Sigma)$ for which $\Lambda(F)$ is continuous is closed in the inductive limit topology. Since that set includes sums of functions of the form $(x, u, H, t) \mapsto \varphi(x)\psi(u, H, t)$ for $\varphi \in C_c(X)$ and $\psi \in C_c(\Sigma)$, the result follows. □

Note that G acts on Σ_0 in the obvious way: let $G * \Sigma_0 = \{ (\sigma, H) \in G \times \Sigma_0 : s(\sigma) = p(H) \}$. If $(\sigma, H) \in G * \Sigma_0$, then we define $\sigma \cdot H = \{ \sigma t \sigma^{-1} : t \in H \}$. It is not hard to check that the action is continous (Ex 3.4.5). The following is a slight variation on [**Ren91**, Lemma 1.6].

LEMMA 3.25. *There is a continuous map* $\omega : G * \Sigma_0 \to (0, \infty)$ *such that*

$$\int_H f(\sigma\gamma\sigma^{-1})\, d\beta^H(\gamma) = \omega(\sigma, H) \int_{\sigma \cdot H} f(\gamma)\, d\beta^{\sigma \cdot H}(\gamma) \tag{3.15}$$

for all $f \in C_c(G)$. *Furthermore, for all* $\sigma, \tau \in G$ *and* $H \in \Sigma_0$, *we have*

$$\omega(\sigma\tau, H) = \omega(\tau, H)\omega(\sigma, \tau \cdot H) \quad \text{and} \quad \omega(\sigma, H)^{-1} = \omega(\sigma^{-1}, \sigma \cdot H). \tag{3.16}$$

REMARK 3.26. The cocycle condition (3.16) above amounts to saying that ω is a groupoid homomorphism from the transformation groupoid $\mathcal{A} = \{ (\sigma \cdot H, \sigma, H) : (\sigma, H) \in G * \Sigma_0 \}$ to the multiplicative positive reals where $(\sigma \cdot H, \sigma, H) \mapsto \omega(\sigma, H)$.

SKETCH OF THE PROOF. The existence of $\omega(\sigma, H)$ follows from the uniqueness of the Haar measure on $\sigma \cdot H$. The fact that ω is continuous follows from the fact that both integrals in (3.15) are continuous with respect to (σ, H). Equation (3.16) is a straightforward computation. □

Looking at elementary examples such as tranformation groups shows that we can rarely expect the *isotropy map* sending $u \in G^{(0)}$ to $G(u)$ to be continuous from $G^{(0)}$ into Σ_0. Nevertheless, we have introduced the Fell topology at least in part to study this map. In Chapter 11, the next observation will be important.

LEMMA 3.27. *The isotropy map $u \mapsto G(u)$ is a Borel map from $G^{(0)}$ to Σ_0.*

PROOF. Recall that a basis for the topology on Σ_0 is given by the sets

$$U(K;\mathcal{F}) = \{\, H \in \Sigma_0 : K \cap H = \emptyset \text{ and } U \cap F \neq \emptyset \text{ for all } F \in \mathcal{F}\,\}$$

where K is compact in G and $\mathcal{F}$ is a finite collection of open sets in G. Let $\mathfrak{S} : G^{(0)} \to \Sigma_0$ be the isotropy map. Since the topology is second countable, it will suffice to see that $\mathfrak{S}^{-1}(U(K,\mathcal{F}))$ is Borel.

For any $S \subset G$, let

$$W(S) = \mathfrak{S}^{-1}\big(\{\, H \in \Sigma_0 : H \cap S \neq \emptyset\,\}\big) = \{\, u \in G^{(0)} : G(u) \cap S \neq \emptyset\,\}.$$

If K is compact, then it is not hard to see that $W(K)$ is closed: let $\{u_n\} \subset W(K)$ be such that $u_n \to u$. Let $\gamma_n \in K$ be such that $\gamma_n \in G(u_n)$. Since K is compact, we can pass to a subsequence, relabel, and assume that $\gamma_n \to \gamma \in K$. Then $\gamma \in G(u)$ and $u \in W(K)$ as claimed.

If U is open, then U is σ-compact and it follows that $W(U)$ is the countable union of closed sets. That is, $W(U)$ is a F_σ set and hence Borel. Since

$$\mathfrak{S}^{-1}(U(K;\mathcal{F})) = W(K)^c \cap \bigcap_{F \in \mathcal{F}} W(U),$$

the result follows. □

We will also need some constructions paralleling those for groups in [**Wil07**, §H.3]. Define an equivalence relation on $G * \Sigma_0$ by $(\gamma, H) \sim (\eta, H)$ if $r(\gamma) = r(\eta)$ and $\gamma^{-1}\eta \in H$. Note that $(\gamma, H) \sim (\eta, H)$ exactly when $\eta = \gamma t$ with $t \in H$.

LEMMA 3.28. *The quotient map $q : G*\Sigma_0 \to G*\Sigma_0/{\sim}$ is open and the quotient space $G * \Sigma_0/{\sim}$ is a second countable locally compact Hausdorff space.*

PROOF. We can let Σ act on the right of $G * \Sigma_0$: $(\gamma, H) \cdot (u, H, t) := (\gamma t, H)$. Then $G * \Sigma_0/{\sim}$ is the orbit space. Using (PA3) of Proposition 2.17 on page 32, it follows that the action is proper. Since Σ has a Haar system, $G*\Sigma_0/{\sim}$ is Hausdorff by Proposition 2.18 on page 33, and the quotient map is open by Proposition 2.12 on page 31. □

LEMMA 3.29. *Suppose that $F \in C_c(X \times G)$. Let $X \times \Sigma_0 * G = \{\, (x, H, \gamma) : p(H) = s(\gamma)\,\}$. Then*

$$\Lambda(F)(x, H, \gamma) = \int_H F(x, \gamma t)\, d\beta^H(t)$$

is continuous on $X \times \Sigma_0 * G$ and defines an element of $C_c(X \times \Sigma_0 * G/{\sim})$.

PROOF. Note that if F is supported in $L \times K$ with $L \subset X$ and $K \subset G$ compact, then $\Lambda(F)$ is supported in the compact set $L \times q(p_0^{-1}(s(K)) \times K)$. Let $F'(x, \gamma, u, H, t) = F(x, \gamma t)$ on the closed set $X \times G * \Sigma$. Since we can assume (x, H, γ) all belong to a compact set, we can assume F' has compact support and extend it to an element of $C_c(X \times G \times \Sigma)$. Now the continuity follows from Lemma 3.24 on the preceding page. □

PROPOSITION 3.30. *Let $q : G * \Sigma_0 \to G * \Sigma_0/\sim$ be the quotient map. There is a full q-system $\{\sigma_{q(\gamma,H)}\}$ of Radon measures on $G * \Sigma_0$ such that*

$$\sigma_{q(\gamma,H)}(f) = \int_H f(\gamma t, H)\, d\beta^H(t).$$

PROOF. The only issue is continuity and that follows from Lemma 3.29 on the preceding page. □

As an immediate corollary, we obtain what is often called an *Bruhat approximate cross section.*

COROLLARY 3.31. *There is a continuous function $b : G * \Sigma_0 \to [0, \infty)$ such that*

(a) *For all $(\gamma, H) \in G * \Sigma_0$,*

$$\int_H b(\gamma t, H)\, d\beta^H(t) = 1,$$

and

(b) *for every compact set $K \subset G$,*

$$\operatorname{supp} b \cap q^{-1}(q(K * \Sigma_0))$$

is compact.

PROOF. By Proposition 3.18 on page 77 there is a $b \in C^+_{c,q}(G * \Sigma_0)$ satisfying (a) and (b). □

EXERCISES

Ex 3.4.1. Show that $p_0 : \Sigma_0 \to G^{(0)}$ is continuous.

—Answer on page 365

Ex 3.4.2. Show that $\Sigma_0 \cup \emptyset$ is closed in $\mathscr{C}(G)$ in the Fell topology. Hence Σ_0 is locally compact.

—Answer on page 365

Ex 3.4.3. Show that if $K \subset G^{(0)}$ is compact, then $p_0^{-1}(K)$ is compact in Σ_0.

—Answer on page 365

Ex 3.4.4. Show that the projection $p : \Sigma \to \Sigma_0$ is both continuous and open.

—Answer on page 366

Ex 3.4.5. Show that the action of G on Σ_0 given by $\gamma \cdot H = \{\, \gamma t \gamma^{-1} : t \in H \,\}$ is continuous.

—Answer on page 366

3.5. Borel Hilbert Bundles

Especially in Section 7, we will need the technology of Borel Hilbert bundles and direct integrals.[7] The material in this section is taken from [**Wil07**, Appendix F.1] which has its origins in Ramsay's treatment in [**Ram76**, p. 264+]. We recall some of the basics here, but rely on [**Wil07**] for the details and many of the proofs.

We start with a collection

$$\mathscr{H} := \{\, \mathcal{H}(x) \,\}_{x\in X}$$

of complex Hilbert spaces. Then the total space is the disjoint union

$$X * \mathscr{H} := \{\, (x,h) : h \in \mathcal{H}(x) \,\},$$

and we let $\pi : X * \mathscr{H} \to X$ be the obvious map.

DEFINITION 3.32. Let $\mathscr{H} = \{\, \mathcal{H}(x) \,\}_{x\in X}$ be a family of Hilbert spaces. Then $(X * \mathscr{H}, \pi)$ is an analytic (standard) Borel Hilbert Bundle if $X * \mathscr{H}$ has an analytic (standard) Borel structure such that

(a) E is a Borel subset of X if and only if $\pi^{-1}(E)$ is Borel in $X * \mathscr{H}$,

(b) there is sequence $\{\, f_n \,\}$ of sections such that

 (i) the maps $\tilde{f}_n : X * \mathscr{H} \to \mathbf{C}$ are each Borel where

$$\tilde{f}_n(x,h) := \big(f_n(x) \mid h\big),$$

 (ii) for each n and m,

$$x \mapsto \big(f_n(x) \mid f_m(x)\big)$$

 is Borel, and

 (iii) the functions $\{\, \tilde{f}_n \,\}$, together with π, separate points of $X * \mathscr{H}$.

EXAMPLE 3.33 (Trivial Bundles). Let $\mathcal{H}$ be any separable Hilbert space with orthonormal basis $\{\, e_n \,\}$, and let X be an analytic (standard) Borel space. If we let $\mathcal{H}(x) := \mathcal{H}$ for all x, then with the product Borel structure on $X * \mathscr{H} = X \times \mathcal{H}$, $(X \times \mathcal{H}, \pi)$ becomes an analytic (standard) Borel Hilbert bundle with respect to the fundamental sequence given by $f_n(x) := e_n$ for all x. More generally, let $X = X_\infty \cup X_1 \cup X_2 \cup \dots$ be a Borel partition of X, and let $\mathcal{H}_d$ be a Hilbert space of dimension d with orthonormal basis $\{\, e_n^d \,\}_{n=1}^{n=d}$ for $d = 1, \dots, \aleph_0$. For each $x \in X_d$, let $\mathcal{H}(x) = \mathcal{H}_d$. Then we can give $X * \mathscr{H} = \coprod_{d=1}^{d=\infty} X_d \times \mathcal{H}_d$ the obvious Borel structure coming from the product Borel structure on each factor.[8] Then $X * \mathscr{H}$ is an analytic (standard) Borel Hilbert bundle with respect to the fundamental sequence

$$f_n(x) = \begin{cases} e_n^d & \text{if } x \in X_d \text{ and } 1 \le n \le d, \text{ and} \\ 0 & \text{otherwise.} \end{cases}$$

[7] Borel Hilbert bundles are the total space of a direct integral of Hilbert spaces as described, for example, in Dixmier [**Dix81**].

[8] If X and Y are Borel spaces, then the disjoint union $X \coprod Y$ is a Borel space where $B \subset X \coprod Y$ is Borel if and only if $B \cap X$ and $B \cap Y$ are Borel. This generalizes in a straightforward way to *countable* disjoint unions.

The following is a summary of [**Wil07**, Proposition F.6 and Remark F.3].

PROPOSITION 3.34. *Let $X * \mathscr{H}$ be an analytic Borel Hilbert Bundle as above with fundamental sequence $\{f_n\}$.*

(a) *A section $f : X \to X * \mathscr{H}$ is a Borel map if and only if $x \mapsto \big(f(x) \mid f_n(x)\big)$ is Borel for all n. The collection of Borel sections is denoted by $B(X * \mathscr{H})$.*
(b) *If $f, g \in B(X * \mathscr{H})$, then $x \mapsto \big(f(x) \mid g(x)\big)$ is Borel. In particular, $x \mapsto \|f(x)\|$ is Borel for all $f \in B(X * \mathscr{H})$.*
(c) *For each $d \in \{1, 2, \dots, \aleph_0\}$,*
$$X_d = \{\, x \in X : \dim \mathcal{H}(x) = d \,\}$$
is Borel.

We say two analytic Borel Hilbert bundles $X * \mathscr{H}$ and $X * \mathscr{K}$ over X are isomorphic if there is a bundle map $\varphi : X * \mathscr{H} \to X * \mathscr{K}$ which is a Borel isomorphism and is such that $(x, h) \mapsto \varphi(x, h)$ is a unitary isomorphism of $\mathcal{H}(x)$ and $\mathcal{K}(x)$.

The flexibility of Borel maps makes the following isomorphism result possible. It should be noted however, that in practice, it is not always natural to invoke it.

PROPOSITION 3.35. *Let $X * \mathscr{H}$ be an analytic Hilbert bundle. Let $X_d = \{\, x \in X : \dim \mathcal{H}(x) = d \,\}$ be as in Proposition 3.34. Let $\mathcal{H}_d$ be a Hilbert space of dimension d. Then $X * \mathscr{H}$ is isomorphic to $\coprod_{d=1}^{d=\infty} X_d \times \mathcal{H}_d$ (as in Example 3.33 on the preceding page).*

PROOF. This is [**Wil07**, Corollary F.9]. □

Given a Borel Hilbert bundle $X * \mathscr{H}$ and a Radon measure μ on X, we can follow [**Wil07**, Appendix F.2] and form the vector space
$$\mathcal{L}^2(X * \mathscr{H}, \mu) = \{\, f \in B(X * \mathscr{H}) : x \mapsto \|f(x)\|^2 \text{ is integrable} \,\}.$$
The usual Cauchy-Schwarz inequality implies that $x \mapsto \big(f(x) \mid g(x)\big)$ is integrable if $f, g \in \mathcal{L}^2$. Hence $\mathcal{L}^2(X * \mathscr{H}, \mu)$ becomes a pre-inner product space in the usual way. Taking the quotient by elements of norm zero gives us a Hilbert space $L^2(X * \mathscr{H}, \mu)$ elements of which can be viewed as almost everywhere equal equivalence classes of functions in $\mathcal{L}^2(X * \mathscr{H}, \mu)$. The Hilbert space $L^2(X * \mathscr{H}, \mu)$ is sometimes called the *direct integral* and denoted by
$$\int_X^{\oplus} \mathcal{H}(x)\, d\mu(x).$$

EXAMPLE 3.36. If $X * \mathscr{H}$ is the trivial bundle $X \times \mathscr{H}$, then as described in [**Wil07**, p. 413], $L^2(X \times \mathcal{H}, \mu)$ is just the vector space $L^2(X, \mu, \mathcal{H}) \cong L^2(X, \mu) \otimes \mathcal{H}$ of equivalences classes of $\mathcal{H}$-valued square integrable functions (see Section 3.6).

DEFINITION 3.37. If $X * \mathscr{H}$ is a Borel Hilbert Bundle, then its *isomorphism groupoid* is the groupoid
$$\operatorname{Iso}(X * \mathscr{H}) := \{\, (x, V, y) : V \in U(\mathcal{H}(y), \mathcal{H}(x)) \,\}$$
with the weakest Borel structure such that
$$(x, V, y) \mapsto \big(V f(y) \mid g(x)\big)$$
is Borel for all $f, g \in B(X * \mathscr{H})$.

PROPOSITION 3.38. *Let* $\operatorname{Iso}(X * \mathscr{H})$ *be the isomorphism groupoid of an analytic (standard) Borel Hilbert bundle over* X. *Then* $\operatorname{Iso}(X * \mathscr{H})$ *is a groupoid with respect to the natural operations:* $(x, V, y)(y, U, z) = (x, VU, z)$ *and* $(x, V, y)^{-1} = (y, V^{-1}, x)$. *Furthermore,* $\operatorname{Iso}(X * \mathscr{H})$ *is an analytic (standard) Borel space, and the groupoid operations are Borel. (We say that* $\operatorname{Iso}(X * \mathscr{H})$ *is an analytic (standard) Borel groupoid.)*

SKETCH OF THE PROOF. The fact that $X * \mathscr{H}$ is a groupoid is routine. To see that it is analytic, we appeal to Proposition 3.35 on the preceding page and assume that $X * \mathscr{H}$ is of the form $\coprod_{d=1}^{d=\infty} X_d \times \mathcal{H}_d$. Then it is not hard to check that $\operatorname{Iso}(\coprod_{d=1}^{d=\infty} X_d \times \mathcal{H}_d)$ is Borel isomorphic to $\coprod_d X_d \times U(\mathcal{H}_d) \times X_d$ where $U(\mathcal{H}_d)$ is given the weak operator topology. Since $U(\mathcal{H}_d)$ is Polish in this topology [**Wil07**, Lemma D.42], $\operatorname{Iso}(X * \mathscr{H})$ is always analytic and will be standard if X is. $\square$

3.6. The Hilbert Space $L^2(X, \mathcal{H})$

Let X be a second countable locally compact Hausdorff space and μ a Radon measure on X. Let $\mathcal{H}$ be a separable Hilbert space. Then any self respecting analyst would want to construct the Hilbert space $L^2(X, \mu, \mathcal{H})$ of square integrable $\mathcal{H}$-valued functions on X.[9]

We say that $f : X \to \mathcal{H}$ is Borel if

$$x \mapsto \big(f(x) \mid v\big) \tag{3.17}$$

is a Borel map for all $v \in \mathcal{H}$. More formally, we are calling f Borel when it is a Borel section for the trivial Borel Hilbert bundle $X \times \mathcal{H}$ as in [**Wil07**, Definition F.1].[10] We let $\mathcal{L}^2(X, \mu, \mathcal{H})$ be the set of Borel functions $h : X \to \mathcal{H}$ such that

$$\int_X \|h(x)\|^2 \, d\mu(x) < \infty. \tag{3.18}$$

The vector space obtained by identifying functions μ-almost everywhere is denoted by $L^2(X, \mu, \mathcal{H})$.[11] Alternatively, we can think of $L^2(X, \mu, \mathcal{H})$ as the direct integral of the trivial Borel Hilbert bundle $X \times \mathcal{H}$ (see [**Wil07**, Example F.5]). General nonsense shows (Ex 3.6.2) that $L^2(X, \mu, \mathcal{H})$ is a Hilbert space with respect to the inner product

$$(h \mid k) = \int_X \big(h(x) \mid k(x)\big) \, d\mu(x). \tag{3.19}$$

If $f \in L^\infty(X)$, then we get a bounded multiplication operator L_f on $L^2(X, \mu, \mathcal{H})$ such that

$$L_f h(x) = f(x)h(x) \quad \text{for } x \in X \text{ and } h \in \mathcal{L}^2(X, \mu, \mathcal{H}).$$

Such operators are called *diagonal* and the algebra of diagonal operators is denoted by $\mathscr{L}$.[12]

---

[9] This material almost exactly parallels that in [**Wil07**, §I.4], but allows for the simplifications possible due our assumptions on separability. In particular, we can work with Borel functions and drop assumptions on "separably-valued" functions.

[10] Since $\mathcal{H}$ is separable, Ex 3.6.1 shows that a Borel section f satisfying (3.17) is actually Borel as a map from X into $\mathcal{H}$ with its metric topology.

[11] The inequality $\|f(x)+g(x)\|^2 \leq 2(\|f(x)\|^2+\|g(x)\|^2)$ is helpful in verifying that $L^2(X, \mu, \mathcal{H})$ is a vector space.

[12] In [**Wil07**, §F.2], $\mathscr{L}$ is denoted by $\Delta(X \times \mathcal{H}, \mu)$. See [**Wil07**, Definition F.13 and Example F.14].

A function $T : X \to B(\mathcal{H})$ is called a *Borel field of operators* if

$$x \mapsto \big(T(x)h(x) \mid k(x)\big)$$

is Borel for all $h, k \in \mathcal{L}^2(X, \mu, \mathcal{H})$. Such a field defines an operator T on $L^2(X, \mu, \mathcal{H})$ determined by

$$Th(x) = T(x)h(x) \quad \text{for } x \in X \text{ and } h \in \mathcal{L}^2(X, \mu, \mathcal{H}).$$

The operator T is often called the *direct integral* of the $\{T(x)\}$, and the situation is often summarized by writing

$$T := \int_X^{\oplus} T(x)\, d\mu(x). \tag{3.20}$$

Note that if $\{v_n\}$ is a countable dense set in the unit ball of $\mathcal{H}$, then

$$\|T(x)\| = \sup_{n,m} \big|\big(T(x)v_n \mid v_m\big)\big|.$$

It follows that $x \mapsto \|T(x)\|$ is Borel. Then we easily check that

$$\|T\| \le \operatorname{ess\,sup}_{x\in X} \|T(x)\|, \tag{3.21}$$

and that $T = 0$ if and only if $T(x) = 0$ for almost all x (Ex 3.6.3).[13] If an operator $T \in B(L^2(X, \mu, \mathcal{H}))$ can be realized as a direct integral as in (3.20), then we call T *decomposible*. We will use without proof the following classical result characterizing decomposible operators. Recall that if $\mathscr{A} \subset B(\mathcal{H})$, then

$$\mathscr{A}' = \{\, S \in B(\mathcal{H}) : SA = AS \text{ for all } A \in \mathcal{A}\,\}$$

is called the commutant of $\mathcal{A}$.

THEOREM 3.39. *An operator $T \in B(L^2(X, \mu, \mathcal{H}))$ is decomposible if and only if it belongs to the commutant, $\mathscr{L}'$, of $\mathcal{L}$.*

This result is proved in [**Arv76**, Theorem 4.2.1] and is a special case of [**Wil07**, Theorem F.21].

We will need the following vector-valued version of [**Wil81**, Lemma 4.15].

PROPOSITION 3.40. *Let X, Y, and Z be second countable locally compact spaces. Suppose that μ_Y and μ_Z are Radon measures on Y and Z, respectively, and that $\mathcal{H}_Y$ and $\mathcal{H}_Z$ are separable Hilbert spaces. Let $i : Y \to X$ and $j : Z \to X$ be continuous injections such that $i(Y) \cap j(Z) = \emptyset$. Define multiplication representations $M_Y : C_0(X) \to B(L^2(Y, \mathcal{H}_Y, \mu_Y))$ and $M_Z : C_0(X) \to B(L^2(Z, \mathcal{H}_Z, \mu_Z))$ by*

$$M_Y(f)h(y) = f(i(y))h(y) \quad \text{and} \quad M_Z(f)k(z) = f(j(z))k(z)$$

for $h \in \mathcal{L}^2(Y, \mathcal{H}_Y, \mu_Y)$ and $k \in \mathcal{L}^2(Z, \mathcal{H}_Z, \mu_Z)$. Then M_Y and M_Z have no equivalent sub-representations. In particular, M_Y and M_Z are not equivalent.

The next lemma will be helpful for the proof of the proposition.

LEMMA 3.41. *Let $\mathscr{L}$ be the diagonal operators in $B(L^2(Y, \mu_Y, \mathcal{H}_Y))$. If $T \in M_Y(C_0(X))'$, then $T \in \mathscr{L}'$.*

[13] Actually, we have equality in (3.21)—see Ex 3.6.4.

PROOF. If $\varphi \in \mathcal{B}^b(Y)$, it will suffice to find $\{f_n\} \subset C_0(X)$ such that $M_Y(f_n) \to L_\varphi$ in the strong operator topology. We can assume that $0 \le \varphi(y) \le 1$ for all y. Let $Y = \bigcup K_n$ with each K_n compact and K_n contained in the interior of K_{n+1}. Since $i(K_n)$ is compact, there is a $f_n \in C_c(X)$ such that $0 \le |f_n(x)| \le 1$ for all $x \in X$ and $f_n(i(y)) = \varphi(y)$ for all $y \in K_n$. Then $|\varphi(y) - f_n(i(y))| \le 1$ and converges to 0 pointwise. Therefore if $h \in L^2(Y, \mathcal{H}_Y, \mu_Y)$, then $M_Y(f_n)h \to L_\varphi h$ by the Dominated Convergence Theorem. □

PROOF OF PROPOSITION 3.40 ON THE FACING PAGE. Let $\mathcal{E}$ be a non-zero invariant subspace in $L^2(Y, \mu_Y, \mathcal{H}_Y)$ for M_Y. If $P_\mathcal{E}$ is the orthogonal projection onto $\mathcal{E}$, then $P_\mathcal{E} \in \mathscr{L}'$ by Lemma 3.41 on the preceding page. Therefore $P_\mathcal{E}$ is given by a Borel field of operators $y \mapsto p(y)$. Since $p(y)$ is determined μ_Y-almost everywhere [**Wil07**, Lemma F.20] and $P_\mathcal{E} = P_\mathcal{E}^2 = P_\mathcal{E}^*$, we modify the $p(y)$ on a null set and assume $p(y) = p(y)^2 = p(y)^*$ for all y. The set $E = \{\, y \in Y : \|p(y)\| = 1 \,\}$ must be such that $\mu_Y(E) > 0$. If B is a non-null Borel subset of E, then

$$P_\mathcal{E}^B = \int_Y^\oplus \mathbb{1}_B(y) p(y)\, d\mu_Y(y)$$

is a nonzero projection in $\mathcal{L}'$ and determines a sub-representation M_Y^B of the sub-representation of $M_Y = M_Y^E$ determined by $P_\mathcal{E}$.

Of course, similar conclusions hold for invariant subspaces $\mathcal{F}$ for M_Z.

Suppose that the sub-representation M_Y^E of M_Y corresponding to $\mathcal{E}$ is equivalent to a sub-representation M_Z^F of M_Z corresponding to $\mathcal{F}$. Since μ_Y is a Radon measure, there is a compact subset $C \subset E$ such that $\mu_Y(C) > 0$. Then M_Y^C is equivalent to a sub-representation $M_Z^{F'}$ with $\mu_Z(F') > 0$. Let $D \subset F'$ be a compact subset such that $\mu_Z(D) > 0$. Then M_Z^D is equivalent to a sub-representation of M_Y^C. Let $f \in C_0(X)$ be such that f is identically one on $i(C)$ and identically zero on $j(D)$. Then $M_Y^C(f)$ is non-zero and $M_Z^D(f)$ is the zero operator. This is a contradiction and completes the proof. □

EXERCISES

Ex 3.6.1. Let X and $\mathcal{H}$ be as above. Show that $f : X \to \mathcal{H}$ is Borel if and only if $f^{-1}(V)$ is Borel for all open sets $V \subset \mathcal{H}$. In other words, our Borel sections are truly Borel functions from X into $\mathcal{H}$ with its metric topology.

—Answer on page 366

Ex 3.6.2. Let $\mathcal{L}^2(X, \mu, \mathcal{H})$ and $L^2(X, \mu, \mathcal{H})$ be as above.

(a) If $h, k \in \mathcal{L}^2(X, \mu, \mathcal{H})$ show that
$$x \mapsto \big(h(x) \mid k(x)\big)$$
is an μ-integrable Borel function on X. In particular, $x \mapsto \|h(x)\|$ is a Borel function (so that (3.18) makes sense).

(b) Show that (3.19) induces a well defined inner product on $L^2(X, \mu, \mathcal{H})$.

(c) Show that $L^2(X, \mu, \mathcal{H})$ is complete in the L^2-norm so that $L^2(X, \mu, \mathcal{H})$ becomes a Hilbert space.

—Answer on page 366

Ex 3.6.3. Let T be a decomposable operator as in (3.20). Show that $T = 0$ if and only if $T(x) = 0$ for almost all x.

—Answer on page 367

Ex 3.6.4. Let T be a decomposable operator as in (3.20). Show that

$$\|T\| = \text{ess sup}\,\|T(x)\|.$$

—Answer on page 367

3.7. The Hilbert Space $L^2_\pi(G_u, \sigma_u, \mathcal{H}_\pi)$

In this section, we fix a groupoid G and work with the space G_u viewed as a right $G(u)$-space. Let $\mathcal{H}$ be a complex Hilbert space. We want to build a special Hilbert space of $\mathcal{H}$-valued functions which will be important when we look at representations of $C^*(G)$ which are induced from unitary representations of $G(u)$. (Induced representations will be defined in Section 5.3. The result in question is Proposition 5.44 on page 130.)

Even though we are fixing u, for convenience we use the Haar measure $\beta^{G(u)}$ coming from our fixed continuous choice of Haar measures $\{\beta^H\}_{H\in\Sigma_0}$ given by Proposition 3.23 on page 82. This will be sufficient for our purposes in view of Remark 5.46 on page 131.

Since G_u is a free and proper right $G(u)$-space, the orbit space $G_u/G(u)$ is a second countable locally compact Hausdorff space, and the orbit map $q : G_u \to G_u/G(u)$ is open. We'll often write $\dot\gamma$ in place of $q(\gamma)$. Thus by Lemma 2.21 on page 33, we get a surjection $\beta : C_c(G_u) \to C_c(G_u/G(u))$ given by

$$\beta(f)(\dot\gamma) = \int_{G_u} f(\gamma h)\, \beta^{G(u)}(h).$$

Since our Haar measures are left-invariant, for each $\dot\gamma \in G_u/G(u)$ we get a well defined Radon measure on G_u determined by

$$\alpha^{\dot\gamma}(f) = \int_{G(u)} f(\gamma h)\, d\beta^{G(u)}(h),$$

and $\alpha = \{\alpha^{\dot\gamma}\}_{\dot\gamma\in G_u/G(u)}$ is a proper q-system of measures. Proposition 3.18 on page 77 implies that there is a $b_u \in C_{c,q}(G_u)$ such that

$$\alpha(f)(\dot\gamma) = \int_{G(u)} b_u(\gamma h)\, d\beta^{G(u)}(h) = 1 \quad \text{for all } \gamma \in G_u.$$

Let $\omega : G * \Sigma_0 \to (0,\infty)$ be as in Lemma 3.25 on page 82 and define

$$\rho_u(\gamma) = \omega(\gamma, G(u)).$$

Notice that ρ_u is continuous and

$$\rho_u(\gamma h) = \Delta_{G(u)}(h)\rho_u(\gamma) \quad \text{if } \gamma \in G_u \text{ and } h \in G(u) \tag{3.22}$$

where $\Delta_{G(u)}$ is the modular function on $G(u)$.[14]

Our next lemma is closely related to Proposition 6.15 on page 148.

LEMMA 3.42. *There is a Radon measure σ_u on $C_c(G_u/G(u))$ such that for all $f \in C_c(G_u)$*

$$\int_{G_u} f(\gamma)\rho_u(\gamma)\, d\lambda_u(\gamma) = \int_{G_u/G(u)} \int_{G(u)} f(\gamma h)\, d\beta^{G(u)}(h)\, d\sigma_u(\dot\gamma). \tag{3.23}$$

[14]In fact, any continuous function $\rho_u : G_u \to (0,\infty)$ satisfying (3.22) would suffice. Our specific choice just makes things a bit less complicated down the road.

REMARK 3.43 (Some Measure Theory). Given σ_u, we can rewrite (3.23) as

$$\nu(f) = \int_{G_u/G(u)} \int_{G_u} f(\gamma)\, d\alpha^{\dot\gamma}(\gamma)\, d\sigma_u(\dot\gamma)$$

where ν is the measure $\rho_u \cdot \lambda_u$. Thus by Proposition 3.6 on page 71, (3.23) makes good sense if $f \in \mathcal{B}^+(G_u)$ (suitably interpreted as in Remark 3.5 on page 71).

PROOF OF LEMMA 3.42 ON THE PRECEDING PAGE. Clearly, we want to define σ_u by

$$\sigma_u(\beta(f)) = \int_{G_u} f(\gamma)\rho_u(\gamma)\, \lambda_u(\gamma). \tag{3.24}$$

To see that this is well-defined, we need to see that $\beta(f) = 0$ implies that the right-hand side of (3.24) is also zero.

So, suppose $f \in C_c(G_u)$ is such that $\beta(f) = 0$. Then if $g \in C_c(G)$ we have

$$\begin{aligned}\int_{G(u)} g * f(h)\, \beta^{G(u)}(h) &= \int_{G(u)} \int_G g(\gamma) f(\gamma^{-1}h)\, d\lambda^u(\gamma)\, d\beta^{G(u)}(h) \\ &= \int_G g(\gamma^{-1}) \int_{G(u)} f(\gamma h)\, d\beta^{G(u)}(h)\, d\lambda_u(\gamma) = 0\end{aligned} \tag{3.25}$$

Since $q(\operatorname{supp} f)$ is compact in $G_u/G(u)$, we can find $\tau \in C_c(G_u/G(u))$ be such that $\tau(\dot\gamma) = 1$ for all $\gamma \in \operatorname{supp} f$. Then

$$\tilde g(\gamma) = \tau(\dot\gamma) b_u(\gamma)\rho_u(\gamma)$$

is in $C_c(G_u)$ since $b_u \in C_{c,q}(G_u)$. Using the Tietze Extension Theorem ([**Wil07**, Lemma 1.42]) we may as well assume $\tilde g \in C_c(G)$. Then we let $g(\gamma) = \tilde g(\gamma^{-1})$ and plug into (3.25). Thus

$$\begin{aligned} 0 &= \int_{G(u)} \int_G g(\gamma) f(\gamma^{-1}h)\, d\lambda^u(\gamma)\, \beta^{G(u)}(h) \\ &= \int_{G(u)} \int_G g(h\gamma) f(\gamma^{-1})\, d\lambda^u(\gamma)\, d\beta^{G(u)}(h) \\ &= \int_G \int_{G(u)} \tilde g(\gamma h^{-1}) f(\gamma)\, d\beta^{G(u)}(h)\, d\lambda_u(\gamma) \\ &= \int_G \int_{G(u)} b_u(\gamma h^{-1}) \rho_u(\gamma h^{-1})\, d\beta^{G(u)}(h) f(\gamma)\, d\lambda_u(\gamma) \\ &= \int_G \int_{G(u)} b_u(\gamma h^{-1}) \Delta_{G(u)}(h)^{-1}\, d\beta^{G(u)}(h) f(\gamma)\rho_u(\gamma)\, d\lambda_u(\gamma) \\ &= \int_G \int_{G(u)} b_u(\gamma h)\, d\beta^{G(u)}(h) f(\gamma)\rho_u(\gamma)\, d\lambda_u(\gamma) \\ &= \int_G f(\gamma)\rho_u(\gamma)\, d\lambda_u(\gamma). \end{aligned}$$

This finishes the proof. □

COROLLARY 3.44. *Let $A = A \cdot G(u)$ be a saturated Borel subset of G_u. Then A is λ_u-null if and only if $q(A)$ is σ_u-null.*

PROOF. This follows immediately by plugging $\mathbb{1}_A$ into (3.23) (see Remark 3.43). □

LEMMA 3.45. *The quotient map $q : G_u \to G_u/G(u)$ has a Borel cross section $c : G_u/G(u) \to G_u$.*

PROOF. Since second countable locally compact Hausdorff spaces are Polish [**Wil07**, Lemma 6.5 and Remark 6.6] and since q is open, this is an immediate consequence of [**Arv76**, Theorem 3.4.1]. $\square$

LEMMA 3.46. *Let $\mathcal{H}$ be a separable Hilbert space. Suppose we have a commutative diagram*

$$\begin{array}{ccc} G_u & \xrightarrow{\quad f \quad} & \\ \downarrow q & & \searrow \\ G_u/G(u) & \xrightarrow[\dot f]{\qquad\qquad} & \mathcal{H}. \end{array}$$

Then f is Borel if and only if $\dot f$ is Borel.

PROOF. This is an immediate consequence of Lemma 3.45 and the observation that the Borel structure on $G_u/G(u)$ is the quotient Borel structure by Lemma 3.48 on the facing page. $\square$

We follow the same lines as in the group case—see [**Wil07**, §5.1]. Let $\pi : G(u) \to U(\mathcal{H}_\pi)$ be a unitary representation of $G(u)$ on a separable Hilbert space. Let $\mathcal{L}^2_\pi(G_u, \sigma_u, \mathcal{H}_\pi)$ be the set of Borel functions $\xi : G_u \to \mathcal{H}_\pi$ such that

$$\xi(\gamma h) = \pi(h)^{-1}\big(\xi(\gamma)\big) \quad \text{for all } \gamma \in G_u \text{ and } h \in G(u), \tag{3.26}$$

and such that

$$\int_{G_u/G(u)} \|\xi(\gamma)\|^2 \, d\sigma_u(\dot\gamma) < \infty.$$

Just as in Section 3.6 on page 87 and Ex 3.6.2, it is standard to show that

$$(\xi_1 \mid \xi_2) := \int_{G_u/G(u)} \big(\xi_1(\gamma) \mid \xi_2(\gamma)\big) \, d\sigma_u(\dot\gamma)$$

is a pre-inner product on $\mathcal{L}^2_\pi(G_u, \sigma_u, \mathcal{H}_\pi)$ that provides an inner product on the λ_u-almost everywhere equivalence classes $L^2_\pi(G_u, \sigma_u, \mathcal{H}_\pi)$ of functions in $\mathcal{L}^2_\pi(G_u, \sigma_u, \mathcal{H}_\pi)$ making $L^2_\pi(G_u, \sigma_u, \mathcal{H}_\pi)$ into a Hilbert space (Ex 3.7.1). If 1 is the trivial representation on $\mathcal{H}$, then using Lemma 3.46, we can identify $L^2_1(G_u, \sigma_u, \mathcal{H})$ with $L^2(G_u/G(u), \sigma_u, \mathcal{H}_\pi)$ defined as one would expect as equivalence classes of square integral Borel functions $\xi : G_u/G(u) \to \mathcal{H}$ in $\mathcal{L}^2(G_u/G(u), \sigma_u, \mathcal{H}_\pi)$. As usual, we will not spend much effort distinguishing between functions and their almost everywhere equivalence classes.

Despite the apparent additional generality, it is always the case that $L^2_\pi(G_u, \sigma_u, \mathcal{H}_\pi)$ and $L^2(G_u/G(u), \sigma_u, \mathcal{H}_\pi)$ are isomorphic. However, the isomorphism in not canonical as we need to choose a Borel cross section $c : G_u/G(u) \to G_u$ using Lemma 3.45. Then we can define a Borel map $b : G_u \to G(u)$ such that for all $\gamma \in G_u$ we have

$$\gamma = c(\dot\gamma)b(\gamma). \tag{3.27}$$

Note that $b(\gamma h) = b(\gamma)h$ for all $h \in G(u)$.

LEMMA 3.47. *Let $c : G_u/G(u) \to G_u$ be a Borel cross section for the quotient map $q : G_u \to G_u/G(u)$ and let $b : G_u \to G(u)$ be as in* (3.27). *Then $\Phi(f)(\gamma) = \pi(b(\gamma))^{-1}\big(f(\dot\gamma)\big)$ defines a unitary from $L^2(G_u/G(u), \sigma_u, \mathcal{H}_\pi)$ onto $L^2_\pi(G_u, \sigma_u, \mathcal{H}_\pi)$ with inverse given by $\Phi^{-1}(g)(\dot\gamma) = g(c(\dot\gamma))$.*

PROOF. It is straightforward to verify that Φ is isometric with the given inverse. □

EXERCISES

Ex 3.7.1. Provide the details showing that $L^2_\pi(G_u, \sigma_u, \mathcal{H}_\pi)$ is a Hilbert space.
—Answer on page 367

3.8. The Quotient Borel Structure

If Y is a topological space, we let $\mathscr{B}(Y)$ be the Borel σ-algebra on Y. If $f : X \to Y$ is a Borel map, then the *quotient Borel structure* on Y is given by the σ-algebra

$$\mathscr{B} = \{\, U \subset Y : f^{-1}(U) \in \mathscr{B}(X)\,\}.$$

LEMMA 3.48. *Let X and Y be Polish spaces and $f : X \to Y$ a Borel surjection with a Borel cross section $c : Y \to X$. Then the quotient Borel structure $\mathscr{B}$ on Y coincides with the usual one.*

PROOF. The map $f : (X, \mathscr{B}(X)) \to (Y, \mathscr{B})$ is a Borel map by definition of $\mathcal{B}$. But $c : (Y, \mathscr{B}) \to (X, \mathscr{B}(X))$ is borel since $\mathscr{B}(Y) \subset \mathscr{B}$. Then [**Arv76**, Proposition 3.4.2] implies that $(Y, \mathscr{B})$ is a standard Borel space. Then $\mathscr{B}(Y) = \mathscr{B}$ by [**Arv76**, Theorem 3.3.5]. □

CHAPTER 4

Proof of the Equivalence Theorem

In this chapter, I want to give proofs of the Equivalence Theorems stated in Section 2.6. Hence in this chapter, G and H will always be second countable locally compact Hausdorff groupoids with Haar systems $\{\lambda^u\}_{u\in G^{(0)}}$ and $\{\beta^v\}_{v\in H^{(0)}}$. We assume Z is a second countable locally compact Hausdorff space implementing an equivalence between G and H. The proof is centered around the linking groupoid, L, with its Haar system $\{\kappa^w\}_{w\in L^{(0)}}$ as in Section 2.5. Since the Haar systems are fixed, we will, as usual, simply write $C^*(L)$ in place of $C^*(L,\kappa)$ and similarly for $C^*(G)$ and $C^*(H)$. Our aim is to identify $C^*(G)$ and $C^*(H)$ as complementary full corners in $C^*(L)$ so that their Morita equivalence follows from the theory of linking algebras as in [**RW98**, Theorem 3.19].

The linking groupoid and linking algebra approach was taken from [**SW12**] where the results are proved for locally Hausdorff, locally compact groupoids.

4.1. The C^*-Algebra of the Linking Groupoid

The linking groupoid $L = G\sqcup Z\sqcup Z^{\mathrm{op}}\sqcup H$ for G and H comes from Lemma 2.66 on page 56, and the Haar system κ from Lemma 2.68 on page 58. Let $V : C_b(L^{(0)}) \to M(C^*(L))$ be the usual map (see Lemma 1.48 on page 19). For the moment, fix any pre-C^*-norm, $\|\cdot\|_\alpha$, on $C_c(L)$ which is continuous in the inductive limit topology.[1] We let the completion be $C^*_\alpha(L)$. (We are only interested in the universal norm $\|\cdot\|$ and the reduced norm $\|\cdot\|_r$—and hence the universal algebra $C^*(L)$ and the reduced algebra $C^*_r(L)$—but the added generality doesn't complicate anything.) Via composition, we get a map of $C_b(L^{(0)})$ into $M(C^*_\alpha(L))$ and the images of the characteristic functions of $G^{(0)}$ and $H^{(0)}$ in $L^{(0)}$ are complementary projections $\mathfrak{p}_G$ and $\mathfrak{p}_H$ in $M(C^*_\alpha(G))$. Recall that a projection in a C^*-algebra is *full* if the ideal it generates is the whole algebra.

PROPOSITION 4.1. *The complementary projections $\mathfrak{p}_G$ and $\mathfrak{p}_H$ are full in $M(C^*_\alpha(L))$.*

The proof of Proposition 4.1 will be easier once we have collected some machinery, so I'll postpone the proof to the end of Section 4.2 (on page 105) so that we can concentrate on the key ideas first.

Given the fullness of $\mathfrak{p}_G$ and $\mathfrak{p}_H$, the easy part of [**RW98**, Theorem 3.19] implies that the corners $\mathfrak{p}_G C^*_\alpha(L)\mathfrak{p}_G$ and $\mathfrak{p}_H C^*_\alpha(L)\mathfrak{p}_H$ are Morita equivalent via $\mathsf{X} = \mathfrak{p}_G C^*_\alpha(L)\mathfrak{p}_H$. (The module actions and inner products being given by multiplication in $C^*_\alpha(L)$.)

[1] That is, if $\{f_i\}\subset C_c(G)$ have supports in a fixed compact set and converge uniformly to f, then $\|f_i - f\|_\alpha \to 0$.

We want to look a bit more closely at this Morita equivalence. For $F \in C_c(L)$, let $F_{11} = F|_G \in C_c(G)$, $F_{12} = F|_Z \in C_c(Z)$, $F_{21} = F|_{Z^{\mathrm{op}}} \in C_c(Z^{\mathrm{op}})$ and $F_{22} = F|_H \in C_c(H)$. We write F as a matrix

$$F = \begin{pmatrix} F_{11} & F_{12} \\ F_{21} & F_{22} \end{pmatrix}.$$

(The matrix representation is motivated by the observation that $C^*(L)$ will turn out to be isomorphic to the linking algebra for X—see Ex 4.2.4.) The involution on $C_c(L)$ is then given by

$$F^* = \begin{pmatrix} F_{11}^* & F_{21}^* \\ F_{12}^* & F_{22}^* \end{pmatrix},$$

where F_{11}^* and F_{22}^* are the images of F_{11} and F_{22} under the standard involutions on $C_c(G)$ and $C_c(H)$, while $F_{12}^*(z) = \overline{F_{21}(\bar{z})}$ and $F_{21}^*(\bar{z}) = \overline{F_{12}(z)}$ for all $z \in Z$.

Since $\|\cdot\|_\alpha$ is assumed to be continuous with respect to the inductive limit topology, we view $C_c(Z) \subset \mathsf{X}$, $C_c(G) \subset \mathfrak{p}_G C_\alpha^*(L)\mathfrak{p}_G$ and $C_c(H) \subset \mathfrak{p}_H C_\alpha^*(L)\mathfrak{p}_H$ as dense subspaces. Thus the convolution operation on $C_c(L)$ induces a $C_c(G)$ – $C_c(H)$-pre-imprimitivity bimodule structure on $C_c(Z)$ where we view $C_c(G)$ as dense subalgebra of $C_\alpha^*(G) := \mathfrak{p}_G C_\alpha^*(L)\mathfrak{p}_G$ and $C_c(H)$ as a dense subalgebra $C_\alpha^*(H) := \mathfrak{p}_H C_\alpha^*(L)\mathfrak{p}_H$ (see [**RW98**, Definition 3.9]). (The notation comes from viewing $C_\alpha^*(G)$ as the completion of the image of $C_c(G)$ in $C_c(L)$ with respect to $\|\cdot\|_\alpha$.) Clearly the imprimitivity bimodule completion of $C_c(Z)$ is $\mathsf{X}_\alpha = \mathfrak{p}_G C_\alpha^*(L)\mathfrak{p}_H$.

If $\varphi, \psi \in C_c(Z)$, then the $C_c(H)$ valued inner product is given by

$$\begin{pmatrix} 0 & 0 \\ 0 & \langle \varphi\,,\psi\rangle_\star \end{pmatrix}(\eta) = \begin{pmatrix} 0 & \varphi \\ 0 & 0 \end{pmatrix}^* * \begin{pmatrix} 0 & \psi \\ 0 & 0 \end{pmatrix}(\eta)$$

To get an explicit formula, let $F_\varphi = \left(\begin{smallmatrix} 0 & \varphi \\ 0 & 0 \end{smallmatrix}\right)$. Then

$$\begin{aligned} F_\varphi^* * F_\psi(\eta) &= \int_L F_\varphi^*(l)F_\psi(l^{-1}\eta)\,d\kappa^{r(\eta)}(l) \\ &= \int_{Z^{\mathrm{op}}} F_\varphi^*(\bar{z})F_\psi((\bar{z})^{-1}\cdot\eta)\,d\sigma_{Z^{\mathrm{op}}}^{r(\eta)}(\bar{z}) \end{aligned}$$

which, provided $r_{Z^{\mathrm{op}}}(\bar{z}) = r(\eta)$, is

$$\begin{aligned} &= \int_G F_\varphi^*(\bar{z}\cdot\gamma)F_\psi((\bar{z}\cdot\gamma)^{-1}\cdot\eta)\,d\lambda^{s_{Z^{\mathrm{op}}}(\bar{z})}(\gamma) \\ &= \int_G F_\varphi^*(\overline{\gamma^{-1}\cdot z})F_\varphi(\gamma^{-1}\cdot z\cdot\eta)\,d\lambda^{r(z)}(\gamma). \end{aligned}$$

From which we obtain the formula (2.6) on page 59 in Section 2.6:

$$\langle \varphi\,,\psi\rangle_\star(\eta) = \int_G \overline{\varphi(\gamma^{-1}\cdot z)}\psi(\gamma^{-1}\cdot z\cdot\eta)\,d\lambda^{r(z)}(\gamma) \tag{4.1}$$

which is valid for any z such that $s_Z(z) = r(\eta)$. We get the $f \in C_c(G)$ action on $\varphi \in C_c(Z)$ via

$$\begin{pmatrix} f\cdot\varphi & 0 \\ 0 & 0 \end{pmatrix} = \begin{pmatrix} f & 0 \\ 0 & 0 \end{pmatrix} * \begin{pmatrix} 0 & \varphi \\ 0 & 0 \end{pmatrix}.$$

Thus we get the formula (2.9):

$$f\cdot\varphi(z) = \int_G f(\gamma)\varphi(\gamma^{-1}\cdot z)\,d\lambda^{r(z)}(\gamma). \tag{4.2}$$

Similarly, the right $C_cH)$-action is given by (2.7):

$$\varphi \cdot b(z) = \int_H \varphi(z \cdot \eta) b(\eta^{-1}) \, d\beta^{s(z)}(\eta), \tag{4.3}$$

and the $C_c(G)$-valued inner product is given by (2.8):

$$_\star\langle \varphi \, , \, \psi \rangle(\gamma) = \int_H \varphi(\gamma \cdot w \cdot \eta) \overline{\psi(w \cdot \eta)} \, d\beta^{s(w)}(\eta) \tag{4.4}$$

with w such that $r(w) = s(\gamma)$.

Modulo a proof of Proposition 4.1 on page 95, we have proved the following lemma.

LEMMA 4.2. *View $C_c(G) \subset C^*_\alpha(G)$ and $C_c(H) \subset C^*_\alpha(H)$ as dense $*$-subalgebras. Then when equipped with the actions* (4.2) *and* (4.3) *and pre-inner products* (4.1) *and* (4.4), *$C_c(Z)$ is a $C_c(G) - C_c(H)$-pre-imprimitivity bimodule.*

REMARK 4.3. It is useful to keep in mind that because all the actions and inner products come from the convolution operation in $C_c(L)$, the actions and inner products are all continuous in the inductive limit topology. For example, if $\varphi_i \to \varphi$ and $\psi_i \to \psi$ in the inductive limit topology in $C_c(Z)$, then $_\star\langle \varphi_i \, , \, \psi_i \rangle \to {}_\star\langle \varphi \, , \, \psi \rangle$ in the inductive limit topology in $C_c(G)$.

Hence to prove the Equivalence Theorem 2.70 on page 60, it will suffice to see that the universal norm on $C_c(L)$ restricts to the universal norm on $C_c(G)$ and $C_c(H)$ viewed as corners in $C_c(L)$. Similarly, to prove the reduced version (Theorem 2.71 on page 60), it will suffice to see that the reduced norm on $C_c(L)$ restricts to the reduced norm on the corners.

REMARK 4.4 (The Linking Algebra Approach). In the original proof of the Equivalence Theorem in [**MRW87**], the approach was to start with the formulas for the pre-inner products and module actions, and derive all the properties required for a pre-imprimitivity bimodule structure. This involves considerable overhead. It has to be verified that the pre-inner products and actions define continuous, compactly supported functions. Then one has to verify all the properties required by [**RW98**, Definition 3.9] for a pre-imprimitivity bimodule: positivity of the pre-inner products, fullness of the pre-inner products, the boundedness of the actions, etc. A great advantage of the linking algebra approach is that all this is automatic from the overlying C^*-structure in $C^*_\alpha(L)$, and the continuity of the norm in the inductive limit topology. But there is still work to do; we have to identify the norms on the corners. By symmetry, it will suffice to consider just the upper right-hand corner.

As it turns out, it is a bit easier to prove the version of the Equivalence Theorem for the reduced norm. Note that the following subsumes Theorem 2.71.

THEOREM 4.5. *Suppose that G and H are second countable locally compact Hausdorff groupoids with Haar systems as above, and suppose that Z is a (G, H)-equivalence. If $f \in C_c(G)$, and*

$$F := \begin{pmatrix} f & 0 \\ 0 & 0 \end{pmatrix} \in C_c(L),$$

then $\|F\|_{C^*_r(L)} = \|f\|_{C^*_r(G)}$. In particular, the completion X_r of $C_c(Z)$ in the norm $\|x\| := \|{}_\star\langle x \, , \, x \rangle\|^{1/2}_{C^*_r(G)}$, equipped with the actions and inner products given in

(2.6)–(2.9) on page 59, is a $C^*_r(G)$–$C^*_r(H)$-imprimitivity bimodule isometrically isomorphic to $\mathfrak{p}_G C^*_r(L)\mathfrak{p}_H$. Hence $C^*_r(G)$ and $C^*_r(H)$ are Morita equivalent.

For the proof of Theorem 4.5, we have to look carefully at the regular representations associated to point masses. For this, it will be convenient to introduce some additional constructs. Recall that $\sigma^v_{Z^{\mathrm{op}}}$ and σ^u_Z are the measures on Z^{op} and Z, respectively, used in the definition of the Haar system κ on L in Lemma 2.68 on page 58. Let ρ^v_Z be the image of $\sigma^v_{Z^{\mathrm{op}}}$ under inversion. It is the Radon measure on Z with support Z_v given on $\varphi \in C_c(Z)$ by

$$\rho^v_Z(\varphi) = \int_G \varphi(\gamma^{-1} \cdot z_0)\, d\lambda^{r(z_0)}(\gamma)$$

for any $z_0 \in Z$ such that $s(z_0) = v$. Similarly, we will use the notation $\rho^u_{Z^{\mathrm{op}}}$ for the Radon measure on Z^{op} which is the image of σ^u_Z on Z under inversion. Although we don't need to describe $\rho^u_{Z^{\mathrm{op}}}$ for the proof of the theorem, for the sake of symmetry, we note that it is the Radon measure on Z^{op} supported on Z^{op}_u such that for all $\psi \in C_c(Z^{\mathrm{op}})$

$$\rho^u_{Z^{\mathrm{op}}}(\psi) = \int_H \psi(\eta^{-1} \cdot \overline{z_0})\, d\beta^{r(\overline{z})}(\eta),$$

for any $\overline{z_0}$ such that $s(\overline{z_0}) = u$. Thus after identifying $H \cdot \overline{z_0}$ with u, $\rho^u_{Z^{\mathrm{op}}}$ is a measure on the free and proper left H-space Z^{op}.

EXAMPLE 4.6 (The Representation $R_{G\cdot z}$). Since the action of G on Z is free and proper, the map $\gamma \to \gamma \cdot z$ is a homeomorphism of $G_{r(z)}$ onto the orbit $G \cdot z$ which is closed in Z (Ex 4.2.5). This induces a unitary from $L^2(G \cdot z, \rho^{s(z)}_Z)$ onto $L^2(G_{r(z)}, \lambda_{r(z)})$. This unitary intertwines $\operatorname{Ind}\delta_{r(z)}$ with a representation $R_{G\cdot z}$ on $L^2(G \cdot z, \rho^{s(z)}_Z)$ given by

$$R_{G\cdot z}(f)h(\gamma \cdot z) = \int_G f(\eta)h(\eta^{-1}\gamma \cdot z)\, d\lambda^{r(\gamma)}(\eta).$$

In particular, $R_{G\cdot z}$ is a representation of $C_c(G)$ and $\|R_{G\cdot z}(f)\| \le \|f\|_r$.

REMARK 4.7 (The κ_w). To discuss regular representations of $C_c(L)$, we will need to use the Radon measures $\{\kappa_w\}_{w\in L^{(0)}}$ on L, where κ_w is the forward image of the measure κ^w under inversion. It is not hard to check that for $F \in C_c(L)$ we have

$$\kappa_w(F) = \begin{cases} \lambda_w(F|_G) + \rho^w_{Z^{\mathrm{op}}}(F|_{Z^{\mathrm{op}}}) & \text{if } w \in G^{(0)}, \text{ and} \\ \rho^w_Z(F|_Z) + \beta_w(F|_H) & \text{if } w \in H^{(0)}, \end{cases}$$

where we have identified $H^{(0)}$ with $G\backslash Z$, and $G^{(0)}$ with Z/H.

PROOF OF THEOREM 4.5 ON THE PREVIOUS PAGE. Fix $f \in C_c(G)$ and let F be the corresponding element of $\mathfrak{p}_G C_c(L)\mathfrak{p}_G \subset C_c(L)$. The theorem follows from Lemma 4.2 on the preceding page once we establish that $\|F\|_{C^*_r(L)} = \|f\|_{C^*_r(G)}$.

For $u \in G^{(0)}$, we have $L_u = G_u \sqcup Z^{\mathrm{op}}_u$, where $Z^{\mathrm{op}}_u := \{\, \overline{z} \in Z^{\mathrm{op}} : s(\overline{z}) = r(z) = u \,\}$. By definition, $\operatorname{Ind}^L \delta_u$ acts on $L^2(L_u, \kappa_u)$. Following Remark 4.7, $L^2(L_u, \kappa_u) = L^2(G, \lambda_u) \oplus L^2(Z^{\mathrm{op}}, \rho^u_{Z^{\mathrm{op}}})$, and with respect to this decomposition,

$(\operatorname{Ind}^L \delta_u)(F) = (\operatorname{Ind}^G \delta_u)(f) \oplus 0$. It follows that

$$\begin{aligned} \|F\|_{C_r^*(L)} &:= \max\Bigl\{ \sup_{u\in G^{(0)}} \|(\operatorname{Ind}^L \delta_u)(F)\|,\ \sup_{v\in H^{(0)}} \|(\operatorname{Ind}^L \delta_v)(F)\| \Bigr\} \\ &= \max\Bigl\{ \|f\|_{C_r^*(G)},\ \sup_{v\in H^{(0)}} \|(\operatorname{Ind}^L \delta_v)(F)\| \Bigr\}. \end{aligned} \tag{4.5}$$

For $v \in H^{(0)}$, let $Z_v = \{ z \in Z : s(z) = v \}$. Then $L_v = Z_v \sqcup H_v$. Furthermore, $L^2(L_v, \kappa_v) = L^2(Z, \rho_Z^v) \oplus L^2(H, \beta_v)$. It is not hard to check that $(\operatorname{Ind}^L \delta_v)(F) = R^Z_{\delta_{G\cdot x_0}}(f) \oplus 0$. By Example 4.6 on the preceding page, we have $\|R^Z_{\delta_{G\cdot x_0}}(f)\| \le \|f\|_{C_r^*(G)}$. It follows from (4.5) that $\|F\|_{C_r^*(L)} = \|f\|_{C_r^*(G)}$. □

To get the corresponding result for the universal norm seems to require Renault's Disintegration Theorem in the form of Theorem 8.2 on page 169. Using this, Theorem 2.70 on page 60 is a consequence of Theorem 4.9.

First, we need to know a bit more than Lemma 4.2 on page 97 gives us. We get for free that ${}_\star\langle \varphi , \varphi\rangle$ is an element of $C_c(G)$ that is positive in the completion of $C_c(G)$ with the norm induced by the universal norm on $C_c(L)$. Since the whole point of the next theorem is that this norm is the universal norm on $C_c(G)$, we can't assume ${}_\star\langle \varphi , \varphi\rangle$ is positive in $C^*(G)$. Fortunately, we can establish this with a fancy approximate unit argument originally due to Green.

LEMMA 4.8. *Suppose that Z is a (G, H)-equivalence and that $C_c(Z)$ is the $C_c(G)$ – $C_c(H)$-pre-imprimitivity bimodule defined in Lemma 4.2 on page 97. Then for all $\varphi \in C_c(Z)$, we have ${}_\star\langle \varphi , \varphi\rangle \ge 0$ in $C^*(G)$ and $\langle \varphi , \varphi\rangle_\star \ge 0$ in $C^*(H)$.*

We'll prove Lemma 4.8 on page 105 of Section 4.2 when we have the right kind of approximate unit in hand.

THEOREM 4.9. *Suppose that G and H are second countable locally compact Hausdorff groupoids with Haar systems as above, and suppose that Z is a (G, H)-equivalence. If $f \in C_c(G)$, and*

$$F := \begin{pmatrix} f & 0 \\ 0 & 0 \end{pmatrix} \in C_c(L),$$

then $\|F\|_{C_r^*(L)} = \|f\|_{C_r^*(G)}$. In particular, the completion X of $C_c(Z)$ in the norm $\|x\| := \|\langle x , x\rangle_\star\|^{1/2}_{C^*(G)}$, equipped with the actions and inner products given in (2.6)–(2.9) on page 59, is a $C^*(G)$ – $C^*(H)$-imprimitivity bimodule isometrically isomorphic to $\mathfrak{p}_G C^*(L)\mathfrak{p}_H$. Hence $C^*(G)$ and $C^*(H)$ are Morita equivalent.

PROOF. Since every representation of $C_c(L)$ restricts to a representation of $C_c(G)$ (possibly on a subspace of the original representation), we certainly have $\|F\|_{C^*(L)} \le \|f\|_{C^*(G)}$.

To obtain the reverse inequality, let π be a faithful representation of $C^*(G)$ on $\mathcal{H}_\pi$. By the universal properties of the tensor product, there is a sesquilinear form $(\cdot \mid \cdot)_\pi$ on the algebraic tensor product $\mathcal{H}_{00} := C_c(L) * \mathfrak{p}_G \odot \mathcal{H}_\pi$ such that for F and K in $C_c(L)$ we have

$$\begin{aligned} (F * \mathfrak{p}_G \otimes \xi \mid K * \mathfrak{p}_G \otimes \zeta)_\pi &= \bigl(\pi(\mathfrak{p}_G * K^* * F * \mathfrak{p}_G)\xi \mid \zeta\bigr) \\ &= \bigl(\pi\bigl(K_{11}^* * F_{11} + {}_\star\langle K_{21}^* , F_{21}^*\rangle\bigr)\xi \mid \zeta\bigr) \end{aligned}$$

(Ex 4.2.1). We want to see that $(\cdot \mid \cdot)_\pi$ is positive. Fix $t = \sum_{i=1}^n F^i \otimes \xi_i \in \mathcal{H}_{00}$. Using Lemma 4.8 on the previous page, [**RW98**, Lemma 2.65] implies that the matrix $M = \big({}_{\star}\langle (F^i_{21})^*\,,\,(F^j_{21})^*\rangle\big)_{ij}$ is positive in $M_n\big(C^*(G)\big)$. Hence $M = D^*D$ for some $D \in M_n(C^*(G))$, so there are elements $d_{ij} \in C^*(G)$ such that

$${}_{\star}\langle (F^i_{21})^*\,,\,(F^j_{21})^*\rangle = \sum_{k=1}^n d^*_{ki} d_{kj}.$$

Then we have

$$\begin{aligned}(t \mid t)_\pi &= \sum_{ij} \big(\pi\big((F^i_{11})^* * F^j_{11} + {}_{\star}\langle (F^i_{21})^*\,,\,(F^j_{21})^*\rangle\big)\xi_j \mid \xi_i\big)\\ &= \sum_{ij}\big(\pi(F^j_{11})\xi_j \mid \pi(F^i_{11})\xi_i\big) + \sum_{ijk}\big(\pi(d_{kj})\xi_j \mid \pi(d_{ki})\xi_i\big)\\ &= \Big(\sum_i \pi(F^i_{11})\xi_i \mid \sum_i \pi(F^i_{11})\xi_i\Big) + \sum_k \Big(\sum_i \pi(d_{ki})\xi_i \mid \sum_i \pi(d_{ki})\xi_i\Big) \geq 0.\end{aligned}$$

Therefore $(\cdot \mid \cdot)_\pi$ is a pre-inner product on $\mathcal{H}_{00}$. Let $\mathcal{N}$ denote the subspace $\{\xi \in \mathcal{H}_{00} : (\xi \mid \xi)_\pi = 0\}$. Then the Cauchy-Schwarz inequality (as in [**Ped89**, §3.1.1]) implies that $(\cdot \mid \cdot)_\pi$ descends to a bona fide inner product on the quotient $\mathcal{H}_0 = \mathcal{H}_{00}/\mathcal{N}$. Furthermore, for each $F \in C_c(L)$, we can define a linear map $R(F) : \mathcal{H}_{00} \to \mathcal{H}_{00}$ such that

$$R(F)(K \otimes \xi) := F * K \otimes \xi.$$

Another application of the Cauchy-Schwarz inequality shows that $R(F)$ defines an operator on $\mathcal{H}_0$. However, it isn't immediately clear that R is bounded with respect to the universal norm. But we can appeal to Renault's Disintergration Theory from Chapter 8.

An easy calculation shows that

$$(R(F)t \mid t')_\pi = (t \mid R(F^*)t')_\pi \quad \text{for } t, t' \in \mathcal{H}_{00}. \tag{4.6}$$

Furthermore, since π is continuous in the inductive limit topology, it is not hard to see that

$$F \mapsto (R(F)t \mid t')_\pi \tag{4.7}$$

is also continuous in the inductive limit topology.

Equations (4.6) and (4.7) imply that $R : C_c(L) \to \mathrm{Lin}(\mathcal{H}_0)$ is a weak-representation of $C_c(L)$ as defined in Definition 8.1 on page 169. By Corollary 8.4 on page 170, R determines a bounded representation of $C^*(L)$ on the completion $\mathcal{H}_R$ of $\mathcal{H}_0$.

Since π is faithful, it suffices to show that

$$\|R(F)\|_{C^*(L)} \geq \|\pi(f)\| = \|f\|_{C^*(G)}. \tag{4.8}$$

Fix $\epsilon \in (0, \|f\|)$ and $\xi \in \mathcal{H}_\pi$ such that $\|\xi\| = 1$ and $\|\pi(f)\xi\|^2 > \|\pi(f)\|^2 - \epsilon$. Let $\{k_\alpha\}$ be an approximate identity in $C_c(G)$ for the inductive limit topology as in Proposition 1.49 on page 20, and let

$$K_\alpha = \begin{pmatrix} k_\alpha & 0 \\ 0 & 0 \end{pmatrix}$$

be the corresponding functions in $C_c(L)$. Then, since π is nondegenerate,

$$\lim_\alpha \|K_\alpha \otimes \xi\|^2 = \lim_\alpha \big(\pi(k^*_\alpha * k_\alpha)\xi \mid \xi\big) = \lim_\alpha \|\pi(k_\alpha)\xi\|^2 = 1.$$

It follows that

$$\begin{aligned}\|R(F)\|^2 &\geq \limsup_\alpha \|R(F)(K_\alpha)\otimes\xi\|^2 = \limsup_\alpha\bigl(\pi\bigl(f^* * k_\alpha^* * k_\alpha * f\bigr)\xi \mid \xi\bigr)\\ &= \lim_\alpha \|\pi(k_\alpha)\pi(f)\xi\|^2 = \|\pi(f)\xi\|^2 > \|\pi(f)\|^2 - \epsilon.\end{aligned}$$

Since ϵ is arbitrary, (4.8) holds. This completes the proof. □

We still need to prove the third part of the Equivalence Theorem: we require a proof of Morita Compatibility as stated in Theorem 2.72 on page 60.

PROOF OF THEOREM 2.72. If $\varphi \in C_c(Z)$, then

$$\|\varphi\|_{\mathsf{X}}^2 = \|\langle \varphi , \varphi\rangle_\star\|_{C^*(H)} \geq \|\langle \varphi , \varphi\rangle_\star\|_{C_r^*(H)} = \|\varphi\|_{\mathsf{X}_r}^2.$$

Therefore the identity map from $C_c(Z)\subset \mathsf{X}$ to $C_c(Z)\subset\mathsf{X}_r$ induces a surjection κ_{X} of X onto X_r. Let $\kappa_G : C^*(G) \to C_r^*(G)$ be the natural map with kernel $I_{C_r^*(G)}$ and define κ_H similarly. Since

$$\kappa_{\mathsf{X}}(f\cdot\varphi\cdot b) = \kappa_G(f)\cdot\kappa_{\mathsf{X}}(\varphi)\cdot\kappa_H(b) \tag{4.9}$$

hold tatologously if $f\in C_c(G)$, $\varphi\in C_c(Z)$, and $b\in C_c(H)$, (4.9) extends to the completions by continuity.

Let Y denote the kernel of κ_{X}. Then it follows from (4.9) that Y is a closed sub-bimodule of X.

The Rieffel correspondence (in the form of [**RW98**, Theorem 3.22] and [**RW98**, Lemma 3.23]) implies that

$$\mathsf{Y} = \mathsf{X}\cdot I = J\cdot\mathsf{X},$$

where I and J are ideals in $C^*(H)$ and $C^*(G)$, respectively, such that $\mathsf{X}\text{–Ind}(I) = J$, and

$$I = \overline{\operatorname{span}}\{\, \langle x , y\rangle_\star : x\in\mathsf{X} \text{ and } y\in\mathsf{Y}\,\} = \overline{\operatorname{span}}\{\,\langle y , y\rangle_\star : y\in Y\,\}.$$

Thus $I\subset I_{C_r^*(H)}$. On the other hand, if $b\in I_{C_r^*(H)}$, then it follows from (4.9) that for all x and y in X, we have $\langle x , y\rangle_\star b = \langle x , y\cdot b\rangle_\star \in I$. Since $\langle\cdot , \cdot\rangle_\star$ is full, it follows that $b\in I$. Therefore $I = I_{C_r^*(H)}$. Similarly, we also must have $J = I_{C_r^*(G)}$.

It follows from [**RW98**, Proposition 3.25] that the quotient Banach space X/Y is a $C_r^*(G)$–$C_r^*(H)$-imprimitivity bimodule. It is not hard to see that the map $z + Y \mapsto \kappa_{\mathsf{X}}(z)$ is an imprimitivity bimodule isomorphism of X/Y onto X_r. □

4.2. Approximate Units

The key to our proof of the Equivalence Theorem, and certainly Proposition 4.1 on page 95, is the following which is based on a technique that goes back to Green's [**Gre78**]. This result was also used to establish the existence of a *two-sided self-adjoint* approximate identity for the inductive limit topology on $C_c(G)$ in Proposition 1.49 on page 20. As far as I am aware, a construction such as given here is necessary to get a self-adjoint left approximate identity—which is then necessarily a two-sided version.

PROPOSITION 4.10. *We suppose that G and H are locally compact Hausdorff groupoids with Haar systems and paracompact unit spaces. Let Z be a (G,H)-equivalence. Then there is a net $\{e_\alpha\}$ in $C_c(G)$ consisting of elements of the form*

$$e_\alpha = \sum_{i=1}^{n_\alpha} {}_\star\langle \varphi_i^\alpha , \varphi_i^\alpha\rangle,$$

which each $\varphi_i^\alpha \in C_c(Z)$*, which is an approximate identity with respect to the inductive limit topology for both the left-action of* $C_c(G)$ *on itself and on* $C_c(Z)$.[2] *Furthermore, we can assume* $\limsup_\alpha \|e_\alpha\|_I \leq 1$ *for all* α*. A similar statement holds for* $C_c(H)$.

We need a number of preliminary results. First of all, a subset U of G is called *conditionally compact* if KU and UK are relatively compact for all compact $K \subset G^{(0)}$. We call U *diagonally compact* if KU and UK are actually compact. Since $UK = U \cap s^{-1}(K)$, we see that U is conditionally (diagonally) compact if and only if both $U \cap s^{-1}(K)$ and $U \cap r^{-1}(K)$ are relatively compact (compact) for all $K \subset G^{(0)}$ compact.

LEMMA 4.11 ([**Ren87**, Proposition 2.1.9]). *Let* G *be a locally compact Hausdorff groupoid with* $G^{(0)}$ *paracompact. Then* G *has a fundamental system* $\mathscr{U}$ *of conditionally (diagonally) compact neighborhoods of* $G^{(0)}$*. That is given any neighborhood* V *of* $G^{(0)}$*, there exists* $U \in \mathscr{U}$ *such that* $U \subset V$.

PROOF. Let V be a neighborhood of $G^{(0)}$ in G. It will suffice to produce a diagonally compact neighborhood of $G^{(0)}$ with $U \subset V$.

Since $G^{(0)}$ is paracompact, it has a locally finite cover $\{V_i\}$ of relatively compact open sets $V_i \subset G^{(0)}$. By the Shrinking Lemma ([**RW98**, Lemma 4.32]), there are compact sets K_i whose interiors cover $G^{(0)}$ and such that $K_i \subset V_i$. Since $\{K_i\}$ is locally finite, every compact subset of $G^{(0)}$ meets at most finitely many K_i.

Let U_i' be a compact neighborhood of K_i in G such that $U_i' \subset V$. Let $U_i = U_i' \cap s^{-1}(K_i) \cap r^{-1}(K_i)$. Note that U_i is compact and its interior (in G) contains the interior (in $G^{(0)}$) of K_i. Therefore

$$U = \bigcup U_i$$

is a neighborhood of $G^{(0)}$.

If $K \subset G^{(0)}$ is compact, then our construction of the U_i implies that

$$U \cap s^{-1}(K) = \bigcup_{K \cap K_i \neq \emptyset} U_i \cap s^{-1}(K).$$

Since the union is finite, $U \cap s^{-1}(K)$ is compact. A similar argument applies to $U \cap r^{-1}(K)$. This completes the proof. □

Later, we'll need the following observation. For groups, it is a straightforward consequence of the continuity of multiplication. For groupoids, one has to be a little more clever.

LEMMA 4.12. *Let* G *be a locally compact groupoid with* $G^{(0)}$ *paracompact. Let* U *be a neighborhood of* $G^{(0)}$ *in* G*. Then there is a diagonally compact neighborhood* V *of* $G^{(0)}$ *such that* $V^2 \subset U$.

PROOF. By Lemma 4.11, we can find a diagonally compact neighborhood W of $G^{(0)}$ such that $W \subset U$. Then, given $u \in G^{(0)}$, I claim there is a neighborhood V_u of u in G such that $V_u W \subset U$. (Note this forces $V_u \subset U$.) If not, then for each neighborhood V of u there is a $\gamma_V \in V$ and $\eta_V \in W$ such that $\gamma_V \eta_V \notin U$. Then $\gamma_V \to u$ while $\gamma_V \eta_V$ is eventually in the compact set AW for a given compact

[2]This means that both $e_\alpha * g \to g$ and $e_\alpha \cdot \varphi \to \varphi$ uniformly with supports eventually contained in a fixed compact set.

neighborhood A of u. Hence we can pass to a subnet, relabel, and assume that $\gamma_V\eta_V \to \eta \in G$. But then $\eta_V \to \eta$ and $\eta \in W$ since W is closed. Since $W \subset U$, we eventually have $\gamma_V\eta_V$ in U which is a contradiction.

Now let $V' = \bigcup_{u\in G^{(0)}} V_u$ with $V_uW \subset U$. Then $V' \subset U$ and $V'W = \bigcup_{u\in G^{(0)}} V_uW \subset U$. Hence $(V')^2 \subset U$. Now let V be a diagonally compact neighborhood of $G^{(0)}$ such that $V \subset V'$. This suffices. □

LEMMA 4.13. *Suppose that for each triple (K, U, ϵ) consisting of a compact subset $K \subset G^{(0)}$, a conditionally compact neighborhood of U of $G^{(0)}$ in G and $\epsilon > 0$, we have $e = e_{(K,U,\epsilon)} \in C_c^+(G)$ such that*

(a) $\operatorname{supp} e \subset U$ *and*

(b) *for all $u \in K$*

$$\Big|\int_G e(\gamma)\,d\lambda^u(\gamma) - 1\Big| < \epsilon.$$

Then the net $\{e_{K,U,\epsilon}\}$ directed by increasing K, and decreasing U and ϵ is an approximate identity for the left-action of $C_c(G)$ on $C_c(Z)$.

PROOF. The proof is left as exercise Ex 4.2.2. □

LEMMA 4.14. *Suppose that G acts freely and properly on Z and that U is a neighborhood of $G^{(0)}$ in G. Then each point $z \in Z$ has a neighborhood V such that $(\gamma \cdot z, z) \in V \times V$ implies $\gamma \in U$.*

PROOF. Suppose the assertion were false. Then given any neighborhood V of z, there would be a $\gamma_V \notin U$ such that $\gamma_V \cdot z \in V$. But then $\{\gamma_V\}$ must have a subnet converging to $\gamma \notin G^{(0)}$. But then $\gamma \cdot z = z$ and this forces $\gamma = r(z) \in G^{(0)}$. This contradiction finishes the proof. □

Let $s_Z : Z \to H^{(0)}$ be our moment map for the right free and proper H-action. Recall from Section 3.2 that a s_Z-system is a family of Radon measures $\{\underline{\beta}^v\}_{v\in H^{(0)}}$ on Z with $\operatorname{supp}\underline{\beta}^v \subset s_Z^{-1}(v)$, and such that

$$v \mapsto \int_Z \varphi(z)\,d\underline{\beta}^v(z)$$

is continuous for all $\varphi \in C_c(Z)$. We call the system full if $\operatorname{supp}\underline{\beta}^v = s_Z^{-1}(v)$ for all v. A r_Z-system is defined similarly.

LEMMA 4.15. *Let Z be a (G, H)-equivalence as above. Then we get a full s_Z-system $\{\underline{\beta}^v\}_{v\in H^{(0)}}$ by*

$$\underline{\beta}^v(\varphi) = \int_G \varphi(\gamma^{-1} \cdot z)\,d\lambda^{r_Z(z)}(\gamma) \tag{4.10}$$

provided $r_Z(z) = v$.

Similarly, we get a full r_Z-system $\{\underline{\lambda}\}_{u\in G^{(0)}}$ by

$$\underline{\lambda}^u(\varphi) = \int_H \varphi(z \cdot \eta)\,d\beta^{s_Z(z)}(\eta)$$

provided $r_Z(z) = u$.

PROOF. The right-hand side of (4.10) defines an element of $C_c(G\backslash Z)$ by part (b) of Lemma 2.21 on page 33. Since Z is an equivalence, we can identify $G\backslash Z$ with $H^{(0)}$. The rest holds by symmetry. □

Proof of Proposition 4.10 on page 101. We will consider only the G-action as the situation is symmetric. Since G is itself a (G, G)-equivalence, it suffices to consider only the action of $C_c(G)$ on $C_c(Z)$.

In view of Lemma 4.13 on the previous page, we can let (K, U, ϵ) be given and produce e as specified in that lemma.

Using Lemma 4.14 on the preceding page, we can find relatively compact open sets $V_1, \dots, V_n$ in Z such that the $r_Z(V_i)$ cover K and such that $(\gamma \cdot z, z) \in V_i \times V_i$ implies that $\gamma \in U$.

Since $G^{(0)}$ is paracompact, we can let $\{b_i\}$ be a partition of unity subordinate to $\{r_Z(V_i)\}$. In particular, $b_i \in C_c^+(G^{(0)})$, $\operatorname{supp} b_i \subset r_Z(V_i)$ and such that

$$\sum_{i=1}^{n} b_i(u) \leq 1$$

for all $u \in G^{(0)}$ with equality whenever $u \in K$.

Let

$$M = \sup_{z \in Z} \sum_{i=1}^{n} \int_H \mathbb{1}_{V_i}(z \cdot \eta)\, d\beta^{s_Z(z)}(\eta). \tag{4.11}$$

Then M is finite by Ex 4.2.3.

Combining Lemma 4.15 on the previous page and Lemma 3.13 on page 73, we can find $\psi_i \in C_c^+(Z)$ with $\operatorname{supp} \psi_i \subset V_i$ such that

$$\int_H \psi_i(z \cdot \eta)\, d\beta^{s_Z(z)}(\eta) = b_i(r_Z(z)),$$

as well as $\varphi_i \in C_c^+(Z)$ with $\operatorname{supp} \varphi_i \subset V_i$ and such that

$$\Big| \psi_i(z) - \varphi_i(z) \int_G \varphi_i(\gamma^{-1} \cdot z)\, d\lambda^{r_Z(z)}(\gamma) \Big| < \frac{\epsilon}{M}.$$

I claim that $e = \sum_{i=1}^{n} {}_{\star}\langle \varphi_i\,,\, \varphi_i \rangle$ satisfies the requirements of Lemma 4.13 on the previous page with respect to (K, U, ϵ). Firstly, we have $\operatorname{supp}(e) \subset U$ by our choice of the V_i. Moreover, letting $w = \gamma^{-1} \cdot z$ in (4.4), we have

$$\begin{aligned}
&\Big| \int_G e(\gamma)\, d\lambda^{r(z)}(\gamma) - \sum_{i=1}^{n} b_i(r(z)) \Big| \\
&\quad = \Big| \sum_{i=1}^{n} \int_G \int_H \varphi_i(z \cdot \eta) \varphi_i(\gamma^{-1} \cdot z \cdot \eta)\, d\beta^{s(z)}(\eta)\, d\lambda^{r(z)}(\gamma) - \sum_{i=1}^{n} b_i(r(z)) \Big| \\
&\quad = \Big| \sum_{i=1}^{n} \int_H \Big[\varphi_i(z \cdot \eta) \int_G \varphi_i(\gamma^{-1} \cdot z \cdot \eta)\, d\lambda^{r(z)}(\gamma) - \psi_i(z \cdot \eta) \Big]\, d\beta^{s(z)}(\eta) \Big| \\
&\quad < \epsilon.
\end{aligned}$$

Thus, if $r_Z(z) \in K$, then $\sum_{i=1}^{n} b_i(r(z)) = 1$ and condition (b) of Lemma 4.13 on the preceding page is satisfied.

On the other hand, for all z, $\sum_{i=1}^{n} b_i(r(z)) \leq 1$ and it follows that

$$\int_G e(\gamma)\, d\lambda^u(\gamma) \leq 1 + \epsilon$$

in all cases. Since e is nonnegative and self-adjoint, $\|e\|_I \leq 1 + \epsilon$. □

Now we obtain Proposition 1.49 on page 20 is an easy consequence.

PROOF OF PROPOSITION 1.49. Proposition 4.10 on page 101 provides a $\|\cdot\|_I$ bounded self-adjoint approximate identity for the left-action of $C_c(G)$ on itself. Since it is self-adjoint, it is also an approximate identity for the right-action. □

Now we can easily complete the proofs of Lemma 4.8 on page 99 as well as Proposition 4.1 on page 95.

PROOF OF LEMMA 4.8. By symmetry, it suffices to consider $\langle \varphi , \varphi \rangle_\star$ in $C_c(H)$. Let $\{e_\alpha\}$ be as in Proposition 4.10 on page 101. Then in view of Remark 4.3 on page 97, we have

$$\begin{aligned} \langle \varphi , \varphi \rangle_\star &= \lim_\alpha \langle e_\alpha \cdot \varphi , \varphi \rangle_\star \\ &= \lim_\alpha \sum_{i=1}^{n_\alpha} \langle {}_\star\langle \varphi_i^\alpha , \varphi_i^\alpha \rangle \cdot \varphi , \varphi \rangle_\star \\ &= \lim_\alpha \langle \varphi_i^\alpha \cdot \langle \varphi_i^\alpha , \varphi \rangle_\star , \varphi \rangle_\star \\ &= \lim_\alpha \langle \varphi_i^\alpha , \varphi \rangle_\star^* \langle \varphi_i^\alpha , \varphi \rangle_\star \end{aligned} \tag{4.12}$$

in the inductive limit topology. Since (4.12) is positive in $C^*(G)$ and since convergence in the inductive limit topology implies convergence in the C^*-norm, it follows that $\langle \varphi , \varphi \rangle_\star \geq 0$ in $C^*(H)$. □

PROOF OF PROPOSITION 4.1. By symmetry, it will suffice to see that $\mathfrak{p}_G$ is full. As in Ex 4.2.1, we have

$$\begin{pmatrix} F_{11} & F_{12} \\ F_{21} & F_{22} \end{pmatrix} * \mathfrak{p}_G * \begin{pmatrix} K_{11} & K_{12} \\ K_{21} & K_{22} \end{pmatrix} = \begin{pmatrix} F_{11} * K_{11} & F_{11} \cdot K_{22} \\ (K_{11}^* \cdot F_{21}^*)^* & \langle F_{12}^* , K_{12} \rangle_\star \end{pmatrix}. \tag{4.13}$$

To see that right-hand side of (4.13) spans a dense subspace of $C_c(L)$ in the inductive limit topology, we can consider each entry separately. Since $C_c(G)$ has an approximate identity in the inductive limit topology for both its action on itself and on $C_c(Z)$ by Proposition 4.10 on page 101, it follows that elements of the form $F_{11} * K_{11}$ are dense in $C_c(G)$ and that elements of the form $F_{11} \cdot K_{12}$ are dense in $C_c(Z)$. By symmetry, elements of the form $(K_{11}^* \cdot F_{21}^*)^*$ also are inductive limit dense in $C_c(Z^{\mathrm{op}})$.

To check the lower right-hand entry, note that if $b \in C_c(H)$ and if $e_\alpha = \sum_i \langle \varphi_i , \varphi_i \rangle_\star$ is as in Proposition 4.10 on page 101 for right action of $C_c(H)$ on itself, then

$$b \cdot e_\alpha = \sum_i \langle \varphi_i \cdot b^* , \varphi_i \rangle_\star .$$

It follows that b is in the inductive limit closure of the span of the inner product $\langle \cdot , \cdot \rangle_\star$. This suffices. □

Exercises

Ex 4.2.1. View each $F \in C_c(L)$ as a matrix $\left(\begin{smallmatrix} F_{11} & F_{12} \\ F_{21} & F_{22} \end{smallmatrix}\right)$ as above.

(a) Show that convolution in $C_c(L)$ is given by

$$F * K = \begin{pmatrix} F_{11} & F_{12} \\ F_{21} & F_{22} \end{pmatrix} \begin{pmatrix} K_{11} & K_{12} \\ K_{21} & K_{22} \end{pmatrix}$$
$$= \begin{pmatrix} F_{11} * K_{11} + {}_{\star}\langle F_{12}\,,\, K_{21}^{*}\rangle & F_{11} \cdot K_{12} + F_{12} \cdot K_{22} \\ (K_{11}^{*} \cdot F_{21}^{*})^{*} + (K_{21}^{*} \cdot F_{22}^{*})^{*} & \langle F_{21}^{*}\,,\, K_{12}\rangle_{\star} + F_{22} * K_{22} \end{pmatrix}.$$

(b) Verify formulas (4.2), (4.3) and (4.4).

—Answer on page 368

Ex 4.2.2. Prove Lemma 4.13.

—Answer on page 369

Ex 4.2.3. Show that the constant M in (4.11) is finite.

—Answer on page 370

Ex 4.2.4. If X is an A–B-imprimitivity bimodule, then the linking algebra $L(\mathsf{X})$ is introduced in [**RW98**, Lemma 3.20]. If G and H are second countable locally compact Hausdorff groupoids with Haar systems and if L is the linking groupoid equipped with its Haar systems as defined here, then show that $C^*(L)$ is isomorphic to the linking algebra $L(\mathsf{X})$ for the $C^*(G)$–$C^*(H)$-imprimitivity bimodule defined in this Chapter.

—Answer on page 370

Ex 4.2.5. Suppose that G acts freely and properly on Z. The orbits $G \cdot z = \{\, \gamma \cdot z : \gamma \in G_{r(z)} \,\}$ are closed in Z and $\gamma \mapsto \gamma \cdot z$ is a homeomorphism of $G_{r(z)}$ onto $G \cdot z$.

—Answer on page 370

CHAPTER 5

Basic Representation Theory

The structure of any C^*-algebra is intimately connected to its representation theory. Like groups, representations of the groupoid C^*-algebra come from unitary representations of the underlying groupoid. Unlike groups, these unitary groupoid representations are a bit unwieldy and mired in measure theory. We'll examine these in detail in Chapter 7. However, these beasts are not to be brought out into the open unless other means fail. In this chapter, we'll set out some of the basics for representations of groupoid C^*-algebras that can be dealt with without using unitary representations and without (much) measure theory. In this chapter, G will always be a second countable locally compact Hausdorff groupoid with a Haar system λ.

The Associated M-Representation. Let L be a (nondegenerate) representation of $C^*(G)$. Then there is an associated representation M_L, or just M if there is no ambiguity, of $C_0(G^{(0)})$ such that

$$M(\varphi)L(f)h = L\big((\varphi \circ r) \cdot f\big)h \quad \text{for all } f \in C_c(G). \tag{5.1}$$

In fact, M is just $\bar{L} \circ V$, were $V : C_0(G^{(0)}) \to M(C^*(G))$ is the canonical map described in Lemma 1.48 on page 19 and $\bar{L}$ the canonical extension of L to $M(C^*(G))$. Then $\ker M$ is an ideal J of $C_0(G^{(0)})$ and we say the *support of* M is the closed set $C \subset G^{(0)}$ such that J is the ideal of functions vanishing on C. One excuse for the use of the word "support" goes as follows. As M is a representation of $C_0(G^{(0)})$, then as in [**Wil07**, Example F.25], there is a Borel Hilbert bundle $G^{(0)} * \mathscr{H}$ and a (Radon) probablity measure μ on $G^{(0)}$ so that M is (equivalent to) the multiplication representation on $L^2(G^{(0)} * \mathscr{H}, \mu)$:

$$M(\varphi)h(u) = \varphi(u)h(u) \quad \text{for } \varphi \in C_0(G^{(0)}) \text{ and } h \in L^2(G^{(0)} * \mathscr{H}). \tag{5.2}$$

It is not hard to check that the support of M is exactly $\operatorname{supp} \mu$ (Ex 5.1.2). We'll come back to the role of supports in Section 5.2.

5.1. Invariant Ideals

If X is a (second countable) locally compact space and if $U \subset X$ is open, then simply by extending functions to be zero on $X \setminus U$, we can view $C_0(U)$ as the ideal of functions in $C_0(X)$ which vanish on the closed set $F := X \setminus U$. Moreover, every ideal in $C_0(X)$ arises in this way (Ex 5.1.3). Restriction to F gives a homomorphism $j : C_0(X) \to C_0(F)$ with kernel $C_0(U)$. The Tietze Extension Theorem implies j is surjective and we have a short exact sequence

$$0 \longrightarrow C_0(U) \xrightarrow{\iota} C_0(X) \xrightarrow{j} C_0(F) \longrightarrow 0. \tag{5.3}$$

(These assertions are still true even if X is not second countable, but some extra work is required. See [**Wil07**, §1.2.2].)

In this section, we want to establish a similar short exact sequence for groupoid C^*-algebras that reduces to (5.3) when G is a topological space viewed as a groupoid.

First, some observations on Haar systems and reductions. We say that $A \subset G^{(0)}$ is *saturated* or *G-invariant* if it is invariant for the usual G-action on $G^{(0)}$. Thus if $u \in A$ and $\gamma \in G_u$, then $r(\gamma) \in A$. If A is G-invariant and if $u \in A$, then $(G|_A)^u$ coincides with G^u. Therefore if A is either a closed or open G-invariant subset of $G^{(0)}$, then $G|_A$ is a locally compact groupoid and any Haar system on G restricts to a Haar system on $G|_A$. In the sequel, we will always assume that the Haar systems on reductions to open or closed G-invariant sets are chosen in this way.

Let U be an open *G-invariant* subset of $G^{(0)}$. Then, with our consistent choice of Haar systems, $C_c(G|_U)$ can be viewed as a $*$-subalgebra of $C_c(G)$; that is, the inclusion map is a $*$-homomorphism. A moments thought reveals that $C_c(G|_U)$ is a two-sided $*$-ideal in $C_c(G)$. Note that $F := G^{(0)} \setminus U$ is a G-invariant closed subset of $G^{(0)}$ and that the restriction map gives a surjective $*$-homomorphism $j : C_c(G) \to C_c(G|_F)$. Especially in analogy with (5.3), the next result is expected. Nevertheless, some work is required to give a proper proof.

THEOREM 5.1. *Let G be a second countable locally compact groupoid with a Haar system. Let $U \subset G^{(0)}$ be an open G-invariant set with complement $F = G^{(0)} \setminus U$. Then there is a short exact sequence of C^*-algebras*

$$0 \longrightarrow C^*(G|_U) \xrightarrow{\ \iota\ } C^*(G) \xrightarrow{\ j\ } C^*(G|_F) \longrightarrow 0,$$

where ι is induced by the inclusion map on $C_c(G|_U)$, and j by the restriction map on $C_c(G)$.

PROOF. Let $\mathrm{Ex}(U)$ be the closure of $C_c(G|_U)$ in $C^*(G)$. Then $\mathrm{Ex}(U)$ is an ideal. Since any representation of $C_c(G)$ restricts to a representation of $C_c(G|_U)$, the inclusion map $\iota : C_c(G|_U) \to C_c(G)$ extends to a homomorphism $\iota : C^*(G|_U) \to C^*(G)$ with range exactly $\mathrm{Ex}(U)$. Since any representation of $C_c(G|_F)$ determines a representation of $C_c(G)$, we get a homomorphism $j : C^*(G) \to C^*(G|_F)$ which is surjective since j maps $C_c(G)$ onto $C_c(G|_F)$ by the Tietze Extension Theorem.

Clearly, $\mathrm{Ex}(U) \subset \ker j$.

Note that if L is a representation of $C_c(G)$ such that $L(f) = 0$ if $j(f) = 0$, then we get a well-defined $*$-homomorphism $L' : C_c(G|_F) \to B(\mathcal{H}_L)$ via $L'(j(f)) = L(f)$. We want to see that L' is a representation; that is, we want to see that

$$\|L'(j(f))\| \le \|j(f)\|_I = \sup_{u \in F}\{\, \lambda(|f|)(u), \lambda(|f^*|)(u) \,\} \tag{5.4}$$

Let $\epsilon > 0$ and fix $f \in C_c(G)$. Since $u \mapsto \lambda(|f|)(u)$ and $u \mapsto \lambda(|f^*|)(u)$ are continuous,

$$F' = \{\, u \in G^{(0)} : \max\{\, \lambda(|f|)(u), \lambda(|f^*|)(u) \,\} \ge \|j(f)\|_I + \epsilon \,\}$$

is a closed set disjoint from F. Let $\varphi \in C^+(G^{(0)})$ be such that $\varphi(u) = 1$ for all $u \in F$ and $\varphi(u) = 0$ if $u \in F'$ and $0 \le \varphi(u) \le 1$ otherwise. Then $f'(\gamma) := \varphi(r(\gamma))f(\gamma)\varphi(s(\gamma))$ defines an element of $C_c(G)$ such that $j(f') = j(f)$. On the other hand,

$$\|L'(j(f))\| = \|L'(j(f'))\| = \|L(f')\| \le \|f'\|_I \le \|j(f)\|_I + \epsilon.$$

Since ϵ is arbitrary, (5.4) holds and L' is a representation. By construction, $L = L' \circ j$ and $\ker j \subset \ker L$.

Suppose that L is a nondegenerate representation of $C^*(G)$ such that $\mathrm{Ex}(U) \subset \ker L$. We want to see that $\ker j \subset \ker L$. We'll do that using the criteria in the previous paragraph. Let C be the support of M_L. Suppose that $C \cap U \neq \emptyset$. Then there is a $\varphi \in C_c^+(G^{(0)})$ such that φ vanishes off U and is not identically zero on C. Thus $M_L(\varphi) \neq 0$ while $V(\varphi)f \in C_c(G|_U) \subset \mathrm{Ex}(U) \subset \ker L$ for all $f \subset C_c(G)$. Then $M_L(\varphi)L(f) = L(V(\varphi)f) = 0$ for all f and $M_L(\varphi)$ must be zero. This is a contradiction and we conclude that $U \cap C = \emptyset$.

Now if $f \in C_c(G)$ is such that $j(f) = 0$, then $K_\epsilon := \{\, \gamma \in G : |f(\gamma)| \geq \epsilon \,\}$ is a compact subset of $G|_U$ and $C_\epsilon = r(K_\epsilon)$ is a compact subset of U. Therefore there is a $\varphi \in C_c^+(U) \subset \ker M_L$ such that $\|V(\varphi)f - f\|_\infty < \epsilon$. It follows that there is a sequence $\{f_i\} \subset \ker L$ such that $f_i \to f$ in the inductive limit topology. Thus $f \in \ker L$. We showed above that this implies $\ker j \subset \ker L$.

Therefore, if $\mathrm{Ex}(U) \subset \ker L$, then $\ker j \subset \ker L$. It follows that we have an exact sequence

$$0 \longrightarrow \mathrm{Ex}(U) \hookrightarrow C^*(G) \xrightarrow{\ j\ } C^*(G|_F) \longrightarrow 0.$$

We can complete the proof by showing the ι is isometric. For this, it will suffice to show that any nondegenerate representation L of $C_c(G|_U)$ is the restriction of a representation of $C_c(G)$. Let $\{e_i\}$ be a $\|\cdot\|_I$-bounded approximate identity for $C_c(G|_U)$ in the inductive limit topology as provided in Proposition 1.49 on page 20. Then if $f \in C_c(G)$, $g_i \in C_c(G|_U)$, and $h_i \in \mathcal{H}_L$,

$$\begin{aligned} \Big\| \sum_i L(f * g_i)h_i \Big\| &= \Big\| \sum_i L(\lim_k (f * (e_k * g_i))h_i \Big\| \\ &\leq \limsup_k \|L(f * e_k)\| \Big\| \sum_i L(g_i)h_i \Big\|. \end{aligned}$$

But, since the I-norm on $C_c(G|_U)$ coincides with the I-norm on $C_c(G)$, we have

$$\|L(f * e_k)\| \leq \|f * e_k\|_I \leq \|f\|_I \|e_k\|_I$$

where the last inequality follows from Ex 1.4.4. Consequently,

$$\Big\| \sum_i L(f * g_i)h_i \Big\| \leq \|f\|_I \limsup_k \|e_k\|_I \Big\| \sum_i L(g_i)h_i \Big\| = \|f\|_I \Big\| \sum_i L(g_i)h_i \Big\|.$$

The above computations shows that we get a well-defined $*$-homomorphism, $\bar{L}$ of $C_c(G)$ on the range of L by setting $\bar{L}(f)L(g)h := L(f*g)h$. The above computation also shows that $\bar{L}$ is $\|\cdot\|_I$-norm bounded. Hence we can extend to all of $\mathcal{H}_L$ since L was assumed nondegenerate. □

It is natural to try to prove a version of Theorem 5.1 on the preceding page for the reduced norm. However, the result is not always true in this case.[1] Hence it is instructive to see where the obvious modifications to the above proof fail. We can let $\mathrm{Ex}_r(U)$ be the closure of $C_c(G|_U)$ in $C_r^*(G)$. If $u \in F$, then $\mathrm{Ind}\,\delta_u(f) = \mathrm{Ind}\,\delta_u(j(f))$ and restriction induces a surjection $j_r : C_r^*(G) \to C_r^*(G|_F)$ with $\mathrm{Ex}_r(U) \subset \ker j_r$. Since $\mathrm{Ind}\,\delta_u(f)$ is the same operator whether we view $f \in C_c(G|_U)$ or as an element of $C_c(G)$, the inclusion map induces an isomorphism of $C_r^*(G|_U)$ and $\mathrm{Ex}_r(U)$. The

---

[1]While there are examples where (5.5) fails to be exact—Renault gives such an example in [**Ren91**, Remark 4.10]—it does not appear to be easy to produce such examples. See Remark 5.3.

problem arises when we try to show $\operatorname{Ex}_r(U) = \ker j_r$. The representation L' of $C^*(G|_F)$ constructed in the proof of Theorem 5.1 on page 108, need not factor through $C^*_r(G|_F)$ and the proof breaks down there. However, if the reduced norm on $C_c(G|_F)$ equals the universal norm, then there is no problem. Thus we have proved the following.

PROPOSITION 5.2. *Let G be a second countable locally compact groupoid with a Haar system. Let $U \subset G^{(0)}$ be an open G-invariant set with complement $F = G^{(0)} \setminus U$. Suppose that that the reduced norm on $C_c(G|_F)$ equals the universal norm. Then there is a short exact sequence of C^*-algebras*

$$0 \longrightarrow C^*_r(G|_U) \xrightarrow{\iota_r} C^*_r(G) \xrightarrow{j_r} C^*_r(G|_F) \longrightarrow 0, \tag{5.5}$$

where ι_r is induced by the inclusion map on $C_c(G|_U)$ and j_r by the restriction map on $C_c(G)$.

REMARK 5.3. In Chapters 9 and 10, we investigate amenability and measurewise amenability for groupoids. We'll show that if $G|_F$ is amenable or even measurewise amenable, then the reduced norm on $C_c(G|_F)$ is equal to the universal norm so that (5.5) is exact in these cases. In general, the issue of the exactness of (5.5) for the reduced norm is very subtle. It is intimately related to the notion of exact groups [**KW99**], and there is an interesting class of nonamenable groupoids for which (5.5) is always exact. Examples include transformation groupoids $G = \mathcal{G} \times X$ where $\mathcal{G}$ is an exact nonamenable group. Such groups are currently a subject of considerable study.

As a corollary of the proof, we note that $\operatorname{Ex}_r(U)$ is always a nontrivial ideal of $C^*_r(G)$ if U is a proper open invariant set: it lies in the kernel of j_r. Recalling that the G-action on $G^{(0)}$ is *minimal* if $G^{(0)}$ has no nontrivial invariant closed sets, we obtain the following observation.

PROPOSITION 5.4. *If $C^*_r(G)$ is simple, then the action of G on $G^{(0)}$ is minimal.*

EXERCISES

Ex 5.1.1. Let $B \subset G^{(0)}$. Show that $A = s(r^{-1}(B))$ is saturated.

—Answer on page 370

Ex 5.1.2. Let M be the representation of $C_0(G^{(0)})$ given in (5.2). Show that the support of M equals $\operatorname{supp}\mu$.

—Answer on page 370

Ex 5.1.3. Let X be a locally compact Hausdorff space. Either assume that X is second countable or use [**Wil07**, §1.2.2]. Show that every ideal in $C_0(X)$ is of the form $J(C) = \{ f \in C_0(X) : f(x) = 0 \text{ for all } x \in C \}$ for some closed set $C \subset X$.

—Answer on page 371

Ex 5.1.4. Let G be a second countable locally compact groupoid with a Haar system λ. Suppose that $U \subset G^{(0)}$ is an open G-invariant set and $F = G^{(0)} \setminus U$.

(a) If $f \in C_c(G|_U)$, show that $\|f\|_{\mathcal{L}^I(G|_U,\lambda|_U)} = \|f\|_{\mathcal{L}^I(G,\lambda)}$.

(b) If $f \in C_c(G)$ and $j(f)$ is the restriction to $G|_F$, then $\|j(f)\|_{\mathcal{L}^I(G|_F,\lambda|_F)} \leq \|f\|_{\mathcal{L}^I(G,\lambda)}$.

(c) Let $u \in U$. Let $\operatorname{Ind}^G \delta_u$ be the representation of $C_c(G)$ and $\operatorname{Ind}^{G|_U} \delta_u$ be the representation of $C_c(G|_U)$. Show that $\operatorname{Ind}^G \delta_u(f) = \operatorname{Ind}^{G|_U} \delta_u(f)$ for any $f \in C_c(G|_U)$.

(d) Let $u \in F$, $f \in C_c(G)$ and j the restriction map. Show that $\operatorname{Ind}^G \delta_u(f) = \operatorname{Ind}^{G|_F} \delta_u(j(f))$.

—Answer on page 371

Ex 5.1.5. Suppose that L and R are nondegenerate representations of $C^*(G)$ with associated representations M_L and M_R of $C_0(G^{(0)})$, respectively. Show that if L and R are equivalent, then so are M_L and M_R.

—Answer on page 371

Ex 5.1.6. Suppose that G has open range and source maps and that A is a saturated subset of $G^{(0)}$. Then its closure, $\overline{A}$, is saturated.

—Answer on page 371

5.2. The Support of a Representation

Recall that if $M : C_0(X) \to B(\mathcal{H})$ is a representation, then the *support of* M is the closed set $C \subset X$ such that

$$\ker M = J_C = \{ f \in C_0(X) : f(x) = 0 \text{ for all } x \in C. \}$$

If L is a nondegenerate representation of $C^*(G)$, then we will also refer to the support of M_L as the *support of* L. If L fails to be nondegenerate, we call the support of L the support of $\operatorname{ess} L$.[2]

PROPOSITION 5.5. *Let L be a nondegenerate representation of $C^*(G)$ and M_L the associated representation of $C_0(G^{(0)})$. Then the support of M_L is a G-invariant closed set F, and L factors through $C^*(G|_F)$.*

First a preliminary result. Recall that if $A \subset G^{(0)}$, then its *saturation* $[A] := G\cdot A$ is the smallest invariant set containing A. We'll need the following observation.

LEMMA 5.6. *Let U be an open set. Then the ideal of $C^*(G)$ generated by $C_c(G|_U)$ is* $\operatorname{Ex}([U])$ *where $[U]$ is the saturation of U.*

PROOF. The ideal E generated by $C_c(G|_U)$ is the closure of

$$E_0 = C_c(G) * C_c(G|_U) * C_c(G) := \operatorname{span}\{ f * g * h : f, h \in C_c(G) \text{ and } g \in C_c(U) \}.$$

Since $C_c(G|_{[U]})$ is an ideal, $E_0 \subset C_c(G|_{[U]})$. We will show that E_0 is dense in $C_c(G|_{[U]})$ in the inductive limit topology.

Fix $f \in C_c(G|_{[U]})$. Let K_0 be a pre-compact open neighborhood of $\operatorname{supp} f$ in $G|_{[U]}$. By the Stone-Weierstrass Theorem [**Ped89**, Corollary 4.3.5], it will suffice to see that functions in E_0 with support in K_0 separate points of K_0 and do not all vanish at some point of K_0. For this, we just need to find, for each $\gamma \in K_0$ and neighborhood V of γ in K_0, a function $F \in E_0$ with $\operatorname{supp} F \subset V$ and $F(\gamma) \neq 0$.

Let $\gamma = \gamma_1\gamma_2\gamma_3$ with $\gamma_2 \in G|_U$. Let V_i be neighborhoods of γ_i such that $V_1V_2V_3 \subset V$ and $V_2 \subset U$. Let $g_i \in C_c^+(G)$ be such that $\operatorname{supp} g_i \subset V_i$ and $g_i(\gamma_i) = 1$. Since we can arrange that $g_2 \in C_c(G|_U)$, it follows that $g_1 * g_2 * g_3 \in E_0$, and has support in $V \subset K_0$. It is a simple matter to check that $g_1 * g_2 * g_3(\gamma) > 0$. □

[2]If $\pi : A \to B(\mathcal{H})$ is possibly degenerate representation, then $\operatorname{ess}\pi$ is the subrepresentation of π assoicated to the subspace generated by $\pi(a)h$ for $a \in A$ and $h \in \mathcal{H}$. Then $\pi = \operatorname{ess}\pi \oplus 0$.

PROOF OF PROPOSITION 5.5. Let $F = \operatorname{supp} M_L$ and $U = G^{(0)} \setminus F$. To see that F is invariant, it will suffice to see that $U = [U]$.

But if $f \in C_c(G|_U)$, then there is a $\varphi \in C_c(U)$ such that $V(\varphi)f = f$. Then $L(f) = M_L(\varphi)L(f)$ and $L(f) = 0$. Hence $C_c(G|_U) \subset \ker L$. By Lemma 5.6 on the previous page, $C_c(G|_{[U]}) \subset \ker L$.

If $[U] \setminus U \neq \emptyset$, then there is a $\varphi \in C_c^+([U])$ which does not vanish on F. Then for all $f \in C_c(G)$, $M_L(\varphi)L(f) = L(V(\varphi)f) = 0$ (since $V(\varphi)f \in C_c(G|_{[U]})$). This forces $M_L(\varphi) = 0$ which is a contradiction.

Hence F is invariant and $\operatorname{Ex}(G^{(0)} \setminus F) \subset \ker L$. This suffices by Theorem 5.1 on page 108. □

EXAMPLE 5.7. Let μ be a Radon measure on $G^{(0)}$ with support C. We claim that if $L = \operatorname{Ind} \mu$, then the support of M_L is the closure of the saturation $[C]$ of C. In particular, the support of $\operatorname{Ind} \delta_u$ is the orbit closure $\overline{[u]}$.

To see this, recall that L acts on $L^2(G, \nu^{-1})$ via convolution: $L(f)h = f * h$. A short calculation shows that $M_L(\varphi)h(\gamma) = \varphi(r(\gamma))h(\gamma)$. Since

$$\|M(\varphi)h\|^2 = \int_{G^{(0)}} \int_G |\varphi(r(\gamma))|^2 |h(\gamma)|^2 \, d\lambda_u(\gamma) \, d\mu(u), \tag{5.6}$$

it is clear that if φ vanishes on $[C] = r(G_C)$, then $M_L(\varphi) = 0$. On the other hand, if φ is nonzero on $[C]$ we can find $h \in C_c(G) \subset L^2(G, \nu^{-1})$ so that $\varphi(r(\gamma))h(\gamma)$ is nonzero for some γ with $s(\gamma) \in C$. Then (5.6) is nonzero and the kernel of M_L is exactly those functions vanishing on $[C]$ and hence on the closure of $[C]$.

A nonempty closed subset F of a topological space X is called *irreducible* if it can't be written as the union of two nonempty proper closed subsets.

LEMMA 5.8. *Let G be a second countable locally compact groupoid with open range and source maps. Then every irreducible subset of $G \backslash G^{(0)}$ is a point closure.*

PROOF. A space is a Baire space if the countable intersection of open dense sets is dense. If every closed subset is Baire, we say the space is totally Baire. Since complete metric spaces are totally Baire, so is every second countable locally compact space (or use [**Wil70**, Corollary 25.4]). Since the orbit map $q : G^{(0)} \to G\backslash G^{(0)}$ is open, $G\backslash G^{(0)}$ is a second countable totally Baire space. Suppose $C \subset G\backslash G^{(0)}$ is irreducible. Since $G\backslash G^{(0)}$ is second countable, C has a countable basis of nonempty relatively open sets U_n, and $C = \bigcup U_n$. Since C is irreducible, each U_n is dense and $\bigcap U_n$ is nonempty. Any point in the intersection meets every open set and is therefore dense. □

Let $F \subset G^{(0)}$ be a G-invariant closed subset which is *G-irreducible* in the sense that it can't be written as the union of two nonempty proper closed G-invariant sets. Then, if $q : G^{(0)} \to G\backslash G^{(0)}$ is the quotient map, $q(F)$ is closed—its complement $q(G^{(0)} \setminus F)$ is open since q is open. If $q(F)$ is not irreducible, then $q(F) = C_1 \cup C_2$ for proper nonempty closed sets C_i, and $F = q^{-1}(C_1) \cup q^{-1}(C_2)$ which contradicts our assumptions. By Lemma 5.8, $q(F) = \overline{\{q(u)\}}$. Hence $F = q^{-1}(q(F)) = \overline{[u]} = \overline{G \cdot u}$ (Ex 2.2.7).

PROPOSITION 5.9. *Suppose that L is an irreducible representation of $C^*(G)$. Then the support of M_L is an orbit closure.*

REMARK 5.10. In our proof of Proposition 5.9, the hypothesis that L is irreducible is only used to guarantee that $\ker L$ is a primitive, and hence prime, ideal. Thus the result holds with the same proof for any representation L with $\ker L$ a prime ideal. Therefore we could equally have well supposed that L was factorial or even a homogeneous representation.

PROOF OF PROPOSITION 5.9. Let F be the support of M_L. By the discussion preceding the proof, it will suffice to show that F can't be written as the union of two nonempty proper closed invariant subsets and still have $\ker L$ a prime ideal. We proceed to see that if $F = C_1 \cup C_2$ with each C_i a nonempty proper closed invariant subset of F, then $\ker L$ is not a prime ideal.

Let $U_i = G^{(0)} \setminus C_i$. Note that $\mathrm{Ex}(U_i) \not\subset \ker L$ by Ex 5.2.2. On the other hand, using Ex 2.2.7,

$$\begin{aligned} C_c(G|_{U_1})C_c(G|_{U_2}) &\subset C_c(G|_{U_1\cap U_2}) \\ &= C_c(G|_{G^{(0)}\setminus(C_1\cup C_2)}) = C_c(G|_{G^{(0)}\setminus F}) = C_c(G|_U). \end{aligned}$$

Thus,

$$\mathrm{Ex}(U_1)\,\mathrm{Ex}(U_2) \subset \ker L.$$

Since $\mathrm{Ex}(U_i) \not\subset \ker L$, this implies $\ker L$ is not prime. □

EXERCISES

Ex 5.2.1. Let $U \subset G^{(0)}$ be open and suppose that $\varphi \in C_c(G^{(0)})$ with $\operatorname{supp}\varphi \subset U$. Show that for all $f \in C_c(G)$, we have $V(\varphi)f \in C_c(G|_{[U]})$.

—Answer on page 372

Ex 5.2.2. Let L be a nondegenerate representation of $C^*(G)$ with $F = \operatorname{supp} M_L$. Suppose that U is an open invariant subset of $G^{(0)}$ such that $U \cap F \neq \emptyset$. Then $\mathrm{Ex}(U) \not\subset \ker L$.

—Answer on page 372

Ex 5.2.3. Show that if U_1 and U_2 are open saturated subsets of $G^{(0)}$, then $C_c(G|_{U_1})C_c(G|_{U_2}) \subset C_c(G|_{U_1\cap U_2})$.

—Answer on page 372

5.3. Inducing Representations

If H is a closed subgroupoid of G, then it is natural to try to induce representations of $C^*(H)$ to representations of $C^*(G)$. In the group case, this is often done at the level of unitary representations. Here we work at the algebra level—therefore both H and G need to have Haar systems. A complete approach using the Rieffel machinery requires the Disintegration Theorem from Chapter 8. For this reason, induced representations were treated in a somewhat *ad hoc* manner in the literature before that result was available. It is also why it is necessary to stick to the second countable case throughout this section. Much of this material is taken from [**IW09b**, §2].

We assume that G is a second countable locally compact Hausdorff groupoid with Haar system $\{\lambda^u\}_{u\in G^{(0)}}$. Let H be a closed subgroupoid of G with Haar system $\{\alpha^u\}_{u\in H^{(0)}}$. Since H is closed in G, we also have $H^{(0)}$ closed in $G^{(0)}$. Then

$G_{H^{(0)}} := s^{-1}(H^{(0)})$ is a locally compact free and proper right H-space, and we can form the imprimitivity groupoid $H^{G_{H^{(0)}}}$ as in Lemma 2.45 on page 47. Specifically,

$$G_{H^{(0)}} *_s G_{H^{(0)}} := \{ (\eta, \gamma) \in G_{H^{(0)}} \times G_{H^{(0)}} : s(\eta) = s(\gamma) \}$$

is a free and proper right H-space for the diagonal action $(\eta, \gamma) \cdot h := (\eta h, \gamma h)$. Consequently, the orbit space

$$H^{G_{H^{(0)}}} := (G_{H^{(0)}} *_s G_{H^{(0)}})/H$$

is a locally compact Hausdorff space. Writing $[\eta, \gamma]$ for the image of (η, γ) in $H^{G_{H^{(0)}}}$, the groupoid structure is given by $[\eta, \gamma][\gamma, \zeta] = [\eta, \zeta]$ and $[\eta, \gamma]^{-1} = [\gamma, \eta]$ (suitably interpreted—see the discussion preceding Lemma 2.45 on page 47). Recall that we can identify $(H^{G_{H^{(0)}}})^{(0)}$ with $G_{H^{(0)}}/H$, and then

$$r([\eta, \gamma]) = \eta \cdot H \quad \text{and} \quad s([\eta, \gamma]) = \gamma \cdot H.$$

Note that the moment map $s_{G_{H^{(0)}}} = s_G|_{G_{G_{H^{(0)}}}}$ for the right H-action on $G_{H^{(0)}}$ is open. To see this, I'll use Fell's criterion. If $u_i \to s_{G_{H^{(0)}}}(\gamma)$ in $H^{(0)}$, then since s_G is open, we can pass to a subnet, relabel, and assume that there are $\gamma_i \in G$ such that $\gamma_i \to s(\gamma)$ and $s(\gamma_i) = u_i$. But then $\gamma_i \in G_{H^{(0)}}$ and $s_{G_{H^{(0)}}}$ must be open. Now Proposition 2.47 on page 48 implies that $G_{H^{(0)}}$ is a $(H^{G_{H^{(0)}}}, H)$-equivalence for the left $H^{G_{H^{(0)}}}$-action

$$[\eta, \gamma] \cdot \gamma = \eta.$$

To get a Haar system on $H^{G_{H^{(0)}}}$, we use Lemma 3.16 on page 75.[3] Since $G_{H^{(0)}}$ is closed in G and since λ is a Haar system on G, it is not hard to check that

$$\beta'(\varphi)(\eta) := \int_G \varphi(\gamma)\, d\lambda_{s(\eta)}(\gamma) \quad \varphi \in C_c(G_{H^{(0)}})$$

is a full equivariant s-system for the map $s : G_{H^{(0)}} \to H^{(0)}$ (as defined in Section 3.2 on page 72). Therefore Lemma 3.16 on page 75 implies that we get a Haar system $\{ \beta^{\eta\cdot H} \}_{\eta\cdot H \in (H^{G_{H^{(0)}}})^{(0)}}$ for $H^{G_{H^{(0)}}}$ via

$$\beta(F)(\eta \cdot H) = \int_{H^{G_{H^{(0)}}}} F([\eta, \gamma])\, d\beta^{\eta\cdot H}([\eta, \gamma]) = \int_G F([\eta, \gamma])\, d\lambda_{s(\eta)}(\gamma).$$

Since both H and $H^{G_{H^{(0)}}}$ have Haar systems, the Equivalence Theorem 4.9 on page 99 implies that $C_c(G_{H^{(0)}})$ is a pre-$C_c(H^{G_{H^{(0)}}}, \beta)$ – $C_c(H, \alpha)$-imprimitivity bimodule with actions and inner products given by

$$F \cdot \varphi(\zeta) = \int_G F([\zeta, \gamma])\varphi(\gamma)\, d\lambda_{s(\zeta)}(\gamma),$$

$$\varphi \cdot g(\zeta) = \int_H \varphi(\zeta h) g(h^{-1})\, d\alpha^{s(\zeta)}(h), \tag{5.7}$$

$$\langle \varphi \,, \psi \rangle_\star (h) = \int_G \overline{\varphi(\gamma)}\psi(\gamma h)\, d\lambda_{r(h)}(\gamma), \quad \text{and} \tag{5.8}$$

$$_\star\langle \varphi \,, \psi \rangle([\eta, \gamma]) = \int_H \varphi(\eta h)\overline{\psi(\gamma h)}\, d\alpha^{s(\eta)}(h).$$

[3] In Section 6.5, we'll show that any groupoid equivalent to a groupoid with a Haar system has a Haar system. But we want an explicit one here, and we have a natural one available.

We will write $\mathsf{X} = \mathsf{X}_H^G$ for the completion of $C_c(G_{H^{(0)}})$ as a $C^*(H_{G_{H^{(0)}}})$–$C^*(H)$-imprimitivity bimodule.

If L is a representation of $C^*(H,\alpha)$, then we write X–$\operatorname{Ind} L$ for the representation of $C^*(H_{G_{H^{(0)}}},\beta)$ induced via X (see the discussion following [**RW98**, Proposition 2.66]). By definition, X–$\operatorname{Ind} L$ acts on the completion $\mathcal{H}_{\operatorname{Ind} L}$ of $C_c(G_{H^{(0)}}) \odot \mathcal{H}_L$ with respect to the pre-inner product given on elementary tensors by

$$(\varphi \otimes h \mid \psi \otimes k) = \big(L\big(\langle \psi \,,\, \varphi\rangle_{\star}\big)h \mid k\big).$$

If $\varphi \otimes_H h$ denotes the class of $\varphi \otimes h$ in $\mathcal{H}_{\operatorname{Ind} L}$, then

$$(\mathsf{X}\text{–}\operatorname{Ind} L)(F)(\varphi \otimes_H h) = F \cdot \varphi \otimes_H h.$$

To get an induced representation of $C^*(G)$ out of this machinery (i.e., using [**RW98**, Proposition 2.66]), we need a nondegenerate homomorphism of $C^*(G)$ into $\mathcal{L}(\mathsf{X})$. If $f \in C_c(G)$ and $\varphi \in C_c(G_{H^{(0)}})$ we can define

$$f \cdot \varphi(\zeta) = \int_G f(\gamma)\varphi(\gamma^{-1}\zeta)\, d\lambda^{r(z)}(\gamma). \tag{5.9}$$

Remark 5.11. Since $G_{H^{(0)}}$ is closed in G, each $\varphi \in C_c(G_{H^{(0)}})$ is the restriction of an element $f_\varphi \in C_c(G)$. Thus we can write

$$\langle \varphi \,,\, \psi\rangle_{\star} = \varphi^* * \psi \quad \text{and} \quad f \cdot \varphi = f * \varphi,$$

where, for example, $\varphi * \psi$ should be interpreted as $f_\varphi * f_\psi$ restricted to $G_{H^{(0)}}$—the point being that the restriction is independent of our choice of f_φ and f_ψ. Similarly, $f * \varphi$ is meant to be the restriction of $f * f_\varphi$ to $G_{H^{(0)}}$.

If ρ is a state on $C^*(H)$, then

$$(\cdot \mid \cdot)_\rho := \rho\big(\langle \cdot \,,\, \cdot\rangle_{\star}\big)$$

is a pre-inner product on $C_c(G_{H^{(0)}})$ with Hilbert space completion denoted by $\mathcal{H}_\rho$. If we let $V(f)\varphi := f \cdot \varphi$, then it follows from direct computation, or by invoking Remark 5.11 above, that

$$\langle V(f)\varphi \,,\, \psi\rangle_{\star} = \langle f \cdot \varphi \,,\, \psi\rangle_{\star} = \langle \varphi \,,\, f^* \cdot \psi\rangle_{\star} = \langle \varphi \,,\, V(f^*)\psi\rangle_{\star}.$$

Thus V induces a map of $C_c(G)$ into the linear operators on the dense image of $C_c(G_{H^{(0)}})$ in $\mathcal{H}_\rho$ which is clearly a weak-representation of $C_c(G)$ as in Definition 8.1 on page 169. Thus by Theorem 8.2 on page 169 we obtain a bona fide representation of $C^*(G)$ on $\mathcal{H}_{\operatorname{Ind} L}$, and it follows that

$$\rho\big(\langle f \cdot \varphi \,,\, f \cdot \varphi\rangle_{\star}\big) \le \|f\|^2_{C^*(G)}\rho\big(\langle \varphi \,,\, \varphi\rangle_{\star}\big).$$

Since this holds for all ρ,

$$\langle f \cdot \varphi \,,\, f \cdot \varphi\rangle_{\star} \le \|f\|^2 \langle \varphi \,,\, \varphi\rangle_{\star},$$

and $V(f)$ is a bounded adjointable operator on X. Therefore we obtain an induced representation $\operatorname{Ind}_H^G L$ of $C^*(G)$ on $\mathcal{H}_{\operatorname{Ind} L}$ such that

$$(\operatorname{Ind}_H^G L)(f)(\varphi \otimes_H h) = f * \varphi \otimes_H h. \tag{5.10}$$

Definition 5.12. Let H be a closed subgroupoid of a second countable locally compact Hausdorff groupoid G. Assume both G and H have Haar systems. If $L : C^*(H) \to B(\mathcal{H}_L)$ is a representation of $C^*(H)$, then the representation $\operatorname{Ind}_H^G L : C^*(G) \to B(\mathcal{H}_{\operatorname{Ind} L})$ satisfying (5.10) is called the *representation of* $C^*(G)$ *induced from* L.

REMARK 5.13. Since $M\bigl(C^*(H_{G_{H^{(0)}}})\bigr) \cong \mathcal{L}(\mathsf{X})$ ([**RW98**, Corollary 2.54 and Proposition 3.8]), it is not hard to see that $\operatorname{Ind}_H^G L$ is the composition of the nondegenerate[4] homomorphism V with the natural extension of X–$\operatorname{Ind} L$ to $\mathcal{L}(\mathsf{X})$. Thus $\operatorname{Ind}_H^G L$ is just the induced representation $\operatorname{Ind}_{C^*(H)}^{C^*(G)}$ defined in [**RW98**, Proposition 2.66] upon viewing $C^*(G)$ as a C^*-algebra of adjointable operators on X. In particular, $\operatorname{Ind}_H^G L$ is nondegenerate if L is.

Remark 5.13 allows us to plug into the Rieffel machinery. For example, it follows that the kernel of an induced representation depends only on the kernel of the initial representation. More precisely, [**RW98**, Corollary 2.37] implies that

$$\ker(\operatorname{Ind}_H^G L) = \operatorname{Ind}_H^G(\ker L), \tag{5.11}$$

where as is standard, we have written Ind_H^G for the map $\operatorname{Ind}_{C^*(H)}^{C^*(G)} : \mathfrak{I}(C^*(H)) \to \mathfrak{I}(C^*(G))$ on ideals given by [**RW98**, Corollary 3.35].

LEMMA 5.14. *Let H, G, V, and X be as in Remark 5.13, and let $\operatorname{Ind}_H^G : \mathfrak{I}(C^*(H)) \to \mathfrak{I}(C^*(G))$ be the map on ideals given by* (5.11). *Then*

$$\operatorname{Ind}_H^G J = \{\, a \in C^*(G) : V(a)x \in \overline{\mathsf{X} \cdot J}\,\}. \tag{5.12}$$

In particular, if $\{\, J_k \,\}_{k\in K}$ is any family of ideals in $C^*(H)$, then

$$\operatorname{Ind}_H^G \bigcap_k J_k = \bigcap_k \operatorname{Ind}_H^G I_k. \tag{5.13}$$

PROOF. This result exactly parallels [**Wil07**, Lemmas 5.16 and 5.19]. In particular, (5.12) follows from [**RW98**, Proposition 3.34] and the observation that X–$\operatorname{Ind} J = \overline{\mathsf{X} \cdot J}$ by [**RW98**, Proposition 3.24]. Then (5.13) is established exactly as in the proof of [**Wil07**, Lemma 5.19]. □

REMARK 5.15. The Cohen Factorization Theorem [**RW98**, Proposition 2.33] implies that

$$\overline{\mathsf{X} \cdot J} := \overline{\operatorname{span}}\{\, x \cdot a : x \in \mathsf{X} \text{ and } a \in J\,\} = \{\, x \cdot a : x \in \mathsf{X} \text{ and } a \in J\,\}$$

where both the closure *and* the span have been eliminated after the second equal sign. Hence we could replace $\overline{\mathsf{X} \cdot J}$ with simply $\mathsf{X} \cdot J$ in (5.12), but I've choosen to proceed with the overlines in place to reduce the overhead.

EXAMPLE 5.16. Let $L : C_0(G^{(0)}) \to B(\mathcal{K})$ be a representation of $C_0(G^{(0)})$. As described in Remark E.15 and Example F.25 of [**Wil07**], we may as well assume that $\mathcal{K}$ is $L^2(G^{(0)} * \mathscr{H}, \mu)$ for a Borel Hilbert bundle $G^{(0)} * \mathscr{H}$ and a finite Radon measure μ, and that L is given by

$$L(\varphi)h(u) = \varphi(u)h(u) \quad \text{for } \varphi \in C_0(G^{(0)}) \text{ and } h \in L^2(G^{(0)} * \mathscr{H}, \mu).$$

We want to examine the induced representation $\operatorname{Ind}_{G^{(0)}}^G L$ on $\mathcal{H}_{\operatorname{Ind}_{G^{(0)}}^G L}$ in more detail.

Let $G * \mathscr{H}_s$ be the pull-back of $G^{(0)} * \mathscr{H}$ via the source map. More precisely, let $\{h_j\}$ be a special orthogonal fundamental sequence for $G^{(0)} * \mathscr{H}$ as in [**Wil07**, Proposition F.6]. For each j, let $\tilde{h}_j(\gamma) = h_j(s(\gamma))$. Then $G * \mathscr{H}_s$ is the Borel Hilbert

[4]Since $C_c(G)$ has an approximate identity for the inductive limit topology (Proposition 1.49 on page 20), it is not hard to check that V is nondegenerate.

bundle built from $\coprod_{\gamma\in G} H(s(\gamma))$ with the fundamental sequence $\{\tilde h_j\}$. As usual we let $\nu = \mu\circ\lambda$ and let ν^{-1} be the push forward under inversion.

Define $U : C_c(G)\odot L^2(G^{(0)} * \mathscr{H}, \mu) \to L^2(G * \mathscr{H}_s, \nu^{-1})$ by

$$U(f\otimes h)(\gamma) = f(\gamma)h(s(\gamma)).$$

Since

$$\big(U(f\otimes h(\gamma) \mid \tilde h_j(\gamma)\big) = f(\gamma)\big(h(s(\gamma)) \mid h_j(s(\gamma))\big),$$

$\gamma\mapsto U(f\otimes h)(\gamma)$ is Borel and $U(f\otimes h)$ is a Borel section of $G * \mathscr{H}_s$. Therefore we can use (5.8) to show that in $\mathcal{H}_{\operatorname{Ind}_{G^{(0)}}^G L}$,

$$\begin{aligned}(f\otimes h \mid g\otimes k) &= \int_{G^{(0)}}\int_G \overline{g(\gamma)}f(\gamma)\big(h(u)\mid k(u)\big)\,d\lambda_u(\gamma)\,d\mu(u)\\ &= \int_G \big(U(f\otimes h)(\gamma) \mid U(g\otimes k)(\gamma)\big)\,d\nu^{-1}(\gamma).\end{aligned}\tag{5.14}$$

It follows that U is an isometry and that $U(f\otimes h)\in L^2(G*\mathscr{H}_s,\nu^{-1})$. Arguing as in [**Wil07**, Lemma F.17], it is not hard to see that the range of U is dense. Hence U extends to a unitary from $\mathcal{H}_{\operatorname{Ind}_{G^{(0)}}^G L}$ onto $L^2(G*\mathscr{H}_s,\nu^{-1})$.

Recall that $(\operatorname{Ind}_{G^{(0)}}^G L)(f)(g\otimes h) = f*g\otimes h$. Then as in (5.14),

$$\begin{aligned}&\big(U(\operatorname{Ind}_{G^{(0)}}^G L)(f)(g\otimes h) \mid g'\otimes h'\big)\\ &= \int_G \overline{g'(\eta)}f*g(\eta)\big(h(s(\eta)) \mid h'(s(\eta))\big)\,d\nu^{-1}(\eta)\\ &= \int_G\int_G f(\gamma)\big(g(\gamma^{-1}\eta)h(s(\eta)) \mid g'(\eta)h'(s(\eta))\big)\,d\lambda^{r(\eta)}(\gamma)\,d\nu^{-1}(\eta)\end{aligned}$$

which, after sending $\gamma\mapsto\eta\gamma$, is

$$\begin{aligned}&= \int_G\int_G f(\eta\gamma)\big(g(\gamma^{-1})h(s(\gamma^{-1})) \mid g'(\eta)h'(s(\eta))\big)\,d\lambda^{s(\eta)}(\gamma)\,d\nu^{-1}(\eta)\\ &= \int_{G^{(0)}}\int_G\int_G f(\eta\gamma)\big(U(g\otimes h)(\gamma^{-1}) \mid U(g'\otimes h')(\eta)\big)\,d\lambda^u(\gamma)\,d\lambda_u(\eta)\,d\mu(u).\end{aligned}$$

Therefore, we have established the following.

LEMMA 5.17. *Let L be the representation of $C_0(G^{(0)})$ on $L^2(G^{(0)}*\mathscr{H},\mu)$ described above. Then $\operatorname{Ind}_{G^{(0)}}^G L$ is equivalent to the representation* $\operatorname{Ind} L$ *on $L^2(G * \mathscr{H}_s,\nu^{-1})$ given by*

$$\big((\operatorname{Ind} L)(f)k \mid k'\big) = \int_{G^{(0)}}\int_G\int_G f(\eta\gamma)\big(k(\gamma^{-1}) \mid k'(\eta)\big)\,d\lambda^u(\gamma)\,d\lambda_u(\eta)\,d\mu(u).$$

EXAMPLE 5.18 ($\operatorname{Ind}\mu$). Let $H = G^{(0)}$. Let μ be a Radon measure on $G^{(0)}$ and $L : C_0(G^{(0)})\to B(L^2(\mu))$ the usual multiplication representation:

$$L(\varphi)h(u) = \varphi(u)h(u)\quad\text{for } \varphi\in C_0(G^{(0)}) \text{ and } h\in L^2(\mu).$$

It is not hard to see that $\operatorname{Ind}_{G^{(0)}}^G L$ is equivalent to the regular representation $\operatorname{Ind}\mu$ defined in (1.7) on page 17. See Ex 5.3.1.

EXAMPLE 5.19 ($\operatorname{Ind}\delta_u$). In Example 5.18, consider the case that μ is the point mass δ_u at $\{u\}$. Then $\operatorname{Ind}_{G^{(0)}}^G L$ is equivalent to the regular representation $\operatorname{Ind}\delta_u$ defined in (1.9).

One advantage of having a formal theory of induction for representations of groupoid C^*-algebras is that we can apply the Rieffel machinery. An example is the following version of induction in stages. The proof, modulo technicalities, is a straightforward modification of Rieffel's original "C^*-version" from [**Rie74**, Theorem 5.9].

THEOREM 5.20 (Induction in Stages). *Suppose that H and K are closed subgroupoids of a second countable locally compact Hausdorff groupoid G with $H \subset K$. Assume that H, K and G have Haar systems. If L is a representation of $C^*(H)$, then*

$$\operatorname{Ind}_H^G L \quad \textit{and} \quad \operatorname{Ind}_K^G\big(\operatorname{Ind}_H^K L\big)$$

are equivalent representations of $C^*(G)$.

We need some preliminaries for the proof. We let λ, β and α be Haar systems on G, K and H, respectively. It will be helpful to notice that the space $\mathcal{H}_{\operatorname{Ind} L}$ of $\operatorname{Ind}_H^G L$ is an internal tensor product $\mathsf{X}_H^G \otimes_H \mathcal{H}_L$ for the appropriate actions of $C^*(H)$.[5] Thus the space of $\operatorname{Ind}_K^G\big(\operatorname{Ind}_H^G L\big)$ is $\mathsf{X}_K^G \otimes_K (\mathsf{X}_H^K \otimes_H \mathcal{H}_L)$. The natural map on the algebraic tensor products induces an isomorphism U of $\mathsf{X}_K^G \otimes_K (\mathsf{X}_H^K \otimes_H \mathcal{H}_L)$ with $(\mathsf{X}_K^G \otimes_K \mathsf{X}_H^K) \otimes_H \mathcal{H}_L$ (see [**Wil07**, Lemma I.6]). We need to combine this with the following observation.

LEMMA 5.21. *The map sending $\varphi \odot \psi \in C_c(G_{K^{(0)}}) \odot C_c(K_{H^{(0)}})$ to $\theta(\varphi \otimes \psi)$ in $C_c(G_{H^{(0)}})$, given by*

$$\theta(\varphi \otimes \psi)(\gamma) := \int_K \varphi(\gamma\kappa)\psi(\kappa^{-1})\, d\beta^{s(\gamma)}(\kappa),$$

induces an isomorphism, also called θ, of $\mathsf{X}_K^G \otimes \mathsf{X}_H^K$ onto X_H^G.

PROOF. The first step is to see that θ is isometric. Notice that we have three sets of actions and inner products. We have not tried to invent notation to distinguish one from another. Instead, we will hope that it is "clear from context" which formula is being employed. In this spirit,

$$\langle \varphi_1 \otimes \psi_1 \,,\, \varphi_2 \otimes \psi_2 \rangle_\star(h) = \langle \psi_1 \,,\, \langle \varphi_1 \,,\, \varphi_2 \rangle_\star \cdot \psi_2 \rangle_\star(h)$$

which, using (5.8) on $C_c(K_{H^{(0)}})$, is

$$= \int_K \overline{\psi_1(\kappa)} \langle \varphi_1 \,,\, \varphi_2 \rangle_\star \cdot \psi_2(\kappa h)\, d\beta_{r(h)}(\kappa)$$

which, using (5.9) for the $C_c(K)$-action on $C_c(K_{H^{(0)}})$, is

$$= \int_K \int_K \overline{\psi_1(\kappa)} \langle \varphi_1 \,,\, \varphi_2 \rangle_\star(\kappa\kappa_1)\psi_2(\kappa_1^{-1}\kappa)\, d\beta^{r(h)}(\kappa_1)\, d\beta_{r(h)}(\kappa)$$

which, using (5.8), is

$$\begin{aligned} &= \int_K \int_K \int_G \overline{\psi_1(\kappa)\varphi_1(\gamma)}\varphi_2(\gamma\kappa\kappa_1)\psi_2(\kappa_1^{-1}\kappa)\, d\lambda_{r(\kappa)}(\gamma)\, d\beta^{r(h)}(\kappa_1)\, d\beta_{r(h)}(\kappa) \\ &= \int_K \int_K \int_G \overline{\psi_1(\kappa)\varphi_1(\gamma\kappa^{-1})}\varphi_2(\gamma\kappa_1)\psi_2(\kappa_1^{-1}\kappa)\, d\lambda_{r(h)}(\gamma)\, d\beta^{r(h)}(\kappa_1)\, d\beta_{r(h)}(\kappa) \end{aligned}$$

[5]Internal tensor products of Hilbert modules are discussed in [**Lan95**] and also [**Wil07**, App. I.1].

which, after using Fubini and sending κ_1 to $h\kappa_1$, is

$$= \int_G \overline{\theta(\varphi_1 \otimes \psi_1)(\gamma)}\theta(\varphi_2 \otimes \psi_2)(\gamma h)\, d\lambda_{r(h)}(\gamma)$$
$$= \langle \theta(\varphi_1 \otimes \psi_1)\, ,\, \theta(\varphi_2 \otimes \psi_2)\rangle_\star .$$

Thus, θ is isometric. We just need to see that it has dense range.

However, notice that $\theta(\varphi \otimes \psi) = \varphi \cdot g_\psi$ for the right action on $C_c(K)$ on $C_c(G_{K^{(0)}})$ with g_ψ *any* extension of ψ to $C_c(K)$ (see (5.7) and Remark 5.11). It follows from Proposition 4.10 on page 101 that there is an approximate identity for $C_c(K)$ such that $\varphi \cdot g_i \to \varphi$ in the inductive limit topology for all $\varphi \in C_c(G_{K^{(0)}})$. This implies that the range of θ is dense, and completes the proof of the lemma. $\square$

PROOF OF THEOREM 5.20. Define a unitary $V : \mathsf{X}^G_K \otimes_K (\mathsf{X}^K_H \otimes_H \mathcal{H}_L) \to \mathsf{X}^G_H \otimes_H \mathcal{H}_L$ by $V = \theta \circ U$ (where U is defined prior to Lemma 5.21). Then on elementary tensors, $V\big(\varphi \otimes (\psi \otimes h)\big) = \theta(\varphi \otimes \psi) \otimes h$. Then on the one hand,

$$V\big(\operatorname{Ind}^G_K\big(\operatorname{Ind}^K_H(L)(f)\big)\big)\big(\varphi \otimes (\psi \otimes h)\big) = \theta(f * \varphi \otimes \psi) \otimes h.$$

On the other hand, $\theta(f * \varphi \otimes \psi) = f * \theta(\varphi \otimes \psi)$. Therefore

$$V(\operatorname{Ind}^G_K\big(\operatorname{Ind}^K_H(L)\big) = (\operatorname{Ind}^G_H L)V.$$

This completes the proof. $\square$

Regular Representations and the Reduced Norm. To calculate the reduced norm, it is important to keep in mind that the kernel of an induced representation depends only on the kernel of the initial representation as in (5.11). As before, we write Ind^G_H for map on ideals from $\mathcal{I}(C^*(H))$ to $\mathcal{I}(C^*(G))$.

Let ev_u be evaluation at $u \in G^{(0)}$ viewed as a representation of $C_0(G^{(0)})$. Hence $\operatorname{Ind}^G_{G^{(0)}} \mathrm{ev}_u$ is equivalent to $\operatorname{Ind} \delta_u$ (see Example 5.19 on page 117 and Ex 5.3.2). Let $L = \bigoplus_{u \in G^{(0)}} \mathrm{ev}_u$. Since Rieffel induction respects arbitrary direct sums,

$$\operatorname{Ind}^G_{G^{(0)}} \bigoplus_{u \in G^{(0)}} \mathrm{ev}_u \cong \bigoplus_{u \in G^{(0)}} \operatorname{Ind}^G_{G^{(0)}} \mathrm{ev}_u \cong \bigoplus_{u \in G^{(0)}} \operatorname{Ind} \delta_u.$$

It follows that for any $f \in C_c(G)$, we have

$$\|f\|_r = \|(\operatorname{Ind}^G_{G^{(0)}} L)(f)\|. \tag{5.15}$$

Now it follows from (5.11), that (5.15) holds if L is any faithful representation of $C_0(G^{(0)})$. We have proved most of the following.

PROPOSITION 5.22. *Let L be a representation of $C_0(G^{(0)})$. Then*

$$\|(\operatorname{Ind}^G_{G^{(0)}} L)(f)\| \leq \|f\|_r \tag{5.16}$$

Furthermore, (5.16) *is an equality for all $f \in C_c(G)$ if L is faithful.*

COROLLARY 5.23. *If μ is a Radon measure on $G^{(0)}$ with full support, then*

$$\|f\|_r = \|(\operatorname{Ind} \mu)(f)\|.$$

REMARK 5.24. In Section 1.4 on page 13, we called representations of the form $\operatorname{Ind} \mu$ with μ a measure on $G^{(0)}$ a regular representation. More generally, if L is any representation of $C_0(G^{(0)})$, then we also call $\operatorname{Ind}^G_{G^{(0)}} L$ a *regular representation.* By Proposition 5.22, all regular representations factor through $C^*_r(G)$.

Exercises

Ex 5.3.1. Prove the equivalence asserted in Example 5.18.

—Answer on page 372

Ex 5.3.2. Prove the equivalence asserted in Example 5.19.

—Answer on page 372

Ex 5.3.3. Suppose that $D \subset G^{(0)}$ is dense. Show that

$$\|f\|_r = \sup_{u \in D} \|(\operatorname{Ind} \delta_u)(f)\| \quad \text{for all } f \in C_c(G).$$

—Answer on page 372

5.4. Supports of Induced Representations

Let F be a saturated closed subset of $G^{(0)}$. Let

$$j_F^G : C^*(G) \to C^*(G|_F)$$

be the natural map from Theorem 5.1 on page 108. It follows from Proposition 5.5 on page 111 that if L is a nondegenerate representation of $C^*(G)$ with support contained in F, then L factors through $C^*(G|_F)$. Thus there is a unique nondegenerate representation L' of $C^*(G|_F)$ such that

$$L = L' \circ j_F^G.$$

The following observation will be convenient. Let $V : C_b(G^{(0)}) \to M(C^*(G))$ be the usual map given in Lemma 1.48 on page 19.

Lemma 5.25. *Suppose that F is a closed saturated subset of $G^{(0)}$ and that L a nondegenerate representation of $C^*(G)$. Let $U = G^{(0)} \setminus F$. Then the support of L is a subset of F if and only if $L\big(V(\varphi)f\big) = 0$ for all $f \in C_c(G)$ and $\varphi \in C_c(U)$.*

Proof. If the support of L is contained in F, then $C_c(U) \subset \ker M_L$. Hence $L(V(\varphi)f) = M_L(\varphi)V(f) = 0$. On the other hand, if the support of L is not contained in F, then there is a $\varphi \in C_c(U)$ such that $M_L(\varphi) \neq 0$. Thus there is a $v \in \mathcal{H}_L$ such that $M_L(\varphi)v \neq 0$. Since L is nondegenerate, there must be a $f \in C_c(G)$ and $v' \in \mathcal{H}_L$ such that $M_L(\varphi)L(f)v' \neq 0$. Hence $L(V(\varphi)f) \neq 0$. □

Corollary 5.26. *Let L be a nondegenerate representation of $C^*(G)$ and F a closed invariant subset of $G^{(0)}$. Then L factors through $C^*(G|_F)$ if and only if the support of L is contained in F.*

Proof. If the support of L is contained in F, then the result follows from Proposition 5.5 on page 111 as noted above. On the other hand, suppose F factors through $C^*(G|_F)$ so that $L = L' \circ j_F^G$. Let $\varphi \in C_c(G^{(0)} \setminus F)$ and $f \in C_c(G)$. Then $V(\varphi)f \in \ker j_F^G$. Hence $L(V(\varphi)f) = 0$ and the assertion follows from Lemma 5.25. □

Now let H be a closed subgroupoid of G with a Haar system $\{\alpha^u\}_{u \in H^{(0)}}$ as in Section 5.3. Then if L is a nondegenerate representation of $C^*(H)$, we can form the induced representation $\operatorname{Ind}_H^G L$ of $C^*(G)$. Note that $\operatorname{Ind}_H^G L$ is nondegenerate if L is (see Remark 5.13 on page 116).

LEMMA 5.27. *Let H and G be as above. Let $L = L' \circ j_E^H$ be a nondegenerate representation of $C^*(H)$ with support contained in the closed H-invariant set $E \subset H^{(0)}$. Let F be a closed G-invariant set containing E. Then $\operatorname{Ind}_H^G L = \operatorname{Ind}_H^G(L' \circ j_E^H)$ is equivalent to $\bigl(\operatorname{Ind}_{H|_E}^{G|_F} L'\bigr) \circ j_F^G$.*

PROOF. The space of $\operatorname{Ind}_H^G L$ is the completion of

$$C_c(G_{H^{(0)}}) \odot \mathcal{H}_L$$

with respect to the inner product

$$\bigl(\xi_1 \otimes v_1 \mid \xi_2 \otimes v_2\bigr) = \bigl(L' \circ j_E^H\bigl(\langle \xi_2 , \xi_1 \rangle_{\star}\bigr) v_1 \mid v_2\bigr)$$

where

$$\langle \xi_2 , \xi_1 \rangle_{\star}(h) = \int_G \overline{\xi_2(\gamma)} \xi_1(\gamma h)\, d\lambda_{s(h)}(\gamma) \quad \text{for } h \in H.$$

On the other hand, the space of $\bigl(\operatorname{Ind}_{H|_E}^{G|_F} L'\bigr) \circ j_F^G$, which is just the space of $\operatorname{Ind}_{H|_E}^{G|_F} L'$, is the completion of

$$C_c(G_E^F) \odot \mathcal{H}_L$$

with respect to the inner product

$$\bigl(\xi_1' \otimes v_1 \mid \xi_2' \otimes v_2\bigr) = \bigl(L'\bigl(\langle \xi_2' , \xi_1' \rangle_{\star}'\bigr) v_1 \mid v_2\bigr)$$

where

$$\langle \xi_2' , \xi_1' \rangle_{\star}'(h) = \int_G \overline{\xi_2'(\gamma)} \xi_1'(\gamma h)\, d\lambda_{s(h)}(\gamma) \quad \text{for } h \in H|_E.$$

We let $r : C_c(G_{H^{(0)}}) \to C_c(G_E^F)$ be the restriction map given by $r(\xi)(\gamma) = \xi(\gamma)$ if $r(\gamma) \in E$. Since F contains the saturation of E, r is surjective. Moreover, if $h \in H|_E$, then

$$\langle \xi_2 , \xi_1 \rangle_{\star}(h) = \langle r(\xi_2) , r(\xi_1) \rangle_{\star}'(h).$$

It follows that the map $\xi \otimes v \mapsto r(\xi) \otimes v$ extends to a unitary U from the space of $\operatorname{Ind}_H^G L$ onto that for $\bigl(\operatorname{Ind}_{H|_E}^{G|_F} L'\bigr) \circ j_F^G$.

If $f \in C_c(G)$, $\xi \in C_c(G_{H^{(0)}})$, and $\gamma \in G_E^F$, then since F is saturated,

$$\begin{aligned} r(f * \xi)(\gamma) &= \int_G f(\eta) \xi(\eta^{-1}\gamma)\, d\lambda^{r(\gamma)}(\eta) \\ &= \int_G j_F^G(f)(\eta) r(\xi)(\eta^{-1}\gamma)\, d\lambda^{r(\gamma)}(\eta) \\ &= j_F^G(f) * r(\xi)(\gamma). \end{aligned}$$

Thus U intertwines the two representations as required. □

COROLLARY 5.28. *Let H and G be as above. Suppose that L is a representation of $C^*(H)$ with support contained in a closed H-invariant set $E \subset H^{(0)}$. Then $\operatorname{Ind}_H^G L$ has support contained in $F = \overline{G \cdot E}$.*

PROOF. By Corollary 5.26 on the preceding page, $L = L' \circ j_E^H$. Hence $\operatorname{Ind}_H^G L$ is equivalent to $(\operatorname{Ind}_{H|_E}^{G|_F} L') \circ j_F^G$ and factors through $C^*(G|_F)$. Thus $\operatorname{Ind}_H^G L$ has support contained in F by Corollary 5.26. □

COROLLARY 5.29. *Let $u \in G^{(0)}$. Suppose that L is a nondegenerate representation of the isotropy group $G(u)$. Then*

$$\operatorname{Ind}_{G(u)}^{G} L \quad \textit{is equivalent to} \quad \operatorname{Ind}_{G(u)}^{G|_{\overline{[u]}}} L \circ j_{\overline{[u]}}^{G}.$$

PROOF. We apply Lemma 5.27 on the preceding page with $H = G(u)$, $E = \{u\}$, and $F = \overline{[u]}$. Then $j_E^H = \mathrm{id}$ and the result is immediate from Lemma 5.27. □

EXAMPLE 5.30 (Regular Representations). Suppose that M is a nondegenerate representation of $C_0(G^{(0)})$ with contained in a closed G-invariant set F. Then $j_F^{G^{(0)}}$ is just the usual map $C_0(G) \to C_0(F)$ induced by restriction, and $M = M' \circ j_F^{G^{(0)}}$ for a nondegenerate representation M' of $C_0(F)$. Since F is G-invariant, Lemma 5.27 on the previous page implies that the regular representation $\operatorname{Ind}_{G^{(0)}}^{G} M$ is equivalent to $\big(\operatorname{Ind}_F^{G|_F} M'\big) \circ j_F^G$. If the support of M is equal to F, then M' is a faithful representation of $C_0(F)$ and $\operatorname{Ind}_F^{G|_F} M'$ factors through a faithful representation of $C_r^*(G|_F)$ by Proposition 5.22 on page 119. Thus if $\kappa_F : C^*(G|_F) \to C_r^*(G|_F)$ is the natural map, then $\ker \operatorname{Ind}_{G^{(0)}}^{G} M$ is equal to the kernel of $\kappa_F \circ j_F^G$.

If U is an open, saturated neighborhood of $H^{(0)}$ in $G^{(0)}$, then $C^*(G|_U)$ can be identified with the ideal $\mathrm{Ex}(U)$ in $C^*(G)$ via the map ι from Theorem 5.1 on page 108. Therefore the representation $\operatorname{Ind}_H^{G|_U} L$ extends naturally to a representation of $C^*(G)$ by taking the canonical extension of the nondegenerate representation $(\operatorname{Ind}_H^{G|_U} L) \circ \iota^{-1}$ of $\mathrm{Ex}(U)$.[6] As with restriction, this process is compatible with induction.

LEMMA 5.31. *Let H, G, U, and ι be as above. If L is a nondegenerate representation of $C^*(H)$, then $\operatorname{Ind}_H^G L$ is the canonical extension of $\big(\operatorname{Ind}_H^{G|_U} L\big) \circ \iota^{-1}$ to $C^*(G)$.*

PROOF. Since U is saturated, $[H^{(0)}] \subset U$ and $G_{H^{(0)}} = (G|_U)_{H^{(0)}}$. Therefore, as in the proof of Lemma 5.27 on the preceding page, the identity map induces an unitary from the space of $\operatorname{Ind}_H^G L$ to the space of $\operatorname{Ind}_H^{G|_U} L$ allowing us to identify the two spaces. If $f \in C_c(G)$ and $g \in C_c(G|_U)$, then it is not hard to check that

$$\big(\operatorname{Ind}_H^G L\big)(f) \circ \big(\operatorname{Ind}_H^{G|_U} L\big)(g)[\xi \otimes v] = \big(\operatorname{Ind}_H^{G|_U} L\big)(f * g)[\xi \otimes v].$$

This suffices. □

EXERCISES

Ex 5.4.1. Let $A \subset G^{(0)}$ and suppose that $[A] \subset B$. Show that $G_A = G_A^B$.

—Answer on page 372

---

[6]If I is an ideal in A and π is a nondegenerate representation of I on $\mathcal{H}$, then π extends to a nondegenerate representation $\bar{\pi}$ if A satisfying $\bar{\pi}(a)(\pi(b)h) = \pi(ab)h$.

5.5. Irreducible Representations

Let G be a second countable, locally compact groupoid with a Haar system $\lambda = \{\lambda^u\}_{u\in G^{(0)}}$. If $u \in G^{(0)}$ and L is a representation of the stability group $G(u)$, then (somewhat tautologously) we say the induced representation $\operatorname{Ind}_{G(u)}^G L$ is a *representation of $C^*(G)$ induced from an isotropy group*. Such representations play a fundamental role in the representation theory of $C^*(G)$.

To see why representations of this type might be important, suppose for the moment that G is transitive. Since we are assuming G has a Haar system, its range and source maps are open, and G is equivalent to any of its isotropy groups $G(u)$ via G_u (Example 2.32 on page 44). Thus the Equivalence Theorem implies that $C^*(G)$ is Morita equivalent to $C^*(G(u))$ via an imprimitivity bimodule $\mathsf{X} = \overline{C_c(G_u)}$. In particular, every nondegenerate representation of $C^*(G)$ is equivalent to one of the form $\mathsf{X}\text{–}\operatorname{Ind} L$ for a nondegenerate representation L of $C^*(G(u))$. Recall that the space of $\mathsf{X}\text{–}\operatorname{Ind} L$ is the completion of $C_c(G_u) \odot \mathcal{H}_L$ with respect to the pre-inner product

$$(\xi_1 \otimes v_1 \mid \xi_2 \otimes v_2) = \big(L(\langle \xi_2 \,,\, \xi_1\rangle_\star)v_1 \mid v_2\big) \tag{5.17}$$

where, using (2.6),

$$\langle \xi_2 \,,\, \xi_1\rangle_\star(h) = \int_G \overline{\xi_2(\gamma)}\xi_1(\gamma h)\,d\lambda_u(\gamma).$$

In this case, we can identify the space of $\mathsf{X}\text{–}\operatorname{Ind} L$ with that of $\operatorname{Ind}_{G(u)}^G L$ as well as the two representations themselves. We have proved most of the following.

LEMMA 5.32. *Let G be a second countable locally compact Hausdorff* transitive *groupoid with $u \in G^{(0)}$. Then every nondegenerate representation of $C^*(G)$ is of the form $\operatorname{Ind}_{G(u)}^G L$ for some nondegenerate representation of $G(u)$. In particular, every irreducible representation of $C^*(G)$ is induced from an isotropy group.*

PROOF. Since $C^*(G)$ is Morita equivalent to $C^*(G(u))$ via an imprimitivity bimodule X, the general theory of Morita Equivalence implies that every nondegenerate representation of $C^*(G)$ is of the form $\mathsf{X}\text{–}\operatorname{Ind} L$ ([**RW98**, Theorem 3.29]). By the above discussion, it follows that every nondegenerate representation of $C^*(G)$ is induced from a stability group. Since $\mathsf{X}\text{–}\operatorname{Ind} L$ is irreducible if and only if L is ([**RW98**, Corollary 3.33], this completes the proof. □

REMARK 5.33 (Modular Warning). There is a subtlety buried in (5.17) that deserves mention as it will impact some of our calculations in the sequel. Our theory of induced representations developed in Section 5.3 is for *groupoids*. This means that when we work with a isotropy *group* $G(u)$, then we must treat $C^*(G(u))$ as a groupoid C^*-algebra. This means that the modular function $\Delta_{G(u)}$ does not appear in the formula for the involution in $C_c(G(u))$. Instead it appears in the integrated form of representations; that is, if L is the integrated form of a unitary representation $\pi : G(u) \to U(\mathcal{H}_\pi)$, then

$$\big(L(b)v \mid w\big) = \int_{G(u)} b(h)\big(\pi(h)v \mid w\big)\Delta_{G(u)}(h)^{-\frac{1}{2}}\,d\beta^{G(u)}(h)$$

for $b \in C_c(G(u))$ and $v, w \in \mathcal{H}_\pi$. As is usual with group representations, in the sequel we will often write π for both the unitary representation and the integrated form.

Lemma 5.32 on the preceding page is more generally applicable than one might think at first blush. Recall that a topological space X is said to satisfy the T_1-axiom of separability if given x and y in X, there is an open set containing x and not y. It is not hard to see that this is equivalent to point sets being closed in X (Ex 5.5.1). In particular, if G is a groupoid, then G-orbits $[u]$ for $u \in G^{(0)}$ are closed in $G^{(0)}$ exactly when the orbit space $G\backslash G^{(0)}$ is a T_1-topological space (Ex 5.5.2).

PROPOSITION 5.34. *Suppose that G is a second countable locally compact Hausdorff groupoid with a Haar system. Suppose that the orbit space $G\backslash G^{(0)}$ is T_1; that is, assume that all the orbits $[u]$ in $G^{(0)}$ are closed. Then every irreducible representation of $C^*(G)$ is induced from a stability group.*

PROOF. By Proposition 5.9 on page 112 and Proposition 5.5 on page 111, every irreducible representation factors through $C^*(G|_{[u]})$ for some $u \in G^{(0)}$. The result now follows from Lemma 5.32 on the previous page and Corollary 5.29 on page 122. □

Recall that an orbit in $G^{(0)}$ is locally closed if it is open in its closure. I want to show that the conclusion of Proposition 5.34 can be extended to the case of locally closed orbits—but not beyond.

THEOREM 5.35. *Let G be a second countable locally compact Hausdorff groupoid with a Haar system. Suppose that the orbits in $G^{(0)}$ are locally closed. Then every irreducible representation of $C^*(G)$ is induced from a stability group.*

Before embarking on the proof, the following example will provide useful motivation.

EXAMPLE 5.36. Suppose that U is an open G-invariant subset of $G^{(0)}$ and the complement $F = G^{(0)} \setminus U$ is such that orbits are closed in F. That is, points are closed $(G|_F)\backslash F$, which can be identified with $q(F)$ in $G\backslash G^{(0)}$ where $q : G^{(0)} \to G\backslash G^{(0)}$ is the orbit map. By Theorem 5.1 on page 108, we have a short exact sequence of C^*-algebras

$$0 \longrightarrow \operatorname{Ex}(U) \hookrightarrow C^*(G) \xrightarrow{j_F^G} C^*(G|_F) \longrightarrow 0.$$

Then any irreducible representation R of $C^*(G)$ with $\operatorname{Ex}(U) \subset \ker R$ factors through an irreducible representation R' of $C^*(G|_F)$. By Proposition 5.34, R' is equivalent to $\operatorname{Ind}_{G(u)}^{G|_F} L$ for an irreducible representation L of the isotropy group $G(u)$ for some $u \in F$. Thus, R is equivalent to $\left(\operatorname{Ind}_{G(u)}^F L\right) \circ j_F^G$. By Lemma 5.27 on page 121, with $H = G(u)$, $E = \{u\}$, and $j_E^H = \mathrm{id}$, we see that R is equivalent to $\operatorname{Ind}_{G(u)}^G L$. Of course, if we also know that every irreducible representation of $C^*(G|_U) \cong \operatorname{Ex}(U)$ is of the form $\operatorname{Ind}_{G(u)}^{G|_U} L$ for an irreducible representation L of an isotropy group $G(u)$ with $u \in \mathscr{U}$, then we can use Lemma 5.31 on page 122 to conclude that every irreducible representation of $C^*(G)$ is induced from a stability group.

PROOF OF THEOREM 5.35. Recall from Mackey-Glimm-Ramsay Dichotomy (Theorem 2.27 on page 38), that having locally closed orbits is not only equivalent to the orbit space $G\backslash G^{(0)}$ satisfying the T_0-axiom of separability, but also equivalent to the orbit space $G\backslash G^{(0)}$ being locally Hausdorff. By [**Wil07**, Theorem 6.3] we have a countable ordinal α_0 such that for each $0 \leq \alpha \leq \alpha_0$ there is an open set V_α in $G\backslash G^{(0)}$ such that

(a) $\alpha < \beta \leq \alpha_0$ implies $V_\alpha \subsetneq V_\beta$,

(b) $\alpha < \alpha_0$ implies that $V_{\alpha+1} \setminus V_\alpha$ is a dense Hausdorff subset of $G^{(0)} \setminus V_\alpha$,

(c) if δ is a limit ordinal, then $V_\delta = \bigcup_{\alpha<\delta} V_\alpha$, and

(d) $V_0 = \emptyset$ and $V_{\alpha_0} = G\backslash G^{(0)}$.

Let $q : G^{(0)} \to G\backslash G^{(0)}$ be the orbit map and let $U_\alpha = q^{-1}(V_\alpha)$. Then U_α is a G-invariant open subset of $G^{(0)}$ and

$$I_\alpha := \operatorname{Ex}(U_\alpha) \cong C^*(G|_{U_\alpha})$$

is an ideal in $C^*(G)$. In fact, $\{I_\alpha\}_{\alpha\leq\alpha_0}$ is what is normally called a composition series in $C^*(G)$ [**Wil07**, Definition 8.11]. The important observation we need here is that every irreducible representation R of $C^*(G)$ must live on a sub-quotient $I_{\alpha+1}/I_\alpha$ in the sense that $I_\alpha \subset \ker R$ and $R(I_{\alpha+1}) \neq \{0\}$ [**Wil07**, Lemma 8.13]. Since R is irreducible, this means that R is the canonical extension of the necessarily irreducible representation $R|_{I_{\alpha+1}}$ to $C^*(G)$. By Theorem 5.1 on page 108, we have commutative diagram of C^*-algebra homomorphisms with exact rows given by

$$\begin{array}{ccccccccc}
0 & \longrightarrow & I_\alpha & \hookrightarrow & I_{\alpha+1} & \xrightarrow{j'_\alpha} & C^*(G|_{U_{\alpha+1}\setminus U_\alpha}) & \longrightarrow & 0 \\
 & & \iota'_\alpha \big\uparrow \cong & & \iota'_{\alpha+1} \big\uparrow \cong & & \big\| & & \\
0 & \longrightarrow & C^*(G|_{U_\alpha}) & \xrightarrow{\iota_\alpha} & C^*(G|_{U_{\alpha+1}}) & \xrightarrow{j_\alpha} & C^*(G|_{U_{\alpha+1}\setminus U_\alpha}) & \longrightarrow & 0
\end{array}$$

where $j_\alpha = j^{G|_{U_{\alpha+1}}}_{U_{\alpha+1}\setminus U_\alpha}$ Since we can identify the orbit space for $G|_{U_{\alpha+1}\setminus U_\alpha}$ with $V_{\alpha+1} \setminus V_\alpha$, it follows that orbits are closed.

Thus if R is an irreducible representation of $C^*(G)$ that lives on $I_{\alpha+1}/I_\alpha$, then just as in Example 5.36 on the facing page, R must be equivalent to an irreducible representation induced from an isotropy group. This completes the proof. □

Exercises

Ex 5.5.1. Show that a topological space X is T_1 if and only if points are closed in X.

—Answer on page 372

Ex 5.5.2. Suppose that G is a groupoid and that $q : G^{(0)} \to G\backslash G^{(0)}$ is the orbit map. Let $[u]$ be the G-orbit of u in $G^{(0)}$. Show that $[u]$ is closed in $G^{(0)}$ if and only if $q(u)$ is closed in $G\backslash G^{(0)}$.

—Answer on page 372

5.6. C^*-Bundles

Of course, if the orbit space $G\backslash G^{(0)}$ is actually Hausdorff, then the Proposition 5.34 on the preceding page applies. But we can say a good deal more. Recall that a C^*-algebra A is called a $C_0(T)$-*algebra* if there is a nondegenerate homomorphism Φ_A of $C_0(T)$ into the center, $ZM(A)$, of the multiplier algebra of A. I'll refer to [**Wil07**, §C.1] for the basic definitions, and only recall some of the basics here. It is standard to suppress the map Φ_A and write $f \cdot a$ in place of $\Phi_A(f)a$. If J_t is the ideal of functions in $C_0(T)$ vanishing at t, then

$$I_t := \overline{\operatorname{span}}\{\, f \cdot a : f \in J_t \text{ and } a \in A \,\}$$

is an ideal in A. We write $A(t)$ for the quotient A/I_t and $a(t)$ for the image of a in $A(t)$. We than think of a as a function $t \mapsto a(t)$ taking values in $\coprod A(t)$. In general the map $t \mapsto \|a(t)\|$ is upper semicontinuous, and we can think of A as being given by continuous sections of an upper semicontinuous bundle over T with fibres $A(t)$.[7] There is a continuous map $\sigma : \operatorname{Prim} A \to T$ characterized by $\sigma(P) = t$ whenever $P \supset I_t$. It turns out that σ is open exactly when $t \mapsto \|a(t)\|$ is continuous for all a. In this case, A is called a *continuous field* over T.

Now we assume that G is a groupoid bundle over a locally compact Hausdorff space T with respect to a continuous map $p : G^{(0)} \to T$ such that $p \circ s = p \circ r$ as in Definition 1.16 on page 7. Then $C^*(G)$ becomes a $C_0(T)$-algebra with respect to the map

$$\Phi : C_0(T) \to ZM(C^*(G)) \tag{5.18}$$

given by $\Phi(\varphi) = V(\varphi \circ p)$ where $V : C_b(G^{(0)}) \to M(C^*(G))$ is given in Lemma 1.48 on page 19. We let $G(t)$ be the reduction $G|_{p^{-1}(t)}$. Note that $p^{-1}(t)$ is a closed saturated subset of $G^{(0)}$ and that $G(t)$ is a closed subgroupoid of G with a Haar system given by restricting the Haar system on G to $p^{-1}(t)$. In particular, if we let $U(t) = G^{(0)} \setminus p^{-1}(t)$, then

$$0 \longrightarrow C^*(G|_{U(t)}) \xrightarrow{\ \iota\ } C^*(G) \xrightarrow{\ j_t^G\ } C^*(G(t)) \longrightarrow 0 \tag{5.19}$$

is a short exact sequence of C^*-algebras by Theorem 5.1 on page 108.

PROPOSITION 5.37. *Suppose that G is a second countable locally compact Hausdorff groupoid bundle with respect to the map $p : G^{(0)} \to T$. Then $C^*(G)$ is a $C_0(T)$-algebra with respect to the map* (5.18). *Furthermore, the fibre $C^*(G)(t)$ over $t \in T$ is (isomorphic to) $C^*(G(t))$.*

PROOF. We just have to show that $C^*(G)(t) = C^*(T)/I_t$ is isomorphic to $C^*(G(t))$. But I_t is the closure of

$$I_t^0 = \operatorname{span}\{\, \Phi(\varphi)f : f \in C_c(G) \text{ and } \varphi \in J_t \,\}. \tag{5.20}$$

Notice that $I_t^0 = C_c(G|_{U(t)})$ (Ex 5.6.1). Therefore $I_t = \operatorname{Ex}(U(t))$ which is the image of $C^*(G|_{U(t)})$ under ι. Thus the assertion follows from (5.19). □

Unrestrained optimism would suggest that we should be able to strengthen Proposition 5.37 to say that $C^*(G)$ is a continuous field over T (rather than just an upper semicontinuous field). However this is too optimistic since it is false even if $G = C_0(G^{(0)})$: $C_0(G^{(0)})$ is a continuous field over T only when the map $p : G^{(0)} \to T$ is open (by [**Wil07**, Proposition C.10(b)]). If we assume that p is open as well as continuous, then while I don't have any counter examples, I can't prove the general result without additional assumptions. To this end, I'll say that $G(t)$ is *metrically amenable* if the reduced norm on $C_c(G(t))$ is equal to the universal norm so that $C^*(G(t)) = C_r^*(G(t))$.

[7] This terminology is formalized in [**Wil07**, §C.2]. Specifically, it is proved [**Wil07**, Theorem C.26] that a $C_0(T)$-algebra A is $C_0(T)$-isomorphic to the C_0-section algebra of a boni fide upper semicontinuous C^*-bundle $\mathcal{A}$.

REMARK 5.38 (Amenability). As the name suggests, "metric amenability" is related to groupoid amenability. In Chapter 9, we'll look extensively at the concept of amenability for groupoids. Since amenability implies equality of the reduced and universal norms (Proposition 9.6 on page 190), amenability of G implies each $G(t)$ is metrically amenable by Proposition 9.81 on page 238. If the orbit space $G\backslash G^{(0)}$ is T_0, then G is amenable if and only if each stability group $G(u)$ is an amenable group (Theorem 9.86 on page 242). The general case is subtle as equality of the reduced and universal norms is known not to be equivalent to amenability of the groupoid [**Wil15**].

COROLLARY 5.39. *Suppose that G is a second countable locally compact Hausdorff groupoid bundle with respect to a continuous and open map $p : G^{(0)} \to T$. Suppose in addition that each $G(t)$ is metrically amenable. Then $C^*(G)$ is a continuous field over T with fibres $C^*(G(t))$.*

PROOF. In view of Proposition 5.37 on the preceding page and its proof, we already know $C^*(G)$ is a $C_0(T)$-algebra. By [**Wil07**, Proposition C.5], we have a continuous map $\sigma : \operatorname{Prim} C^*(G) \to T$ such that $\sigma(P) = t$ if and only if $P \supset I_t := \operatorname{Ex}(U(t))$. Since (5.19) is exact, $I_t = \ker j_t^G$. To prove the corollary, we just need to see that σ is an open map ([**Wil07**, Proposition C.10]).

Let J_t be the ideal of functions in $C_0(G^{(0)})$ that vanish on $p^{-1}(t)$. If M_t is a representation of $C_0(G^{(0)})$ with kernel equal to J_t, then M_t factors through a faithful representation of $C_0(p^{-1}(t))$. Thus $M_t = M_t' \circ j_{p^{-1}(t)}^{G^{(0)}}$ and $\operatorname{Ind}_{p^{-1}(t)}^{G(t)} M_t'$ is a faithful representation of $C_r^*(G(t))$ by Proposition 5.22 on page 119. By Lemma 5.27 on page 121 (together with Example 5.30 on page 122), we have $I_t = \operatorname{Ind}_{G^{(0)}}^G J_t$.

To see that σ is open, it will suffice to see that if $U \subset \operatorname{Prim} C^*(G)$ is open and if $t_n \to t$ in T with $t_n \notin \sigma(U)$ for all n, then $t \notin \sigma(U)$.

Suppose to the contrary that for some U and $t_n \to t$ as above, we have $t \in \sigma(U)$. Then there is a $P \in \operatorname{Prim} C^*(G)$ such that $P \supset I_t$ and an ideal J of $C^*(G)$ such that

$$P \in \mathcal{O}_J = \{\, Q \in \operatorname{Prim} C^*(G) : Q \not\supset J \,\} \subset U.$$

Since p is open, it follows (Ex 5.6.2) that

$$J_t \supset \bigcap_n J_{t_n}. \tag{5.21}$$

It follows from Lemma 5.14 on page 116 that

$$I_t = \operatorname{Ind}_{G^{(0)}}^G J_t \supset \bigcap \operatorname{Ind}_{G^{(0)}}^G J_{t_n} = \bigcap I_{t_n}.$$

Since $P \not\supset J$, there is a n such that $I_{t_n} \not\supset J$. Since every ideal in a C^*-algebra is the intersection of those primitive ideals that contain it, there is a $Q \in \operatorname{Prim} C^*(G)$ such that $Q \supset I_{t_n}$ and $Q \not\supset J$. Then $Q \in \mathcal{O}_J \subset U$ and $\sigma(Q) = t_n$. This is a contradiction and completes the proof. □

EXERCISES

Ex 5.6.1. Let I_t^0 be as in (5.20). Show that $I_t^0 = C_c(G|_{U(t)})$ as claimed in the text.
—Answer on page 372

Ex 5.6.2. Establish (5.21)
—Answer on page 372

5.7. Type Structure

A C^*-algebra can be very simple object such as a matrix algebra whose properties can be described completely, or it can be mystifyingly complex. As a result, it is common to break the collection of all C^*-algebras into certain "types". Perhaps the most fundamental distinction is between those C^*-algebras that have, in a technical sense, a tractable representation theory and those that don't. (This is closely related to the Mackey-Glimm-Ramsay dichotomy discussed in Section 2.2.) The standard references are [**Mur90**, §5.6] and [**Ped79**, Chap. 6]. For our purposes, we can sweep most of the (highly nontrivial) theory under the rug, and simply recall that a seperable C^*-algebra is GCR if every irreducible representation π of A on $\mathcal{H}_\pi$ satisfies $\pi(A) \supset \mathcal{K}(\mathcal{H}_\pi)$ where $\mathcal{K}(\mathcal{H}_\pi)$ are the compact operators on $\mathcal{H}_\pi$. While GCR algebras are those with a good representation theory, I also want to look at an especially well-behaved subclass: we say that A is CCR if $\pi(A) = \mathcal{K}(\mathcal{H}_\pi)$ for all $\pi \in \hat{A}$. It is easy to see that if A is CCR and $I \in \mathfrak{I}(A)$, then I and A/I are also CCR. (The converse is false.) However, A is GCR if and only if given $I \in \mathfrak{I}(A)$, both I and A/I are GCR. Since the compacts have only one equivalence class of irreducible representation, it follows that if A is GCR, then irreducible representations π and ρ are equivalent if and only if $\ker \pi = \ker \rho$. Since the compacts are simple, it follows that if A is CCR, then irreducible representations π and ρ are equivalent if and only if $\ker \pi \subset \ker \rho$.[8]

As is standard, we say that the group $G(u)$ is CCR or GCR whenever $C^*(G(u))$ enjoys the same property.

THEOREM 5.40 (Clark & van Wyk). *Let G be a second countable locally compact Hausdorff groupoid with a Haar system. Then $C^*(G)$ is GCR if and only if $G\backslash G^{(0)}$ is T_0 and $G(u)$ is GCR for all $u \in G^{(0)}$.*

REMARK 5.41. This result was originally proved by Clark [**Cla07**, Theorem 7.1] under some mild amenability assumptions. These assumptions were shown to be unnecessary by van Wyk [**vW18b**, Theorem 4.2]. The proof given here, as is van Wyk's, is essentially Clark's together with van Wyk's Theorem 5.56 on page 139. This allows us to eliminate some of the complicated machinery from Clark's original proof.

PROOF. Suppose that $G\backslash G^{(0)}$ is T_0 and $G(u)$ is GCR for all $u \in G^{(0)}$. Let R be an irreducible representation of $C^*(G)$. Then by Theorem 5.35 on page 124 we may was well assume that $R = \operatorname{Ind}_{G(u)}^{G} L$ for an irreducible representation L of $G(u)$. Let $F = \overline{[u]}$. Thus R factors through $C^*(G|_F)$. Since $G\backslash G^{(0)}$ is T_0, the orbits are locally closed (Theorem 2.27 on page 38) and $[u]$ is open in F. Hence by Lemma 5.31 on page 122, R is the canonical extension of $\operatorname{Ind}_{G(u)}^{G|_{[u]}} L$ to $C^*(G|_F)$ (where I have supressed the identification of $C^*(G|_{[u]})$ as an ideal in $C^*(G|_F)$). But $G|_{[u]}$ is equivalent to $G(u)$ and $\operatorname{Ind}_{G(u)}^{G|_{[u]}} L$ is the same as $\mathsf{X}\text{–}\operatorname{Ind} L$ where $\mathsf{X} = \mathsf{X}_{G(u)}^{G|_{[u]}}$ is a $C^*(G|_{[u]})$ – $C^*(G(u))$-imprimitivity bimodule. Since we're assuming $G(u)$ is GCR, then $C^*(G|_{[u]})$ is GCR by [**Wil07**, Proposition I.44] and $\mathsf{X}\text{–}\operatorname{Ind} L$ is irreducible by

[8]Our definition of a separable GCR C^*-algebra is equivalent to a host of other conditions as listed in [**Ped79**, Theorem 6.8.7]. (For non-separable algebras, consult [**Ped79**, §6.8.8].) In place of GCR, many authors prefer type I, postliminal, or postliminary. Similarly, CCR is often replaced by liminal or liminary.

[**RW98**, Corollary 3.32]. This means the image of $\operatorname{Ind}_{G(u)}^{G|_{[u]}} L$ contains the compacts. Hence, so does the image of R. Since R was an arbitrary irreducible representation, $C^*(G)$ must be GCR.

Now assume that $C^*(G)$ is GCR. Then van Wyk's Theorem 5.56 on page 139 implies that $G\backslash G^{(0)}$ is T_0. If $u \in G^{(0)}$ and $F = \overline{[u]}$, then $C^*(G|_F)$ is (isomorphic to) a quotient of $C^*(G)$ and must be GCR. Since $G\backslash G^{(0)}$ is T_0, $[u]$ is locally closed and therefore open in F. Then $C^*(G|_{[u]})$ is (isomorphic to) an ideal in $C^*(G|_F)$. Hence it is GCR. Since $C^*(G|_{[u]})$ is Morita equivalent to $C^*(G(u))$, we must have $G(u)$ GCR by [**Wil07**, Proposition I.44]. □

A similar result is true for the CCR case.

THEOREM 5.42 (Clark & van Wyk). *Let G be a second countable locally compact Hausdorff groupoid with a Haar system. Then $C^*(G)$ is CCR if and only if $G\backslash G^{(0)}$ is T_1 and $G(u)$ is CCR for all $u \in G^{(0)}$.*

REMARK 5.43. Clark proved this result in [**Cla07**, Theorem 6.1] with amenability conditions and machinery that was circumvented by van Wyk. The proof given here depends on van Wyk's Proposition 5.51 on page 134 which will require some additional work.

PROOF. Suppose that $G\backslash G^{(0)}$ is T_1 and each $G(u)$ is CCR. If R is an irreducible representation of $C^*(G)$, then as above we can assume that $R = \operatorname{Ind}_{G(u)}^{G} L$ for some $u \in G^{(0)}$ and an irreducible representation L of $G(u)$. Then G must factor through some $C^*(G|_{[u]})$. But $C^*(G|_{[u]})$ is Morita Equivalent to $C^*(G(u))$. Hence $C^*(G|_{[u]})$ is CCR by [**Wil07**, Proposition I.43]. It follows that R must map onto the compacts.

Suppose that $C^*(G)$ is CCR. By van Wyk's Proposition 5.51 on page 134, all the orbits in $G\backslash G^{(0)}$ are closed. Assuming this, we just need to see that each $G(u)$ is CCR. But since $[u]$ is closed, $C^*(G|_{[u]})$ is a quotient of $C^*(G)$ and hence CCR. But $C^*(G|_{[u]})$ is Morita equivalent to $C^*(G(u))$, hence $G(u)$ is CCR by [**Wil07**, Proposition I.43]. □

5.8. Closed Orbits

In this section, we want to establish the key result, Proposition 5.51, used in the proof of Theorem 5.42. Namely that whenever $C^*(G)$ is CCR, orbits in $G^{(0)}$ must be closed. For this, we will need to look closely at representations induced from isotropy groups. As is the case for induced representations of groups (or even crossed products), it is convenient to have a realization of such an induced representation on a concrete Hilbert space as in [**Wil07**, Proposition 5.4]. Specifically, if $\pi : G(u) \to U(\mathcal{H}_\pi)$ is a unitary representation of $G(u)$, we want to realize $\operatorname{Ind}_{G(u)}^{G} \pi$ on the Hilbert space $L^2_\pi(G_u, \sigma_u, \mathcal{H}_\pi)$ constructed in Section 3.7. Recall that this entails the Radon measure σ_u on $G_u/G(u)$ from Lemma 3.42 on page 90 and a continuous function $\rho_u : G_u \to (0, \infty)$ such that $\rho_u(\gamma h) = \Delta_{G(u)}(h)\rho_u(\gamma)$ for all $\gamma \in G_u$ and $h \in G(u)$. Then $L^2_\pi(G_u, \sigma_u, \mathcal{H}_\pi)$ is the vector space of all λ_u-almost everywhere equivalence classes of Borel maps $\xi : G_u \to \mathcal{H}_\pi$ which satisfy

$$\xi(\gamma h) = \pi(h)^{-1}\big(\xi(\gamma)\big) \quad \text{for all } \gamma \in G_u \text{ and } h \in G(u), \tag{5.22}$$

and which are such that $\dot{\gamma} \mapsto \|\xi(\gamma)\|^2$ is σ_u-integrable.

PROPOSITION 5.44. *Let $u \in G^{(0)}$ and let π be a representation of $G(u)$ on $\mathcal{H}$. The space $\mathcal{V}_c$ of continuous functions $\xi : G_u \to \mathcal{H}$ satisfying (5.22) and such that $\dot\gamma \mapsto \|\xi(\gamma)\|$ has compact support in $G_u/G(u)$ is dense in $L^2_\pi(G_u, \sigma_u, \mathcal{H}_\pi)$. Furthermore, $\operatorname{Ind}_{G(u)}^G \pi$ is equivalent to the representation $L^{u,\pi}$ on $L^2_\pi(G_u, \sigma_u, \mathcal{H}_\pi)$ such that for all $g \in C_c(G)$ and $\xi \in \mathcal{V}_c$ we have*

$$L^{u,\pi}(g)\xi(\gamma) = \int_G g(\eta)\xi(\eta^{-1}\gamma)\Big(\frac{\rho_u(\eta^{-1}\gamma)}{\rho_u(\gamma)}\Big)^{\frac12}\, d\lambda^{r(\gamma)}(\eta). \tag{5.23}$$

PROOF. The induced representation $\operatorname{Ind}_{G(u)}^G \pi$ acts on the completion of $C_c(G_u) \odot \mathcal{H}_\pi$ with respect to the pre-inner product[9]

$$\begin{aligned}
(f_1 \otimes v_1 \mid f_2 \otimes v_2) &= \big(\pi\big(\langle f_2\,,\, f_1\rangle_\star\big)v_1 \mid v_2\big)\\
&= \int_{G(u)} \langle f_2\,,\, f_1\rangle_\star(h)\big(\pi(h)v_1 \mid v_2\big)\Delta_{G(u)}(h)^{-\frac12}\, d\beta^{G(u)}(h)\\
&= \int_{G(u)}\int_G \overline{f_2(\gamma)}f_1(\gamma h)\, d\lambda_u(\gamma)\big(\pi(h)v_1 \mid v_2\big)\Delta_{G(u)}(h)^{-\frac12}\, d\beta^{G(u)}(h)
\end{aligned}$$

which, by definition of σ_u, is

$$= \int_{G(u)}\int_{G_u/G(u)}\int_{G(u)} \overline{f_2(\gamma t)}f_1(\gamma t h)\rho_u(\gamma t)^{-1}\Delta_{G(u)}(h)^{-\frac12}\big(\pi(h)v_1 \mid v_2\big)\, d\beta^{G(u)}(t)\, d\sigma_u(\dot\gamma)\, d\beta^{G(u)}(h)$$

which, by left-invariance and Fubini, is

$$= \int_{G_u/G(u)}\int_{G(u)}\int_{G(u)} \overline{f_2(\gamma t)}f_1(\gamma h)\rho_u(\gamma t)^{-1}\Delta_{G(u)}(t^{-1}h)^{-\frac12}\big(\pi(h)v_1 \mid \pi(t)v_2\big)\, d\sigma_u(\dot\gamma)\, d\beta^{G(u)}(t)\, d\beta^{G(u)}(h)$$

which, since $\rho_u(\gamma t)^{-1}\Delta_{G(u)}(t^{-1}h)^{-\frac12} = \rho_u(\gamma t)^{-\frac12}\rho_u(\gamma h)^{-\frac12}$, is

$$= \int_{G_u/G(u)} \big(W(f_1 \otimes v_1)(\gamma) \mid W(f_2 \otimes v_2)(\gamma)\big)\, d\sigma_u(\dot\gamma)$$

where

$$W(f \otimes v)(\gamma) = \int_{G(u)} \rho_u(\gamma h)^{-\frac12} f(\gamma h)\pi(h)v\, d\beta^{G(u)}(h).$$

Clearly, the image of W is a subspace of $\mathcal{V}_c$, and W defines an isometry of the space of $\operatorname{Ind}_{G(u)}^G \pi$ onto a closed subspace in $\overline{\mathcal{V}_c} \subset L^2_\pi(G_u, \sigma_u, \mathcal{H}_\pi)$. To see that W is a unitary, and that $\mathcal{V}_c$ is dense, it suffices to see that if $\xi \in \mathcal{L}^2_\pi(G_u, \sigma_u, \mathcal{H}_\pi)$ and

$$\big(W(f \otimes v) \mid \xi\big) = 0$$

for all $f \otimes v \in C_c(G_u) \odot \mathcal{H}$ imples that $\xi = 0$ almost everywhere λ_u. But

$$\begin{aligned}
\big(W(f \otimes v) \mid \xi\big) &= \int_{G_u/G(u)} \big(W(f \otimes v)(\gamma) \mid \xi(\gamma)\big)\, d\sigma_u(\dot\gamma)\\
&= \int_{G_u/G(u)}\int_{G(u)} \rho_u(\gamma h)^{-\frac12} f(\gamma h)\big(\pi(h)v \mid \xi(\gamma)\big)\, d\beta^{G(u)}(h)\, d\sigma_u(\dot\gamma)
\end{aligned}$$

[9]The modular function appearing in the second line is explained in Remark 5.33 on page 123.

which, using (3.26), is

$$= \int_{G_u/G(u)} \int_{G(u)} \rho_u(\gamma h)^{-\frac{1}{2}} f(\gamma h)\big(v \mid \xi(\gamma h)\big)\, d\beta^{G(u)}(h)\, d\sigma_u(\dot\gamma)$$

which, using (3.23), is

$$= \int_G f(\gamma)\big(v \mid \xi(\gamma)\big)\rho_u(\gamma)^{\frac{1}{2}}\, d\lambda_u(\gamma).$$

Since given v, this holds for all $f \in C_c(G_u)$, we must have $\big(v \mid \xi(\gamma)\big) = 0$ off a λ_u-null set that depends on v. Since $\mathcal{H}$ is separable, we must have $\xi = 0$ for λ_u-almost all γ as required.

It only remains to establish (5.23). That follows from

$$\begin{aligned} W(g * f \otimes v)(\gamma) &= \int_{G(u)} \rho_u(\gamma h)^{-\frac{1}{2}} g * f(\gamma h)\pi(h)v\, d\beta^{G(u)}(h) \\ &= \int_G g(\eta) \int_{G(u)} \rho_u(\gamma h)^{-\frac{1}{2}} f(\eta^{-1}\gamma h)\pi(h)v\, d\beta^{G(u)}(h)\, d\lambda^{r(\gamma)}(\eta) \end{aligned}$$

which, using (3.22), is

$$\begin{aligned} &= \int_G \rho_u(\eta^{-1}\gamma)^{\frac{1}{2}}\rho_u(\gamma)^{-\frac{1}{2}} \int_{G(u)} \rho_u(\eta^{-1}\gamma h)^{-\frac{1}{2}} f(\eta^{-1}\gamma h)\pi(h)v \\ &\qquad\qquad d\beta^{G(u)}(h)\, d\lambda^{r(\gamma)}(\eta) \\ &= \int_G g(\eta) W(f \otimes v)(\eta^{-1}\gamma)\rho_u(\eta^{-1}\gamma)^{\frac{1}{2}}\rho_u(\gamma)^{-\frac{1}{2}}\, d\lambda^{r(\gamma)}(\eta). \qquad \square \end{aligned}$$

Another crucial ingredient is how representations induced from $G(u)$ vary as u varies. Let Σ_0 be the locally compact Hausdorff space of closed subgroups of G. Note that Σ_0 is a bundle over $G^{(0)}$ where $p : \Sigma_0 \to G^{(0)}$ is given by $p(H) = u$ if $r(H) = s(H) = \{u\}$. We give the set $\mathfrak{I}(A)$ of closed two-sided ideals in A the topology with subbasis

$$\mathcal{O}_J = \{\, I \in \mathfrak{I}(A) : I \not\supset J \,\}$$

where J varies over $\mathfrak{I}(A)$. One of the few respectable things about this topology is that the relative topology on $\operatorname{Prim} A$, viewed as a subset of $\mathfrak{I}(A)$, is the Jacobson topology. It also has uses in the theory of induced representations. For example, Lemma 5.14 on page 116 and [**Wil07**, Lemma 5.18] imply that $\operatorname{Ind}_H^G : \mathfrak{I}(C^*(H)) \to \mathfrak{I}(C^*(G))$ is continuous. We also have the following.

LEMMA 5.45. *For each $H \in \Sigma_0$, let 1_H be the trivial representation of H. Then $H \mapsto \ker\big(\operatorname{Ind}_H^G 1_H\big)$ is continuous from Σ_0 to $\mathfrak{I}(C^*(G))$.*

REMARK 5.46 (Choice of Haar Measure). Since our usual notation for a groupoid C^*-algebra supresses the choice of Haar system, we have to admit that indulging this convenience has obscured an issue here. If H is a closed subgroupoid of G, then the definition of $\operatorname{Ind}_H^G L$ depends on the choice of a Haar system for H. Hence the statement above would seem to require that we choose a Haar measure for each $H \in \Sigma_0$ first. But in the case of a group, this choice is unique up to a positive scalar. Hence different choices of Haar measures give equivalent induced representations which fortunately have the same kernel. The most reasonable way to proceed here is to continue using the continuous family $\{\, \beta^H \,\}_{H \in \Sigma_0}$ guaranteed by Proposition 3.23 on page 82.

REMARK 5.47 (Trivial Representations). As noted in Remark 5.33 on page 123, the integrated form of unitary representations of a sub*group* H of G must involve the modular function Δ_H of H. Thus if 1_H is the trivial representation and $b \in C_c(H)$, then

$$1_H(b) = \int_H b(h)\Delta_H(h)^{-\frac{1}{2}}\, d\beta^H(h).$$

If $\omega : G * \Sigma_0 \to (0,\infty)$ is the continuous function given in Lemma 3.25 on page 82, then $\Delta_H(h) = \omega(h,H)$. Thus $(h,H) \mapsto \Delta_H(h)$ is continuous for our choice of Haar measures.

PROOF OF LEMMA 5.45. We fix a continuous system of Haar measures on Σ_0 as in Remark 5.46 on the preceding page. Suppose that $H_i \to H_0$ in Σ_0. Let $L_i := \operatorname{Ind}_{H_i}^G 1_{H_i}$, and $J_i = \ker L_i$. Then it will suffice to see that $J_i \to J_0$ in $\mathfrak{I}(C^*(G))$.

If $J \in \mathfrak{I}(C^*(G))$ and $J_0 \in \mathcal{O}_J$, it will suffice to see that we eventually have $J_i \in \mathcal{O}_J$. Since $J_0 = \ker L_0 \not\supset J$, there is a positive element $a \in J$ such that $\|L_0(a)\| > 1$. It will suffice to see that we eventually have $L_i(a) \neq 0$.

Let $\{u_i\} = r(H_i)$. The space $\mathcal{H}_i$ of L_i is the completion of $C_c(G_{u_i})$ with respect to the pre-inner product

$$(f_1 \mid f_2)_i = \int_{H_i}\int_G \overline{f_2(\gamma)} f_1(\gamma h)\, d\lambda_{u_i}(\gamma)\Delta_{H_i}(h)^{-\frac{1}{2}}\, d\beta^{H_i}(h). \tag{5.24}$$

The Tietze Extension Theorem (see [**Wil07**, Lemma 1.42]) implies that every $f \in C_c(G_{u_i})$ the restriction of an element of $C_c(G)$. Hence we can replace each f_i in (5.24) with such an extension and view $\mathcal{H}_i$ as the completion of $C_c(G)$. Then

$$(f_1 \mid f_2)_i = \int_{H_i} f_2^* * f_1(h)\Delta_{H_i}(h)^{-\frac{1}{2}}\, d\beta^{H_i}(h). \tag{5.25}$$

If $f \in C_c(G)$, let $[f]_i$ be the image of f in $\mathcal{H}_i$. Because of our choice of the β^{H_i}, (5.25), and the continuity of $(h,H) \mapsto \Delta_H(h)$, it follows that for any $f_i \in C_c(G)$

$$(f_1 \mid f_2)_i \to (f_1 \mid f_2)_0.$$

In particular, if $\|[f]_i\|_i$ is the norm of $[f]_i$ in $\mathcal{H}_i$, then $\|[f]_i\|_i \to \|[f]_0\|_0$. Fix $f \in C_c(G)$ such that

$$\|[f]_0\|_0 = 1 \quad\text{and}\quad \big(L_0(a)[f]_0 \mid [f]_0\big) > 1.$$

Let $g \in C_c(G)$ be such that

$$\|a - g\| < \frac{1}{4} \quad\text{and}\quad \big(L_0(g)[f]_0 \mid [f]_0\big) > 1.$$

Notice that

$$\big(L_i(g)[f]_i \mid [f]_i\big)_i = \int_{H_i} f^* * g * f(h)\, d\beta^{H_i}(h). \tag{5.26}$$

Since the integrand in (5.26) is in $C_c(G)$, we eventually have

$$\|[f]_i\|_i < 2 \quad\text{and}\quad \big(L_i(g)[f]_i \mid [f]_i\big)_i > 1.$$

Then eventually

$$\big(L_i(a)[f]_i \mid [f]_i\big)_i \geq \big|\big(L_i(g)[f]_i \mid [f]_i\big)_i\big| - \big|\big(L(a-g)[f_i]_i \mid [f_i]\big)_i\big| > 0.$$

Thus we eventually have $L_i(a) \neq 0$. This completes the proof. □

We need the following technical result. The proof is routine—especially with our choice of Haar measures on Σ_0.

LEMMA 5.48. *Let π be a unitary representation of $G(u)$ and let $\eta\cdot\pi$ be the representation of $G(r(\eta))$ given by $\eta\cdot\pi(h) = \pi(\eta^{-1}h\eta)$. Then $\operatorname{Ind}_{G(u)}^{G}\pi$ and $\operatorname{Ind}_{G(r(\eta))}^{G}\eta\cdot\pi$ are equivalent.*

SKETCH OF THE PROOF. Define $V : C_c(G_u) \to C_c(G_{\eta)})$ by

$$V(f)(\gamma) = \omega\bigl(\eta, G(u)\bigr)^{\frac{1}{2}} f(\gamma\eta).$$

Then

$$\begin{aligned}\pi\bigl(\langle f\,,g\rangle_{\star}\bigr) &= \int_{G(u)} \langle f\,,g\rangle_{\star}(h)\pi(h)\Delta_{G(u)}(h)^{-\frac{1}{2}}\,d\beta^{G(u)}(h)\\ &= \int_{G(u)}\int_G \overline{f(\gamma)}g(\gamma h)\pi(h)\Delta_{G(u)}(h)^{-\frac{1}{2}}\,d\lambda_u(\gamma)\,d\beta^{G(u)}(h)\end{aligned}$$

which, after $\gamma \mapsto \gamma\eta$, is

$$= \int_{G(u)}\int_G \overline{f(\gamma\eta)}g(\gamma\eta h)\pi(h)\Delta_{G(u)}(h)^{-\frac{1}{2}}\,\lambda_{r(\eta)}(\gamma)\,d\beta^{G(u)}(h)$$

which, in view of Lemma 3.25, is

$$= \omega(\eta, G(u))\int_{G(r(\eta))}\int_G \overline{f(\gamma\eta)}g(\gamma h\eta)\eta\cdot\pi(h)\Delta_{G(u)}(\eta^{-1}h\eta)^{-\frac{1}{2}} \qquad d\lambda_{r(\eta)}(\gamma)\,d\beta^{G(r(\eta))}(h). \tag{5.27}$$

After observing that (3.16) implies

$$\begin{aligned}\omega(\eta, G(u))\Delta_{G(u)}(\eta^{-1}h\eta)^{-1} &= \omega(\eta, G(u))\Delta_{G(u)}(\eta^{-1}h^{-1}\eta)\\ &= \omega(h^{-1}\eta, G(u))\\ &= \omega(\eta, G(u))\omega(h^{-1}, G(r(\eta)))\\ &= \omega(\eta, G(u))\Delta_{G(r(\eta))}(h)^{-1},\end{aligned}$$

it follows that (5.27) equals

$$\begin{aligned}\int_{G(G(u))}\int_G &\overline{V(f)(\gamma)}V(g)(\gamma h)\,\lambda_{r(\eta)}(\gamma)\Delta_{G(r(\eta))}(h)^{-\frac{1}{2}}\,d\beta^{G(r(\eta))}(h)\\ &= \eta\cdot\pi\bigl(\langle V(f)\,,V(g)\rangle_{\star}\bigr).\end{aligned}$$

Since V is clearly onto, $f \otimes h \mapsto V(f) \otimes h$ extends to a unitary intertwining the two representations. $\square$

REMARK 5.49. If $[u]$ is locally closed with $F = \overline{[u]}$, then $\operatorname{Ind}_{G(u)}^{G}\pi$ factors through the representation $\operatorname{Ind}_{G(u)}^{G|_F}\pi$ which is the canonical lift of $\operatorname{Ind}_{G(u)}^{G|_{[u]}}\pi$. The latter is equivalent to X–$\operatorname{Ind}\pi$ where X is an equivalence between $C^*(G|_{[u]})$ and $C^*(G(u))$. It is not hard to deduce from this that $\operatorname{Ind}_{G(u)}^{G}\pi$ is equivalent to $\operatorname{Ind}_{G(\eta\cdot u)}^{G}\tau$ if and only if τ is equivalent to $\eta\cdot\pi$ as in Lemma 5.48. However, if u and v are not in the same orbit, the situation is less clear. It is especially murky if the orbits aren't even locally closed. Fortunately, we have the following general observation.

PROPOSITION 5.50. *Let $u, v \in G^{(0)}$. Suppose that π and τ are unitary representations of $G(u)$ and $G(v)$, respectively. If $v \notin [u]$, then $\operatorname{Ind}_{G(u)}^G \pi$ and $\operatorname{Ind}_{G(v)}^G \tau$ are not equivalent.*

PROOF. Suppose to the contrary, that $\operatorname{Ind}_{G(u)}^G \pi$ and $\operatorname{Ind}_{G(v)}^G \tau$ are equivalent. Using Proposition 5.44 on page 130, we get equivalent representations $L^{u,\pi}$ and $L^{v,\tau}$ on $L^2_\pi(G_u, \sigma_u, \mathcal{H}_\pi)$ and $L^2_\tau(G_v, \sigma_v, \mathcal{H}_\tau)$, respectively. As in Ex 5.1.5, the associated representations, $M^{u,\pi}$ and $M^{v,\tau}$ of $C_0(G^{(0)})$ are equivalent. The isomorphism of $L^2_\pi(G_u, \sigma_u, \mathcal{H}_\pi)$ and $L^2(G_u/G(u), \sigma_u, \mathcal{H}_\pi)$ given by Lemma 3.47 on page 93 intertwines $M^{u,\pi}$ with a representation $N^{u,\pi}$. Since $M^{u,\pi}(\varphi)\xi(\gamma) = \varphi(r(\gamma))\xi(\gamma)$, it is clear that $N^{u,\pi}(\varphi)\eta(\dot\gamma) = \varphi(i(\dot\gamma))\eta(\dot\gamma)$ where $i : G_u/G(u) \to G^{(0)}$ is the continuous injection given by $i(\dot\gamma) = r(\gamma)$. Hence $N^{u,\pi}$ is equivalent to a representation $N^{v,\tau}$ given by $N^{v,\tau}(\varphi)\eta(\dot\gamma) = \varphi(j(\dot\gamma))\eta(\dot\gamma)$ where $j : G_v/G(v) \to G^{(0)}$ is the continuous injection given by $j(\dot\gamma) = r(\gamma)$. Since the range of i is $[u]$ and the range of j is $[v]$ and $[u] \cap [v] = \emptyset$ by assumption, $N^{u,\pi}$ and $N^{v,\tau}$ are not equivalent by Proposition 3.40 on page 88. This contradiction finishes the proof. □

The next result is key to the characterization of CCR groupoid C^*-algebras. It was first proved by Clark [**Cla07**, Theorem 6.1] with some amenability assumptions. van Wyk was able to remove the amenability assumptions in his thesis [**vW18a**, Theorem 3.5]. A version of his argument is given here.

PROPOSITION 5.51 (Clark & van Wyk). *Suppose that G is a second countable locally compact Hausdorff groupoid with a Haar system. If $C^*(G)$ is CCR, then $G\backslash G^{(0)}$ is T_1.*

REMARK 5.52. It will be crucial in the proof to note that, since $C^*(G)$ is CCR, if L and R are irreducible representations of $C^*(G)$ with $\ker L \subset \ker R$, then L and R are equivalent. We will also use the observation that $\operatorname{Ind}_{G(u)}^G \pi$ is irreducible if π is. This certainly will follow from Theorem 5.53 on the next page. But we could also invoke Theorem 5.42 on page 129 to assume that the orbits are locally closed. Then the argument is relatively straightforward.

PROOF. Suppose to the contrary that there is a $u \in G^{(0)}$ such that $[u]$ is not closed. Then there is a $v \notin [u]$ and η_n such that $\eta_n \cdot u \to v$ in $G^{(0)}$. Since we can assume that the $r(\eta_n)$ all lie in a compact set K and $p^{-1}(K) \cap \Sigma_0$ is compact (Ex 3.4.3), we can pass to a subnet, relabel, and assume $G(r(\eta_n)) \to C$ for some $C \in \Sigma_0$ with $C \subset G(v)$. Let $J_n = \ker\bigl(\operatorname{Ind}_{G(r(\eta_n))}^G 1_{G(r(\eta_n))}\bigr)$ where 1_H denotes the trivial representation of a group H. Hence, by Lemma 5.45 on page 131, $J_n \to K$ where $K = \ker\bigl(\operatorname{Ind}_C^G 1_C\bigr)$. Since Lemma 5.48 on the preceding page implies each $\operatorname{Ind}_{G(r(\eta_n))}^G 1_{G(r(\eta_n))}$ is equivalent to $\operatorname{Ind}_{G(u)}^G 1_{G(u)}$, we have $\ker \operatorname{Ind}_{G(u)}^G 1_{G(u)} = J = J_n$ for all n. Since $\{ I \in \mathfrak{I}(C^*(G)) : J \subset I \}$ is the complement of $\mathcal{O}_J = \{ I \in \mathfrak{I}(C^*(G)) : I \not\supset J \}$, the latter set is closed and we must have $J \subset K$. Since every ideal in a C^*-algebra is the intersection of those primitive ideals that contain it, there is an irreducible representation π of $G(u)$ such that $\ker\bigl(\operatorname{Ind}_C^{G(u)} 1_C\bigr) \subset \ker \pi$. However, Induction in Stages (Theorem 5.20 on page 118) implies that $\operatorname{Ind}_{G(u)}^G \operatorname{Ind}_C^{G(u)} 1_C$ is equivalent to $\operatorname{Ind}_C^G 1_C$. Since inducing ideals preserves containment,

$$\ker\bigl(\operatorname{Ind}_{G(u)}^G 1_{G(u)}\bigr) = J \subset \ker\bigl(\operatorname{Ind}_C^G 1_C\bigr) \subset \ker\bigl(\operatorname{Ind}_{G(v)}^G \pi\bigr). \tag{5.28}$$

Since $\operatorname{Ind}_{G(u)}^{G} 1_{G(u)}$ and $\operatorname{Ind}_{G(v)}^{G} \pi$ are irreducible by Theorem 5.53 and $C^*(G)$ is CCR, (5.28) implies that they are equivalent. This contradicts Proposition 5.50 on the preceding page. This completes the proof. □

Exercises

Ex 5.8.1. Let X be a locally compact Hausdorff space and $A = C_0(X)$. If F is closed in X, let I_F be the ideal in $\mathfrak{I}(A)$ of functions that vanish on F. Show that a net $\{I_{F_i}\}$ converges to I_F in $\mathfrak{I}(A)$ if and only if given $t \in F$ every subnet of $\{I_{F_i}\}$ has a subnet $\{F_{i_j}\}$ such that there is a $t_j \in F_{i_j}$ such that $t_j \to t$. (Compare with Lemma 3.22 on page 81 and [**Wil81**, Lemma 2.4].)

—Answer on page 373

5.9. Inducing Irreducible Representations

The central theme above is to establish general conditions in which every irreducible representation of $C^*(G)$ is induced from a stability group. A careful look at what we've proved shows that we never verify the converse—is $\operatorname{Ind}_{G(u)}^{G} L$ irreducible whenever L is? While we can easily use the techniques above to prove this is the case when $G\backslash G^{(0)}$ is locally Hausdorff, it is true in full generality.

THEOREM 5.53 ([**IW09b**, Theorem 5]). *Let G be a second countable locally compact groupoid with Haar system $\{\lambda^u\}_{u\in G^{(0)}}$. Suppose that L is an irreducible representation of the stability group $G(u)$ at $u \in G^{(0)}$. Then $\operatorname{Ind}_{G(u)}^{G} L$ is an irreducible representation of $C^*(G)$.*

The idea of the proof is straightforward. Let L be an irreducible representation of $C^*\big(G(u)\big)$. Let $\mathsf{X} = \mathsf{X}_{G(u)}^{G}$ be the $C^*(G(u)_{G_u})$ – $C^*(H)$-imprimitivity bimodule built in Section 5.3 in order to define $\operatorname{Ind}_{G(u)}^{G} L$ in Definition 5.12 on page 115. Then [**RW98**, Corollary 3.32] implies that X–$\operatorname{Ind} L$ is an irreducible representation of $C^*\big(G(u)_{G_u}\big)$ where $G(u)_{G(u)}$ is the imprimitivity groupoid $(G_u *_r G_u)/G(u)$. To see that $\operatorname{Ind}_{G(u)}^{G} L$ is irreducible, it will suffice to see that any T in the commutant of $\operatorname{Ind}_{G(u)}^{G} L$ is a scalar multiple of the identity. To show this, it will suffice to see that any such T commutes with $(\mathsf{X}\text{–}\operatorname{Ind} L)(F)$ for all $F \in C_c\big(G(u)_{G_u}\big)$. Our proof will consist in producing, given F, a net $\{f_i\}$ in $C_c(G)$ such that

$$(\operatorname{Ind}_{G(u)}^{G} L)(f_i) \to (\mathsf{X}\text{–}\operatorname{Ind} L)(F)$$

in the weak operator topology. For convenience, we will write $\varphi \otimes_{G(u)} h$ for the image of $\varphi \otimes h$ in the space $\mathcal{H}_{\operatorname{Ind} L}$ of $\operatorname{Ind}_{G(u)}^{G} L$ (which is the completion of $C_c(G_u) \odot \mathcal{H}_L$). Since we will also arrange that this net is uniformly bounded in the $\|\cdot\|_I$-norm on $C_c(G)$—so that the net $\{(\operatorname{Ind}_{G(u)}^{G} L)(f_i)\}$ is uniformly bounded in $B\big(\mathcal{H}_{\operatorname{Ind} L}\big)$—we just have to arrange that

$$\big((\operatorname{Ind}_{G(u)}^{G} L)(f_i)(\varphi \otimes_{G(u)} h) \mid \psi \otimes_{G(u)} k\big) \to \big((\mathsf{X}\text{–}\operatorname{Ind} L)(F)(\varphi \otimes_{G(u)} h) \mid \psi \otimes_{G(u)} k\big)$$

for all $\varphi, \psi \in C_c(G_u)$ and $h, k \in \mathcal{H}_L$.

The next lemma is the essential ingredient to our proof.

LEMMA 5.54. *Suppose that* $F \in C_c\bigl(G(u)_{G_u}\bigr)$. *Then there is compact set* C_F *in* G *such that for each compact set* $K \subset G_u$ *there is a* $f_K \in C_c(G)$ *such that*

(a) $f_K(zy^{-1}) = F\bigl([z,y]\bigr)$ *for all* $(z,y) \in K \times K$,

(b) $\operatorname{supp} f_K \subset C_F$ *and*

(c) $\|f_K\|_I \le \|F\|_I + 1$.

The proof of Lemma 5.54 is a bit technical, so we'll postpone the proof for a bit, and show that the lemma allows us to prove Theorem 5.53 on the preceding page.

PROOF OF THEOREM 5.53. For each $K \subset G_u$, let f_K be as in Lemma 5.54. Then $\{\, f_K \,\}$ and $\{\, (\operatorname{Ind}_{G(u)}^G L)(f_K) \,\}$ are nets indexed by increasing K. Notice that

$$\begin{aligned}(5.29)\quad \bigl((\operatorname{Ind}_{G(u)}^G L)(f_K)(\varphi \otimes_{G(u)} h) \mid \psi \otimes_{G(u)} k\bigr) - \\ \bigl((\mathsf{X}\text{–}\operatorname{Ind} L)(F)(\varphi \otimes_{G(u)} h) \mid \psi \otimes_{G(u)} k\bigr) \\ = \bigl(L\bigl(\langle \psi \,,\, f_K * \varphi - F \cdot \varphi\rangle_{\star}\bigr)h \mid k\bigr)\end{aligned}$$

Furthermore, using the invariance of the Haar system on G, we can compute as follows:

$$\begin{aligned}(5.30)\qquad \langle \psi \,,\, f_K * \varphi\rangle_{\star}(s) &= \int_G \overline{\psi(x)} f_K * \varphi(xs)\, d\lambda_u(x) \\ &= \int_G\int_G \overline{\psi(x)} f_K(xz^{-1})\varphi(zs)\, d\lambda_u(z)\, d\lambda_u(x),\end{aligned}$$

while on the other hand,

$$\begin{aligned}(5.31)\qquad \langle \psi \,,\, F \cdot \varphi\rangle_{\star}(s) &= \int_G \overline{\psi(x)} F \cdot \varphi(xs)\, d\lambda_u(x) \\ &= \int_G\int_G \overline{\psi(x)} F\bigl([xs,z]\bigr)\varphi(z)\, d\lambda_u(z)\, d\lambda_u(x) \\ &= \int_G\int_G \overline{\psi(x)} F\bigl([x,zs^{-1}]\bigr)\varphi(z)\, d\lambda_u(z)\, d\lambda_u(x) \\ &= \int_G\int_G \overline{\psi(x)} F\bigl([x,z]\bigr)\varphi(z)\, d\lambda_u(zs)\, d\lambda_u(x)\end{aligned}$$

Notice that $\operatorname{supp}\langle \psi \,,\, \varphi\rangle_{\star} \subset (\operatorname{supp}\psi)(\operatorname{supp}\varphi)$. Since $\operatorname{supp} f_K \subset C_F$ for all K, we have

$$\operatorname{supp} f_K * \varphi \subset (\operatorname{supp} f_K)(\operatorname{supp}\varphi) \subset C_F(\operatorname{supp}\varphi).$$

Therefore if (5.30) does not vanish, then we must have $s \in (\operatorname{supp}\psi)C_F(\operatorname{supp}\varphi)$. Therefore there is a compact set K_0—which does *not* depend on K—such that both (5.30) and (5.31) vanish if $s \notin K_0$. Thus if $s \in K_0$ and if $K \supset (\operatorname{supp}\psi) \cup (\operatorname{supp}\varphi)K_0^{-1}$, then the integrand in (5.30) is either zero or we must have $(x,z) \in K \times K$. Therefore we can replace $f_K(xz^{-1})$ by $F\bigl([x,z]\bigr)$ and $f_K * \varphi - F \cdot \varphi$ is the zero function whenever K contains $(\operatorname{supp}\psi) \cup (\operatorname{supp}\varphi)K_0^{-1}$. Therefore the left-hand side of (5.29) is eventually zero, and the theorem follows. □

We still need to prove Lemma 5.54, and to do that, we need some preliminaries. Recall that if S is a Borel subset of G, then

$$\int_S f(x)\,d\lambda^u(x) := \int_G \mathbb{1}_S(x)f(x)\,d\lambda^u(x)$$

where $\mathbb{1}_S$ is the characteristic function of S.

LEMMA 5.55. *Suppose that $f \in C_c^+(G)$ and that $K \subset G$ is a compact set such that*

$$\int_K f(x)\,d\lambda^u(x) \leq M \quad \textit{for all } u \in G^{(0)}.$$

There there is a neighborhood V of K such that

$$\int_V f(x)\,d\lambda^u(x) \leq M+1 \quad \textit{for all } u \in G^{(0)}.$$

PROOF. Let K_1 be a compact neighborhood of K. Since G is second countable, we can find a countable fundamental system $\{ V_n \}$ of neighborhoods of K in K_1; thus, given any neighborhood V of K, there is a n such that $V_n \subset V$ and $K = \bigcap V_n$. Certainly, we can assume that $V_{n+1} \subset V_n$.

If no V as prescribed in the lemma exists, then for each n we can find $u_n \in G^{(0)}$ such that

$$\int_{V_n} f(x)\,d\lambda^{u_n}(x) \geq M+1.$$

Since we must have each $u_n \in r(K_1)$, we can pass to a subsequence, relabel, and assume that $u_n \to u_0$. Since $\mathbb{1}_{V_n} \to \mathbb{1}_K$ pointwise, the dominated convergence theorem implies that

$$\int_{V_n} f(x)\,d\lambda^{u_0}(x) \to \int_K f(x)\,d\lambda^{u_0}(x).$$

In particular, there is a n_1 such that

$$\int_{V_{n_1}} f(x)\,d\lambda^{u_0}(x) \leq M + \frac{1}{2}$$

Let W be an open set such that $K \subset W \subset \overline{W} \subset V_{n_1}$, and let $f_0 \in C_c^+(G)$ be such that $f_0|_{\overline{W}} = f$, $f_0 \leq f$ and $\operatorname{supp} f_0 \subset V_{n_1}$. Then

$$\int_G f_0(x)\,d\lambda^{u_0}(x) \leq M + \frac{1}{2}.$$

However, since $\{ \lambda^u \}$ is a Haar system,

$$\int_G f_0(x)\,d\lambda^{u_n}(x) \to \int_G f_0(x)\,d\lambda^{u_0}(x) \leq M + \frac{1}{2}.$$

But for large n, we have $V_n \subset W$ and therefore

$$\int_G f_0(x)\,d\lambda^{u_n}(x) \geq \int_{V_n} f_0(x)\,d\lambda^{u_n}(x) = \int_{V_n} f(x)\,d\lambda^{u_n}(x) \geq M+1.$$

This leads to a contradiction and completes the proof of the lemma. □

PROOF OF LEMMA 5.54. The map $(z, y) \mapsto zy^{-1}$ is certainly continuous on $G_u \times G_u$ and factors through the orbit map $\pi : G_u \times G_u \to G(u)_{G_u}$. In fact, if $zy^{-1} = xw^{-1}$, then we must have $z = x(w^{-1}y)$ and $y = w(x^{-1}z)$. But $w^{-1}y = x^{-1}z$

and lies in $G(u)$. Therefore, we have a well-defined injection $\Theta : G(u)_{G_u} \to G$ sending $[z, y]$ to zy^{-1}. We let C_F be a compact neighborhood of $\Theta(\operatorname{supp} F)$.

Fix a compact set $K \subset G_u$. The restriction of Θ to the compact set $\pi(K \times K)$ is a homeomorphism so we can find a function $\tilde{f}_K \in C_c(G)$ such that $\operatorname{supp} \tilde{f}_K \subset C_F$ and such that $\tilde{f}_K(zy^{-1}) = F\big([z, y]\big)$ for all $(z, y) \in K \times K$.

Let $K_G := \pi(K \times K)$. If

$$\int_{K_G} |\tilde{f}_K(y)| \, d\lambda^w(y) \neq 0,$$

then $K_G \cap G^w \neq \emptyset$. Thus there is a $z \in K$ such that $r(z) = w$ (and $s(z) = u$). Then by left invariance

$$\begin{aligned}
\int_{K_G} |\tilde{f}_K(y)| \, d\lambda^w(y) &= \int_G \mathbb{1}_{K_G}(zy) |\tilde{f}_K(zy)| \, d\lambda^u(y) \\
&= \int_G \mathbb{1}_{K_G}(zy^{-1}) |\tilde{f}_K(zy^{-1})| \, d\lambda_u(y) \\
&= \int_G \mathbb{1}_{K_G}(zy^{-1}) |F\big([z, y]\big)| \, d\lambda_u(y) \\
&\leq \|F\|_I.
\end{aligned}$$

Similarly, if

$$\int_{K_G} |\tilde{f}_K(y^{-1})| \, d\lambda^w(y) \neq 0,$$

then as before there is a $z \in K$ such that $r(z) = w$ and

$$\begin{aligned}
\int_{K_G} |\tilde{f}_K(y^{-1})| \, d\lambda^w(y) &= \int_G \mathbb{1}_{K_G}(zy) |\tilde{f}_K(y^{-1}z^{-1})| \, d\lambda^u(y) \\
&= \int_G \mathbb{1}_{K_G}(zy^{-1}) |\tilde{f}_K(yz^{-1})| \, d\lambda_u(y)
\end{aligned}$$

which, since $K_G^{-1} = K_G$, is

$$\begin{aligned}
&= \int_G \mathbb{1}_{K_G}(yz^{-1}) |\tilde{f}_K(yz^{-1})| \, d\lambda_u(y) \\
&\leq \int_G |F\big([y, z]\big)| \, d\lambda_u(y) \\
&= \int_{G(u)^G} |F\big([z, y]^{-1}\big) \, d\beta^{z \cdot G(u)}\big([z, y]\big) \\
&\leq \|F\|_I.
\end{aligned}$$

Using Lemma 5.55, we can find a neighborhood V of K_G contained in C_F such that both

$$\int_V |\tilde{f}_K(x)| \, d\lambda^w(x) \quad \text{and} \quad \int_V |\tilde{f}_K(x^{-1})| \, d\lambda^w(x)$$

are bounded by $\|F\|_I + 1$ for all $w \in G^{(0)}$. Since K_G is symmetric, we can assume that $V = V^{-1}$ as well. We can now let f_K be any element of $C_c(G)$ such that $f_K = \tilde{f}_K$ on K_G, $\operatorname{supp} f_K \subset V$ and $f_K \leq \tilde{f}_K$ everywhere. Then $\|f\|_I \leq \|F\| + 1$, $\operatorname{supp} f_K \subset C_F$ and $f_K(yz^{-1}) = F\big([z, y]\big)$ for all $(z, y) \in K \times K$. This completes the proof of the lemma. $\square$

5.10. The Non-Smooth Case

If $C^*(G)$ is GCR, we have the following result.

THEOREM 5.56 (van Wyk). *If G is a second countable locally compact groupoid with a Haar system and if $C^*(G)$ is GCR, then $G\backslash G^{(0)}$ is T_0.*

This result if from van Wyk's [**vW18b**, Theorem 4.2]. This was proved in the case of a transformation groupoid by Gootman in [**Goo73**] using the machinery of direct integrals. Clark gave a proof for groupoids that does not require this overhead in [**Cla07**], but at the expense of an amenability assumption. Van Wyk's proof reprises Gootman's technique and eliminated the amenability assumption. I have decided not to include the necessary background here and settle for the reference.

If $C^*(G)$ fails to be GCR, then there is good reason to believe that we can't describe the irreducible representations of $C^*(G)$ in a reasonable way.[10] In particular, when $C^*(G)$ is not GCR, then the orbit space $G\backslash G^{(0)}$ is not going to be T_0 and none of the nice properties listed in Theorem 2.27 on page 38 are going to hold. Therefore it no longer seems reasonable to expect all irreducibles to be induced from stability groups.

The upshot of this is that looking at the non-smooth case leaves what I would call "basic representation theory". Nevertheless, the isotropy groups still play a fundamental role. Here I'll settle some general comments on the "Effros-Hahn theory" with few if any details.

In 1967, Effros and Hahn conjectured that if $G = \mathcal{G} \rtimes X$ is a (second countable) transformation groupoid with the group $\mathcal{G}$ amenable, then every primitive ideal $K \in \operatorname{Prim} C^*(G)$ is induced from an isotropy group in that $K = \operatorname{Ind}_{G(u)}^G J$ for a primitive ideal $J \in \operatorname{Prim} C^*(G(u))$ for some isotropy group $G(u)$ (see [**EH67**]). This conjecture was verified (as well as its natural generalization to crossed products) by Gootman and Rosenberg in [**GR79**] building on work of Sauvageot. (For further references, discussion and proof, see [**Wil07**, Theorem 8.21].) Since transformation groups are a motivating example for groupoids—perhaps *the* motivating example—it is natural to conjecture that the Gootman-Rosenberg-Sauvageot result generalizes to groupoids. Marius Ionescu and I verified this in 2009.

THEOREM 5.57 ([**IW09a**]). *If G is an amenable second countable locally compact groupoid with a Haar system and $K \in \operatorname{Prim} C^*(G)$, then there is a $u \in G^{(0)}$ and a $J \in \operatorname{Prim} C^*(G(u))$ such that $K = \operatorname{Ind}_{G(u)}^G J$.*

The proof is somewhat involved and uses techniques from the theory of groupoid crossed products. Hence it requires too much background to present here.

[10]This assertion is not meant to be precise. This is partly because the precise statements are a bit involved. The set of all irreducible representations of $C^*(G)$ on a fixed infinite-dimensional Hilbert space $\mathcal{H}$ is a Polish space in a natural way. It turns out that $C^*(G)$ is GCR exactly when we can find a Borel set of irreducible representations that contains one and only element of each equivalence class of irreducible representations. When such a Borel set does not exist, many would argue that this is evidence that "one can't describe the irreducible representations in a reasonable way".

CHAPTER 6

The Existence and Uniqueness of Haar Systems

If $\mathcal{G}$ is a locally compact group viewed as a groupoid, then a Haar system for $\mathcal{G}$ is a single Radon measure μ with full support such that μ is left invariant. Such a measure is called a (left) Haar measure on $\mathcal{G}$, and it is well known that *every* locally compact group has a Haar measure. Haar measures automatically have full support, and are unique up to a positive scalar multiple. For example, see [**Fol95**, Theorems 2.10 and 2.20]. The question of existence and uniqueness for Haar systems on groupoids is considerably more complicated.

6.1. First Steps

We can see immediately that we can't expect existence and uniqueness in the general setting. Recall that if G has a Haar system, then the range map must be open (Proposition 1.23 on page 10). But there exist groupoids where this fails.

EXAMPLE 6.1 ([**Sed86**, §3]). Let H be the group bundle $[0,1]\times \mathbf{T}$ viewed as a subset of $\mathbf{R}^3 = \mathbf{R}\times\mathbf{C}$. Of course, the fibre $H(x)$ over $x\in[0,1]$ can be identified with the circle group $\mathbf{T}$. We let G be the closed subgroupoid $G = \{\,(x,z)\in[0,1]\times\mathbf{T} : z=1 \text{ if } x>\frac12\,\}$ where $H(x)$ has been replaced by the trivial group if $x>\frac12$. Since the projection map $p=r=s$ on G is not open, G does not admit a Haar system.

EXAMPLE 6.2. Let $R\subset\mathbf{R}^2$ be the graph of the equivalence relation in $\mathbf{R}$ where $x\sim y$ if either $x=y$ or both x and y have absolute value less that or equal to 1. Then R is closed in $\mathbf{R}^2$, hence locally compact, but the range map is not open. Hence R does not admit a Haar system.

Since Examples 6.1 and 6.2 show groupoids need not have open range and source maps, the most we can hope for is a positive answer to the following.

QUESTION. Does every locally compact Hausdorff groupoid with open range and source maps admit a Haar system?

Recently, Anton Deitmar informed me that even the above question is too much to expect. Note that if X is a locally compact Hausdorff space, then we can form the *pair groupoid* $G = X\times X$—that is, the full equivalence relation on X. Then G has a Haar system exactly when there is a full Radon measure μ on X—see Ex 1.3.8. But, as the next example shows, there are spaces which do not admit full Radon measures.

EXAMPLE 6.3 ([**Dei18**, Proposition 3.2]). Let D be an uncountable discrete set and let X be the one-point compactification of D. If $d\in D$, then $\{d\}$ is open in X. If μ were a full Radon measure on X, then $\mu(\{d\})>0$. But $\mu(X)<\infty$. Since

D is uncountable, this leads to a contradiction.[1] Therefore the Pair groupoid on X does not admit a Haar system.

Any disappointment coming from the previous example is ameliorated by the observation that the space X is pathological compared to the second countable spaces we usually restrict attention to. In fact, if X is second countable, then it is easy to see that there are always full Radon measures on X. So a more realistic quest might be the following.

QUESTION 6.4. Does every second countable locally compact Hausdorff groupoid with open range and source maps admit a Haar system?

As of this writing, I don't know of any examples giving a negative answer to Question 6.4. However, no experts I've talked to are willing to go on record conjecturing a positive answer to Question 6.4.

EXAMPLE 6.5 (No Uniqueness). Observe that as *any* full Radon measure on X produces a Haar system on $X \times X$, it is clear that there will little in the way of uniqueness for Haar systems. For example, let G be the pair groupoid on $[0, 1]$. They any probability measure on $[0, 1]$ with full support gives us a Haar system for G. In particular, let μ_1 be Lebesgue measure, and let μ_2 be given by

$$\mu_2(f) = \sum_{n=1}^{\infty} f(d_n)$$

where $D = \{ d_n \}$ is any countable dense subset of $[0, 1]$. Then μ_1 and μ_2 are mutually singular measures, yet both give Haar systems on G.

But the lack of uniqueness in Example 6.5 is not as crushing as it first appears. Note that in the case of the pair groupoid $G = [0, 1]^2$, $C^*(G, \mu_i) \cong \mathcal{K}(L^2(\mu_i))$ and hence

$$C^*(G, \mu_1) \cong C^*(G, \mu_2).$$

In fact, $C^*(G)$ is well-defined up to isomorphism for any choice of Haar system on G (Ex 2.8.1).

As we observed in Section 2.7, a fairly strong form of uniqueness is provided by the Equivalence Theorem. If λ and β are two Haar systems on a second countable locally compact groupoid, then since G is equivalent to itself, the C^*-algebras $C^*(G, \lambda)$ and $C^*(G, \beta)$ are Morita equivalent by Proposition 2.74 on page 61. However, as far as I am aware, the following question could have a positive answer.

QUESTION 6.6. If λ and β are Haar systems on a second countable locally compact groupoid G, must $C^*(G, \lambda)$ and $C^*(G, \beta)$ be isomorphic?

In the positive direction, we have the following.

EXAMPLE 6.7. Let G be a second countable *transitive* groupoid. If λ and β are Haar systems on G, then $C^*(G, \lambda) \cong C^*(G, \beta)$. This follows from Theorem 2.78 on page 62 and the fact that the isomorphism class of $\mathcal{K}(L^2(G^{(0)}, \nu))$ is unchanged if ν is replaced by another full Radon measure on $G^{(0)}$.

[1]The sets $X_n = \{ d \in D : \mu(\{d\}) \geq \frac{1}{n} \}$ can't all be finite.

6.2. Group Bundles

Recall that a groupoid G is called a group bundle if the range and source maps coincide on G. Then $G = \coprod_{u\in G^{(0)}} G(u)$ where as usual $G(u)$ is the isotropy group $\{\,\gamma \in G : s(\gamma) = u = r(\gamma)\,\}$. If we let $p = r = s$, then $p : G \to G^{(0)}$ is a bundle map whose fibres are the groups $G(u)$.

EXAMPLE 6.8. Let G be any groupoid and let

$$G' = \{\,\gamma \in G : r(\gamma) = s(\gamma)\,\}$$

be the *isotropy groupoid*. Note that the isotroy groupoid is a closed subgroupoid of G which is clearly a group bundle over $G^{(0)}$.

Notice that in the topological setting, not every bundle $p : B \to X$ for which each fibre is a group is a groupoid. Given a topological bundle p, to construct a groupoid bundle we would have to fix an isomorphism of each fibre with a group in order to define the multiplication. Then we need to identify X with the set of units $B^{(0)}$. In particular, this gives a continuous section for p. This is not always possible. For example, if $p : B \to X$ is a principal circle bundle, then it is groupoid group bundle only when p has a section. But if a principal bundle has a section, then it is trivial.

THEOREM 6.9 (Renault). *Let G be a second countable locally compact (groupoid) group bundle. Then G has a Haar system if and only if the projection $p = r = s$ is open.*

The proof here comes from [**Ren91**, Lemma 1.3], and is based on a technique from Glimm's [**Gli62**, p. 908]. It requires the notion of a *generalized limit*. A generalized limit is any state Λ on ℓ^∞ that extends the limit functional on $\mathfrak{c} = \{\,x \in \ell^\infty : \lim_n x_n \text{ exists}\,\}$. Notice that any norm one extension of the limit functional on $\mathfrak{c}$ to all of ℓ^∞ is a functional on ℓ^∞ attaining its norm at the identity. Hence it is a positive linear functional, and hence a state, by [**RW98**, Lemma A.9]. Furthermore, if $\Lambda(x) = a$ for all generalized limits Λ, then $x \in \mathfrak{c}$ and $\lim_n x_n = a$ (Ex 6.2.1).

PROOF. We already know that if G has a Haar system, then r, and hence p, is open (Proposition 1.23 on page 10).

So assume p is open. Let N be a diagonally compact neighborhood of $G^{(0)}$ in G. Since G is normal, there is a continuous non-negative function f_0 on G such that $f_0(u) = 1$ for all $u \in G^{(0)}$ and such that $\operatorname{supp} f_0 \subset N$. Hence $f_0 \in C^+_{c,p}(G)$. Let λ^u be the Haar measure on $G(u)$ such that

$$\int_{G(u)} f_0(t)\, d\lambda^u(t) = 1.$$

Since Haar measures always have full support, it will suffice to see that

$$u \mapsto \int_{G(u)} f(t)\, d\lambda^u(t)$$

is continuous for all $f \in C_c(G)$.

Let $K \subset G$ be compact. We want to see that $u \mapsto \lambda^u(K)$ is bounded. Let $U = \{\,\gamma \in G : f_0(\gamma) > \frac{1}{2}\,\}$. Then U is an open neighborhood of $G^{(0)}$ and

$$\lambda^u(U) \le 2 \int_U f_0(t)\, d\lambda^u(t) \le 2 \int_G f_0(t)\, d\lambda^u(t) = 2.$$

If $\gamma \in K$, then $\gamma^{-1}\gamma \in U$, and since multiplication is continuous, there is an open neighborhood V of γ such that $V^{-1}V \subset U$. Hence we can over K with open sets $V_1, \dots, V_n$ such that $V_i^{-1}V_i \subset U$. If $G(u) \cap V_i = \emptyset$, then $\lambda^u(V_i) = 0$. Otherwise if $t \in G(u) \cap V_i$, then

$$\lambda^u(V_i) = \lambda^u(tt^{-1}V_i) \le \lambda^u(tV_i^{-1}V_i) \le \lambda^u(tU) = \lambda^u(U) \le 2.$$

Thus $\lambda^u(K) \le 2n$.

Let Λ be a generalized limit. Suppose that $\{u_n\}$ a sequence converging to u, and that $f \in C_c(G)$. By the above

$$n \mapsto \tilde{f}(n) := \int_{G(u_n)} f(t)\,d\lambda^{u_n}(t)$$

defines an element $\tilde{f}$ in ℓ^∞. Since Λ is arbitrary, it will suffice to see that

$$\Lambda(\tilde{f}) = \int_{G(u)} f(t)\,d\lambda^u(t).$$

Notice that $f \mapsto \mu(f) := \Lambda(\tilde{f})$ is a positive linear functional on $C_c(G)$. Hence μ is given by a Radon measure on G also called μ. We want to see that μ is supported on $G(u)$. Let f and g be elements of $C_c(G)$ that agree on $G(u)$. Let K be a compact set containing both $\operatorname{supp} f$ and $\operatorname{supp} g$. A compactness argument (Ex 6.2.4) implies

$$\lim_n \sup_{t \in G(u_n)} |f(t) - g(t)| = 0. \tag{6.1}$$

Moreover,

$$\begin{aligned} \lim_n |\tilde{f}(n) - \tilde{g}(n)| &\le \lim_n \int_{G(u_n)} |f(t) - g(t)|\,d\lambda^{u_n}(t) \\ &\le \lim_n \sup_{t \in G(u_n)} |f(t) - g(t)| \lambda^{u_n}(K) = 0 \end{aligned}$$

Thus $\mu(f) = \mu(g)$, and we've shown that μ is supported in $G(u)$. Hence we can view μ as a positive linear functional on $C_c(G(u))$.

We want to see that μ is left-invariant and hence a Haar measure on $G(u)$. Let $s \in G(u)$, and let $g \in C_c(G)$ be such that $g(t) = f(st)$ for all $t \in G(u)$. It will suffice to see that $\Lambda(\tilde{g}) = \Lambda(\tilde{f})$.

Let N be a compact neighborhood of s in G such that $\{u_n\} \subset p(N)$. Let $K \subset G$ be a compact set containing both $\operatorname{supp} g$ and $N^{-1} \cdot \operatorname{supp} f$. As we saw above, there is a $M > 0$ such that $\lambda^u(K) \le M$ for all $u \in G^{(0)}$. Fix $\epsilon > 0$. I claim there is a neighborhood $U \subset N$ of s in G such that $s' \in U$ implies that

$$|g(t) - f(s't)| < \epsilon/M \quad \text{for all } t \in G(p(s')).$$

If the claim were not true, then for all $U \subset N$, there are $s'_U \in U$ and $t_U \in K \cap G(p(s'_U))$ such that

$$|g(t_U) - f(s'_U t_U)| \ge \epsilon/M > 0.$$

Hence $s'_U \to s$ and we can pass to a subnet, relabel, and assume $t_U \to t \in K \cap G(u)$. But this eventually contradicts our choice of g. This establishes the claim.

Let n be large enough so that $u_n \in p(U)$. Then $U \cap G(u_n) \neq \emptyset$ and we can find $s_n \in U \cap G(u_n)$. Then

$$\begin{aligned}|\tilde{g}(n) - \tilde{f}(n)| &= \Big|\int_{G(u_n)} g(t)\,d\lambda^{u_n}(t) - \int_{G(u_n)} f(t)\,d\lambda^{u_n}(t)\Big| \\ &= \Big|\int_{G(u_n)} \big(g(t) - f(s_n t)\big)\,d\lambda^{u_n}(t) + \int_{G(u_n)} \big(f(s_n t) - f(t)\big)\,d\lambda^{u_n}(t)\Big| \\ &\leq \epsilon + 0 = \epsilon.\end{aligned}$$

It follows that

$$\lim_n |\tilde{g}(n) - \tilde{f}(n)| \leq \limsup_n |\tilde{g}(n) - \tilde{f}(n)| = 0.$$

Therefore $\Lambda(\tilde{f}) = \Lambda(\tilde{g})$. Hence $\mu(f) = \mu(g)$, and μ is a Haar measure on $G(u)$. Since $\tilde{f}_0(n) = 1$ for all n, $\mu(f_0) = \lambda^u(f_0)$. The uniqueness of Haar measure insures that $\mu = \lambda^u$.

This completes the proof. □

Exercises

Ex 6.2.1. Let $x \in \ell^\infty$ be such that $\Lambda(x) = a$ for all generalized limits Λ. Then $x \in \mathfrak{c}$ and $\lim_n x_n = a$. (Hint: if x has a subsequence $\{x_{n_k}\}$ with limit a, then there is a Λ such that $\Lambda(x) = a$.)

—Answer on page 373

Ex 6.2.2. Let Λ be a generalized limit on ℓ^∞. Show that $|\Lambda(x)| \leq \Lambda(|x|)$ where $|x| = \big(|x_n|\big)$.

—Answer on page 373

Ex 6.2.3. If Λ is a generalized limit on ℓ^∞, show that $|\Lambda(x)| \leq \limsup_n |x_n|$.

—Answer on page 373

Ex 6.2.4. Establish (6.1).

—Answer on page 373

6.3. The Isotropy Groupoid and the Isotropy Map

The map $u \mapsto G(u)$ from $G^{(0)}$ to Σ_0 is called the *isotropy map*. Examples show that it is rarely continuous with respect to the Fell Topology on Σ_0. In fact, we'll see in Theorem 6.12 on the next page that its continuity is intimately related to the existence of Haar systems on the *isotropy groupoid* $G' := \{\, \gamma \in G : r(\gamma) = s(\gamma)\,\}$ (see Example 6.8 on page 143). We spend some time exploring this and some related issues. But first a few more general observations.

LEMMA 6.10. *Suppose that $G(u_0) = \{u_0\}$, then the isotropy map $u \mapsto G(u)$ is continuous at u_0.*

PROOF. Let $u_i \to u_0$. It will suffice to see that every subnet of $\{G(u_i)\}$ has a subnet converging to $\{u_0\}$. Since we can assume all the u_i are in a compact neighborhood of u_0, any subnet of $\{G(u_i)\}$ will have a subnet converging to some $C \in \Sigma_0$. It will suffice to see that C is always trivial. We may as well assume that $G(u_i) \to C$. But if $\gamma \in C$, then we may as well assume that there are $\gamma_i \in G(u_i)$ converging to γ. But then $r(\gamma) = \lim r(\gamma_i) = u_0$. Similarly, $s(\gamma) = u_0$ and $\gamma \in G(u_0) = \{u_0\}$. Thus $C = \{u_0\}$. □

We say that a map $f : X \to Y$ is open at $x \in X$ if $f(V)$ is a neighborhood of $f(x)$ for every neighborhood V of x in X.

PROPOSITION 6.11 ([**Ren91**, Lemma 1.1]). *Let G be a locally compact Hausdorff groupoid. Let $r' : G' \to G^{(0)}$ be the restriction of the range map to the isotropy groupoid G'. Then the isotropy map $u \mapsto G(u)$ is continuous at v if and only if r' is open at v.*

PROOF. Suppose that r' is open at v. Let $u_n \to v$. We need to verify that $G(u_n) \to G(v)$. We'll use the criteria in Lemma 3.22 on page 81. Suppose that $\gamma_n \in G(u_n)$ and that $\gamma_n \to \gamma$. Then $u_n = r(\gamma_n) \to r(\gamma)$. Since $G^{(0)}$ is Hausdorff, $r(\gamma) = u$. Similarly, we have $s(\gamma) = v$ and $\gamma \in G(v)$. Since this argument applies to any subnet, we have established property (a) of Lemma 3.22.

Now let $\gamma \in G(v)$. Let V be any neighborhood of γ in G. By assumption, $r'(V \cap G')$ is a neighborhood of v in $G^{(0)}$. Hence given any n_0, there is a $n \geq n_0$ such that $u_n \in r'(V \cap G')$. Hence there is a $\gamma_{n,V} \in G(u_n) \cap V$. It is not hard to check that if $\mathscr{N}(\gamma)$ is the set of neighborhoods of γ in G directed by reverse containment, then

$$D = \{ (n, V) \in \mathbf{N} \times \mathscr{N}(\gamma) : u_n \in r'(V \cap G') \}$$

is directed. Letting $u_{(n,V)} = u_n$ gives us a subnet $\{u_d\}$ such that $\{\gamma_d\}$ is a net in G converging to γ with $\gamma_d \in G(u_d)$. Since we could repeat this argument for any subset, we have established property (b) of Lemma 3.22 on page 81. Thus $G(u_n) \to G(v)$ as required.

Conversely, assume $u \mapsto G(u)$ is continuous at v. Let V be a neighborhood of $\gamma \in G(v)$ in G. We want to see that $r'(V \cap G')$ is a neighborhood of v. Suppose that $u_n \to v$ in $G^{(0)}$. It will suffice to see that $\{u_n\}$ has a subnet eventually in $r'(V \cap G')$. By assumption, $G(u_n) \to G(v)$. We can pass to a subnet, and relabel, so that there are $\gamma_n \in G(u_n)$ such that $\gamma_n \to \gamma$ in G. But $\{\gamma_n\}$ must eventually be in $V \cap G'$. Hence u_n is eventually in $r'(V \cap G')$. This completes the proof. □

Combining Proposition 6.11 with Theorem 6.9 on page 143 gives the following useful observation.

THEOREM 6.12. *Let G be a locally compact groupoid with isotropy groupoid G'. The following are equivalent.*

(a) *G' admits a Haar system.*

(b) *The restriction of the range map r to G' is open.*

(c) *The isotropy map $u \mapsto G(u)$ is continuous from $G^{(0)}$ to Σ_0.*

Examples show that asking for a Haar system on G' is a strong requirement. Nevertheless, if we do have such a Haar system, there are some useful consequences regarding Haar systems on G that we explore here. Most of these ideas are due to Anton Deitmar [**Dei18**].

Let $\pi : G \to G^{(0)} \times G^{(0)}$ we the map given by $\pi(\gamma) = \big(r(\gamma), s(\gamma)\big)$. Let $R \subset G^{(0)} \times G^{(0)}$ be the range of π. Clearly R is a topological groupoid in the relative product topology, but as we have seen before, R need not be a nice subset of $G^{(0)} \times G^{(0)}$. In particular, R will generally not be a locally compact groupoid in this topology. Usually we agree to give R the quotient topology $\tau_q = \{ U \subset R : \pi^{-1}(U) \text{ is open in } G \}$. Even so, R is still a bit troublesome.

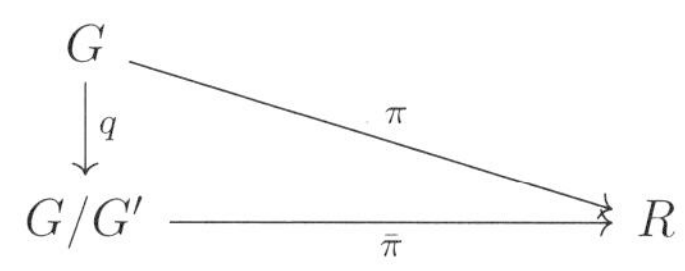

FIGURE 6.1. The map $\bar{\pi}$

Of course, π is certainly continuous into R with the quotient topology by definition), and π factors through G/G' for the natural right action of G' on G by multiplication: if $\pi(\gamma) = \pi(\eta)$, then $\eta = \gamma(\gamma^{-1}\eta)$. Hence if q is the quotient map, then we get a bijection $\bar{\pi}$ so that Figure 6.1 commutes.

LEMMA 6.13. *Let π, q and R be as above. Suppose that G' has a Haar system. Then*

(a) *π is continuous and open,*

(b) *R is a locally compact groupoid, and*

(c) *$\bar{\pi}$ is a homeomorphism.*

PROOF. Since G' has a Haar system, the quotient map q is open by Proposition 2.12 on page 31, and G/G' is locally compact Hausdorff by Proposition 2.18 on page 33. If V is open in G, then

$$\pi^{-1}(\pi(V)) = VG' = q^{-1}(q(V)). \tag{6.2}$$

Since q is open, VG' is open and it follows that π is open. It then follows from Figure 6.1 that $\bar{\pi}$ is a homeomorphism. It only remains to see that R is a topological groupoid. But if $\pi(\gamma_i) \to \pi(\gamma)$ and $\pi(\eta_i) \to \pi(\eta)$, it will suffice to show that every subnet of $\{\pi(\gamma_i\eta_i)\}$ has a subnet converging to $\pi(\gamma\eta)$ and that every subnet of $\{\pi(\gamma_i^{-1})\}$ has a subnet converging to $\pi(\gamma^{-1})$. After replacing everything in sight with a subnet, and relabeling, we can pass to another subnet, relabel, and assume that $\gamma_i h_i \to \gamma$ and $\eta_i k_i \to \eta$ in G for appropriate $h_i, k_i \in G'$. But then

$$\gamma_i h_i \eta_i k_i \to \gamma\eta,$$

and $\pi(\gamma_i h_i \eta_i k_i) = \pi(\gamma_i\eta_i) \to \pi(\gamma\eta)$. Since $\pi(h_i^{-1}\gamma_i^{-1}) = \pi(\gamma_i^{-1})$, the argument for the inverse is similar. □

For the rest of this section, we are going to identify the unit spaces of both R and G' with $G^{(0)}$.

COROLLARY 6.14. *If G' has a Haar system $\beta = \{\beta^u\}_{u\in G^{(0)}}$, then we get a surjection $\beta : C_c(G) \to C_c(R)$ given by*

$$\beta(f)\big(\pi(\gamma)\big) = \int_{G(s(\gamma))} f(\gamma t)\, d\beta^{s(\gamma)}(t) \tag{6.3}$$

PROOF. The right hand side of (6.3) is in $C_c(G/G')$ by Lemma 2.21 on page 33 which also imples that every element of $C_c(G/G')$ is of this form. The rest follows from Lemma 6.13(c). □

PROPOSITION 6.15 (Deitmar). *Suppose that G is a second countable locally compact Hausdorff groupoid. Let G' be the isotropy groupoid and R the associated equivalence relation in $G^{(0)} \times G^{(0)}$ with the quotient topology. Suppose that $\beta = \{\beta^u\}_{u\in G^{(0)}}$ is a Haar system for G', and we define $\beta : C_c(G) \to C_c(R)$ as in (6.3).*

(a) *If $m = \{m^u\}_{u\in G^{(0)}}$ is a Haar system for R, then $\lambda = \{\lambda^u\}_{u\in G^{(0)}}$ is a Haar system for G where*

$$\lambda^u(f) = \int_R \beta(f)(\pi(\gamma))\, dm^u(\pi(\gamma)). \tag{6.4}$$

(b) *If $\lambda = \{\lambda^u\}_{u\in G^{(0)}}$ is a Haar system for G, then there is a Haar system $m = \{m^u\}_{u\in G^{(0)}}$ for R such that (6.4) holds.*

For the proof, of the part (b), we need the following observation.

LEMMA 6.16. *Let G, and β be as in the Proposition. Suppose that $f, g \in C_c(G)$ and that $\lambda = \{\lambda^u\})_{u\in G^{(0)}}$ is a Haar system on G. Then*

$$\lambda^u\big(g\beta(f)\big) = \lambda^u\big(\beta(g)f\big) \quad \textit{for all } u \in G^{(0)}.$$

PROOF. We have

$$\begin{aligned}\lambda^u\big(g\beta(f)\big) &= \int_G g(\gamma)\beta(f)(\pi(\gamma))\, d\lambda^u(\gamma)\\ &= \int_G\int_{G(s(\gamma))} g(\gamma)f(\gamma t)\, d\beta^{s(\gamma)}(t)\, d\lambda^u(\gamma)\end{aligned}$$

which, invoking Lemma 3.25, is

$$= \int_G\int_{G(u)} \omega(\gamma, G(s(\gamma)))g(\gamma)f(t\gamma)\, \beta^u(t)\, d\lambda^u(\gamma)$$

which, after using Fubini and $\gamma \mapsto t^{-1}\gamma$, is

$$= \int_G\int_{G(u)} \omega(t^{-1}\gamma, G(s(\gamma)))g(t^{-1}\gamma)f(\gamma)\, d\beta^u(t)\, d\lambda^u(\gamma)$$

which, using (3.16) and the observation that $\omega(t^{-1}, G(u)) = \Delta_{G(u)}(t)^{-1}$, is

$$\begin{aligned}&= \int_G\int_{G(u)} \omega(\gamma, G(s(\gamma)))\Delta_{G(u)}(t)^{-1}g(t^{-1}\gamma)f(\gamma)\, d\beta^u(t)\, d\lambda^u(\gamma)\\ &= \int_G\int_{G(u)} \omega(\gamma, G(s(\gamma)))g(t\gamma)f(\gamma)\, d\beta^u(t)\, d\lambda^u(\gamma)\\ &= \int_G\int_{G(s(\gamma))} g(\gamma t)f(\gamma)\, d\beta^{s(\gamma)}(t)\, d\lambda^u(\gamma)\\ &= \lambda^u(\beta(g)f).\end{aligned}$$

□

PROOF OF PROPOSITION 6.15. (a) Clearly, $u \mapsto \lambda^u(f) = m^u(\beta(f))$ is continuous. If $\gamma \in G^u$ and $f \in C_c^+(G)$ is such that $f(\gamma) > 0$, then $\beta(f)(\pi(\gamma)) > 0$. Since m^u has support R^u, it follows that $\lambda^u(f) > 0$. Hence $\operatorname{supp} \lambda^u = G^u$. We also have

$$\begin{aligned}\int_G f(\eta\gamma)\, d\lambda^{s(\eta)}(\gamma) &= \int_R \beta(f)(\pi(\gamma\eta))\, dm^{s(\eta)}(\pi(\gamma))\\ &= \int_R \beta(f)\big(\pi(\eta)\pi(\gamma)\big)\, dm^{s(\eta)}(\pi(\gamma))\end{aligned}$$

which, since m is a Haar system, is

$$= \int_R \beta(f)(\pi(\gamma))\,dm^{r(\eta)}(\pi(\gamma))$$
$$= \int_G f(\gamma)\,d\lambda^{r(\eta)}(\gamma).$$

Hence λ is a Haar system for G.

(b) Suppose that $\beta(f) = \beta(g)$. Since $\beta : C_c(G) \to C_c(R)$ is surjective, we can find $h \in C_c^+(G)$ such that $\beta(h)$ is identically one on $\operatorname{supp}\beta(f) = \operatorname{supp}\beta(g)$. It follows that for all γ, $f(\gamma) = f(\gamma)\beta(h)(\pi(\gamma))$ and similarly for g. Hence using Lemma 6.16 on the preceding page, we have

$$\lambda^u(f) = \lambda^u(f\beta(h)) = \lambda^u(\beta(f)h) = \lambda^u(\beta(g)h) = \lambda^u(g).$$

Hence we get a well-defined positive linear functional on $C_c(R)$ via

$$m^u(\beta(f)) := \lambda^u(f).$$

It is clear that $u \mapsto m^u(\beta(f))$ is continuous, and it is not hard to see that $\operatorname{supp} m^u = R^u$. Clearly, (6.4) holds. For left-invariance, check

$$\begin{aligned}\int_R \beta(f)\big(\pi(\eta)\pi(\gamma)\big)\,dm^{s(\eta)}(\pi(\gamma)) &= \int_R \beta(f)\big(\pi(\eta\gamma)\big)\,dm^{s(\eta)}(\pi(\gamma))\\ &= \int_G f(\eta\gamma)\,d\lambda^{s(\eta)}(\gamma)\\ &= \int_G f(\gamma)\,d\lambda^{r(\eta)}(\gamma)\\ &= \int_R \beta(f)(\pi(\gamma))\,dm^{r(\eta)}(\pi(\gamma)).\end{aligned}$$

Hence m is the desired Haar system on R. □

REMARK 6.17. Proposition 6.15 on the facing page depends crucially on the existence of a Haar system on G'. Nevertheless, we will prove a Borel version of part (b) in Proposition 11.5 on page 282. These proofs are all motivated by [**Ren91**, Lemma 1.7].

COROLLARY 6.18. *Suppose that G is a second countable locally compact Hausdorff groupoid such that the isotropy groupoid G' has a Haar system. Let R be the associated equivalence relation with the the quotient topology as above. If R is a proper groupoid with open range and source maps, then G has a Haar system.*

PROOF. We will see that every proper principal groupoid with open range and source maps necessarily has a Haar system (see Proposition 6.24 on page 153). Hence G has a Haar system by Proposition 6.15(a). □

EXERCISES

Ex 6.3.1. Suppose that H is a closed subgroupoid of a locally compact groupoid G. Then $G_{H^{(0)}}$ is a free and proper right H-space for the natural action by right multiplication.

—Answer on page 373

Ex 6.3.2. Let G' be the isotropy groupoid for G and let π, $\bar{\pi}$, q and R be as in Figure 6.1 on page 147.

(a) Show that π is open if and only if q is open.

(b) Suppose that G has open range and source maps. Show that q is open if and only if the restriction r' of r to G' is open.

—Answer on page 373

6.4. Haar Systems on Étale Groupoids

Recall that a groupoid is called *r-discrete* if $G^{(0)}$ is open in G. A r-discrete groupoid is called *étale* if the range and source maps are local homeomorphisms. (See Definitions 1.24 and 1.30.)

PROPOSITION 6.19. *A r-discrete groupoid has a Haar system if and only if it is étale. If G is an étale groupoid and λ and β are Haar systems on G, then $C^*(G,\lambda) \cong C^*(G,\beta)$.*

PROOF. Suppose that G is r-discrete. If G has a Haar system, then r is open by Proposition 1.23 on page 10. Then G is étale by Proposition 1.29 on page 11. If G is étale, then counting measures give a Haar system by Proposition 1.29 on page 11.

Let G be an étale groupoid and let $\sigma = \{\sigma^u\}_{u\in G^{(0)}}$ be the Haar system of counting measures on G. Let $\lambda = \{\lambda^u\}_{u\in G^{(0)}}$ be another Haar system on G. By Ex 1.3.4, there is a strictly positive continuous function α on $G^{(0)}$ such that $\lambda^u(\{\gamma\}) = \alpha(s(\gamma))$ for all $\gamma \in G^u$. Clearly, it will suffice to see that $C^*(G,\lambda)$ and $C^*(G,\sigma)$ are isomorphic.

Define $\Phi : C_c(G,\lambda) \to C_c(G,\sigma)$ by

$$\Phi(f)(\gamma) = \alpha(r(\gamma))^{\frac{1}{2}} f(\gamma)\alpha(s(\gamma))^{\frac{1}{2}}. \tag{6.5}$$

It is not hard to see that Φ is a $*$-homomorphism—see Ex 6.4.1.

To see that Φ extends to an isomorphism of the C^*-completions seems to be non-trivial. The proof given here uses Renault's Disintegration Theorem in the form of Corollary 8.4 on page 170. In particular, if L is a representation of $C^*(G,\sigma)$, then $L' := L \circ \Phi$ is a weak representation of $C_c(G,\lambda)$ as defined in Definition 8.1 on page 169, and hence bounded for the C^*-norm on $C^*(G,\alpha)$. Hence

$$\|\Phi(f)\|_\lambda \le \|f\|_\sigma.$$

We get the reverse inequality by employing Φ^{-1}. Hence Φ is isometric for the C^*-norms and extends to the required isomorphism. $\square$

Exercises

Ex 6.4.1. Show that Φ in (6.5) is a $*$-homomorphism.

—Answer on page 374

6.5. Haar Systems on Equivalent Groupoids

Most of the time we find a groupoid in the wild, it comes with an "obvious" choice of Haar system. For example, if G acts on a space X, and if G has a Haar system, then so does the action groupoid—see Ex 2.1.7. An important situation where this paradigm proved difficult to verify is the case of a groupoid equivalent to a groupoid that is known to have a Haar system. Fortunately it is possible to show that even if the choice is not necessarily "obvious", there is nevertheless a choice of Haar system.

THEOREM 6.20 ([**Wil16**, Theorem 2.1]). *Suppose that G is a second countable, locally compact Hausdorff groupoid with a Haar system $\{\lambda^u\}_{u\in G^{(0)}}$. If H is a second countable, locally compact groupoid which is equivalent to G, then H has a Haar system.*

The conclusion in Theorem 6.20 is relatively new. A first step in its proof comes from [**KMRW98**, Proposition 5.2] which was the best we had for some time.

LEMMA 6.21. *Let Z be a (G,H) equivalence. Then H has a Haar system if and only if there is a full equivariant r-system for the moment map $r : Z \to G^{(0)}$ for the left G-action.*

PROOF. Suppose that $\alpha = \{\alpha^z\}_{z\in Z}$ is a full equivariant r-system. Since can identify H with the imprimitivity groupoid $G^Z = G\backslash(Z *_r Z)$ by Lemma 2.44 on page 47, it will suffice to produce a Haar system on G_Z. Therefore the result follows from Lemma 3.16 on page 75. □

In view of Lemma 6.21, to prove Theorem 6.20 it will suffice to see that if Z is a (G,H) equivalence and G has a Haar system, then the moment map $r : Z \to G^{(0)}$ has a full equivariant r-system.

In fact, we can do considerably better than that. The next proposition shows that every proper G-space has a full equivariant r-system for the moment map whether or not the action is free.[2]

PROPOSITION 6.22. *Let G be a second countable locally compact Hausdorff groupoid with a Haar system $\{\lambda^u\}_{u\in G^{(0)}}$. Suppose that Z is a proper G-space with open momment map $r_Z : Z \to G^{(0)}$. Then there is a full equivariant r_Z-system $\{\nu^u\}_{u\in G^{(0)}}$.*

Blanchard's Lemma 3.20 on page 77 implies that there is a full r_Z-system $\beta = \{\beta^u\}_{u\in G^{(0)}}$. The idea of the proof of Proposition 6.22 is to use the Haar system on G to average β to create an equivariant system. Some of the technicalities are provided by the next lemma. Notice that since G acts properly, the orbit map $q : Z \to G\backslash Z$ is a continuous open surjection between locally compact Hausdorff spaces.

LEMMA 6.23. *Let G, Z, λ and β be as above.*

(a) *If $F \in C_c(G \times Z)$, then*

$$\Phi(F)(g,u) = \int_Z F(g,z)\,d\beta^u(z)$$

defines an element of $C_c(G \times G^{(0)})$.

---

[2]We've noted before that the literature is inconsistent on whether the moment map for a groupoid action need be open. In this section, it is critical that moment maps to be open.

(b) *If* $f \in C_c(Z)$ *and* $\varphi \in C_{c,q}(Z)$, *then*

$$\Psi_\varphi(f)(g) = \int_Z f(g \cdot z)\varphi(z)\, d\beta^{s(g)}(z)$$

defines an element of $C_c(G)$.

PROOF. (a) This is straightforward if $F(g, z) = f(g)\varphi(z)$ with $f \in C_c(G)$ and $\varphi \in C_c(Z)$. But we can approximate F properly in the inductive limit topology with sums of such functions.

(b) Let $L = \operatorname{supp} \varphi \cap q^{-1}\big(q(\operatorname{supp} f)\big)$. By assumption on φ, L is compact. Since G acts properly on Z, the set

$$P(\operatorname{supp} f, L) = \{\, g \in G : g \cdot L \cap \operatorname{supp} f \neq \emptyset \,\}$$

is compact. It follows that

$$F(g, z) = f(g \cdot z)\varphi(z)$$

defines an element of $C_c(G \times Z)$. Then

$$\Psi_\varphi(f)(g) = \Phi(F)(g, s(g)).$$

The assertion follows. □

PROOF OF PROPOSITION 6.22. Using Lemma 3.19 on page 77, we fix $\varphi \in C^+_{c,q}(Z)$ such that $q\big(\{\, z : \varphi(z) > 0 \,\}\big) = G\backslash Z$. Then we define a Radon measure on $C_c(Z)$ by

$$\nu^u(f) = \int_G \int_Z f(g \cdot z)\varphi(z)\, d\beta^{s(g)}(z)\, d\lambda^u(g) = \int_G \Psi_\varphi(f)(g)\, d\lambda^u(g).$$

Since λ is a Haar system and $\Psi_\varphi(f) \in C_c(G)$ by our technical Lemma 6.23 on the preceding page, we see immediately that

$$u \mapsto \nu^u(f)$$

is continuous.

Clearly, $\operatorname{supp} \nu^u \subset r_Z^{-1}(u)$. Suppose $r_Z(w) = u$ and $f \in C_c^+(Z)$ is such that $f(w) > 0$. Then there is a $z' \in \{\, z : \varphi(z) > 0 \,\}$ such that $q(z') = q(w)$. Hence there is a $g \in G$ such that $g \cdot z' = w$. Note that $r_Z(z') = s(g)$ and $r(g) = r_Z(g \cdot z) = r_Z(w) = u$. Since $\beta^{s(g)}$ has full support and since everything in sight is continuous and non-negative,

$$\Psi_\varphi(g) = \int_Z f(g \cdot z)\varphi(z)\, \beta^{s(g)}(z) > 0.$$

Hence $\nu^u(f) > 0$ and we have

$$\operatorname{supp} \nu^u = r_Z^{-1}(u).$$

Hence to complete the proof of the theorem, we just need to establish equivariance. But

$$\begin{aligned} \int_Z f(g' \cdot z)\, d\nu^{s(g')}(z) &= \int_G \int_Z f(g'g \cdot z)\varphi(z)\, d\beta^{s(g)}(z)\, d\lambda^{s(g')}(g) \\ &= \int_G \Psi_\varphi(f)(g'g)\, d\lambda^{s(g')}(g) \end{aligned}$$

which, since λ is a Haar system on G, is

$$= \int_G \Psi_\varphi(g)\,d\lambda^{r(g')}(g)$$

$$= \int_G \int_Z f(g\cdot z)\varphi(z)\,d\beta^{s(g)}(z)\,d\lambda^{r(g')}(g)$$

which, since $g'\cdot s(g') = r(g')$, is

$$= \int_Z f(z)\,d\nu^{g'\cdot s(g')}(z).$$

This completes the proof. □

6.6. Some Applications

Using Theorem 6.20 on page 151, we can add to the list of interesting groupoids that necessarily have Haar systems.

Proper Principal Groupoids. Recall that G is called principal if the natural action of G on $G^{(0)}$ given by $x\cdot s(x) = r(x)$ is free. We call G proper if this action is proper in that $(x, s(x)) \mapsto (r(x), s(x))$ is proper from $G\times G^{(0)} \to G^{(0)}\times G^{(0)}$. If G is a proper principal groupoid with open range and source maps, then the orbit space $G\backslash G^{(0)}$ is locally compact Hausdorff by Proposition 2.18 on page 33, and it is straightforward to check that $G^{(0)}$ implements an equivalence between G and the orbit space $G\backslash G^{(0)}$ (see Example 2.35 on page 45). Since the orbit space trivially has a Haar system, the following is a simple corollary of Theorem 6.20.

PROPOSITION 6.24 (Blanchard). *Every second countable proper principal groupoid with open range and source maps has a Haar system.*

REMARK 6.25. If G is a second countable proper principal groupoid with open range and source maps, then we can explicitly describe a Haar system. The orbit map $q : G^{(0)} \to G\backslash G^{(0)}$ sending u to $\dot u := G\cdot u$ is continuous and open. Hence Blanchard's Lemma 3.20 on page 77 implies there is a full q-system $\{\beta^{\dot u}\}_{\dot u\in G\backslash G^{(0)}}$. It is not hard to check that $\lambda^u = \delta_u \times \beta^{\dot u}$ is a Haar system for

$$G_q := \{\, (u, v)\in G^{(0)}\times G^{(0)} : \dot u = \dot v \,\}.$$

Since $x\mapsto (r(x), s(x))$ is a groupoid isomorphism of G and G_q, we get an elementary description for a Haar system on G.

Note that Example 6.2 on page 141 is an example of a principal proper groupoid that does not have an open range map. Hence the hypothesis that the range and source maps be open is required in Proposition 6.24.

Transitive Groupoids. We have already observed in Example 6.7 on page 142 that if G is a second countable transitive groupoid, then the isomorphism class of $C^*(G)$ is independent of choice of Haar system. On the other hand, we know that a second countable transitive groupoid is equivalent to any of its isotropy groups provided its range and source maps are open (see Example 2.32 on page 44).[3] Since locally compact groups always have a Haar measure, the following is an immediate consequence of Theorem 6.20. (Similar assertions can be found in [**Sed76**].)

[3]It should be noted that in both [**Ram90**] and [**MRW87**] openness of the range and source maps on a topological groupoid is a standing assumption.

PROPOSITION 6.26 (Seda). *If G is a second countable, locally compact transitive groupoid with open range and source maps, then G has a Haar system.*

I had asked in [**Wil16**] whether it could be the case that second countable, locally compact, *transitive* groupoids always had open range and source maps. As I was preparing this book for publication, M. Buneci produced an example where this is not the case [**Bun19**]. Hence the hypothesis in Proposition 6.26 that G has open range and source maps is a necessary one.

Blowing Up the Unit Space. While there are myriad ways groupoid equivalences arise in applications, one standard technique deserves special mention (for example, see Theorem 2.52 on page 49 and Proposition 2.64 on page 55). Suppose that G is a second countable locally compact groupoid with a Haar system (or at least open range and source maps). Let $f : Z \to G^{(0)}$ be a continuous and open map. Then we can form the blow-up or pull-back groupoid as in Example 2.37 on page 45:

$$G[Z] = \{ (z, g, w) \in Z \times G \times Z : f(z) = r(g) \text{ and } s(g) = f(w) \}.$$

The idea being that we use f to "blow-up" the unit space of G to all of Z. We saw in Example 2.37 that the blow up $G[Z]$ is equivalent to G.[4]

Assuming G has a Haar system and Z is second countable, Theorem 6.20 implies $G[Z]$ has a Haar system. However in this case we can do a bit better and write down a tidy formula for the Haar system on the blow-up. We still require Blanchard's Lemma 3.20 on page 77 to assume that there is a full f-system for any continuous open map $f : Z \to G^{(0)}$ (provided that Z is second countable).

PROPOSITION 6.27. *Suppose that G is a locally compact Hausdorff groupoid with a Haar system $\{\lambda^u\}_{u\in G^{(0)}}$, and that Z is second countable. Let $f : Z \to G^{(0)}$ be a continuous open map, and let*

$$G[Z] = \{ (w, g, z) \in Z \times G \times Z : f(w) = r(g) \textit{ and } s(g) = f(z) \}$$

be the "blow-up" of G by f. If $\{\beta^u\}_{u\in G^{(0)}}$ is a full f-system, then we get a Haar system $\{\kappa^z\}_{z\in Z}$ on $G[Z]$ given by

$$\kappa^z(f) = \int_G \int_Z f(z, g, w)\, d\beta^{s(g)}(w)\, d\lambda^{f(z)}(g).$$

The proof is left as an exercise (Ex 6.6.2).

Imprimitivity Groupoids. If X is a free and proper right G-space, then assuming G has open range and source maps, we can form the imprimitivity groupoid G^Z. If the moment map $r_Z : X \to G^{(0)}$ is open, then X implements an equivalence between G^Z and G (see Proposition 2.47 on page 48). Then we can apply Theorem 6.20 (or its proof).

PROPOSITION 6.28. *Suppose that G is a second countable, locally compact Hausdorff groupoid with a Haar system. Let X be a free and proper right G-space with open moment map. Then the imprimitivity groupoid G^Z has a Haar system.*

---

[4] It is worth repeating that this assertion requires f to be open and G to have open range and source maps.

Wide Subgroupoids. Suppose that G has a Haar system $\{\lambda^u\}$. Then any open subgroupoid H has a Haar system—just take the restriction $\{\lambda^u\}_{u\in H^{(0)}}$. The situation for closed subgroupoids is more subtle. As we have seen in Theorem 6.12 on page 146, the isotropy groupoid has a Haar system if and only if the restriction of the range map to G' is open. More generally, we have the following for subgroupoids H with $H^{(0)} = G^{(0)}$ (such subgroupoids are called *wide*). This result is due to Scott LaLonde [**LaL18**, Theorem 4.2].

PROPOSITION 6.29. *Suppose that G is a second countable locally compact groupoid with a Haar system and that H is closed subgroupoid with $H^{(0)} = G^{(0)}$. If the $r|_H$ is open, then H has a Haar system.*

PROOF. If H has an open range map, then H is equivalent to the action groupoid $G \rtimes G/H$ by Corollary 2.50 on page 48. If G has a Haar system, then so does $G \rtimes G/H$ (Ex 2.1.7). Hence H has a Haar system by Theorem 6.20. □

EXERCISES

Ex 6.6.1. Suppose that G acts on the left of X. Show that the corresponding action groupoid has a Haar system if and only if there is a family of Radon measures $\{\sigma^x\}_{x\in X}$ such that $\operatorname{supp}\sigma^x = G^{r_X(x)}$, $\gamma\cdot\sigma^x = \sigma^{\gamma\cdot x}$, and for all $f \in C_c(G)$ we have

$$x \mapsto \int_G f(\gamma)\, d\sigma^x(\gamma)$$

continuous.

—Answer on page 374

Ex 6.6.2. Prove Proposition 6.27.

—Answer on page 374

CHAPTER 7

Unitary Representations

Unlike the situation for groups, unitary representations of groupoids are, at least in my opinion, more of tool than an end in themselves. We will work quite hard in Chapter 8 to show that every $*$-homomorphism $L : C_c(G) \to B(\mathcal{H}_L)$ is the integrated form of a unitary representation of G (to be defined below) with only modest assumptions on L. This will be most interesting as a means to show that such $*$-homomorphisms are bounded and extend to $C^*(G)$. Another difference from the group case is that unitary representations of groupoids are Borel functions. Hence we will need some measure theoretic constructs as well as Borel Hilbert bundles together with direct integrals such as treated in [**Wil07**, Appendix F]. There is a brief review of the sort of direct integrals required in Section 3.5.

7.1. Quasi-invariant Measures

To get from a Haar system on G to a Radon measure on G, we need to supply a Radon measure μ on $G^{(0)}$: given a Haar system $\{\lambda^u\}_{u\in G^{(0)}}$ and a Radon measure μ on $G^{(0)}$, we get a Radon measure $\nu = \mu \circ \lambda$ on G via

$$\nu(f) = \int_{G^{(0)}} \int_G f(\gamma)\, d\lambda^u(\gamma)\, d\mu(u) \quad \text{for all } f \in C_c(G). \tag{7.1}$$

(Measures of this sort are considered in Section 3.1.) As we will see, there is no canonical choice for μ. But we do want our measure ν to play nice with the involution and hence with inversion. To this end, we define ν^{-1} to be the push-forward of ν under inversion. Hence,

$$\nu^{-1}(f) = \int_{G^{(0)}} \int_G f(\gamma)\, d\lambda_u(\gamma)\, d\mu(u) \quad \text{for all } f \in C_c(G), \tag{7.2}$$

where as usual λ_u is the push-forward of λ^u under inversion.

Definition 7.1. Let G be a locally compact Hausdorff groupoid with Haar system $\{\lambda^u\}_{u\in G^{(0)}}$. Then a Radon measure μ on $G^{(0)}$ is called *quasi-invariant* if the measures ν and ν^{-1} defined in (7.1) and (7.2), respectively, are equivalent. In that case, the Radon-Nykodym derivative $\Delta := d\nu/d\nu^{-1}$ is called the *modular function* for μ.

Hence by definition, if μ is quasi-invariant with modular function Δ, then

$$\int_{G^{(0)}} \int_G f(\gamma)\Delta(\gamma)\, d\lambda_u(\gamma)\, d\mu(u) = \int_{G^{(0)}} \int_G f(\gamma)\, d\lambda^u(\gamma)\, d\mu(u) \tag{7.3}$$

for all $f \in C_c(G)$.

It is hardly obvious that quasi-invariant measures should exist at all, let alone that they should play an important role in the theory. It might be comforting to notice that if G is a group, then ν is simply a left-Haar measure for G. Then

Δ is just the usual modular function on a locally compact group—see [**Wil07**, Lemma 1.67]. As it turns out, we have already seen examples of modular functions in the transformation group setting.

EXAMPLE 7.2 (Transformation Groups). Let $G = G(\mathcal{G}, X)$ be a transformation group groupoid as in Section 1.4.2. Suppose that (M, π) is a covariant representation of $(\mathcal{G}, X)$ and that M is equivalent to the multiplication representation of $C_0(X)$ on $L^2(X * \mathscr{H}, \mu)$ for a Borel Hilbert bundle $X * \mathscr{H}$ and a finite Radon measure μ as in [**Wil07**, Example F.25]. Corollary 1.59 on page 25 implies that μ is quasi-invariant for G with modular function $(s, x) \mapsto \Delta(s)d(s, x)^{-1}$ where d is given by Lemma 1.58 on page 24 and Δ is the modular function on $\mathcal{G}$.

To see that there are lots of quasi-invariant measures, we use the next lemma. Since we are assuming everything in sight is second countable, a Radon measure is necessarily σ-finite. In particular, every Radon measure on a second countable space is equivalent to a probability measure. Recall that if $\tau : X \to Y$ is a Borel map and ν is a Borel measure on X, then the *push-forward measure* on Y is the Borel measure $\tau_*\mu(E) := \mu\big(\tau^{-1}(E)\big)$ and

$$\int_Y f(y)\, d\tau_*\mu(x) = \int_X f(\tau(x))\, d\mu(x) \tag{7.4}$$

for suitable f—see [**Wil07**, Lemma H.13].

LEMMA 7.3. *We assume that G is a second countable locally compact Hausdorff groupoid with Haar system* $\{\lambda^u\}_{u \in G^{(0)}}$. *Suppose that* μ_0 *is any Radon measure on* $G^{(0)}$. *Let* ν_0 *be the Radon measure* $\nu_0 = \mu_0 \circ \lambda$ *as above. Then* ν_0 *is* σ*-finite. If* ν *is a probability measure equivalent to* ν_0 *and if* $\mu := s_*\nu$ *is the push-forward by* s*, then* μ *is quasi-invariant and is called the* saturation *of* μ_0*. Furthermore, if* μ_0 *is quasi-invariant to begin with, then* μ *and* μ_0 *are equivalent.*

REMARK 7.4. Note that "the" saturation of a measure is only defined up to equivalence of measures. This is unfortunate, but nevertheless, standard terminology.

Recall that *a priori*, the formula

$$\nu_0(f) = \int_{G^{(0)}} \int_G f(\gamma)\, d\lambda^u(\gamma)\, d\mu_0(u)$$

holds only for $f \in C_c(G)$. However as usual, we take advantage of the observation that it holds for all $f \in \mathcal{B}^+(G)$ by Corollary 3.9 on page 72. Note that by definition, $\mu(E) = \nu\big(s^{-1}(E)\big)$ and by (7.4)

$$\int_{G^{(0)}} f(u)\, d\mu(u) = \int_G f\big(s(\gamma)\big)\, d\nu(\gamma).$$

For convenience, we let φ be a representative of the Radon-Nikodym derivative $d\nu/d\nu_0$ such that $\varphi(\gamma) \in (0, \infty)$ for all $\gamma \in G$.

PROOF OF LEMMA 7.3. We let $\nu' := \int_G \lambda^u\, d\mu(u)$. For the first part of the proof, it will suffice to show that if f is a bounded non-negative Borel function on G such that

$$\int_G f(\gamma)\, d\nu'(\gamma) = 0,$$

then

$$\int_G \tilde{f}(\gamma)\,d\nu'(\gamma) = 0,$$

where $\tilde{f}(\gamma) := f(\gamma^{-1})$.

Now we compute as follows.

$$\begin{aligned} 0 &= \int_G f(\gamma)\,d\nu'(\gamma) = \int_{G^{(0)}} \int_G f(\gamma)\,d\lambda^u(\gamma)\,d\mu(u) \\ &= \int_G \int_G f(\gamma)\,d\lambda^{s(\eta)}(\gamma)\,d\nu(\eta) \end{aligned}$$

which, using the definition of φ, is

$$\begin{aligned} &= \int_G \int_G f(\gamma)\,d\lambda^{s(\eta)}(\gamma)\varphi(\eta)\,d\nu_0(\gamma) \\ &= \int_{G^{(0)}} \int_G \int_G f(\gamma)\varphi(\eta)\,d\lambda^{s(\eta)}(\gamma)\,d\lambda^u(\eta)\,d\mu_0(u) \end{aligned}$$

which, after sending $\gamma \mapsto \eta^{-1}\gamma$ and noting that $r(\eta) = u$, is

$$= \int_{G^{(0)}} \int_G \int_G f(\eta^{-1}\gamma)\varphi(\eta)\,d\lambda^u(\gamma)\,d\lambda^u(\eta)\,d\mu_0(u)$$

which, by Fubini/Tonelli, is

$$= \int_{G^{(0)}} \int_G \int_G f(\eta^{-1}\gamma)\varphi(\eta)\,d\lambda^u(\eta)\,d\lambda^u(\gamma)\,d\mu_0(u)$$

which, by sending $\eta \mapsto \gamma\eta$, is

$$\begin{aligned} &= \int_{G^{(0)}} \int_G \int_G f(\eta^{-1})\varphi(\gamma\eta)\,d\lambda^{s(\gamma)}(\eta)\,d\lambda^u(\gamma)\,d\mu_0(u) \\ &= \int_G \int_G \tilde{f}(\eta)\varphi(\gamma\eta)\,d\lambda^{s(\gamma)}(\eta)\,d\nu_0(\gamma) \end{aligned}$$

which, using the fact that $\varphi^{-1} = d\nu_0/d\nu$, is

$$= \int_G \int_G \tilde{f}(\eta)\varphi(\gamma\eta)\varphi(\gamma)^{-1}\,d\lambda^{s(\gamma)}(\eta)\,d\nu(\gamma). \tag{7.5}$$

Notice that since $\varphi(\gamma) > 0$ for all γ, we have

$$\int_G \tilde{f}(\eta)\varphi(\gamma\eta)\varphi(\gamma)^{-1}\,d\lambda^{s(\gamma)}(\eta) = 0 \tag{7.6}$$

if and only if

$$\int_G \tilde{f}(\eta)\,d\lambda^{s(\gamma)}(\eta) = 0. \tag{7.7}$$

But if (7.5) is zero, then there is a set ν-null set N such that $\gamma \notin N$ implies that (7.6) vanishes. Clearly, N is s-saturated: $s^{-1}\big(s(N)\big) = N$. Thus, $s^{-1}(N)$ is μ-null and, since (7.6) vanishes if and only if (7.7) does, we've established that

$$\int_G \tilde{f}(\eta)\,d\lambda^u(\eta) = 0 \quad \text{for } \mu\text{-almost all } u.$$

But then

$$\int_{G^{(0)}} \int_G \tilde{f}(\eta)\,d\lambda^u(\eta)\,d\mu(u) = 0 = \int_G \tilde{f}(\eta)\,d\nu'(\eta).$$

This is what we wanted to show.

For the second assertion, suppose that μ_0 is quasi-invariant. Then $\nu_0 \sim \nu_0^{-1}$ and we can let ψ be the Radon-Nikodym derivative $d\nu_0^{-1}/d\nu_0$. Again, we can assume that $\psi(\gamma) \in (0, \infty)$ for all γ. But then if f is a bounded non-negative Borel function on $G^{(0)}$,

$$\begin{aligned}\mu(f) &= \int_G f\big(s(\gamma)\big)\, d\nu(\gamma) \\ &= \int_G f\big(s(\gamma)\big)\varphi(\gamma)\, d\nu_0(\gamma)\end{aligned}$$

which, using the quasi-invariance of μ_0, is

$$\begin{aligned}&= \int_G f\big(r(\gamma)\big)\varphi(\gamma^{-1})\psi(\gamma^{-1})\, d\nu_0(\gamma) \\ &= \int_{G^{(0)}}\int_G f\big(r(\gamma)\big)\varphi(\gamma^{-1})\psi(\gamma^{-1})\, d\lambda^u(\gamma)\, d\mu_0(u) \\ &= \int_{G^{(0)}} f(u)\Big(\int_G \varphi(\gamma^{-1})\psi(\gamma^{-1})\, d\lambda^u(\gamma)\Big)\, d\mu_0(u).\end{aligned}$$

Since

$$\alpha(u) := \int_G \varphi(\gamma^{-1})\psi(\gamma)\, d\lambda^u(\gamma)$$

is a non-negative (extended) real-valued function, it follows that $\mu \ll \mu_0$.

By Proposition 3.18 on page 77, there is a continuous nonnegative function $b \in C^+_{s,c}(G)$ such that

$$\int_G b(\gamma)\, d\lambda_u(\gamma) = 1 \quad \text{for all } u \in G^{(0)}.$$

Then for any nonnegative Borel function on $G^{(0)}$,

$$\begin{aligned}\mu_0(f) &= \int_{G^{(0)}} f(u)\, \mu_0(u) \\ &= \int_{G^{(0)}} f(u) \int_G b(\gamma)\, d\lambda_u(\gamma)\, d\mu_0(u) \\ &= \int_G f(s(\gamma))b(\gamma)\, d\nu_0^{-1}(\gamma)\end{aligned}$$

which, since $\psi = d\nu_0^{-1}/d\nu_0$, is

$$= \int_G f(s(\gamma))b(\gamma)\psi(\gamma)\, d\nu_0(\gamma)$$

which, since $\varphi = d\nu/d\nu_0$, is

$$= \int_G f(s(\gamma))b(\gamma)\psi(\gamma)\varphi(\gamma)^{-1}\, d\nu(\gamma). \tag{7.8}$$

However, if $\mu(f) = 0$, then $f(\gamma) = 0$ for ν-almost all γ. Then if follows from (7.8) that $\mu_0(f) = 0$. Therefore $\mu_0 \ll \mu$, and μ is equivalent to μ_0. □

In the case of groups, the modular function Δ is a continuous homomorphism. For a groupoid and a given quasi-invariant measure, Δ is only defined ν-almost everywhere. We eventually want to see that the modular function can be chosen to be multiplicative—see Proposition 7.6. The first step is to establish the following which comes from [**Ren80**, Proposition I.3.3].[1] For this, we will need the Radon measure $\nu^{(2)}$ on the space of composible paris $G^{(2)}$ given on $g \in C_c(G^{(2)})$ by

$$\nu^{(2)}(g) = \int_{G^{(0)}} \int_G \int_G g(\gamma, \eta)\, d\lambda_u(\gamma)\, d\lambda^u(\eta)\, d\mu(u). \tag{7.9}$$

LEMMA 7.5. *Suppose that G is a second countable locally compact groupoid with a Haar system $\{\lambda^u\}_{u\in G^{(0)}}$ and that μ is a quasi-invariant measure on $G^{(0)}$ with modular function Δ. Then*

$$\Delta(\gamma\eta) = \Delta(\gamma)\Delta(\eta)$$

for $\nu^{(2)}$-almost all $(\gamma, \eta) \in G^{(2)}$. Furthermore,

$$\Delta(\gamma) = \Delta(\gamma)^{-1} \quad \text{for } \nu\text{-almost all } \gamma. \tag{7.10}$$

PROOF. We'll use (7.3) as well as the left-invariance property of Haar systems at will. Let $\rho \in C_c(G^{(2)})$. Then on the one hand,

$$\begin{aligned}
\int_{G^{(2)}} \rho(\gamma\eta, \eta^{-1})\, d\nu^{(2)}(\gamma, \eta) &= \int_{G^{(0)}} \int_G \int_G \rho(\gamma\eta, \eta^{-1})\, d\lambda_{r(\eta)}(\gamma)\, d\lambda^u(\eta)\, d\mu(u) \\
&= \int_{G^{(0)}} \int_G \int_G \rho(\gamma, \eta^{-1})\, d\lambda_{s(\eta)}(\gamma)\, d\lambda^u(\eta)\, d\mu(u) \\
&= \int_{G^{(0)}} \int_G \int_G \varphi(\gamma, \eta)\, d\lambda_{r(\eta)}(\gamma)\, d\lambda_u(\eta)\, d\mu(u) \\
&= \int_{G^{(0)}} \int_G \int_G \varphi(\gamma, \eta)\, d\lambda_u(\gamma)\Delta(\eta)^{-1}\, d\lambda^u(\eta)\, d\mu(u) \\
&= \int_{G^{(0)}} \int_G \int_G \rho(\gamma, \eta)\Delta(\eta)^{-1}\, d\nu^{(2)}(\gamma, \eta).
\end{aligned}$$

On the other hand, we can use Fubini in (7.9) to see that

$$\begin{aligned}
\int_{G^{(2)}} \rho(\gamma\eta, \eta^{-1})\, d\nu^{(2)}(\gamma, \eta) &= \int_{G^{(0)}} \int_G \int_G \rho(\gamma\eta, \eta^{-1})\, d\lambda^{s(\gamma)}(\eta)\, d\lambda_u(\gamma)\, d\mu(u) \\
&= \int_{G^{(0)}} \int_G \int_G \rho(\eta, \eta^{-1}\gamma)\, d\lambda^{r(\gamma)}(\eta)\, d\lambda_u(\gamma)\, d\mu(u) \\
&= \int_{G^{(0)}} \int_G \int_G \rho(\eta, \eta^{-1}\gamma)\Delta(\gamma)^{-1}\, d\lambda^{r(\gamma)}(\eta)\, d\lambda^u(\gamma)\, d\mu(u) \\
&= \int_{G^{(0)}} \int_G \int_G \rho(\eta, \gamma)\Delta(\eta\gamma)^{-1}\, d\lambda^{s(\eta)}(\gamma)\, d\lambda^u(\eta)\, d\mu(u) \\
&= \int_{G^{(0)}} \int_G \int_G \rho(\eta, \gamma)\Delta(\eta\gamma)^{-1}\Delta(\eta)\, d\lambda^u(\gamma)\, d\lambda_u(\eta)\, d\mu(u)
\end{aligned}$$

which, after interchanging η and γ, is

$$\begin{aligned}
&= \int_{G^{(0)}} \int_G \int_G \rho(\gamma, \eta)\Delta(\gamma\eta)^{-1}\Delta(\gamma)\, d\lambda_u(\gamma)\, d\lambda^u(\eta)\, d\mu(u) \\
&= \int_{G^{(2)}} \rho(\gamma, \eta)\Delta(\gamma\eta)^{-1}\Delta(\gamma)\, d\nu^{(2)}(\gamma, \eta).
\end{aligned}$$

[1] The idea is that $G^{(2)}$ has a groupoid structure with $(\gamma, \eta)(\gamma\eta, \zeta) = (\gamma, \eta\zeta)$ and $(\gamma, \eta)^{-1} = (\gamma\eta, \eta^{-1})$. Then the argument given here just checks that if $D(\gamma, \eta) = d\nu^{(2)}/(d\nu^{(2)})^{-1}$, then on the one hand, $D(\gamma, \eta) = \Delta(\eta)$, while on the other hand $D(\gamma, \eta) = \Delta(\gamma\eta)\Delta(\gamma)^{-1}$.

It follows that $\Delta(\eta) = \Delta(\gamma\eta)\Delta(\gamma)^{-1}$ for $\nu^{(2)}$-almost all (γ, η).

The proof of (7.10) is left as an exercise (Ex 7.1.1). □

Some clever work of Ramsay's from Appendix D allows us to spruce up Lemma 7.5 considerably. Let $\mathbf{R}^{+,\times}$ be the multiplicative group of positive reals.

PROPOSITION 7.6. *Let G be a second countable locally compact groupoid with a Haar system. Suppose μ is a quasi-invariant measure on $G^{(0)}$. Then we can choose the modular function Δ so that $\Delta : G \to \mathbf{R}^{+,\times}$ is a Borel homomorphism.*

PROOF. Lemma 7.5 asserts that any choice of Δ is an almost everywhere homomorphism of G into the group $\mathbf{R}^{+,\times}$ as defined in Lemma D.3 on page 347. Then Theorem D.8 on page 350 implies that we can find a homomorphism Δ_1 on G into $\mathbf{R}^{+,\times}$ that equals Δ ν-almost everywhere. □

EXERCISES

Ex 7.1.1. Prove (7.10).

—Answer on page 375

7.2. Unitary Representations

With the preliminaries in hand, we have the machinery to make the basic definition for the analogue of a unitary representation of a group. Note that we must fix a Haar system λ in order to make sense of quasi-invariant measures. If μ is a quasi-invariant measure on $G^{(0)}$, then we write ν for $\mu \circ \lambda$ as per usual. The basics of Borel Hilbert bundles are outlined in Section 3.5.

DEFINITION 7.7. A *unitary representation* of a groupoid G with Haar system $\lambda = \{\lambda^u\}_{u\in G^{(0)}}$ is a triple $(\mu, G^{(0)} * \mathscr{H}, \hat{L})$ consisting of a quasi-invariant measure μ on $G^{(0)}$, a Borel Hilbert bundle $G^{(0)} * \mathscr{H}$ over $G^{(0)}$, and a Borel homomorphism $\hat{L} : G \to \operatorname{Iso}(G^{(0)} * \mathscr{H})$ such that

$$\hat{L}(\gamma) = \big(r(\gamma), L_\gamma, s(\gamma)\big).$$

Definition 7.7 is the ideal form of a unitary representation of a groupoid. But in practice, we can get away with a lot less. So here, I'm often going to work with weaker alternatives to be described shortly. Naturally, it is the weaker forms whose existence is easiest to establish in applications.

Recall that a subset U of a measure space (X, μ) is *conull* if $\mu(X \setminus U) = 0$.

REMARK 7.8 (Inessential Reductions). Suppose that μ is a quasi-invariant measure on $G^{(0)}$ and $U \subset G^{(0)}$ is μ-conull. Then $r^{-1}(U)$ is clearly ν-conull. Thus $s^{-1}(U)$ is ν^{-1}-conull. Since ν and ν^{-1} are equivalent, the intersection $r^{-1}(U) \cap s^{-1}(U) = UGU = G|_U$ is ν-conull. In [**Muh99**], Muhly calls such reductions *inessential*.

DEFINITION 7.9. An *almost everywhere unitary representation* of a groupoid G with Haar system $\{\lambda^u\}_{u\in G^{(0)}}$ is a quadruple $(\mu, G^{(0)} * \mathscr{H}, U, \hat{L})$ consisting of a quasi-invariant measure μ on $G^{(0)}$, a Borel Hilbert bundle $G^{(0)} * \mathscr{H}$ over $G^{(0)}$, a μ-conull set U, and a Borel map $\hat{L} : G|_U \to \operatorname{Iso}(G^{(0)} * \mathscr{H})$ of the form $\hat{L}(\gamma) = \big(r(\gamma), L_\gamma, s(\gamma)\big)$ such that

(a) $L_u = I_{\mathcal{H}(u)}$ for μ-almost all $u \in G^{(0)}$,

(b) $L_{\gamma\eta} = L_\gamma L_\eta$ for ν^2-almost all (γ, η), and
(c) $L_\gamma^{\ 1} = L_{\gamma^{-1}}$ for ν-almost all γ.

For technical reasons, or just elegance, we will sometimes want to work with real homomorphisms. Then we upgrade the previous definition as follows.

DEFINITION 7.10. A *weak unitary representation* or sometimes an *essential unitary representation* of a groupoid G with Haar system $\{\lambda^u\}_{u\in G^{(0)}}$ is a quadruple $(\mu, G^{(0)} * \mathscr{H}, U, \hat{L})$ consisting of a quasi-invariant measure μ on $G^{(0)}$, a Borel Hilbert bundle $G^{(0)} * \mathscr{H}$ over $G^{(0)}$, a μ-conull set U, and a Borel homomorphism $\hat{L} : G|_U \to \operatorname{Iso}(G^{(0)} * \mathscr{H})$ of the form $\hat{L}(\gamma) = \big(r(\gamma), L_\gamma, s(\gamma)\big)$.

Of course a weak unitary representation $(\mu, G^{(0)} * \mathscr{H}, U, \hat{L})$ with $U = G^{(0)}$ is simply a *unitary representation* $(\mu, G^{(0)} * \mathscr{H}, \hat{L})$ of G as above. Notice that in all cases—whether $(\mu, G^{(0)} * \mathscr{H}, U, \hat{L})$ is an almost everywhere unitary representation, a weak unitary representation, or even a unitary representation—we always have $L_\gamma \in U(\mathcal{H}(s(\gamma)), \mathcal{H}(r(\gamma)))$ so that $L_\gamma^{-1} = L_\gamma^*$.

REMARK 7.11. Of course, every weak unitary representation or every unitary representation is an almost everywhere unitary representation. The extra generality in working with the almost everywhere variety makes the following results a bit more flexible. A remarkable result due to Ramsay (see Corollary D.9 on page 350) says that an almost everywhere unitary representation of G is is equal almost everywhere to an weak unitary representation.

PROPOSITION 7.12. *Let $(\mu, G^{(0)} * \mathscr{H}, U, \hat{L})$ be a almost everywhere unitary representation of G. Then there is an I-norm bounded representation L of $C_c(G)$ on $L^2(X * \mathscr{H}, \mu)$ characterized by*

$$\big(L(f)h \mid k\big) = \int_{G|_U} f(\gamma)\big(L_\gamma(h(s(\gamma))) \mid k(r(\gamma)\big)\Delta(\gamma)^{-\frac{1}{2}}\, d\nu(\gamma). \tag{7.11}$$

REMARK 7.13. Since $G|_U$ is ν-conull, there is no harm in computing the right-hand side of (7.11) if we simply define $L_\gamma = 0$ if $\gamma \notin G|_U$. We can then write (7.11) as

$$\int_{G^{(0)}} \int_G f(\gamma)\big(L_\gamma(h(s(\gamma))) \mid k(r(\gamma)\big)\Delta(\gamma)^{-\frac{1}{2}}\, d\lambda^u(\gamma)\, d\mu(u).$$

With the same convention on L_γ for $\gamma \notin G|_U$, L can also be defined via the vector-valued integral

$$L(f)h(u) = \int_G f(\gamma)L_\gamma\big(h(s(\gamma))\big)\Delta(\gamma)^{-\frac{1}{2}}\, d\lambda^u(\gamma).$$

These sorts of integrals are discussed in [**Wil07**, §1.5].

PROOF OF PROPOSITION 7.12. Using Remark 7.13, we have

$$\begin{aligned}
\big(L(f*g)h \mid k\big) &= \int_{G^{(0)}} \int_G L(f*g)(\gamma)\big(L_\gamma(h(s(\gamma))) \mid k(r(\gamma))\big)\Delta(\gamma)^{-\frac{1}{2}}\, \lambda^u(\gamma)\, d\mu(u) \\
&= \int_{G^{(0)}} \int_G \int_G f(\eta)g(\eta^{-1}\gamma)\big(L_\gamma(h(s(\gamma))) \mid k(r(\gamma))\big)\Delta(\gamma)^{-\frac{1}{2}}\, d\lambda^u(\eta)\, \lambda^u(\gamma)\, d\mu(u) \\
&= \int_{G^{(2)}} f(\eta^{-1})g(\eta\gamma)\big(L_\gamma(h(s(\gamma))) \mid k(r(\gamma))\big)\Delta(\gamma)^{-\frac{1}{2}}\, d\nu^{(2)}(\eta, \gamma)
\end{aligned}$$

which, since $L_\gamma = L^*_\eta L_{\eta\gamma}$ ν^2-almost everywhere, is

$$= \int_{G^{(0)}} \int_G \int_G f(\eta) g(\eta^{-1}\gamma) \big(L^*_{\eta^{-1}} L_{\eta^{-1}\gamma}(h(s(\gamma))) \mid k(r(\gamma)) \big) \Delta(\gamma)^{-\frac{1}{2}} \\ d\lambda^u(\eta)\, d\lambda^u(\gamma)\, d\mu(u)$$

which, after Fubini and letting $\gamma \mapsto \eta\gamma$, is

$$= \int_{G^{(0)}} \int_G \int_G f(\eta) g(\gamma) \big(L^*_{\eta^{-1}} L_\gamma(h(s(\gamma))) \mid k(r(\eta)) \big) \Delta(\eta\gamma)^{-\frac{1}{2}} \\ d\lambda^{s(\eta)}(\gamma)\, d\lambda^u(\eta)\, d\mu(u)$$

which, since we can take Δ multiplicative by Proposition 7.6, is

$$= \int_{G^{(0)}} \int_G f(\eta) \Big(L^*_{\eta^{-1}} \big(\int_G g(\gamma) L_\gamma(h(s(\gamma))) \Delta(\gamma)^{-\frac{1}{2}}\, d\lambda^{s(\eta)}(\gamma) \big) \Bigm| k(r(\eta)) \Big) \\ \Delta(\eta)^{-\frac{1}{2}}\, d\lambda^u(\eta)\, d\mu(u)$$

which, in view of Remark 7.13, is

$$= \int_G f(\eta) \big(L^*_{\eta^{-1}} L(g)(h)(s(\eta)) \mid k(r(\eta))\, \Delta(\eta)^{-\frac{1}{2}}\, d\nu(\eta)$$

which, since $L_{\eta^{-1}} = L^*_\eta$ for ν-almost all η, is

$$= \big(L(f)L(g)h \mid k \big).$$

Hence L is multiplicative.

To see that L is $*$-preserving, we use the fact that $\Delta(\gamma)^{-\frac{1}{2}}\, d\nu(\gamma)$ is invariant under inversion. Thus

$$\begin{aligned} \big(h \mid L(f^*)k \big) &= \int_G f(\gamma^{-1}) \big(h(r(\gamma)) \mid L_\gamma(k(s(\gamma))) \big) \Delta(\gamma)^{-\frac{1}{2}}\, d\nu(\gamma). \\ &= \int_G f(\gamma) \big(h(s(\gamma)) \mid L^{-1}_\gamma(k(r(\gamma))) \big) \Delta(\gamma)^{-\frac{1}{2}}\, d\nu(\gamma) \end{aligned}$$

which, since $L_{\gamma^{-1}} = L^*_\gamma$ for ν-almost all γ, is

$$= \big(L(f)h \mid k \big).$$

It only remains to see that L is I-norm bounded. But using the Cauchy-Schwarz inequality in $L^2(\nu)$, we have

$$\begin{aligned} \big| \big(L(f)h \mid k \big) \big|^2 &\le \Big(\int_G |f(\gamma)| \|h(s(\gamma))\| \|k(r(\gamma)\| \Delta(\gamma)^{-\frac{1}{2}}\, d\nu(\gamma) \Big)^2 \\ &\le \Big(\int_G |f(\gamma)| \|h(s(\gamma))\|^2 \Delta(\gamma)^{-1}\, d\nu(\gamma) \Big) \Big(\int_G |f(\gamma)| \|k(r(\gamma)\|^2\, d\nu(\gamma) \Big) \\ &\le \big(\|f\|_{I,s} \|h\|_2^2 \big) \big(\|f\|_{I,r} \|k\|_2^2 \big) \\ &\le \|f\|_I^2 \|h\|_2^2 \|k\|_2^2. \end{aligned}$$

□

Definition 7.14. If $(\mu, G^{(0)} * \mathscr{H}, U, \hat{L})$ is an almost everywhere unitary representation of G or a weak unitary representation of G, then we call the representation L of $C_c(G)$ defined in Proposition 7.12 on the previous page the *integrated form* of $(\mu, G^{(0)} * \mathscr{H}, U, \hat{L})$.

EXAMPLE 7.15 (Transformation Groups). We have—almost anyway—seen an example of a weak unitary representation in Section 1.4.2 for the transformation group groupoid $G = G(\mathcal{G}, X)$ associated to a transformation group $(\mathcal{G}, X)$. Recall that the unit space $G^{(0)}$ is identified with X. There, given a covariant pair (M, π), we found a Hilbert bundle $X * \mathscr{H}$ and a quasi-invariant measure μ as well as unitaries $u(t,x) : \mathcal{H}(s(t,x)) \to \mathcal{H}(r(t,x))$. Had we worked a bit harder—say by following [**Wil07**, Proposition G.24]—we could have produced a conull set $U \subset X$ such that $u(tv,x) = u(t,x)u(v,t^{-1}\cdot x)$ provided x,$t^{-1}\cdot x$, and $v^{-1}t^{-1}\cdot x$ are in U. Then $(\mu, X * \mathscr{H}, U, \hat{u})$ would be weak unitary representation of $G = G(\mathcal{G}, X)$. As in Proposition 1.60 on page 25, the integrated form as in Proposition 7.12 on page 163 is the representation L corresponding the the covariant pair (M, π).

7.3. An Example: Regular Representations

If nothing else, Example 7.15 shows that getting our hands on a bonafide (weak) unitary representation is a bit daunting. After all, transformation groups are supposed to be *the* basic example and we omitted many of the details of the construction. As another example, consider one of the most fundamental examples of representations of $C_c(G)$—a regular representation $L := \operatorname{Ind}\mu$ as in Proposition 1.41 on page 18. Recall that L is defined as follows. As usual, let ν be the Radon measure on G given by

$$\nu(f) = \int_{G^{(0)}} \int_G f(\gamma)\, d\lambda^u(\gamma)\, d\mu(u) \quad \text{for } f \in C_c(G),$$

and write ν^{-1} for the forward image of ν under inversion. Then $L(f)$ acts on $L^2(G, \nu^{-1})$ via convolution:

$$L(f)\xi(\gamma) = \int_G f(\eta)h(\eta^{-1}\gamma)\, d\lambda^{r(\gamma)}(\eta) := f * h(\gamma).$$

The whole point of introducing unitary representations of groupoids is that we aim to prove (in Chapter 8) that every representation is the integrated form of a unitary representation. As motivation, we want to see that $L = \operatorname{Ind}\mu$ is (equivalent to) the integrated form of a weak unitary representation given any Radon measure μ on $G^{(0)}$. This material is based on [**Muh99**, Example 3.28].

A hint that doing this might be non-trivial is provided by the observation that μ is not required to be quasi-invariant. So let ρ be a probability measure on G which is equivalent to ν. As in Lemma 7.3 on page 158, we get a quasi-invariant measure, $[\mu]$, called the saturation of μ, by letting $[\mu] := s_*\rho$, where s is the source map. We write $[\nu]$ for the measure determined by

$$[\nu](f) = \int_{G^{(0)}} \int_G f(\gamma)\, d\lambda^u(\gamma)\, d[\mu](u).$$

We write Δ for a choice of Radon-Nikodym derivative for $[\nu]$ with respect to $[\nu]^{-1}$ which is a bona fide Borel homomorphism from G into $\mathbf{R}^{+,\times}$ (Proposition 7.6 on page 162).

Then, using [**Wil07**, Theorem I.5] for example, we can disintegrate ρ with respect to the map $s : G \to G^{(0)}$ to obtain probability measures $\bar{\nu}_u$ on G with $\operatorname{supp}\bar{\nu}_u \subset s^{-1}(u) = G_u$ and such that

$$\rho(f) = \int_{G^{(0)}} \int_G f(\gamma)\, d\bar{\nu}_u(\gamma)\, d[\mu](u).$$

Let $\varphi = \frac{d\nu}{d\rho}$ be a Radon-Nikodym derivative. Then $\nu = \varphi \cdot \rho$, and we get Radon measures $\nu_u = \varphi \cdot \bar{\nu}_u$ such that

$$\nu(f) = \int_{G^{(0)}} \int_G f(\gamma)\, d\nu_u(\gamma)\, d[\mu](u). \tag{7.12}$$

We will write ν^u for the Radon measure on G with $\operatorname{supp} \nu^u \subset G^u$ which is the forward image of ν_u under inversion. When convenient, we will view ν^u as a Radon measure on G^u. Of course, the ν_u, and hence the ν^u, are only determined up to a $[\mu]$-null set.

We need to examine the decomposition in (7.12) in a bit more detail. (We work with the ν^u rather than the ν_u as the former will be more relevant in the sequel.) Recall that $\eta \cdot \nu^{s(\eta)} = \Delta(\eta)\nu^{r(\eta)}$ simply means that $\Delta(\eta)\nu^{r(\eta)}(E) = \nu^{s(\eta)}(\eta^{-1}E)$ for all Borel sets $E \subset G^{r(\gamma)}$. This is equivalent to the assertion that

$$\Delta(\eta) \int_G \varphi(\gamma)\, d\nu^{r(\eta)}(\gamma) = \int_G \varphi(\eta\gamma)\, d\nu^{s(\eta)}(\gamma) \quad \text{for all } \varphi \in C_c(G). \tag{7.13}$$

A key step is the following which ultimately requires some subtle gymnastics with measures on groupoids due to Ramsay from Appendix D.

LEMMA 7.16. *There is a $[\mu]$-conull set U such that*

$$\eta \cdot \nu^{s(\eta)} = \Delta(\eta)\nu^{r(\eta)} \quad \textit{for all } \eta \in G|_U.$$

PROOF. Observe that

$$\begin{aligned} \big(L(f)h \mid k\big) &= \int_G L(f)(h)(\gamma)\overline{k(\gamma)}\, d\nu^{-1}(\gamma) \\ &= \int_G L(f)(h)(\gamma^{-1})\overline{k(\gamma^{-1})}\, d\nu(\gamma) \end{aligned} \tag{7.14}$$

which, using (7.12), recalling the definition of ν^u, and using Fubini, is

$$= \int_{G^{(0)}} \int_G L(f)(h)(\gamma)k(\gamma)\, d\nu^u(\gamma)\, d[\mu](u)$$

$$= \int_{G^{(0)}} \int_G f(\eta)\Big(\int_G h(\eta^{-1}\gamma)\overline{k(\gamma)}\, d\nu^{r(\eta)}(\gamma)\Big)\, d\lambda^u(\eta)\, d[\mu](u). \tag{7.15}$$

On the other hand, we also have $\big(L(f)h \mid k\big) = \overline{\big(L(f^*)k \mid h\big)}$. Replacing f by f^* in (7.15) and taking conjugates implies that (7.14) is also equal to

$$\begin{aligned} &= \int_{G^{(0)}} \int_G \overline{f^*(\eta)}\Big(\int_G \overline{k(\eta^{-1}\gamma)}h(\gamma)\, d\nu^{r(\eta)}(\gamma)\Big)\, d\lambda^u(\eta)\, d[\mu](u) \\ &= \int_{G^{(0)}} \int_G f(\eta^{-1})\Big(\int_G \overline{k(\eta^{-1}\gamma)}h(\gamma)\, d\nu^{r(\eta)}(\gamma)\Big)\, \lambda^u(\eta)\, d[\mu](u) \\ &= \int_{G^{(0)}} \int_G f(\eta)\Big(\int_G \overline{k(\eta\gamma)}h(\gamma)\, d\nu^{s(\eta)}(\gamma)\Big)\, \lambda_u(\eta)\, d[\mu](u) \end{aligned}$$

which, by definition of $\Delta = d([\mu] \circ \lambda)/d([\mu] \circ \lambda)^{-1}$, is

$$= \int_{G^{(0)}} \int_G f(\eta)\Big(\Delta(\eta)^{-1} \int_G h(\gamma)\overline{k(\eta\gamma)}\, d\nu^{s(\eta)}(\gamma)\Big)\, d\lambda^u(\eta)\, d[\mu](u).$$

It follows that there is a $[\nu]$-null set $N(h,k)$ such that $\eta \notin N(h,k)$ implies that

$$\Delta(\eta) \int_G h(\eta^{-1}\gamma)\overline{k(\gamma)}\, d\nu^{r(\eta)}(\gamma) = \int_G h(\gamma)\overline{k(\eta\gamma)}\, d\nu^{s(\eta)}(\gamma) \tag{7.16}$$

Let $D \subset C_c(G)$ be a countable set which is dense in the inductive limit topology as in Proposition C.3 on page 344. Hence there is a $[\nu]$-null set N such that (7.16) holds provided $h, k \in D$. Now fix $\eta \notin N$ and $\varphi \in C_c(G)$. We can find h_i and k_i in D such that the functions $\gamma \mapsto h_i(\eta^{-1}\gamma)\overline{k_i(\gamma)}$ converge to $\gamma \mapsto \varphi(\gamma)$ properly in the inductive limit topology on $C_c(G^{r(\eta)})$ while the functions $\gamma \mapsto h_i(\gamma)\overline{k_i(\eta\gamma)}$ converge properly to $\gamma \mapsto \varphi(\eta\gamma)$ in the inductive limit topology on $C_c(G^{s(\eta)})$ (see Ex 7.3.1). Since (7.16) holds for each pair h_i and k_i, and since $\sup_{u\in G^{(0)}} \nu^u(K)$ is finite for any compact set in G, it follows that (7.13) holds for φ and η.

We have shown that

$$\Sigma := \{\, \eta \in G : \eta \cdot \nu^{s(\eta)} = \Delta(\eta)\nu^{r(\eta)} \,\}$$

is $[\nu]$-conull. Since we can assume that Δ is a homomorphism, we easily check that if $\gamma, \eta \in \Sigma$ and $(\gamma, \eta) \in G^{(2)}$, then $\gamma\eta \in \Sigma$. Now we can employ Ramsay's Proposition D.2 on page 346, and conclude that there is $[\mu]$-conull set U such that the inessential reduction $G|_U \subset \Sigma$.[2] □

Next we want to view $L^2(G, \nu^{-1})$ as the sections of a Borel Hilbert bundle. Note that we can use (7.12) to decompose ν^{-1}:

$$\nu^{-1}(f) = \int_{G^{(0)}} \int_G f(\gamma)\, d\nu^u(\gamma)\, d[\mu](u).$$

Let $\mathcal{H}(u) := L^2(G, \nu^u) = L^2(G^u, \nu^u)$ and view $h \in \mathcal{L}^2(G, \nu^{-1})$ as a section $u \mapsto h(u)$ where $h(u) = h|_{G^u}$. Then the norm of $u \mapsto h(u)$ in $L^2(G^{(0)} * \mathscr{H}, [\mu])$ is the same as that of $h \in L^2(G, \nu^{-1})$. (For more details, see [**Wil07**, Example F.19].)

If $\eta \in G|_U$, then we can define a unitary $R_\eta : \mathcal{H}(s(\eta)) \to \mathcal{H}(r(\eta))$ by

$$R_\eta(h(s(\eta))(\gamma) = \Delta(\eta)^{\frac{1}{2}} h(\eta^{-1}\gamma).$$

Then $(\hat{R}, G^{(0)} * \mathscr{H}, [\mu])$ is a unitary representation of $G|_U$ which, since U is conull, we can view as a weak unitary representation of G.

Now we compute:

$$\begin{aligned}
\big(L(f)h \mid k\big) &= \int_G L(f)(h)(\gamma)\overline{k(\gamma)}\, d\nu^{-1}(\gamma) \\
&= \int_{G^{(0)}} \int_G L(f)(h)(\gamma)\overline{k(\gamma)}\, d\nu^u(\gamma)\, d[\mu](u) \\
&= \int_{G^{(0)}} \int_G \int_G f(\eta)h(\eta^{-1}\gamma)\overline{k(\gamma)}\, d\lambda^u(\eta)\, d\nu^u(\gamma)\, d[\mu](u) \\
&= \int_{G^{(0)}} \int_G f(\eta)\Big(\int_G \Delta(\eta)^{\frac{1}{2}} h(\eta^{-1}\gamma)\overline{k(\gamma)}\, d\nu^u(\gamma)\Big)\Delta(\eta)^{-\frac{1}{2}}\, d\lambda^u(\eta)\, d[\mu](u) \\
&= \int_{G^{(0)}} \int_G f(\eta)\Big(\int_G R_\eta(h(s(\eta))(\gamma)\overline{k(u)(\gamma)}\nu^u(\gamma)\Big)\Delta(\eta)^{-\frac{1}{2}}\, d\lambda^u(\eta)\, d[\mu](u) \\
&= \int_{G^{(0)}} \int_G f(\eta)\big(R_\eta(h(s(\eta)) \mid k(r(\eta))\big)\Delta(\eta)^{-\frac{1}{2}}\, d\lambda^u(\eta)\, d[\mu](u) \\
&= \big(R(f)h \mid k\big).
\end{aligned}$$

Thus L is the integrated form of $(\hat{R}, G^{(0)} * \mathscr{H}, U, [\mu])$ as claimed.

[2] Ramsay's result is a generalization of [**Wil07**, Lemma D.36] to groupoids.

Exercises

Ex 7.3.1. Let $\varphi \in C_c(G)$. Suppose $D \subset C_c(G)$ is dense in the inductive limit topology. Show that for each $\eta \in G$ there are $h_i, k_i \in D$ such that the functions $\gamma \mapsto h_i(\eta^{-1}\gamma)\overline{k_i(\gamma)}$ converge to $\gamma \mapsto \varphi(\gamma)$ properly in the inductive limit topology on $C_c(G^{r(\eta)})$ while at the same time the functions $\gamma \mapsto h_i(\gamma)\overline{k_i(\eta\gamma)}$ converge properly to the function $\gamma \mapsto \varphi(\eta\gamma)$ on $C_c(G^{s(\eta)})$.

—Answer on page 375

CHAPTER 8

Renault's Disintegration Theorem

In Chapter 7, we introduced the notion of a unitary representation of a groupoid. While these objects clearly parallel the corresponding notion for groups, they are unwieldy and not of much interest in their own right. While it might technically be an overstatement that their sole importance is that they only serve to give rise to *bounded* representations of the convolution algebra $C_c(G)$, it would not be terribly misleading. Moreover, unlike the case for locally compact groups, unitary representations of groupoids do not seem to arise in nature. They have to be extracted by brute force. Consider the example of regular representations outlined in Section 7.3.

The fundamental importance of the Disintegration Theorem is that it shows that weak representations of $C_c(G)$ (as defined below) are necessarily given by integration of a unitary representation and hence are bounded for the universal norm. One of the key applications of this automatic boundedness of weak representations is that it is exactly what is needed for the Equivalence Theorem.

8.1. The Statement

DEFINITION 8.1. Suppose that G is a locally compact Hausdorff groupoid with a Haar system. Let $\mathcal{H}_0$ be a dense subspace of a Hilbert space $\mathcal{H}$ and $\operatorname{Lin}(\mathcal{H}_0)$ the algebra of linear operators on $\mathcal{H}_0$. A *weak-representation* of $C_c(G)$ is a homomorphism $L : C_c(G) \to \operatorname{Lin}(\mathcal{H}_0)$ such that

(a) for each $f \in C_c(G)$ and $h, k \in \mathcal{H}_0$ we have $\big(L(f)h \mid k\big) = \big(h \mid L(f^*)k\big)$, and

(b) for each $h, k \in \mathcal{H}_0$, $f \mapsto \big(L(f)h \mid k\big)$ is continuous in the inductive limit topology.

We say that L is *nondegenerate* if in addition

(c) $\operatorname{span}\{\, L(f)h : f \in C_c(G) \text{ and } h \in \mathcal{H}_0 \,\}$ is dense in $\mathcal{H}$.

If $L' : C_c(G) \to \operatorname{Lin}(\mathcal{H}_0')$ is another weak-representation, then we say L' is equivalent to L if there is a unitary $U : \mathcal{H} \to \mathcal{H}'$ mapping $\mathcal{H}_0$ onto $\mathcal{H}_0'$ and intertwining L and L'.

Weak-representations are indeed weak. Since $\mathcal{H}_0$ is a arbitrary dense subspace, it is not complete and is merely a vector space. Thus linear operators $T \in \operatorname{Lin}(\mathcal{H}_0)$ don't necessarily have adjoints nor need they be bounded. Hence it is not at all clear how they could be extended to all of $\mathcal{H}$. It is remarkable that we can say anything at all about weak-representations.

THEOREM 8.2 (Renault's Disintegration Theorem). *Suppose that G is a second countable locally compact Hausdorff groupoid with a Haar system. Suppose that $L : C_c(G) \to \operatorname{Lin}(\mathcal{H}_0)$ is a nondegenerate weak-representation of $C_c(G)$ where $\mathcal{H}_0$ is a dense subspace of a separable Hilbert space $\mathcal{H}$. Then for each $f \in C_c(G)$, $L(f)$ is* *bounded and extends to an opertor* $\bar{L}(f)$ *on* $\mathcal{H}$ *of norm at most* $\|f\|_I$. *Hence* $\bar{L}$ *is a representation of* $C_c(G)$ *on* $\mathcal{H}$. *Moreover there is a unitary representation* $(\mu, G^{(0)} * \mathscr{H}, \hat{U})$ *of* G *such that* $\bar{L}$ *is equivalent to the integrated form of* $(\mu, G^{(0)} * \mathscr{H}, \hat{U})$ *on* $L^2(X * \mathscr{H}, \mu)$.

REMARK 8.3. An essential conclusion of Theorem 8.2 is that weak-representations L are automatically bounded and hence extend to bona fide representations. The assumption that the Hilbert space of L be separable is essential to exhibiting L as the integrated form of a (weak) unitary representation. However, if L maps into operators on a Hilbert space, we can dispense with the separability assumption.

COROLLARY 8.4. *Suppose that* G *is a second countable locally compact groupoid with a Haar system and that* $L : C_c(G) \to \operatorname{Lin}(\mathcal{H}_0)$ *is a weak-representation. Then* L *is* I*-norm bounded. In particular,* L *extends to a representation if it is nondegenerate.*

PROOF OF COROLLARY 8.4. It suffices to see that for all nonzero $h \in \mathcal{H}$ and $f \in C_c(G)$, we have $\|L(f)h\| \leq \|f\|_I\|h\|$. Let $\mathcal{V}$ be the closure of the cyclic subspace of $\mathcal{H}$

$$\mathcal{V}_0 := \{\, L(f)h : f \in C_c(G) \,\}$$

determined by h. By Proposition C.3 on page 344, there is a countable set $D \subset C_c(G)$ that given $f \in C_c(G)$ there is a sequence $\{f_i\} \subset D$ such that $f_i \to f$ uniformly with supports contained in a fixed compact set. Then

$$\begin{aligned}\|L(f_i)h - L(f)h\|^2 \\ &= \big(L(f_i^* * f_i)h \mid h\big) - \big(L(f_i^* * f)h \mid h\big) - \big(L(f^* * f_i)h \mid h\big) + \big(L(f)h \mid L(f)h\big)\end{aligned}$$

tends to zero. Hence $\mathcal{V}_0$ is separable as is $\mathcal{V}$. But the restriction of L to $\mathcal{V}$ is now a nondegenerate weak-representation on a separable Hilbert space. Hence $L|_{\mathcal{V}}$ is bounded and $\|L(f)h\| \leq \|f\|_I\|h\|$ as required. □

At the risk of being redundant, I feel it is important to emphasize the application of Corollary 8.4 to bona fide representations. (This should be compared with the corresponding result for crossed products, [**Wil07**, Lemma 2.45] which is considerably easier to prove.)

COROLLARY 8.5. *Let* G *be a second countable locally compact groupoid with a Haar system. Let* $L : C_c(G) \to B(\mathcal{H})$ *be a* $*$*-homomorphism that is continuous in the inductive limit topology. Then* L *is a representation and hence* $\|L(f)\| \leq \|f\|$ *for all* $f \in C_c(G)$.

With this corollary in hand, we can deal with our "modular worries" mentioned in Remark 1.37 on page 15 as well as Section 1.4.1 very efficiently. Let $G = G(\mathcal{G}, X) = G \rtimes X$ be the action groupoid associated to a locally compact transformation group $(\mathcal{G}, X)$. Let $C_c(\mathcal{G}, X)$ be $C_c(\mathcal{G} \times X)$ viewed as a subalgebra of the crossed product $C_0(X) \rtimes_{\mathrm{lt}} \mathcal{G}$. Then we get a $*$-algebra isomorphism $\Phi : C_c(G) \to C_c(\mathcal{G}, X)$ given by

$$\Phi(f)(g, x) = \Delta(g)^{-\frac{1}{2}} f(g, x). \tag{8.1}$$

We saw in Ex 1.4.11 that Φ is isometric with respect to the reduced norms on $C_c(G)$ and $C_c(\mathcal{G}, X)$, but we need Corollary 8.5 to get the same conclusion for the universal norms.

PROPOSITION 8.6. *Let $G = G(\mathcal{G}, X)$ as above. The map Φ defined in* (8.1) *is isometric for the both the reduced and universal C^*-norms. Hence Φ extends to both an isomorphism of $C^*(G)$ onto $C_0(X) \rtimes_{\mathrm{lt}} \mathcal{G}$ and an isomorphism of $C_r^*(G)$ onto $C_0(X) \rtimes_{\mathrm{lt},r} \mathcal{G}$.*

PROOF. As remarked above, we only have to deal with the universal norms. Let R be an isometric representation of $C_0(X) \rtimes_{\mathrm{lt}} \mathcal{G}$. Then $R \circ L$ is continuous in the inductive limit topology on $C_c(G)$. Hence

$$\|\Phi(f)\|_{C_0(X)\rtimes_{\mathrm{lt}}\mathcal{G}} = \|R(\Phi(f))\| \leq \|f\|_{C^*(G)}.$$

But representations of $C_c(\mathcal{G}, X)$ that are continuous in the inductive limit topology are necessarily bounded representations of $C_0(X) \rtimes_{\mathrm{lt}} \mathcal{G}$ by [**Wil07**, Lemma 2.45]. So we can reverse the argument by taking an isometric representation L of $C^*(G)$:

$$\|f\|_{C^*(G)} = \|L(f)\| = \|L \circ \Phi^{-1}(\Phi(f)\| \leq \|\Phi(f)\|_{C_0(X)\rtimes_{\mathrm{lt}}\mathcal{G}}.$$

This suffices. □

An Example. Before proceeding with the proof, it might be instructive to see what the Disintegration Theorem says about transformation group groupoids $G = G(\mathcal{G}, X)$ and compare with Section 1.4.2. If $L : C_c(\mathcal{G} \times X) \to B(\mathcal{H})$ is a representation of the the groupoid $G = \mathcal{G} \times X$, then we'll prove in the next section that L is the integrated form of a unitary representation $(\mu, X * \mathscr{H}, \hat{u})$. Recall that the modular function on G is given by $\Delta_G(s, x) = d(s, x)^{-1}\Delta(s)$ where Δ is the usual modular function on the group $\mathcal{G}$ and d is given by Corollary 1.59 on page 25. Therefore we may as well assume that $\mathcal{H} = L^2(X * \mathscr{H}, \mu)$ and hence

$$\begin{aligned}&\big(L(f)h \mid k\big)\\ &\quad= \int_X \int_{\mathcal{G}} f(s, x)\big(u(s, x)h(s^{-1} \cdot x) \mid k(x)\big)\, d\mu_{\mathcal{G}}(s) d(s, x)^{\frac{1}{2}} \Delta(s)^{-\frac{1}{2}}\, d\mu(x).\end{aligned}$$

We can let M be the multiplication representation of $C_0(X)$ on $\mathcal{H}$. Clearly, $x \mapsto d(s, x)^{\frac{1}{2}} u(s, x)h(s^{-1} \cdot x)$ is a Borel section of $X * \mathscr{H}$ and

$$\int_X \|d(s, x)^{\frac{1}{2}} u(s, x)h(s^{-1} \cdot x)\|^2\, d\mu(x) = \|h\|^2.$$

Hence we can define an isometry $\pi(s)$ on $L^2(X * \mathscr{H}, \mu)$ by

$$\pi(s)h(x) = d(s, x)^{\frac{1}{2}} u(s, x)h(s^{-1} \cdot x).$$

Since $d(sr, x) = d(s, x)d(r, s^{-1} \cdot x)$ for almost all x by Lemma 1.58 on page 24, we see that $\pi(sr) = \pi(s)\pi(r)$ and that $\pi(s^{-1}) = \pi(s)^{-1}$. Hence $\pi(s)$ is a unitary on $\mathcal{H}$. Since $s \mapsto \big(\pi(s)h \mid k\big)$ is Borel by Fubini, π is a unitary representation of G on $\mathcal{H}$ by [**Wil07**, Lemma D.43 and Theorem D.11]. It is straightforward to see that (M, π) is a covariant representation of $(C_0(X), G)$. Hence using Fubini,

$$\big(L(f)h \mid k\big) = \int_{\mathcal{G}} \big(M(f(s, \cdot)\pi(s)h \mid k\big)\Delta(s)^{-\frac{1}{2}}\, d\mu(s),$$

and we recover the $*$-homomorphism from (1.14) corresponding to (M, π). This justifies (and more) the assertions made in Section 1.4.2 where we showed "directly" that (1.14) defined a $\|\cdot\|_I$-bounded representation. It also shows that the arguments in Section 1.4.2 are essentially the same as invoking Proposition 8.6.

8.2. The Proof

Let L, $\mathcal{H}$, and $\mathcal{H}_0$ be as in the statement of Theorem 8.2 on page 169. Let

$$\mathcal{H}_{00} = \operatorname{span}\{\, L(f)h : f \in C_c(G) \text{ and } h \in \mathcal{H}_0 \,\}.$$

The measure μ. The first step in the proof will be to produce the measure μ that appears in the unitary representation. This is straightforward and is done in the next proposition. The real work will be to show that the measure is quasi-invariant.

PROPOSITION 8.7. *Suppose that $L : C_c(G) \to \operatorname{Lin}(\mathcal{H}_0)$ is a weak-representation as above. Then there is a representation $M : C_0(G^{(0)}) \to B(\mathcal{H})$ such that for all $\varphi \in C_0(G^{(0)})$, $f \in C_c(G)$ and $h \in \mathcal{H}_0$ we have*

$$M(\varphi)L(f)h = L\big((\varphi \circ r) \cdot f\big)h. \tag{8.2}$$

In particular, after replacing L by an equivalent weak representation, we may assume that $\mathcal{H} = L^2(G^{(0)} * \mathscr{V}, \mu)$ for a Borel Hilbert bundle $G^{(0)} * \mathscr{V}$ and a finite Radon measure μ on $G^{(0)}$ such that

$$M(\varphi)\xi(u) = \varphi(u)h(u) \quad \textit{for all } \varphi \in C_0(G^{(0)}) \textit{ and } h \in L^2(G^{(0)} * \mathscr{V}, \mu).$$

PROOF. We can easily make sense of $(\varphi \circ r) \cdot f$ for $\varphi \in C_b(G^{(0)})$. Furthermore, we can compute that

$$\big(L((\varphi \circ r) \cdot f)h \mid L(g)k\big) = \big(L(f)h \mid L((\bar{\varphi} \circ r) \cdot g)k\big).$$

Then, if $\psi \in C_b^+(G^{(0)})$ is such that

$$\|\varphi\|_\infty^2 1 - |\varphi|^2 = \psi^2,$$

we can compute that

$$\begin{aligned}
\|\varphi\|_\infty^2 \Big\|\sum_{i=1}^n L(f_i)h_i\Big\|^2 - \Big\|\sum_{i=1}^n L\big((\varphi \circ r) \cdot f_i\big)h_i\Big\|^2 \\
&= \sum_{ij} \big(L\big(\big((\|\varphi\|_\infty^2 1 - |\varphi|^2\big) \circ r\big) \cdot f_i\big)h_i | L(f_j)h_j\big) \\
&= \Big\|\sum_{i=1}^n L\big((\psi \circ r) \cdot f_i\big)h_i\Big\|^2 \\
&\geq 0
\end{aligned}$$

Since $\mathcal{H}_{00}$ is dense in $\mathcal{H}$, it follows that there is a well-defined bounded operator $M(\varphi)$ on all of $\mathcal{H}$ satisfying (8.2). It is not hard to see that M is a $*$-homomorphism. To see that M is a representation, by convention, we must also see that M is nondegenerate. But if $f \in C_c(G)$, then $r\big(\operatorname{supp} f\big)$ is compact in $G^{(0)}$. Hence there is a $\varphi \in C_0^+(G^{(0)})$ such that $M(\varphi)f = f$. From this, it is straightforward to see that M is nondegenerate and therefore a representation.

Since M is a representation of $C_0(G^{(0)})$, it is equivalent to a multiplication representation on $L^2(G^{(0)} * \mathscr{V}, \mu)$ for an appropriate Borel Hilbert bundle $G^{(0)} * \mathscr{V}$ and finite Radon measure μ—for example, see [**Wil07**, Example F.25]. The second assertion follows, and this completes the proof. □

LEMMA 8.8. *If $\mathcal{H}'_{00}$ is a dense subspace of $\mathcal{H}_{00}$, then*

$$\operatorname{span}\{\, L(f)h : f \in C_c(G) \text{ and } h \in \mathcal{H}'_{00} \,\}$$

is dense in $\mathcal{H}$.

PROOF. By Proposition 1.49 on page 20, there is a self-adjoint approximate identity $\{\, e_i \,\}$ for $C_c(G)$ in the inductive limit topology. Then if $L(f)h \in \mathcal{H}_{00}$, we see that

$$\|L(e_i)L(f)h - L(f)h\|^2 = \\ \big(L(f^* * e_i * e_i * f)h \mid h\big) - 2\operatorname{Re}\big(L(f^* * e_i * f)h \mid h\big) + \big(L(f^* * f)h \mid h\big),$$

which tends to zero since L is continuous in the inductive limit topology (by part (b) of Definition 8.1 on page 169). It follows that $\mathcal{H}'_{00} \subset \overline{\operatorname{span}}\{\, L(f)h : h \in \mathcal{H}'_{00} \text{ and } f \in C_c(G) \,\}$. Since $\mathcal{H}'_{00}$ is dense, the result follows. □

Forms on $C_c(G) \odot \mathcal{H}_0$. The key to Renault's proof, which we are following here, is to realize $\mathcal{H}$ as the completion of (a quotient of) the algebraic tensor product $C_c(G) \odot \mathcal{H}_0$. Then we take advantage of the natural fibring of $C_c(G) \odot \mathcal{H}_0$ over $G^{(0)}$.

LEMMA 8.9. *There is a positive sesquilinear form $\langle \cdot \,, \cdot \rangle$ on $C_c(G) \odot \mathcal{H}_0$ such that*

$$\langle f \otimes h \,, g \otimes k \rangle = \big(L(g^* * f)h \mid k\big). \tag{8.3}$$

Furthermore, the Hilbert space completion $\mathcal{K}$ of $C_c(G) \odot \mathcal{H}_0$ with respect to this form is isomorphic to $\mathcal{H}$. If $[f \otimes h]$ is the class of $f \otimes h$ in $\mathcal{K}$, then $[f \otimes h] \mapsto L(f)h$ is well-defined and induces an isomorphism of $\mathcal{K}$ with $\mathcal{H}$ which maps the quotient $C_c(G) \odot \mathcal{H}_0/\mathcal{N}$ onto $\mathcal{H}_{00}$ where $\mathcal{N}$ is the subspace $\mathcal{N} = \{\, \sum_i f_i \otimes h_i : \sum_i L(f_i)h_i = 0 \,\}$ of vectors in $C_c(G) \odot \mathcal{H}_0$ of length zero.

PROOF. Using the universal properties of the *algebraic* tensor product, as in the proof of [**RW98**, Proposition 2.64] for example, it is not hard to see that there is a unique sesquilinear form on $C_c(G) \odot \mathcal{H}_0$ satisfying (8.3) (Ex 8.2.1). Thus to see that $\langle \cdot \,, \cdot \rangle$ is a pre-inner product, we just have to see that it is positive. But

$$\begin{aligned} \Big\langle \sum_i f_i \otimes h_i \,, \sum_i f_i \otimes h_i \Big\rangle &= \sum_{ij} \big(L(f_j^* * f_i)h_i \mid h_j\big) \\ &= \sum_{ij} \big(L(f_i)h_i \mid L(f_j)h_i\big) \\ &= \Big\|\sum_i L(f_i)h_i\Big\|^2. \end{aligned} \tag{8.4}$$

As in [**RW98**, Lemma 2.16], $\langle \cdot \,, \cdot \rangle$ defines an inner-product on $C_c(G) \odot \mathcal{H}_0/\mathcal{N}$, and $[f_i \otimes h] \mapsto L(f_i)h$ is well-defined in view of (8.4). Since this map has range $\mathcal{H}_{00}$ and since $\mathcal{H}_{00}$ is dense in $\mathcal{H}$ by assumption, the map extends to an isomorphism of $\mathcal{K}$ onto $\mathcal{H}$ as claimed. □

Using Lemma 8.9, we will normally identify $\mathcal{H}$ with $\mathcal{K}$, and $\mathcal{H}_{00}$ with $C_c(G) \odot \mathcal{H}_0/\mathcal{N}$. Thus we will interpret $[f \otimes h]$ as a vector in $\mathcal{H}_{00} \subset \mathcal{H}_0 \subset \mathcal{H}$. Then we have

$$L(g)[f \otimes h] = [g * f \otimes h] \quad \text{and} \tag{8.5}$$

$$M(\varphi)[f \otimes h] = [(\varphi \circ r) \cdot f \otimes h], \tag{8.6}$$

where M is the representation of $C_0(G^{(0)})$ defined in Proposition 8.7 on page 172, $g \in C_c(G)$ and $\varphi \in C_0(G^{(0)})$.

REMARK 8.10. Using Proposition 8.7 on page 172, we can extend M to a $*$-homomorphism of $\mathcal{B}_c^b(G)$ into $B(\mathcal{H})$ such that $M(h) = 0$ if $h(u) = 0$ for μ-almost all u (where μ is the measure defined in that proposition). However, at this point, we can not assert that (8.6) holds for any $\varphi \notin C_0(G^{(0)})$.

Enlarging the domain of L. Showing that μ is quasi-invariant requires that we extend the operators L and M in equations (8.5) and (8.6) to a larger class of functions. A major obstruction is that we don't as yet know that each $L(f)$ is bounded in any sense. Furthermore, $\mathcal{H}_0$ is not complete. We'll introduce only those functions we absolutely need.

DEFINITION 8.11. Let $\mathcal{B}_c^1(G)$ be the collection of bounded Borel functions on G which are the pointwise limit of a uniformly bounded sequence $\{ f_n \} \subset C_c(G)$ such that there is a compact set K such that $\operatorname{supp} f_n \subset K$ for large n.

Note that $\mathcal{B}_c^1(G)$ is not a very robust class of functions on G. For example, it is generally not closed under the type of convergence used in its definition. Nevertheless, its elements are all integrable with respect to any Radon measure on G, and the following lemma is an easy consequence of the dominated convergence theorem applied to the total variation measure. (Complex Radon measures are discussed in Section 3.3.)

LEMMA 8.12. *Suppose that σ is a complex Radon measure on $C_c(G)$ such that*

$$\sigma(f) = \int_G f(\gamma)\rho(\gamma)\, d|\sigma|(\gamma) \tag{8.7}$$

for a unimodular function ρ and total variation $|\sigma|$. Then σ extends to a linear functional on $\mathcal{B}_c^1(G)$ such that (8.7) holds and such that if $\{ f_n \}$ is a uniformly bounded sequence in $\mathcal{B}_c^1(G)$ converging pointwise to $f \in \mathcal{B}_c^1(G)$ with supports eventually contained in a fixed compact set, then $\sigma(f_n) \to \sigma(f)$.

SKETCH OF THE PROOF. Since $|\sigma|$ is a Radon measure, (8.7) makes good sense for any $f \in \mathcal{B}_c^1(G)$. Thus σ extends as claimed. The rest is an easy consequence of the dominated convergence theorem applied to $|\sigma|$. □

COROLLARY 8.13. *If $f, g \in \mathcal{B}_c^1(G)$, then*

$$f * g(\gamma) := \int_G f(\eta)g(\eta^{-1}\gamma)\, d\lambda^{r(\gamma)}(\eta)$$

defines an element $f * g$ of $\mathcal{B}_c^1(G)$.

PROOF. Let $\{ f_n \}$ and $\{ g_n \}$ be uniformly bounded sequences in $C_c(G)$ such that $f_n \to f$ and $g_n \to g$ pointwise and with supports contained in a fixed compact set. Since

$$|f_n * g_n(\gamma)| \le \|f_n\|_\infty \|g_n\|_\infty \sup_{u \in G^{(0)}} \lambda^u\big((\operatorname{supp} f_n)(\operatorname{supp} g_n)\big),$$

it follows that $\{ f_n * g_n \}$ is a uniformly bounded sequence in $C_c(G)$, all of whose supports are in a fixed compact set, converging pointwise to $f * g$. Thus $f * g \in \mathcal{B}_c^1(G)$. □

In view of the continuity assumption on L, we can define a complex Radon measure $L_{h,k}$ on $C_c(G)$ via

$$L_{h,k}(f) := \big(L(f)h \mid k\big) \tag{8.8}$$

for each h and k in $\mathcal{H}_0$. By Lemma 8.12 on the preceding page we can extend each $L_{h,k}$ to a linear functional on all of $\mathcal{B}^1_c(G)$.

LEMMA 8.14. *There is a positive sesquilinear form on $\mathcal{B}^1_c(G) \odot \mathcal{H}_0$, extending that on $C_c(G) \odot \mathcal{H}_0$, such that*

$$\langle f \otimes h \,,\, g \otimes k \rangle = L_{h,k}(g^* * f) \quad \text{for all } f, g \in \mathcal{B}^1_c(G) \text{ and } h, k \in \mathcal{H}_0.$$

In particular, if

$$\mathcal{N}_b := \Big\{ \sum_i f_i \otimes h_i \in \mathcal{B}^1_c(G) \odot \mathcal{H}_0 : \big\langle \sum_i f_i \otimes h_i \,,\, \sum_i f_i \otimes h_i \big\rangle = 0 \Big\}$$

is the subspace of vectors of zero length, then the quotient $(\mathcal{B}^1_c(G) \odot \mathcal{H}_0)/\mathcal{N}_b$ can be identified with a subspace of $\mathcal{H}$ containing $\mathcal{H}_{00} := C_c(G) \odot \mathcal{H}_0/\mathcal{N}$.

REMARK 8.15. As before, we will write $[f \otimes h]$ for the class of $f \otimes h$ in the quotient $(\mathcal{B}^1_c(G) \odot \mathcal{H}_0)/\mathcal{N}_b \subset \mathcal{H}$.

PROOF. Just as in Lemma 8.9 on page 173, there is a well-defined sesquilinear form on $\mathcal{B}^1_c(G) \odot \mathcal{H}_0$ extending that on $C_c(G) \odot \mathcal{H}_0$. (Note that the right-hand side of (8.3) can be rewritten as $L_{h,k}(g^* * f)$.) In particular, we have

$$\big\langle \sum_i f_i \otimes h_i \,,\, \sum_j g_j \otimes k_j \big\rangle = \sum_{ij} L_{h_i,k_j}(g_j^* * f_i).$$

We need to see that the form is positive. Let $\alpha := \sum_i f_i \otimes h_i$, and let $\{ f_{i,n} \}$ be a uniformly bounded sequence in $C_c(G)$ converging pointwise to f_i with all the supports contained in a fixed compact set. Then for each i and j, $f_{j,n}^* * f_{i,n} \to f_j^* * f_i$ in the appropriate sense. In particular, Lemma 8.12 on the preceding page implies that

$$\begin{aligned} \langle \alpha \,,\, \alpha \rangle &= \sum_{ij} L_{h_i,h_j}(f_j^* * f_i) \\ &= \lim_n \sum_{ij} L_{h_i,h_j}(f_{j,n}^* * f_{i,n}) \\ &= \lim_n \langle \alpha_n \,,\, \alpha_n \rangle, \end{aligned}$$

where $\alpha_n := \sum_i f_{i,n} \otimes h_i$. Since $\langle \cdot \,,\, \cdot \rangle$ is positive on $C_c(G) \odot \mathcal{H}_0$ by Lemma 8.9 on page 173, we have $\langle \alpha_n \,,\, \alpha_n \rangle \geq 0$, and we've shown that $\langle \cdot \,,\, \cdot \rangle$ is still positive on $\mathcal{B}^1_c(G) \odot \mathcal{H}_0$.

Clearly the map sending the class $f \otimes h + \mathcal{N}$ to $f \otimes h + \mathcal{N}_b$ is isometric and therefore extends to an isometric embedding of $\mathcal{H}$ into the Hilbert space completion $\mathcal{H}_b$ of $\mathcal{B}^1_c(G) \odot \mathcal{H}_0$ with respect to $\langle \cdot \,,\, \cdot \rangle$. However if $g \otimes h \in \mathcal{B}^1_c(G) \odot \mathcal{H}_0$ and if $\{ g_n \}$ is a sequence in $C_c(G)$ such that $g_n \to g$ in the usual way, then

$$\|(g_n \otimes h + \mathcal{N}_b) - (g \otimes h + \mathcal{N}_b)\|^2 = L_{h,h}(g_n^* * g_n - g_n^* * g - g * g_n^* + g^* * g),$$

and this tends to zero by Lemma 8.12 on the preceding page. Thus the image of $\mathcal{H}$ in $\mathcal{H}_b$ is all of $\mathcal{H}_b$. Consequently, we can identify the completion of $\mathcal{B}^1_c(G) \odot \mathcal{H}_0$ with $\mathcal{H}$ and $\mathcal{B}^1_c(G) \odot \mathcal{H}_0/\mathcal{N}_b$ with a subspace of $\mathcal{H}$ containing $\mathcal{H}_{00}$. □

The "extra" vectors provided by $\mathcal{B}_c^1(G) \odot \mathcal{H}_0/\mathcal{N}_b$ are just enough to allow us to use a bit of general nonsense about unbounded operators to extend the domain of each $L(f)$. More precisely, for $f \in C_c(G)$, we can view $L(f)$ as an operator in $\mathcal{H}$ with domain $\mathcal{D}(L(f)) = \mathcal{H}_{00}$. Then using part (a) of Definition 8.1 on page 169, we see that

$$L(f^*) \subset L(f)^*.$$

This implies that $L(f)^*$ is a densely defined operator. Hence $L(f)$ is closable [**Con85**, Proposition X.1.6]. Consequently, the closure of the graph of $L(f)$ in $\mathcal{H} \times \mathcal{H}$ is the graph of the closure $\overline{L(f)}$ of $L(f)$ [**Con85**, Proposition X.1.4].

Suppose that $g \in \mathcal{B}_c^1(G)$. Let $\{ g_n \}$ be a uniformly bounded sequence in $C_c(G)$ all with supports in a fixed compact set such that $g_n \to g$ pointwise. Then

$$\|[g_n \otimes h] - [g \otimes h]\|^2 = L_{h,h}(g_n^* * g_n - g^* * g_n - g_n^* * g + g * g). \tag{8.9}$$

However $\{ g_n^* * g_n - g^* * g_n - g_n^* * g + g * g \}$ is uniformly bounded and converges pointwise to zero. Since the supports are all contained in a fixed compact set, the left-hand side of (8.9) tends to zero by Lemma 8.12 on page 174. Similarly,

$$\|[f * g_n \otimes h] - [f * g \otimes h]\|^2 \to 0.$$

If follows that

$$\big([g_n \otimes h], L(f)[g_n \otimes h]\big) \to \big([g \otimes h], [f * g \otimes h]\big)$$

in $(\mathcal{B}_c^1(G) \odot \mathcal{H}_0/\mathcal{N}_b) \times (\mathcal{B}_c^1(G) \odot \mathcal{H}_0/\mathcal{N}_b) \subset \mathcal{H} \times \mathcal{H}$. Therefore $[g \otimes h] \in \mathcal{D}\big(\overline{L(f)}\big)$ and $\overline{L(f)}[g \otimes h] = [f * g \otimes h]$. We have proved the following.

LEMMA 8.16. *For each $f \in C_c(G)$, let $L(f)$ be the operator in $\mathcal{H}$ with domain $\mathcal{D}(L(f)) = \mathcal{H}_{00} = C_c(G) \odot \mathcal{H}_0/\mathcal{N}$. Then $L(f)$ is closable, $\mathcal{B}_c^1(G) \odot \mathcal{H}_0/\mathcal{N}_b \subset \mathcal{D}\big(\overline{L(f)}\big)$, and*

$$\overline{L(f)}[g \otimes h] = [f * g \otimes h] \quad \text{for all } f \in C_c(G),\ g \in \mathcal{B}_c^1(G) \text{ and } h \in \mathcal{H}_0.$$

Now can extend L a bit.

LEMMA 8.17. *For each $f \in \mathcal{B}_c^1(G)$, there is a well-defined operator $L_b(f) \in \operatorname{Lin}(\mathcal{B}_c^1(G) \odot \mathcal{H}_0)/\mathcal{N}_b)$ such that*

$$L_b(f)[g \otimes h] = [f * g \otimes h]. \tag{8.10}$$

If $f \in C_c(G)$, then $L_b(f) \subset \overline{L(f)}$.

PROOF. To see that (8.10) determines a well-defined operator, we need to see that

$$\sum_i [g_i \otimes h_i] = 0 \quad \text{implies} \quad \sum_i [f * g_i \otimes h_i] = 0. \tag{8.11}$$

However,

$$\Big\|\sum_i [f * g_i \otimes h_i]\Big\|^2 = \sum_{ij} L_{h_i,h_j}(g_j^* * f^* * f * g_i). \tag{8.12}$$

Since $f \in \mathcal{B}_c^1(G)$, we can approximate the right-hand side of (8.12) by sums of the form

$$\sum_{ij} L_{h_i,h_j}(g_j^* * f_0^* * f_0 * g_i), \tag{8.13}$$

where $f_0 \in C_c(G)$. But (8.13) equals

$$\left\|\overline{L(f_0)}\left(\sum_i [g_i \otimes h_i]\right)\right\|^2$$

which is zero if the left-hand side of (8.11) is zero. Hence the right-hand side of (8.11) is also zero and $L_b(f)$ is well-defined.

If $f \in C_c(G)$, then $L_b(f) \subset \overline{L(f)}$ by Lemma 8.16 on the facing page. □

Technicalities. The previous gymnastics have allowed us to produce some additional vectors in the domain of each $L(f)$ and to extend slightly the domain of L. The next lemma provides the technical assurances that, despite the subtle definitions above, our new operators act via the formulas we expect.

We let $\mathcal{B}_c^1(G^{(0)})$ be the bounded Borel functions on $G^{(0)}$ which are the pointwise limit of a uniformly bounded sequence from $C_c(G^{(0)})$.

LEMMA 8.18. *Suppose that $f \in \mathcal{B}_c^1(G)$ and that $\psi \in \mathcal{B}_c^1(G^{(0)})$. Then for all $g_1, g_2 \in C_c(G)$ and $h, k \in \mathcal{H}_0$, we have the following.*

(a)
$$\begin{aligned}\big(L_b(f)[g_1 \otimes h] \mid [g_2 \otimes k]\big) &= \big([f * g_1 \otimes h] \mid [g_2 \otimes k]\big)\\ &= L_{h,k}(g_2^* * f * g_1)\\ &= L_{[g_1 \otimes h],[g_2 \otimes k]}(f)\end{aligned}$$

(b)
$$\begin{aligned}\big(M(\psi)[g_1 \otimes h] | [g_2 \otimes k]\big) &= L_{h,k}(g_2^* * ((\psi \circ r) \cdot g_1))\\ &= \big([(\psi \circ r) \cdot g_1 \otimes h] \mid [g_2 \otimes k]\big)\\ &= \big(M(\psi)L(g_1)h \mid L(g_2)k\big)\end{aligned}$$

(c)
$$\big(M(\psi)L_b(f)[g_1 \otimes h] \mid [g_2 \otimes k]\big) = \big(L_b((\psi \circ r) \cdot f)[g_1 \otimes h] \mid [g_2 \otimes k]\big).$$

PROOF. We start with part (a). The first equality is just the definition of $L_b(f)$. The second follows from the definition of the inner product on $\mathcal{B}_c^1(G) \odot \mathcal{H}_0/\mathcal{N}_b$. If f is in $C_c(G)$, then the third equation holds just by untangling the definition of the complex Radon measure $L_{h,k}$. The third equality holds for $f \in \mathcal{B}_c^1(G)$ by applying the continuity assertion in Lemma 8.12 on page 174.

Part (b) is proved similarly. The first equation holds if $\psi \in C_0(G^{(0)})$ by definition of $M(\psi)$ and $L_{h,k}$. If $\{\psi_n\} \subset C_0(G^{(0)})$ is a bounded sequence converging pointwise to ψ, then $M(\psi_k) \to M(\psi)$ in the weak operator topology by the dominated convergence theorem. On the other hand $g_2^* * (\psi_n \circ r) \cdot g_1 \to g_2^* * (\psi \circ r) g_1$ in the required way. Thus $L_{h,k}(g_2^* * (\psi_n \circ r) \cdot g_1) \to L_{h,k}(g_2^* * (\psi \circ r) g_1)$ by Lemma 8.12 on page 174. Thus the first equality is valid. The second equality is clear if $\psi \in C_0(G^{(0)})$ and passes to the limit as above. The third equality is simply our identification of $[g \otimes h]$ with $L(g)h$ as in Lemma 8.9 on page 173.

For part (c), first note that if $f_n \to f$ and $\psi_n \to \psi$ are uniformly bounded sequences converging pointwise with supports in fixed compact sets independent of n, then $(\psi \circ r) \cdot f = \lim_n (\psi_n \circ r) \cdot f_n$. It follows that $(\psi \circ r) \cdot f \in \mathcal{B}_c^1(G)$. Also, $[f \otimes h] = \lim [f_n \otimes h]$, and since $M(\psi)$ is bounded, part (b) implies that

$$\begin{aligned}M(\psi)[f \otimes h] &= \lim_n M(\psi)[f_n \otimes h]\\ &= \lim_n [(\psi \circ r) \cdot f_n \otimes h]\\ &= [(\psi \circ r) \cdot f \otimes h].\end{aligned}$$

Since it is not hard to verify that $M(\psi)^*[f \otimes h] = (\bar{\psi} \circ r) \cdot f \otimes h]$, we can compute that

$$\begin{aligned}\big(M(\psi)L_b(f)[g_1 \otimes h] \mid [g_2 \otimes k]\big) &= \big([f * g_1 \otimes h] \mid (\bar{\psi} \circ r) \cdot g_2 \otimes k]\big) \\ &= \big([\psi \circ r) \cdot (f * g_1) \otimes h] \mid [g_2 \otimes k]\big) \\ &= \big([((\psi \circ r) \cdot f) * g_1 \otimes h] \mid [g_2 \otimes k]\big) \\ &= \big(L_b((\psi \circ r) \cdot f)[g_1 \otimes h] \mid [g_2 \otimes k]\big). \qquad \square\end{aligned}$$

The quasi-invariance of μ.

PROPOSITION 8.19. *Let μ be the Radon measure on $G^{(0)}$ associated to L by Proposition 8.7 on page 172. Then μ is quasi-invariant.*

PROOF. We need to show that measures $\nu = \mu \circ \lambda$ and and its push forward under inversion ν^{-1} (defined in (7.1) and (7.2), respectively) are equivalent. Therefore, we have to show that if A is pre-compact in G, then $\nu(A) = 0$ if and only if $\nu(A^{-1}) = 0$. Since $(A^{-1})^{-1} = A$, it's enough to show that A ν-null implies that A^{-1} is too. Since ν is regular, we may as well assume that A is a G_δ-set. Let $\tilde{f}(x) = f(x^{-1})$. We need to show that $\tilde{f}(x) = 0$ for ν-almost all x. Since A is a G_δ, we can find a sequence $\{ f_n \} \subset C_c^+(G)$ such that $f_n \searrow f$. In particular, $f := \mathbb{1}_A$ is in $\mathcal{B}_c^1(G)$.

If d is any function in $C_c(G)$, then $df\bar{d} = |d|^2 f \in \mathcal{B}_c^1(G)$ and vanishes ν-almost everywhere. By the monotone convergence theorem,

$$\lambda(df\bar{d})(u) := \int_G |d(\gamma)|^2 f(\gamma)\, d\lambda^u(\gamma)$$

defines a function in $\mathcal{B}_c^1(G^{(0)})$ which is equal to 0 for μ-almost all u. In particular, $M(\lambda(df\bar{d})) = 0$.

It then follows from part (b) of Lemma 8.18 that

$$0 = \big(M\big(\lambda(df\bar{d})\big)L(g)h \mid L(g)h\big) = L_{h,h}\big(g^* * \big(\lambda(df\bar{d}) \circ r\big) \cdot g\big) \tag{8.14}$$

for all $g, d \in C_c(G)$ and $h \in \mathcal{H}_0$. On the other hand, if (8.14) holds with f replaced by $\tilde{f}$ for all $g, d \in C_c(G)$ and $h \in \mathcal{H}_0$, then we must have $M\big(\lambda(d\tilde{f}\bar{d})\big) = 0$ for all $d \in C_c(G)$. Since $\tilde{f}(\gamma) \geq 0$ everywhere, this forces $|d(\gamma)|^2 \tilde{f}(\gamma) = 0$ for ν-almost all γ. Since d is arbitrary, we conclude that $\tilde{f}(\gamma) = 0$ for ν-almost all γ. Therefore it will suffice to show that

$$L_{h,h}\big(g^* * \big(\lambda(d\tilde{f}\bar{d}) \circ r\big) \cdot g\big) = 0 \quad \text{for all } g, d \in C_c(G) \text{ and } h \in \mathcal{H}_0. \tag{8.15}$$

First, we compute that with f in (8.14) we have

$$\begin{aligned}g^* * \big(\lambda(df\bar{d}) \circ r\big) \cdot g(\sigma) &= \int_G \overline{g(\gamma^{-1})}\big(\lambda(df\bar{d}) \circ r\big) \cdot g(\gamma^{-1}\sigma)\, d\lambda^{r(\sigma)}(\gamma) \\ &= \int_G \overline{g(\gamma^{-1})}\lambda(df\bar{d})\big(s(\gamma)\big)g(\gamma^{-1}\sigma)\, d\lambda^{r(\sigma)}(\gamma) \\ &= \int_G \int_G \overline{g(\gamma^{-1})d(\eta)}f(\eta)d(\eta)g(\gamma^{-1}\sigma)\, d\lambda^{s(\gamma)}(\eta)\, d\lambda^{r(\sigma)}(\gamma)\end{aligned}$$

which, after sending $\eta \mapsto \gamma^{-1}\eta$ and using left-invariance of the Haar system, is

$$= \int_G \int_G \overline{g(\gamma^{-1})d(\gamma^{-1}\eta)}f(\gamma^{-1}\eta)d(\gamma^{-1}\eta)g(\gamma^{-1}\sigma)\, d\lambda^{r(\sigma)}(\eta)\, d\lambda^{r(\sigma)}(\gamma)$$

which, after defining $F(\gamma,\eta) := d(\gamma^{-1}\eta)g(\gamma^{-1})$ and $f\cdot F(\gamma,\eta) := f(\gamma^{-1}\eta)F(\gamma,\eta)$ for $(\gamma,\eta)\in G *_r G$, is

$$(8.16)\qquad = \int_G\int_G \overline{F(\gamma,\eta)}f\cdot F(\sigma^{-1}\gamma,\sigma^{-1}\eta)\,d\lambda^{r(\sigma)}(\eta)\,d\lambda^{r(\sigma)}(\gamma).$$

We will have to look at integrals of the form (8.16) in some detail. Note that if $g,d\in C_c(G)$, then $F(\gamma,\eta) := d(\gamma^{-1}\eta)g(\gamma^{-1})$ defines an element $F\in C_c(G *_r G)$.

LEMMA 8.20. *Suppose that $F_1,F_2\in C_c(G *_r G)$. Then*

$$\sigma\mapsto \int_G\int_G \overline{F_1(\gamma,\eta)}F_2(\sigma^{-1}\gamma,\sigma^{-1}\eta)\,d\lambda^{r(\sigma)}(\eta)\,d\lambda^{r(\sigma)}(\gamma)$$

defines an element of $C_c(G)$ which we denote by $\overline{F}_1 *_{\lambda*\lambda} F_2$.

PROOF. Note that

$$\|\overline{F}_1 *_{\lambda*\lambda} F_2\|_\infty \le \|F_1\|_\infty\|F_2\|_\infty \sup_{u\in G^{(0)}} \lambda^u(K_1)\lambda^u(K_2)$$

whenever $\operatorname{supp} F_1\subset K_1 *_r K_2$. Thus to see that $\overline{F}_1 *_{\lambda*\lambda} F_2\in C_c(G *_r G)$, it will suffice to consider only those F_i in a subset that spans a dense subspace in the inductive limit topology. Hence we may as well assume that $F_i(\gamma,\eta) = d_i(\eta)g_i(\gamma^{-1})$ with d_i and g_i in $C_c(G)$. But then

$$\overline{F}_1 *_{\lambda*\lambda} F_2(\sigma) = \overline{d}_1 * \tilde{d}_2(\sigma)g_1^* * g_2(\sigma),$$

and the result follows. □

LEMMA 8.21. *Functions of the form*

$$(\gamma,\eta)\mapsto d(\gamma^{-1}\eta)g(\gamma^{-1})\quad\text{with } d,g\in C_c(G)$$

span a dense subspace of $C_c(G *_r G)$ in the inductive limit topology.

PROOF. This follows from the Stone-Weierstrass Theorem. □

Let $\mathcal{A}\subset C_c(G *_r G)$ be the dense subspace of functions of the form considered in Lemma 8.21. We continue to reserve f for the characteristic function of our fixed pre-compact, ν-null set. Then we know from (8.14) that

$$(8.17)\qquad L_{h,h}\big(\overline{F} *_{\lambda*\lambda} (f\cdot F)\big) = 0\quad\text{for all } F\in\mathcal{A}.$$

It is not hard to check that, if $f'\in\mathcal{B}_c^1(G)$, then $F *_{\lambda*\lambda} (f'\cdot F)\in\mathcal{B}_c^1(G)$ and that if $F_n\to F$ in the inductive limit topology in $C_c(G *_r G)$, then $\{\,\overline{F}_n *_{\lambda*\lambda} (f'\cdot F_n)\,\}$ is uniformly bounded and converges pointwise to $\overline{F} *_{\lambda*\lambda} (f'\cdot F)$. In particular the continuity of the $L_{h,h}$ (see Lemma 8.12 on page 174) implies that (8.17) holds for all $F\in C_c(G *_r G)$. But if we define $\tilde{F}(x,y) := F(y,x)$, then we see from the definition that

$$\overline{\tilde{F}} *_{\lambda*\lambda} (f\cdot\tilde{F}) = \overline{F} *_{\lambda*\lambda} (\tilde{f}\cdot F)),$$

where we recall that $\tilde{f}(x) := f(x^{-1})$. Thus

$$L_{h,h}\big(\overline{F} *_{\lambda*\lambda} (\tilde{f}\cdot F))\big) = 0\quad\text{for all } F\in C_c(G *_r G).$$

Since the above holds in particular for $F\in\mathcal{A}$, this implies (8.15). Hence μ is quasi-invariant. □

The Borel Hilbert Bundle. Recall from Section 3.3 that a Complex Radon measure α is absolutely continuous with respect to a Radon measure ν if whenever $\nu(B) = 0$ for a Borel set B, then B is locally α-null. That is, $\alpha(E) = \alpha(\mathbb{1}_E) = 0$ for every pre-compact Borel subset $E \subset B$. To define the Borel Hilbert bundle for the unitary representation we need to construct, we will need to see that the complex Radon measures $L_{h,k}$ defined in (8.8) are absolutely continuous with respect to the measure $\nu = \mu \circ \lambda$. In order to prove this, we need to restrict h and k to lie in $\mathcal{H}_{00}$ and to employ Lemma 8.18 on page 177.

LEMMA 8.22. *Let a and b be vectors in $\mathcal{H}_{00}$ (identified with $C_c(G) \odot \mathcal{H}_0/\mathcal{N}$). Let $L_{a,b}$ be the complex Radon measure given by*

$$L_{a,b}(f) := \big(L(f)a \mid b\big) \quad \text{for } f \in C_c(G).$$

Then $L_{a,b}$ is absolutely continuous with respect to the measure $\nu = \mu \circ \lambda$ defined in (7.1).

PROOF. It is enough to show that if M is a pre-compact ν-null set and if $f := \mathbb{1}_M$, then $L_{a,b}(f) = 0$. Since ν is a Radon measure, and therefore regular, we may as well assume that M is a G_δ-set. Then $f \in \mathcal{B}_c^1(G)$.

Since M is ν-null,

$$0 = \int_{G^{(0)}} \int_G f(\gamma)\, d\lambda^u(\gamma)\, d\mu(u),$$

and there is a μ-null set $N \subset G^{(0)}$ such that $\lambda^u(M \cap G^u) = 0$ if $u \notin N$. As above, we can assume that N is a G_δ set. Then for any $g \in C_c(G)$, we have

$$f * g(\gamma) = \int_G f(\eta) g(\eta^{-1}\gamma)\, d\lambda^{r(\gamma)}(\eta) = 0$$

whenever $r(\gamma) \notin N$. Since $\operatorname{supp} \lambda^{r(\gamma)} = G^{r(\gamma)}$, it follows that for all $\gamma \in G$ (without exception),

$$f * g(\gamma) = \mathbb{1}_N\big(r(\gamma)\big) f * g(\gamma) = \big((\mathbb{1}_N \circ r) \cdot f\big) * g(\gamma). \tag{8.18}$$

Since $a, b \in \mathcal{H}_{00}$, it suffices to consider $a = [g \otimes h]$ and $b = [d \otimes k]$ (with $g, d \in C_c(G)$ and $h, k \in \mathcal{H}_0$). Note that f and $\mathbb{1}_N$ satisfy the hypotheses of Lemma 8.18 on page 177. Therefore, by part (a) of that lemma,

$$L_{[g\otimes h, d\otimes k]}(f) = \big([f * g \otimes h] \mid [d \otimes k]\big)$$

which, by (8.18), is

$$= \big([((\mathbb{1}_N \circ r) \cdot f) * g \otimes h] \mid [d \otimes k]\big)$$

which, by part (a) of Lemma 8.18, is

$$= \big(L_b((\mathbb{1}_N \circ r) \cdot f)[g \otimes h] \mid [d \otimes k]\big)$$

which, by part (c) of Lemma 8.18, is

$$= \big(M(\mathbb{1}_N) L_b(f)[g \otimes h] \mid [d \otimes k]\big).$$

Since $M(\mathbb{1}_N) = 0$, the last inner product is zero as desired. This completes the proof. □

Since Δ is nowhere zero, the measures ν and $\nu_0 := \Delta(\gamma)^{\frac{1}{2}}\nu$ are equivalent. Hence for each $h, k \in \mathcal{H}_{00}$, we can, in view of Lemma 8.22 on the preceding page, let $\rho_{h,k}$ be the Radon-Nikodym derivative of $L_{h,k}$ with respect to ν_0. Then for each $h, k \in \mathcal{H}_{00}$ and $f \in C_c(G)$, we have

$$\begin{aligned}\big(L(f)h \mid k\big) = L_{h,k}(f) &= \int_G f(\gamma)\, dL_{h,k}(\gamma)\\ &= \int_G f(\gamma)\rho_{h,k}(\gamma)\Delta(\gamma)^{-\frac{1}{2}}\, d\nu(\gamma)\\ &= \int_{G^{(0)}}\int_G f(\gamma)\rho_{h,k}(\gamma)\Delta(\gamma)^{-\frac{1}{2}}\, d\lambda^u(\gamma)\, d\mu(u).\end{aligned}$$

To motivate the construction in Lemma 8.23, consider $h, k \in \mathcal{H}_{00}$ and $f, g \in C_c(G)$. Then we can apply Lemma 8.22 on the preceding page and compute that

$$\begin{aligned}\big(L(f)h \mid L(g)k\big) &= \big(L(g^* * f)h \mid k\big) = L_{h,k}(g^* * f)\\ &= \int_{G^{(0)}}\int_G g^* * f(\gamma)\rho_{h,k}(\gamma)\Delta(\gamma)^{-\frac{1}{2}}\, d\lambda^u(\gamma)\, d\mu(u)\\ &= \int_{G^{(0)}}\int_G\int_G \overline{g(\eta^{-1})}f(\eta^{-1}\gamma)\rho_{h,k}(\gamma)\Delta(\gamma)^{-\frac{1}{2}}\, d\lambda^u(u)\, d\lambda^u(\gamma)\, d\mu(u)\end{aligned}$$

which, by Fubini and sending $\gamma \mapsto \eta\gamma$, is

$$= \int_{G^{(0)}}\int_G\int_G \overline{g(\eta^{-1})}f(\gamma)\rho_{h,k}(\eta\gamma)\Delta(\eta\gamma)^{-\frac{1}{2}}\, d\lambda^{s(\eta)}(\gamma)\, d\lambda^u(\eta)\, d\mu(u)$$

which, after sending $\eta \mapsto \eta^{-1}$, and using the symmetry of ν_0, is

$$= \int_{G^{(0)}}\int_G\int_G \overline{g(\eta)}f(\gamma)\rho_{h,k}(\eta^{-1}\gamma)\Delta(\eta)^{-\frac{1}{2}}\Delta(\gamma)^{-\frac{1}{2}}\, d\lambda^u(\gamma)\, d\lambda^u(\eta)\, d\mu(u).$$

Since it is not clear to what extent $\rho_{h,k}$ is a sesquilinear function of (h, k), we fix once and for all a countable orthonormal basis $\{ z_i \}$ for $\mathcal{H}$ in $\mathcal{H}_{00}$ (Ex 8.2.2). We let

$$\mathcal{H}_{00}' := \operatorname{span}\{ z_i \}.$$

To make the subsequent formulas a bit easier to read, we will write ρ_{ij} in place of the Radon-Nikodym derivative ρ_{z_i,z_j}. The linear independence of the z_i guarantees that each $\alpha \in C_c(G) \odot \mathcal{H}_{00}'$ can be written *uniquely* as

$$\alpha = \sum_i f_i \otimes z_i$$

where all but finitely many f_i are zero.

LEMMA 8.23. *For each $u \in G^{(0)}$, there is a sesquilinear form $\langle \cdot\, , \cdot \rangle_u$ on $C_c(G) \odot \mathcal{H}_{00}'$ such that*

$$\langle f \otimes z_i\, , g \otimes z_j \rangle_u = \int_G\int_G \overline{g(\eta)}f(\gamma)\rho_{ij}(\eta^{-1}\gamma)\Delta(\eta\gamma)^{-\frac{1}{2}}\, d\lambda^u(\gamma)\, d\lambda^u(\eta). \tag{8.19}$$

Furthermore, there is a μ-conull set $F \subset G^{(0)}$ such that $\langle \cdot\, , \cdot \rangle_u$ is a pre-inner product for all $u \in F$.

PROOF. Given $\alpha = \sum_i f_i \otimes z_i$ and $\beta = \sum_j g_j \otimes z_j$, the usual arguments give a well-defined sesquilinear form on $C_c(G) \odot \mathcal{H}'_{00}$ via

$$\langle \alpha , \beta \rangle_u = \sum_{ij} \int_G \int_G \overline{g_j(\eta)} f_i(\gamma) \rho_{ij}(\eta^{-1}\gamma) \Delta(\eta\gamma)^{-\frac{1}{2}} \, d\lambda^u(\gamma) \, d\lambda^u(\eta).$$

It only remains to provide a conull Borel set F such that $\langle \cdot , \cdot \rangle_u$ is positive for all $u \in F$.

However, (8.19) was inspired by the calculation preceding the lemma. Hence if $\alpha := \sum_i f_i \otimes z_i$, then

$$\begin{aligned} \Big\| \sum L(f_i) z_i \Big\|^2 &= \sum_{ij} \big(L(f_i) z_i \mid L(f_j) z_j \big) \\ &= \sum_{ij} \big(L(f_j^* * f_i) z_i \mid z_j \big) \\ &= \sum_{ij} \int_{G^{(0)}} \int_G \int_G \overline{f_j(\eta)} f_i(\gamma) \rho_{ij}(\eta^{-1}\gamma) \Delta(\gamma\eta)^{-\frac{1}{2}} \\ &\qquad\qquad d\lambda^u(\gamma) \, d\lambda^u(\eta) \, d\mu(u) \\ &= \sum_{ij} \int_{G^{(0)}} \langle f_i \otimes z_i , f_j \otimes z_j \rangle_u \, d\mu(u) \\ &= \int_{G^{(0)}} \langle \alpha , \alpha \rangle_u \, d\mu(u). \end{aligned} \tag{8.20}$$

Thus, for μ-almost all u, we have $\langle \alpha , \alpha \rangle_u \geq 0$. The difficulty is that the exceptional null set depends on α. However, by Proposition C.3 on page 344, there is a sequence $\{ f_i \} \subset C_c(G)$ which is dense in $C_c(G)$ in the inductive limit topology. Let $\mathcal{A}$ be the rational vector space spanned by the countable set $\{ f_i \otimes z_j \}_{i,j}$. Since $\mathcal{A}$ is countable, there is a μ-conull set F such that $\langle \cdot , \cdot \rangle_u$ is a positive $\mathbf{Q}$-sesquilinear form on $\mathcal{A}$. Using Proposition C.3 on page 344 we can find sequences from our dense set such that $g_i \to g$ and $h_i \to h$ uniformly with supports in a fixed compact set, then, since $\lambda^u \times \lambda^u$ is a Radon measure on $G^u \times G^u$, we have $\langle g_i \otimes z_j , h_i \otimes z_k \rangle_u \to \langle g \otimes z_j , h \otimes z_k \rangle_u$. It follows that for all $u \in F$, $\langle \cdot , \cdot \rangle_u$ is a positive sesquilinear form (over $\mathbf{C}$) on the complex vector space generated by

$$\{ f \otimes z_i : f \in C_c(G) \}.$$

However, as that is all of $C_c(G) \odot \mathcal{H}'_{00}$, the proof is complete. □

The following is a straightforward application of the Tietze Extension Theorem (see [**Wil07**, Lemma 1.42]).

LEMMA 8.24. *Suppose that $g \in C_c(G^u)$ for some $u \in G^{(0)}$. Then there is a $f \in C_c(G)$ such that $f|_{G^u} = g$.*

Note that for *any* $u \in G^{(0)}$, the value of $\langle f \otimes z_i , g \otimes z_j \rangle_u$ depends only on $f|_{G^u}$ and $g|_{G^u}$. Using the Tietze Extension result above, we can view $\langle \cdot , \cdot \rangle_u$ as a sesquilinear form on $C_c(G^u)$. In particular, if $f \in C_c(G)$ and $\sigma \in G$, then we let $\check{u}(\sigma) f$ be any element of $C_c(G)$ such that

$$(\check{u}(\sigma) f)(\gamma) = \Delta(\sigma)^{\frac{1}{2}} f(\sigma^{-1}\gamma) \quad \text{for all } \gamma \in G^{r(\sigma)}.$$

Of course, $\check{u}(\sigma) f$ is only well-defined on $G^{r(\sigma)}$.

The next computation is critical to what follows. We have

$$\big\langle \check{u}(\sigma)f \otimes z_i \,,\, g \otimes z_j \big\rangle_{r(\sigma)} = \int_G \int_G \overline{g(\eta)} f(\sigma^{-1}\gamma)\rho_{ij}(\gamma^{-1}\eta)\Delta(\sigma^{-1}\gamma\eta)^{-\frac{1}{2}} \, d\lambda^{r(\sigma)}(\eta)\, d\lambda^{r(\sigma)}(\gamma)$$

which, after sending $\gamma \mapsto \sigma\gamma$, is

$$= \int_G \int_G \overline{g(\eta)} f(\gamma)\rho_{ij}(\gamma^{-1}\sigma^{-1}\eta)\Delta(\gamma\eta)^{-\frac{1}{2}} \, d\lambda^{r(\sigma)}(\eta)\, d\lambda^{s(\sigma)}(\gamma)$$

which, after sending $\eta \mapsto \sigma\eta$, is

$$= \int_G \int_G \overline{g(\sigma\eta)} f(\gamma)\rho_{ij}(\gamma^{-1}\eta)\Delta(\sigma)^{-\frac{1}{2}}\Delta(\gamma\eta)^{-\frac{1}{2}} \, d\lambda^{s(\sigma)}(\eta)\, d\lambda^{s(\sigma)}(\gamma)$$

$$= \int_G \int_G \big(\check{u}(\sigma^{-1})g\big)(\eta) f(\gamma)\rho_{ij}(\gamma^{-1}\eta)\Delta(\gamma\eta)^{-\frac{1}{2}} \, d\lambda^{s(\sigma)}(\eta)\, d\lambda^{s(\sigma)}(\gamma)$$

$$= \big\langle f \otimes z_i \,,\, \check{u}(\sigma^{-1})g \otimes \eta \big\rangle_{s(\sigma)}.$$

Recall that G acts continuously on the left of $G^{(0)}$: $\gamma \cdot s(\gamma) = r(\gamma)$. In particular, if C is compact in G and if K is compact in $G^{(0)}$, then

$$C \cdot K = \{\, \gamma \cdot u : (\gamma, u) \in (G * G^{(0)}) \cap (C \times K) \,\}$$

is compact. If $U \subset G^{(0)}$, then we say that U is *saturated* if U is G-invariant. More simply, U is saturated if $s(\gamma) \in U$ implies $r(\gamma)$ is in U. If $V \subset G^{(0)}$, then its *saturation* is the set $[V] = G \cdot V$ which is the smallest saturated set containing V.

The next result is a key technical step in our proof and takes the place of the Ramsay selection theorem (Theorem D.8 on page 350) used in Muhly's and Renault's proof.

LEMMA 8.25. *We can choose the μ-conull Borel set $F \subset G^{(0)}$ in Lemma 8.23 to be saturated for the G-action on $G^{(0)}$.*

PROOF. Let F be the Borel set from Lemma 8.23 on page 181. We want to see that $\langle \cdot \,,\, \cdot \rangle_v$ is positive for all v in the saturation of F. To this end, suppose that $u \in F$ and that $\sigma \in G$ is such that $s(\sigma) = u$ and $r(\sigma) = v$. Then

$$\gamma \mapsto \Delta(\sigma)^{\frac{1}{2}} f(\sigma^{-1}\gamma)$$

is in $C_c(G^v)$, and such functions span a dense subspace of $C_c(G^v)$ in the inductive limit topology. Moreover, as we observed at the end of the proof of Lemma 8.23 on page 181,

$$\langle f_i \otimes z_j \,,\, g_i \otimes z_k \rangle_v \to \langle f \otimes z_j \,,\, g \otimes z_k \rangle_v$$

provided $f_i \to f$ and $g_i \to g$ properly in the inductive limit topology in $C_c(G^v)$. Therefore, to show that $\langle \cdot \,,\, \cdot \rangle_v$ is positive, it will suffice to check on vectors of the form $\alpha := \sum_i \check{u}(\sigma)(f_i) \otimes z_i$. Then using the key calculation preceding Lemma 8.25, we have

$$
\begin{aligned}
\langle \alpha \,,\, \alpha\rangle_v &= \sum_{ij} \bigl\langle \check{u}(\sigma^{-1}\sigma) f_i \otimes z_i \,,\, f_j \otimes z_j \bigr\rangle_u. \\
&= \sum_{ij} \bigl\langle f_i \otimes z_i \,,\, f_j \otimes z_j \bigr\rangle_u. \\
&= \bigl\langle \sum_i f_i \otimes z_i \,,\, \sum_i f_i \otimes z_i \bigr\rangle_u
\end{aligned}
\tag{8.21}
$$

which is positive since $u \in F$.

It only remains to verify that the saturation of F is Borel. Since μ is a Radon measure—and therefore regular—we can shrink F a bit, if necessary, and assume it is σ-compact. Say $F = \bigcup K_n$. On the other hand, G is second countable and therefore σ-compact. If $G = \bigcup C_m$, then $[F] = \bigcup C_m \cdot K_n$. Since each $C_m \cdot K_n$ is compact, $[F]$ is σ-compact and therefore Borel. This completes the proof. □

From here on, we will assume that F is saturated. In view of Lemma 8.23 on page 181, for each $u \in F$ we can define $\mathcal{H}(u)$ to be the Hilbert space completion of $C_c(G) \odot \mathcal{H}_{00}'$ with respect to $\langle \cdot \,,\, \cdot \rangle_u$. We will denote the image of $f \otimes z_i$ in $\mathcal{H}(u)$ by $f \otimes_u z_i$. Since the complement of F is μ-null and also saturated, what we do off F has little consequence. In particular, G is the disjoint union of $G|_F$ and the ν-null set $G|_{G^{(0)} \setminus F}$.[1] Nevertheless, for the sake of nicety, we let $\mathcal{V}$ be a Hilbert space with an orthonormal basis $\{ e_{ij} \}$ doubly indexed by the same index sets as for $\{ f_i \}$ and $\{ z_j \}$, and set $\mathcal{H}(u) = \mathcal{V}$ if $u \in G^{(0)} \setminus F$. We then let

$$G^{(0)} * \mathscr{H} = \{ (u, h) : u \in G^{(0)} \text{ and } h \in \mathcal{H}(u) \},$$

and define $\Phi_{ij} : G^{(0)} \to G^{(0)} * \mathscr{H}$ by

$$\Phi_{ij}(u) := \begin{cases} f_i \otimes_u z_j & \text{if } u \in F \text{ and} \\ e_{ij} & \text{if } u \notin F. \end{cases}$$

(Technically, $\Phi_{ij}(u) = (u, f_i \otimes_u z_j)$—at least for $u \in F$—but we have agreed to obscure this subtlety.) Then [**Wil07**, Proposition F.8] implies that we can make $G^{(0)} * \mathscr{H}$ into a Borel Hilbert bundle over $G^{(0)}$ in such a way that the $\{ \Phi_{ij} \}$ form a fundamental sequence (see [**Wil07**, Definition F.1]).

A unitary representation and the rest of the proof. Note that if $f \otimes z_i \in C_c(G) \odot \mathcal{H}_{00}'$ and if $\Phi(u) := f \otimes_u z_i$, then

$$u \mapsto \bigl\langle \Phi(u) \,,\, \Phi_{ij}(u) \bigr\rangle_u$$

is Borel on F.[2] It follows that Φ is a Borel section of $G^{(0)} * \mathscr{H}$ and defines a class in $L^2(G^{(0)} * \mathscr{H}, \mu)$.

Furthermore, (8.21) shows that for each $\sigma \in G|_F$, there is a unitary $U_\sigma : \mathcal{H}\bigl(s(\sigma)\bigr) \to \mathcal{H}\bigl(r(\sigma)\bigr)$ characterized by

$$U_\sigma(f \otimes_{s(\sigma)} z_i) = \check{u}(\sigma) f \otimes_{r(\sigma)} z_i.$$

[1] The saturation of F is critical to what follows. If F is not saturated, then in general G is not the union of $G|_F$ and $G|_{G^{(0)} \setminus F}$. But as F *is* saturated, note that a homomorphism $\varphi : G|_F \to H$ can be trivially extended to a homomorphism on all of G by letting φ be suitably trivial on $G|_{G^{(0)} \setminus F}$. This is certainly not the case if F is not saturated.

[2] We can define $\Phi(u)$ to be zero off F. We are going to continue to pay as little attention as possible to the null complement of F in the sequel.

If $\sigma \notin G|_F$, then $\mathcal{H}\big(s(\sigma)\big) = \mathcal{H}\big(r(\sigma)\big) = \mathcal{V}$, and we can let U_σ be the identity operator.

LEMMA 8.26. *The map $\hat{U}$ from G to $\operatorname{Iso}(G^{(0)} * \mathscr{H})$ defined by $\hat{U}(\sigma) := \big(r(\sigma), U_\sigma, s(\sigma)\big)$ is a Borel homomorphism. Hence $(\mu, G^{(0)} * \mathscr{H}, \hat{U})$ is a unitary representation of G on $L^2(G^{(0)} * \mathscr{H}, \mu)$.*

PROOF. If $\sigma \in G|_F$, then

$$\big(U_\sigma \Phi_{ij}\big(s(\sigma)\big) \mid \Phi_{kl}\big(r(\sigma)\big)\big) = \int_G \int_G \overline{f_k(\eta)} f_i(\sigma^{-1}\gamma)\rho_{jl}(\eta^{-1}\gamma)\Delta(\sigma^{-1}\gamma\eta)^{-\frac{1}{2}}\, d\lambda^{r(\sigma)}(\gamma)\, d\lambda^{r(\sigma)}(\eta).$$

Thus $\sigma \mapsto \big(U_\sigma \Phi_{ij}\big(s(\sigma)\big) \mid \Phi_{kl}\big(r(\sigma)\big)\big)$ is Borel on $G|_F$ by a couple of applications of Corollary 3.10 on page 72. Since it is clearly Borel on the complement, $\hat{U}$ is Borel. The algebraic properties are straightforward. For example, assuming that $\gamma \in G^{r(\sigma)}$, we have on the one hand,

$$\big(\check{u}(\sigma\eta)f\big)(\gamma) = \Delta(\sigma\eta)^{\frac{1}{2}} f\big((\sigma\eta)^{-1}\gamma\big),$$

while

$$\begin{aligned}\big(\check{u}(\sigma)\check{u}(\eta)f\big)(\gamma) &= \Delta(\sigma)^{\frac{1}{2}} \big(\check{u}(\eta)f\big)(\sigma^{-1}\gamma) \\ &= \Delta(\sigma\eta)^{\frac{1}{2}} f\big(\eta^{-1}\sigma^{-1}\gamma\big).\end{aligned}$$

It follows that $\hat{U}$ is multiplicative on $G|_F$. Of course, it is clearly multiplicative on the complement (which is $G|_{G^{(0)}\setminus F}$ since F is saturated). □

LEMMA 8.27. *Each $f \otimes z_i \in C_c(G) \odot \mathcal{H}'_{00}$ determines a Borel section $\Phi(u) := f \otimes_u z_i$ in $\mathcal{L}^2(G^{(0)} * \mathscr{H}, \mu)$ whose class in $L^2(G^{(0)} * \mathscr{H}, \mu)$ depends only on the class of $[f \otimes z_i] \in C_c(G) \odot \mathcal{H}'_{00}/\mathcal{N} \subset C_c(G) \odot \mathcal{H}_0/\mathcal{N} = \mathcal{H}_{00}$. Furthermore, there is a unitary isomorphism V of $\mathcal{H}$ onto $L^2(G^{(0)} * \mathscr{H}, \mu)$ such that $V(L(f)z_i) = [\Phi]$.*

PROOF. We have already seen that Φ is in $\mathcal{L}^2(F * \mathscr{H}, \mu)$. More generally, the computation (8.20) in the proof of Lemma 8.23 on page 181 shows that if $\alpha = \sum_i f_i \otimes z_i$ and $\Psi(u) := \sum_i f_i \otimes_u z_i$, then

$$\|\Psi\|_2^2 = \Big\|\sum_i L(f_i)z_i\Big\|^2$$

Thus there is a well defined isometric map V as in the statement of lemma mapping $\operatorname{span}\{\, L(f)z_i : f \in C_c(G) \,\}$ onto a dense subspace of $L^2(F * \mathscr{H}, \mu)$. Since $\mathcal{H}'_{00}$ is dense in $\mathcal{H}_{00}$, and therefore in $\mathcal{H}$, the result follows by Lemma 8.8 on page 173. □

The proof of Theorem 8.2 on page 169 now follows almost immediately from the next proposition.

PROPOSITION 8.28. *The unitary V defined in Lemma 8.27 intertwines L with a representation L' which in the integrated form of the unitary representation $(\mu, G^{(0)} * \mathscr{H}, \hat{U})$ from Lemma 8.26.*

PROOF. We have $L'(f_1) = VL(f_1)V^*$. On the one hand,

$$\begin{aligned}\big(L(f_1)[f \otimes z_i] \mid [g \otimes z_j]\big)_{\mathcal{H}} &= \big(VL(f_1)[f \otimes z_i] \mid V[g \otimes z_j]\big) \\ &= \big(L'(f_1)V[f \otimes z_1] \mid V[g \otimes z_j]\big).\end{aligned}$$

But the left-hand side is

$$\begin{aligned}
\big(L(f_1 * f)z_i \mid L(g)z_j\big) &= L_{z_i,z_j}(g^* * f_1 * f) \\
&= \int_{G^{(0)}} \int_G \int_G \overline{g(\eta)} f_1 * f(\gamma) \rho_{ij}(\eta^{-1}\gamma)\Delta(\eta\gamma)^{-\frac12}\, d\lambda^u(\gamma)\, d\lambda^u(\eta)\, d\mu(u) \\
&= \int_{G^{(0)}} \int_G \int_G \int_G \overline{g(\eta)} f_1(\sigma) f(\sigma^{-1}\gamma)\rho_{ij}(\eta^{-1}\gamma)\Delta(\eta\gamma)^{-\frac12} \\
&\qquad\qquad d\lambda^u(\sigma)\, d\lambda^u(\gamma)\, d\lambda^u(\eta)\, d\mu(u) \\
&= \int_{G^{(0)}} \int_G \int_G \int_G f_1(\sigma)\overline{g(\eta)}\big(\check u(\sigma) f\big)(\gamma)\rho_{ij}(\eta^{-1}\gamma)\Delta(\eta\gamma)^{-\frac12}\Delta(\sigma)^{-\frac12} \\
&\qquad\qquad d\lambda^u(\sigma)\, d\lambda^u(\gamma)\, d\lambda^u(\eta)\, d\mu(u) \\
&= \int_F \int_G f_1(\sigma)\big\langle \check u(\sigma) f \otimes z_i\, ,\, g \otimes z_j \big\rangle_u \Delta(\sigma)^{-\frac12}\, d\lambda^u(\sigma)\, d\mu(u) \\
&= \int_F \int_G f_1(\gamma)\big\langle U_\gamma(f \otimes_{s(\gamma)} z_i)\, ,\, (g \otimes_u z_j\big\rangle_u \Delta(\sigma)^{-\frac12}\, d\lambda^u(\sigma)\, d\mu(u) \\
&= \int_G f_1(\sigma)\big\langle U_\sigma V[f \otimes z_i]\big(s(\sigma)\big)\, ,\, V[g \otimes z_j]\big(r(\sigma)\big)\big\rangle_{r(\sigma)} \Delta(\sigma)^{-\frac12}\, d\nu(\sigma).
\end{aligned}$$

Thus L' is the integrated form as claimed. □

Exercises

Ex 8.2.1. Show that (8.3) defined in Lemma 8.9 is a well-defined sesquilinear form on $C_c(G) \odot \mathcal{H}_0$.

—Answer on page 375

Ex 8.2.2. Suppose that V is a dense subspace of a separable Hilbert space $\mathcal{H}$. Then V contains an orthonormal basis for $\mathcal{H}$.

—Answer on page 375

CHAPTER 9

Amenability for Groupoids

Unlike the situation for groups, it is not so clear what the right definition of amenable groupoid is. As we shall see, there are (at least) two notions which may or may not be distinct. In the group case, from the perspective of operator algebras, perhaps the most important property of amenability is that it is *equivalent* to the reduced norm on $C_c(\mathcal{G})$ being equal to the universal norm. All current notions of amenability for groupoids also imply the equality of these norms, but we now know that the converse can fail [**Wil15**]. On the plus side, all the common notions of amenability for groupoids reduce to the usual notion when the groupoid in question is a group.

Just about everything to be discussed here, and quite a bit more, can be found in Anantharaman-Delaroche and Renault's monograph [**ADR00**] together with Renault's [**Ren15**]. My excuse for repeating many of these results here is to keep this monograph self contained, provide some additional details here and there, and to incorporate the results in [**Ren15**] to those in [**ADR00**] in a useful way. Also, the focus in [**ADR00**] is on Borel groupoids so concentrating on the continuous case here might be helpful in extracting results and provide some useful tools for reseachers.

I want to emphasize that in this chapter—as it has been throughout—it is a standing assumption that our groupoids and spaces are *second countable* as well as locally compact Hausdorff. We also assume that our groupoids have open range and source maps.

9.1. Some Comments on the Group Case

For motivation, I want to review some of the fundamental results for amenable *groups*. Much more on this venerable subject can be found in [**Pat88**] and a short summary is available in [**Wil07**, §A.2]. The concept of amenability for groups dates to back to von Neumann's work in 1929 while investigating the consequences of the Banach-Tarski paradox. The term “amenable” originated from a purposeful mis-translation of the German word “messbar” by Day: amenability involves the existence of a mean so that a group is a-mean-able if it has a such mean. One definition of a mean is a finitely additive probability measure that is left-invariant: $m(E) = m(aE)$ for every Borel set E in G and every $a \in G$. This is equivalent to the existence of a norm one linear functional $m : L^\infty(G) \to C$ such that $m(1) = 1$ and $m(\lambda_r f) = m(f)$ where $\lambda_r(f)(s) = f(r^{-1}s)$.[1] Thus a mean on $L^\infty(G)$ is just a left-invariant state.

[1]Of course, if G is not discrete, we have to recall that elements of $L^\infty(G)$ are equivalence classes. Fortunately, $\lambda_r(f)$ is well defined on equivalence classes. In fact, since we have a running assumption of second countability, we are dealing with ordinary almost everywhere equivalence here. But the same ideas work in general. See [**Wil07**, p. 319] for the necessary alterations in the general case.

We see immediately that compact groups are amenable—Haar measure provides a countably additive mean. But it turns out that abelian groups are also amenable—this is proved in [**Gre69**, Theorem 1.2.1] in the discrete case and the general case follows from this [**Wil07**, Theorem A.13]. If H is a closed subgroup of an amenable group G, then H is amenable as is the quotient G/H if H is normal [**Gre69**, Theorems 2.3.1 and 2.3.2]. If N is normal and if N and G/N are amenable, then so is G [**Gre69**, Theorem 2.3.3]. Thus all solvable and nilpotent groups are amenable.

The canonical examples of non-amenable groups are the free groups.

PROPOSITION 9.1. *The free group* $\mathbf{F}_2$ *on two generators in not amenable.*

PROOF. If $\mathbf{F}_2 = \langle a, b\rangle$ were amenable, then there would be a left invariant mean m on $\ell^\infty(\mathbf{F}_2)$ and a left invariant finitely additive probability measure μ on the subsets of $\mathbf{F}_2$. Let

$$\begin{aligned} A_0 &= \{\, v \in \mathbf{F}_2 : v = a^r w \text{ for } r \in 2\mathbf{Z}\,\} \quad \text{and} \\ A_1 &= \{\, v \in \mathbf{F}_2 : v = a^{r+1} w \text{ for } r \in 2\mathbf{Z}\,\}. \end{aligned}$$

(Of course, in the above, w is meant to be either trivial or a reduced word beginning with b or b^{-1}.) Notice that $\mathbf{F}_2$ is the disjoint union of A_0 and A_1 and that $A_1 = aA_0$. Since $\mu(\mathbf{F}_2) = 1$, we must have

$$\mu(A_0) = \frac{1}{2} = \mu(A_1).$$

On the other hand, for $i = 0, 1, 2$, we can define

$$B_i = \{\, v \in \mathbf{F}_2 : v = b^{r+i} w \text{ for } r \in 3\mathbf{Z}\,\}.$$

Then, just as above, $\mathbf{F}_2$ is the disjoint union of B_0, B_1 and B_2. Furthermore $B_i = b^i B_0$. Therefore

$$\mu(B_0) = \mu(B_1) = \mu(B_2) = \frac{1}{3}.$$

On the other hand, B_1 and B_2 are disjoint subsets of A_0. But a finitely additive measure must be monotonic, and hence,

$$\frac{2}{3} = \mu(B_1 \cup B_2) \le \mu(A_0) = \frac{1}{3}.$$

This contradiction means that $\mathbf{F}_2$ can't be amenable. □

Amenability is closely tied to the behavior of *positive definite functions* on G. Recall that $f : G \to \mathbf{C}$ is called positive definite if for all finite subsets $\{\, s_1, \dots, s_n \,\} \subset G$, the $n \times n$-matrix $\bigl(f(s_i^{-1} s_j)\bigr)$ is positive in M_n. (For equivalent definitions, see [**Wil07**, Proposition A.1].) In particular, f is positive definite on G if given complex scalars $c_1, \dots, c_n$ and $s_1, \dots, s_n$ in G, then

$$\sum_{i,j} f(s_i^{-1} s_j) \overline{c_i} c_j \ge 0.$$

If $\varphi \in \mathcal{L}^2(G)$, then $f(r) = (\varphi \mid \lambda_r \varphi)_{L^2(G)} = \varphi * \tilde{\varphi}(r)$ is positive definite where $\tilde{\varphi}(r) = \overline{\varphi(r^{-1})}$. If f is continuous with compact support, then such a φ always exists [**Wil07**, Lemma A.9]. Also note that if $f \in C_c(G)$ and $\varphi \in \mathcal{L}^\infty(G)$, then

$$f * \varphi(s) = \int_G f(r) \varphi(r^{-1} s) \, d\mu_G(r)$$

is a bounded continuous function on G (see [**Wil07**, p. 319]).

We can combine a number of classical results in one statement for later reference.

THEOREM 9.2. *Let G be a locally compact group. The following are equivalent.*

(a) *G is amenable.*

(b) *There is a mean m on $L^\infty(G)$ such that*
$$m(f * \varphi) = \int_G f(s)\, d\mu_G(s) \cdot m(\varphi)$$
for all $f \in C_c(G)$ and $\varphi \in \mathcal{L}^\infty(G)$.

(c) *There is a net $\{f_i\}$ of compactly supported positive definite functions of the form $f_i = g_i * \tilde{g}_i$ with $g_i \in C_c^+(G)$ and $\|g_i\|_2 = 1$ such that $f_i \to 1$ uniformly on compact subsets of G.*

(d) *The universal and reduced norms agree on $C_c(G)$.*

PROOF. The equivalence of (a) and (b) can be found in [**Wil07**, Theorem A.13]. The equivalence of (a) and (c) is in [**Wil07**, Proposition A.7]. The equivalence of (a) and (d) is [**Wil07**, Theorem A.18]. □

9.2. First Definitions

The quest to find a workable notion of amenability for groupoids has it origins in Zimmer's work on amenable group actions in the 1970s. The extension to more general settings came shortly thereafter. I think it is fair to say that groupoid amenability was originally introduced in a *ad hoc* manner using various generalizations of definitions given for groups. While in the group case, all these definitions are equivalent, the situation turns out to be more complicated for groupoids. At this point in time, the fundamental source for all things amenable in the groupoid world is Anantharaman-Delaroche & Renault's [**ADR00**]. For good reasons, the definitions in [**ADR00**] don't seem, *a priori*, to match up with those in Renault's original [**Ren80**] and the twenty years of work leading up to [**ADR00**]. Fortunately, they do coincide, but it will take us some work to see this.

As discussed in Section 9.1, one of the basic characterizations of amenability for groups involves positive definite functions. On a groupoid, a function h on G is said to be of *positive type* if for all $u \in G^{(0)}$, all $n \geq 1$, all $\gamma_1, \dots, \gamma_n \in G^u$ and all $c_1, \dots, c_n \in \mathbf{C}$, the sum
$$\sum_{i,j} h(\gamma_i^{-1}\gamma_j)\overline{c_i}c_j \geq 0.$$
As shown in Ex 9.2.1, functions of the form $f * f^*$ with $f \in C_c(G)$ are of positive type on G. Thus in analogy with the group case (see Theorem 9.2(c)), a reasonable first attempt at a Definition is the following.

DEFINITION 9.3. A locally compact groupoid with a Haar system $\{\lambda^u\}_{u\in G^{(0)}}$ is *topologically amenable* if there is a net $\{ f_i \} \subset C_c(G)$ such that

(a) the functions $u \mapsto \lambda(|f_i|^2)(u) = \int_G |f_i(\gamma)|^2\, d\lambda^u(\gamma) = f_i * f_i^*(u)$ are uniformly bounded, and

(b) the functions $x \mapsto f_i * f_i^*(\gamma)$ converge to the constant function 1 uniformly on compacta in G.

Definition 9.3 was Renault's choice in [**Ren80**, p. 92]. Although Renault stresses in [**Ren80**] that he "has no real justification for" this definition, the definition has stood the test of time. However, it will take some work in the next sections to see that it is a robust choice and that it coincides with the definition of amenability used in [**ADR00**] (see Definition 9.14 on page 199). In fact, I'll continue to use the term "topologically amenable" in order to distinguish it from the notion of "amenable" we'll define in Section 9.3. Once we manage to prove that the two notions are equivalent, we can drop the "topological" (see Corollary 9.46 on page 216).

REMARK 9.4 (Normalizing). Condition (a) in Definition 9.3 on the preceding page requiring that the f_i are "normalized" is useful in applications, but an annoyance to check. If we have a net $\{f_i\}$ in $C_c(G)$ satisfying only (b), then we can let

$$\varphi_i(u) = \begin{cases} 1 & \text{if } \lambda(|f_i|^2)(u) \leq 1, \text{ and} \\ \frac{1}{\lambda(|f_i|^2)(u)} & \text{otherwise.} \end{cases}$$

Notice that as $\lambda(|f_i|^2)(u) = f_i * f_i^*(u)$, we have $\varphi_i \to 1$ uniformly on compacta in $G^{(0)}$. Let

$$g_i(x) := \varphi_i(r(x))^{\frac{1}{2}} f_i(x),$$

and observe that

$$g_i * g_i^*(x) = \varphi_i(r(x))^{\frac{1}{2}} \varphi_i(s(x))^{\frac{1}{2}} f_i * f_i^*(x).$$

Thus if $\{f_i\}$ satisfies only condition (b) of Definition 9.3, then there are $\{g_i\} \subset C_c(G)$ such that

(a)′ $\int_G |g_i(\gamma)|^2 \, d\lambda^u(\gamma) \leq 1$ for all $u \in G^{(0)}$ and

(b)′ the functions $x \mapsto g_i * g_i^*(\gamma)$ converge to the constant function 1 uniformly on compacta in G.

Thus we have proved the following.

LEMMA 9.5. *A locally compact groupoid G with Haar system is topologically amenable if and only if there is a net $\{f_i\} \subset C_c(G)$ such that the functions $x \mapsto f_i * f_i^*(x)$ converge to the constant function 1 uniformly on compacta in G.*

Some good evidence that Definition 9.3 is a useful one is provided by the following. (However, see Ex 9.2.5.)

PROPOSITION 9.6. *Suppose that G is a second countable locally compact groupoid with a Haar system. If G is topologically amenable, then the reduced norm on $C_c(G)$ coincides with the universal norm.*

PROOF. Let $I_{C_r^*(G)}$ be the kernel of the natural map $\iota_r : C^*(G) \to C_r^*(G)$. Let $a \in I_{C_r^*(G)}$. We want to see that a must be zero. To that end, let L be a representation of $C^*(G)$. In view of the Disintegration Theorem 8.2 on page 169, we can assume that L is the integrated form of a unitary representation $(\mu, G^{(0)} * \mathscr{H}, \hat{L})$ as in Definition 7.7 on page 162. Let $h \in L^2(G^{(0)} * \mathscr{H}, \mu)$ be a vector of norm 1, and let ω_h be the corresponding vector state: $\omega_h(b) = (L(b)h \mid h)$. It will suffice to see that $\omega_h(a) = 0$.

Recall that if $f \in C_c(G)$, then

$$\omega_h(f) = \int_{G^{(0)}} \int_G f(\gamma)\big(L_\gamma(h(s(\gamma)) \mid h(u)\big)\Delta(\gamma)^{-\frac{1}{2}} \, d\lambda^u(\gamma) \, d\mu(u).$$

Since G is topologically amenable, there is a net $\{f_i\} \subset C_c(G)$ satisfying assertions (a) and (b) of Definition 9.3 on page 189. For convenience, let

$$e_i(\gamma) = f_i * f_i^*(\gamma) = \int_G f_i(\eta)\overline{f_i(\gamma^{-1}\eta)}\, d\lambda^{r(\gamma)}(\eta).$$

By assumption (b) of Definition 9.3, the pointwise product $e_i f$ converges to f in the inductive limit topology and hence in the C^*-norm. Therefore $\omega_h(e_i f) \to \omega_h(f)$. But

$$\begin{aligned}
\omega_h(e_i f) &= \int_{G^{(0)}} \int_G e_i(\gamma) f(\gamma) \big(L_\gamma(h(s(\gamma))) \mid h(r(\gamma))\big) \Delta(\gamma)^{-\frac{1}{2}} \\
&\qquad\qquad d\lambda^u(\gamma)\, d\mu(u) \\
&= \int_{G^{(0)}} \int_G \int_G f(\gamma) f_i(\eta) \overline{f_i(\gamma^{-1}\eta)} \big(L_\gamma(h(s(\gamma))) \mid h(r(\gamma))\big) \\
&\qquad\qquad \Delta(\gamma)^{-\frac{1}{2}}\, d\lambda^u(\eta)\, d\lambda^u(\gamma)\, d\mu(u)
\end{aligned} \tag{9.1}$$

which, after using Fubini and sending $\gamma \mapsto \eta\gamma$, is

$$\begin{aligned}
&= \int_{G^{(0)}} \int_G \int_G f(\eta\gamma) f_i(\eta) \overline{f_i(\gamma^{-1})} \big(L_\gamma(h(s(\gamma))) \mid L_\eta^{-1}(h(r(\eta)))\big) \\
&\qquad\qquad \Delta(\eta\gamma)^{-\frac{1}{2}}\, d\lambda^{s(\eta)}(\gamma)\, d\lambda^u(\eta)\, d\mu(u) \\
&= \int_G \int_G f(\eta\gamma) f_i(\eta) \overline{f_i(\gamma^{-1})} \big(L_\gamma(h(s(\gamma))) \mid L_\eta^{-1}(h(r(\eta)))\big) \\
&\qquad\qquad \Delta(\eta\gamma)^{-\frac{1}{2}}\, d\lambda^{s(\eta)}(\gamma)\, d\nu(\eta)
\end{aligned}$$

which, since $\nu = \Delta \cdot \nu^{-1}$,

$$\begin{aligned}
&= \int_G \int_G f(\eta\gamma) f_i(\eta) \overline{f_i(\gamma^{-1})} \big(L_\gamma(h(s(\gamma))) \mid L_\eta^{-1}(h(r(\eta)))\big) \\
&\qquad\qquad \Delta(\eta\gamma)^{-\frac{1}{2}} \Delta(\eta)\, d\lambda^{s(\eta)}(\gamma)\, d\nu^{-1}(\eta) \\
&= \int_{G^{(0)}} \int_G \int_G f(\eta\gamma) f_i(\eta) \overline{f_i(\gamma^{-1})} \big(L_\gamma(h(s(\gamma))) \mid L_\eta^{-1}(h(r(\eta)))\big) \\
&\qquad\qquad \Delta(\eta)^{\frac{1}{2}} \Delta(\gamma)^{-\frac{1}{2}}\, d\lambda^u(\gamma)\, d\lambda_u(\eta)\, d\mu(u)
\end{aligned}$$

which, if we let $k_i(\gamma) = \Delta(\gamma)^{\frac{1}{2}} \overline{f_i(\gamma)} L_\gamma^{-1}(h(r(\gamma)))$, is

$$\begin{aligned}
&= \int_{G^{(0)}} \int_G \int_G f(\eta\gamma) \big(k_i(\gamma^{-1}) \mid k_i(\eta)\big)\, d\lambda^u(\gamma)\, d\lambda_u(\eta)\, d\mu(u) \\
&= \big((\operatorname{Ind} M)(f) k_i \mid k_i\big),
\end{aligned}$$

where $\operatorname{Ind} M$ is the representation of $C^*(G)$ as in Lemma 5.17 on page 117. Recall that $\operatorname{Ind} M$ is equivalent to the regular representation induced by the usual multiplication representation of $C_0(G^{(0)})$ on $L^2(G^{(0)} * \mathscr{H}, \mu)$. In particular, $\|\operatorname{Ind} M(f)\| \leq \|f\|_r$, and

$$|\omega_h(e_i f)| \leq \|f\|_r \|k_i\|_2^2 \quad \text{for all } i$$

On the other hand,

$$\begin{aligned}\|k_i\|_2^2 &= \int_G |k_i(\gamma)|^2 \, d\nu^{-1}(\gamma) \\ &= \int_G |f_i(\gamma)|^2 \|h(r(\gamma))\|^2 \Delta(\gamma) \, d\nu^{-1}(\gamma) \\ &= \int_G |f_i(\gamma)|^2 \|h(r(\gamma))\|^2 \, d\nu(\gamma) \\ &= \int_{G^{(0)}} \Big(\int_G |f(\gamma)|^2 \, d\lambda^u(\gamma) \Big) \|h(u)\|^2 \, d\mu(u) \\ &= \int_{G^{(0)}} e_i(u) \|h(u)\|^2 \, d\mu(u)\end{aligned}$$

which, by assumption (a) of Definition 9.3, is

$$\leq \|h\|^2 = 1.$$

Taking limits we have

$$|\omega_h(f)| \leq \|f\|_r \quad \text{for all } f \in C_c(G).$$

It now follows easily that $\omega_h(a) = 0$. This completes the proof. $\square$

Since Definition 9.3 on page 189 was inspired by the role of positive definite functions in the group case, it seems reasonable to wonder where did the "mean" in "a-mean-able" go? The first problem is to find a substitute for $L^\infty(G)$: even if G has a Haar system λ, this doesn't give us a measure on G. As usual, to get a measure on G from λ we nominate a measure μ on $G^{(0)}$. Then we can form the L^∞-spaces $L^\infty(G^{(0)}, \mu)$ and $L^\infty(G, \nu)$ where $\nu = \mu \circ \lambda$. If $h \in \mathcal{B}^b(G^{(0)})$ and $\varphi \in \mathcal{B}^b(G)$, then we get a bounded Borel function $h \cdot \varphi$ on G via

$$h \cdot \varphi(\gamma) = h(r(\gamma))\varphi(\gamma). \tag{9.2}$$

Since $r^{-1}(N)$ is ν-null if N is μ-null, it follows that (9.2) extends to an action of $L^\infty(G^{(0)})$ on $L^\infty(G)$. But we need μ to be quasi-invariant to get actions of $C_c(G)$ on $L^\infty(G^{(0)})$ and $L^\infty(G)$ as follows (see Ex 9.2.8):

$$f * h(u) = \int_G f(\gamma) h(\gamma^{-1} \cdot u) \, d\lambda^u(\gamma) = \int_G f(\gamma) h(s(\gamma)) \, d\lambda^u(\gamma) \quad \text{and} \tag{9.3}$$

$$f * \varphi(\gamma) = \int_G f(\eta) \varphi(\eta^{-1}\gamma) \, d\lambda^{r(\gamma)}(\eta). \tag{9.4}$$

Then, again with no other justification than analogy with the group case, we can make the following definitions (see Theorem 9.2(b) on page 189).

Definition 9.7. Let G be a second countable locally compact groupoid with a Haar system λ. Let μ be a quasi-invariant measure on $G^{(0)}$ and let $\nu = \mu \circ \lambda$. We call μ *amenable* if there is a positive, unital $L^\infty(G^{(0)}, \mu)$-linear map $m : L^\infty(G, \nu) \to L^\infty(G^{(0)}, \mu)$ such that

$$m(f * \varphi)(u) = f * m(\varphi)(u) \quad \text{for all } u \in G^{(0)},\ f \in C_c(G), \text{ and } \varphi \in L^\infty(G). \tag{9.5}$$

We say that G is *measurewise amenable* if every quasi-invariant measure on $G^{(0)}$ is amenable.[2]

EXAMPLE 9.8. If $G = \mathcal{G}$ is a locally compact group, then $\mathcal{G}$ is measurewise amenable exactly when Theorem 9.2(b) on page 189 holds. Hence $\mathcal{G}$ is amenable in this case.

A positive unital $L^\infty(G^{(0)}, \mu)$-linear map $m : L^\infty(G, \nu) \to L^\infty(G^{(0)}, \mu)$ is called a *mean*. A mean is called *invariant* if it satisfies (9.5). Notice that if G is a group, then an invariant mean on G is a left-invariant mean on G by (the proof of) [**Wil07**, Theorem A.13].

REMARK 9.9. In [**Ren80**], measurewise amenability is defined in terms of Borel functions of positive type—see [**Ren80**, Definitions II.3.1 and II.3.6]. The definition given here is perhaps more natural, but it is not clearly equivalent to Renault's original. As observed in Remark 10.51 on page 275 and the references given there, the two notions are equivalent but I won't pursue the matter here.

Looking Forward. The remainder of this chapter, as well as all of Chapter 10 on measurewise amenability, is devoted to a self-contained summary of the current state of the art of amenability for groupoids taken primarily from [**ADR00**] and [**Ren15**]. It is a complicated and intertwined story, so it seems only decent to provide an outline and a summary.

A key idea is to work with maps $\pi : Y \to X$ between G-spaces and π-systems rather than the groupoid itself and its Haar system. For example, in Section 9.3, we will define a continuous equivariant map $\pi : Y \to X$ to be *properly amenable* if there is an equivariant π-system $\{ m^x \}_{x\in X}$ of probability measures (so that by definition $\gamma^{-1}\cdot m^x = m^{\gamma^{-1}\cdot x}$ for all $(\gamma, x) \in G *_r X$). Then we say that G is properly amenable if $r : G \to G^{(0)}$ is properly amenable. Proper amenability is quite restrictive; we show in Corollary 9.26 on page 202 that G is properly amenable exactly when it is a proper groupoid. A group $\mathcal{G}$ is properly amenable as a groupoid exactly when it is compact.

To get a notion that is less restrictive—and more in line with the goal of generalizing amenability for groups—we look for π-systems of probability measures that are only "approximately" equivariant. Therefore we define $\pi : Y \to X$ to be *amenable* (see Definition 9.13 on page 199) if there is a net $\{ m_i = \{m_i^x\}_{x\in X} \}$ of π-systems so that $\gamma^{-1} \cdot m_i^x$ converges to m_i^x in a suitable sense. Such a net is called a *approximately invariant continuous mean*. Then we define G to be *amenable* if $r : G \to G^{(0)}$ is an amenable map, and we define a G-space X to be an *amenable G-space* if the action groupoid $G \rtimes X$ is amenable.

Having dealt with these new definitions in Section 9.3, we look at the fundamental properties and calculus of amenable maps in Section 9.4. We see that a G-space is automatically amenable whenever G is amenable (Corollary 9.30). We can show that if Z is a free and proper G-space such that the moment map is an amenable map, then G must be amenable. This turns out to be a critical characterization in the sequel. To ease some of the proofs in later sections, Section 9.4.1 is devoted to some technical variations on approximately invariant continuous means.

[2]It would have been more appropriate to define the pair (G, λ) to be measurewise amenable as the Haar system clearly plays a role. However, we will show in Theorem 10.41 on page 266 that measurewise amenability is preserved by equivalence. Therefore measurewise amenability does not depend on the choice of Haar system.

Then we can use our technology to specialize to $r : G \to G^{(0)}$ and derive a number of alternate characterizations of amenability of G in terms the existence of nets of functions on G with desirable properties (see Theorem 9.43 on page 211). As a consequence in Corollary 9.46 on page 216, we see that our new notion of amenability of G is exactly what we defined as topological amenability earlier. Hence we can simply use the term "amenable" in the sequel.

In Section 9.5 we use the calculus for approximately invariant continuous means from Section 9.4.1 to give a purely topological proof that equivalence of groupoids preserves amenability (Theorem 9.48 on page 217). This result will also be obtained as a special case of the main result in Section 9.7 where we see that amenability is preserved under a much weaker Borel version of equivalence.

This weaker notion of equivalence is useful in light of recent work of Renault's [**Ren15**] that shows that the amenability of a locally compact groupoid depends only on its underlying Borel structure. We start exposing this in Section 9.6 where we start trying to generalize the concept of a mean from the classical group setting. We let $\mathcal{E} := C_0(G^{(0)}, L^1(G,\lambda))$ be the completion of $C_c(G)$ with respect to the norm

$$\|f\|_{\lambda,1} = \sup_{u\in G^{(0)}} \int_G |f(\gamma)|\, d\lambda^u(\gamma).$$

Note that if $\mathcal{G}$ is a group, then $\mathcal{E}$ is simply $L^1(\mathcal{G})$, so that $\mathcal{G}$ is amenable as a group if there is a left-invariant mean $m \in L^\infty(\mathcal{G})^* = \mathcal{E}^{**}$ as in Theorem 9.2 on page 189. Motivated by this, it is natural to consider $\mathcal{E}^{**} = \big(C_0(G^{(0)}, L^1(G,\lambda))\big)^{**}$. Notice that $C_0(G^{(0)})$ acts on $\mathcal{E}$ by pointwise multiplication. Thus if $\varphi \in C_0(G^{(0)})$ and $\nu \in \mathcal{E}^*$ we have $\varphi \cdot \nu(g) = \nu(\varphi g)$. Similarly, if $m \in \mathcal{E}^{**}$, then $\varphi \cdot m(\nu) = m(\varphi\nu)$. If we denote the double transpose of left-convolution by $f \in C_c(G)$ on $\mathcal{E}$ by $f \diamond m$ for $m \in \mathcal{E}^{**}$, then we say that a suitably normalized positive element of $m \in \mathcal{E}^{**}$ is a *topological invariant mean* on G if $f \diamond m = (\lambda(f)\circ r)\cdot m$. It has been known for some time that the amenability of G implies the existence of a topological invariant mean. Renault uses a subtle convexity argument in [**Ren15**] to demonstrate the converse. In the course of the proof we verify that it is the underlying Borel structure of G that matters. This is all done in Theorem 9.58 on page 222.

In Section 9.7, we fully, but temporarily, embrace the Borel world by defining Borel groupoids, Borel amenability for a Borel groupoid, and Borel equivalence between Borel groupoids. It should be kept in mind that the only Borel groupoids we have more than passing interest in here are those coming from second countable, locally compact Hausdorff groupoids with their underlying Borel structure. Using the characterizations in the previous section, we can show that a locally compact groupoid is amenable if and only if it is amenable as a Borel groupoid (Theorem 9.69 on page 230). We also show that Borel amenability is preserved by a notion of Borel equivalence that is considerably weaker than topological equivalence (Theorem 9.74 on page 232). Combining these observations, we establish a stronger version of the main result from Section 9.5 showing that amenability of locally compact groupoids is preserved by the much weaker requirement that they be Borel equivalent (Theorem 9.75 on page 232).

In Section 9.8, we look at some examples and important consequences that can be extracted from the previous sections. In particular, we see that closed and open subgroupoids of an amenable groupoid are amenable—at least when these subgroupoids have open source and range maps (see Proposition 9.77 on page 235).

The corresponding result for quotients is more complicated. However, we can prove the following. If $\pi : G \to H$ is a continuous open surjection *that restricts to a homeomorphism of* $G^{(0)}$ *onto* $H^{(0)}$, then H is amenable if G is. If both $\ker \pi$ and H are amenable, then so is G. These results are proved in Theorem 9.80 on page 236. We also show that if the orbit space $G\backslash G^{(0)}$ is T_0, then G is amenable if and only if all its isotropy groups are amenable groups (see Theorem 9.86 on page 242).

The concept of measurewise amenability is examined in Chapter 10. Accepting the message from [**ADR00**] that there is much to be gained by working with a calculus of maps, we replace $r : G \to G^{(0)}$ with equivariant maps $\pi : Y \to X$ and the Haar system on G with an equivariant π-system α. We also require a measure μ on X which is suitably "quasi-invariant". This gives us measures $\mu \circ \alpha$ on Y as well as μ on X. Therefore in Section 10.1 we can generalize the notion of mean $m : L^\infty(G) \to L^\infty(G^{(0)})$ to general means which are defined be positive unital $L^\infty(X)$-linear maps from $L^\infty(Y)$ to $L^\infty(X)$. The quasi-invariance of μ is needed to define $C_c(G)$-actions on $L^\infty(Y)$ and $L^\infty(X)$. Then a mean m is *invariant* if $m(f * \varphi) = f * m(\varphi)$ (Definition 10.5 on page 244). We show that m is invariant exactly when we can find a suitable net of Borel functions on Y called a *approximate weakly invariant mean* (Proposition 10.12 on page 248).

In Section 9.6, we saw that G was amenable exactly when there was a *topological invariant mean* $m \in \mathcal{E}^{**}$ (Theorem 9.58 on page 222). In Section 10.2, we note that in the measurewise context, we only need the key invariance property of a topological invariant mean to hold on the subspace $\mathcal{E}^*_G$ of $\mathcal{E}^*$ corresponding to quasi-invariant measures. A mean satisfying this weaker requirement is called a *measurewise invariant mean* and the existence of such a mean is equivalent to G being measurewise amenable (Proposition 10.20 on page 253). Then it is clear that if G is amenable, then it must also be measurewise amenable (Theorem 10.21 on page 253). Since we can show that $\mathcal{E}^*_G = \mathcal{E}^*$ when G is étale, we see that in the case that G is étale, the concepts of amenability and measurewise amenability coincide (Theorem 10.22 on page 253). As of this writing, it is possible that the two concepts agree in general. At least, I know of no counterexamples.

In Section 10.3 we investigate a number of invariance properties for means. We fix a *quasi-invariant* measure on X and say that $\pi : Y \to X$ is *amenable with respect to* (λ, μ, α) when there is an invariant mean $m : L^\infty(Y) \to L^\infty(X)$ (the only role λ plays is in defining the $C_c(G)$-actions). The fundamental construction we need in the sequel involves an *auxillary* locally compact G-space T equipped with an equivariant $r : T \to G^{(0)}$ system β. Then we can equip $X * T$ with the measure $\mu \circ \beta$ and $Y * T$ with the measure $(\mu \circ \alpha) \circ \beta$. Then if $m : L^\infty(Y) \to L^\infty(X)$ is a mean, we can, with some work, define mean $m \otimes 1_T : L^\infty(Y * T) \to L^\infty(X * T)$ by "intergrating out the T dependence". The key result is that $m \otimes 1_T$ is invariant if m is (Lemma 10.26 on page 257). An important choice of auxillary space is G itself. Then if X is a G-space we recall that $X * G$ can be realized as a transformation groupoid with inverse map $\iota_X(x, \gamma) := (\gamma^{-1} \cdot x, \gamma^{-1})$. This induces a map $\mathfrak{I}_X$ on $L^\infty(X * G)$. A mean $m : L^\infty(Y) \to L^\infty(X)$ is called *completely invariant* if $I_X \circ (m \otimes 1_G) = (m \otimes 1_G) \circ \mathfrak{I}_Y$. As the name suggests, a completely invariant mean is invariant (Proposition 10.33 on page 260), and the existence of a completely invariant means implies the existence of approximately completely invariant means with stronger properties called *approximately completely invariant means*.

Section 10.4 is devoted to showing that equivalence preserves measurewise amenability (Theorem 10.41 on page 266. Among other things, this affirms that G being measurewise amenable does not depend on the choice of a Haar system on G. As a corollary of the proof, we see that if G admits an invariant mean $m : L^\infty(G) \to L^\infty(G^{(0)})$ for a given quasi-invariant measure μ, then it also admits a completely invariant mean (Corollary 10.46 on page 273).

We use Section 10.5 we specialize to the case of $r : G \to G^{(0)}$. We call a quasi-invariant measure on $G^{(0)}$ *amenable* if there is an invariant mean $m : L^\infty(G) \to L^\infty(G^{(0)})$ (so that G is measurewise amenable exactly when every quasi-invariant measure is amenable). Then we can summarize a number of equivalent characterizations of an amenable measure μ on $G^{(0)}$ derived in the earlier sections as Theorem 10.49 on page 274. This gives us the opportunity to show that measurewise amenability implies that the reduced norm coincides with the universal norm on $C_c(G)$ (Corollary 10.50 on page 275). This is sometimes called *metric amenability*. It should be noted that examples show that there are metrically amenable groupoids that are not even measurewise amenable (see [**Wil15**]). But in the special case that G is transitive, then the three flavors of amenability do coincide—see Corollary 10.54 on page 277.

We saw that measurewise amenability implied amenability at least for étale groupoids in Section 10.2 and wondered how general such a result might be. In Section 10.6 we observe that if $G\backslash G^{(0)}$ is T_0, then measurewise amenabilty of G implies that G is amenable just as in the étale case. Suffice it to say, the complete story of the relationship between amenability, measurewise amenability, and metric amenability is mysterious.

Unlike the situation for amenability, we don't have a lot of results about measurewise amenability passing to subgroupoids and open images. A few technical results—well beyond the scope of this effort—are given in [**ADR00**, Section 5.3].

Exercises

Ex 9.2.1. Let G be a locally compact groupoid with a Haar system $\lambda = \{\lambda^u\}_{u\in G^{(0)}}$. Suppose that $f \in C_c(G)$. Show that $h := f*f^*$ is a continuous compactly supported function of positive type on G.

—Answer on page 376

Ex 9.2.2. Suppose that h is a continuous function on G of positive type and that φ is continuous on $G^{(0)}$. Show that

$$\tilde{h}(\gamma) = \overline{\varphi(r(\gamma)}h(\gamma)\varphi(s(\gamma))$$

is also of positive type on G.

—Answer on page 376

Ex 9.2.3. Suppose that $\{h_i\}$ is a net or sequence of continuous compactly supported functions of positive type on G converging to 1 uniformly on compact subsets of G. Show that there is a net or sequence $\{h_i'\}$ (indexed by the same set) converging uniformly to 1 on compact subsets of G such that $h_i'(u) \le 1$ for all $u \in G^{(0)}$.

—Answer on page 376

Ex 9.2.4. Show that if k is a compactly supported continuous function of positive type on G, then $k = k^*$ and for all $f \in C_c(G)$ we have

$$\int_G \int_G k(\gamma^{-1}\eta)\overline{f(\gamma)}f(\eta)\,d\lambda^u(\eta)\,d\lambda^u(\gamma) \geq 0. \tag{9.6}$$

(Hint: see [**Wil07**, Lemma A.4 and Proposition A.1].)

—Answer on page 376

Ex 9.2.5. Let G be a locally compact groupoid with a Haar system λ. Suppose that for every quasi-invariant measure μ on $G^{(0)}$ there is a net of Borel functions $\{f_i\}$ on G such that

$$e_i(\gamma) = \int_G f_i(\eta)\overline{f_i(\gamma^{-1}\eta)}\,d\lambda^{r(\gamma)}(\eta)$$

defines a Borel function on G such that

(a)″ The net $\{e_i|_{G^{(0)}}\}$ of restrictions converges to 1 in the weak-$*$ topology on $L^\infty(G^{(0)},\mu)$, and

(b)″ The net $\{e_i\}$ converges to 1 in the weak-$*$ topology on $L^\infty(G,\mu\circ\lambda)$.

By modifying the proof of Proposition 9.6, show that the reduced norm on $C_c(G)$ coincides with the universal norm.

—Answer on page 377

Ex 9.2.6. Let X be a locally compact Hausdorff space. Show that when viewed as a groupoid, X is topologically amenable.

—Answer on page 377

Ex 9.2.7. Let $G = X \times X$ be the trivial or pair groupoid on the locally compact Hausdorff space X. Let μ be a Radon measure on X with full support so that $\lambda^x = \delta_x \times \mu$ give a Haar system on G. Show that G is topologically amenable.

—Answer on page 377

Ex 9.2.8. Let μ be a quasi-invariant measure on $G^{(0)}$ and $\nu = \mu\circ\lambda$. Suppose that $f \in C_c(G)$, $h \in \mathcal{B}^b(G^{(0)})$, and $\varphi \in \mathcal{B}^b(G)$. Let $f * h$ and $f * \varphi$ be defined as in (9.3) and (9.4), respectively. Show that, with respect to the essential supremum norms on $L^\infty(G^{(0)},\mu)$ and $L^\infty(G,\nu)$, we have $\|f * h\|_\infty \leq \|f\|_I\|h\|_\infty$ and $\|f * \varphi\|_\infty \leq \|f\|_I\|\varphi\|_\infty$. Hence we get actions of $C_c(G)$ on $L^\infty(G^{(0)})$ and $L^\infty(G)$ as claimed prior to Definition 9.7.

—Answer on page 377

9.3. Amenable Groupoids

Recall that if Y is a (second countable) locally compact Hausdorff space, then $M(Y)$ will denote the Banach space of complex measures on Y. Recall that we can identify $M(Y)$ with the norm dual of $C_0(Y)$ with respect to the total variation norm $\|\cdot\|_1$ [**Fol99**, Theorem 7.17]. Thus if $m \in M(Y)$, then

$$\|m\|_1 = \sup_{\substack{f\in C_0(Y)\\ \|f\|_\infty=1}} |m(f)| = \sup_{\substack{f\in C_0(Y)\\ \|f\|_\infty=1}} \Bigl|\int_Y f(y)\,dm(y)\Bigr|.$$

Note that a (positive) Radon measure on Y is in $M(Y)$ only when it is a finite measure. Furthermore, there is a unique Radon measure $|m| \in M(Y)$, called the *total variation measure* of m, such that

$$\int_Y f(y)\,dm(y) = \int_Y f(y)\rho(y)\,d|m|(y) \quad \text{for all } f \in C_0(Y)$$

for a unimodular function ρ determined $|m|$-almost everywhere. In particular,

$$\Big|\int_Y f(y)\,dm(y)\Big| \leq \int_Y |f(y)|\,d|m|(y) \quad \text{for all } f \in C_0(Y).$$

Suppose that $\pi : Y \to X$ is a continuous equivariant surjection between second countable locally compact Hausdorff spaces, and that $m^x \in M(Y)$ is such that $\operatorname{supp} m^x \subset \pi^{-1}(x)$. Since the Tietze Extension Theorem implies every $f \in C_0(\pi^{-1}(x))$ is the restriction of a function in $C_0(X)$, we have

$$\|m^x\|_1 = \sup_{\substack{f \in C_0(\pi^{-1}(x)) \\ \|f\|_\infty = 1}} |m^x(f)|.$$

If $r_G(\gamma) = r_X(x)$, then we can define a measure $\gamma^{-1} \cdot m^x \in M(Y)$ with support in $\pi^{-1}(\gamma^{-1} \cdot x) = \gamma^{-1} \cdot \pi^{-1}(x)$ satisfying

$$\int_Y f(y)\,d(\gamma^{-1} \cdot m^x)(y) = \int_Y f(\gamma^{-1} \cdot y)\,dm^x(y)$$

for $f \in C_0(\gamma^{-1} \cdot \pi^{-1}(x))$. Since $y \mapsto \gamma^{-1} \cdot y$ is a homeomorphism of $\pi^{-1}(x)$ onto $\gamma^{-1} \cdot \pi^{-1}(x)$, we have

$$\|m^x\|_1 = \|\gamma^{-1} \cdot m^x\|_1. \tag{9.7}$$

DEFINITION 9.10 ([**ADR00**, Definition 2.2.1]). Suppose that $\pi : Y \to X$ is a continuous, equivariant surjection between second countable locally compact G-spaces. Then an *approximately invariant continuous mean* for π is a net $\{m_i\}$ of π-systems[3] $m_i = \{m_i^x\}_{x \in X}$ of probability measures on Y such that the total variation norm $\|\gamma^{-1} \cdot m_i^x - m_i^{\gamma^{-1} \cdot x}\|_1$ tends to zero uniformly on compact subsets of the action groupoid $G \ltimes X$.

REMARK 9.11 ($G \ltimes X$). As observed in Remark 1.13 on page 6, we can view the action groupoid $G \ltimes X$ as either triples in $X \times G \times X$ or as pairs (γ, x) in $G \times_r X$ since the map $(x, \gamma, \gamma^{-1} \cdot x) \mapsto (\gamma, x)$ is a homeomorphism. In this chapter, we will identify the unit space of $G \ltimes X$ with X and usually use the pair description for $G \ltimes X$ with the groupoid structure induced by the above homeomorphism. Thus $r((\gamma, x)) = x$ and $s((\gamma, x)) = \gamma^{-1} \cdot x$.

REMARK 9.12 (Terminology). It could be argued that an approximately invariant continuous mean should be called an "approximately *equivariant* continuous mean" since having $\gamma^{-1} \cdot m_i^x = m_i^{\gamma^{-1} \cdot x}$ would result in what we called an equivariant π-system in Section 3.2. But since the definition of "approximately invariant continuous mean"s comes from [**ADR00**] and plays a key role in their theory, it would be confusing to change the terminology here. So in the Chapter at least, we will agree to use the terms "equivariant π-system" and "invariant π-system" interchangeably.

---

[3]Recall that π-system are introduced in Section 3.2.

DEFINITION 9.13 ([**ADR00**, Definition 2.2.2]). Suppose that $\pi : Y \to X$ is a continuous, equivariant surjection between second countable locally compact G-spaces. Then we say that π is *amenable* if it admits an approximately invariant continuous mean as in Definition 9.10 on the preceding page.

DEFINITION 9.14. A second countable, locally compact groupoid G is *amenable* if $r : G \to G^{(0)}$ is amenable. A locally compact G-space X is called *amenable* if the corresponding action groupoid $G \rtimes X$ is amenable.

REMARK 9.15 (Dueling Notions). The question immediately arises as to the relationship between amenability of a groupoid defined above versus to naive notions of topological amenability and measurewise amenability introduced in Section 9.2. In due course, we'll prove that topological amenability and amenability as defined here are the same (see Corollary 9.46 on page 216).

EXAMPLE 9.16. Since we can identify $G \rtimes G^{(0)}$ with G, it follows that $G^{(0)}$ is an amenable G-space exactly when G is amenable.

REMARK 9.17 (Amenable G-spaces). Suppose that X is a G-space. View $G \rtimes X$ as pairs as in Remark 9.11 on the preceding page. Let $m = \{m^x\}_{x\in X}$ be a continuous r-family of measures for the structure map $r : G \rtimes X \to X$. Since each m^x is supported in $r^{-1}(x) = \{ (\gamma, x) : \gamma \in G^{r(x)} \}$, we can view m^x as a Radon measure on G supported in $G^{r(x)}$. The continuity assertion boils down to the continuity of

$$x \mapsto m^x(f) \quad \text{for all } f \in C_c(G). \tag{9.8}$$

(To see this, it suffices to check on $F \in C_c(G \rtimes X)$ of the form $f \otimes \varphi$ with $f \in C_c(G)$ and $\varphi \in C_c(X)$.) It is not hard to see that $\|\gamma^{-1} \cdot m^x - m^{\gamma^{-1}\cdot x}\|_1$ is the same whether we view $\gamma^{-1} \cdot m^x - m^{\gamma^{-1}\cdot x}$ as a measure on $G \rtimes X$ with support $r^{-1}(\gamma^{-1} \cdot x) = \gamma^{-1} \cdot r^{-1}(x)$, or as measures on G. Thus an approximately invariant continuous mean for $r : G \rtimes X \to X$ can be viewed as a collection of probability measures $m_i = \{m_i^x\}$ on G satisfying (9.8) such that $\operatorname{supp} m_i^x \subset G^{r(x)}$ and such that

$$\|\gamma^{-1} \cdot m_i^x - m_i^{\gamma^{-1}\cdot x}\|_1 \to 0$$

uniformly on compact subsets of $G \rtimes X$.

Properly Amenable Groupoids. An excellent source of motivation for amenable maps $\pi : Y \to X$ can be drawn from the special case where the net $\{m_i\}$ of nearly invariant π-systems is replaced by a single invariant π-system $m = \{m^x\}_{x\in X}$ so that $\gamma^{-1} \cdot m^x = m^{\gamma^{-1}\cdot x}$ (see Remark 9.12 on the facing page).

DEFINITION 9.18 ([**ADR00**, Definition 2.1.13]). Suppose that X and Y are second countable locally compact G-spaces. A continuous equivariant map $\pi : Y \to X$ is called *properly amenable* if there is a equivariant, continuous π-system of probability measures. A second countable locally compact groupoid G is called *properly amenable* if $r : G \to G^{(0)}$ is properly amenable. A second countable locally compact G-space X is called properly amenable if the map $r : G \rtimes X \to X$ is properly amenable.

REMARK 9.19. The definition of a properly amenable groupoid above differs slightly from that in [**ADR00**, Definition 2.1.13]. There it is required that the equivariant r-system be full—that is, G is required to have a Haar system of probability measures. This is not required here. Hence X is a properly amenable G-space exactly when $G \rtimes X$ is a properly amenable groupoid.

REMARK 9.20 (Properly Amenable G-spaces). As in Remark 9.17 on the previous page, X is a properly amenable G-space if and only if there are probability measures $\{ m^x \}_{x\in X}$ on G such that $\operatorname{supp} m^x \subset G^{r(x)}$ and $\gamma^{-1} \cdot m^x = m^{\gamma^{-1}\cdot x}$.

REMARK 9.21. Note that a group is properly amenable if and only if it has a finite Haar measure. Thus a group is properly amenable if and only if it is compact [**Fol99**, Proposition 11.4d].

LEMMA 9.22 ([**ADR00**, Proposition 2.1.14]). *Suppose that $\pi : Y \to X$ is a properly amenable (continuous equivariant) map between G-spaces and that Y is a proper G-space. Then X is a proper G-space. (Compare with Lemma 2.20 on page 33.)*

PROOF. Let $\alpha = \{\alpha^x\}$ be an invariant π-system of probability measures. Let A and B be compact sets in X. By Proposition 2.17 on page 32, it will suffice to see that

$$P(A, B) = \{ \gamma \in G : A \cap \gamma \cdot B \neq \emptyset \}$$

is compact. Let $f, g \in C_c(Y)$ be such that $0 \leq f(y), g(y) \leq 1$ for all $y \in Y$ and such that $\alpha(f)(x) > \frac{1}{2}$ for $x \in A$ and $\alpha(g)(x) > \frac{1}{2}$ for all $x \in B$ (Ex 9.3.1). Let $A' = \operatorname{supp} f$ and $B' = \operatorname{supp} g$. Suppose that

$$A' \cap \gamma B' = \emptyset.$$

Then, provided $r(x) = r(\gamma)$, the functions $y \mapsto f(y)$ and $y \mapsto g(\gamma^{-1}y)$ have disjoint supports C and D, respectively, on $\pi^{-1}(x)$ Since the invariance property implies $\gamma \cdot \alpha^{\gamma^{-1}\cdot x} = \alpha^x$, we have

$$\alpha(f)(x) + \alpha(g)(\gamma^{-1} \cdot x) \leq \alpha^x(C \cup D) \leq 1.$$

But this means that $A \cap \gamma \cdot B = \emptyset$. It follows that

$$P(A, B) \subset P(A', B').$$

Since the later is compact, so is $P(A, B)$. □

The proof of the next proposition also implies a slight strengthening of Proposition 3.14(a) on page 75.

PROPOSITION 9.23 ([**ADR00**, Proposition 2.1.15]). *Suppose that G has open range and source maps. Let $\pi : Y \to X$ be a continuous G-equivariant surjection which is properly amenable and so admits a* equivariant *π-system $\beta = \{\beta^x\}_{x\in X}$. Suppose that X is a proper G-space.*

(a) *There is a G-*invariant *$k \in C^+_{\pi,c}(Y)$ such that $\beta(k)(x) = 1$ for all $x \in X$.*
(b) *π is properly amenable.*

PROOF. By Lemma 2.20 on page 33, Y is a proper G-space. In particular, both $G\backslash Y$ and $G\backslash X$ are locally compact Hausdorff spaces (Proposition 2.18 on page 33). Let $\dot\pi : G\backslash Y \to G\backslash X$ be the induced map. Define $\varphi : Y \to X * (G\backslash Y) := \{ (x, G\cdot y) : G \cdot x = G \cdot \pi(y) \}$ by $\varphi(y) = (\pi(y), G \cdot y)$. Clearly, φ is a continuous surjection. I claim it is proper. To establish this, let $K \subset X$ and $L \subset G\backslash Y$ be compact subsets. Let $\{y_i\} \subset \varphi^{-1}(K \times L)$. It will suffice to see that $\{y_i\}$ has a convergent subnet.

Passing to a subnet, and relabeling, we can assume $\pi(y_i) \to x$ and $G \cdot y_i \to G \cdot z$. Since the quotient map $q : Y \to G\backslash Y$ is open, we can pass to a subnet, relabel, and assume that there are $\gamma_i \in G$ such that $\gamma_i \cdot y_i \to z$. But then $\gamma_i \cdot \pi(y_i) \to \pi(z)$. Since X is a proper G-space, Proposition 2.17 on page 32 implies that $\{\gamma_i\}$ has a convergent subnet. Passing to a subnet, and relabeling, allows us to assume that $\gamma_i \to \gamma$. But then $y_i \to \gamma^{-1} \cdot z$. This proves the claim about φ.

Now assume that $x_0 \in X$ and that $f \in C_c(G\backslash Y)$. Let K be a compact neighborhood of x_0. Since φ is proper, there is a $\psi \in C_c^+(Y)$ such that $\psi(y) = 1$ for all $y \in \varphi^{-1}(K \times \operatorname{supp} f)$. Then $g(y) = \psi(y)f(G \cdot y)$ is in $C_c(Y)$. Then for all $x \in K$, we have

$$\beta^x(f \circ q) = \beta^x(g). \tag{9.9}$$

Since the left-hand side of (9.9) depends only on $G \cdot x$, we can define a Radon measure $\dot\beta^{G\cdot x}$ on $G\backslash Y$ by $\dot\beta^{G\cdot x}(f) = \beta^x(f \circ q)$. To see that $G \cdot x \mapsto \dot\beta^{G\cdot x}(f)$ is continuous (and hence that $\dot\beta = \{\dot\beta^{G\cdot x}\}_{G\cdot x \in G\backslash X}$ is a $\dot\pi$-system), suppose that $G \cdot x_i \to G \cdot x_0$. It will suffice to see that every subnet of $\{\dot\beta^{G\cdot x_i}(f)\}$ has a subnet converging to $\dot\beta^{G\cdot x}(f)$. After passing to a couple of subnets, we may as well assume that $\gamma_i \cdot x_i \to x$ for suitable γ_i. But then with g as above, for large i,

$$\dot\beta^{G\cdot x_i}(f) = \beta^{\gamma_i\cdot x}(g) \to \beta^x(g) = \dot\beta^{G\cdot x}(f)$$

since β is a π-system.

By Proposition 3.18 on page 77, there is a $g \in C_{c,\dot\pi}^+(G\backslash Y)$ such that $\dot\beta(g)(G\cdot x) = 1$ for all $x \in X$. Then $k = g \circ q$ suffices for part (a).

Let g be as above and let α^x be the probability measure on Y given by

$$\alpha^x(f) = \beta^x(f \cdot (g \circ q)).$$

Since $f \cdot (g \circ q) \in C_c(Y)$, $x \mapsto \alpha^x(f)$ is continuous, and α is a π-system consisting of probability measures. It remains only see that α is equivariant. But, since $\gamma \cdot \beta^x = \beta^{\gamma\cdot x}$,

$$\begin{aligned}\int_G f(\gamma \cdot y)\, d\alpha^x(y) &= \int_G f(\gamma \cdot y) g(q(y))\, d\beta^x(y) \\ &= \int_G f(\gamma \cdot y) g(q(\gamma \cdot y)\, d\beta^x(y) \\ &= \int_G f(y) g(q(y))\, d\beta^{\gamma\cdot x}(y) \\ &= \int_G f(y)\, d\alpha^{\gamma\cdot x}(y). \qquad \square\end{aligned}$$

If we are willing to assume that X is a free G space, we can use Blanchard's Lemma 3.20 on page 77 to drop the assumption in Proposition 9.23 on the preceding page that π have an equivariant system.

LEMMA 9.24. *Suppose that G has open range and source maps and that $\pi : Y \to X$ is a continuous, open, equivariant surjection. Suppose that X is a free and proper G-space. Then π is properly amenable.*

PROOF. We retain the notations from the proof of Proposition 9.23 on the facing page. Because the action is free, the map

$$\varphi : Y \to X * (G\backslash Y) \tag{9.10}$$

is a homeomorphism of G-spaces (Ex 9.3.2). Hence we can replace Y with $X*(G\backslash Y)$ and π with the projection onto the first factor. Since the orbit maps are open, $\dot{\pi} : G\backslash Y \to G\backslash X$ is open and there is a full $\dot{\pi}$-system $m = \{m^{G\cdot x}\}_{G\cdot x\in G\backslash X}$ of probability measures by Blanchard's Lemma 3.20. If $f \in C_c(X * (G\backslash Y))$, then we define a probability measure by

$$n^x(f) = \int_{G\backslash Y} f(x, G\cdot y)\, dm^{G\cdot x}(G\cdot y).$$

Then $n = \{n^x\}_{x\in X}$ is a π-system of probability measures. Moreover,

$$n^{\gamma\cdot x}(f) = \int_{G\backslash Y} f(\gamma\cdot x, G\cdot y)\, dm^{G\cdot x}(G\cdot y) = \gamma\cdot n^x(f).$$

Hence n is equivariant and π is properly amenable. □

COROLLARY 9.25. *Suppose that G has open range and source maps.*

(a) *If X is is a properly amenable G-space, then X is a proper G-space.*
(b) *If X is a free and proper G-space, then X is a properly amenable G-space.*
(c) *If X is a proper G-space and G has a Haar system λ, then X is a properly amenable G-space.*

PROOF. (a) The action $\eta\cdot(\gamma, x) = (\eta\gamma, \eta\cdot x)$ is a proper action on $G \rtimes X$ and $r : G\rtimes X \to X$ is equivariant. If X is properly amenable, then r is properly amenable by definition. Thus X is a proper G-space by Lemma 9.22 on page 200.

(b) By definition, X is properly amenable as a G-space when $r : G\rtimes X \to X$ is properly amenable. The openness of $r_G : G \to G^{(0)}$ implies that $r : G\rtimes X \to X$ is open.[4] Thus if X is a free and proper G-space, then (b) follows from Lemma 9.24 on the preceding page.

(c) Suppose X is a proper G-space. Since G has a Haar system λ, then $\beta = \{\lambda^{r(x)} \times \delta_x\}$ is a full equivariant r-system. Hence r is properly amenable by Proposition 9.23(b) on page 200. □

COROLLARY 9.26. *Let G be a locally compact group with a Haar system. Then G is properly amenable if and only if G is a proper groupoid.*

PROOF. The groupoid G is proper if and only if $G^{(0)}$ is a proper G-space. But $G^{(0)}$ is a proper G-space if and only if $G^{(0)}$ is a properly amenable G-space. The latter is equivalent to G being properly amenable. □

EXERCISES

Ex 9.3.1. Suppose that $\pi : Y \to X$ is a continuous surjection, that $A \subset X$ is compact, and that $\alpha = \{\alpha^x\}_{x\in X}$ is a π-system of probability measures. Show that there is a $f \in C_c^+(Y)$ such that $0 \le f(y) \le 1$ for all y and $\alpha(f)(x) > \frac{1}{2}$ for all $x \in A$.

—Answer on page 378

Ex 9.3.2. Show that the map φ in (9.10) is a homeomorphism.

—Answer on page 378

[4] Suppose that $x_i \to x$ and that $(\gamma, x) \in G\rtimes X$. Then $r(x_i) \to r(x) = r_G(\gamma)$. After passing to a subnet and relabeling, we can find $\gamma_i \to \gamma$ such that $r_G(\gamma_i) = r(x_i)$. Then $(\gamma_i, x_i) \to (\gamma, x)$.

9.4. Amenable Maps

The following calculus for amenable maps will be used frequently in the sequel.

PROPOSITION 9.27 ([**ADR00**, Proposition 2.2.4]). *Suppose that $\pi : Y \to X$ and $\rho : Z \to Y$ are continuous G-equivariant surjections.*

(a) *If ρ and π are amenable, then so is $\pi \circ \rho$.*

(b) *If $\pi \circ \rho$ is amenable, then so is π.*

For the proof we need a bit more measure theory:

LEMMA 9.28 ([**ADR00**, Lemma 2.2.3]). *Suppose that $m = \{m^x\}_{x\in X}$ is a continuous $\pi : Y \to X$ system of probability measures. Given $\epsilon > 0$ and a compact set $L \subset X$, there is a compact set $K \subset Y$ such that $m^x(Y \setminus K) < \epsilon$ for all $x \in L$.*

PROOF. Since Y is σ-compact, for each $x \in X$, there is a $f_x \in C_c(Y)$ such that $0 \le f_x \le 1$ and $m^x(f_x) > 1 - \epsilon$. Hence there is a neighborhood U_x of x such that $z \in U_x$ implies $m^z(f_x) > 1 - \epsilon$. Since L is compact, there are $x_1, \dots, x_n$ such that $\bigcup U_{x_i}$ cover L. Now let K be the union of the supports of the f_{x_i}. □

PROOF OF PROPOSITION 9.27. (a) We need to see that for any compact subset T of $G \rtimes X$ and $\epsilon > 0$ that there is a $(\pi \circ \rho)$-system $m = \{m^x\}_{x\in X}$ of probability measures on Z such that

$$\|\gamma^{-1} \cdot m^x - m^{\gamma^{-1}\cdot x}\|_1 \le \epsilon. \tag{9.11}$$

Then our net will be directed by increasing T and decreasing ϵ.

To establish (9.11), it suffices to see that for any $f \in C_0(Z)$ with $\|f\|_\infty \le 1$ and any $(\gamma, x) \in T$, then we have

$$|\gamma^{-1} \cdot m^x(f) - m^{\gamma^{-1}\cdot x}(f)| = \Big|\int_Z f(\gamma^{-1} \cdot z)\, dm^x(x) - \int_Z f(z)\, dm^{\gamma^{-1}\cdot x}(x)\Big| < \epsilon.$$

We can assume $T = C * L$ with C compact in G and L compact in X. Since π is amenable, there is a π-system of probability measures $\alpha = \{\alpha^x\}_{x\in X}$ such that

$$\|\gamma^{-1} \cdot \alpha^x - \alpha^{\gamma^{-1}\cdot x}\|_1 < \frac{\epsilon}{4} \tag{9.12}$$

if $(\gamma, x) \in C * L$. Using Lemma 9.28, we can find a compact set K in Y such that $\alpha^x(Y \setminus K) < \epsilon/4$ for all $x \in L$. Since ρ is amenable, we have a ρ-system $\beta = \{\beta^y\}_{y\in Y}$ of probability measures such that

$$\|\gamma^{-1} \cdot \beta^y - \beta^{\gamma^{-1}\cdot y}\|_1 < \frac{\epsilon}{4} \tag{9.13}$$

provided $(\gamma, y) \in C * K \subset G \rtimes Y$. Then we can define a probability measure on X via

$$m^x(f) = \int_Y \beta^y(f)\, d\alpha^x(y) = \int_Y \int_Z f(z)\, d\beta^y(z)\, d\alpha^x(y).$$

Clearly, $\operatorname{supp} m^x \subset (\pi \circ \rho)^{-1}(x)$ and m is a $\pi \circ \rho$-system. Furthermore, if $(\gamma, x) \in C * L$ and $f \in C_c(Z)$ with $\|f\|_\infty \leq 1$, then

$$\begin{aligned}
|\gamma^{-1} \cdot m^x(f) - m^{\gamma^{-1}\cdot x}(f)| \\
&= \Bigl| \int_Y \gamma^{-1} \cdot \beta^y(f)\, d\alpha^x(y) - \int_Y \beta^y(f)\, d\alpha^{\gamma^{-1}\cdot x}(y) \Bigr| \\
&\leq \Bigl| \int_Y \bigl(\gamma^{-1} \cdot \beta^y(f) - \beta^{\gamma^{-1}\cdot y}(f)\bigr)\, d\alpha^x(y) \Bigr| \\
&\qquad + \Bigl| \int_Y \beta^{\gamma^{-1}\cdot y}(f)\, d\alpha^x(y) - \int_Y \beta^y(f) \alpha^{\gamma^{-1}\cdot x}(y) \Bigr|
\end{aligned}$$

which, since $\|\gamma^{-1} \cdot \beta^y(f) - \beta^{\gamma^{-1}\cdot y}(f)\|_\infty \leq 2$ and our choice of K, is

$$\begin{aligned}
&\leq \frac{\epsilon}{2} + \Bigl| \int_K \bigl(\gamma^{-1} \cdot \beta^y(f) - \beta^{\gamma^{-1}\cdot y}(f)\bigr)\, d\alpha^x(y) \Bigr| \\
&\qquad + \Bigl| \int_Y \beta^y(f)\, d(\gamma^{-1} \cdot \alpha^x)(y) - \int_Y \beta^y(f) \alpha^{\gamma^{-1}\cdot x}(y) \Bigr|
\end{aligned}$$

which, using (9.13) and (9.12), is

$$< \frac{\epsilon}{2} + \frac{\epsilon}{4} + \frac{\epsilon}{4} = \epsilon.$$

This proves part (i).

For part (b), note that if (p_i) is an approximately invariant continuous mean for $\pi \circ \rho$, then we can define an approximately invariant continuous mean for π via

$$m_i^x(g) = \rho_* \cdot p_i^x(g) := p_i^x(g \circ \rho) \tag{9.14}$$

for $g \in C_c(Y)$. (We call $\rho_* \cdot p_i^x$ the *push-forward* of p_i by ρ.) The details are left to Ex 9.4.1. □

The next result should be compared with the corresponding results (and their proofs) in the properly amenable case—see Lemma 9.22 and Proposition 9.23.

PROPOSITION 9.29 ([**ADR00**, Proposition 2.2.9]). *Let $\pi : Y \to X$ be a continuous, G-equivariant surjection.*

(a) *If X is an amenable G-space, then Y is an amenable G-space.*
(b) *Suppose, in addition, that π is open. Then X is an amenable G space if and only if*
    (i) *Y is an amenable G space and*
    (ii) *π is an amenable map.*

PROOF. (a) Let $\{m_i\}$ be an approximately invariant continuous mean for $r : G \rtimes X \to X$ which we view as measures on G (as in Remark 9.17). Define a measure on n_i^y on $G \rtimes Y$ by

$$n_i^y(f) = \int_G f(\gamma, y)\, dm_i^{\pi(y)}(\gamma).$$

It is not hard to see that $\{n_i\}$ is an approximately invariant continuous mean for $r : G \rtimes Y \to Y$. This proves (a).

(b) Since we are now assuming π is open, we can use Lemma 3.20 on page 77 to produce a continuous π-system $\alpha = \{\alpha^x\}$ of probability measures. Let $\mathrm{id} \otimes \pi : G \rtimes Y \to G \rtimes X$ be the obvious map (which is well-defined since π is equivariant).

Then $\mathrm{id} \otimes \pi$ is equivariant for the diagonal G-actions on $G \ltimes Y$ and $G \ltimes X$. For $(\gamma, x) \in G \ltimes X$, let $n^{(\gamma,x)}$ be the measure on $G \ltimes Y$ given by

$$n^{(\gamma,x)}(f) = \int_Y f(\gamma, \gamma \cdot y)\, d\alpha^{\gamma^{-1}\cdot x}(y).$$

Clearly, $n^{(\gamma,x)}$ has support in $(\mathrm{id} \otimes \pi)^{-1}(\gamma, x) = \{\, (\gamma, y) : \pi(y) = x \,\}$. Further it is not hard to see that $(\gamma, x) \mapsto n^{(\gamma,x)}(f)$ is continuous: we have

$$n^{(\gamma,x)}(f) = \int \tilde{f}(\gamma^{-1}, y)\, d\alpha^{\gamma^{-1}\cdot x}(y),$$

where $\tilde{f}(\gamma, y) := f((\gamma, y)^{-1}) = f(\gamma^{-1}, \gamma^{-1} \cdot y)$. Then we can assume $\tilde{f}$ is of the form $g \otimes \varphi$ and approximate.

Then observe that

$$\eta^{-1} \cdot n^{(\gamma,x)}(f) = \int f(\eta^{-1}\gamma, \eta^{-1}\gamma \cdot y)\, d\alpha^{\gamma^{-1}\cdot x}(y) = n^{(\eta^{-1}\gamma, \eta^{-1}\cdot x)}(f).$$

Thus $\{n^{(\gamma,x)}\}_{(\gamma,x)\in G\ltimes X}$ is invariant and $\mathrm{id} \otimes \pi$ is not only amenable, but even properly amenable.

Since

$$\begin{array}{ccc} G \ltimes Y & \xrightarrow{\quad \mathrm{id} \otimes \pi \quad} & G \ltimes X \\ \big\downarrow{\scriptstyle r_{G\ltimes Y}} & & \big\downarrow{\scriptstyle r_{G\ltimes X}} \\ Y & \xrightarrow{\quad \pi \quad} & X \end{array}$$

is a commuting diagram of G-spaces with $\mathrm{id} \otimes \pi$ amenable, $r_{G\ltimes X}$ amenable implies $\pi \circ r_{G\ltimes Y}$ is amenable by Proposition 9.27(a) on page 203. Then π is amenable by Proposition 9.27(b). This together with part (a) implies the ($\Longrightarrow$) part of part (b).

For the converse, if $r_{G\ltimes Y}$ and π are both amenable, then $r_{G\ltimes X} \circ \mathrm{id} \otimes \pi$ is amenable by Proposition 9.27(a), and $r_{G\ltimes X}$ is amenable by Proposition 9.27(b). □

COROLLARY 9.30. *Suppose that G is amenable. Then every G-space X is amenable. In fact, every open G-equivariant surjection is amenable.*

PROOF. Since G is amenable, $G^{(0)}$ is an amenable G-space (Example 9.16 on page 199). Now we can apply Proposition 9.29(a) to the structure map $r_X : X \to G^{(0)}$ to get the first assertion. The second follows from the first and Proposition 9.29(b). □

And now a sort of converse—which will be crucial down the road: specifically in the proof that equivalence of groupoids preserves amenability (Theorem 9.48 on page 217).

PROPOSITION 9.31. *Suppose that G is a second countable locally compact Hausdorff groupoid with open range and source maps. Suppose that Z is a free and proper G-space such that the moment map $r_Z : Z \to G^{(0)}$ is amenable. Then G is amenable. If G has a Haar system, then it suffices for Z to be merely a proper G-space.*

REMARK 9.32. Note that if G is amenable, then G is a (free and) proper G-space such that $r : G \to G^{(0)}$ is amenable.

PROOF. If Z is a free and proper G-space, it is a properly amenable G-space by Corollary 9.25(b) on page 202. If Z is a proper G-space and G has a Haar system, Z is properly amenable by Corollary 9.25(c). In either case, Z is *a fortiori* an amenable G-space. Since $r_Z : Z \to G^{(0)}$ is amenable by assumption, $G^{(0)}$ is an amenable G-space by Proposition 9.29(b). But this just means that G is amenable (Example 9.16 on page 199). □

9.4.1. Variations on Continuous Means. In applications, it will be convenient to add additional requirements to the approximately invariant continuous means in play.

DEFINITION 9.33. A continuous $\pi : Y \to X$ system $\alpha = \{\alpha^x\}$ is *conditionally compactly supported* if given a compact set $L \subset X$, there is a compact set $K \subset Y$ such that $\operatorname{supp} \alpha^x \subset K$ for all $x \in L$. We say that an approximately invariant continuous mean $m = \{m_i\}$ is conditionally compactly supported if each m_i is.

PROPOSITION 9.34. *If $\pi : Y \to X$ is an amenable map, then π admits a conditionally compactly supported approximately invariant continuous mean.*

For the proof of the Proposition, we require the following technical lemma.

LEMMA 9.35. *Suppose that $\{n^x\}_{x\in X}$ is a continuous $\pi : Y \to X$ system of probability measures. If $L \subset X$ is compact and $\epsilon > 0$, then there is a continuous π-system $m = \{m^x\}_{x\in X}$ of probability measures with conditionally compact support such that $\|n^x - m^x\|_1 < \epsilon$ for all $x \in L$.*

PROOF. Fix L and $\epsilon > 0$ as in the statement of the lemma. For convenience, let $0 < \delta < \min\{\,\epsilon/3, \frac{1}{4}\,\}$. Using Proposition 3.18 on page 77, we can find $g \in C^+_{c,\pi}(Y)$ such that $n(g)(x) = 1$ for all x. By Lemma 9.28 on page 203, there is a compact set $K \subset Y$ such that $n^x(Y \setminus K) < \delta$ for all $x \in L$. Let $\varphi_0 \in C_c^+(Y)$ be such that $0 \le \varphi_0 \le 1$ and φ_0 equals 1 on K. Since $a \mapsto n(\varphi_0)(x)$ is continuous and $n(\varphi_0)(x) > 1 - \delta$ if $x \in L$, there is a pre-compact open set $U \supset L$ such that $n(\varphi_0)(x) > \frac{1}{2}$ if $x \in U$. Let $\rho \in C^+(X)$ be such that $0 \le \rho \le 1$, ρ is 1 off U and 0 on L. Let

$$\varphi(y) = \varphi_0(y) + \rho(\pi(y))g(y).$$

Then $\varphi \in C^+_{c,\pi}(Y)$ and we get a conditionally compactly supported π-system $p := \varphi \cdot n$ (Ex 9.4.2). Since $p^x(Y) = n(\varphi)(x)$, $x \mapsto p^x(Y)$ is continuous and nonzero by our choices of φ_0, ρ, and g. Thus $m^x = p^x(Y)^{-1}p^x$ is a π-system of probability measures on Y.

Fix $x \in L$ and $f \in C_0(Y)$ with $\|f\|_\infty \le 1$. Then

$$\begin{aligned} \Big|\int_Y f(y)\,dn^x(y) - \int_Y f(y)\,dm^x(y)\Big| &= \Big|\int_Y f(y)\big(1 - n(\varphi_0)(x)^{-1}\varphi_0(y)\big)\,dn^x(y)\Big| \\ &\le \Big|\int_{Y\setminus K} f(y)\big(1 - n(\varphi_0)(x)^{-1}\varphi_0(y)\big)\,dn^x(y)\Big| \\ &\qquad + \Big|\int_K f(y)\big(1 - n(\varphi_0)(x)^{-1}\big)\,dn^x(y)\Big| \end{aligned}$$

which, since $|1 - n(\varphi_0)(x)^{-1}\varphi_0(y)| \le 1$ and $|1 - n(\varphi_0)(x)^{-1}| < 2\delta$, is

$$\le \delta + 2\delta = \epsilon.$$

Since this holds for all such f, $\|n^x - m^x\|_1 < \epsilon$ if $x \in L$ as required. □

PROOF OF PROPOSITION 9.34. Let $\{m_i\}$ be an approximately invariant continuous mean for π. For each compact set $L \subset X$ and $\epsilon > 0$, use Lemma 9.35 to produce $m_{i,L,\epsilon} = \{m^x_{i,L,\epsilon}\}$ which is a conditionally compactly supported approximately invariant continuous mean such that $\|m^x_{i,L,\epsilon} - m^x_i\|_1 < \epsilon$ for all $x \in L$.

Let K be a compact subset of $G \rtimes X$ and let $\epsilon_0 > 0$. We can assume $K \subset A * B$ with A compact in G and B compact in X. Let i_0 be such that $i \geq i_0$ implies that

$$\|\gamma^{-1} \cdot m^x_i - m^{\gamma^{-1}\cdot x}_i\|_1 < \epsilon_0$$

provided $(\gamma, x) \in K$. Let L_0 be the compact set $B \cup A^{-1} \cdot B \subset X$. If $(i, L, \epsilon) \geq (i_0, L_0, \epsilon_0)$ and $(\gamma, x) \in K$, then (9.7) on page 198 implies that $\|\gamma^{-1} \cdot (m_{i,L,\epsilon} - m^x_i)\|_1 = \|m_{i,L,\epsilon} - m^x_i\|_1$. Hence, we have

$$\begin{aligned}\|\gamma^{-1}&\cdot m^x_{i,L,\epsilon} - m^{\gamma^{-1}\cdot x}_{i,L,\epsilon}\|_1 \\ &\leq \|\gamma^{-1} \cdot (m_{i,L,\epsilon} - m^x_i)\|_1 + \|\gamma^{-1} \cdot m^x_i - m^{\gamma^{-1}\cdot x}_i\|_1 + \|m^{\gamma^{-1}\cdot x}_i - m^{\gamma^{-1}\cdot x}_{i,L,\epsilon}\|_1 \\ &\leq 2\epsilon + \epsilon_0 \leq 3\epsilon_0.\end{aligned}$$

This suffices. □

We are now going to drop our standing requirement that π-systems be proper.

DEFINITION 9.36. Let $m = \{m^x\}$ be a (not necessarily proper) $\pi : Y \to X$ system.

(a) We say m is *compactly supported* if there is a compact set $K \subset Y$ such that $\operatorname{supp} m^x \subset K$ for all $x \in X$.
(b) We say m is *finite* if m^x is a finite measure for all $x \in X$.

DEFINITION 9.37. Let $\pi : Y \to X$ be a continuous, equivariant surjection. Then a *weak approximately invariant continuous mean* for π is a net of compactly supported (not necessarily proper) π-systems $\{m_i\}$ such that

(a) $m^x_i(Y) \leq 1$ for all x and i,
(b) $\lim_i m^x_i(Y) = 1$ uniformly on compact subsets of X, and
(c) $\|\gamma^{-1} \cdot m^x_i - m^{\gamma^{-1}\cdot x}_i\|_1 \to 0$ uniformly on compact subsets of $G \rtimes X$.

PROPOSITION 9.38. *Let $\pi : Y \to X$ be a continuous equivariant surjection. Suppose that either π is open or that π admits a continuous π-system of probability measures. Then π is amenable if and only if π admits a* weak *approximately invariant continuous mean.*

Note that π will necessarily be open if it admits a *full* π-system by Lemma 3.20 on page 77.

COROLLARY 9.39. *Suppose that G is a second countable locally compact groupoid with open range and source maps. Then G is amenable if and only if there is a weak approximately invariant continuous mean for the range map $r : G \to G^{(0)}$.*

PROOF. Recall that G is amenable if and only if $r : G \to G^{(0)}$ is amenable. □

Before proceeding with the proof of the proposition, we need some preliminary results.

LEMMA 9.40. *Let $m = \{m^x\}$ be a continuous, conditionally compactly supported finite $\pi : Y \to X$ system (proper or not). Then $x \mapsto m^x(Y)$ is continuous.*

PROOF. Fix $x_0 \in X$ and let L be a compact neighborhood of x_0. By assumption, there is a compact set $K \subset Y$ such that $\operatorname{supp} m^x \subset K$ for all $x \in L$. Let $f \in C_c^+(Y)$ be such that f is 1 on K. Then for all $x \in L$, $m^x(Y) = m^x(f) = m(f)(x)$. Since $x \mapsto m(f)(x)$ is continuous, this suffices. $\square$

LEMMA 9.41. *Let $\pi : Y \to X$ be a continuous equivariant map. Suppose that $\{m_i\}$ is a net of continuous, conditionally compactly supported π-systems such that*

(a) $0 < m_i^x(Y) \leq 1$ *for all i and all x,*

(b) $\lim_i m_i^x(Y) = 1$ *uniformly on compact subsets of X, and*

(c) $\|\gamma^{-1} \cdot m_i^x - m_i^{\gamma^{-1}\cdot x}\|_1 \to 0$ *uniformly on compact subsets of $G \rtimes X$.*

Then π is amenable.

PROOF. Since we've assumed $m_i^x(Y) > 0$, we can let $n_i^x := m_i^x(Y)^{-1} m_i^x$. By Lemma 9.40, $x \mapsto n_i^x$ is a continuous π-system of probability measures. Let $K \subset G \rtimes X$ be compact and let $\epsilon > 0$. We can find i_0 such that $i \geq i_0$ implies that for all $(\gamma, x) \in K$,

(i) $|1 - m_i^x(Y)^{-1}| < \epsilon/3$,

(ii) $|1 - m_i^{\gamma^{-1}\cdot x}(Y)^{-1}| < \epsilon/3$, and

(iii) $\|\gamma^{-1} \cdot m_i^x - m_i^{\gamma^{-1}\cdot x}\|_1 < \epsilon/3$.

Let $f \in C_0(Y)$ with $\|f\|_\infty \leq 1$. If $i \geq i_0$ and $(\gamma, x) \in K$, then

$$\begin{aligned}
&|\gamma^{-1} \cdot n_i^x(f) - n_i^{\gamma^{-1}\cdot x}(f)| \\
&\quad = \Big| \int_Y m_i^x(Y)^{-1} f(\gamma^{-1}\cdot y)\, dm_i^x(y) - \int_Y m_i^{\gamma^{-1}\cdot x}(Y)^{-1} f(y)\, dm^{\gamma^{-1}\cdot x}(y) \Big| \\
&\quad \leq \Big| \int_Y \big(m_i^x(Y)^{-1} - 1\big) f(\gamma^{-1}\cdot y)\, dm_i^x(y) \Big| \\
&\qquad\qquad + \Big| \int_Y f(\gamma^{-1}\cdot y)\, dm_i^x(y) - f(y)\, dm_i^{\gamma^{-1}\cdot x}(y) \Big| \\
&\qquad\qquad\qquad + \Big| \int_Y \big(1 - m_i^{\gamma^{-1}\cdot x}(Y)^{-1}\big) f(y)\, dm_i^{\gamma^{-1}\cdot x}(y) \Big| \\
&\quad \leq |m_i^x(Y)^{-1} - 1| + \|\gamma^{-1}\cdot m_i^x - m_i^{\gamma^{-1}\cdot x}\|_1 + |1 - m_i^{\gamma^{-1}\cdot x}(Y)^{-1}| \\
&\quad < \epsilon.
\end{aligned}$$

Since f is arbitrary, this suffices. $\square$

PROOF OF PROPOSITION 9.38. Let $\{m_i\}$ be a weak approximately invariant continuous mean for π. Either by assumption or by invoking Lemma 3.20 on page 77, we can fix a continuous π-system $\{n^x\}_{x\in X}$ of probability measures. Let $X = \bigcup_{n=1}^\infty V_n$ with each V_n compact and contained in the interior of V_{n+1}. Note that every compact subset of X is eventually contained in some V_n.

Let

$$J := \Big\{ (i, n, m) \in I \times \mathbf{N} \times \mathbf{N} : |m_i^x(Y) - 1| < \frac{1}{m+1} \text{ if } x \in V_n \Big\}.$$

Note that for all $(n, m) \in \mathbf{N}\times\mathbf{N}$, there is an i_0 such that $i \geq i_0$ implies $(i, n, m) \in J$. In fact, it is not hard to see that J is a directed set. Furthermore, if $(i, n, m) \in J$, then

$$m_i^x(Y) \geq \frac{m}{m+1} \quad \text{if } x \in V_n.$$

Since $x \mapsto m_i^x(Y)$ is continuous (by Lemma 9.40), $(i,n,m) \in J$ implies that

$$A_n = \{\, x \in X : m_i^x(Y) \leq \frac{m}{2(m+1)} \,\}$$

is a closed set disjoint from V_n. Thus there is a $\varphi_{i,n} \in C^+(X)$ such that $0 \leq \varphi_{i,n} \leq 1$, $\varphi_{i,n}$ vanishes on V_n and $\varphi_{i,n}(x)$ is equal to 1 whenever $m_i^x(Y) = 0$. Then $\varphi_{i,n}(x) + m_i^x(Y) > 0$ for all x. We define

$$p^x_{(i,n,m)} = (1 - m_i^x(Y))\varphi_{i,n}(x) \cdot n^x + m_i^x.$$

Note that $p^x_{(i,n,m)} = m_i^x$ if $x \in V_n$.

Then each $x \mapsto p^x_{(i,n,m)}$ is a π-system such that $\{p_{(i,n,m}\}_{(i,n,m)\in J}$ satisfies hypotheses (a) and (b) of Lemma 9.41. Note that if $K \subset G \rtimes X$ is compact, then there is a compact set V in X such that $(\gamma, x) \in K$ implies both x and $\gamma^{-1} \cdot x$ are in V. Thus hypothesis (c) of Lemma 9.41 is also satisfied. Hence π is amenable.

Conversely, if π is amenable, then π admits a conditionally compactly supported approximately invariant continuous mean $\{m_i\}$ by Proposition 9.34. Let $X = \bigcup V_n$ with each V_n compact and contained in the interior of V_{n+1}. Let $\varphi_n \in C_c^+(X)$ be such that $0 \leq \varphi_n \leq 1$ and such that φ_n is equal to 1 on V_n. Then we can let $p^x_{n,i} = \varphi_n(x) \cdot m_i^x$. It is not hard to see that $\{p_{n,i}\}$ is a weak approximately invariant continuous mean. □

9.4.2. Topological Amenability. In this section we want to develop a number of useful criteria for amenability. As a consequence, we'll verify that a groupoid is amenable if and only if it is topologically amenable and therefore connect up with our naïve approach in Section 9.2.

PROPOSITION 9.42. *Let $\alpha = \{\alpha^x\}_{x\in X}$ be a continuous, equivariant $\pi : Y \to X$ system of measures. Suppose in addition that Y is a proper G-space. Then the following are equivalent.*

(a) *π is amenable.*

(b) *There is a net $\{g_i\} \subset C^+_{c,\pi}(Y)$ such that*

  (i) *$\alpha(g_i)(x) = 1$ for all i and x and*

  (ii) *$\int_Y |g_i(\gamma^{-1} \cdot y) - g_i(y)|\, d\alpha^x(y)$ tends to zero uniformly on compact subsets of $G \rtimes X$.*

(c) *There is a net $\{h_i\} \subset C_c^+(Y)$ such that*

  (i) *$\alpha(h_i)(x) \leq 1$ for all i and x,*

  (ii) *$\lim_i \alpha(h_i)(x) = 1$ uniformly on compact subsets of X, and*

  (iii) *$\int_Y |h_i(\gamma^{-1} \cdot y) - h_i(y)|\, d\alpha^x(y)$ tends to zero uniformly on compact subsets of $G \rtimes X$.*

PROOF. (a)$\Longrightarrow$(b): Suppose that π is amenable. Then by Proposition 9.34 on page 206, π admits a conditionally compactly supported approximately invariant continuous mean $\{m_i\}$. Let

$$Y *_\pi Y = \{\, (y, z) \in Y \times Y : \pi(y) = \pi(z) \,\},$$

and let $s : Y *_\pi Y \to Y$ be the projection onto the second factor. Define a measure on $Y *_\pi Y$ by

$$\beta^z(f) = \int_Y f(y,z)\,d\alpha^{\pi(z)}(y).$$

Then

$$\begin{aligned}\gamma \cdot \beta^z(f) &= \int_Y f(\gamma \cdot y, \gamma \cdot z)\,d\alpha^{\pi(z)}(y)\\ &= \int_Y f(y, \gamma \cdot z)\,d\alpha^{\gamma\cdot\pi(z)}(y)\\ &= \beta^{\gamma\cdot z}(f).\end{aligned}$$

Thus β is an equivariant, continuous s-system. Since Y is a proper G-space, so is $Y *_\pi Y$ and Proposition 9.23(a) on page 200 implies there in a G-invariant $f \in C_{c,s}^+(Y *_\pi Y)$ such that

$$\beta(f)(z) = \int_Y f(y,z)\,d\alpha^{\pi(z)}(y) = 1 \quad \text{for all } y \in Y.$$

The idea now is to define $g_i \in C_{c,\pi}^+(Y)$ by

$$g_i(y) = \int_Y f(y,z)\,dm_i^{\pi(y)}(z).$$

To see that this will work, consider a compact set $L \subset X$. By assumption, there is a compact set $K \subset Y$ such that $\operatorname{supp} m_i^x \subset K$ if $x \in L$. Thus if $y \in \pi^{-1}(L)$, we have

$$g_i(y) = \int_Y f(y,z)\,dm_i^{\pi(y)}(z) = \int_K f(y,z)\,dm_i^{\pi(y)}(z) < \infty, \tag{9.15}$$

and g_i has support in the projection onto the first factor of the compact set $\operatorname{supp} f \cap (Y \times K) = \operatorname{supp} f \cap s^{-1}(K)$. Using this, it is not hard to see that g_i is continuous (Ex 9.4.3). (For example, we can let L be neighborhood of $\pi(y)$ for a given $y \in Y$.) Hence $g_i \in C_{c,\pi}^+(Y)$ as required.

Using Fubini,

$$\begin{aligned}\alpha(g_i)(w) &= \int_Y g_i(y)\,d\alpha^w(y)\\ &= \int_Y\int_Y f(y,z)\,dm_i^w(z)\,d\alpha^w(y)\\ &= \int_Y\int_Y f(y,z)\,d\alpha^w(y)\,dm_i^w(z)\\ &= \int_Y dm_i^w(y) = 1.\end{aligned}$$

Thus (b)(i) holds.

But if $(\gamma, x) \in G \ltimes X$, we have

$$\begin{aligned}&\int_Y |g_i(\gamma^{-1}\cdot y) - g_i(y)|\,d\alpha^x(y)\\ &\quad= \int_Y \Bigl|\int_Y f(\gamma^{-1}\cdot y, z)\,dm_i^{\gamma^{-1}\cdot x}(z) - \int_Y f(y,z)\,dm_i^x(z)\Bigr|\,d\alpha^x(y)\end{aligned}$$

which, since f is invariant, is

$$= \int_Y \Big| \int_Y f(\gamma^{-1}\cdot y, z)\, dm_i^{\gamma^{-1}\cdot x}(z) - \int_Y f(\gamma^{-1}\cdot y, \gamma^{-1}\cdot z)\, dm_i^x(z) \Big|\, d\alpha^x(y)$$

$$= \int_Y \Big| \int_Y f(\gamma^{-1}\cdot y, z)\, dm_i^{\gamma^{-1}\cdot x}(z) - \int_Y f(\gamma^{-1}\cdot y, z)\, d(\gamma^{-1}\cdot m_i^x)(z) \Big|\, d\alpha^x(y)$$

which, using the properties of the total variation measure, is

$$\leq \int_Y \int_Y f(\gamma^{-1}\cdot y, z)\, d|m_i^{\gamma^{-1}\cdot x} - \gamma^{-1}\cdot m_i^x|(z)\, d\alpha^x(y)$$

which, by Fubini, is

$$= \int_Y \int_Y f(\gamma^{-1}\cdot y, z)\, d\alpha^x(y)\, d|m_i^{\gamma^{-1}\cdot x} - \gamma^{-1}\cdot m_i^x|(z)$$

which, by invariance of f, is

$$= \int_Y \int_Y f(y, \gamma\cdot z)\, d\alpha^x(y)\, d|m_i^{\gamma^{-1}\cdot x} - \gamma^{-1}\cdot m_i^x|(z)$$

which, since we can assume $\pi(z) = \gamma^{-1}\cdot x$ and hence that the inner integral is 1, is

$$= \|m_i^{\gamma^{-1}\cdot x} - \gamma^{-1}\cdot m_i^x\|_1.$$

Thus (b)(ii) holds, and we've proved that (a)$\Longrightarrow$(b).

(b)$\Longrightarrow$(c): This follows easily by cutting down the g_i as in last part of the proof of Proposition 9.38.

(c)$\Longrightarrow$(a): Let $\{h_i\}$ be as in part (c). Define $m_i^x = h_i \cdot \alpha^x$. It follows that $x \mapsto m_i^x$ is a (not necessarily proper) π-system of measures satisfying (a) and (b) of the definition of weak approximately invariant continuous mean for π (see Definition 9.37 on page 207). But if $f \in C_c(Y)$, then if $(\gamma, x) \in G \ltimes X$,

$$\begin{aligned} &\gamma^{-1}\cdot m_i^x(f) - m_i^{\gamma^{-1}\cdot x}(f) \\ &\quad = \int_Y f(\gamma^{-1}\cdot y)\, dm_i^x(y) - \int_Y f(y)\, dm_i^{\gamma^{-1}\cdot x}(y) \\ &\quad = \int_Y f(\gamma^{-1}\cdot y) h_i(y)\, d\alpha^x(y) - \int_Y f(y) h_i(y)\, d\alpha^{\gamma^{-1}\cdot x}(y) \end{aligned}$$

which, since α is invariant, is

$$= \int_Y f(\gamma^{-1}\cdot y)\big(h_i(y) - h_i(\gamma^{-1}\cdot y)\big)\, d\alpha^x(y).$$

Therefore if $\|f\|_\infty \leq 1$, then

$$\big|\gamma^{-1}\cdot m_i^x(f) - m_i^{\gamma^{-1}\cdot x}(f)\big| \leq \int_Y \big|h_i(y) - h_i(\gamma^{-1}\cdot y)\big|\, d\alpha^x(y).$$

It follows that $\{m_i\}$ is a weak approximately invariant continuous mean. Thus π is amenable by Proposition 9.38 on page 207. □

For the sake of clarity, we now replace a general equivariant map $\pi : Y \to X$ with equivariant π system α with the map $r : G \to G^{(0)}$ and a Haar system λ on G.

THEOREM 9.43. *Let G be a second countable locally compact Hausdorff groupoid with Haar system λ. Then the following are equivalent.*

(a) *G is amenable.*

(b) *There is a net* $\{g_i\} \subset C_c^+(G)$ *such that*

(i) $\lambda(g_i)(u) \leq 1$ *for all* i *and all* $u \in G^{(0)}$,

(ii) $\lim_i \lambda(g_i)(u) = 1$ *uniformly on compact subsets of* $G^{(0)}$, *and*

(iii) $\lim_i \int_G |g_i(\gamma^{-1} \cdot \eta) - g_i(\eta)| \, d\lambda^{r(\gamma)}(\eta) \to 0$ *uniformly on compact subsets of* G.

(c) *There is a net* $\{\xi_i\} \subset C_c(G)$ *such that*

(i) $\int_G |\xi_i(\gamma)|^2 \, d\lambda^u(\gamma) \leq 1$ *for all* i *and all* $u \in G^{(0)}$,

(ii) $\lim_i \int_G |\xi_i(\gamma)|^2 \, d\lambda^u(\gamma) \to 1$ *uniformly on compact subsets of* $G^{(0)}$, *and*

(iii) $\lim_i \int_G |\xi_i(\gamma^{-1}\eta) - \xi_i(\eta)|^2 \, d\lambda^{r(\gamma)}(\gamma) \to 0$ *uniformly on compact subsets of* G.

(d) *There is a net* $\{\xi_i\} \subset C_c(G)$ *such that*

(i) $\xi_i * \xi_i^*(u) \leq 1$ *for all* i *and all* $u \in G^{(0)}$ *and*

(ii) $\lim_i \xi_i * \xi_i^*(\gamma) \to 1$ *uniformly on compact subsets of* G.

(e) *There is a net* $\{h_i\}$ *of positive type functions in* $C_c(G)$ *such that*

(i) $h_i(u) \leq 1$ *for all* i *and all* $u \in G^{(0)}$ *and*

(ii) $\lim_i h_i(\gamma) = 1$ *uniformly on compact subsets of* G.

Moreover, we can replace the nets in (b), (c), (d) *and* (e) *with sequences. Also we can drop the normalization conditions specified by item* (i) *in each of* (b), (c), (d) *and* (e).

REMARK 9.44. It is important in Theorem 9.43 that we can use *either* sequences or nets. It can be much easier in trying to establish amenability to have the added flexibility to produce a net with the given properties rather than a sequence. But given amenability, a sequence with the given property might be easier to apply in practice. Using recent remarkable work of Renault's in [**Ren15**], we can lower the bar still further and introduce Borel versions of part (b) and part (c). We'll take this up in Section 9.6 (see Theorem 9.58 on page 222 and Ex 9.6.2). But in the Borel category, using sequences will be critical.

Before we proceed with the proof, we need some comments about functions of positive type on G and a nontrivial result generalizing a well known result of Godement [**Dix77**, Theorem 13.8.6]. This material is taken from [**ADR00**, §2.2b].

If $f, g \in C_c(G)$, then we define

$$(f, g)(\gamma) = \int_G \overline{f(\eta)} g(\gamma^{-1}\eta) \, d\lambda^{r(\gamma)}(\eta) = g * f^*(\gamma^{-1}). \tag{9.16}$$

As in Ex 9.2.1 on page 196, a straightforward computation shows that (f, f) is a continuous compactly supported function of positive type for any $f \in C_c(G)$.

We can make $C_c(G)$ into a pre-Hilbert $C_0(G^{(0)})$-module via

$$\langle f \, , g \rangle_\star(u) = \int_G \overline{f(\gamma)} g(\gamma) \, d\lambda^u(\gamma) = (g \mid f)_{L^2(G^u, \lambda^u)}.$$

We complete with respect to the induced norm

$$\|f\|_{r,2} = \|\langle f\,,\,f\rangle_{\star}\|_\infty = \|\lambda(|f|^2)\|_\infty^{\frac{1}{2}}$$

to get a Hilbert $C_0(G^{(0)})$-module we denote $C_0(G^{(0)}, L^2(G,\lambda))$.

Notice that

$$|(\xi,\zeta)(\gamma)| \le \lambda(|\xi|^2(r(\gamma))^{\frac{1}{2}}\lambda(|\zeta|^2)(s(\gamma))^{\frac{1}{2}} \le \|\xi\|_{r,2}\|\zeta\|_{r,2}. \tag{9.17}$$

It follows that if $\{f_i\}$ and $\{g_i\}$ are sequences in $C_c(G)$ converging to h and k, respectively, in $C_0(G^{(0)}, L^2(G,\lambda))$, then $\{(f_i, g_i)\}$ is uniformly Cauchy. Hence it converges to a continuous function we denote by (h, k). It particular, (h, h) is a continuous function of positive type on G for all $h \in C_0(G^{(0)}, L^2(G,\lambda))$.

LEMMA 9.45 ([**ADR00**, Lemma 2.2.12]). *Let G be a second countable locally compact groupoid with a Haar system λ. If k is a compactly supported continuous function on G of positive type, then there is a $h \in C_0(G^{(0)}, L^2(G,\lambda))$ such that $k = (h, h)$.*

PROOF. Define an operator $\rho(k)$ on $C_c(G)$ via

$$\rho(k)f(\gamma) = \int_G k(\gamma^{-1}\eta)f(\eta)\,d\lambda^{r(\gamma)}(\eta).$$

Let $U : L^2(G_u, \lambda_u) \to L^2(G^u, \lambda^u)$ be the unitary given by $U(f)(\gamma) = f(\gamma^{-1})$. Let L^u be the representation of $C^*(G)$ on $L^2(G^u, \lambda^u)$ given by $L^u(f) = U(\operatorname{Ind}\delta_u)U^*$. Check that

$$L^u(f)h(\gamma) = \int_G f(\gamma^{-1}\eta)h(\eta)\,d\lambda^{r(\gamma)}(\eta).$$

Therefore,

$$\begin{aligned} \big|\langle\rho(k)f\,,\,g\rangle_{\star}(u)\big| &= \big|(g \mid L^u(k)f)\big| \\ &\le \|k\|\|g\|_{L^2(G^u)}\|f\|_{L^2(G^u)} \\ &= \|k\|\lambda(|g|^2(u)^{\frac{1}{2}}\lambda(|f|^2)(u)^{\frac{1}{2}}. \end{aligned} \tag{9.18}$$

Hence

$$\|\langle\rho(k)f\,,\,g\rangle_{\star}\|_\infty \le \|k\|\|f\|_{r,2}\|g\|_{r,2},$$

and $\rho(k)$ is a bounded operator on $C_0(G^{(0)}, L^2(G,\lambda))$ with adjoint $\rho(k^*)$.[5] Moreover,

$$\langle\rho(k)f\,,\,f\rangle_{\star}(u) = \int_G\int_G \overline{k(\gamma^{-1}\eta)f(\eta)}f(\gamma)\,d\lambda^u(\eta)\,d\lambda^u(\gamma)$$

which is non-negative by Ex 9.2.4.

Hence $\rho(k)$ is a positive operator in $\mathcal{L}(C_0(G^{(0)}, L^2(G,\lambda)))$. Hence $\rho(k)$ has a positive square root $\rho(k)^{\frac{1}{2}}$ that is the norm limit of polynomials in $\rho(k)$. Let $\{e_n\}$ be an self-adjoint approximate identity for $C_c(G)$ in the inductive limit topology with $\|e_n\|_I \le 1$ for all n as in Proposition 1.49 on page 20. Let $f_n = \rho(k)^{\frac{1}{2}}e_n \in C_0(G^{(0)}, L^2(G,\lambda))$. Note that

$$\langle f_n\,,\,f_m\rangle_{\star}(u) = \langle\rho(k)e_n\,,\,e_m\rangle_{\star}(u)$$

[5] It is not hard to see that since k is of positive type, $k = k^*$. See Ex 9.2.4.

which, as in (9.18), is

$$= \big(e_m \mid L^u(k)e_n\big)$$

which, since $e_n^* = e_n$, works out to be

$$= e_m * \overline{k} * e_n(u).$$

Which in view of Ex 1.4.12, converges uniformly to $\overline{k(u)}$. Since

$$\begin{aligned}&\langle f_n - f_m \,, f_n - f_m\rangle_\star(u)\\ &\qquad = \langle f_n \,, f_n\rangle_\star(u) - \langle f_n \,, f_m\rangle_\star(u) - \langle f_m \,, f_n\rangle_\star(u) + \langle f_m \,, f_m\rangle_\star(u),\end{aligned}$$

it follows that $\{f_n\}$ is Cauchy in $C_0(G^{(0)}, L^2(G,\lambda))$ and converges to an $h \in C_0(G^{(0)}, L^2(G,\lambda))$.

For each $\gamma \in G$, we get a unitary operator $\tau_\gamma : L^2(G^{r(\gamma)}) \to L^2(G^{s(\gamma)})$ via $\tau_\gamma h(\eta) = h(\gamma^{-1}\eta)$. Furthermore,

$$(f,g)(\gamma) = \langle f \,, \tau_\gamma g\rangle_\star(r(\gamma)).$$

We can compute that

$$\langle f \,, \tau_\gamma(\rho(k)g)\rangle_\star(r(\gamma)) = \langle f \,, \rho(k)(\tau_\gamma(g))\rangle_\star(r(\gamma)). \tag{9.19}$$

Since $\rho(k)^{\frac{1}{2}}$ is the norm limit of polynomials in $\rho(k)$, we can conclude from (9.19) that

$$\langle f \,, \tau_\gamma(\rho(k)^{\frac{1}{2}}g)\rangle_\star(r(\gamma)) = \langle f \,, \rho(k)^{\frac{1}{2}}(\tau_\gamma(g))\rangle_\star(r(\gamma)). \tag{9.20}$$

Hence

$$\begin{aligned}(h,h)(\gamma) &= \langle h \,, \tau_\gamma(h)\rangle_\star(r(\gamma))\\ &= \lim_n \langle f_n \,, \tau_\gamma(f_n)\rangle_\star(r(\gamma))\end{aligned}$$

which, in view of (9.20), is

$$\begin{aligned}&= \lim_n \langle \rho(k)(e_n) \,, \tau_\gamma(e_n)\rangle_\star\\ &= \lim_n (\tau_\gamma(e_n) | L^{r(\gamma)}(k)(e_n))\end{aligned}$$

which, after a straightforward computation, is

$$= \lim_n e_n * \overline{k} * e_n(\gamma^{-1})$$

which, using view of Ex 1.4.12 and Ex 9.2.4, is

$$= \overline{k(\gamma^{-1})} = k(\gamma).$$

This completes the proof of the lemma. □

Proof of Theorem 9.43 on page 211. Since G is a proper G-space which is amenable precisely when $r : G \to G^{(0)}$ is amenable, the equivalence of (a) and (b) follows immediately from Proposition 9.42 on page 209.

We next show that (b), (c), (d) and (e) are equivalent. At the same time we show that (b)′, (c)′, (d)′ and (e)′ are equivalent where the prime versions are without the normailizing condition (i) in each case.

(c)$\Longrightarrow$(b): Let $\{\xi_i\}$ be as in (c)$'$. Let $g_i(\gamma) = |\xi_i(\gamma)|^2$. Thus we get (b)(ii) immediately. Since $\big||a| - |b|\big| \le |a - b|$, it follows that $\big||a|^2 - |b|^2\big| \le \big(|a| + |b|\big)|a - b|$. Hence

$$\int_G |g_i(\gamma^{-1}\eta) - g_i(\eta)|\, d\lambda^{r(\gamma)}(\eta)$$
$$\le \int_G \big(|\xi_i(\gamma^{-1}\eta)| + |\xi_i(\eta)|\big)\big|\xi_i(\gamma^{-1}\eta) - \xi_i(\eta)\big|\, d\lambda^{r(\gamma)}(\eta)$$

which, by Cauchy-Schwarz, is

$$\le \Big(\int_G (|\xi_i(\gamma^{-1}\eta)| + |\xi_i(\eta)|)^2\, d\lambda^{r(\gamma)}(\eta)\Big)^{\frac12} \Big(\int_G |\xi_i(\gamma^{-1}\eta) - \xi_i(\eta)|^2\, d\lambda^{r(\gamma)}(\eta)\Big)^{\frac12}$$
$$\le 2\Big(\int_G |\xi_i(\gamma^{-1}\eta) - \xi_i(\eta)|^2\, d\lambda^{r(\gamma)}(\eta)\Big)^{\frac12}.$$

Thus, (c)$'$ $\Longrightarrow$(b)$'$. Since b(i) is an immediate consequence of c(i), we also have (c)$\Longrightarrow$(b).

(b)$\Longrightarrow$(d): Given $\{g_i\}$ as in (b)$'$, let $\xi_i(\gamma) = \sqrt{g_i(\gamma)}$. Now let $h_i(\gamma) = (\xi_i, \xi_i)(\gamma)$ as in (9.16). To establish (d)$'$, it will suffice to show that $\{h_i\}$ satisfies (ii) of (e). But on the one hand,

$$\int_G |\xi_i(\gamma^{-1}\eta) - \xi_i(\eta)|^2\, d\lambda^{r(\gamma)}(\eta)$$
$$= \int_G \big(|\xi_i(\gamma^{-1}\eta)|^2 + |\xi_i(\eta)|^2\big)\, d\lambda^{r(\gamma)}(\eta) - 2h_i(\gamma)$$
$$= \alpha(g_i)(s(\gamma)) + \alpha(g_i)(r(\gamma)) - 2h_i(\gamma).$$

And on the other hand, since $a, b \ge 0$ implies $|a - b|^2 \le |a^2 - b^2|$,

$$\int_G |\xi(\gamma^{-1}\eta) - \xi_i(\eta)|^2\, d\lambda^{r(\gamma)}(\eta) \le \int_G |g_i(\gamma^{-1}\eta) - g_i(\eta)|\, d\lambda^{r(\gamma)}(\eta).$$

Thus $\alpha(g_i)(s(\gamma)) + \alpha(g_i)(r(\gamma)) - 2h_i(\gamma)$ goes to zero uniformly on compact subsets of G. Since $\alpha(g_i)(u) \to 1$ uniformly on compact subsets of $G^{(0)}$, we have shown that (b)$'$ $\Longrightarrow$(d)$'$, and then (b)$\Longrightarrow$(d) is clear.

Of course, (d)$'$ $\Longrightarrow$(e)$'$ and (d)$\Longrightarrow$(e) are immediate, so we are left with (e)$\Longrightarrow$(c). Let $\{h_i\}$ be as in (e)$'$. Then using Ex 9.2.3, we can assume $h_i(u) \le 1$ for all $u \in G^{(0)}$. Lemma 9.45 on page 213 implies that each h_i is of the form (ξ_i, ξ_i) with $\xi_i \in C_0(G^{(0)}, L^2(G, \lambda))$. As in Ex 9.4.4, we can use (9.17) to replace $\{\, h_i \,\}$ with a new net, with possibly a different index set, such that each h_i is of the form (ξ_i, ξ_i) with $\xi_i \in C_c(G)$. Furthermore, if $\{\, h_i \,\}$ was a sequence to begin with, we can take the new net to be a sequence as well.

Now we observe that

$$\int_G |\xi_i(\gamma^{-1}\eta) - \xi_i(\eta)|^2\, d\lambda^{r(\gamma)}(\eta) = h_i(s(\gamma)) + h_i(r(\gamma)) - 2h_i(\gamma).$$

This shows (e)$'$ $\Longrightarrow$(c) and (c) trivially implies (c)$'$.

Since G is second countable, it is the increasing union of compact sets each contained in the interior of the next. Thus if $\{h_i\}$ is a net satisfying (e) above, it can be replaced with a sequence. And as observed above, that sequence can be of the form $h_i = (\xi_i, \xi_i)$ with each $\xi_i \in C_c(G)$. Now it follows from the above arguments that we get sequences in (b), (c) and (d) as required.

As for dropping the normalization condition (i), we have already observed in Lemma 9.5 on page 190 that (d) and (d)′ are equivalent. Hence so are the other prime versions. □

With Theorem 9.43 on page 211 in hand, we no longer need to distinguish between amenable groupoids and topologically amenable groupoids.

COROLLARY 9.46. *Let G be a second countable locally compact Hausdorff groupoid. Then G is amenable if and only if G is topologically amenable.*

PROOF. Condition (d) in Theorem 9.43 says exactly that G is topologically amenable. □

By definition an amenable groupoid has an approximately invariant continuous mean. It is worth remarking that we can insist that the approximately invariant continuous mean be given by a *sequence* of r-systems.

COROLLARY 9.47. *Suppose that G is an amenable second countable locally compact Hausdorff groupoid. Then G has an approximately invariant continuous mean m which is a* sequence *$\{m_i\}_{i=1}^{\infty}$ of r-systems.*

PROOF. Since G is amenable, it has a sequence $\{h_i\} \subset C_c^+(G)$ satisfying Theorem 9.43(b) on page 211, and hence also Proposition 9.42(c) on page 209. The proof of (c) implies (a) in Proposition 9.42 yields the desired approximately invariant continuous mean. □

EXERCISES

Ex 9.4.1. Show that (9.14) defines an approximately invariant continuous mean for π as claimed in the proof of Proposition 9.27.

—Answer on page 378

Ex 9.4.2. Show that the π-system p constructed in the proof of Lemma 9.35 is conditionally compactly supported.

—Answer on page 378

Ex 9.4.3. Show that g_i as defined in (9.15) is continuous on Y.

—Answer on page 379

Ex 9.4.4. Show that the net $h_i = (\xi_i, \xi_i)$ in the (e)$\Longrightarrow$(c) portion of the proof of Theorem 9.43 can be replaced with $\xi_i \in C_c(G)$ with a possibly different index set. Observe that if $\{h_i\}$ is a sequence, the new net can be taken to be a sequence as well.

—Answer on page 379

9.5. Amenability and Equivalence

In this section, I want to present Anantharaman-Delaroche and Renault's proof that amenability is preserved under equivalence. The proof does not require the existence of Haar systems. Nevertheless, recall that we are assuming that all our groupoids have open range and source maps. If G and H have Haar systems, then we will give a different proof in Section 9.7 (Theorem 9.75 on page 232) where a much weaker form of equivalence is used—it is only required that G and H be Borel equivalent as in Definition 9.72 on page 231. I've included the topological proof here for completeness.

THEOREM 9.48 ([**ADR00**, Theorem 2.2.17]). *Suppose that G and H are equivalent (second countable locally compact Hausdorff) groupoids. If H is amenable, then G is amenable.*

PROOF. Let Z be a (G, H)-equivalence so that we can identify $(Z *_s Z)/H$ with G (see Lemma 2.42 on page 46). Assume that H is amenable. By Corollary 9.39 on page 207, we can find a weak approximately invariant continuous mean $m = \{m_i\}$ for $r_H : H \to H^{(0)}$. Proposition 9.31 on page 205 implies that it will suffice to see that that the quotient map $q : Z \to Z/H$ is amenable: it is the moment map for the $G := (Z *_s Z)/H$-action.

Since q is open, Lemma 3.20 on page 77 implies that there is a full q-system $\{n^{[z]}\}_{[z]\in Z/H}$ of probability measures. For a given i and z, define a measure p_i^z on Z via

$$p_i^z(f) = \int_H f(z \cdot \eta)\, dm_i^{s(z)}(\eta) \quad \text{for } f \in C_c(Z).$$

Note that $\operatorname{supp} p_i^z \subset Z^{[z]} = q^{-1}([z]) = z \cdot H$. But, as p_i^z depends on z and not just $[z]$, it isn't necessarily q-system. Since the $\{m_i\}$ are a weak approximately invariant continuous mean, for a fixed i, the m_i^v have supports contained in a common compact subset of H. Hence if $f \in C_c(Z)$, then

$$y \mapsto \int_H f(y \cdot \eta)\, dm_i^{s(y)}(\eta) \tag{9.21}$$

has compact support. If K is a compact neighborhood of z in Z, the properness of the action implies $F(y, \eta) = f(y \cdot \eta)$ is in $C_c(K * H)$, and then a standard argument implies that (9.21) is continuous.[6]

Therefore we can define measures on Z by

$$w_i^{[z]}(f) = \int_Z \int_H f(y \cdot \eta)\, dm_i^{s(y)}(\eta)\, dn^{[z]}(y) = \int_Z p_i^y(f)\, dn^{[z]}(y).$$

Since n is a q-system and $y \mapsto p_i^y(f)$ is in $C_c(Z)$, it follows that $w_i = \{w_i^{[z]}\}_{[z]\in Z/H}$ is a continuous q-system. It only remains to see that $\{w_i\}$ is a weak approximately invariant continuous mean. (This suffices by Proposition 9.38 on page 207.)

But the $\{w_i\}$ clearly satisfy axiom (a) (of Definition 9.37 on page 207). Since $y \mapsto m_i^{s(y)}(H)$ is continuous by Lemma 9.40 on page 207, we have

$$w_i^{[z]}(Z) = \int_Z m_i^{s(y)}(H)\, n^{[z]}(y).$$

Since $y \mapsto m_i^{s(y)}(H)$ converges uniformly to 1 on Z, it is not hard to see that axiom (b) holds as well. So it only remains to show the approximate invariance (a.k.a. axiom (c)).

Suppose that K is compact in $G \rtimes Z/H$. Since the quotient map from $Z *_s Z \to G$ is open, there is a compact set L in Z such that $K \subset K' := \{ ([y, x], [y]) \in G \rtimes Z/H : x, y \in L \}$. Using Lemma 9.28 on page 203, we can enlarge L if necessary such that $n^{[z]}(L) > 1 - \frac{\epsilon}{8}$ if $z \in L$. Since the right H-action is proper,

$$M = \{ \eta \in H : L \cap L \cdot \eta^{-1} \neq \emptyset \}$$

[6]Approximate F by functions of the form $g(y)h(\eta)$, etc.

is compact. Let i_0 be such that $i \geq i_0$ and $\eta \in M$ implies that

$$\|\eta \cdot m_i^{s(\eta)} - m_i^{r(\eta)}\|_1 < \epsilon/4. \tag{9.22}$$

Now let $x_1, y_1 \in L$ with $s(x_1) = s(y_1)$ (so that $([y_1, x_1], [y_1]) \in K'$). If $x \in L$ satisfies $[x] = [x_1]$, then $x_1 = x \cdot \eta$ for some $\eta \in M$, and it follows that $\|p_i^x - p_i^{x_1}\|_1 \leq \epsilon/4$ (Ex 9.5.1). Similarly, if $[y] = [y_1]$ with $y \in L$, then $\|p_i^y - p_i^{y_1}\|_1 \leq \epsilon/4$. Since $n^{[z]}(Z \setminus L) < \epsilon/8$ for $z \in L$, we can also show that $\|w_i^{[x_1]} - p_i^{x_1}\|_1 \leq \epsilon/2$ and $\|w_i^{[y_1]} - p_i^{y_1}\|_1 \leq \epsilon/2$ (Ex 9.5.2). But then,

$$\begin{aligned}&\|[x_1, y_1] \cdot w_i^{[y_1]} - w_i^{[x_1]}\|_1 \\ &\quad \leq \|[x_1, y_1] \cdot w_i^{[y_1]} - [x_1, y_1] \cdot p_i^{y_1}\|_1 + \|[x_1, y_1] \cdot p_i^{y_1} - p_i^{x_1}\|_1 + \|p_i^{x_1} - w_i^{[x_1]}\|_1\end{aligned}$$

which, in view of (9.7) on page 198, is

$$= \|w_i^{[y_1]} - p_i^{y_1}\|_1 + \|[x_1, y_1] \cdot p_i^{y_1} - p_i^{x_1}\|_1 + \|p_i^{x_1} - w_i^{[x_1]}\|_1.$$

Since

$$\begin{aligned}[x_1, y_1] \cdot p_i^{y_1}(f) &= \int_H f([x_1, y_1] \cdot y_1 \cdot \eta)\, dm_i^{s(x_1)}(\eta) \\ &= \int_H f(x_1 \cdot \eta)\, dm_i^{s(x_1)}(\eta) \\ &= p_i^{x_1}(f),\end{aligned}$$

we have $\|[x_1, y_1] \cdot p_i^{y_1} - p_i^{x_1}\|_1 = 0$, and hence

$$\|[x_1, y_1] \cdot w_i^{[y_1]} - w_i^{[x_1]}\|_1 \leq \epsilon.$$

This suffices. □

Exercises

Ex 9.5.1. Show that $\|p_i^x - p_i^{x_1}\|_1 \leq \epsilon/4$ as in the proof of Theorem 9.48.
—Answer on page 379

Ex 9.5.2. Show that $\|w_i^{[x_1]} - p_i^{x_1}\|_1 \leq \epsilon/2$ as in the proof of Theorem 9.48.
—Answer on page 379

9.6. Topological Invariant Means

In this section, we fix a Haar system λ on a second countable locally compact Hausdorff groupoid G. We let $\mathcal{E} = C_0(G^{(0)}, L^1(G, \lambda))$ be the completion of $C_c(G)$ with respect to the norm[7]

$$\|f\|_{\lambda,1} = \sup_{u \in G^{(0)}} \lambda(|f|) = \sup_{u \in G^{(0)}} \int_G |f(\gamma)|\, d\lambda^u(\gamma).$$

Since it is not hard to check that

$$\|f * g\|_{\lambda,1} \leq \|f\|_{\lambda,1} \|g\|_{\lambda,1}, \tag{9.23}$$

[7]Of course, $\|f\|_{\lambda,1}$ is the same as $\|f\|_{I,r}$ defined in Section 1.4. However, this notation motivates that in Appendix A.3.

it follows that $\mathcal{E}$ is a Banach algebra with respect to convolution.[8] Since $C_c(G)$ is dense in $\mathcal{E}$ and since $C_c(G)$ is separable in the inductive limit topology (Lemma C.2 on page 344), it follows that $\mathcal{E}$ is separable.

Complex Radon measures, as opposed to complex measures, are discussed in Section 3.3. If μ_1 and μ_2 are (positive) measures, then we write $\mu_1 \leq \mu_2$ if $\mu_1(E) \leq \mu_2(E)$ for all Borel sets E (Ex 9.6.1).

DEFINITION 9.49. A complex Radon measure on G is λ-*bounded* if there is a finite (positive) measure μ on $G^{(0)}$ such that $|\nu| \leq \mu \circ \lambda$. We set

$$\|\nu\| = \inf\{\, \|\mu\|_1 : |\nu| \leq \mu \circ \lambda \,\}$$

where $\|\mu\|_1 = \mu(G^{(0)})$. The collection of λ-bounded measures is denoted by $M(G, \lambda)$.

Using Ex 9.6.1, it is not hard to see that $M(G, \lambda)$ is a normed vector space. If $\nu \in M(G, \lambda)$ and $|\nu| \leq \mu \circ \lambda$, then for all $f \in C_c(G)$ we have

$$|\nu(f)| \leq |\nu|(|f|) \leq \mu \circ \lambda(|f|) \leq \|\mu\|_1 \|f\|_{\lambda,1}.$$

It follows that $|\nu(f)| \leq \|\nu\| \|f\|_{\lambda,1}$. Hence ν defines a linear functional on $\mathcal{E}$ of norm at most $\|\nu\|$. The following is a special case of Proposition A.10 on page 319.

PROPOSITION 9.50. *The correspondence given above allows us to identify* $M(G, \lambda)$ *with the norm dual of* $\mathcal{E}$.

REMARK 9.51. If ν is a complex Radon measure in $M(G, \lambda)$, then $\nu = \psi \cdot |\nu|$ for a unimodular function ψ. Hence every $\nu \in M(G, \lambda)$ is of the form $\varphi(\mu \circ \lambda)$ for a finite measure μ on $G^{(0)}$ and a $\varphi \in L^\infty(G, \mu \circ \lambda)$ where $\varphi = \psi \cdot \frac{d|\nu|}{d(\mu \circ \lambda)}$. Moreover, by Proposition A.15 on page 324, $\|\nu\| = \inf\{\, \|\mu\|_1 \|\varphi\|_\infty \,\}$ taken over all such representations.

We write $f \mapsto m_f$ for the standard embedding of $\mathcal{E}$ into its double dual $\mathcal{E}^{**}$. Let $L(f)$ denote the operator on $\mathcal{E}$ acting by left convolution by $f \in C_c(G)$. If $f \in C_c(G)$ and $m \in \mathcal{E}^{**}$ we let $f \diamond m = {}^{tt}L(f)$ where ${}^{tt}L(f)$ is the double transposition acting on $\mathcal{E}^{**}$. Thus if $g \in C_c(G)$,

$$\langle f \diamond m_g \,,\, \nu \rangle = \nu(f * g) = \langle m_{f*g} \,,\, \nu \rangle,$$

where $\langle \cdot \,,\, \cdot \rangle$ is the natural pairing of $\mathcal{E}^{**}$ acting on $\mathcal{E}^*$. We also write

$$\langle \varphi \cdot \nu \,,\, g \rangle = \nu(\varphi g) \quad \text{and} \quad \langle \varphi \cdot m \,,\, \nu \rangle = m(\varphi \cdot \nu)$$

for $\varphi \in C_0(G), m \in \mathcal{E}^{**}$ and $\nu \in \mathcal{E}^*$.

DEFINITION 9.52 ([**ADR00**, Definition 3.3.4]). Let G be a locally compact groupoid endowed with a Haar system λ. A *topological invariant mean* on G with respect to λ is an element $m \in \mathcal{E}^{**}$ such that

(a) $m(\nu) \geq 0$ if $\nu \geq 0$,
(b) for any probability measure μ on $G^{(0)}$, we have $m(\mu \circ \lambda) = 1$ and
(c) for any $f \in C_c(G)$, we have $f \diamond m = \big(\lambda(f) \circ r\big) \cdot m$ as functionals on $\mathcal{E}^*$.

[8] Equation (9.23) only applies to $f, g \in C_c(G)$. The point is that we can extend the convolution product to a product on the completion. At least at this stage of the game, we are not claiming that $\mathcal{E}$ is a set of equivalence classes of functions on G, nor are we trying to write down a formula for the product of these equivalence classes.

The concept of a topological invariant mean plays an important role in [**ADR00**, §3.3] (and hence here). The concept has become even more important after Renault's [**Ren15**, Theorem 2.14] connecting Borel approximate invariant means and amenability. In this section we'll connect up with Renault's work in [**Ren15**] and prove Renault's result.[9]

Note that in the Borel context, it is necessary to work strictly with sequences rather than nets.

DEFINITION 9.53 ([**Ren15**, Definition 2.3]). Let G be a locally compact groupoid endowed with a continuous Haar system λ. A *sequence* of nonnegative Borel functions $\{g_i\} \subset \mathcal{B}^+(G)$ on G will be called a *Borel approximate invariant density* if

(a) $\lambda(g_i)(u) \leq 1$ for all $u \in G^{(0)}$,

(b) $\lambda(g_i)(u) \to 1$ for all $u \in G^{(0)}$, and

(c) for all $\gamma \in G$ we have

$$\int_G |g_i(\gamma^{-1}\eta) - g_i(\eta)|\, d\lambda^{r(\gamma)}(\eta) \to 0.$$

Recall that by Definition 3.1 on page 70, a Radon family of measures $\rho = \{\rho^u\}_{u\in G^{(0)}}$ on G is a collection of Radon measures ρ^u such that $u \mapsto \rho(f)(u) = \int_G f(\gamma)\, d\rho^u(\gamma)$ is Borel for all $f \in C_c(G)$.

DEFINITION 9.54 ([**Ren15**, Definition 2.1]). If G is a locally compact groupoid, then a Borel approximate invariant mean is *sequence* $m = \{m_m\}$ of Radon families of measures $m_m = \{m_n^u\}_{u\in G^{(0)}}$ such that $\operatorname{supp} m_n^u \subset G^u$ for all $u \in G^{(0)}$ for all u, and such that

(a) $m_n^u(G) \leq 1$,

(b) $m_n^u(G) \to 1$ for all $u \in G^{(0)}$, and

(c) $\|\gamma \cdot m_n^{s(\gamma)} - m_n^{r(\gamma)}\|_1 \to 0$ for all $\gamma \in G$.

REMARK 9.55. Since each m_n^u is a finite measure, Proposition 3.3 on page 70 implies that

$$u \mapsto \int_G f(\gamma)\, dm_n^u(\gamma)$$

is Borel for any bounded Borel function f on G. Hence Definition 9.54 is the same as Renault's in [**Ren15**].

In analogy with Proposition 9.42 on page 209, we have the following.

PROPOSITION 9.56. *Suppose G is a second countable locally compact Hausdorff groupoid with a Haar system λ. Then G has a Borel approximate invariant density if and only if it has a Borel approximate invariant mean.*

PROOF. Let $\{g_n\} \subset \mathcal{B}^+(G)$ be a Borel approximate invariant density. Define $m_n^u = g_n \cdot \lambda^u$. We have $m_n^u(G) = \lambda(g_n)(u)$. Hence to see that $m = \{m_n\}$ is a Borel

[9] The related concept of a *measurewise invariant mean* will be introduced in Chapter 10 in Definition 10.18 on page 252.

approximate invariant mean, we just have to verify (c). But

$$\begin{aligned}(\gamma\cdot m_n^{s(\gamma)}-m_n^{r(\gamma)})(f)&=\int_G f(\gamma\eta)\,dm_n^{s(\gamma)}(\eta)-\int_G f(\eta)\,dm_n^{r(\gamma)}(\eta)\\&=\int_G f(\gamma\eta)g_n(\eta)\,d\lambda^{s(\gamma)}(\eta)-\int_G f(\eta)g_n(\eta)\,d\lambda^{r(\gamma)}(\eta)\\&=\int_G f(\eta)\big(g_n(\gamma^{-1}\eta)-g_n(\eta)\big)\,d\lambda^{r(\gamma)}(\eta).\end{aligned}$$

Hence

$$\|\gamma\cdot m_n^{s(\gamma)}-m_n^{r(\gamma)}\|_1\le\int_G|g_n(\gamma^{-1}\eta)-g_n(\eta)|\,d\lambda^{r(\gamma)}(\eta).$$

Thus m is a Borel approximate invariant mean.

Now suppose we have a Borel approximate invariant mean $m=\{m_n\}$. Using Proposition 3.18 on page 77, we have a $f\in C^+_{r,c}(G)$ such that $\lambda(f)(u)=1$ for all u. Define

$$g_n(\gamma)=\int_G f(\eta^{-1}\gamma)\,dm_n^{r(\gamma)}(\eta).$$

Since everything in sight is nonnegative,

$$\begin{aligned}\lambda(g_n)(u)&=\int_G\int_G f(\eta^{-1}\gamma)\,dm_n^u(\eta)\,d\lambda^u(\gamma)=\int_G\int_G f(\gamma)\,d\lambda^{s(\eta)}(\gamma)\,dm_n^u(\eta)\\&=m_n^u(G).\end{aligned}$$

Moreover,

$$\begin{aligned}&\int_G|g_n(\gamma^{-1}\eta)-g_n(\eta)|\,d\lambda^{r(\gamma)}(\eta)\\&\qquad=\int_G\Big|\int_G f(\zeta^{-1}\gamma^{-1}\eta)\,dm_n^{s(\gamma)}(\zeta)-\int_G f(\zeta^{-1}\eta)\,dm^{r(\gamma)}(\zeta)\Big|\,d\lambda^{r(\gamma)}(\eta)\\&\qquad=\int_G\Big|\int_G f(\zeta^{-1}\eta)\,d(\gamma\cdot m_n^{s(\gamma)})(\zeta)-\int_G f(\zeta^{-1}\eta)\,dm^{r(\gamma)}(\zeta)\Big|\,d\lambda^{r(\gamma)}(\eta)\end{aligned}$$

which, using the properties of the total variation measure, is

$$\le\int_G\int_G f(\zeta^{-1}\eta)\,d|\gamma\cdot m_n^{s(\gamma)}-m_n^{r(\gamma)}|(\zeta)\,d\lambda^{r(\gamma)}(\eta)$$

which, after using Fubini and the properties of f, is

$$\le\|\gamma\cdot m_n^{s(\gamma)}-m_n^{r(\gamma)}\|_1.$$

It now follows easily that $\{g_n\}$ is a Borel approximate invariant density. □

REMARK 9.57. Clearly, producing a Borel approximate invariant mean is a much less demanding requirement that that of a approximately invariant continuous mean. Similarly, the concept of a Borel approximate invariant density should be contrasted with the sort of functions required in Theorem 9.43(b) on page 211. We'll call a sequence of functions as in Theorem 9.43(b) a *topological approximate invariant density* since it sounds nicer. It should be noted however, that a Borel approximate invariant density and topological approximate invariant density are required to be a *sequences*. Nevertheless, we have the following result from [**Ren15**, Theorem 2.14].[10]

THEOREM 9.58 (Renault). *Let G be a second countable locally compact groupoid with continuous Haar system λ. The following are equivalent.*

(a) *G is amenable.*

(b) *G admits a Borel approximate invariant density.*

(c) *G admits a Borel approximate invariant mean.*

(d) *G admits a topological invariant mean.*

(e) *G admits a topological approximate invariant density.*

Untangling the notation and comparing with Theorem 9.43 on page 211, we see that G is amenable if and only if it has a topological approximate invariant density: that is, (a)$\Longleftrightarrow$(e). And (b)$\Longleftrightarrow$(c) by Proposition 9.56 on page 220. Of course, (e)$\Longrightarrow$(b) is immediate. Hence we can complete the proof by showing just (b)$\Longrightarrow$(d) and (d)$\Longrightarrow$(e). The heavy lifting in this proof is the implication (d)$\Longrightarrow$(e).

PROOF. We start with the implication (b)$\Longrightarrow$(d). Let $\{g_i\}$ be a Borel approximate invariant density as in Definition 9.53 on page 220. Since Remark 9.51 on page 219 allows us to write a typical element of $\mathcal{E}^* = M(Y,\alpha)$ as $\varphi(\mu\circ\lambda)$ with $\|\nu\| \le \|\mu\|_1\|\varphi\|_\infty$, we can define $m_i := m_{g_i} \in \mathcal{E}^{**}$. Then if $\nu = \varphi(\mu\circ\alpha)$,

$$\langle m_i\,,\varphi(\mu\circ\lambda)\rangle = \int_{G^{(0)}}\int_G g_i(\gamma)\varphi(\gamma)\,d\lambda^u(\gamma)\,d\mu(u) = \langle\varphi(\mu\circ\lambda)\,,g_i\rangle.$$

Since

$$|\langle\varphi(\mu\circ\lambda)\,,g_i\rangle| \le \|\varphi(\mu\circ\lambda)\|\|g_i\|_{\lambda,1},$$

we see that $\|m_i\| \le 1$ in view of Definition 9.53(a) on page 220.

Let m be a weak-$*$ cluster point of $\{m_i\}$. To see that m is a topological invariant mean we need to check axioms (a), (b) and (c) of Definition 9.52 on page 219.

Clearly, axiom (a) holds. Furthermore, if μ is a probability measure on $G^{(0)}$, then since $\{m_i\}$ is a sequence, the dominated convergence theorem applies and

$$\lim_i m_i(\mu\circ\lambda) = \lim_i \int_{G^{(0)}}\int_G g_i(\gamma)\,d\lambda^u(\gamma)\,d\mu(u) = 1. \tag{9.24}$$

But then (9.24) must hold for any subnet $\{m_{i_j}\}$ converging to m.[11] Hence, axiom (b) holds as well.

[10] Renault observes that his approach is based on the corresponding result for groups: [**BdlHV08**, Theorem G.3.1].

[11] Note that some care was required here. A subnet $\{m_{i_j}\}$ might not be a sequence and the Dominated Convergence Theorem would not apply.

To see that axiom (c) holds, consider $f \in C_c(G)$. Then for any $\nu = \varphi(\mu \circ \lambda) \in \mathcal{E}^*$, we have

$$\begin{aligned}\big\langle f \diamond m_i - \big(\lambda(f) \circ r\big) \cdot m_i \,,\, \nu \big\rangle &= \big\langle \nu \,,\, f * g_i - \big(\lambda(f) \circ r\big) g_i \big\rangle \\ &= \int_{G^{(0)}} \int_G \varphi(\gamma)\big(f * g_i(\gamma) - \lambda(f)(u) g_i(\gamma)\big)\, d\lambda^u(\gamma)\, d\mu(u) \\ &= \int_{G^{(0)}} \int_G \int_G \varphi(\gamma) f(\eta)(g_i(\eta^{-1}\gamma) - g_i(\gamma))\, d\lambda^u(\eta)\, d\lambda^u(\gamma)\, d\mu(u) \\ &= \int_{G^{(0)}} \int_G f(\eta)\Big(\int_G \varphi(\gamma)(g_i(\eta^{-1}\gamma) - g_i(\gamma))\, d\lambda^u(\gamma)\Big)\, d\lambda^u(\eta)\, d\mu(u).\end{aligned}$$

But

$$(9.25)\quad \Big|\int_G \varphi(\gamma)(g_i(\eta^{-1}\gamma) - g_i(\gamma))\, d\lambda^{r(\eta)}(\gamma)\Big| \le \|\varphi\|_\infty \int_G |g_i(\eta^{-1}\gamma) - g_i(\gamma)|\, d\lambda^{r(\eta)}(\gamma).$$

Hence the left-hand side of (9.25) tends to zero for all η (in view of Definition 9.53(c)). On the other hand, the left-hand side of (9.25) is bounded by

$$\|\varphi\|_\infty \Big(\int_G |g_i(y^{-1}\gamma)|\, d\lambda^{r(y)}(\gamma) + \int_G |g_i(\gamma)|\, \lambda^{r(y)}(\gamma)\Big) \le 2\|\varphi\|_\infty$$

by left invariance of λ and Definition 9.53(a). It now follows from the Dominated Convergence Theorem (applied in $L^1(\mu \circ \lambda)$) that

$$\big\langle f \diamond m_i - \big(\lambda(f) \circ r\big) \cdot m_i \,,\, \nu \big\rangle \to 0 \quad \text{for all } \nu.$$

As in above, this must also hold for any subnet $\{m_{i_j}\}$ converging to m. Therefore $f \diamond m = \big(\lambda(f) \circ r)\big) \cdot m$ on $\mathcal{E}^*$ for all $f \in C_c(G)$. Thus m is a topological invariant mean and we have shown that (b)$\Longrightarrow$(d).

As discussed above, we can finish the proof by establishing (d)$\Longrightarrow$(e). We let m be a topological invariant mean in $\mathcal{E}^{**}$. To establish (e), I claim it will suffice to produced a net satisfying condition (b) in Theorem 9.43 on page 211—it then follows from Theorem 9.43 that we can replace that net by a sequence and obtain a topological approximate invariant density as required.

Let $C_c^+(G)_{\lambda,1}$ be those elements of $f \in C_c^+(G)$ such that $\|f\|_{\lambda,1} \le 1$. The natural inclusion $g \mapsto m_g$ of $C_c(G)$ into $\mathcal{E}^{**}$ allows us to identify $C_c^+(G)_{\lambda,1}$ as a subspace of the unit ball of $\mathcal{E}^{**}$. We'll need the observation that $C_c^+(G)_{\lambda,1}$ is weak-$*$ dense in the unit ball of $\mathcal{E}^{**}_+ = \{\, m \in \mathcal{E}^{**} : m(\nu) \ge 0 \text{ if } \nu \ge 0.\,\}$. The proof is given in Proposition A.20 on page 329. Using this, we can find a net $\{g_i\} \subset C_c^+(G)_{\lambda,1}$ such that m is the weak-$*$ limit of $m_i := m_{g_i}$. Our net satisfies the criterion in Theorem 9.43(b)(i) on page 211 by construction. I claim it also satisfies

(ii)′ $\lambda(g_i) \to 1$ in the weak topology on $C_0(G^{(0)})$.

To verify this, recall that the norm dual of $C_0(G^{(0)})$ can be identified with the set of complex measures $M(G^{(0)})$ on $G^{(0)}$ with respect to the total variation norm.[12] Moreover, if μ is a probability measure on $G^{(0)}$, then by assumption

$$\int_{G^{(0)}} \lambda(g_i)(u)\, d\mu(u) = m_i(\mu \circ \lambda) \to m(\mu \circ \lambda) = 1.$$

[12]See Section A.1.

If $\varphi \in L^\infty(\mu)$, then

$$\Bigl|\int_{G^{(0)}} \varphi(u)\,d\mu(u) - \int_{G^{(0)}} \varphi(u)\lambda(g_i)(u)\,d\mu(u)\Bigr| \leq \|\varphi\|_\infty \int_{G^{(0)}} \bigl|1-\lambda(g_i)(u)\bigr|\,d\mu(u)$$

which, since $\lambda(g_i)(u) \leq 1$, is

$$= \|\varphi\|_\infty\Bigl(1 - \int_{G^{(0)}} \lambda(g_i)(u)\,d\mu(u)\Bigr)$$

which tends to 0. Since every complex measure on $G^{(0)}$ is of the form $\varphi\cdot\mu$ for some probability measure μ, (ii)′ holds.

Next, I claim that

(iii)′ For every $f \in C_c(G)$, we have $f * g_i - \bigl(\lambda(f)\circ r)g_i\bigr) \to 0$ in the weak topology on $\mathcal{E}$.

To see this, let $\nu \in \mathcal{E}^*$. Then

$$\bigl\langle \nu\ ,\ f * g_i - \bigl(\lambda(f)\circ r)\bigr)g_i\bigr\rangle = \bigl\langle f\diamond m_i - \bigl(\lambda(f)\circ r\bigr)\cdot m_i\ ,\ \nu\bigr\rangle,$$

and the latter tends to $\bigl\langle f\diamond m - \bigl(\lambda(f)\circ r\bigr)\cdot m\ ,\ \nu\bigr\rangle$ which is zero since m is a topological invariant mean. Thus (iii)′ holds as claimed.

We need to "upgrade" properties (ii)′ and (iii)′ to criteria (ii) and (iii) of Theorem 9.43(b) on page 211. In order to do this, we will need to replace our net $\{g_i\}$ with a new net. Doing so will require a convexity argument. To do this, let $\mathscr{E}_0$ be the topological vector space $C_b(G^{(0)})$ equipped with the strict topology (viewed as the multiplier algebra of $C_0(G^{(0)})$). For each $f \in C_c(G)$, let $\mathcal{E}_f$ be $\mathcal{E}$ with its norm topology. Then form the topological vector space

$$\mathcal{F} = \mathscr{E}_0 \times \prod_{f\in C_c(G)} \mathcal{E}_f$$

(with the product topology). We'll write $\mathcal{E}_{f,w}$ for $\mathcal{E}$ equipped with the weak-topology $\sigma(\mathcal{E},\mathcal{E}^*)$ coming from the norm dual $\mathcal{E}^*$. On the other hand, we can use Theorem A.1 on page 315 to realize the strict dual of $C_b(G^{(0)})$ as $M(G^{(0)})$, and we let $\mathscr{E}_{0,w}$ be $C_b(G^{(0)})$ equipped with the corresponding weak topology $\sigma(C_b(G^{(0)}), M(G^{(0)}))$.[13] Then we can also form the topological vector space

$$\mathcal{F}_w := \mathscr{E}_{0,w} \times \prod_{f\in C_c(G)} \mathcal{E}_{f,w}.$$

It is a consequence of Proposition A.5 on page 317 that the product topology on $\mathcal{F}_w$ is just the weak topology on $\mathcal{F}$ induced by the continuous linear functionals on $\mathcal{F}$. In particular, $\mathcal{F}$ and $\mathcal{F}_w$ have the same closed convex sets by Corollary A.6 on page 317. Properties (ii)′ and (iii)′ allow us to assert that $(1,(0)_{f\in C_c(G)})$ belongs to the closure of the convex set

$$C := \bigl\{\,\bigl(\lambda(g), \bigl(f*g - \bigl(\lambda(f)\circ r\bigr)g\bigr)_{f\in C_c(G)}\ :\ g \in C_c^+(G)_{\lambda,1}\,\bigr\}$$

in $\mathcal{F}_w$. Hence $(1,(0)_{f\in C_c(G)})$ must also be in the closure of C in $\mathcal{F}$. Therefore we can replace $\{g_i\}$ with a net in $C_c^+(G)_{\lambda,1}$, which we still call $\{g_i\}$, such that

[13]Weak topologies of this type are discussed in [**Ped89**, §2.3]. In particular, $f_i \to f$ in $C_b(G^{(0)})$ if and only if $m(f_i) \to m(f)$ for all $m \in M(G^{(0)})$.

$\lambda(g_i) \to 1$ strictly in $C_b(G^{(0)})$ and such that for all $f \in C_c(G)$,

$$\|f * g_i - (\lambda(f) \circ r)g_i\|_{\lambda,1} = \sup_{u \in G^{(0)}} \int_G |f * g_i(\gamma) - \lambda(f)(u)g_i(\gamma)| \, d\lambda^u(\gamma) \tag{9.26}$$

goes to zero. In particular, since the strict and compact-open topologies agree on bounded sets of $C_b(G^{(0)})$, we see immediately that the first of these properties implies that

(ii)″ $\lambda(g_i) \to 1$ uniformly on compact subsets of $G^{(0)}$.

Note that we can reformulate (9.26) to say that

$$\int_G \Big| \int_G f(\gamma)(g_i(\gamma^{-1}\eta) - g_i(\eta))) \, d\lambda^u(\eta) \Big| \, d\lambda^u(\gamma) \to 0 \quad \text{uniformly on } G^{(0)}. \tag{9.27}$$

In fact, I claim that for all $F \in C_c(G *_r G)$ we have

$$\int_G \Big| \int_G F(\zeta, \gamma)(g_i(\gamma^{-1}\eta) - g_i(\eta))) \, d\lambda^{r(\zeta)}(\eta) \Big| \, d\lambda^{r(\zeta)}(\gamma) \to 0 \quad \text{uniformly on } G. \tag{9.28}$$

However if $F = f_1 \otimes f_2$ (so that $F(\zeta, \gamma) = f_1(\zeta)f_2(\gamma)$) with $f_i \in C_c(G)$, then the left-hand side of (9.28) is dominated by

$$\|f_1\|_\infty \int_G \Big| \int_G f_2(\gamma)(g_i(\gamma^{-1}\eta) - g_i(\eta)) \, d\lambda^{r(\zeta)}(\eta) \Big| \, d\lambda^{r(\zeta)}(y),$$

and it follows from (9.27) that (9.28) goes to zero uniformly on compact subsets of G as claimed. But if $F \in C_c(G *_r G)$ has $\operatorname{supp} F \subset K \times K$ with $K \subset G$ compact, then for any $\epsilon > 0$, there is a $\underline{F} = \sum_k f_{1,k} \otimes f_{2,k}$ such that $\operatorname{supp} \underline{F} \subset K \times K$ and $\|F - \underline{F}\|_\infty < \epsilon$. On the other hand, the modulus of the left-hand side of (9.28) is bounded by

$$\begin{aligned} \|F\|_\infty \int_K \int_G &|g_i(\gamma^{-1}\eta) - g_i(\eta)| \, d\lambda^{r(z)}(\eta) \, d\lambda^{r(\zeta)}(\gamma) \\ &\le \|F\|_\infty \int_K \int_G \big(g_i(\gamma^{-1}\eta) + g_i(\eta)\big) \, d\lambda^{r(\zeta)}(\eta) \, d\lambda^{r(\zeta)}(\gamma) \\ &\le 2M\|F\|_\infty, \end{aligned}$$

where M is an upper bound for $\lambda^u(K)$. The full claim now follows easily.

If $f \in C_c(G)$, then $F_f(\zeta, \gamma) := f(\zeta^{-1}\gamma)$ is not necessarily compactly supported. However, if K is compact in G and $g \in C_c^+(G)$ is such that $g(\zeta) = 1$ for all $\zeta \in K$, then

$$F(\zeta, \gamma) := g(\zeta)f(\zeta^{-1}\gamma)$$

does have compact support and

$$\int_G \Big| \int_G F(\zeta, \gamma)(g_i(\gamma^{-1}\eta) - g_i(\eta)) \, d\lambda^{r(\zeta)}(\eta) \Big| \, d\lambda^{r(\zeta)}(\gamma)$$

is equal to

$$\int_G \Big| \int_G f(\zeta^{-1}\gamma)(g_i(\gamma^{-1}\eta) - g_i(\eta)) \, d\lambda^{r(\zeta)}(\eta) \Big| \, d\lambda^{r(\zeta)}(\gamma)$$

for all $\zeta \in K$. We have established the following:

(iii)″ For all $f \in C_c(G)$,

$$\int_G \Big| \int_G f(\zeta^{-1}\gamma)(g_i(\gamma^{-1}\eta) - g_i(\eta)) \, d\lambda^{r(\zeta)}(\eta) \Big| \, d\lambda^{r(\zeta)}(\gamma) \to 0.$$

uniformly on compact subsets of G.

Now to finish the proof of (d)$\Longrightarrow$(e), it will suffice to produce for all $\epsilon > 0$ and each pair of compact sets $L \subset G^{(0)}$ and $K \subset G$ a function $g = g_{L,K,\epsilon} \in C_c^+(G)_{\lambda,1}$ such that

(I) $1 - \lambda(g)(u) \le \epsilon$ for all $u \in L$ and

(II) for all $\gamma \in K$

$$\int_G |g(\gamma^{-1}\eta) - g(\eta)|\, d\lambda^{r(\gamma)}(\eta) \le \epsilon.$$

After enlarging K and L if necessary, we can assume that $r(K) \subset K$ and $K \subset G_L^L$. We can also fix $f \in C_c^+(G)_{\lambda,1}$ such that $\lambda(f)(u) = 1$ for all $u \in L$. By (ii)$''$, there is an i_0 such that $i \ge i_0$ implies that

$$\lambda(g_i)(u) \ge 1 - \epsilon \quad \text{for all } u \in s(\operatorname{supp} f). \tag{9.29}$$

In view of (iii)$''$, we can find $i \ge i_0$ such that for all $\zeta \in K$,

$$\int_G \Bigl| \int_G f(\zeta^{-1}\gamma)(g_i(\gamma^{-1}\eta) - g_i(\eta))\, d\lambda^{r(\zeta)} \Bigr|\, d\lambda^{r(\zeta)}(\gamma) < \frac{\epsilon}{2}. \tag{9.30}$$

Then we fix i and define

$$g(\eta) = \int_G f(\gamma) g_i(\gamma^{-1}\eta)\, d\lambda^{r(\eta)}(\gamma).$$

It is not hard to see that $g \in C_c^+(G)_{\lambda,1}$. In fact,

$$\begin{aligned} \lambda(g)(u) &= \int_G g(\eta)\, d\lambda^u(\eta) \\ &= \int_G \int_G f(\gamma) g_i(\gamma^{-1}\eta)\, d\lambda^u(\gamma)\, d\lambda^u(\eta) \\ &= \int_G f(\gamma) \int_G g_i(\gamma^{-1}\eta)\, d\lambda^u(\eta)\, d\lambda^u(\gamma) \\ &= \int_G f(\gamma)\lambda(g_i)(s(\gamma))\, d\lambda^u(\gamma) \\ &\le \int_G f(\gamma)\, d\lambda^u(\gamma) \le 1. \end{aligned} \tag{9.31}$$

On the other hand, we can apply (9.29) to (9.31) to conclude that

$$\lambda(g)(u) \ge (1-\epsilon) \int_G f(\gamma)\, d\lambda^u(\gamma) \ge 1 - \epsilon.$$

Hence g also satisfued condition (I).

On the other hand, if $\zeta \in G_L^L$, then

$$\begin{aligned} &g(\zeta^{-1}\eta) - g(\eta) \\ &\quad = \int_G f(\gamma) g_i(\gamma^{-1}\zeta^{-1}\eta)\, d\lambda^{s(\zeta)}(\gamma) - \int_G f(\gamma) g_i(\gamma^{-1}\eta)\, d\lambda^{r(\eta)}(\gamma) \\ &\quad = \int_G f(\zeta^{-1}\gamma) g_i(\gamma^{-1}\eta)\, d\lambda^{r(\zeta)}(\gamma) - \int_G f(\gamma) g_i(\gamma^{-1}\eta)\, d\lambda^{r(\zeta)}(\gamma). \end{aligned}$$

Since $\zeta \in G_L^L$, our assumptions on f guarantee that

$$g_i(\eta) \int_G f(\zeta^{-1}\gamma)\, d\lambda^{r(\zeta)}(\gamma) = g_i(\eta) = g_i(\eta) \int_G f(\gamma)\, d\lambda^{r(\zeta)}(\gamma).$$

Thus we can add and substract $g_i(\eta)$ from our expression

$$g(\zeta^{-1}\eta) - g(\eta) =$$
$$\int_G f(\zeta^{-1}\gamma)(g_i(\gamma^{-1}\eta) - g_i(\eta))\,d\lambda^{r(\zeta)}(\gamma) - \int_G f(\gamma)(g_i(\gamma^{-1}\eta) - g_i(\eta))\,d\lambda^{r(\zeta)}(\gamma).$$

But if $\zeta \in K$, then $K \subset G^L_L$ and we can apply (9.30)—once with ζ and once with $r(\zeta) \in K$—to conclude that

$$\int_G |g(\zeta^{-1}\eta) - g(\eta)|\,d\lambda^{r(\zeta)}(\eta) < \epsilon.$$

This completes the proof that (d)$\Longrightarrow$(e). □

Exercises

Ex 9.6.1. If μ_1 and μ_2 are Radon measures on Y, then we write $\mu_1 \leq \mu_2$ if $\mu_1(E) \leq \mu_2(E)$ for all Borel sets E.

(a) Show that $\mu_1 \leq \mu_2$ if and only if $\mu_1(f) \leq \mu_2(f)$ for all $f \in C_c^+(Y)$.

(b) Suppose that ν_1 and ν_2 are in $M(Y)$. Show that $|\nu_1 + \nu_2| \leq |\nu_1| + |\nu_2|$.

—Answer on page 379

Ex 9.6.2. Using the proof of Theorem 9.43 and the observation that for $a, b \geq 0$ we have $|a - b|^2 \leq |a^2 - b^2|$, show that a locally compact groupoid G with a Haar system λ is amenable if and only if there is a sequence $\{\xi_i\}$ of Borel functions on G such that

(i) $\lambda(|\xi_i|^2)(u) \leq 1$ for all $u \in G^{(0)}$,

(ii) $\lambda(|\xi_i|^2)(u) \to 1$ for all $u \in G^{(0)}$, and

(iii) for all $\gamma \in G$, we have $\lim_i \int_G |\xi_i(\gamma^{-1}\eta) - \xi_i(\eta)|^2\,d\lambda^{r(\gamma)}(\eta) = 0$.

—Answer on page 380

9.7. Borel Equivalence

Renault's work connecting amenability to the Borel structure on G in [**Ren15**] allows a powerful strengthening of Theorem 9.48 on page 217 which implies that groupoid equivalence preserves amenability. In this section we will prove that amenability is preserved under the weaker requirement of equivalence as Borel groupoids—see Theorem 9.75 on page 232. Although our primary interest is Borel groupoids whose Borel structure comes from and underlying locally compact topology, it is more natural to phrase our approach in a completely Borel setting. Therefore we will work with Borel groupoids as in [**ADR00**, §2.1a] (although the treatment here is closer to that given in [**RW17**, §§2-3]). To avoid pathologies, we always assume that our Borel spaces are analytic Borel spaces. The product Borel structure on $X \times Y$ is that generated by the Borel rectangles.

Definition 9.59. A *Borel groupoid* is a groupoid G equipped with an analytic Borel structure such that the range, source, inverse, and multiplication maps are Borel when $G^{(2)}$ is given the relative Borel stucture coming from the product Borel structure on $G \times G$ and $G^{(0)}$ has the relative Borel stucture coming from G. By a *Borel Haar system* on G we mean a collection $\{\lambda^u\}_{u\in G^{(0)}}$ of nonzero measures on

G such that $G \setminus G^u$ is λ^u-null for each u, $\gamma \cdot \lambda^{s(\gamma)} = \lambda^{r(\gamma)}$ and $u \mapsto \int f \, d\lambda^u$ is a Borel function (into $[0, \infty]$) for all $f \in \mathcal{B}^+(G)$. We insist that $\{\lambda^u\}_{u \in G^{(0)}}$ is *proper* in that G is the increasing union of Borel sets A_n such that $u \mapsto \lambda^u(A_n)$ is bounded.[14]

EXAMPLE 9.60. Observe that if G is a second countable, locally compact Hausdorff groupoid, then the Borel structure on $G^{(2)}$ is the same as the Borel structure generated by the relative product topology on $G^{(2)}$. Hence G is a Borel groupoid. Furthermore, any Haar system on G is a Borel Haar system in view of Proposition 3.3 on page 70.

We proceed in nearly perfect analogy with the topological setting. For example, a G-space X is called a Borel G-space if the moment map is Borel and the map $(\gamma, x) \mapsto \gamma \cdot x$ is Borel when $G * X = \{\, (\gamma, x) \in G \times X : s(\gamma) = r(x) \,\}$ is given the relative Borel structure. If X and Y are Borel spaces and $\pi : Y \to X$ is Borel surjection, then a *Borel π-system* is a family of measures $m = \{m^x\}_{x \in X}$ such that each m^x is supported in $\pi^{-1}(x)$ and such that

$$x \mapsto \int_Y f(y) \, dm^x(y)$$

is Borel for any nonnegative Borel function f on Y. We say that π is proper if each m^x is nonzero, and finite if each m^x is a finite measure. If G is a Borel groupoid, X and Y are Borel G-spaces and π is G-invariant, then we have that m is *invariant*[15] if $\gamma \cdot m^x = m^{\gamma \cdot x}$ whenever $s(\gamma) = r(y)$. (By definition, $\gamma \cdot m^x(E) = m^x(\gamma^{-1} \cdot E)$.)

We define Borel amenability and proper Borel amenability in analogy with the topological versions in Section 9.3.

DEFINITION 9.61 ([**ADR00**, Definition 2.1.1(b)]). Suppose that G is a Borel groupoid and that X and Y are Borel G-spaces. A surjective Borel G-map $\pi : Y \to X$ is *properly Borel amenable* if there is a invariant Borel π-system $m = \{m^x\}_{x \in X}$ of probability measures on Y. Then we say that m is an *invariant mean for π*.

REMARK 9.62. Notice that if $\pi : Y \to X$ is a Borel G-map, then the induced map $\dot\pi : G \backslash Y \to G \backslash X$ is Borel with respect to the quotient Borel structures. Suppose that π is G-properly amenable and that $\{\lambda^x\}$ is an invariant mean for π. Let $p : Y \to G \backslash Y$ be the quotient map. Then the push-forward $p_* \lambda^x$ is a probability measure supported on $\dot\pi^{-1}(\dot x)$. By invariance, it depends only on $\dot x$ and we can denote this measure by $\dot\lambda^{\dot x}$. If b is a bounded Borel function on $G \backslash Y$, then

$$\int_{G \backslash Y} b(\dot z) \, d\dot\lambda^{\dot x}(\dot z) = \int_Y b(p(y)) \, d\lambda^x(y).$$

Hence $\{\dot\lambda^{\dot x}\}_{\dot x \in G \backslash X}$ is a Borel $\dot\pi$-system of probability measures on $G \backslash Y$.

DEFINITION 9.63 ([**ADR00**, Defintion 2.1.2]). A Borel groupoid G is *proper* if the range map $r : G \to G^{(0)}$ is properly Borel amenable. A Borel G-space X is *proper* if the projection $p : X * G \to X$ is properly Borel amenable.

---

[14] In the Borel context, we can no longer talk about Radon measures or even regular measures. Properness implies that each λ^u is σ-finite. The definition given above is taken from [**Ren15**] and is slightly stronger than that given in [**ADR00**] where the measures λ^u are only assumed to be σ-finite and nonzero.

[15] See Remark 9.12 on page 198.

REMARK 9.64. Thus X is a proper G-space if and only if there is a family $\{m^x\}_{x\in X}$ of probability measures m^x on G supported in $G^{r(x)}$ such that $x \mapsto m^x(f)$ is Borel for all non-negative Borel functions on G and such that $\gamma \cdot m^x = m^{\gamma\cdot x}$.

If X and Y are (left) G-spaces, then the fibered product $X * Y$ is a G-space with respect to the diagonal action. If X and Y are Borel G-spaces, then we give $X * Y$ the Borel structure as a subset of $X \times Y$ with Borel structure generated by the Borel rectangles. In particular, if X is a G-space, then $X * G = \{ (x, \gamma) \in X \times G : r(x) = r(\gamma) \}$ is a G-space with respect to the diagonal action.[16]

Of course, if G is a second countable locally compact groupoid and X is a second countable properly (topological) amenable G-space, then X is a properly Borel amenable G-space. But the bar is not so high in the Borel category. Recall that by assumption, all of our Borel spaces, and our Borel groupoids in particular, are analytic Borel spaces.

LEMMA 9.65. *Let G be a Borel groupoid acting freely on a Borel space X such that the quotient map $q : X \to G\backslash X$ has a Borel cross section. Then X is a proper Borel G-space.*

PROOF. Let $G \rtimes X$ be the transformation groupoid for the right G-action.[17] The map $(\gamma, x)) \mapsto (x, \gamma^{-1}\cdot x)$ is a Borel as well as an algebraic isomorphism of $X\rtimes G$ onto the equivalence relation groupoid $X *_q X = \{ (x, y) \in X \times X : q(x) = q(y) \}$. Since $X *_q X$ and $X \rtimes G$ are both Borel subsets of the corresponding product spaces, they are themselves analytic Borel spaces. Hence $X \rtimes G$ and $X *_q X$ are Borel isomorphic by Corollary 2 of [**Arv76**, Theorem 3.3.4]. Hence it suffices to see that $X *_q X$ is proper.

We can identify X with the unit space of $X *_q X$. For each $x \in X$, let $m^x = \delta_x \times \delta_{c(q(x))}$ where $c : G\backslash X \to X$ is our Borel cross section for q. Since $c(q(x)) = c(q(y))$ if $(x, y) \in X *_q X$, we have $(y, x)\cdot m^x = m^y$. Hence X is a proper Borel G-space. □

DEFINITION 9.66. Suppose that G is a Borel groupoid and that X and Y are Borel G-spaces. A surjective equivariant Borel map $\pi : Y \to X$ is *Borel amenable* if there is a *sequence* $\{m_n\}_{n\in\mathbf{N}}$ of Borel π-systems $m_n = \{m_n^x\}_{x\in X}$ of probability measures on Y which is approximately invariant in that for all $(\gamma, x) \in G * X$, $\|\gamma \cdot m_n^x - m_n^{\gamma\cdot x}\|_1$ converges to 0, where $\|\cdot\|_1$ is the total variation norm. A Borel groupoid G is (Borel) *amenable* if the range map $r : G \to G^{(0)}$ is G-amenable. A Borel G-space X is *Borel amenable* if the projection $p : X * G \to X$ is G-amenable.

If $\pi : Y \to X$ admits a nontrivial π-system, then we can weaken the requirement for the amenability of π slightly.

LEMMA 9.67. *Suppose that $\pi : Y \to X$ is an surjective equivariant Borel map admitting a proper, finite Borel π-system $p = \{p^x\}$. It there is a sequence $\{m_n\}_{n\in\mathbf{N}}$ of Borel π-systems $p_n = \{p_n^x\}_{x\in X}$ such that*

(a) $p_n^x(Y) \le 1$ *for all n and x,*

(b) *for all x, $\|p_n^x\|_1 \to 1$, and*

[16]In fact, we can identify $X * G$ with the groupoid $G \rtimes X$ as in Remark 9.11 on page 198. After identifying its unit space with X, the range map $(x, \gamma) \mapsto x$ is clearly G-equivariant. Hence, in Defintion 9.63, we defined a G-space X to be properly amenable exactly when the groupoid $G \rtimes X$ is properly amenable.

[17]Thus $G\rtimes X$ is the set of pairs (γ, x) with multiplication defined by $(\gamma, x)(\eta, \gamma^{-1}x) = (\gamma\eta, x)$.

(c) *for all* $\gamma \in G$, $\|\gamma \cdot p_n^x - p_n^{\gamma\cdot x}\|_1 \to 0$,

then π *is Borel amenable.*

PROOF. We can replace p^x by $\|p^x\|_1^{-1}p^x$, and assume that p is a π-system of probability measures. For all n and x, let

$$m_n^x = \begin{cases} \|p_n^x\|_1^{-1}p_n^x & \text{if } \|p_n^x\|_1 \neq 0, \text{ and} \\ p^x & \text{otherwise.} \end{cases}$$

It is not hard to see that $m = \{m_n\}$ is a Borel approximate invariant mean for π as specified in Definition 9.66. □

We define a Borel approximate invariant density for a Borel groupoid G with a Borel Haar system $\{\lambda^u\}_{u\in G^{(0)}}$ just as in Definition 9.53 in the topological case: a Borel approximate invariant density on G is a sequence $\{g_i\}$ of non-negative Borel functions on G satisfying conditions (a), (b), and (c) of Definition 9.53 on page 220.

We can also define a Borel approximate invariant mean on a Borel groupoid exactly as in Definition 9.54: a Borel approximate invariant mean on G is a sequence of Borel $r : G \to G^{(0)}$-systems $\{m_n\}$ such that conditions (a), (b), and (c) of Definition 9.54 on page 220 hold.

LEMMA 9.68. *Let* G *be a Borel groupoid with a Borel Haar system. Then the following are equivalent.*

(a) G *is Borel Amenable.*

(b) G *has a Borel approximate invariant mean.*

(c) G *has a Borel approximate invariant density.*

PROOF. We have (a)$\Longrightarrow$(b) by definition. If G has a Borel approximate invariant mean and a Haar system, then G is Borel amenable in view of Ex 9.7.1 and Lemma 9.67 on the preceding page. Hence, (b)$\Longrightarrow$(a).

Then we can see that (b) and (c) are equivalent just as in the topological case following the proof of Proposition 9.56 on page 220 *mutatis mutandis* (see also [**Ren15**, Proposition 2.4]). □

Our primary interest in Borel amenability stems from the following observation which is an immediate corollary of Theorem 9.58 on page 222 and Lemma 9.68.

THEOREM 9.69. *Let* G *be a second countable locally compact group with a Haar system. Then* G *is amenable if and only if it is Borel amenable as a Borel groupoid.*

A key result for Borel amenable maps is the following which is the Borel analogue of Proposition 9.31 on page 205.

LEMMA 9.70. *Let* G *be a Borel groupoid. If there is a proper Borel* G*-space* X *such that the moment map* $r_X : X \to G^{(0)}$ *is properly Borel amenable (resp., Borel amenable), then* G *is properly Borel amenable (resp., Borel amenable).*

PROOF. Since, by assumption, the map $\pi_X : X * G \to X$ is properly amenable, there is an invariant system $m = \{m^x\}_{x\in X}$ of probability measures for π_X. We can view each m^x as a measure on $G^{r_X(x)}$ such that $\gamma \cdot m^x = m^{\gamma\cdot x}$. Suppose that r_X is Borel amenable. Then there is an approximate invariant mean $\{\mu_n\}$ as in

Definition 9.66 on page 229. Then μ_n^u is a probability measure on $r_X^{-1}(u)$, and for all $\gamma \in G$, $\lim_n \|\gamma \cdot \mu_n^{s(\gamma)} - \mu_n^{r(\gamma)}\|_1 = 0$. Define

$$\lambda_n^u = \int_X m^x \, d\mu_n^u(x).$$

Then λ_n^u is a probability measure supported in G^u. The system $\lambda_n = \{\lambda_n^u\}_{u \in G^{(0)}}$ is certainly Borel, and

$$\begin{aligned}\|\gamma \cdot \lambda_n^{s(\gamma)} - \lambda_n^{r(\gamma)}\|_1 &= \Big\| \int_X \gamma \cdot m^y \, d\mu_n^{s(\gamma)}(y) - \int_X m^x \, \mu_n^{r(\gamma)}(x) \Big\| \\ &= \Big\| \int_X m^{\gamma \cdot x} \, d\mu_n^{s(\gamma)}(y) - \int_X m^x \, d\mu_n^{r(\gamma)}(x) \Big\| \\ &= \Big\| \int_X m^x \, d(\gamma \cdot \mu_n^{s(\gamma)} - \mu_n^{r(\gamma)})(x) \Big\| \\ &\leq \|\gamma \cdot \mu_n^{s(\gamma)} - \mu_n^{r(\gamma)}\|_1.\end{aligned}$$

It follows that λ is an approximate invariant mean for $r : G \to G^{(0)}$. Thus G is Borel amenable.

If r_X is properly Borel amenable, then the argument is similar, but easier as we can take μ to be invariant. □

Of course, a proper Borel groupoid is Borel amenable. Our next lemma should be compared with [**ADR00**, Proposition 5.3.37].

LEMMA 9.71 ([**RW17**, Lemma 2.8]). *Suppose that G is a Borel groupoid with a Haar system which is the increasing union of a sequence of proper Borel subgroupoids G_n. Then G is Borel amenable.*

PROOF. By assumption, we can find an system $m_m = (m_n^u)_{u \in G_n^{(0)}}$ of invariant probability measures on G_n; that is, $\gamma \cdot m_n^{s(\gamma)} = m_n^{r(\gamma)}$ for all $\gamma \in G_n$. By adding zero measures if necessary, we can view m_n as a Borel r-system for $r : G \to G^{(0)}$. Then that $m = (m_n)$ is an Borel approximate invariant mean for $r : G \to G^{(0)}$ and G is Borel amenable by Lemma 9.68 on the facing page. □

The definition of equivalence for Borel groupoids is given in the appendix of [**ADR00**]. In light of [**Ren15**], it turns out to be a much more significant notion than originally thought.

DEFINITION 9.72 ([**ADR00**, Definition A.1.11]). Let G and H be Borel groupoids. A (G, H)-*Borel equivalence* is a Borel space Z such that

(a) Z is a free and proper left Borel G-space,
(b) Z is a free and proper right Borel H-space,
(c) The G- and H-actions commute,
(d) $r : Z \to G^{(0)}$ induces a Borel isomorphism between Z/H and $G^{(0)}$, and
(e) $s : Z \to H^{(0)}$ induces a Borel isomorphism between $G\backslash Z$ and $H^{(0)}$.

In this case, we say that G and H are *Borel equivalent*.

EXAMPLE 9.73. Suppose that G and H are second countable locally compact Hausdorff groupoids and that Z is a (G, H) equivalence. Then Z is a (G, H)-Borel equivalence (Ex 9.7.2).

We saw in Theorem 9.48 on page 217 that equivalence of locally compact groupoids preserves (topological) amenability. Here we show a similar result holds in the Borel case using virtually the same argument.

THEOREM 9.74. *Suppose that G and H are equivalent Borel groupoids. If H is Borel amenable (resp., properly amenable), then so is G.*

Since Borel equivalence is much weaker than equivalence for topological groupoids, Theorem 9.74 provides a dramatic strengthening of Theorem 9.48 on page 217 when the groupoids in question have Haar systems.

THEOREM 9.75. *Suppose that G and H are second countable locally compact Hausdorff groupoids with Haar systems. If G is amenable and Borel equivalent to H, then H is amenable.*

PROOF. If G is amenable, then it is Borel amenable by the easy half of Theorem 9.69 on page 230. Hence H is Borel amenable by Theorem 9.74 and hence amenable by the other half of Theorem 9.69. □

PROOF OF THEOREM 9.74. Let Z be a (G, H)-equivalence. Consider the commutative diagram

$$\begin{array}{ccc} G & \xrightarrow{\quad r \quad} & G^{(0)} \\ {\scriptstyle p_G}\uparrow & & \uparrow{\scriptstyle r} \\ Z * G & \xrightarrow{\quad p_Z \quad} & Z \\ {\scriptstyle p}\downarrow & & \downarrow{\scriptstyle q} \\ Z & \xrightarrow{\quad q \quad} & G\backslash Z, \end{array}$$

where $p(z, \gamma) := \gamma^{-1} \cdot z$.

First, assume that G is properly Borel amenable. To see that H is properly Borel amenable it will suffice, Lemma 9.70 on page 230 to see that that the moment map $s : Z \to H^{(0)}$ is properly Borel amenable.

Let $\lambda = \{\lambda^u\}_{u \in G^{(0)}}$ be an invariant mean for $r : G \to G^{(0)}$. We first build an invariant mean λ_Z for $p_Z : Z * G \to Z$: let λ_Z^z be given by

$$\lambda_Z^z(f) = \int_G f(z, \gamma)\, d\lambda^{r_Z(z)}(\gamma).$$

Consider the measure on Z obtained by the push-forward $p_* \lambda_Z^z$. Since invariance implies $\eta^{-1} \cdot \lambda^{r(\eta)} = \lambda^{s(\eta)}$ for all $\eta \in G$, it follows that

$$\begin{aligned} p_* \lambda_Z^{\eta \cdot z}(b) &= \int_G b(p(\eta \cdot z, \gamma))\, d\lambda_Z^{\eta \cdot z}(\gamma) \\ &= \int_G b(\gamma^{-1}\eta \cdot z)\, d\lambda^{r_Z(\eta \cdot z)}(\gamma) = \int_G b(\gamma^{-1} \cdot z)\, d(\eta^{-1} \cdot \lambda^{r(\eta)})(\gamma) \\ &= p_* \lambda_Z^z(b). \end{aligned}$$

Hence the push-forward $p_* \lambda_Z^z$ depends only on $\dot{z}$ in $G\backslash Z$. Since Z is a Borel equivalence, we can identify the quotient map $q : Z \to G\backslash Z$ with the moment map $s : Z \to H^{(0)}$. Thus we get an s-system of measures $\nu = \{\nu^v\}_{v \in H^{(0)}}$ where $\nu^v = q_* \lambda_Z^z$ for any z with $s(z) = v$. It will suffice to see that ν is invariant. But if

$s(z) = r(h)$, then

$$\begin{aligned}\int_Z b(z)\,d(\nu^{r(h)}\cdot h)(z) &= \int_G b((\gamma^{-1}\cdot z)\cdot h)\,d\lambda^{r(z)}(\gamma)\\ &= \int_G b(\gamma^{-1}\cdot(z\cdot h))\,d\lambda^{r(z\cdot h)}(\gamma)\\ &= \int_Z b(z)\,d\nu^{s(h)}(z).\end{aligned}$$

This completes the proof for proper Borel amenability.

Now we assume that G is Borel amenable and that $\lambda = \{\lambda_n\}$ is a sequence of Borel r-systems for $r : G \to G^{(0)}$ with $\lambda_n = \{\lambda_n^u\}_{u\in G^{(0)}}$ consisting probability measures as in Definition 9.66 on page 229. By Lemma 9.70, it will suffice to show that the moment map $s : Z \to H^{(0)}$ is Borel amenable. The argument parallels that above argument for proper Borel amenability.

We begin by lifting each λ_n to a system $\lambda_{n,Z} = \{\lambda_{n,Z}^z\}_{z\in Z}$ of probability measures for $p_Z : Z * G \to Z$:

$$\lambda_{n,Z}^z(f) = \int_G f(z,\gamma)\,d\lambda_n^{r(z)}(\gamma).$$

We let $\mu_n^z = p_*\lambda_{n,Z}^z$. Since $\lambda_{n,Z}$ is not necessarily invariant, we can't assert that μ_n^z depends only on $\dot z$. However

$$\int_Z b(w)\,d\mu_n^z(w) = \int_G b(\gamma^{-1}\cdot z)\,d\lambda_n^{r(z)}(\gamma),$$

and μ_n^z is supported in $q^{-1}(z) = G\cdot z$.

Since Z is a proper Borel G-space, by definition there is a G-invariant system of probability measures $\rho = \{\rho^z\}_{z\in Z}$ for the Borel G-map $p_Z : Z*G \to Z$ which we view as a family of measures on G such that $\operatorname{supp}\rho^z \subset G^{r(z)}$. As in Remark 9.62, ρ drops to a system $\dot\rho = \{\dot\rho^{\dot z}\}_{\dot z\in G\backslash Z}$ for the quotient map $q : G\backslash(Z*G) \to G\backslash Z$. Since $(z,\gamma)\mapsto \gamma^{-1}\cdot z$ identifies $G\backslash(Z*G)$ with Z, we can view these as measures on Z. Explicitly, $\dot\rho^{\dot z} = p_*\rho^z$ and

$$\dot\rho^{\dot z}(b) = \int_G b(\gamma^{-1}\cdot z)\,d\rho^z(\gamma).$$

Using the invariance of ρ we get measures depending only on $\dot z$ supported on $q^{-1}(\dot z)$ by averaging the μ_n^z with respect to $\dot\rho^{\dot z}$:

$$\nu_n^{\dot z} = \int_Z \mu_n^z\,d\dot\rho^{\dot z}(z) = \int_G \mu_n^{\gamma^{-1}\cdot z}\,d\rho^z(\gamma). \tag{9.32}$$

As in the first part of the proof, we use the fact that Z implements an equivalence to identify the quotient map $q : Z \to G\backslash Z$ with the moment map $s : Z \to H^{(0)}$. Thus we get an s-system $\{\nu^v\}_{v\in H^{(0)}}$ where $\nu^v = \nu^{\dot z}$ for any z such that $s(z) = v$. We will complete the proof by showing that ν is an approximately invariant mean for s. Therefore we need to see that for all $h\in H$, $\|\nu_n^{r(h)}\cdot h - \nu_n^{s(h)}\|_1$ tends to zero with n. Notice that if $(z,\gamma)\in Z*G$, then

$$\|\mu_n^z - \mu_n^{\gamma^{-1}\cdot z}\|_1 \le \|\lambda_n^{r(\gamma)} - \gamma\cdot\lambda_n^{s(\gamma)}\|_1.$$

Next we claim that for fixed $z \in Z$, $\lim_n \|\nu^{\dot{z}} - \mu_n^z\|_1 = 0$. To see this we employ (9.32), view ρ^z as a measure on $G^{r(z)}$, and deduce that

$$\|\nu_n^{\dot{z}} - \mu_n^z\|_1 \leq \int_G \|\mu_n^{\gamma^{-1}\cdot z} - \mu_n^z\|_1 \, d\rho^z(\gamma) \leq \int_G \|\lambda_n^{r(\gamma)} - \gamma \cdot \lambda_n^{s(\gamma)}\|_1 \, d\rho^z(\gamma). \tag{9.33}$$

Since for each γ, $\lim_n \|\lambda_n^{r(\gamma)} - \gamma \cdot \lambda_n^{s(\gamma)}\|_1 = 0$, we see that (9.33) goes to zero by the Lebesgue Dominated Convergence Theorem.

Now fix $h \in H$ and $\epsilon > 0$. Let $z \in Z$ be such that $s(z) = r(h)$ and let $z' = z \cdot h$ and observe that $\mu_n^z \cdot h = \mu_n^{z'}$. Let M be such that $n \geq M$ implies that

$$\|\nu_n^{\dot{z}'} - \mu_n^{z'}\|_1 < \frac{\epsilon}{2}.$$

Then for $n \geq M$, we have

$$\begin{aligned} \|\nu_n^{r(h)} \cdot h - \nu_n^{s(h)}\|_1 &\leq \|\nu_n^{r(h)} \cdot h - \mu_n^z \cdot h\|_1 + \|\mu_n^z \cdot h - \mu_n^{\dot{z}}\|_1 + \|\mu_n^{\dot{z}} - \nu_n^{s(h)}\|_1 \\ &= \|\nu_n^{\dot{z}'} - \mu_n^{z'}\|_1 + 0 + \|\mu_n^{z'} - \nu_n^{\dot{z}'}\|_1 < \epsilon. \end{aligned}$$

Thus $s : Z \to H^{(0)}$ is Borel amenable. This completes the proof. □

LEMMA 9.76. *Let G be a Borel groupoid with a Borel Haar system $\{\lambda^u\}_{u \in G^{(0)}}$. Let $\{Y_i\}_{i=1}^\infty$ be a countable cover of $G^{(0)}$ by invariant Borel subsets such that each $G|_{Y_i}$ is Borel amenable. Then G is Borel amenable.*

PROOF. Since each Y_i is invariant, $u \in Y_i$ implies that $(G|_{Y_i})^u = G^u$. Hence λ restricts to a Borel Haar system on $G|_{Y_i}$. Hence by Lemma 9.68 on page 230, the Borel amenability of $G|_{Y_i}$ implies that there is a Borel approximate invariant density), $\{g_n^i\}_{n=1}^\infty$, on $G|_{Y_i}$.

Let $B_1 = Y_1$ and if $i \geq 2$, let $B_i = Y_i \setminus \bigcup_{j=1}^{i-1} Y_j$. Then the $\{B_i\}$ are a pairwise disjoint cover of $G^{(0)}$ by invariant Borel sets (some of which might be empty). Let b^i be the characteristic function of B_i. Then each b^i is a Borel function on $G^{(0)}$ (taking the values 0 and 1) such that b_i vanishes off Y_i and for each $u \in G^{(0)}$, there is one and only one i such that $b^i(u) = 1$. In particular, we trivially have

$$\sum_{i=1}^\infty b^i(u) = 1 \quad \text{for all } u \in G^{(0)}.$$

For each n, define g_n on G by

$$g_n(\gamma) = \sum_i g_n^i \cdot b^i,$$

where $g_n^i \cdot b^i(\gamma) = g_n^i(\gamma) b^i(s(\gamma))$.

By invariance,

$$\int_G g_n(\gamma) \, d\lambda^u(\gamma) = \int_G g_n^i(\gamma) \, d\lambda^u(\gamma),$$

where i is such that $b^i(u) = 1$. Now it is clear that $\{g_n\}$ is a Borel approximate invariant density for G. Hence G is Borel amenable as claimed. □

Exercises

Ex 9.7.1. Suppose that G is a Borel groupoid with a Borel Haar system $\lambda = \{\lambda^u\}_{u\in G^{(0)}}$. Show that $r : G \to G^{(0)}$ admits a Borel r-system consisting of probability measures.

—Answer on page 380

Ex 9.7.2. Verify that a topological groupoid equivalence is also a Borel groupoid equivalence as asserted in Example 9.73.

—Answer on page 380

9.8. Applications and Examples

It is well known that closed subgroups and quotients of amenable groups by closed subgroups are amenable. The *closed* part is important: for example, the orthogonal group $O(3)$ is compact, hence amenable, but it contains the free group $\mathbf{F}_2$ as a non-closed subgroup. For locally compact groups, this ends the discussion: any subgroup which is locally compact in the relative topology is locally closed [**Wil07**, Lemma 1.26], and any locally closed subgroup of a locally compact group is necessarily closed [**Wil07**, Lemma 1.34].

Naturally, things are not quite so straightforward for groupoids. But it is still true that any subgroupoid that is locally compact in the relative topology is still locally closed [**Wil07**, Lemma 1.26]. Hence the following proposition covers both closed subgroupoids and open subgroupoids. Note that $G^{(0)} = \{\, \gamma \in G : \gamma = \gamma^{-1} = \gamma^2 \,\}$. Hence if H is a subgroupoid of G, then $H^{(0)} \subset G^{(0)}$. Note that the range map on H is just the restriction of the range map from G. But even assuming $r : G \to G^{(0)}$ is open, if H is not open in G, then the restriction $r|_H$ need not be open.

Proposition 9.77 ([**ADR00**, Proposition 5.1.1]). *Let H be a locally compact subgroupoid of a second countable locally compact amenable groupoid G. If H has open range and source maps, then H is amenable.*

Proof. The space $G^{H^{(0)}} = r^{-1}(H^{(0)})$ is a locally compact free and proper left H-space. If $m = \{m_i\}$ is an approximately invariant continuous mean for $r : G \to G^{(0)}$, then we restrict each m_i to $H^{(0)}$: $m_i' = \{m_i^u\}_{u\in H^{(0)}}$. Then $m' = \{m_i'\}$ forms an approximately invariant continuous mean for $r : G^{H^{(0)}} \to H^{(0)}$. Since H is assumed to have open range and source maps, the result follows from Proposition 9.31 on page 205. □

Corollary 9.78. *If G is an amenable groupoid, then all of its isotropy groups $G(u) = G_u^u$ are amenable.*

For transitive groupoids, the Equivalence Theorem implies the converse.

Proposition 9.79. *Let G be a second countable* transitive *groupoid with open range and source maps. Then G is amenable if any, and hence all, of its isotropy groups $G(u) = G_u^u$ are amenable.*

Proof. Since all the isotropy groups are isomorphic, if one is amenable, they all are.

Fix $u \in G^{(0)}$. If G is amenable, then $G(u)$ is by Proposition 9.77 on the previous page. If $G(u)$ is amenable then G is since G is equivalent to $G(u)$ and equivalence preserves amenability by Theorem 9.48 on page 217. □

Next we turn to quotients. As usual, for groupoids, we best say exactly what we mean. Let $\pi : G \to H$ be a continuous *open* surjective groupoid homomorphism that *induces a homeomorphism of $G^{(0)}$ onto $H^{(0)}$*. Let $N = \{\, \gamma \in G : \pi(\gamma) \in H^{(0)} \,\}$. We can identify $H^{(0)}$ and $N^{(0)}$ with $G^{(0)}$. Then we view G as a *unit fixing groupoid extension* of N by H, and write

(9.34)

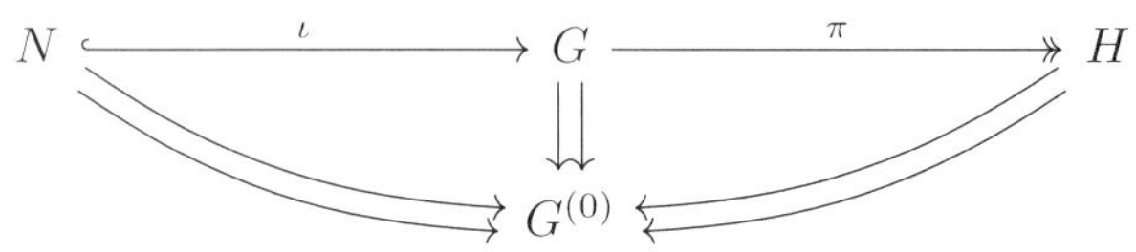

to summarize the situation. Taking care that all these hypotheses are satisfied, we have the usual "two out of three" result just as for groups. This result is taken from [**ADR00**, Proposition 5.1.2].

THEOREM 9.80. *Let G be an extension of N by H as summarized in* (9.34). *If G is amenable, then H is amenable as is N if it has open range and source maps. If N and H are amenable, so is G.*

PROOF. Suppose that G is amenable. Then N is amenable by Proposition 9.77 provided its range and source maps are open. Consider the commutative diagram

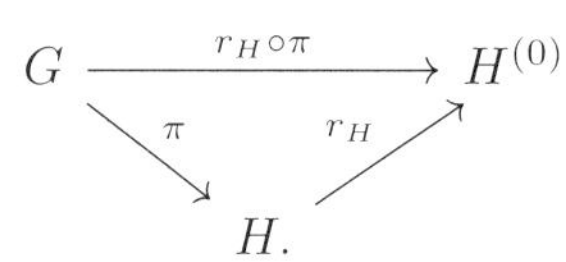

The composition is G-equivariant (since we have identified $G^{(0)}$ and $H^{(0)}$ via $\pi|_{G^{(0)}}$). Since G is amenable, $r_H \circ \pi$ is amenable by Corollary 9.30 on page 205. Hence $r_H : H \to H^{(0)}$ is amenable by Proposition 9.27 on page 203. Thus H is amenable by definition.

The proof of the converse is more complicated. Suppose that N and H are amenable. We need to produce an approximately invariant continuous mean for $r_G : G \to G^{(0)}$. For this, it will suffice, given $\epsilon > 0$ and a compact set $K_1 \subset G$, to produce a r_G-system α of probability measures such that

$$\|\gamma_1 \cdot \alpha^{s(\gamma_1)} - \alpha^{r(\gamma_1)}\|_1 \leq 4\epsilon \quad \text{for all } \gamma_1 \in K_1. \tag{9.35}$$

Since π is open, Lemma 3.20 on page 77 implies that there is a (full) π-system $\beta = \{\beta^h\}_{h\in H}$ of probability measures on G. By Proposition 3.18 on page 77, there is a $g \in C^+_{\pi,c}(G)$ such that $\beta(g)(h) = 1$ for all $h \in H$. Let $m^h = g \cdot \beta^h$. Then $m = \{m^h\}_{h\in H}$ is a π-system of probability measures such that given any compact set $L \subset H$, then $K = \operatorname{supp} g \cap \pi^{-1}(L)$ is a compact set in G such that $m^h(K) = 1$ for all $h \in L$.

Since we have assumed $r_H : H \to H^{(0)} = G^{(0)}$ is amenable, there is a r_H-system $\delta = \{\delta^u\}_{u\in G^{(0)}}$ of probability measures such that

$$\|h \cdot \delta^{s(h)} - \delta^{r(h)}\|_1 \leq \epsilon \quad \text{for all } h \in \pi(K_1). \tag{9.36}$$

Using Lemma 9.28 on page 203, we can find a compact set $L \subset H$ such that

$$\delta^u(L) > 1 - \epsilon \quad \text{for all } u \in s(K_1). \tag{9.37}$$

By our construction of m, we can also find a compact set $K \subset G$ such that

$$m^h(K) = 1 \quad \text{if } h \in L \cup \pi(K_1)L.$$

Let $M \subset G$ be a compact set containing $K^{-1}K$ and $K^{-1}K_1K$.

Since $r_N : N \to N^{(0)} = G^{(0)}$ is amenable, we can find a r_N-system $\sigma = \{\sigma^u\}_{u\in G^{(0)}}$ of probability measures such that

$$\|n \cdot \sigma^{s(n)} - \sigma^{r(n)}\|_1 \le \frac{\epsilon}{2} \quad \text{for all } n \in M \cap N. \tag{9.38}$$

Using this data, we define a probability measure α^u on G via

$$\alpha^u(f) = \int_H \int_G \int_N f(\gamma n)\, d\sigma^{s(\gamma)}(n)\, dm^h(\gamma)\, d\delta^u(h) \quad \text{for } f \in C_c(G).$$

It is not so hard to check that $u \mapsto \alpha^u(f)$ is continuous and that α^u is supported in G^u. Thus α is a $r : G \to G^{(0)}$ system of probability measures. Thus it will suffice to establish (9.35). Fix $f \in C_c(G)$ with $\|f\|_\infty \le 1$. It will suffice to show that

$$\Big| \int_G f(\gamma_1\gamma)\, d\alpha^{s(\gamma_1)}(\gamma) - \int_G f(\gamma)\, d\alpha^{r(\gamma_1)}(\gamma) \Big| \le 4\epsilon \quad \text{if } \gamma_1 \in K_1. \tag{9.39}$$

To ease the notational burden, let

$$\underline{\sigma}(f)(\gamma) = \int_N f(\gamma n)\, d\sigma^{s(\gamma)}(n).$$

I claim that if $\gamma', \gamma \in G$ are such that $\pi(\gamma) = \pi(\gamma')$ and $\gamma^{-1}\gamma' \in M$ (and therefore in $M \cap N$), we have

$$\big|\underline{\sigma}(f)(\gamma) - \underline{\sigma}(f)(\gamma')\big| \le \frac{\epsilon}{2}. \tag{9.40}$$

To establish the claim, note that if $g \in C_c(N)$ with $\|g\|_\infty \le 1$, then (9.38) implies that

$$\Big| \int_N g(p \cdot n)\, d\sigma^{s(p)}(n) - \int_N g(n)\, d\sigma^{r(p)}(n) \Big| \le \frac{\epsilon}{2} \quad \text{if } p \in M \cap N.$$

We verify (9.40) by letting $g(n) = f(\gamma n)$ and $p = \gamma^{-1}\gamma'$.

Now if $\gamma_1 \in K_1$, $h \in L$ is such that $r(h) = s(\gamma_1)$, and $\gamma' \in K$ with $\pi(\gamma') = h$, then our choice of K implies

$$\begin{aligned} &\Big| \int_G \underline{\sigma}(f)(\gamma_1\gamma)\, dm^h(\gamma) - \underline{\sigma}(f)(\gamma_1\gamma') \Big| \\ &\qquad = \Big| \int_{K\cap\pi^{-1}(h)} \big(\underline{\sigma}(f)(\gamma_1\gamma) - \underline{\sigma}(f)(\gamma_1\gamma')\big)\, dm^h(\gamma) \Big| \end{aligned} \tag{9.41}$$

which, since $\gamma^{-1}\gamma' \in K^{-1}K \cap N \subset M \cap N$, is

$$\le \frac{\epsilon}{2}.$$

Similarly,

$$\Big| \int_G \underline{\sigma}(f)(\gamma)\, dm^{\pi(\gamma_1)h}(\gamma) - \underline{\sigma}(f)(\gamma_1\gamma') \Big| \le \frac{\epsilon}{2}. \tag{9.42}$$

Combining (9.41) and (9.42) gives

$$\Bigl|\int_G \underline{\sigma}(f)(\gamma_1\gamma)\,dm^h(\gamma) - \int_G \underline{\sigma}(f)(\gamma)\,dm^{\pi(\gamma_1)h}(\gamma)\Bigr| \le \epsilon. \tag{9.43}$$

Now we consider (9.39). Notice that on the one hand,

$$\int_G f(\gamma_1\gamma)\,d\alpha^{s(\gamma_1)}(\gamma) = \int_H\int_G \underline{\sigma}(f)(\gamma_1\gamma)\,dm^h(\gamma)\,d\delta^{s(\gamma_1)}(h),$$

while

$$\int_G f(\gamma)\,d\alpha^{r(\gamma_1)}(\gamma) = \int_H\int_G \underline{\sigma}(f)(\gamma)\,dm^h(\gamma)\,d\delta^{r(\gamma_1)}(h) = \int_H \psi(h)\,d\delta^{r(\gamma_1)}(h),$$

where

$$\psi(h) = \int_G \underline{\sigma}(f)(\gamma)\,dm^h(\gamma).$$

The point is that (9.39) is bounded by $A+B$ where

$$A = \Bigl|\int_H\Bigl(\int_G \underline{\sigma}(f)(\gamma_1\gamma')\,dm^h(\gamma') - \int_G \underline{\sigma}(f)(\gamma')\,dm^{\pi(\gamma_1)h}(\gamma')\Bigr)\,d\delta^{s(\gamma_1)}(h)\Bigr|,$$

and

$$B = \Bigl|\int_H \psi(\pi(\gamma_1)h)\,d\delta^{s(\gamma_1)}(h) - \int_H \psi(h)\,d\delta^{r(\gamma_1)}(h)\Bigr|.$$

Since $\|\psi\|_\infty \le 1$ and $\pi(\gamma_1)\in\pi(K_1)$, we can use (9.36) to show that

$$B \le \|\pi(\gamma_1)\cdot\delta^{s(\gamma_1)} - \delta^{r(\gamma_1)}\|_1\|\psi\|_\infty \le \epsilon.$$

On the other hand,

$$\begin{aligned} A \le {} & \Bigl|\int_L\Bigl(\int_G \underline{\sigma}(f)(\gamma_1\gamma')\,dm^h(\gamma') - \int_G \underline{\sigma}(f)(\gamma')\,dm^{\pi(\gamma_1)h}(\gamma')\Bigr)\,d\delta^{s(\gamma_1)}(h)\Bigr| \\ & + \Bigl|\int_{H\setminus L}\Bigl(\int_G \underline{\sigma}(f)(\gamma_1\gamma')\,dm^h(\gamma') - \int_G \underline{\sigma}(f)(\gamma')\,dm^{\pi(\gamma_1)h}(\gamma')\Bigr)\,d\delta^{s(\gamma_1)}(h)\Bigr|. \end{aligned} \tag{9.44}$$

Then (9.43) applies and the first term in (9.44) is bounded by ϵ. However, (9.37) implies that $\delta^{s(\gamma_1)}(H\setminus L) < \epsilon$. Hence the second term in (9.44) is bounded by 2ϵ. Combining, we establish the inequality in (9.39) as required. □

Another application involves groupoid bundles as defined in Definition 1.16 on page 7. Recall that if G is a groupoid bundle over T with respect to $p : G^{(0)} \to T$, then G is a bundle over T with respect to the map $p\circ r = p\circ s$. The fibre over $t\in T$ is the groupoid $G(t)$ which is the reduction of G to the saturated set $p^{-1}(t)\subset G^{(0)}$. Notice that if $u\in p^{-1}(t)$, then $G^u\subset G(t)$. It follows that if G has a Haar system $\lambda = \{\lambda^u\}_{u\in G^{(0)}}$, then $\lambda^t = \{\lambda^u\}_{u\in p^{-1}(t)}$ is a Haar system for the fibre groupoid $G(t)$.

PROPOSITION 9.81 ([**Ren15**, Theorem 3.5]). *Suppose that G is a second countable locally compact groupoid with a Haar system λ. Suppose that G is a groupoid bundle with respect to $p : G^{(0)} \to T$. Then G is amenable if and only if each $G(t)$ is amenable.*

PROOF. If G is amenable, then each $G(t)$ is amenable by Proposition 9.77 on page 235.[18]

Now assume each $G(t)$ is amenable. To see that G is amenable it suffices to see that G has a net $\{g_i\} \subset C_c^+(G)$ as in Theorem 9.43(b) on page 211. Thus given a compact set $K \subset G$, a compact set $L \subset G^{(0)}$ and $\epsilon > 0$, it will suffice to produce a $g = g_{K,L} \in C_c^+(G)$ such that

(A) $\lambda(g)(u) \leq 1$ for all $u \in G^{(0)}$,

(B) $\lambda(g)(u) \geq 1 - \epsilon$ for all $u \in L$, and

(C) for all $\gamma \in K$, we have

$$\int_G |g(\gamma^{-1}\eta) - g(\eta)| \, d\lambda^{r(\gamma)}(\eta) < \epsilon.$$

Fix $t \in p(L)$. Since $G(t)$ is amenable, there is a $g^t \in C_c^+(G(t))$ such that

(a) $\lambda(g^t)(u) \leq 1$ for all $u \in p^{-1}(t)$,

(b) $\lambda(g^t)(u) \geq 1 - \epsilon$ for all $u \in L \cap p^{-1}(t)$, and

(c) for all $\gamma \in K \cap G(t)$, we have

$$\int_G |g^t(\gamma^{-1}\eta) - g^t(\eta)| \, d\lambda^{r(\gamma)}(\eta) \leq \epsilon.$$

By the Tietze Extension Theorem, we can extend g^t to an element of $C_c^+(G)$. Since $u \mapsto \lambda(g^t)(u)$ is continuous,

$$W = \{ u \in G^{(0)} : \lambda(g^t)(u) > 0 \}$$

is an open neighborhood of $L \cap p^{-1}(t)$. Hence we can add an element of $C_c^+(G)$, which vanishes off $G(t)$, to g^t and assume that

$$\lambda(g^t)(u) > 0 \quad \text{for all } u \in L.$$

Again using the continuity of $u \mapsto \lambda(g)(u)$ we can modify g^t off $G(t)$ to ensure that

(A)′ $\lambda(g^t)(u) \leq 1$ for all $u \in G^{(0)}$ and

(B)′ $\lambda(g^t)(u) \geq 1 - \epsilon$ for all $u \in L$.

The function

$$\gamma \mapsto \int_G |g^t(\gamma^{-1}\eta) - g^t(\eta)| \, d\lambda^{r(\gamma)}(\eta) \tag{9.45}$$

is continuous. Hence the set U of $\gamma \in G$ such that (9.45) is less than ϵ is open. Thus $K \setminus U$ is compact as is $p(r(K \setminus U))$. Moreover $p(r(K \setminus U))$ does not contain t. Hence there is an open neighborhood V_t of t in T such that

$$\int_G |g^t(\gamma^{-1}\eta) - g^t(\eta)| \, d\lambda^{r(\gamma)}(\eta) < \epsilon \quad \text{for all } \gamma \in K \cap (p \circ r)^{-1}(V_t).$$

Since $p(L)$ is compact, there are $t_1, \dots, t_n \in T$ such that $\{ V_{t_1}, \dots, V_{t_n} \}$ cover $p(L)$. Let $\{ h_0, h_1, \dots, h_n \} \subset C_c^+(T)$ be a partition of unity subordinate to $\{ p(L)^c, V_{t_1}, \dots, V_{t_n} \}$. Thus $\operatorname{supp} h_i \subset V_{t_i}$ for $1 \leq i \leq n$ and

$$\sum_{i=1}^n h_i(t) = 1 \quad \text{for all } t \in p(L) \quad \text{and} \quad \sum_{i=1}^n h_i(t) \leq 1 \quad \text{for all } t \in T.$$

[18] The existence of a Haar system on $G(t)$ means that the range and source maps on $G(t)$ are open.

Define $g \in C_c^+(G)$ by

$$g(\gamma) = \sum_{i=1}^{n} h_i(p(r(\gamma)))g^{t_i}(\gamma).$$

Clearly g satisfies (A) and (B). But since $p \circ r = p \circ s$,

$$\int_G |g(\gamma^{-1}\eta) - g(\eta)|\, d\lambda^{r(\gamma)}(\eta)$$
$$\leq \sum_{i=1}^{n} h_i(p(r(\gamma))) \int_{(p\circ r)^{-1}(V_{t_i})} |g^{t_i}(\gamma^{-1}\eta) - g^{t_i}(\eta)|\, d\lambda^{r(\gamma)}(\eta)$$

which is less than ϵ if $t \in K$. This suffices. □

EXAMPLE 9.82. Suppose that G is a second countable locally compact groupoid with a Haar system such that the orbit space $G\backslash G^{(0)}$ is Hausdorff. Then G is a groupoid bundle over $G\backslash G^{(0)}$ with fiber over the orbit $[u]$ equal to the restriction $G([u])$ of G to $[u]$. Note that $G([u])$ is a transitive groupoid with open range and source maps. Of course, the isotropy group $G([u])(u)$ is equal to $G(u)$. Hence combining Proposition 9.79 on page 235 and Proposition 9.81 on page 238, we see that G is amenable if and only if every stablity group $G(u)$ is amenable. (See Theorem 9.86 on page 242.)

Let G be a second countable locally compact groupoid with a Haar system $\lambda = \{\lambda^u\}_{u\in G^{(0)}}$. Let U be an open saturated subset of $G^{(0)}$ with complement $F = G^{(0)} \setminus U$. Let $G(U)$ and $G(F)$ be the reductions of G to U and F, respectively. Note that λ restricts to a Haar system for both $G(U)$ and $G(F)$. In Theorem 5.1 on page 108 we saw that we could identify $C^*(U)$ with an ideal in $C^*(G)$ with quotient isomorphic to $C^*(F)$. This result might predict the following.

PROPOSITION 9.83. *Let G, U, F, and λ be as above. Then G is amenable if and only if both $G(U)$ and $G(F)$ are amenable.*

PROOF. Since $G(U)$ and $G(F)$ are locally closed subgroupoids of G with Haar systems, they are amenable when G is by Proposition 9.77 on page 235.

Assume $G(U)$ and $G(F)$ are amenable. By either Theorem 9.58 on page 222 or Theorem 9.43 on page 211, there are *sequences* $\{g_n^U\}$ and $\{g_n^F\}$ that form topological approximate invariant densities for $G(U)$ and $G(F)$, respectively. Since $G = G(U) \cup G(F)$, we can define Borel functions on G by

$$g_n(\gamma) = \begin{cases} g_n^U(\gamma) & \text{if } \gamma \in G(U), \text{ and} \\ g_n^F(\gamma) & \text{if } \gamma \in G(F). \end{cases}$$

Then $\{g_n\}$ is a Borel approximate invariant density for G. Thus G is amenable by Theorem 9.58 on page 222. □

Our next result should be compared with Lemma 9.71 on page 231.

PROPOSITION 9.84. *Let G be a second countable locally compact Hausdorff groupoid with a Haar system. Suppose $G = \bigcup G_n$ where each G_n is a locally closed amenable subgroupoid such that $G_n \subset G_{n+1}$ and such that every compact subset of G is eventually contained in some G_n. Then G is amenable.*

REMARK 9.85. The hypotheses that every compact set be contained in some G_n is a trifle annoying. But I can't proceed without it. Fortunately, it is satisfied in the case each G_n is open. We'll take advantage of this in Theorem 9.86 on the next page.

PROOF. Note that a complex measure on G_n can be viewed as a measure on G. Furthermore, then total variation norm is independent of whether the measure is viewed as a measure on G_n or G.[19] Let $m_n = \{m_{n,i}\}$ be a approximately invariant continuous mean on G_n consisting of a sequence of $r|_{G_n}$-systems (see Corollary 9.47 on page 216). By setting $m^u_{n,i} = 0$ if $u \notin G^{(0)}_n$, we view each $m_{n,i} = \{m^u_{n,i}\}_{u\in G^{(0)}}$ as a Radon family of measures on G such that $\operatorname{supp} m^u_{i,n} \subset G^u$ and such that $m^u_{i,n}(G) = 1$ if $u \in G^{(0)}_n$. Also we have

$$\lim_i \|\gamma \cdot m^{s(\gamma)}_{n,i} - m^{r(\gamma)}_{n,i}\|_1 = 0$$

uniformly on compact subsets of G_n.

Since G is second countable, we can let $G = \bigcup K_p$ with each K_p compact and contained in the interior of K_{p+1}. Hence every compact subset of G is contained in some K_m, and the every K_m is contained in some G_n (by assumption).

Let D be the set of 4-tuples (p, n, i, k) of positive integers such that

$$\|\gamma \cdot m^{s(\gamma)}_{n,i} - m^{r(\gamma)}_{n,i}\|_1 \leq \frac{1}{k} \quad \text{for all } \gamma \in K_p.$$

Clearly D is nonempty. To see that D is directed, let (p_1, n_1, i_1, k_1) and (p_2, n_2, i_2, k_2) be in D. Let $p = \max\{p_1, p_2\}$. By assumption, there is a n such that $K_p \subset G_n$. We can assume that $n \geq n_i$ for $i = 1, 2$. Let $k = \max\{k_1, k_2\}$. Since by assumption,

$$\|\gamma \cdot m^{s(\gamma)}_{n,i} - m^{r(\gamma)}_{n,i}\|_1 \to 0$$

uniformly on K_p, there is an i dominating i_1 and i_2 such that

$$\|\gamma \cdot m^{s(\gamma)}_{n,i} - m^{r(\gamma)}_{n,i}\|_1 < \frac{1}{k} \quad \text{for all } \gamma \in K_p.$$

Thus $(m, n, i, k) \in D$ and dominates both (p_1, n_1, i_1, k_1) and (p_2, n_2, i_2, k_2). The point is that $(p, n, i, k) \mapsto m_{n,i}$ a net of Radon families on G directed by D. We want to show that it has all the properties of a Borel approximate invariant mean with the exception of being a sequence.

If $u \in G^{(0)}$, then there is a p_0 such that $u \in K_{p_0}$ and an n_0 such that $K_{p_0} \subset G_{n_0}$. Then there is i_0 such that $(p_0, n_0, i_0, 1) \in D$. Then if $(p, n, i, k) \geq (p_0, n_0, i_0, 1)$, we have $m^u_{n,i}(G) = 1$. Similarly, if $\gamma \in G$, then as above, we can assume $\gamma \in K_{p_0} \subset G_{n_0}$. The for any $\epsilon > 0$ there are i_0 and k_0 such that $(p, n, i, k) \geq (p_0, n_0, i_0, k_0)$ implies

$$\|\gamma \cdot m^{s(\gamma)}_{n,i} - m^{r(\gamma)}_{n,i}\|_1 < \epsilon$$

which is what we claimed.

Since D is countable, we can extract a subnet which is a sequence $m' = \{m'_r\}_{r=1}^\infty$ (Ex 9.8.1) . It follows that m' is a Borel approximate invariant mean. Hence G is amenable by Theorem 9.58 on page 222. □

---

[19] Recall that $\nu = \int \varphi \, d|\nu|$ for a unique measure $|\nu|$ and a unimodular function φ uniquely defined almost everywhere. Then $\|\nu\|_1 = |\nu|(X)$.

With these tools in hand, we can prove a converse to Corollary 9.78 on page 235 in the case that G-orbits are locally closed in $G^{(0)}$.

THEOREM 9.86. *Suppose that G is a second countable locally compact groupoid with a Haar system. Suppose that $G\backslash G^{(0)}$ is T_0. Then G is amenable if and only if every isotropy group $G(u)$ is amenable.*

PROOF. If G is amenable, then each $G(u)$ is amenable by Corollary 9.78 on page 235. For the converse, assume each $G(u)$ is amenable. Then we need to see that G is amenable.

Since $G\backslash G^{(0)}$ is T_0, $G\backslash G^{(0)}$ is locally Hausdorff by Theorem 2.27 on page 38. By [**Wil07**, Lemma 6.3], there is a family $\{ V_\alpha : 0 \le \alpha \le \alpha_0 \}$ of open sets indexed by ordinals with α_0 countable such that $V_0 = \emptyset$, $V_{\alpha_0} = G\backslash G^{(0)}$, and such that

(a) $\alpha < \alpha' \le \alpha_0$ implies $V_\alpha \subsetneq V_{\alpha'}$,
(b) $V_{\alpha+1} \setminus V_\alpha$ is a dense Hausdorff subset of $(G\backslash G^{(0)}) \setminus V_\alpha$, and
(c) if α is a limit ordinal, then

$$V_\alpha = \bigcup_{\alpha'<\alpha} V_{\alpha'}.$$

Let $U_\alpha = q^{-1}(V_\alpha)$ where $q : G^{(0)} \to G\backslash G^{(0)}$ is the orbit map. Then each U_α is an open and saturated subset of $G^{(0)}$. Furthermore, we have $G(U_1)$ amenable in view of Example 9.82 on page 240.

Now we can proceed by transfinite induction. Assume $G(U_{\alpha'})$ is amenable for all $\alpha' < \alpha$. If $\alpha = \beta + 1$, then $G(U_{\beta+1} \setminus U_\beta)$ is amenable again as in Example 9.82. Since $G(U_\beta)$ is amenable by the inductive hypotheses, we have $G(U_\alpha) = G(U_{\beta+1})$ amenable by Proposition 9.83 on page 240.

If α is a limit ordinal, then $\{ \alpha' : \alpha' < \alpha \}$ is countable. Hence we can enumerate it as $\{\alpha'_n\}_{n=1}^{\infty}$. Inductively choose β_{n+1} such that $\beta_{n+1} \ge \alpha'_{n+1}$ and $\beta_{n+1} \ge \beta_n$. Then

$$G(U_\alpha) = \bigcup_{n=1}^{\infty} G(U_{\beta_n}).$$

The amenability of $G(U_\alpha)$ follows from Proposition 9.84 on page 240 (see Remark 9.85 on the preceding page).[20] □

EXERCISES

Ex 9.8.1. Let $x = \{x_d\}_{d\in D}$ be a net directed by a countable set D. Show that x has a subnet indexed by $\mathbf{N}$. Thus, x has a subnet which is a sequence.

—Answer on page 380

Ex 9.8.2. Suppose that F is a locally closed subset of $G^{(0)}$ and let $G(F)$ be the reduction of G to F. Show that if G has open range and source maps then so does $G(F)$.

—Answer on page 381

[20]We could also apply Lemma 9.76 on page 234 and Theorem 9.69 on page 230.

CHAPTER 10

Measurewise Amenability for Groupoids

In the previous chapter, we pursued a notion of amenability for groupoids primarily motivated by the characterization of amenable groups via positive definite functions. Now it is time to concentrate on the mean in a*mean*able as in Definition 9.7 on page 192. Since we are still following Anantharaman-Delaroche & Renault, we want to develop a calculus for means which requires set-ups requiring more general G-spaces than just G and $G^{(0)}$. Throughout G will be a second countable locally compact Hausdorff groupoid with a Haar system λ.

10.1. Means

Our standard set-up will include an equivariant map $\pi : Y \to X$ between second countable locally compact Hausdorff G-spaces and a proper *equivariant* π-system α. We also assume we have a finite measure μ on X and equip Y with the measure $\mu \circ \alpha$. We'll usually write $L^\infty(X)$ in place of $L^\infty(X, \mu)$. Similarly, we write $L^\infty(Y)$ in place of $L^\infty(Y, \mu \circ \alpha)$. Notice that $L^\infty(Y)$ is naturally a $L^\infty(X)$ module: $\varphi \cdot f(y) = \varphi(\pi(y))f(y)$. We write $B_{L^\infty(X)}\big(L^\infty(Y), L^\infty(X)\big)$ for the bounded $L^\infty(X)$-linear maps, and say $m \in B_{L^\infty(X)}\big(L^\infty(Y), L^\infty(X)\big)^+$ if $m(L^\infty(Y)^+) \subset L^\infty(X)^+$.

Quasi-Invariant Measures. Since our first goal is to generalize the notion of a mean on $L^\infty(G, \mu \circ \lambda)$—as in Definition 9.7 on page 192—we are going to need actions of $C_c(G)$ on both $L^\infty(Y)$ and $L^\infty(X)$. For this we are going to need the measure μ on X to be suitably "quasi-invariant" with respect to the G-actions on Y and X. For this we need a measure on the action groupoid $G \rtimes X$.

LEMMA 10.1. *We'll write $\mu \circ \lambda$ for the measure on $G \rtimes X$ given by*

$$\mu \circ \lambda(f) = \int_{G \rtimes X} f(\gamma, x)\, d(\mu \circ \lambda)(\gamma, x) = \int_X \int_G f(\gamma, x)\, d\alpha^{r(x)}(\gamma)\, d\mu(x).$$

Then the collection of functions

$$f \otimes h(\gamma, x) = f(\gamma)h(x) \quad \text{for } f \in C_c(G) \text{ and } h \in C_c(X)$$

span a dense subspace of $L^1(G \rtimes X, \mu \circ \lambda)$.

PROOF. Such functions span a dense subspace of $C_c(G \rtimes X)$ in the inductive limit topology. This suffices. □

DEFINITION 10.2. We say that μ is a *quasi-invariant* measure on the G-space X if the measure class of $\mu \circ \lambda$ is invariant under the inversion map: $I_X : G \rtimes X \to G \rtimes X$ given by $I_X(\gamma, x) = (\gamma^{-1}, \gamma^{-1} \cdot x)$. We let δ be the Raydon-Nikodym derivative

$$\delta(\gamma, x) = \frac{d(\mu \circ \lambda)}{d(\mu \circ \lambda)^{-1}}(\gamma, x)$$

where we write $(\mu \circ \lambda)^{-1}$ for the push-forward of $\mu \circ \alpha$ by I_X.

REMARK 10.3. If we view $G \ltimes X$ as a groupoid with unit space (identified with) X and equipped with the usual Haar system $\underline{\lambda} = \{\delta_x \times \lambda^{r(x)}\}_{x\in X}$, then the measure $\mu \circ \lambda$ from Lemma 10.1 on the previous page is just the "usual" measure $\underline{\nu} = \mu \circ \underline{\lambda}$ on $G \ltimes X$. With this in mind, saying that μ is a quasi-invariant measure on the G-space X is equivalent to saying μ is $G \ltimes X$-quasi-invariant on $(G \ltimes X)^{(0)} = X$.

REMARK 10.4. For future reference, note that if μ is quasi-invariant, then

$$\int_X \int_G f(\gamma^{-1}, \gamma^{-1} \cdot x)\delta(\gamma^{-1}, \gamma^{-1} \cdot x)\, d\lambda^{r(x)}(\gamma)\, d\mu(x) = \int_X \int_G f(\gamma, x)\, d\lambda^{r(x)}(\gamma)\, d\mu(x). \tag{10.1}$$

As in Section 9.2, quasi-invariance is key to defining a $C_c(G)$-action on $L^\infty(Y)$ and $L^\infty(X)$. If X is a G-space, $f \in C_c(G)$, and $\psi \in \mathcal{B}^b(X)$, then Proposition 3.3 on page 70 implies that we can define a bounded Borel function on X by

$$f * \psi(x) = \int_G f(\gamma)\psi(\gamma^{-1} \cdot x)\, d\lambda^{r(x)}(\gamma). \tag{10.2}$$

Clearly $|f * \psi(x)| \leq \|f\|_I \cdot \sup_{x\in X} |\psi(x)|$. If μ is quasi-invariant and $\psi(x) = 0$ for μ-almost all x, then $f * \psi(x) = 0$ for μ-almost all x (Ex 10.1.1). Since $f * (\psi + \psi') = f * \psi + f * \psi'$, it follows that $\|f * \psi\|_\infty \leq \|f\|_I \|\psi\|_\infty$ for the essential supremum norm on $L^\infty(X)$. Therefore (10.2) defines an action of $C_c(G)$ on $L^\infty(X)$. Similarly, if $\varphi(y) = 0$ for $\mu \circ \alpha$-almost all y, then $f * \varphi(y) = 0$ for $\mu \circ \alpha$-almost all y, and we also have an action of $C_c(G)$ on $L^\infty(Y)$ (Ex 10.1.1).

DEFINITION 10.5. Let $\pi : Y \to X$ be a continuous equivariant surjection. A *mean* for (π, α, μ) is a unital, positive, $L^\infty(X)$-linear map $m : L^\infty(Y) \to L^\infty(X)$. That is, $m \in B_{L^\infty(X)}\big(L^\infty(Y), L^\infty(X)\big)^+$ with $m(1) = 1$. We say m is *invariant* if

$$m(f * \varphi) = f * m(\varphi) \quad \text{in } L^\infty(X) \text{ for all } f \in C_c(G), \text{ and } \varphi \in L^\infty(Y).$$

REMARK 10.6. Notice that as a positive unital map, a mean necessarily has norm equal to one. In particular, if X is reduced to a point, then a mean is a state. We'll write $B_{L^\infty(X)}\big(L^\infty(Y), L^\infty(X)\big)_1^+$ for the positive part of the unit ball.

EXAMPLE 10.7. A mean as defined in Definition 9.7 on page 192, is exactly a mean for $r : G \to G^{(0)}$ with respect to μ and λ.

Duals. We need some technical results about some dual spaces generalizing the constructs in Section 9.6. I'll summarize the key facts we need here and refer to Appendices A.3, A.4, and A.5 for the full details. Recall that a linear functional $\nu : C_c(Y) \to \mathbf{C}$ which is continuous in the inductive limit topology is called a *Radon measure* (see Section 3.3). We say that ν is *α-bounded* if there is a finite measure μ on X such that $|\nu| \leq \mu \circ \alpha$ and write $M(Y, \alpha)$ for the collection of all α-bounded complex Radon measures on Y. If we let

$$\|f\|_{\alpha,1} = \sup_{x\in X} \lambda(|f|)(x) = \sup_{x\in X} \int_Y |f(y)|\, d\alpha^x(y),$$

then $\|\cdot\|_{\alpha,1}$ is norm on $C_c(Y)$ and if $|\nu| \leq \mu \circ \alpha$ we have

$$|\nu(f)| \leq \|\mu\|_1 \|f\|_{\alpha,1}.$$

Hence ν extends to a bounded linear functional on the completion $C_0(X, L^1(Y,\alpha))$ of $C_c(Y)$ with respect to $\|\cdot\|_{\alpha,1}$. Proposition A.10 on page 319 implies that we can identify the dual of $C_0(X, L^1(Y,\alpha))$ with $M(Y,\alpha)$ where

$$\|\nu\| = \inf\{\, \|\mu\|_1 : |\nu| \le \mu \circ \alpha \,\}.$$

If we fix a measure μ on X, then as usual we equip Y with the measure $\mu\circ\alpha$. If $\varphi \in L^\infty(Y)$ and $h \in L^1(X)$, then we get a complex Radon measure $\nu = \varphi(h\cdot\mu\circ\alpha)$ in $M(Y,\alpha)$ given by[1]

$$\nu(f) = \int_X \int_Y f(y)\varphi(y)\, d\alpha^x(y) h(x)\, d\mu(x).$$

This map factors through the balanced tensor product $L^\infty(Y) \odot_{L^\infty(X)} L^1(X)$ into $M(Y,\alpha)$. By Corollary A.12 on page 322 the range is exactly

$$M(Y,[\mu]\circ\alpha) = \{\, \nu \in M(Y,\alpha) : |\nu| \le \mu_1 \circ \alpha \text{ where } \mu_1 \ll \mu \,\}.$$

Moreover this map is isometric if we view $L^\infty(Y) \odot_{L^\infty(X)} L^1(X)$ as a dense subspace of the balanced projective tensor product $L^\infty(Y) \mathbin{\widehat{\otimes}}_{L^\infty(X)} L^1(X)$ (Corollary A.14 on page 324). The usual theory of projective tensor products (Appendix A.2) implies that the closure $X([\mu])$ of $M(Y,[\mu]\circ\alpha)$ in $M(Y,\alpha)$ has dual $B_{L^\infty(X)}\big(L^\infty(Y), L^\infty(X)\big)$ where $m \in B_{L^\infty(X)}\big(L^\infty(Y), L^\infty(X)\big)$ acts on $\nu = \varphi\otimes h$ by

$$m(\varphi\otimes h) = \big\langle m(\varphi)\,,\, h\big\rangle,$$

where the pairing on the right is the usual one between $L^\infty(X)$ and $L^1(X)$.

Let $C_c^+(Y)_{\alpha,1} = \{\, f \in C_c^+(Y) : \alpha(f)(x) \le 1 \text{ for all } x \,\}$, and let

$$C_0(X, L^1(Y,\alpha))^{**}_+ = M(Y,\alpha)^*_+ = \{\, m \in M(Y,\alpha)^* : m(\nu) \ge 0 \text{ if } \nu \ge 0 \,\}.$$

If $f \in C_c^+(Y)_{\alpha,1}$ and $f \mapsto m_f$ the usual embedding—so that $m_f(\nu) = \nu(f)$—then m_f is in the unit ball of $C_0(X, L^1(Y,\alpha))^{**}_+$. If we restrict m_f to $M(Y,[\mu]\circ\alpha)$ and identify $M(Y,[\mu]\circ\alpha)$ with $L^\infty(Y) \odot_{L^\infty(X)} L^1(X)$, then we can view $m_f \in B_{L^\infty(X)}\big(L^\infty(Y), L^\infty(X)\big)^+_1$ with $m_f(\varphi\otimes h) = \big\langle \alpha(\varphi f)\,,\, h\big\rangle$. We will need to know that the image of $C_c^+(Y)_{\alpha,1}$ is weak-$*$ dense in the unit balls of both $B_{L^\infty(X)}\big(L^\infty(Y), L^\infty(X)\big)^+$ (Lemma A.18 on page 326) and $C_0(X, L^1(Y,\alpha))^{**}_+$ (Proposition A.19 on page 328).

In Lemma A.13 on page 322, we note that any $\nu \in L^\infty(Y) \odot_{L^\infty(X)} L^1(X)$ extends isometrically to a linear functional on the vector space $\mathcal{L}^\infty\big(X, L^1(Y,\alpha)\big)$ of Borel functions f such that $x \mapsto \alpha(f)(x)$ is bounded. This allows us to view $L^\infty(Y) \odot_{L^\infty(X)} L^1(X)$ as a subspace of $\mathcal{L}^\infty\big(X, L^1(Y,\alpha)\big)^*$. If $M(f)$ is the map of $\mathcal{L}^\infty\big(X, L^1(Y,\alpha)\big)$ to itself under convolution, then the transpose ${}^tM(f)$ maps $L^\infty(Y) \odot_{L^\infty(X)} L^1(X)$ to itself (Lemma A.17 on page 325). Since $B_{L^\infty(X)}\big(L^\infty(Y), L^\infty(X)\big)$ is the dual of $L^\infty(Y) \odot_{L^\infty(X)} L^1(X) \subset L^\infty(Y) \mathbin{\widehat{\otimes}}_{L^\infty(X)} L^1(X)$, we can abuse notation a bit and let ${}^{tt}M(f)$ be the operator on $B_{L^\infty(X)}\big(L^\infty(Y), L^\infty(X)\big)$ given by $\big\langle {}^{tt}M(m)\,,\, \psi\big\rangle = \big\langle m\,,\, {}^tM(\psi)\big\rangle$ for ψ in the image of $L^\infty(Y) \odot_{L^\infty(X)} L^1(X) \subset \mathcal{L}^\infty\big(X, L^1(Y,\alpha)\big)^*$, where the later pairing is that between $B_{L^\infty(X)}\big(L^\infty(Y), L^\infty(X)\big)$ and $L^\infty(Y) \odot_{L^\infty(X)} L^1(X)$. Since this pseudo

[1] $|\nu| \le \mu_1 \circ \alpha$ with $\mu_1 = \|\varphi\|_\infty |h|\cdot\mu$.

double transpose plays a key role when working with means, we introduce notation similar to that used in [**ADR00**]:

$$f \star m := {}^{tt}M(f)m \quad \text{for } f \in C_c(G) \text{ and } m \in B_{L^\infty(X)}\big(L^\infty(Y), L^\infty(X)\big).$$

For example, notice that if $\psi \in L^\infty(Y) \odot_{L^\infty(X)} L^1(X)$ and ${}^tM(f)\psi = \psi' \otimes h' \in L^\infty(Y) \odot_{L^\infty(X)} L^1(X)$, then using (A.9), we have

$$\begin{aligned}\langle m_g\,,\,{}^tM(f)\psi\rangle &= \langle \alpha(\psi' g)\,,\,h'\rangle = \langle {}^tM(f)\psi\,,\,g\rangle = \langle \psi\,,\,f*g\rangle \\ &= \langle m_{f*g}\,,\,\psi\rangle.\end{aligned}$$

Thus

$$f \star m_g = m_{f*g} \quad \text{for } f \in C_c(G) \text{ and } g \in \mathcal{L}^\infty\big(X, L^1(Y,\alpha)\big).$$

REMARK 10.8. We are most interested in the case where $C_0(X, L^1(Y,\alpha)) = C_0(G^{(0)}, L^1(G,\lambda))$—we denoted the later by $\mathcal{E}$ in Section 9.6. The action $f \star m$ should be compared with $f \diamond m$ in where $m \in \mathcal{E}^{**}$. They are clearly close cousins, but not quite the same. But we do have the following.

LEMMA 10.9. *Suppose that μ is a quasi-invariant measure on $G^{(0)}$, that $\nu = \varphi \otimes h \in L^\infty(G) \odot_{L^\infty(G^{(0)})} L^1(G^{(0)}) \equiv M(G,[\mu]\circ\lambda)$, and that $m \in \mathcal{E}^{**}$. Let m' be the restriction of m to $M(G,[\mu]\circ\lambda)$. If $f \in C_c(G)$ then $f \star m'$ is the restriction of $f \diamond m$ to $M(G,[\mu]\circ\lambda)^* \equiv B_{L^\infty(G^{(0)})}(L^\infty(G), L^\infty(G^{(0)}))$.*

PROOF. As above, we let $M(f)$ be the operator on $L^\infty(G^{(0)}, L^1(G,\lambda))$ and $L(f)$ the restriction of $M(f)$ to $\mathcal{E} \subset L^\infty(G^{(0)}, L^1(G,\lambda))$. Using Lemma A.17 on page 325, we have ${}^tM(f)\nu = \nu' \in M(G,[\mu]\circ\lambda)$. Then if $g \in \mathcal{E}$,

$$\big\langle \nu\,,\,L(f)g\big\rangle = \big\langle \nu\,,\,M(f)g\big\rangle = \big\langle {}^tM(f)\,,\,g\big\rangle = \big\langle \nu'\,,\,g\big\rangle.$$

On the other hand, $\big\langle \nu\,,\,L(f)g\big\rangle = \big\langle {}^tL(f)\nu\,,\,g\big\rangle$. Hence ${}^tL(f)\nu$ is given by $\nu' = {}^tM(f)\nu$. Thus

$$\begin{aligned}\big\langle {}^{tt}M(f)m'\,,\,\nu\big\rangle &= \big\langle m'\,,\,{}^tM(f)\nu\big\rangle \\ &= \big\langle m\,,\,\nu'\big\rangle = \big\langle m\,,\,{}^tL(f)\nu\big\rangle = \big\langle {}^{tt}L(f)m\,,\,\nu\big\rangle.\end{aligned}$$

Thus on $M(G,[\mu]\circ\lambda)$, $f \star m' = {}^{tt}M(f)m'$ agrees with $f \diamond m = {}^{tt}L(f)m$ as claimed. □

Approximate Means. The key idea here is that if m is an invariant mean in $B_{L^\infty(X)}\big(L^\infty(Y), L^\infty(X)\big)_1^+$—or for that matter any mean with special properties—the density of $C_c^+(Y)_{\alpha,1}$ in $B_{L^\infty(X)}\big(L^\infty(Y), L^\infty(X)\big)_1^+$ means that there is a net $\{g_i\} \subset C_c^+(Y)_{\alpha,1}$ such that $m_{g_i} \to m$ in the weak-$*$ topology. This forces the net $\{g_i\}$ to have certain important properties related to the given properties of m. Conversely, the existence of such nets will force its accumulation points—recall the unit ball is weak-$*$ compact—to be means with desirable properties. For flexibility down the road, we work with the larger space $\mathcal{L}^\infty\big(X, L^1(Y,\alpha)\big)_1^+ = \{\, f \in \mathcal{L}^\infty\big(X, L^1(Y,\alpha)\big)^+ : \alpha(f)(x) \le 1 \text{ for all } x \,\}$ in place of $C_c^+(Y)_{\alpha,1}$.

LEMMA 10.10. *Let* $\{g_i\} \subset \mathcal{L}^\infty\big(X, L^1(Y,\alpha)\big)_1^+$. *Then the following are equivalent.*

(a) *For all* $\varphi \in L^\infty(Y)$, $h \in C_c(X)$ *and* $f \in C_c(G)$,

$$\lim_i \int_X \int_Y \int_G f(\gamma)h(x)\varphi(y)\big(g_i(\gamma^{-1}\cdot y) - g_i(y)\big)\, d\lambda^{r(x)}(\gamma)\, d\alpha^x(y)\, d\mu(x) = 0.$$

(b) *For all* $\varphi \in L^\infty(Y)$, $h \in C_c(X)$ *and* $f \in C_c(G)$,

$$\lim_i \int_X \int_Y \int_G f(\gamma)h(x)\varphi(\gamma^{-1}\cdot y)\big(g_i(\gamma^{-1}\cdot y) - g_i(y)\big)\, d\lambda^{r(x)}(\gamma)\, d\alpha^x(y)\, d\mu(x) = 0.$$

(c) *For all* $\varphi \in L^\infty(Y)$ *and* $F \in L^1(G \rtimes X)$,

$$\lim_i \int_X \int_Y \int_G F(\gamma, x)\varphi(y)(g_i(\gamma^{-1}\cdot y) - g_i(y))\, d\lambda^{r(x)}(y)\, d\alpha^x(y)\, d\mu(x) = 0.$$

PROOF. Since $C_c(G) \odot C_c(X)$ is dense in $L^1(G \rtimes X)$ by Lemma 10.1, (a) and (b) are equivalent to, respectively,

(a)′ $$\lim_i \int_X \int_Y \int_G f(\gamma, x)\varphi(y)\big(g_i(\gamma^{-1}\cdot y) - g_i(y)\big)\, d\lambda^{r(x)}(\gamma)\, d\alpha^x(y)\, d\mu(x) = 0$$
and

(b)′ $$\lim_i \int_X \int_Y \int_G f(\gamma, x)\varphi(\gamma^{-1}\cdot y)\big(g_i(\gamma^{-1}\cdot y) - g_i(y)\big)\, d\lambda^{r(x)}(\gamma)\, d\alpha^x(y)\, d\mu(x) = 0$$

for $f \in L^1(G \rtimes X)$ and $\varphi \in L^\infty(Y)$.

Hence (a) is equivalent to (c). On the other hand, using Fubini and quasi-invariance in the form of (10.1), the integral in (a)′ is equal to

$$\begin{aligned}\int_X \int_G \int_Y & f(\gamma, x)\varphi(y)\big(g_i(\gamma^{-1}\cdot y) - g_i(y)\big)\, d\alpha^x(y)\, d\lambda^{r(x)}(\gamma)\, d\mu(x) \\ &= \int_X \int_G \int_Y f(\gamma^{-1}, \gamma^{-1}\cdot x)\varphi(y)\big(g_i(\gamma\cdot y) - g_i(y)\big)\, d\alpha^{\gamma^{-1}\cdot x}(y) \\ &\qquad\qquad \delta(\gamma^{-1}, \gamma^{-1}\cdot x)\, d\lambda^{r(x)}(\gamma)\, d\mu(x)\end{aligned}$$

which, since α is equivariant, is

$$\begin{aligned}&= \int_X \int_G \int_Y f(\gamma^{-1}, \gamma^{-1}\cdot x)\delta(\gamma^{-1}, \gamma^{-1}\cdot x)\varphi(\gamma^{-1}\cdot y)\big(g_i(y) - g_i(\gamma^{-1}\cdot y)\big) \\ &\qquad\qquad d\alpha^x(y)\, d\lambda^{r(x)}(\gamma)\, d\mu(x).\end{aligned}$$

Since $(\gamma, x) \mapsto f(\gamma^{-1}, \gamma^{-1}\cdot x)\delta(\gamma^{-1}, \gamma^{-1}\cdot x)$ is in $L^1(G \rtimes X)$, it follows that (a)′ implies (b)′. Similarly, (b)′ implies (a)′. □

Lemma 10.10 doesn't force interesting properties on $\{g_i\}$ unless we require some sort of normalization condition: note that the zero net would suffice.

DEFINITION 10.11. A net in $\mathcal{L}^\infty\big(X, L^1(Y,\alpha)\big)_1^+$ is called a *approximate weakly invariant mean* if it satisfies the equivalent conditions in Lemma 10.10 and if $\{\alpha(g_i)\}$ converges to 1 in the weak-$*$ topology on $L^\infty(X)$. That is, for all $h \in L^1(X)$ we have

$$\lim_i \int_X h(x)\alpha(g_i)(x)\, d\mu(x) = \int_X h(x)\, d\mu(x). \tag{10.3}$$

The following is a prototype for other results to follow. Given a mean m with a fundamental property such as invariance, we get an approximate version of that property using the compactness of the unit ball in the weak-$*$ topology. Since

$$C_c^+(Y)_{\alpha,1} \subset C_{\pi,c}^+(Y) \cap \mathcal{L}^\infty\big(X, L^1(Y,\alpha)\big)_1 \subset \mathcal{L}^\infty\big(X, L^1(Y,\alpha)\big)_1^+, \tag{10.4}$$

and since Lemma A.18 on page 326 implies $C_c^+(Y)_{\alpha,1}$ is dense weak-$*$ in the dual space $B_{L^\infty(X)}\big(L^\infty(Y), L^\infty(X)\big)_1^+$, we can get approximate versions from any of the three spaces in (10.4).

If $\varphi \in C_0(Y)$ and $m \in B_{L^\infty(X)}\big(L^\infty(Y), L^\infty(X)\big)$, we let $\varphi \cdot m(\psi) := m(\varphi\psi)$. We are most interested in this when $\varphi(y) = \lambda(f)(r(y)) = \lambda(f)(r(\pi(y)))$. Since m is $L^\infty(X)$-linear, we have $(\lambda(f) \circ r) \cdot m(\varphi)(x) = \lambda(f)(r(x))m(\varphi)(x)$.

PROPOSITION 10.12. *Let $m \in B_{L^\infty(X)}\big(L^\infty(Y), L^\infty(X)\big)_1^+$ be a mean.*

(a) *Then m is invariant if and only if $f \star m = \big(\lambda(f) \circ r\big) \cdot m$ for all $f \in C_c(G)$.*

(b) *If $\{g_i\}$ is a net in $\mathcal{L}^\infty\big(X, L^1(Y,\alpha)\big)_1^+$ such that $m = \lim_i m_{g_i}$ in the weak-$*$ topology, then m is invariant if and only if $\{g_i\}$ is an approximate weakly invariant mean for π.*

(c) *If m is invariant, then there is an approximate weakly invariant mean $\{g_i\}$ in $C_c^+(Y)_{\alpha,1}$ such that $m = \lim_i m_{g_i}$ in the weak-$*$ topology.*

REMARK 10.13. Recall that m is invariant when $f * m(\varphi) = m(f * \varphi)$ for all $f \in C_c(G)$ and $\varphi \in L^\infty(Y)$. It will be handy to keep in mind that $f \star m_g = m_{f*g}$, and that $m_g(\varphi) := \alpha(g\varphi)$.

PROOF. Let $\{g_i\}$ be a net in $\mathcal{L}^\infty\big(X, L^1(Y,\alpha)\big)_1^+$. If $f \in C_c(G)$, $h \in C_c(X)$ and $\varphi \in L^\infty(Y)$), then since α is equivariant,

$$\begin{aligned}\big\langle f * (m_{g_i}(\varphi))\,,\, h\big\rangle - \big\langle m_{g_i}(f * \varphi)\,,\, h\big\rangle \qquad \qquad & \\ = \int_X \int_Y \int_G f(\gamma)h(x)\varphi(\gamma^{-1}\cdot y)\big(g_i(\gamma^{-1}\cdot y) - g_i(y)\big)\, d\lambda^{r(x)}(\gamma)\, d\alpha^x(y)\, d\mu(x).&\end{aligned} \tag{10.5}$$

Assume that $\{g_i\}$ is such that $m = \lim_i m_{g_i}$ in the weak-$*$ topology. If $\{g_i\}$ is an approximate weakly invariant mean, then Lemma 10.10(b) and (10.5) imply that m is invariant. On the other hand, if m is invariant, then (10.5) implies that Lemma 10.10(b) holds. But we also have $m(1) = 1$ so that

$$\big\langle m_{g_i}(1)\,,\, h\big\rangle \to \langle 1\,,\, h\rangle.$$

That is (10.3) holds and $\{g_i\}$ is an approximate weakly invariant mean. This proves (b). Part (c) follows from part (b) and the density of $C_c^+(Y)_{\alpha,1}$ in $B_{L^\infty(X)}\big(L^\infty(Y), L^\infty(X)\big)_1^+$ (Lemma A.18 on page 326).

Next compute that

$$\begin{aligned}\big\langle (f \star m_{g_i})(\varphi)\,,\, h\big\rangle &- \big\langle (\lambda(f) \circ r) \cdot m_{g_i}(\varphi)\,,\, h\big\rangle \\ &= \big\langle m_{f*g_i}(\varphi)\,,\, h\big\rangle - \big\langle (\lambda(f) \circ r) \cdot m_{g_i}(\varphi)\,,\, h\big\rangle \\ &= \int_X \int_Y \int_G f(\gamma)h(x)\varphi(y)\big(g_i(\gamma^{-1}\cdot y) - g_i(y)\big)\, d\lambda^{r(x)}(\gamma)\, d\alpha^x(y)\, d\mu(x)\end{aligned} \tag{10.6}$$

But if $m = \lim_i m_{g_i}$, then

$$\begin{aligned}\big\langle f \star m_{g_i}(\varphi)\,,\,\psi\big\rangle &= \big\langle {}^{tt}M(f) m_{g_i}(\varphi)\,,\,\psi\big\rangle \\ &= \big\langle m_{g_i}(\varphi)\,,\,{}^{t}M(f)\psi\big\rangle \\ &\to \big\langle m(\varphi)\,,\,{}^{t}M(f)\psi\big\rangle \\ &= \big\langle f\star m(\varphi)\,,\,\psi\big\rangle.\end{aligned}$$

On the other hand,

$$\big\langle(\varphi(f)\circ r)\cdot m_{g_i}(\varphi)\,,\,h\big\rangle \to \big\langle(\varphi(f)\circ r)\cdot m(\varphi)\,,\,h\big\rangle.$$

Thus part (a) follows from part (b) and (10.6). □

Using a convexity argument attributed to Day & Namioka (see [**Gre69**, §2.4]), we get a slightly stronger convergence property for approximate weakly invariant means. Here, and later, it will be convenient to swap variables and replace $G \rtimes X$ with $X * G$ and still equip it with the measure $\mu\circ\alpha$ as in Lemma 10.1 on page 243.

PROPOSITION 10.14 ([**ADR00**, Proposition 3.1.8]). *In our standard set-up, the following are equivalent.*

(a) *There is an invariant mean* $m : L^\infty(Y) \to L^\infty(X)$.
(b) *There is an approximate weakly invariant mean for* $\pi : Y \to X$.
(c) *There is a net* $\{g_i\}$ *in* $\mathcal{L}^\infty\big(X, L^1(Y,\alpha)\big)_1^+$ *such that*
   (i) $\{\alpha(g_i)\}$ *converges to* 1 *in the weak-$*$ topology of* $L^\infty(X)$, *and*
   (ii) *for all* $f \in C_c(G)$ *and* $h \in C_c(X)$, *we have*

$$\lim_i \int_X h(x)\int_Y \Big| f * g_i(y) - \lambda(f)(r(y))g_i(y)\Big|\,d\alpha^x(y)\,d\mu(x) = 0.$$

If we wish in part (c), *we can take* $\{g_i\}$ *to be a* sequence *in* $C_c^+(Y)_{\alpha,1}$.

PROOF. We've just observed that (a) and (b) are equivalent and it is straightforward to check that (c) implies (b). We just need to see that (b) implies the seemingly stronger property (c).

Assume that $\{g_i\} \subset C_c^+(Y)_{\alpha,1}$ is a approximate weakly invariant mean. For each $d \in L^1(X * G)$ and $g \in C_c^+(Y)_{\alpha,1}$ let

$$b(d,g)(y) = \int_G d(\pi(y),\gamma)\big(g(\gamma^{-1}\cdot y) - g(\gamma)\big)\,d\lambda^{r(y)}(\lambda).$$

Note that with respect to $\mu\circ\alpha$,

$$\begin{aligned}\|b(d,g)\|_1 &= \int_X\int_Y \big|b(d,g)(y)\big|\,d\alpha^x(y)\,d\mu(x) \\ &\le \int_X\int_Y\int_G \big|d(x,\gamma)\big(g(\gamma^{-1}\cdot y) - g(y)\big)\big|\,d\alpha^x(y)\,d\lambda^{r(x)}(\gamma)\,d\mu(x) \\ &\le 2\|d\|_1.\end{aligned} \tag{10.7}$$

Furthermore if $\varphi \in L^\infty(Y)$,

$$\begin{aligned}&\lim_i \int_X\int_Y b(d,g_i)\varphi(y)\,d\alpha^x(y)\,d\mu(x) = \\ &\qquad \lim_i \int_X\int_Y\int_G \varphi(y)d(x,\gamma)\big(g(\gamma^{-1}\cdot y) - g_i(y)\big)\,d\lambda^{r(x)}(\gamma)\,d\alpha^x(y)\,d\mu(x)\end{aligned} \tag{10.8}$$

which tends to zero by Lemma 10.10(c) on page 247.

Fix $d_1, \dots d_k$ in $L^1(X * G)$. Let

$$L = L^\infty(X) \times \prod_{j=1}^{k} L^1(Y)$$

where we give $L^\infty(X)$ the weak-$*$ topology and $L^1(Y)$ the norm topology. Let L_w be the same set where we replace the norm topology on $L^1(Y)$ with the weak topology. Since the weak-$*$ continuous linear functionals on $L^\infty(X)$ are exactly those coming from $L^1(X)$, Corollary A.6 on page 317 implies that L and L_w have the same closed convex sets.

Let C be the image in L_w of $C_c^+(Y)_{\alpha,1}$ under the map

$$g \mapsto \big(\alpha(g), b(d_1, g), \dots, b(d_k, g)\big).$$

Then C is convex and has $(1, 0, \dots, 0)$ in its closure in L_w in view of (10.8). Hence $(1, 0, \dots, 0)$ is in the closure in L. Hence there is a net $\{g_i'\} \subset C_c^+(Y)_{\alpha,1}$ such that $\alpha(g_i')$ tends to 1 in the weak-$*$ topology, and for each j

$$\lim_i \|b(d_j, g_i')\|_1 = \lim_i \int_X \int_Y \Big| \int_G d_j(x,\gamma)\big(g_i'(\gamma^{-1}\cdot y) - g_i'(y)\big)\, d\lambda^{r(x)}(\gamma) \Big| \, d\alpha^x(y)\, d\mu(x) = 0.$$

Let $D = \{\, d_j \,\}$ be a countable dense subset of $L^1(X * G)$ and $D' = \{\, h_i \,\}$ be a countable dense subset of $L^1(X)$. For each $n \in \mathbf{N}$, let $g_n \in C_c^+(Y)_{\alpha,1}$ be such that

$$\|b(d_j, g_n)\|_1 < \frac{1}{n} \quad \text{for all } 1 \le j \le n, \text{ and}$$

$$\Big| \int_X \alpha(g_n)(x) h_j(x)\, d\mu(x) - \int_X h_j(x)\, d\mu(x) \Big| < \frac{1}{n} \quad \text{for all } 1 \le j \le n.$$

Thus

$$\lim_n \|b(d, g_n)\|_1 = \lim_n \int_X \int_Y \Big| \int_G d(x,\gamma)\big(g_n(\gamma^{-1}\cdot y) - g_n(y)\big)\, d\lambda^{r(x)}(\gamma) \Big| \, d\alpha^x(y)\, d\mu(x) = 0$$

for all $d \in D$ and hence all $d \in L^1(X * G)$ in view of (10.7). Similarly,

$$\lim_n \int_X \alpha(g_n)(x) h(x)\, d\mu(x) = \int_X h(x)\, d\mu(x)$$

for all $h \in L^1(X)$.

If $d(x,\gamma) = h(x) f(\gamma)$ with $h \in C_x(X)$ and $f \in C_c(G)$, then we get

$$\lim_n \int_X |h(x)| \int_Y \big| f * g_n(y) - \alpha(f)(r(x)) g_n(y) \big| \, d\alpha^x(y)\, d\mu(x).$$

The desired result follows. □

In order to get tidy descriptions of measurewise amenability down the road, it will be convenient to observe that we can normalize our approximate weakly invariant mean as described in our next result.

LEMMA 10.15. *Let $m : L^\infty(Y) \to L^\infty(X)$ be a mean.*

(a) *Suppose that $\{g_i\}$ is a net in $\mathcal{L}^\infty\big(X, L^1(Y,\alpha)\big)_1^+$ is such that $\lim m_{g_i} = m$ in the weak-$*$ topology. Then there is a net $\{h_j\} \subset \mathcal{L}^\infty\big(X, L^1(Y,\alpha)\big)_1^+$ such that $\{m_{h_j}\}$ converges to m and such that such that $\alpha(h_j) = 1$ for all j. If $\{g_i\} \subset C^+_{\pi,c}(Y) \cap \mathcal{L}^\infty\big(X, L^1(Y,\alpha)\big)_1$, then we can take $\{\, h_j \,\} \subset C^+_{\pi,c}(Y) \cap \mathcal{L}^\infty\big(X, L^1(Y,\alpha)\big)_1$.*

(b) *Suppose that $\{g_i\}$ in $C^+_c(Y)_{\alpha,1}$ is such that $\lim m_{g_i} = m$ in the weak-$*$ topology. Then there is a net $\{h_j\} \subset C^+_c(Y)_{\alpha,1}$ such that $\{m_{h_j}\}$ converges to m and such that $\alpha(g_i)$ converges to 1 uniformly on compact subsets of X.*

In both cases, if $\{g_i\}$ is a sequence, we can take $\{h_j\}$ to be a sequence.

PROOF. Let $m = \lim m_{g_i}$ with $g_i \in \mathcal{L}^\infty\big(X, L^1(Y,\alpha)\big)_1^+$. Let $g \in C^+_{\pi,c}(Y)$ be such that $\alpha(g) = 1$ (see Proposition 3.18 on page 77). Define

$$h_{i,n}(y) := \frac{g_i(y) + \frac{1}{n} g(y)}{\alpha(g)(\pi(y)) + \frac{1}{n}}.$$

Then clearly $h_{i,n} \in \mathcal{L}^\infty\big(X, L^1(Y,\alpha)\big)_1$ and $\alpha(h_{i,n}) = 1$. To prove (a), it will suffice to see that $m = \lim_{i,n} m_{h_{i,n}}$. Thus for fixed $\varphi \in L^\infty(Y)$ and $h \in L^1(X)$ we need to show that

$$\lim_{i,n} \langle m_{h_{i,n}}(\varphi)\, ,\, h\rangle = \langle m(\varphi)\, ,\, h\rangle.$$

There is no harm in assuming $\|\varphi\|_\infty \le 1$.

Observe, that by assumption, $0 \le \alpha(g_i) \le 1$ and that

$$\lim_i \int_X (1 - \alpha(g_i)(x)) |h(x)| \, d\mu(x) = 0.$$

It follows that $\alpha(g_i) \cdot h \to h$ in norm in $L^1(X)$. Hence it will suffice to see that

$$\langle m_{h_{i,n}}(\varphi)\, ,\, \alpha(g_i) h\rangle \to \langle m(\varphi)\, ,\, h\rangle.$$

Since

$$m_{h_{i,n}}(\varphi)(x) = \frac{1}{\alpha(g_i)(x) + \frac{1}{n}} \big(m_{g_i}(\varphi)(x) + \frac{1}{n} m_g(\varphi)(x)\big)$$

and

$$\begin{aligned} \Big|\Big\langle \frac{1}{\alpha(g_i) + \frac{1}{n}} \cdot (\frac{1}{n} m_g(\varphi))\, ,\, \alpha(g_i) \cdot h\Big\rangle\Big| &= \Big|\frac{1}{n}\Big\langle m_g(\varphi)\, ,\, \frac{\alpha(g_i)}{\alpha(g_i) + \frac{1}{n}} \cdot h\Big\rangle\Big| \\ &\le \frac{1}{n}\|h\|_1, \end{aligned}$$

it will suffice to check that

$$\Big\langle \frac{1}{\alpha(g_i) + \frac{1}{n}} \cdot m_{g_i}(\varphi)\, ,\, \alpha(g_i) \cdot h\Big\rangle \to \langle m(\varphi)\, ,\, h\rangle.$$

But $\frac{1}{n} h \to 0$ in $L^1(X)$, so we may as well check that

$$\Big\langle \frac{1}{\alpha(g_i) + \frac{1}{n}} \cdot m_{g_i}(\varphi)\, ,\, (\alpha(g_i) + \frac{1}{n}) \cdot h\Big\rangle \to \langle m(\varphi)\, ,\, h\rangle.$$

But that simplifies to $\langle m_{g_i}(\varphi)\, ,\, h\rangle \to \langle m(\varphi)\, ,\, h\rangle$ which holds by assumption.

Note that if $\{g_i\} \subset C^+_{\pi,c}(Y) \cap \mathcal{L}^\infty\big(X, L^1(Y,\alpha)\big)_1$, then so is $\{h_j\}$. This proves (a).

For part (b), repeat the above construction to produce $\{h_j\} \subset C^+_{\pi,c}(Y) \cap \mathcal{L}^\infty\big(X, L^1(Y,\alpha)\big)_1$ converging to m. Let $X = \bigcup V_n$ with each V_n compact and with V_n contained in the interior of V_{n+1}. Let $\varphi_n \in C^+_c(X)$ be such that $0 \le \varphi_n \le 1$, $\varphi_n = 1$ on V_n and vanishes off V_{n+1}. Define $g_{j,n} = \varphi_n \cdot h_j$. Then it suffices to check that $\langle m_{g_{j,n}}(\varphi), h\rangle \to \langle m(\varphi), h\rangle$. But this follows from the similar property for the $\{h_j\}$ and the fact that $\varphi_n \cdot h \to h$ in $L^1(X)$.

The statement about sequences follows from a straightforward diagonalization. □

DEFINITION 10.16. An approximate weakly invariant mean $\{g_i\}$ in $C^+_{\pi,c}(Y) \cap \mathcal{L}^\infty\big(X, L^1(Y,\alpha)\big)_1$ is called *normalized* if $\alpha(g_i) = 1$ for all i. An approximate weakly invariant mean $\{g_i\}$ in $C^+_c(Y)_{\alpha,1}$ is called *normalized* if $\alpha(g_i) \to 1$ uniformly on compact subsets of X.

COROLLARY 10.17. *Suppose there is an invariant mean $m : L^\infty(Y) \to L^\infty(X)$ for (π,α,μ). Then there is a normalized approximate weakly invariant mean $\{h_j\}$ for π in either $C^+_{\pi,c}(Y) \cap \mathcal{L}^\infty\big(X, L^1(Y,\alpha)\big)_1$ or $C^+_c(Y)_{\alpha,1}$. Moreover, we can take $\{h_j\}$ to be a sequence.*

PROOF. By Proposition 10.14 on page 249 there is an approximate weakly invariant mean in $C^+_c(Y)_{\alpha,1}$ which is a sequence. Now we can apply Lemma 10.15 on the preceding page. □

EXERCISES

Ex 10.1.1. In our standard set-up, suppose that $\psi(x) = 0$ for μ-almost all x, and that $\varphi(y) = 0$ for $\mu \circ \alpha$ almost all y. Show that for any $f \in C_c(G)$, we have $f * \psi(x) = 0$ for μ-almost all x, and that $f * \varphi(y) = 0$ for $\mu \circ \alpha$-almost all y. (The functions $f * \psi$ and $f * \varphi$ are defined in (10.2).)

—Answer on page 381

10.2. Measurewise Invariant Means

In this section we work in the special case where $\pi : Y \to X$ in our usual set-up is simply $r : G \to G^{(0)}$ with the usual G-actions on G and $G^{(0)}$, and our equivariant r-system is the given Haar system on G. As in Section 9.6, we let $\mathcal{E}$ is the space $C_0(G^{(0)}, L^1(G,\lambda))$. Let $\mathcal{E}^*_G$ be the subset of the dual $\mathcal{E}^*$ consisting of functionals

$$\nu \in M(G, [\mu] \circ \lambda) = \{\, \nu \in M(G,\lambda) : |\nu| \le \mu_1 \circ \lambda \text{ where } \mu_1 \ll \mu \,\}$$

for some *quasi-invariant* measure μ. Since the sum of quasi-invariant measures is quasi-invariant, $\mathcal{E}^*_G$ is a subspace of $\mathcal{E}^*$.

DEFINITION 10.18. Let G be a locally compact groupoid with a Haar system λ. A *measurewise invariant mean* on G with respect to λ is an element $m \in \mathcal{E}^{**}$ such that

(a) $m(\nu) \ge 0$ if $\nu \ge 0$,

(b) for any probability measure μ on $G^{(0)}$, we have $m(\mu \circ \lambda) = 1$ and

(c) for any $f \in C_c(G)$, we have $f \diamond m = \big(\lambda(f) \circ r\big) \cdot m$ on $\mathcal{E}^*_G$

REMARK 10.19. This definition should be compared to that of a topological invariant mean as given in Definition 9.52 on page 219: the only difference is that (c) only need hold for $\nu \in \mathcal{E}_G^*$. In general, $\mathcal{E}_G^*$ can be very small compared to $\mathcal{E}^*$. For example if G is the transformation groupoid for $\mathcal{G}$ acting on a quotient $\mathcal{G}/\mathcal{H}$, then there is a unique measure class of quasi-invariant measures on $\mathcal{G}/\mathcal{H}$. Consequently, the property of having a measurewise invariant mean is *a priori* much weaker than the corresponding notion of a topological invariant mean.

PROPOSITION 10.20 ([**ADR00**, Proposition 3.3.5]). *Let G be a locally compact groupoid endowed with a Haar system λ. Then G is measurewise amenable if and only if G admits a measurewise amenable mean.*

PROOF. Let $m \in \mathcal{E}^{**}$ be a measurewise amenable mean. If μ is quasi-invariant, then m restricts to a linear functional on $X(\mu) = \overline{M(G, \mu \circ \lambda)}$. Since $X(\mu)$ is isomorphic to $L^\infty(G, \mu \circ \lambda) \widehat{\otimes}_{L^\infty(G^{(0)}, \mu)} L^1(G^{(0)}, \mu)$, this restriction gives us a mean

$$m' : L^\infty(G, \mu \circ \lambda) \to L^\infty(G^{(0)}, \mu)$$

in $B_{L^\infty(X)}(L^\infty(G), L^\infty(G^{(0)}))_1^+$. By Lemma 10.9 on page 246, $f \star m'$ is the restriction of $f \diamond m = (\lambda(f) \circ r) \cdot m$. Thus m' is invariant by part (a) of Proposition 10.12 on page 248. It follows that G is measurewise amenable as in Definition 9.7 on page 192.

Now suppose that G is measurewise amenable. Suppose $\epsilon > 0$, $F_1 = \{ f_1, \dots, f_n \} \subset C_c(G)$, and $F_2 = \{ \nu_1, \dots, \nu_m \} \subset \mathcal{E}_G$. Then there are quasi-invariant measures μ_i such that $\nu_k \in M(G, [\mu_k] \circ \lambda)$. Since $\mu_k \ll \mu := \mu_1 + \cdots + \mu_m$, we have $\nu_k \in M(G, [\mu] \circ \lambda)$ for all k. By Proposition 10.12 on page 248 together with Corollary 10.17 (Ex 10.2.1), there is a $g \in C_{r,c}^+(G)$ such that $\lambda(g)(u) = 1$ for all u and such that

$$\big|(f_j \star m_g)(\nu_k) - \big(\lambda(f_j) \circ r\big) m_g(\nu_k)\big| < \epsilon \quad \text{for } 1 \le j \le n \text{ and } 1 \le k \le m.$$

In this way, we obtain a normalized net $\{g_i\} \subset C_{r,c}^+(G)$, indexed by increasing finite sets F_j and decreasing $\epsilon > 0$, such that

$$\lim_i (f \star m_{g_i})(\nu) - \big(\lambda(f) \circ r\big) m_{g_i}(\nu) = 0 \quad \text{for all } \nu \in \mathcal{E}_G \text{ and } f \in C_c(G).$$

But $\{m_{g_i}\}$ must have an accumulation point m in the unit ball of $\mathcal{E}^{**}$. Using Lemma 10.9 on page 246, it is not hard to see that m is a measurewise invariant mean. This completes the proof. □

As a corollary, we obtain the following.

THEOREM 10.21. *Suppose that G is a second countable locally compact groupoid with a Haar system λ. If G is amenable, then G is measurewise amenable.*

PROOF. If G is amenable, then it has a topological invariant mean by Theorem 9.58 on page 222. But a topological invariant mean on G is also a measurewise invariant mean. Hence G is measurewise amenable by Proposition 10.20. □

As observed in [**Ren15**, Corollary 2.16], we can also give an "improved" proof of one of the main results in [**ADR00**]: namely that measurewise amenability coincides with topological amenability in the case of étale groupoids.

THEOREM 10.22 ([**ADR00**, Theorem 3.3.7]). *Let G be a second countable étale groupoid. Then G is amenable if and only if G is measurewise amenable.*

We need the following lemma.

LEMMA 10.23. *Let G be an étale groupoid equipped with counting measures for a Haar system. Then every measure μ on $G^{(0)}$ is absolutely continuous with respect to a quasi-invariant measure ν on $G^{(0)}$.*

PROOF. Let ν be a saturation of μ as in Lemma 7.3 on page 158. Then

$$\nu(\varphi) = \int_{G^{(0)}} \rho^u(\varphi)\, d\mu(u),$$

where

$$\rho^u(\varphi) = \sum_{r(\gamma)=u} \varphi(s(\gamma))\delta(\gamma),$$

and δ is an appropriate everywhere positive function on G.

It follows that if $E \subset G^{(0)}$ is Borel, then $\rho^u(E) \neq 0$ if and only if $\rho^v(E) \neq 0$ for all v in the orbit of u. Hence if $\nu(E) = 0$, then $\mu([E]) = 0$ where $[E]$ is the saturation of E. Since $E \subset [E]$, $\mu(E) = 0$ and $\mu \ll \nu$, and ν is quasi-invariant by Lemma 7.3. □

PROOF OF THEOREM 10.22. We know that amenability always implies measurewise amenability by Theorem 10.21 on the preceding page.

For the other direction, since G is étale, Lemma 10.23 implies every bounded measure on $G^{(0)}$ is absolutely continuous with respect to a quasi-invariant measure. Hence $\mathcal{E}^*_G = \mathcal{E}^*$. But then a measurewise invariant mean is also a topological invariant mean. The result now follows from Theorem 9.58 on page 222. □

REMARK 10.24. We can replace étale in the statement of Theorem 10.22 on the previous page with the assumption that $G^{(0)}$ has countable orbits: it is not hard to modify the proof of Lemma 10.23 to hold in this case.[2] Then since both forms of amenability are invariant under equivalence (see Theorem 9.48 on page 217 and Theorem 10.41 on page 266), the conclusion of the theorem holds for any groupoid equivalent to one with countable orbits.

EXERCISES

Ex 10.2.1. Suppose that m is an invariant mean in $B_{L^\infty(X)}\big(L^\infty(Y), L^\infty(X)\big)_1^+$. Let $\{ f_1, \dots, f_n \} \subset C_c(G)$ and $\nu_1, \dots, \nu_m \in M(G, [\mu] \circ \lambda)$. Show that for any $\epsilon > 0$ there is a $g \in C_{r,c}(G)^+$ such that $\lambda(g)(u) = 1$ for all $u \in G^{(0)}$ and

$$|f_j \star m_g(\nu_k) - (\lambda(f_j) \circ r) \cdot m_g(\nu_k)| < \epsilon$$

for $1 \leq j \leq n$ and $1 \leq k \leq m$.

—Answer on page 381

[2]Once just needs to check that if $G^{(0)}$ has countable orbits and λ is a Haar system, then $\lambda^u(\{\gamma\}) > 0$ for $\gamma \in G^u$ as in Ex 1.3.4 on page 12.

10.3. Fun with Means

Recall our standard set-up: (G, λ) is a locally compact Hausdorff groupoid with a Haar system λ, X and Y are locally compact G-spaces with a continuous equivariant map $\pi : Y \to X$, and an equivariant π-system α. We also fix a quasi-invariant measure μ on X. We say that $\pi : Y \to X$ is *amenable with respect to* (λ, μ, α) if there is an invariant mean $m : L^\infty(Y) \to L^\infty(X)$.[3]

We want to experiment with other properties of means and a critical concept is the notion of an *auxiliary G-space* as in [**ADR00**]. Hence we assume that T is a locally compact G-space equipped with a continuous *equivariant* $r : T \to G^{(0)}$ system β. Then we equip $X * T = \{ (x, t) : r(x) = r(t) \}$ with the measure $\mu \circ \beta$ given by

$$\mu \circ \beta(f) = \int_X \int_T f(x, t) \, d\beta^{r(x)}(t) \, d\mu(x).$$

Similarly, we equip $Y * T$ with the measure $(\mu \circ \alpha) \circ \beta$.

Note that $Y * T$ and $X * T$ are themselves G-spaces with respect to the diagonal G-action. If we let $\pi \otimes 1 : Y * T \to X * T$ be the obvious map, we end up with another standard set-up: for our $\pi \otimes 1$-system, we take $\underline{\alpha} = \{\underline{\alpha}^{(x,t)}\}$ with $\underline{\alpha}^{(x,t)} = \alpha^x \times \delta_{\{t\}}$. It is easy to see that it is equivariant:

$$\begin{aligned}\int_{Y*T} f(\gamma^{-1} \cdot y, \gamma^{-1} \cdot s) \, d\underline{\alpha}^{(x,t)}(y, s) &= \int_Y f(\gamma^{-1} \cdot y, \gamma^{-1} \cdot t) \, \alpha^x(y) \\ &= \int_Y f(y, \gamma^{-1} \cdot t) \, d\alpha^{\gamma^{-1} \cdot x}(y) \\ &= \int_{Y*T} f(y, s) \, d\underline{\alpha}^{(\gamma^{-1} \cdot x, \gamma^{-1} \cdot t)}(y, s).\end{aligned}$$

Moreover, $\mu \circ \beta$ is quasi-invariant with Radon-Nikodym derivative $\underline{\delta}(\gamma, (x, t)) = \delta(\gamma, x)$: suppose $F \in C_c(G \ltimes (X * T))$. Then

$$\begin{aligned}&\int_{X*T} \int_G F(\gamma^{-1}, (\gamma^{-1} \cdot x, \gamma^{-1} \cdot t)) \delta(\gamma^{-1}, \gamma^{-1} \cdot x) \, d\lambda^{r(x)}(\gamma) \, d(\mu \circ \beta)(x, t) \\ &\quad = \int_X \int_T \int_G F(\gamma^{-1}, (\gamma^{-1} \cdot x, \gamma^{-1} \cdot t)) \delta(\gamma^{-1}, \gamma^{-1} \cdot x) \, d\lambda^{r(x)}(\gamma) \, d\beta^{r(x)}(t) \, d\mu(x)\end{aligned}$$

which, by Fubini and equivariance of β, is

$$= \int_X \int_G \int_T F(\gamma^{-1}, (\gamma^{-1} \cdot x, t)) \, d\beta^{\gamma^{-1} \cdot x}(t) \delta(\gamma^{-1}, \gamma^{-1} \cdot x) \, d\lambda^{r(x)}(\gamma) \, d\mu(x)$$

which, by definition of δ, is

$$\begin{aligned}&= \int_X \int_G \int_G F(\gamma, (x, t)) \, d\beta^x(t) \, d\lambda^{r(x)}(\gamma) \, \mu(x) \\ &= \int_{X*T} \int_G F(\gamma, (x, t)) \, d\lambda^{r(x)}(\gamma) \, d(\mu \circ \beta)(x, t).\end{aligned}$$

Now we want to see how to upgrade a mean m from $L^\infty(Y)$ to $L^\infty(X)$ to a mean $m \otimes 1_T$ from $L^\infty(Y * X)$ to $L^\infty(X * T)$. Suppose that $\varphi \in \mathcal{B}^b(Y * T)$ and

[3] The only role the "λ" plays is that invariance depends on the choice of the Haar system λ as it is used to define convolution!

that h is a $\mu\circ\alpha$-integrable Borel function on $X*T$. Then using Ex 10.3.1, we can show that

$$\Theta(\varphi,h)(y)=\int_T \varphi(y,t)h(\pi(y),t)\,d\beta^{r(y)}(t) \tag{10.9}$$

is well-defined $(\mu\circ\alpha)\circ\beta$-almost everywhere. Furthermore, since Θ is bilinear, the $(\mu\circ\alpha)\circ\beta$-almost everywhere equivalence class of $\Theta(\varphi,h)$ depends only on the class of $\varphi\otimes h$ in $L^\infty(Y*T)\odot_{L^\infty(X*T)} L^1(X*T)$. Let

$$\beta(|h|)(x)=\begin{cases}\displaystyle\int_T |h(x,t)|\,d\beta^{r(x)}(t) & \text{if } \int_T |h(x,t)|\,d\beta^{r(x)}(t)<\infty \text{, and}\\ 0 & \text{otherwise.}\end{cases}$$

and

$$\zeta(y,t)=\begin{cases}\dfrac{h(\pi(y),t)}{\beta(|h|)(\pi(y))} & \text{if } \beta(|h|)(\pi(y))\neq 0\\ 0 & \text{otherwise.}\end{cases}$$

Then if

$$\psi(y)=\int_T \varphi(y,t)\zeta(y,t)\,d\beta^{r(y)}(t),$$

we have $\|\beta(|h|)\|_1=\|h\|_1$ and $\|\psi\|_\infty\le\|\varphi\|_\infty$. The point is that

$$\Theta(\varphi,h)(y)=\beta(|h|)(\pi(y))\psi(y)=\big(\psi\otimes\beta(|h|)\big)(y)\quad\text{for } \mu\circ\alpha\text{-almost all } y,$$

and with respect to the balanced projective tensor product norms $\|\Theta(\varphi,h)\|\le\|\varphi\otimes h\|$. Since the map sending $\psi\otimes k\in L^\infty(Y)\odot_{L^\infty(X)} L^1(X)$ to $y\mapsto\psi(y)k(\pi(y))$ is an injection when we indentify functions $\mu\circ\alpha$-almost everywhere, Θ extends to a contraction

$$c_T: L^\infty(Y*T)\mathbin{\widehat{\otimes}}_{L^\infty(X*T)} L^1(X*T)\to L^\infty(Y)\mathbin{\widehat{\otimes}}_{L^\infty(X)} L^1(X).$$

Note that if we view an elements of $L^\infty(Y*T)\odot_{L^\infty(X*T)}L^1(X*T)$ and $L^\infty(Y)\odot_{L^\infty(X)} L^1(X)$ as a functions on $Y*T$ and Y, respectively, then

$$c_T(\psi)(y)=\int_T \psi(y,t)\,d\beta^{r(y)}(t).$$

By transposition, we get a map $m\mapsto m\otimes 1_T$ from

$$B_{L^\infty(X)}\big(L^\infty(Y),L^\infty(X)\big)\quad\text{to}\quad B_{L^\infty(X*T)}\big(L^\infty(Y*T),L^\infty(X*T)\big)$$

such that $\langle m\otimes 1_T\,,\,\nu\rangle=\big\langle m\,,\,c_T(\nu)\big\rangle$ for $\nu\in L^\infty(Y*T)\mathbin{\widehat{\otimes}}_{L^\infty(X*T)} L^1(X*T)\equiv X*T([\mu\circ\alpha])$.

If $\varphi\in L^\infty(Y*T)$ and $g\in C_c(T)$, then define

$$[\![\varphi,g]\!](y)=\int_T \varphi(y,t)g(t)\,d\beta^{r(y)}(t), \tag{10.10}$$

with a similar formula if $\varphi\in L^\infty(X*T)$. Then we get the following characterization of $m\otimes 1_T$ from [**ADR00**, Lemma 1.3.3].

LEMMA 10.25. *Let $m\in B_{L^\infty(X)}\big(L^\infty(Y),L^\infty(X)\big)$ and $n\in B_{L^\infty(X*T)}\big(L^\infty(Y*T),L^\infty(X*T)\big)$. Then $n=m\otimes 1_T$ if and only if*

$$\big[\![n(\varphi),g]\!\big]=m\big([\![\varphi,g]\!]\big)\quad\textit{for all } \varphi\in L^\infty(X*T) \textit{ and } g\in C_c(T).$$

PROOF. Let $h \in L^1(X)$ and $g \in C_c(T)$. Let

$$\nu = \varphi \otimes (h \otimes g) \in L^\infty(Y * T) \odot_{L^\infty(X*T)} L^1(X * T).$$

Then check that $c_T(\nu) = [\![\varphi, g]\!] \otimes h \in L^\infty(Y) \odot_{L^\infty(X)} L^1(X)$. Now compute that

$$\begin{aligned}\big\langle [\![(m \otimes 1_T)(\varphi), g]\!] \,,\, h\big\rangle &= \big\langle m \otimes 1_T(\varphi) \,,\, h \otimes g\big\rangle \\ &= \big\langle m \,,\, c_T(\nu)\big\rangle \\ &= \big\langle m([\![\varphi, g]\!]) \,,\, h\big\rangle.\end{aligned}$$

Since $L^1(X) \odot C_c(T)$ is dense in $L^1(X * T)$, this suffices. □

Let $g \in \mathcal{L}^\infty\big(X, L^1(Y, \alpha)\big)$. Then $g \otimes 1 \in \mathcal{L}^\infty(X * T, L^1(Y * T, \underline{\alpha})$. As usual, let $m_g \in B_{L^\infty(X)}\big(L^\infty(Y), L^\infty(X)\big)$ so that $m_q \otimes 1_T \in B_{L^\infty(X*T)}\big(L^\infty(Y * T), L^\infty(X * T)\big)$. Then, if ψ is a function on $Y * T$ associated to an element of $L^\infty(Y * T) \odot_{L^\infty(X*T)} L^1(X * T)$, we have

$$\begin{aligned}\langle m_g \otimes 1_T \,,\, \psi\rangle &= \langle m_g \,,\, c_T(\psi)\rangle \\ &= \int_X \int_Y g(y) c_T(\psi)(y)\, d\alpha^x(y)\, d\mu(x) \\ &= \int_X \int_Y \int_T g(y)\psi(y, t)\, d\beta^{r(x)}(t)\, d\alpha^x(y)\, d\mu(x) \\ &= \int_{X*T} \int_Y g(y)\psi(y, s)\, d\underline{\alpha}^{(x,t)}(y, s)\, d(\mu \circ \beta)(x, t) \\ &= \langle m_{g\otimes 1} \,,\, \psi\rangle. \end{aligned} \tag{10.11}$$

LEMMA 10.26. *Suppose that* $m \in B_{L^\infty(X)}\big(L^\infty(Y), L^\infty(X)\big)$ *is an invariant mean. Then* $m \otimes 1_T$ *is invariant for any auxiliary space* T.

PROOF. By Proposition 10.12 on page 248, we need to see that $f \star (m \otimes 1_T) = (\lambda(f) \circ r) \cdot (m \otimes 1_T)$. For this, it suffices to see that $f \star (m \otimes 1_T) = (f \star m) \otimes 1_T$. Let $\{g_i\} \subset C_c^+(Y)_{\alpha,1}$ be a approximate weakly invariant mean such that $\lim_i m_{g_i} = m$ as in Proposition 10.12. Since transposition is certainly weak-$*$ continuous, $\{m_{g_i} \otimes 1_T\}$ converges to $m \otimes 1_T$. Similarly, $f \star (m_{g_i} \otimes 1_T) \to f \star (m \otimes 1_T)$. Using (10.11) we have, $f \star (m_{g_i} \otimes 1_T) = f \star m_{g_i \otimes 1} = m_{f*(g_i \otimes 1)} = m_{(f*g_i)\otimes 1} = m_{f*g_i} \otimes 1_T = (f \star m_{g_i}) \otimes 1_T$. But $(f \star m_{g_i}) \otimes 1_T \to (f \star m) \otimes 1_T$. □

EXAMPLE 10.27. Of course *the* basic example of an auxiliary space is G itself together with its Haar system λ. It is important to notice that in this event, $X * G$ is modulo the flip, simply $G \rtimes X$.

A function in $\varphi \in L^\infty(X)$ is called G-*invariant* if for all $f \in C_c(G)$,

$$f * \varphi = (\lambda(f) \circ r) \cdot \varphi \quad \text{in } L^\infty(X). \tag{10.12}$$

We'll write $L^\infty(X)^G$ for the G-invariant elements of $L^\infty(X)$.

COROLLARY 10.28. *Let* $m : L^\infty(Y) \to L^\infty(X)$ *be an invariant mean. Then for any auxilary space* T,

$$m \otimes 1_T\big(L^\infty(Y * T)^G\big) \subset L^\infty(X * T)^G.$$

PROOF. If $\varphi \in L^\infty(Y * T)^G$, then $f * \varphi = (\lambda(f) \circ r) \cdot \varphi$ by (10.12). But $m \otimes 1_T$ is invariant by Lemma 10.26 on the preceding page. Hence

$$\begin{aligned} f * \big(m \otimes 1_T(\varphi)\big) &= (m \otimes 1_T)(f * \varphi) \\ &= (m \otimes 1_T)\big((\lambda(f) \circ r) \cdot \varphi\big) \\ &= (\lambda(f) \circ r) \cdot (m \otimes 1_T)(\varphi). \end{aligned}$$

Thus, by definition, $m \otimes 1_T(\varphi) \in L^\infty(X * T)^G$. □

One advantage of defining G-invariance in terms of (10.12) is that it is clearly well-defined on L^∞. However, in the sequel, we'll need to examine spaces such as $L^\infty(X * G)^G$ a bit more carefully. An important tool is the following from [**ADR00**, Lemma 3.1.12]:

PROPOSITION 10.29. *Suppose that X is a G-space equipped with a quasi-invariant measure μ. Let $\varphi \in \mathcal{B}^b(X)$. Then the following are equivalent.*

(a) *$\varphi \in L^\infty(X)^G$.*
(b) *$\varphi(\lambda^{-1} \cdot x) = \varphi(x)$ for $\mu \circ \lambda$-almost all (x, γ).*
(c) *There is a Borel function ψ on $X * G$ such that $\psi = \varphi$ almost everywhere and $\psi(\gamma^{-1} \cdot x) = \psi(x)$ for all $(x, \gamma) \in X * G$.*

Although seemingly straightforward, the proof requires a technical result from [**ADR00**, Lemma 3.1.11] based on [**Zim84**, Lemma 2.2.16].

LEMMA 10.30. *Suppose that G is a locally compact Hausdorff groupoid with a Haar system λ and let μ be a quasi-invariant measure on $G^{(0)}$. Let $\nu = \mu \circ \lambda$. Suppose $\varphi : G^{(0)} \to \mathbf{C}$ is a Borel function. Then $\varphi \circ r = \varphi \circ s$ for ν-almost all γ if and only if there is a Borel function $\psi : G^{(0)} \to \mathbf{C}$ such that $\psi \circ r = \psi \circ s$ for all γ and such that $\varphi = \psi$ for μ-almost all $u \in G^{(0)}$.*

PROOF. If ψ is as in the statement, let N be a μ-null set such that $\varphi = \psi$ off N. Then $r^{-1}(N)$ is clearly $\mu \circ \lambda$ null. Since μ is quasi-invariant, so is $s^{-1}(N)$. But then off the null set $r^{-1}(N) \cup s^{-1}(N)$, we have $\varphi \circ r = \psi \circ r = \psi \circ s = \varphi \circ s$.

So suppose that φ is as in the statement. Since $\mathbf{C}$ is Borel isomorphic to $[0, 1]$, we can assume φ takes values in $[0, 1]$. We can let $g \in C^+(G)$ be a nonvanishing function such that $\lambda(g)(u) = 1$ for all $u \in G^{(0)}$ (see Lemma 3.19 on page 77).

Let $G_0^{(0)}$ be the set of $u \in G^{(0)}$ such that $\gamma \mapsto \varphi(s(\gamma))$ is essentially constant on G^u. If $\gamma \mapsto \varphi(s(\gamma))$ is constant off a $\lambda^{s(\eta)}$-null set N, then $\eta\gamma \mapsto \varphi(s(\eta\gamma))$ is constant off ηN. Since $\lambda^{r(\eta)}(\eta N) = \lambda^{s(\eta)}(N)$, we see that $s(\eta) \in G_0^{(0)}$ implies $r(\eta) \in G_0^{(0)}$. Thus, $G_0^{(0)}$ is saturated. Define

$$\varphi_1(u) = \int_G \varphi(s(\gamma)) g(\gamma) \, d\lambda^u(\gamma).$$

Then φ_1 is everywhere finite and Borel on $G^{(0)}$. Also let

$$\rho(u) = \int_G \big|\varphi(s(\gamma)) - \varphi_1(u)\big| g(\gamma) \, d\lambda^u(\gamma).$$

Then ρ is also finite-valued and Borel. Moreover, if $\rho(u) = 0$, then $\gamma \mapsto \varphi(s(\gamma))$ is equal to $\varphi_1(u)$ for almost all $\gamma \in G^u$ and $u \in G_0^{(0)}$. On the other hand, if $u \in G_0^{(0)}$, then $\gamma \mapsto \varphi(s(\gamma))$ is essentially constant and that constant must be $\varphi_1(u)$. Hence $\rho^{-1}(\{0\}) = G_0^{(0)}$. In particular, $G_0^{(0)}$ is Borel.

But

$$\int_{G^{(0)}} |\varphi_1(u) - \varphi(u)|\, d\mu(u)$$
$$\leq \int_{G^{(0)}} \Big(\int_G \big|\varphi(s(\gamma)) - \varphi(r(\gamma))\big| g(\gamma)\, d\lambda^u(\gamma) \Big)\, d\mu(u) = 0.$$

Hence $\varphi = \varphi_1$ for μ-almost all u. Furthermore,

$$\rho_1(u) = \int_G \big|\varphi(s(\gamma)) - \varphi(u)\big|\, d\lambda^u(\gamma)$$

is zero almost everywhere and almost everywhere equal to ρ. This implies that $G_0^{(0)}$ is a *co*null set. Define

$$\psi(u) = \begin{cases} \varphi_1(u) & \text{if } u \in G_0^{(0)} \text{ and} \\ 0 & \text{otherwise.} \end{cases}$$

Clearly ψ is Borel and $\psi = \varphi_1$ almost everywhere. Hence $\psi = \varphi$ almost everywhere. Let $\gamma \in G$. Since $G_0^{(0)}$ is saturated, we either have $s(\gamma)$ and $r(\gamma)$ both in $G_0^{(0)}$ or both not in $G_0^{(0)}$. In the later case, $\psi(r(\gamma)) = 0 = \psi(s(\gamma))$. Otherwise, $\eta \mapsto \varphi(s(\eta))$ is essentially constant on both $G^{s(\gamma)}$ and $G^{r(\gamma)}$. We have

$$\begin{aligned} \psi(r(\gamma)) &= \varphi_1(r(\gamma)) \\ &= \int_G \varphi(s(\eta)) g(\eta)\, d\lambda^{r(\gamma)}(\eta) \end{aligned}$$

which, by left invariance, is

$$= \int_G \varphi(s(\eta)) g(\gamma\eta)\, d\lambda^{s(\gamma)}(\eta)$$

which, since $\eta \mapsto \varphi(s(\eta))$ is essentially equal to $\varphi_1(s(\gamma))$ on $G^{s(\gamma)}$, is

$$\begin{aligned} &= \varphi_1(s(\gamma)) \int_G g(\gamma\eta)\, d\lambda^{s(\gamma)}(\eta) \\ &= \varphi_1(s(\gamma)) = \psi(s(\gamma)). \end{aligned}$$

Thus ψ is the desired function. □

PROOF OF PROPOSITION 10.29 ON THE FACING PAGE. The equivalence of (a) and (b) follows almost immediately from

$$f * \varphi(x) - (\varphi(f) \circ r)\varphi(x) = \int_G f(\gamma)\big(\varphi(\gamma^{-1} \cdot x) - \varphi(x)\big)\, d\lambda^{r(x)}(\gamma).$$

The equivalence of (b) and (c) follows by applying Lemma 10.30 to the groupoid $X * G$ with Haar system

$$\lambda^x(f) = \int_G f(x, \gamma)\, d\lambda^{r(x)}(\gamma)$$

(after identifying $(X * G)^{(0)}$ with X). □

REMARK 10.31. In general, it is not so easy to realize $L^\infty(X)^G$ with a space of functions on the orbit space $G\backslash X$.[4] But if the action of G on X is proper and $G\backslash X$ is Hausdorff, then [**Arv76**, Theorem 3.4.1] implies that the orbit map has a

[4] See the discussion prior to [**ADR00**, Definition 3.1.13].

Borel cross section and then, using Proposition 10.29 on page 258, we can easily identify $L^\infty(X)^G$ with $L^\infty(G\backslash X)$ (with respect to a σ-fintie measure equivalent to the push-forward of μ).[5]

Recall that if X is a G-space, then

$$X * G = \{ (x, \gamma) : r(x) = r(\gamma) \}$$

can, modulo the flip, be identified with the transformation groupoid $G \rtimes X$. In particular, we will want to consider the inversion map ι_X from $X * G$ to itself given by

$$\iota_X(x, \gamma) = (\gamma^{-1} \cdot x, \gamma^{-1}).$$

We will be even more interested in the induced map $\mathfrak{I}_X$ on $L^\infty(X * G)$:

$$\mathfrak{I}_X(\varphi) = \varphi \circ \iota_X.$$

DEFINITION 10.32. A mean $m : L^\infty(Y) \to L^\infty(X)$ is called *completely invariant* if

$$\mathfrak{I}_X \circ (m \otimes 1_G) = (m \otimes 1_G) \circ \mathfrak{I}_Y.$$

This terminology is partially justified by the following observation.

PROPOSITION 10.33. *A completely invariant mean is invariant.*

PROOF. Let $m : L^\infty(Y) \to L^\infty(X)$ be completely invariant. Recall from Lemma 10.25 on page 256 that $m \otimes 1_G$ is characterized by

$$\big[\!\big[(m \otimes 1_G)(\varphi), f\big]\!\big] = m\big([\![\varphi, f]\!]\big) \quad \varphi \in L^\infty(Y * G) \text{ and } f \in C_c(G),$$

and $[\![\cdot, \cdot]\!]$ is defined in (10.10). In particular,

$$\big[\!\big[\varphi(\iota_Y(\cdot)), f\big]\!\big](y) = \int_G f(\gamma)\varphi(\gamma^{-1} \cdot y, \gamma^{-1})\, d\lambda^{r(y)}(\gamma).$$

Thus if $\varphi = \psi \otimes 1_G$,

$$\big[\!\big[\varphi(\iota_Y(\cdot)), f\big]\!\big](y) = f * \psi(y).$$

Furthermore, for $\varphi = \psi \otimes 1_G$,

$$(m \otimes 1_G)(\varphi) = m(\psi) \otimes 1_G.$$

However, since m is completely invariant, we have

$$(m \otimes 1_G)\big(\varphi(\iota_Y(\cdot))\big) = (m \otimes 1_G)\big(\varphi\big)(\iota_X(\cdot))$$

Thus, with φ as above,

$$\big[\!\big[(m \otimes 1_G)\big(\varphi(\iota_Y(\cdot))\big), f\big]\!\big] = m\big([\![\varphi(\iota_Y(\cdot)), f]\!]\big) = m(f * \psi). \tag{10.13}$$

But the left-hand side of (10.13) is equal to

$$\big[\!\big[(m \otimes 1_G)\big(\varphi\big)(\iota_X(\cdot)), f\big]\!\big] = \big[\!\big[(m(\psi) \otimes 1_G)(\iota_X(\cdot)), f\big]\!\big] = f * m(\psi).$$

Hence m is invariant. □

[5] In fact, it is enough for $G\backslash X$ to be T_0. This follows from part of the Mackey-Glimm-Ramsay Dichotomy that I did not include in Section 2.2 on page 38. Namely, when the orbit space is nice, the quotient topology generates the quotient Borel structure. Hence Borel functions on the orbit space lift to Borel functions which are constant on orbits and conversely. The details are in [**Ram90**, Theorem 2.1(6)].

Of course, the properties of a completely invariant mean equate to properties of its approximations. As in [**ADR00**, Proposition 3.1.21], this leads to the following definition.

DEFINITION 10.34. We say that a net $\{g_i\}$ in $\mathcal{L}^\infty\big(X, L^1(Y,\alpha)\big)^+_1$ is an *approximately completely invariant mean* if

(a) $\alpha(g_i) \to 1$ in the weak-$*$ topology on $L^\infty(X)$ and
(b) for all $f \in L^1(X * G)$ and $\varphi \in L^\infty(Y * G)$ we have

$$\lim_i \int_X \int_Y \int_G f(x,\gamma)\varphi(y,\gamma)\big(g_i(\gamma^{-1}\cdot y) - g_i(y)\big)\, d\lambda^{r(x)}(y)\, d\alpha^x(y)\, d\mu(x) = 0.$$

This definition should be compared with Lemma 10.10(c) on page 247 where the $\varphi \in L^\infty(Y)$ has been replaced with $\varphi \in L^\infty(Y * T)$.

PROPOSITION 10.35. *Let $m : L^\infty(Y) \to L^\infty(X)$ be a mean. If $\{\, g_i \,\}$ is a net in $\mathcal{L}^\infty\big(X, L^1(Y,\alpha)\big)^+_1$ such that $m = \lim_i m_{g_i}$ in the weak-$*$ topology, then m is completely invariant if and only if $\{g_i\}$ is an approximately completely invariant mean.*

PROOF. Recall that $m_g(\psi) = \alpha(g\psi)$ and that by (10.11) we have $m_g \otimes 1_G = m_{g\otimes 1}$. Thus if $\varphi \in L^\infty(Y * G)$ and $f \in L^1(X * G)$, we have

$$\begin{aligned}\big\langle \mathfrak{I}_X \circ (m_{g_i} \otimes 1_G)(\varphi)\,,\, f\big\rangle &= \int_X \int_G (m_{g_i\otimes 1}(\varphi)(\gamma^{-1}\cdot x, \gamma^{-1}) f(x,\gamma)\, d\lambda^{r(x)}(\gamma)\, d\mu(x) \\ &= \int_X \int_G \int_Y g_i(y)\varphi(y,\gamma^{-1}) f(x,\gamma)\, d\alpha^{\gamma^{-1}\cdot x}(y)\, d\lambda^{r(x)}(\gamma)\, d\mu(x)\end{aligned}$$

which, since α is equivariant, is

$$= \int_X \int_G \int_Y g_i(\gamma^{-1}\cdot y)\varphi(\gamma^{-1}\cdot y, \gamma^{-1}) f(x,\gamma) \\ d\alpha^x(y)\, d\lambda^{r(x)}(\gamma)\, d\mu(x)$$

Similarly,

$$\big\langle (m_{g_i} \otimes 1_G)\circ \mathfrak{I}_Y(\varphi)\,,\, f\big\rangle \\ = \int_X \int_G \int_Y g_i(y)\varphi(\gamma^{-1}\cdot y, \gamma^{-1}) f(x,\gamma)\, d\alpha^x(y)\, d\lambda^{r(x)}(\gamma)\, d\mu(x).$$

Hence, applying Fubini, we have

$$\big\langle \mathfrak{I}_X \circ (m_{g_i} \otimes 1_G)(\varphi)\,,\, f\big\rangle - \big\langle (m_{g_i} \otimes 1_G)\circ \mathfrak{I}_Y(\varphi)\,,\, f\big\rangle \\ = \int_X \int_Y \int_G f(x,\gamma)\varphi(\gamma^{-1}\cdot y, \gamma^{-1})\big(g_i(\gamma^{-1}\cdot y) - g_i(y)\big)\, d\lambda^{r(x)}(\gamma)\, d\alpha^x(y)\, d\mu(x).$$

The result follows as $\varphi \in L^\infty$ if and only if $\varphi \circ \mathfrak{I}_Y$ is, and since

$$\lim_i\big(\big\langle \mathfrak{I}_X \circ (m_{g_i} \otimes 1_G)(\varphi)\,,\, f\big\rangle - \big\langle (m_{g_i} \otimes 1_G)\circ \mathfrak{I}_Y(\varphi)\,,\, f\big\rangle\big) \\ = \big\langle \mathfrak{I}_X \circ (m \otimes 1_G)(\varphi)\,,\, f\big\rangle - \big\langle (m \otimes 1_G)\circ \mathfrak{I}_Y(\varphi)\,,\, f\big\rangle$$

as in Ex 10.3.2. □

REMARK 10.36. Just as in Lemma 10.15 on page 251, we can normalize an approximately completely invariant mean in $\mathcal{L}^\infty\big(X, L^1(Y,\alpha)\big)_1^+$ and assume that $\alpha(g_i) = 1$ for all i. If the approximately completely invariant mean is in $C_c^+(Y)_{\alpha,1}$, then we can insist that $\alpha(g_i) \to 1$ uniformly on compact subsets of X.

PROPOSITION 10.37 ([**ADR00**, Proposition 3.1.22]). *The following are equivalent.*

(a) *There is a completely invariant mean* $m : L^\infty(Y) \to L^\infty(X)$.

(b) *There exists an approximately completely invariant mean.*

(c) *There exists a normalized approximately completely invariant mean.*

(d) *There is a* sequence $\{\, g_n \,\} \subset \mathcal{L}^\infty\big(X, L^1(Y,\alpha)\big)_1^+$ *such that*

  (i) $\alpha(g_n)(x) = 1$ *for all* $x \in X$, *and*

  (ii) *for all* $f \in L^1(X * G)$,

$$\lim_n \int_X \int_Y \int_G f(x,\gamma)\big|g_n(\gamma^{-1}\cdot y) - g_n(y)\big|\, d\lambda^{r(x)}(\gamma)\, d\alpha^x(y)\, d\mu(x) = 0.$$

(e) *There is a* sequence $\{\, g_n \,\} \subset C_c^+(Y)_{\alpha,1}$ *such that*

  (i) $\alpha(g_n) \to 1$ *uniformly on compact subsets of* X, *and*

  (ii) *for all* $f \in L^1(X * G)$,

$$\lim_n \int_X \int_Y \int_G f(x,\gamma)\big|g_n(\gamma^{-1}\cdot y) - g_n(y)\big|\, d\lambda^{r(x)}(\gamma)\, d\alpha^x(y)\, d\mu(x) = 0.$$

If we wish, we can pick $\{g_n\} \subset C_{\pi,c}^+(Y) \cap \mathcal{L}^\infty\big(X, L^1(Y,\alpha)\big)_1$ *in part* (d).

PROOF. Proposition 10.35 implies that (a) and (b) are equivalent. Remark 10.36 implies that (b) and (c) are equivalent. We also easily see (e) implies (b). So we just have to see that (a) implies (d), and that (d) implies (e).

If (a) holds, then by Proposition 10.35 on the preceding page and since $C_c^+(Y)_{\alpha,1}$ is weak-$*$ dense in $B_{L^\infty(X)}\big(L^\infty(Y), L^\infty(X)\big)_1^+$ (by Lemma A.18 on page 326), we can pick a approximately completely invariant mean $\{g_i\} \subset C_{\pi,c}^+(Y) \cap \mathcal{L}^\infty\big(X, L^1(Y,\alpha)\big)_1$. By Remark 10.36, we can assume $\alpha(g_i) = 1$ for all i. For each $g \in C_{\pi,c}(Y)$, let $b_g(y,\gamma) = g(\gamma^{-1}\cdot y) - g(y)$. Then let

$$C = \{\, b_g \in L^\infty(X * G, L^1(Y * G, \alpha\otimes 1)) : g \in C_{\pi,c}^+(Y) \text{ and } \alpha(g) = 1 \,\}.$$

Then C is convex.

Equip $\mathcal{L}^\infty(X * G, L^1(Y * G, \alpha\otimes 1))$ with the τ_w-topology as in Section A.6 on page 329. By assumption, 0 is in the τ_w-closure of C. By Corollary A.22 on page 330, there is a net $\{b_{g_i}\} \subset C$ such that

$$\lim_i \int_X \int_Y \int_G f(x,\gamma)|g_i(\gamma\cdot y) - g_i(y)|\, d\lambda^{r(x)}(\gamma)\, d\alpha^x(y)\, d\mu(x) = 0$$

for all $f \in L^1(X * G)$.

To get a sequence takes some work. Let $D \subset L^1(X * G)$ be a countable dense subset. Then the collection $\mathcal{F}$ of finite subsets of D is countable and we can form the countable directed set $\Lambda = \mathcal{F} \times \mathbf{N}$. Abusing the notation in Section A.6, we can write

$$\|f\|_g := \int_X \int_Y \int_G f(x,y)\big|g(\gamma\cdot y) - g(y)\big|\, d\lambda^{r(x)}(\gamma)\, d\alpha^x(y)\, d\mu(x).$$

(Note that $\|\cdot\|_g$ is only a seminorm when f is nonnegative.) For each $(F,n)\in\Lambda$ there is a $g_{F,n}\in\{g_i\}$ such that

$$\bigl|\|d\|_{g_{F,n}}\bigr| < \frac{1}{n} \quad \text{for all } d\in F.$$

Clearly, $\|d\|_{g_\lambda}\to 0$ for all $d\in D$. Since $\alpha(g_\lambda)=1$ for all λ, $\bigl|\|f\|_{g_\lambda}\bigr|\le 2\|f\|_1$ for any $f\in L^1(G*X)$. It follows easily that

$$\lim_\lambda \|f\|_{g_\lambda}=0 \quad \text{for all } f\in L^1(X*G).$$

Since Λ is countable, we can extract the desired sequence as in Ex 9.8.1. Hence (a) implies (d).

To see that (d) implies (e), note that it suffices to consider $f\in C_c(X*G)$. Let $X=\bigcup K_n$ with each K_n compact and contained in the interior of K_{n+1}. Let $\varphi_n\in C_c(X)$ be such that $\varphi_n(x)=1$ for all $x\in K_n$ and $0\le\varphi_n(x)<1$ for all x. Then $g_n'(y)=\varphi_n(\pi(y))g_n(y)$ defines a function in $C_c^+(Y)_{\alpha,1}$. If $\operatorname{supp} f\subset K\times C$ with K compact in X and C compact in G, then there is an n_f such that $n\ge n_f$ implies $C^{-1}\cdot K\cup K\subset K_n$. But then if $n\ge n_f$

$$\int_X\int_Y\int_G f(x,\gamma)\bigl|g_n'(\gamma^{-1}\cdot y)-g_n'(y)\bigr|\,d\lambda^{r(x)}(\gamma)\,d\alpha^x(y)\,d\mu(x)=$$
$$\int_X\int_Y\int_G f(x,\gamma)\bigl|g_n(\gamma^{-1}\cdot y)-g_n(y)\bigr|\,d\lambda^{r(x)}(\gamma)\,d\alpha^x(y)\,d\mu(x)$$

Thus $\{g_n'\}$ is the desired sequence. □

Let $\mathcal{L}^\infty(X,L^2(Y,\alpha))$ be the set of Borel functions f on Y such that $\alpha(|f|^2)$ is bounded. We let $\mathcal{L}^\infty(X,L^2(Y,\alpha))_1$ be those $f\in\mathcal{L}^\infty(X,L^2(Y,\alpha))$ such that $\alpha(|f|^2)(x)\le 1$ for all x. Then we have the following from [**ADR00**, Proposition 3.1.25].

PROPOSITION 10.38. *In our standard set-up, the following are equivalent.*

(a) *There is a completely invariant mean $m: L^\infty(Y)\to L^\infty(X)$.*

(b) *There is a net $\{\xi_i\}\in\mathcal{L}^\infty(X,L^2(Y,\alpha))_1$ such that, if we let*

$$e_i(x,\gamma)=\int_Y\overline{\xi_i(y)}\xi_i(\gamma^{-1}\cdot y)\,d\alpha^x(y),$$

and $e_i^{(0)}(x)=e_i(x,r(x))$, then

(i) *$e_i^{(0)}\to 1$ in the weak-$*$ topology on $L^\infty(X)$ and*

(ii) *$\lim_n e_i=1$ in the weak-$*$ topology on $L^\infty(X*G)$.*

(c) *There is a* sequence *$\{\xi_n\}\subset C_c^+(Y)\cap\mathcal{L}^\infty(X,L^2(Y,\alpha))_1$ such that if we define e_n and $e_n^{(0)}$ as in part* (b), *then*

(i) *$e_n^{(0)}(x)\le 1$ for all x,*

(ii) *$e_n\to 1$ uniformly on compact subsets of X, and*

(iii) *$e_n\to 1$ in the weak-$*$ topology on $L^1(X*G)$.*

Before proceeding with the proof, we need some preliminaries. Recall that $X*G$ carries the measure $\mu\circ\lambda$:

$$\mu\circ\lambda(f)=\int_X\int_G f(x,\gamma)\,d\lambda^{r(x)}(\gamma)\,d\mu(x). \tag{10.14}$$

Since μ is quasi-invariant,

$$\int_X \int_G f(\gamma^{-1}\cdot x, \gamma^{-1})\delta(\gamma^{-1}\cdot x, \gamma^{-1})\, d\lambda^{r(x)}(\gamma)\, d\mu(x) = \int_X \int_G f(x,\gamma)\, d\lambda^{r(x)}(\gamma)\, d\mu(x) \tag{10.15}$$

using the Radon-Nikodym derivative δ as in Definition 10.2 on page 243. Recall that Proposition 3.6 on page 71 implies that (10.14) and (10.15) hold for nonnegative Borel functions and can be suitably interpreted for $f \in L^1(X * G)$. More to the point, if $f \in \mathcal{L}^1(X * G)$, then (10.14) means that

$$\mu \circ \lambda(f) = \int_X \lambda(f)(x)\, d\mu(x),$$

where

$$\lambda(f)(x) := \begin{cases} 0 & \text{if } (\gamma \mapsto f(x,\gamma)) \notin L^1(\lambda^{r(x)}), \text{ and} \\ \displaystyle\int_X f(x,\gamma)\, d\lambda^{r(x)}(\gamma) & \text{otherwise.} \end{cases}$$

In particular, $\lambda(f) \in L^1(X)$ as is

$$\lambda(\tilde{f\delta})(x) := \int_G f(\gamma^{-1}\cdot x, \gamma^{-1})\delta(\gamma^{-1}\cdot x, \gamma^{-1})\, d\lambda^{r(x)}(\gamma).$$

The next two lemmas are left as Ex 10.3.3 and Ex 10.3.4.

Lemma 10.39. *Suppose that $\{h_i\} \subset L^\infty(X)$ is a net converging to 1 in the weak-$*$ topology on $L^\infty(X)$. Then for all $f \in L^1(X * G)$ we have*

$$\lim_i \int_X \int_G h_i(x) f(x,\gamma)\, d\lambda^x(\gamma)\, d\mu(x) = \int_X \int_G f(x,\gamma)\, d\lambda^x(\gamma)\, d\mu(x), \quad \textit{and}$$

$$\lim_i \int_X \int_G h_i(\gamma^{-1}\cdot x) f(x,\gamma)\, d\lambda^x(\gamma)\, d\mu(x) = \int_X \int_G f(x,\gamma)\, d\lambda^x(\gamma)\, d\mu(x).$$

Lemma 10.40. *Suppose that $\{e_i\} \subset L^\infty(Z)$ is a net converging to 1 in the weak-$*$ topology. Then $\{\operatorname{Re}(e_i)\}$ does too.*

Proof of Proposition 10.38. Assume e_i and ξ_i are as in part (b) of the proposition. Let $g_i(y) := |\xi_i(y)|^2$. Note that $g_i \in \mathcal{L}^\infty\big(X, L^1(Y,\alpha)\big)^+_1$. Hence to see that (b)$\Longrightarrow$(a), it will suffice to show that the g_i form a approximately completely invariant mean as in Proposition 10.37(b) on page 262. I claim that for all $f \in L^1(X * G)$ we have

$$\lim_i \int_X \int_Y \int_G |\xi_i(\gamma^{-1}\cdot y) - \xi_i(y)|^2 f(x,\gamma)\, d\lambda^{r(x)}(\gamma)\, d\alpha^x(y)\, d\mu(x) = 0. \tag{10.16}$$

But the left-hand side of (10.16) simplifies to

$$\int_X \int_G \big(e_i^{(0)}(\gamma^{-1}\cdot x) - 2\operatorname{Re} e_i(x,\gamma) + e_i^{(0)}(x)\big) f(x,\gamma)\, d\lambda^{r(x)}(\gamma)\, d\mu(x).$$

Thus the claim follows from our assumptions on the e_i together with Lemmas 10.39 and 10.40.

But if $f \in L^1(X * G)$, then the inequality $\big||a|^2 - |b|^2\big| \le (|a| + |b|)|a - b|$ implies that

$$\begin{aligned}\Big|\int_X \int_Y \int_G f(x,\gamma)|g_i(\gamma^{-1}\cdot y) - g_i(y)|\,d\lambda^{r(x)}(\gamma)\,d\alpha^x(y)\,d\mu(x)\Big| \\ \le \int_X \int_Y \int_G |f(x,\gamma)|\Big(|\xi_i(\gamma^{-1}\cdot y)| + |\xi_i(y)|\Big)\Big|\xi_i(\gamma^{-1}\cdot y) - \xi_i(y)\Big| \\ d\lambda^{r(x)}(\gamma)\,d\alpha^x(y)\,d\mu(x).\end{aligned}$$

which by Cauchy-Schwarz, is

$$\le A\cdot B,$$

where

$$A = \Big(\int_X \int_Y \int_G |f(x,\gamma)|\Big(|\xi_i(\gamma^{-1}\cdot y)| + |\xi_i(y)|\Big)^2\,d\lambda^{r(x)}(\gamma)\,d\alpha^x(y)\,d\mu(x)\Big)^{\frac12}$$

and

$$B = \Big(\int_X \int_Y \int_G |f(x,\gamma)|\big|\xi_i(\gamma^{-1}\cdot y) - \xi_i(y)\big|^2\,d\lambda^{r(x)}(\gamma)\,d\alpha^x(y)\,d\mu(x)\Big)^{\frac12}.$$

But B tends to zero by (10.16). So it only remains to see that A is eventually bounded. But $(|a| + |b|)^2 \le 2(|a|^2 + |b|^2)$ implies that

$$\begin{aligned}A^2 &\le 2\int_X \int_Y \int_G |f(x,\gamma)|\big(|\xi_i(\gamma^{-1}\cdot y)|^2 + |\xi_i(y)|^2\big)\,d\lambda^{r(x)}(\gamma)\,d\alpha^x(y)\,d\mu(x) \\ &= 2\int_X \int_G |f(x,\gamma)|\big(e_i^{(0)}(\gamma^{-1}\cdot x) + e_i^{(0)}(x)\big)\,d\lambda^x(\gamma)\,d\mu(x)\end{aligned}$$

which converges to $4\|f\|_1$. Thus (b)$\Longrightarrow$(a).

Since (c)$\Longrightarrow$(b), it suffices to see that (a)$\Longrightarrow$(c). Assuming (a), we can find a *sequence* $\{g_i\} \subset C_c^+(Y)_{\alpha,1}$ as in Proposition 10.37(d) on page 262. Let $\xi_i := \sqrt{g_i}$ and

$$e_i(x,\gamma) = \int_Y \xi_i(y)\xi_i(\gamma^{-1}\cdot y)\,d\alpha^x(y).$$

Since $e_i^{(0)}(x) = \alpha(g_i)(x)$ and since $\alpha(g_i) \to 1$ uniformly on compact subsets of X, we just have to show that for all $f \in L^1(X * G)^+$,

$$\int_X \int_G e_i(x,\gamma)f(x,\gamma)\,d\lambda^x(\gamma)\,d\mu(x) \to \int_X \int_G f(x,\gamma)\,d\lambda^x(\gamma)\,d\mu(x).$$

But $\big||a| - |b|\big|^2 \le \big||a|^2 - |b|^2\big|$ allows us to write

$$\begin{aligned}&\int |g_i(\gamma^{-1}\cdot y) - g_i(y)|f(x,\gamma)\,d\lambda^{r(x)}(\gamma)\,d\alpha^x(y)\,d\mu(x) \\ &\qquad \ge \int |\xi_i(\gamma^{-1}\cdot y) - \xi_i(y)|^2 f(x,y)\,d\lambda^{r(x)}(\gamma)\,d\alpha^x(y)\,d\mu(x) \\ &\qquad = \int \big(\alpha(g_i)(\gamma^{-1}\cdot x) - 2e_i(x,\gamma) + \alpha(g_i)(x)\big)f(x,\gamma)\,d\lambda^x(\gamma),\,d\mu(x). \qquad (10.17)\end{aligned}$$

Thus our assumptions on the g_i force (10.17) to tend to zero. But since Lemma 10.39 on page 264 (and our assumptions on the g_i) imply that

$$\int_X \int_G \big(\alpha(g_i)(x)f(x,\gamma) - f(x,\gamma)\big)\,d\lambda^x(\gamma)\,d\mu(x) = 0 \quad \text{and}$$
$$\int_X \int_G \big(\alpha(g_i)(\gamma^{-1}\cdot x)f(x,\gamma) - f(x,\gamma)\big)\,d\lambda^x(\gamma)\,d\mu(x) = 0,$$

the result follows. □

Exercises

Ex 10.3.1. Show the function $\Theta(\varphi, h)$ given by (10.9) is well-defined $(\mu \circ \alpha) \circ \beta$-almost everywhere.

—Answer on page 381

Ex 10.3.2. Let m and g_i be as in Proposition 10.35. Show that we also have

(a) $\lim_i (m_{g_i} \otimes 1_G) = m \otimes 1_G$,
(b) $\lim_i \mathfrak{I}_X \circ (m_{g_i} \otimes 1_G) = I_X \circ (m \otimes 1_G)$, and
(c) $(m_{g_i} \otimes 1_G) \circ \mathfrak{I}_Y = (m \otimes 1_G) \circ \mathfrak{I}_Y$

in the weak-$*$ topology on $B_{L^\infty(X)}\big(L^\infty(Y), L^\infty(X)\big)$.

—Answer on page 381

Ex 10.3.3. Prove Lemma 10.39.

—Answer on page 382

Ex 10.4. Prove Lemma 10.40.

—Answer on page 382

10.4. Measurewise Amenability and Equivalence

Recall that (G, λ) is called measurewise amenable if every quasi-invariant measure μ on $G^{(0)}$ is amenable in that there is an invariant mean $m : L^\infty(G, \mu \circ \lambda) \to L^\infty(G^{(0)}, \mu)$. Clearly, the amenability of μ depends only on its measure class. Our goal in this section is to see that measurewise amenability is preserved by groupoid equivalence. Since every groupoid is equivalent to itself, this implies that measurewise amenability does not depend on the choice of Haar system. Most of this material is taken from [**ADR00**, §3.2(b)].

Theorem 10.41. *Suppose that G and H are equivalent locally compact Hausdorff groupoids with Haar systems λ_G and λ_H, respectively. Then (G, λ_G) is measurewise amenable if and only if (H, λ_H) is.*

Throughout this section, Z will be a (G, H)-equivalence, and μ a finite quasi-invariant measure on $G^{(0)}$. We can use the Haar system on H to equip Z with an full equivariant r-system $\beta = \{\beta^u\}_{u \in G^{(0)}}$ where

$$\beta^u(g) = \int_Z g(z)\,d\beta^u(z) = \int_H g(z_u \cdot \eta)\,d\lambda_H^{s(z_u)}(\eta), \tag{10.18}$$

where we are free to choose $z_u \in r_Z^{-1}(u)$ as we please. Of course, $\operatorname{supp} \beta^u = z_u \cdot H = r_Z^{-1}(u)$ as required.

REMARK 10.42. Note that β is the r-system which allows us to equip the imprimitivity groupoid $\underline{G} := G\backslash(Z * Z)$ with a Haar system $\underline{\lambda}$ such that the natural identification $[z, z\cdot\eta] \mapsto \eta$ of $\underline{G}$ with H is measure preserving (see Lemma 6.21 on page 151). Recall that

$$\underline{\lambda}(f)(G\cdot z) = \int_H f([z, z\cdot\eta])\,\lambda_H^{s(z)}(\eta) = \int_Z f([z,w])\,d\beta^{r(z)}(w).$$

We view (Z, β) as an auxiliary G-space so that $G^{(0)} *_r Z$ acquires the measure $\mu\circ\beta$ given by

$$\mu\circ\beta(g) = \int_{G^{(0)}}\int_Z g(u,z)\,d\beta^u(z)\,d\mu(u) = \int_{G^{(0)}}\int_H g(u, z_u\cdot\eta)\,d\lambda_H^{s(z_u)}(\eta)\,d\mu(u).$$

Of course, $G^{(0)} *_r Z$ can be identified with Z and $G\backslash(G^{(0)} *_r Z)$ with $H^{(0)}$. We'll write $\underline{\mu}$ for a pseudo-image[6] of $\mu\circ\beta$ on $H^{(0)}$. Thus there is a $\mu\circ\beta$-integrable Borel function κ taking values in $(0,\infty)$ such that

$$\begin{aligned}\underline{\mu}(\varphi) &= \int_{G^{(0)}}\int_Z \varphi(s(z))\kappa(u,z)\,d\beta^u(z)\,d\mu(u)\\ &= \int_{G^{(0)}}\int_H \varphi(s(\eta))\kappa(u, z_u\cdot\eta)\,d\lambda_H^{s(z_u)}(\eta)\,d\mu(u).\end{aligned}\tag{10.19}$$

As always with pseudo-images, $\underline{\mu}$ is only defined up to equivalence.

LEMMA 10.43. *The measure $\underline{\mu}$ on $H^{(0)}$ constructed above is quasi-invariant for the H-action on $H^{(0)}$.*

PROOF. Let $\underline{\nu} := \underline{\mu}\circ\lambda_H$ and suppose that $f\in\mathcal{B}^+(H)$ is such that $\underline{\nu}(f) = 0$. Then we need to see that $\underline{\nu}(\tilde f) = 0$ where $\tilde f(\eta) = f(\eta^{-1})$. But

$$\begin{aligned}0 &= \underline{\nu}(f)\\ &= \int_{H^{(0)}}\int_H f(\eta)\,d\lambda_H^v(\eta)\,d\underline{\mu}(v)\\ &= \int_{G^{(0)}}\int_H\int_H f(\eta)\kappa(u, z_u\cdot\zeta)\,d\lambda_H^{s(\zeta)}(\eta)\,d\lambda_H^{s(z_u)}(\zeta)\,d\mu(u)\end{aligned}$$

which, after sending $\eta\mapsto\zeta^{-1}\eta$, is

$$= \int_{G^{(0)}}\int_H\int_H f(\zeta^{-1}\eta)\kappa(u, z_u\cdot\zeta)\,d\lambda_H^{s(z_u)}(\eta)\,d\lambda_H^{s(z_u)}(\zeta)\,d\mu(u)$$

which, after Fubini and $\zeta\mapsto\eta\zeta$, is

$$\begin{aligned}&= \int_{G^{(0)}}\int_H\int_H f(\zeta^{-1})\kappa(u, z_u\cdot\eta\zeta)\,d\lambda_H^{s(\eta)}(\zeta)\,d\lambda_H^{s(z_u)}(\eta)\,d\mu(u)\\ &= \int_{G^{(0)}}\int_Z\int_H f(\zeta^{-1})\kappa(u, z\cdot\zeta)\,d\lambda_H^{s(z)}(\zeta)\,d\beta^u(z)\,d\mu(u).\end{aligned}$$

It follows that, for $\mu\circ\beta$-almost all (u,z), we have

$$\int_H f(\zeta^{-1})\kappa(u, z\cdot\zeta)\,d\lambda_H^{s(z)}(\zeta) = 0.$$

[6]If $f: Y\to X$ is a Borel map and μ is a (σ-finite) measure on Y, then by definition of *pseudo-image of* μ is the push forward by f of a probability measure equivalent to μ.

Since κ is nowhere vanishing, it follows that for $\mu \circ \beta$-almost all (u, z), we have

$$\int_H f(\zeta^{-1})\, d\lambda_H^{s(z)}(\zeta) = 0.$$

However, in that case,

$$\begin{aligned} 0 &= \int_{G^{(0)}} \int_Z \int_H \tilde{f}(\zeta)\, d\lambda_H^{s(z)}(\zeta)\, d\beta^u(z)\, d\mu(u) \\ &= \int_{H^{(0)}} \tilde{f}(\eta)\, d\lambda_H^v(\eta)\, d\underline{\mu}(v) = \underline{\nu}(\tilde{f}). \end{aligned}$$

Thus, $\underline{\mu}$ is quasi-invariant as claimed. □

PROPOSITION 10.44. *Every quasi-invariant measure on $H^{(0)}$ is equivalent to one of the form $\underline{\mu}$ constructed as above from a quasi-invariant measure μ on $G^{(0)}$.*

PROOF. Suppose that μ is a quasi-invariant measure on $G^{(0)}$ with $\underline{\mu}$ the corresponding quasi-invariant measure on $H^{(0)}$. By symmetry, we can construct a quasi-invariant measure $\underline{\underline{\mu}}$ on $G^{(0)}$ using $\underline{\mu}$ and the opposite equivalence Z^{op}. Then, by symmetry, it will suffice to see that $\underline{\underline{\mu}}$ is equivalent to μ.

Note that

$$\underline{\underline{\mu}}(\psi) = \int_{H^{(0)}} \int_Z \psi(r(z))\underline{\kappa}(v, z)\, d\underline{\beta}^v(z)\, d\underline{\mu}(v),$$

for a suitable nowhere vanishing function $\underline{\kappa}$ and where $\underline{\beta}^v$ is the measure supported on $G \cdot z_v$ for any z_v such that $s(z_v) = v$ and

$$\underline{\beta}^v(g) = \int_G g(\gamma^{-1} \cdot z_v)\, d\lambda_G^{r(z_v)}(\gamma).$$

If only to preserve sanity, we momentarily let

$$\alpha(v) = \int_Z \psi(r(z))\underline{\kappa}(v, z)\, d\underline{\beta}^v(z),$$

then

$$\begin{aligned} \underline{\underline{\mu}}(\psi) &= \int_{H^{(0)}} \alpha(v)\, d\underline{\mu}(v) \\ &= \int_{G^{(0)}} \int_Z \alpha(s(z))\kappa(u, z)\, d\beta^u(z)\, d\mu(u) \end{aligned}$$

which, for any $z_u \in r_Z^{-1}(u)$, is

$$\begin{aligned} &= \int_{G^{(0)}} \int_H \alpha(s(\eta))\kappa(u, z_u \cdot \eta)\, d\lambda_H^{s(z_u)}(\eta)\, d\mu(u) \\ &= \int_{G^{(0)}} \int_H \int_Z \psi(r(w))\underline{\kappa}(s(\eta), w)\, d\underline{\beta}^{s(\eta)}(w)\kappa(u, z_u \cdot \eta)\, d\lambda_H^{s(z_u)}(\eta)\, d\mu(u) \end{aligned}$$

which, since $s(z_u \cdot \eta) = s(\eta)$, is

$$= \int_{G^{(0)}} \int_H \int_G \psi(s(\gamma))\underline{\kappa}(s(\eta), \gamma^{-1} \cdot z_u \cdot \eta)\kappa(u, z_u \cdot \eta)\, d\lambda_G^u(\gamma)\, d\lambda_H^{s(z_u)}(\eta)\, d\mu(u).$$

It follows that

$$\underline{\underline{\mu}}(\psi) = \int_{G^{(0)}} \int_G \underline{\underline{\kappa}}(\gamma, z)\, d\lambda_G^u(\gamma)\, d\mu(u),$$

where

$$\underline{\underline{\kappa}}(\gamma, z) = \int_H \underline{\kappa}(s(\eta), \gamma^{-1} \cdot z \cdot \eta)\kappa(u, z \cdot \eta)\, d\lambda_H^{s(z)}(\eta).$$

Since $\underline{\underline{\mu}}$ is finite, $\underline{\underline{\kappa}}$ must be finite $\nu = \mu \circ \lambda_G$-almost everywhere. Since κ and $\underline{\kappa}$ are nowhere vanishing and since λ_G^u has full support, $\underline{\underline{\kappa}}$ is also nowhere vanishing. Hence $\underline{\underline{\mu}}$ is a pseudo-image of the push forward of μ via s_G. Since μ was quasi-invariant to begin with, $\underline{\underline{\mu}}$ and μ are equivalent: see Lemma 7.3 on page 158. □

PROPOSITION 10.45 ([**ADR00**, Proposition 3.2.13]). *Suppose that Z is a (G, H)-equivalence and that (G, λ_G, μ) is amenable for a quasi-invariant measure μ on $G^{(0)}$ (see Definition 9.7 on page 192). Then $r : Z \to G^{(0)}$ is amenable with respect to (λ_G, μ, β), where β is the r-system given in* (10.18). *In other words, there is an invariant mean*

$$p : L^\infty(Z, \mu \circ \beta) \to L^\infty(G^{(0)}, \mu).$$

PROOF. By Lemma 3.19 on page 77, there is a $g \in C(Z)$ such that $\beta(g)(u) = 1$ for all $u \in G^{(0)}$ and such that $g(z) > 0$ for all $z \in Z$. Let m^u be the probability measure $g{\cdot}\beta^u$ on Z with support $z_u{\cdot}H$ (for any $z_u \in r_Z^{-1}(u)$). Let $\mathrm{pr}_2 : Z{*}G \to G$ be the projection on the second factor. Then we can define a pr_2-system $n = \{n^\gamma\}_{\gamma\in G}$ by

$$\begin{aligned} n^\gamma(f) &= \int_Z f(\gamma \cdot z, \gamma)\, dm^{s(\gamma)}(z) = \int_Z f(\gamma \cdot z, \gamma) g(z)\, d\beta^{s(z)}(z) \\ &= \int_H f(\gamma \cdot z_{s(\gamma)} \cdot \eta, \gamma) g(z_{s(\gamma)} \cdot \eta)\, d\lambda_H(\eta). \end{aligned}$$

Since n is a Radon family of measures, we can extend our formulas to bounded Borel functions in the obvious way (see Proposition 3.6 on page 71). Moreover, we can equip $Z * G$ with the measure $\nu_G \circ n = (\mu \circ \lambda) \circ n$ and define a mean

$$\underline{n} : L^\infty(Z * G, \nu_G \circ n) \to L^\infty(G, \nu_G)$$

by

$$\underline{n}(\varphi)(\gamma) = n^\gamma(\varphi).$$

Furthermore, n is an equivariant system:

$$\eta \cdot n^\gamma(f) = \int_Z f(\eta\gamma \cdot z, \eta\gamma)\, dm^{s(\gamma)}(z) = n^{\eta\gamma}(f),$$

as the last equality is clear from $s(\eta\gamma) = s(\gamma)$. Then we can compute that

$$\begin{aligned} \underline{n}(f * \varphi)(\gamma) &= \int_Z f * \varphi(\gamma \cdot z, \gamma)\, dm^{s(\gamma)}(z) \\ &= \int_Z \int_G f(\eta)\varphi(\eta^{-1}\gamma \cdot z, \eta^{-1}\gamma)\, d\lambda_G^{r(\gamma)}(\eta)\, dm^{s(\gamma)}(z) \\ &= \int_G f(\eta)\underline{n}(\varphi)(\eta^{-1}\gamma)\, d\lambda_G^{r(\gamma)}(\eta) \\ &= f * \underline{n}(\varphi)(\gamma). \end{aligned}$$

Thus, $\underline{n}$ is an invariant mean. However, the computation

$$\begin{aligned}\nu_G \circ n(f) &= \int_G \int_Z f(\gamma \cdot z, \gamma)\, dm^{s(\gamma)}(z)\, d\nu_G(\gamma) \\ &= \int_G \int_Z f(\gamma \cdot z, \gamma) g(z)\, d\beta^{s(\gamma)}(z)\, d\nu_G(\gamma) \\ &= \int_G \int_H f(\gamma \cdot z_{s(\gamma)} \cdot \eta, \gamma) g(z_{s(\gamma)} \cdot \eta)\, d\lambda_H^{s(z)}(\eta)\, d\nu_G(\gamma)\end{aligned}$$

which, since $z \cdot \eta = \gamma^{-1}\gamma \cdot z \cdot \eta$ and we can let $z_{r(\gamma)} = \gamma \cdot z_{s(\gamma)}$, is

$$= \int_G \int_Z f(z, \gamma) g(\gamma^{-1} \cdot z)\, d\beta^{r(\gamma)}(z)\, d\nu_G(\gamma)$$

shows that $\nu_G \circ n$ is equivalent to $\nu_G \circ \beta$ on $Z * G$ (since g never vanishes). Hence we can view $\underline{n}$ as a mean on $L^\infty(Z * G, \nu_G \circ \beta)$.

If φ is a bounded Borel function on Z, then define $\underline{\mathrm{pr}}_1(\varphi)$ on $Z*G$ by $\underline{\mathrm{pr}}_1(\varphi)(z, \gamma) = \varphi(z)$. Also, if $\varphi \in \mathcal{B}^+(Z)$, then

$$\nu_G \circ \beta(\underline{\mathrm{pr}}_1(\varphi)) = \int_{G^{(0)}} \int_G \int_Z \varphi(z)\, d\beta^u(z)\, d\lambda_G^u(\gamma)\, d\mu(u).$$

Note that

$$\int_G \int_Z \varphi(z)\, d\beta^u(z)\, d\lambda_G^u(\gamma) = 0 \quad \text{whenever} \quad \int_Z \varphi(z)\, d\beta^u(z) = 0.$$

It follows that $\underline{\mathrm{pr}}_1(\varphi) = 0$ for $\nu_G \circ \beta$-almost all (z, γ) whenever $\varphi = 0$ for $\mu \circ \beta$-almost all z. Thus we get a well-defined invariant mean

$$\underline{\mathrm{pr}}_1 : L^\infty(Z, \mu \circ \beta) \to L^\infty(Z * G, \nu_G \circ \beta).$$

However, by hypothesis, we have an invariant mean

$$m : L^\infty(G, \nu_G) \to L^\infty(G^{(0)}, \mu).$$

We can then let $p = m \circ \underline{n} \circ \underline{\mathrm{pr}}_1$. It is not hard to check that this is the invariant mean we are looking for. This completes the proof of Proposition 10.45. □

Proof of Theorem 10.41 on page 266. In view of Proposition 10.44 on page 268, it will suffice to prove that if (G, λ_G, μ) is amenable, then so is $(H, \lambda_H, \underline{\mu})$. For this, we follow the proof of [**ADR00**, Theorem 3.2.16]. It will be convenient to work with $(\underline{G}, \underline{\lambda})$ in place of (H, λ_H) so that $H^{(0)}$ is identified with $G\backslash Z = \underline{G}^{(0)}$. Recall that $\underline{\lambda}$ is given in Remark 10.42 on page 267, and we can view $\underline{\mu}$ as a measure on $G\backslash Z$ via an appropriate modification of (10.19).

We let $p : L^\infty(Z) \to L^\infty(G^{(0)})$ be our invariant mean from Proposition 10.45 on the previous page. We treat Z as an axillary G-space and form the spaces $Z * Z$ and $G^{(0)} * Z$. Then $G^{(0)} * Z$ is equipped with the measure $\mu \circ \beta$ and $Z * Z$ with the measure $\underline{\beta} \circ (\mu \circ \beta)$ where by definition

$$\begin{aligned}\underline{\beta} \circ (\mu \circ \beta)(f) &= \int_Z \int_Z f(z, w)\, d\beta^{r(z)}(w)\, d(\mu \circ \beta)(z) \\ &= \int_{G^{(0)}} \int_Z \int_Z f(z, w)\, d\beta^u(z)\, d\beta^u(w)\, d\mu(u).\end{aligned}$$

Hence in the sequel, we'll write $\mu \circ (\beta \otimes \beta)$ in place of $\underline{\beta} \circ (\mu \circ \beta)$ for clarity. Then our standard construction for an auxiliary space yeilds a mean

$$p \otimes 1_Z : L^\infty(Z * Z, \mu \circ \beta \otimes \beta) \to L^\infty(G^{(0)} * Z, \mu \circ \beta).$$

Let $\underline{G} = G\backslash(Z * Z)$ as before, and identify $\underline{G}^{(0)}$ with $G\backslash(G^{(0)} * Z)$. Since p is invariant and the G-actions on $Z * Z$ and $G^{(0)} * Z$ are proper, Corollary 10.28 on page 257 together with Remark 10.31 on page 259 imply that $p \otimes 1_Z$ factors through a mean

$$\underline{p} : L^\infty(\underline{G}, \mu \circ (\beta \otimes \beta)) \to L^\infty(\underline{G}^{(0)}, \mu \circ \beta) \tag{10.20}$$

by viewing functions on the quotient spaces and invariant functions upstairs. Note that we can replace both measures in (10.20) with pseudo-images via the quotient maps. In particular, we can replace $\mu \circ \beta$ by $\underline{\mu}$ using the appropriate function κ as in (10.19).

We get a measure ρ on $Z * Z$ equivalent to $\mu \circ (\beta \otimes \beta)$ using the same function κ:

$$\begin{aligned}\rho(f) &= \int_Z \int_Z f(z, w)\, d\beta^{r(z)}(w) \kappa(r(z), z)\, d(\mu \circ \beta)(z) \\ &= \int_{G^{(0)}} \int_Z \int_Z f(z, w)\, d\beta^u(w) \kappa(u, z)\, d\beta^u(z)\, d\mu(u).\end{aligned}$$

Since we are free to evaluate the inner integral using any $z \in r_Z^{-1}(u)$,

$$\rho(f) = \int_{G^{(0)}} \int_Z \int_H f(z, z \cdot \eta)\, d\lambda_H^{s(z)}(\eta) \kappa(u, z)\, d\beta^u(z)\, d\mu(u). \tag{10.21}$$

Since Z is an equivalence, it follows that if f is G-invariant—so that $f(\gamma \cdot z, \gamma \cdot w) = f(z, w)$—the inner integral in (10.21) depends only on $s(z)$. Thus for invariant f, (10.21) becomes

$$\int_{H^{(0)}} \int_H f(z_v, z_v \cdot \eta)\, d\underline{\lambda}^v(\eta)\, d\underline{\mu}(v),$$

where z_v is chosen such that $s(z_v) = v$.[7] Thus on the subalgebra of G-invariant functions, ρ is equivalent to $\underline{\mu} \circ \underline{\lambda}$. Thus

$$L^\infty(G\backslash(Z * Z), \mu \circ (\beta \circ \beta)) \cong L^\infty(G\backslash(Z * Z), \underline{\mu} \circ \underline{\lambda}) \cong L^\infty(H, \nu_H).$$

Hence it will suffice to see that $\underline{p}$ is invariant.

For this we view $\underline{G}$ as an auxiliary $\underline{G}$-space with the natural left $\underline{G}$-action. Observe that $\underline{\lambda}$ is an invariant $\underline{r}$-system for the natural left $\underline{G}$-actions on $\underline{G}$ and $\underline{G}^{(0)}$. Furthermore $\underline{\mu}$ is quasi-invariant by Lemma 10.43 on page 267. To see that $\underline{p}$ is invariant, it will suffice to prove something *a priori* stronger: namely that $\underline{p}$ is completely invariant as in Definition 10.32 on page 260. Therefore we need show that the mean

$$\underline{p} \otimes 1_{\underline{G}} : L^\infty(\underline{G} * \underline{G}, \underline{\mu} \circ \underline{\lambda} \circ \underline{\lambda}) \to L^\infty(\underline{G}^{(0)} * \underline{G}, \underline{\mu} \circ \underline{\lambda})$$

satisfies

$$(\underline{p} \otimes 1_{\underline{G}}) \circ \mathfrak{I}_{\underline{G}} = \mathfrak{I}_{\underline{G}^{(0)}} \circ (\underline{p} \otimes 1_{\underline{G}}) \tag{10.22}$$

for the appropriate inversion maps.[8]

[7] See Remark 10.42 on page 267 for the definition of $\underline{\lambda}$.

[8] For $\underline{G}$ as an auxiliary space paired with X, the inversion map $\iota_X : X * \underline{G} \to X * \underline{G}$ is given by $\iota_X(x, [z, w]) = ([w, z] \cdot x, [w, z])$.

To establish (10.22), we view $(Z * Z, \beta \otimes \beta)$ as an auxiliary G-space and first form

$$p \otimes 1_{Z*Z} : L^\infty(Z * Z * Z, \mu \circ \beta \circ (\beta \otimes \beta)) \to L^\infty(G^{(0)} * Z * Z, \mu \circ (\beta \otimes \beta)).$$

Let $I_Z : Z * Z * Z \to Z * Z * Z$ be given by $(z_1, z_2, z_3) \mapsto (z_1, z_3, z_2)$ and $I_{G^{(0)}} : G^{(0)} * Z * Z \to G^{(0)} * Z * Z$ be given by $(u, z, w) \mapsto (u, w, z)$. Using Lemma 10.25 on page 256 (see Ex 10.4.1), it follows that

$$(p \otimes 1_{Z*Z}) \circ I_Z = I_{G^{(0)}} \circ (p \otimes 1_{Z*Z}). \tag{10.23}$$

Since p is invariant, $p \otimes 1_{Z*Z}$ is invariant by Lemma 10.26 on page 257. As with $p \otimes 1_Z$ and $\underline{p}$, we can pass to invariant functions so that $p \otimes 1_{Z*Z}$ defines a mean

$$\underline{p \otimes 1_{Z*Z}} : L^\infty(G\backslash(Z * Z * Z)) \to L^\infty(G\backslash(G^{(0)} * Z * Z)).$$

The idea is to show that $\underline{p \otimes 1_{Z*Z}}$ essentially the same as $\underline{p} \otimes 1_{\underline{G}}$ and so complete the proof. Unfortunately, checking the details is annoying.

Observe that we have a homeomorphism of $\mathfrak{h} : \underline{G} * \underline{G} \to G\backslash(Z * Z * Z)$ given by $\big([z, w], [z, w']\big) \mapsto [z, w, w']$. Similarly, we get a homeomorphism of $\mathfrak{k} : \underline{G}^{(0)} * \underline{G} \to G\backslash(G^{(0)} * Z * Z)$ given by $\big([r(w), w], [w, w']\big) \mapsto [r(w), w, w']$. These maps induce isomorphisms

$$\Phi_{\mathfrak{h}} : L^\infty(G\backslash(Z * Z * Z), \mu \circ \beta \circ (\beta \times \beta)) \to L^\infty(\underline{G} * \underline{G}, \underline{\mu} \circ \underline{\lambda} \circ \underline{\lambda})$$

and

$$\Psi_{\mathfrak{k}} : L^\infty(G\backslash(G^{(0)} * Z * Z), \mu \circ \beta \times \beta) \to L^\infty(\underline{G}^{(0)} * \underline{G}, \underline{\mu} \circ \underline{\lambda}).$$

Moreover, we can choose suitable pseudo images of our measures such that $\mathfrak{k}$ is an isometric isomorphism of $L^1(G\backslash(G^{(0)} * Z * Z)$ onto $L^1(\underline{G}^{(0)} * \underline{G})$.

Let $h \in L^1(G\backslash(G^{(0)} * Z * Z))$ and $\varphi \in L^\infty(G\backslash(Z * Z * Z)$. It is not hard to verify that

$$\Phi_{\mathfrak{h}}(h \cdot \varphi) = \Psi_{\mathfrak{k}}(h) \cdot \Phi_{\mathfrak{h}}(\varphi).$$

Alternatively, viewing elements of $L^\infty(\underline{G} * \underline{G}) \odot_{L^\infty(\underline{G}^{(0)} * \underline{G})} L^1(\underline{G}^{(0)} * \underline{G})$ as functions on $\underline{G} * \underline{G}$,

$$\Phi_{\mathfrak{h}}(\varphi \otimes h) = \Phi_{\mathfrak{h}}(\varphi) \otimes \Psi_{\mathfrak{k}}(h).$$

Now if

$$m \in B_{L^\infty(\underline{G}^{(0)} * \underline{G})}(L^\infty(\underline{G} * \underline{G}), L^\infty(\underline{G}^{(0)} * \underline{G})),$$

then

$$\Psi_{\mathfrak{k}}^{-1} \circ m \circ \Phi_{\mathfrak{h}} \in B_{L^\infty(G\backslash(G^{(0)} * Z * Z)}(L^\infty(G\backslash(Z * Z * Z)), L^\infty(G\backslash(G^{(0)} * Z * Z))).$$

Thus as a linear functional, where we employ Ex 10.4.2, we have

$$\begin{aligned} \big\langle \Psi_{\mathfrak{k}}^{-1} \circ m \circ \Phi_{\mathfrak{h}} \, , \, \varphi \otimes h \big\rangle &= \big\langle m(\Phi_{\mathfrak{h}}(\varphi)) \, , \, \Psi_{\mathfrak{k}}(h) \big\rangle \\ &= \big\langle m \, , \, \Phi_{\mathfrak{h}}(\varphi) \otimes \Psi_{\mathfrak{k}}(h) \big\rangle \\ &= \big\langle m \, , \, \Phi_{\mathfrak{h}}(\varphi \otimes h) \big\rangle. \end{aligned}$$

Now we employ this with $m = \underline{p} \otimes 1_{\underline{G}}$:

$$\big\langle \underline{p} \otimes 1_{\underline{G}} \, , \, \Phi_{\mathfrak{h}}(\varphi) \big\rangle = \big\langle \underline{p} \, , \, c_{\underline{G}}(\Phi_{\mathfrak{h}}(\varphi)) \big\rangle,$$

where

$$\begin{aligned} c_{\underline{G}}(\Phi_{\mathfrak{h}}(\varphi))([z,w]) &= \int_Z \Phi_{\mathfrak{h}}(\varphi)([z,w],[x,w'])\,d\beta^{r(z)}(w') \\ &= \int_Z \varphi([z,w,w'])\,d\beta^{r(z)}(w'). \end{aligned}$$

But then

$$\langle \underline{p}\,,\, c_{\underline{G}}(\Phi_{\mathfrak{h}}(\varphi))\rangle = \langle p\,,\, c_Z(c_{\underline{G}}(\Phi_{\mathfrak{h}}(\varphi)))\rangle$$

where

$$\begin{aligned} c_Z(c_{\underline{G}}(\Phi_{\mathfrak{h}}(\varphi)))(z) &= \int_Z c_{\underline{G}}(\Phi_{\mathfrak{h}}(\varphi))([z,w])\,d\beta^{r(z)}(w) \\ &= \int_Z\int_Z \varphi([z,w,w'])\,d\beta^{r(z)}(w)\,d\beta^{r(z)}(w'). \end{aligned}$$

On the other hand,

$$\langle \underline{p\otimes 1_{Z*Z}}\,,\,\varphi\rangle = \langle p\,,\, c_{Z*Z}(\varphi)\rangle$$

and it is not hard to check that

$$c_{Z*Z}(\varphi) = c_Z\big(c_{\underline{G}}(\Phi_{\mathfrak{h}}(\varphi))\big).$$

It follows that

$$\Psi_{\mathfrak{k}}\circ(\underline{p\otimes 1_{Z*Z}})\circ\Phi_{\mathfrak{h}}^{-1} = \underline{p}\circ 1_{\underline{G}}.$$

Now we can verify (10.22):

$$\begin{aligned} (\underline{p}\otimes 1_{\underline{G}})\circ\mathfrak{I}_{\underline{G}} &= \Psi_{\mathfrak{k}}\circ(\underline{p\otimes 1_{Z*Z}})\circ(\Phi_{\mathfrak{h}}^{-1}\circ\mathfrak{I}_{\underline{G}}) \\ &= \Psi_{\mathfrak{k}}\circ(\underline{p\otimes 1_{Z*Z}})\circ I_Z\circ\Phi_{\mathfrak{h}}^{-1} \\ &= (\Psi_{\mathfrak{k}}\circ I_{G^{(0)}})\circ(\underline{p\otimes 1_{Z*Z}})\circ\Phi_{\mathfrak{h}}^{-1} \\ &= \mathfrak{I}_{\underline{G}^{(0)}}\circ\Psi_{\mathfrak{k}}\circ(\underline{p\otimes 1_{Z*Z}})\circ\Phi_{\mathfrak{h}}^{-1} \\ &= \mathfrak{I}_{\underline{G}^{(0)}}\circ(\underline{p}\otimes 1_{\underline{G}}). \end{aligned}$$

□

Since the above proof actually produces a completely invariant mean on $(\underline{G},\underline{\lambda})\cong(H,\lambda_H)$, we obtain the following as a corollary of the proof: if every quasi-invariant measure on G is amenable (that is, G is measurewise amenable), then every quasi-invariant measure has a completely invariant mean. As a result, we obtain the following.

Corollary 10.46. *Let G be a locally compact groupoid with Haar system λ. Suppose that μ is a quasi-invariant measure on $G^{(0)}$ such that $r : G\to G^{(0)}$ is amenable with respect to (λ,μ,λ) so that there is an invariant mean $n : L^\infty(G)\to L^\infty(G^{(0)})$. Then there is a completely invariant mean $\underline{m} : L^\infty(G)\to L^\infty(G^{(0)})$.*

Proof. We simply repeat the proof of Theorem 10.41 on page 266 with $(Z,\beta) = (G,\lambda)$. After identifying $\underline{G}$ with G and untangling the measures, $\underline{p} : L^\infty(\underline{G})\to L^\infty(\underline{G}^{(0)})$ is what we want. □

Remark 10.47. Carefully examining the proof of Theorem 10.41 on page 266 reveals that we did not use the invariance of m. We only require that $m\otimes 1_T$ satisfies the conclusion of Corollary 10.28 on page 257 for any auxiliary space T: namely that $m\otimes 1_T$ maps G-invariant functions to G-invariant functions. (This what is called (INV1) in [**ADR00**].) This is made good use of in [**ADR00**], but won't be pursued here.

EXERCISES

Ex 10.4.1. Verify (10.23).

—Answer on page 382

Ex 10.4.2. Let $\mathfrak{h} : X \to Y$ be a homeomorphism. If f is a function on Y, let $\Phi_{\mathfrak{h}}(f)(x) = f(\mathfrak{h}(x))$.

(a) If μ is a finite measure on X, show that $\Phi_{\mathfrak{h}}$ defines an isometric isomorphism of $L^1(Y, \mathfrak{h}_*\mu)$ onto $L^1(X, \mu)$.

(b) Show that the transpose ${}^t\Phi_{\mathfrak{h}} : L^\infty(X, \mu) \to L^\infty(Y, \mathfrak{h}_*\mu)$ satisfies

$$\langle {}^t\Phi_{\mathfrak{h}}(\varphi)\,,\, f\rangle = \langle \Phi_{\mathfrak{h}}^{-1}(\varphi)\,,\, f\rangle.$$

—Answer on page 383

10.5. Amenable Measures

In this section, we specialize our standard set up to $r : G \to G^{(0)}$.

DEFINITION 10.48. A quasi-invariant measure μ on $G^{(0)}$ is called *amenable* if there is an invariant mean $m : L^\infty(G) \to L^\infty(G^{(0)})$. Equivalently, μ is amenable if $r : G \to G^{(0)}$ is amenable with respect to (λ, μ, λ).

Notice that (G, λ) is measurewise amenable precisely when every quasi-invariant measure on $G^{(0)}$ is amenable.

Now we can summarize some of the important properties of amenable measures from the proceeding sections. We write $\mathcal{L}^\infty(G^{(0)}, L^1(G, \lambda))$ for the set of Borel functions f on G such that $\lambda(f)$ is bounded, and $\mathcal{L}^\infty(G^{(0)}, L^1(G, \lambda))_1^+$ for those $f \in \mathcal{L}^\infty(G^{(0)}, L^1(G, \lambda))$ that are nonnegative with $\lambda(f) \le 1$.

THEOREM 10.49 ([**ADR00**, Proposition 3.2.14]). *Let μ be a quasi-invariant measure on (G, λ). The following are equivalent.*

(a) *μ is amenable.*

(b) *There is a net $\{g_i\} \subset \mathcal{L}^\infty(G^{(0)}, L^1(G, \lambda))_1^+$ such that*

  (i) *$\lambda(g_i) \to 1$ in the weak-$*$ topology on $L^\infty(G^{(0)})$, and*

  (ii) *for all $f \in C_c(G)$ and $\varphi \in L^\infty(G)$ we have*

$$\lim_i \int_{G^{(0)}} \int_G \int_G f(\gamma)\varphi(\eta)\big(g(\gamma^{-1}\eta) - g_i(\eta)\big)\, d\lambda^u(\gamma)\, d\lambda^u(\eta)\, d\mu(u) = 0.$$

(c) *There is a net $\{g_i\} \subset \mathcal{L}^\infty(G^{(0)}, L^1(G, \lambda))_1^+$ such that*

  (i) *$\lambda(g_i) \to 1$ in the weak-$*$ topology on $L^\infty(G^{(0)})$, and*

  (ii) *for all $h \in C_c(G^{(0)})$ and $f \in C_c(G)$ we have*

$$\lim_i \int_{G^{(0)}} h(u) \int_G |f * g_i(\gamma) - \lambda(f)(u) g_i(\gamma)|\, d\lambda^u(\gamma)\, d\mu(u) = 0.$$

(d) *There is a net $\{g_n\} \subset \mathcal{L}^\infty(G^{(0)}, L^1(G, \lambda))_1^+$ such that*

  (i) *$\lambda(g_n) \to 1$ in the weak-$*$ topology on $L^\infty(G^{(0)})$, and*

(ii) *for all* $f \in C_c(G)$ *we have*

$$\int_{G^{(0)}} \int_G \int_G f(\gamma)|g_n(\gamma^{-1}\eta) - g_n(\eta)|\, d\lambda^u(\gamma)\, d\lambda^u(\eta)\, d\mu(u) = 0.$$

(e) *There is a net* $\{\xi_n\} \subset \mathcal{L}^\infty(G^{(0)}, L^2(G,\lambda))_1$ *such that if we let*

$$e_n(\gamma) = \int_G \overline{\xi_n(\eta)}\xi_n(\gamma^{-1}\eta)\, d\lambda^{r(\gamma)}(\eta),$$

then

(i) $e_n^{(0)} \to 1$ *in the weak-$*$ topology on* $L^\infty(G^{(0)})$ *and*

(ii) $e_n \to 1$ *in the weak-$*$ topology on* $L^\infty(G)$.

Also equivalent are conditions (b)$'$, (c)$'$, (d)$'$, *and* (e)$'$ *where "net" is replaced by "sequence". Furthermore, if we wish, we can replace each* $\{g_i\}$ *with functions in* $C_c^+(G)_{\lambda,1}$ *and* $\{\xi_i\}$ *by functions in* $C_c(G)$. *Then we can normalize so that each condition* (i) *above reads "converges to one uniformly on* $G^{(0)}$*".*

PROOF. By Definition 10.48, (a) is equivalent to the existence of an invariant mean. Hence we have

$$\text{(a)} \iff \text{(b)} \iff \text{(c)}$$

by Proposition 10.12 on page 248 and Proposition 10.14 on page 249.

But (a) is equivalent to the existence of a complete invariant mean by Proposition 10.33 on page 260 and Corollary 10.46 on page 273. Hence

$$\text{(a)} \iff \text{(d)} \iff \text{(d)}' \iff \text{(e)} \iff \text{(e)}'$$

by Proposition 10.37 on page 262 and Proposition 10.38 on page 263. Simple computations show that

$$\text{(d)} \implies \text{(c)} \implies \text{(b)} \quad \text{and} \quad \text{(d)}' \implies \text{(c)}' \implies \text{(b)}'.$$

Since we trivially have (b)$' \implies$ (b), we have the equivalences claimed in the Theorem.

The statements about normalization follow from Lemma 10.15 on page 251 as in Corollary 10.17 on page 252 and Proposition 10.38 on page 263. □

COROLLARY 10.50. *Let* G *be a second countable locally compact groupoid with a Haar system which is measurewise amenable. Then the reduced norm and the universal norm coincide on* $C_c(G)$.

PROOF. Since every quasi-invariant measure on G is amenable by assumption, we can employ part (e) of the theorem and Ex 9.2.5. □

REMARK 10.51 (Legacy). In [**Ren80**, Definition II.3.1], Renault defines a measure μ on $G^{(0)}$ to be *amenable* if there is a net $\{f_i\} \subset C_c(G)$ such that conditions (i) and (ii) of part (e) of Theorem 10.49 on the preceding page hold. But in [**Ren80**] it *is not* required that $\lambda(e_i)$ be bounded. So Theorem 10.49 on the facing page does not quite tie up with the original [**Ren80**, Definition II.3.1]. But it is pointed out in [**ADR00**] that [**ADR00**, Theorem 6.1.4(v)] implies that [**Ren80**, Definition II.3.1] suffices to show μ is amenable as defined here.

10.6. An Application

We have already seen that amenability and measurewise amenability coincide for groupoids equivalent to étale groupoids and noted that it remains possible that the notion coincide in general. Here we at least want to see that the notions coincide whenever the orbit space is reasonable. This result slightly extends [**SW13**, Theorem 4].

THEOREM 10.52. *Let G be a second countable locally compact groupoid with a Haar system $\lambda = \{\lambda^u\}_{u\in G^{(0)}}$ such that $G\backslash G^{(0)}$ is T_0. The the following are equivalent.*

(a) *G is amenable.*
(b) *G is measurewise amenable.*
(c) *Every stability group $G(u)$ is amenable.*

First we need an observation.

LEMMA 10.53. *Suppose that G is a measurewise amenable second countable locally compact Hausdorff groupoid with a Haar system $\lambda = \{\lambda^u\}_{u\in G^{(0)}}$. If F is a locally closed saturated subset of $G^{(0)}$, then $G(F)$ is measurewise amenable.*

PROOF. Since F is locally closed, it is locally compact. Hence $F = C \cap O$ with C closed and O open. I claim every $f \in C_c(F)$ is the restriction of some $\tilde{f} \in C_c(G)$. To see this, note that f extends to a function $\tilde{f} \in C_c(O)$ by the usual Tietze extension theorem [**Wil07**, Lemma 1.42]. Then we can view $\tilde{f}$ as an element of $C_c(G)$.

Let μ be a quasi-invariant measure on F. We need to see that $(G(F), \lambda^F, \mu)$ is amenable as in Definition 10.48 on page 274 where λ^F is the restriction of λ to F. By Theorem 10.49(b) on page 274 it suffices to produce Borel functions $\{g_i\} \subset L^\infty(F, L^1(G(F), \lambda^F))_1^+$ such that

(a) $\lambda^F(g_i) \to 1$ in the weak-$*$ topology on $L^\infty(F, \mu)$ and
(b) for all $f \in C_c(F)$ and $\varphi \in L^\infty(G(F), \nu^F)$ we have

$$\lim_i \int_F \int_{G(F)} \int_{G(F)} f(\gamma)\varphi(\eta)\big(g_i(\gamma^{-1}\cdot\eta) - g(\eta)\big)\, d\lambda^u(\gamma)\, d\lambda^u(\eta)\, d\mu(u) = 0.$$

But we can view μ as a measure on $G^{(0)}$ and $\nu^F = \nu \circ \lambda^F$ as a measure on all of G. Then $\nu = \nu^F$ where as usual, $\nu = \mu \circ \lambda$. Since $(\nu \circ \lambda^F)^{-1}$ is equivalent to $\nu \circ \lambda^F$, it follows that μ is quasi-invariant as a measure on $G^{(0)}$. Hence there is a net $\{g_i'\}$ satisfying (a) and (b) above with $G^{(0)}$ in place of F. We can identify $L^\infty(G^{(0)})$ and $L^\infty(F, \mu)$ as well as $L^\infty(G(F), \nu^F)$ and $L^\infty(G, \nu)$. Thus we obtain the result by letting g_i be the restriction of g_i' to $G(F)$ and recalling that every $f \in C_c(F)$ is the restriction of an element in $C_c(G)$. □

PROOF OF THEOREM 10.52. Theorem 10.21 on page 253 shows that (a) implies (b). We showed that (c) implies (a) in Theorem 9.86 on page 242. So we just have to prove that (b) implies (c).

Since $G\backslash G^{(0)}$ is T_0, each orbit $[u]$ is locally closed by Theorem 2.27 on page 38. By Lemma 10.53, $G([u])$ is measurewise amenable. Since $[u]$ is saturated and $r : G \to G^{(0)}$ is open, $r|_{G([u])} : G([u]) \to [u]$ is open. Hence $G([u])$ is equivalent to $G(u)$. Hence $G(u)$ is measurewise amenable by Theorem 10.41 on page 266. Hence $G(u)$ is amenable (since $G(u)$ is a group). This completes the proof. □

Recall that we call G metrically amenable if the reduced norm and the universal norm coincide on $C_c(G)$. In general, examples show that, even for étale groupoids, metric amenability does not imply even measurewise amenability [**Wil15**]. However, for transitive groupoids, we get the best possible result.

COROLLARY 10.54. *Suppose that G is a second countable* transitive *groupoid with a Haar system. Then the following are equivalent.*

(a) *G is amenable.*

(b) *G is measurewise amenable.*

(c) *G is metrically amenable.*

(d) *For any $u \in G^{(0)}$, the isotropy group $G(u)$ is amenable.*

PROOF. Let $u \in G^{(0)}$ and recall that G is equivalent to the isotropy group $G(u)$. By Proposition 2.75 on page 61, G is metrically amenable if and only if $G(u)$ is an amenable group. Since u was arbitrary, the rest follows from Theorem 10.52 on the preceding page. □

CHAPTER 11

Comments on Simplicity

A fundamental structural question for any C^*-algebra is whether or not it is simple in the sense that it has no nontrivial (closed two-sided) ideals. For a group C^*-algebra $C^*(G)$, the issue is uninteresting: $C^*(G)$ is simple if and only if G is the trivial group (Ex 11.1.1). However, the question of whether the reduced group C^*-algebra $C^*_r(G)$ is simple is very subtle and is of considerable interest (see, for example, [**dlH07**]). Such groups are called *G-simple*. As usual, the case for locally compact groupoids is far more intricate. But we can collect some low hanging fruit. If the reduced norm and universal norm do not coincide on $C_c(G)$, then the natural map $C^*(G)$ onto $C^*_r(G)$ has a nontrivial kernel and $C^*(G)$ is not simple. Therefore, just as for groups, it makes sense to concentrate on $C^*_r(G)$. Also, if the action of G on its unit space $G^{(0)}$ is not minimal, then by definition, there is a nontrivial closed invariant subset of $G^{(0)}$ and $C^*_r(G)$ is not simple (see Proposition 5.4 on page 110).

In order to talk about either $C^*(G)$ or $C^*_r(G)$, we need to assume G has a Haar system. Hence throughout this chapter, G is going to be a second countable locally compact Hausdorff groupoid with a Haar system.[1] In particular, the range and source maps of G are open. Thus in the special case that G is transitive, it will be equivalent to any of its isotropy groups $H = G(u)$ for some $u \in G^{(0)}$. Hence by Theorem 2.71 on page 60, $C^*_r(G)$ will be Morita equivalent to $C^*_r(H)$. Thus $C^*_r(G)$ will be simple if and only if H is G-simple.[2] Since the theory of nontrivial G-simple groups is well outside the scope of this book, here we'll concentrate on the extreme case where H is the trivial group. This will still give us plenty to work on. Our main goal will be to describe two very interesting results of Renault's from [**Ren91**]. They are easiest to state in the case where G is étale.

THEOREM 11.1. *Suppose that G is a second countable locally compact* étale *groupoid acting minimally on its unit space $G^{(0)}$ and such that there is a $u \in G^{(0)}$ such that $G(u)$ is trivial. Then $C^*_r(G)$ is simple.*

If the action of G on $G^{(0)}$ is not minimal, we could still hope that the only ideals in $C^*_r(G)$ are the ones corresponding to open G-invariant subsets of $G^{(0)}$. Our freeness condition then becomes that the action of G on $G^{(0)}$ is *essentially free*; that is we require that the set of points u where $G(u)$ is trivial is dense in every closed G-invariant subset F of $G^{(0)}$. Then we have the following.

THEOREM 11.2. *Let G be an* amenable *second countable locally* étale *groupoid acting essentially freely on its unit space $G^{(0)}$. Then the lattice $\mathfrak{I}(C^*(G))$ of ideals in $C^*(G)$ is lattice isomorphic to the lattice of open G-invariant subsets of $G^{(0)}$.*

[1] The results in [**Ren91**] only require that G be second countable and locally Hausdorff.

[2] An admitted over use of the symbol G.

We'll obtain Theorem 11.1 as a special case of Theorem 11.24 on page 302. Theorem 11.2 is a special case of Proposition 11.35 on page 311 together with Remark 11.36 on page 312. It should be mentioned that Theorem 11.2 on the previous page is unlikely to pass to the reduced algebra if we drop the assumption of amenability. This is due to the potential lack of exactness of the exact sequence in Proposition 5.2 on page 110.

Renault's method of proof is motivated by the Effros-Hahn conjecture which suggests that—in the amenable case—the primitive ideals of $C^*(G)$ should be induced in some way from primitive ideals of the isotropy subgroups. Renault's approach from [**Ren91**] is as follows. Given a representation $\underline{R}$ of $C^*(G)$, one can restrict the underlying unitary representation of $\underline{R}$ to a unitary representation r of the isotropy groupoid $G' = \{\, \gamma \in G : r(\gamma) = s(\gamma) \,\}$. Then, in spite of the fact that G' generally doesn't have a Haar system—so that our usual theory of induced representations does not apply—one can still form a sort of induced representation $\operatorname{Ind} r$ of $C^*(G)$. The key result is that one always has $\ker \underline{R} \subset \ker(\operatorname{Ind} r)$ (Theorem 11.17 on page 296). This is what is needed to prove Theorem 11.24—and obtain Theorem 11.1 as a corollary. However, in the spirit of the Effros-Hahn conjecture, we can show that if G is suitably amenable, then we get $\ker \underline{R} = \ker(\operatorname{Ind} r)$ (Theorem 11.25 on page 302) which can be viewed as confirmation of the Effros-Hahn conjecture for groupoids. This result allows us to prove Proposition 11.35—and hence Theorem 11.2. Although it is beyond the scope of this book, Theorem 11.25 can also be used to prove a version of the Effros-Hahn conjecture for second countable amenable groupoids asserting that every primitive ideal K of $C^*(G)$ is of the form $\operatorname{Ind}_{G(u)}^G J$ for a primitive ideal J of $C^*(G(u))$ for some $u \in G^{(0)}$ ([**IW09b**, Theorem 2.1]).

11.1. The Auxillary Groupoids

The example of transitive groupoids suggests that a lot of information about the structure of a groupoid is contained in the isotropy groups. We made good on this paradigm in Section 5.5 at least when the orbit space $G\backslash G^{(0)}$ is nice. The main result of this chapter, Theorem 11.25 on page 302, is another version of this rubric. First we are going to need a number of technical observations.

As in Section 3.4, we are going to let Σ_0 be the locally compact space of closed subgroups of G, and let Σ be the associated group bundle over Σ_0. Using Proposition 3.23 on page 82, we can make a continuous choice of Haar measures β^H on each $H \in \Sigma_0$ so that $\beta = \{\beta^H\}_{H\in\Sigma_0}$ is a Haar system for Σ.

The idea of "restricting to the isotropy" will involve the isotropy groupoid

$$G' := \{\, \gamma \in G : r(\gamma) = s(\gamma) \,\}$$

introduced in Section 6.3. Clearly G' is a closed subgroupoid of G and a group bundle over $G^{(0)}$. In addition to the isotropy groupoid, we need another auxillary groupoid called the *associated equivalence relation* R. We define R to be the image in $G^{(0)} \times G^{(0)}$ of the continuous groupoid homomorphism

$$\pi : G \to G^{(0)} \times G^{(0)}$$

given by $\pi(\gamma) = \big(r(\gamma), s(\gamma)\big)$.

Both G' and R will play a key role in the remainder of this chapter. But we are going to have to deal with the unfortunate truth that they are, in general, not very well behaved. The isotropy groupoid G' has a Haar system only when the isotropy map is continuous (Theorem 6.9 on page 143). Much worse is the fact that R is rarely a nice subset of $G^{(0)} \times G^{(0)}$ let alone locally compact. We can improve things a bit by equipping R with its quotient topology $\tau_q = \{ U \subset R : \pi^{-1}(U) \text{ is open in } G \}$. This topology is often finer than the relative topology coming from $G^{(0)} \times G^{(0)}$. Unfortunately, there is nothing guaranteeing that π is an open map with respect to the quotient topology on R. As a result, there is no way to guarantee that even the quotient topology is locally compact.

The situation for these groupoids is much nicer in the Borel context. Recall that Borel groupoids and Borel Haar systems were defined in Section 9.7. Note that every second countable locally compact (Hausdorff) groupoid G is an example of a Borel groupoid, and that a Haar system on G is a Borel Haar system (Example 9.60 on page 228).

Since G' is a closed subgroupoid of G, it is always a Borel groupoid. Since the isotropy map $u \mapsto G(u)$ is always Borel by Lemma 3.27 on page 83, $\beta = \{\beta^{G(u)}\}_{u\in G^{(0)}}$ is a Radon family on G', and β is a Borel Haar system for G' which a Haar system exactly when the isotropy map is continuous.[3] Unlike the topological case, R is always reasonable in a Borel sense. Since R is σ-compact, it is a Borel subset of $G^{(0)} \times G^{(0)}$ and hence a standard Borel space in its quotient Borel structure. Furthermore, our next observation shows that there is no difference between the relative Borel structure and the quotient Borel structure.

LEMMA 11.3. *The relative and quotient Borel stuctures on R coincide. In particular, a function $\varphi : R \to X$ is Borel if and only if $\bar\varphi : G \to X$, given by $\bar\varphi(\gamma) = \varphi(\pi(\gamma))$, is Borel.*

PROOF. Since π is continuous, to see that the Borel stuctures coincide we just need to see that $B \subset G^{(0)} \times G^{(0)}$ is Borel if $\pi^{-1}(B)$ is Borel in G. But every open set U in G is σ-compact. Hence $\pi(U)$ is σ-compact and therefore Borel. Clearly the inverse image of points is closed, so [**Arv76**, Theorem 3.4.1] implies there is a Borel cross section $s : R \to G$ for π; that is, $\pi \circ s = \mathrm{id}_R$. But then $B = s^{-1}\big(\pi^{-1}(B)\big)$ is Borel.

The final assertion is straightforward (Ex 11.1.2). □

LEMMA 11.4. *Let $f \in \mathcal{B}^+(G)$, then*

$$\beta(f)(\pi(\gamma)) = \int_{G(s(\gamma))} f(\gamma t)\, d\beta^{s(\gamma)}(t) \tag{11.1}$$

is in $\mathcal{B}^{+,e}(R)$.

PROOF. The right-hand side of (11.1) clearly depends only on $\pi(\gamma)$. Hence it suffices to see that the right-hand side in Borel in γ. But $\{\beta^H\}$ is a Haar system for the group bundle Σ. Hence for any $F \in \mathcal{B}^+(\Sigma \times G)$,

$$\gamma \mapsto \int F(s(\gamma), G(s(\gamma)), t)\, d\beta^{s(\gamma)}(t)$$

is Borel by Proposition 3.3 on page 70 together with the observation that $\gamma \mapsto G(s(\gamma))$ is Borel by Lemma 3.27 on page 83. Now let $F(p(H), H, t, \gamma) = f(\gamma t)$ if $p(H) = s(\gamma)$ and zero otherwise. □

The next result is crucial. It says that the isotropy groupoid and the induced equivalence relation play well together at the Borel level. It's taken from [**Ren91**, Lemma 1.12].

[3] We will sometimes write β^u in place of $\beta^{G(u)}$ to make some formulae easier to digest.

PROPOSITION 11.5 (Renault). *Suppose that G is a second countable locally compact groupoid with Haar system $\{\lambda^u\}_{u\in G^{(0)}}$ and let R be the associated equivalence relation. Then R is a Borel groupoid such that we can identify $R^{(0)}$ with $G^{(0)}$. Furthermore R admits a Borel Haar system $\{\alpha^u\}_{u\in G^{(0)}}$ such that for all $f \in \mathcal{B}^+(G)$*

$$\int_G f(\gamma)\,d\lambda^u(\gamma) = \int_R \int_{G(s(\gamma))} f(\gamma t)\,d\beta^{s(\gamma)}(t)\,d\alpha^u(\pi(\gamma)). \tag{11.2}$$

PROOF. If $f \in \mathcal{B}^+(G)$, then we define $\beta(f) \in \mathcal{B}^{+,e}(R)$ as in Lemma 11.4. Thus the right-hand side of (11.2) is just

$$\int_R \beta(f)(\pi(\gamma))\,d\alpha^u(\pi(\gamma))$$

which is well defined once we specify α^u.

Let $g \in \mathcal{B}^+(R)$. Define

$$f(\gamma) = b(\gamma, G(s(\gamma)))g(\pi(\gamma))$$

where b is a Bruhat approximate cross section as in Corollary 3.31 on page 84. Then $\beta(f) = g$. Hence every element of $\mathcal{B}^+(R)$ is of the form $\beta(f)$ for some $f \in \mathcal{B}^+(G)$.

It is not hard to check that $\lambda(f\beta(g)) = \lambda(\beta(f)g)$ (Ex 11.1.3). Also as $R^u = \pi(G^u)$ is σ-compact, it is Borel and there is a Borel function $F \in \mathcal{B}^+(R)$ such that $F(\pi(\gamma)) = 1$ for all $\pi(\gamma) \in R^u$. Let $g \in \mathcal{B}^+(G)$ be such that $\beta(g) = F$. Then

$$\lambda(f)(u) = \lambda(f\beta(g))(u) = \lambda(\beta(f)g).$$

Hence $\lambda(f)(u)$ depends only on $\beta(f)$ and we can define

$$\alpha^u : \mathcal{B}^+(R) \to [0, \infty]$$

by

$$\alpha^u(\beta(f)) = \lambda(f)(u).$$

Multiple instances of the Monotone Convergence Theorem show that

$$\begin{aligned}\alpha^u\Big(\sum \beta(f_i)\Big) &= \alpha^u\Big(\beta\Big(\sum f_i\Big)\Big)\\ &= \lambda^u\Big(\sum f_i\Big)\\ &= \sum \lambda^u(f_i)\\ &= \sum \alpha^u(\beta(f_i)).\end{aligned} \tag{11.3}$$

It now follows that α^u defines a measure on R (sending E to $\alpha^u(\mathbb{1}_E)$) also denoted α^u. Since

$$\alpha^u(g) = \int_R g(x, y)\,d\alpha^u(x, y) \tag{11.4}$$

is clear for simple functions, (11.4) holds in general by (11.3) and the Monotone Convergence Theorem.

Now (11.2) holds by construction.

Since $\alpha(\beta(f))(u) = \lambda(f)(u)$, it is clear that $u \mapsto \alpha(h)(u)$ is Borel for all $h \in \mathcal{B}^+(R)$.

If $(u,v)=\pi(\eta)$, then we get left-invariance via

$$\begin{aligned}\int_R \beta(f)(\pi(\gamma))\,d\alpha^u(\pi(\gamma)) &= \int_G f(\gamma)\,d\lambda^u(\gamma)\\ &= \int_G f(\eta\gamma)\,d\lambda^v(\gamma)\\ &= \int_R \beta(f)(\pi(\eta)\pi(\gamma))\,d\alpha^v(\pi(\gamma)),\end{aligned}$$

which suffices since $\beta(f)$ is a typical element of $\mathcal{B}^+(R)$.

Clearly the complement of R^u is α^u-null. If $K\subset G$ is compact and if b is a Bruhat approximate cross section, then $f(\gamma)=b(\gamma,G(s(\gamma)))\mathbb{1}_{\pi(K)}(\pi(\gamma))$ has compact support. Since b is continuous, f is also bounded. Hence $u\mapsto\alpha^u(\pi(K))=\lambda^u(f)$ is bounded. Therefore α is proper and we've established that $\{\alpha^u\}$ forms a Borel Haar system as claimed. □

Note that $s:R^u\to[u]$ is a Borel bijection with Borel inverse $v\mapsto(u,v)$. Hence we get a measure $\underline{\alpha}^u$ on $[u]$ via

$$\int_{[u]} f(y)\,d\underline{\alpha}^u(y)=\int_R f(u\cdot\gamma)\,d\alpha^u(\pi(\gamma)). \tag{11.5}$$

But if $(u,v)=\pi(\eta)\in R$, then left-invariance gives

$$\int_R f(u\cdot\gamma)\,d\alpha^u(\pi(\gamma))=\int_R f(v\cdot\gamma)\,d\alpha^v(\pi(\gamma)).$$

Hence $\underline{\alpha}^u=\underline{\alpha}^v$ and we can write $\underline{\alpha}^{[u]}$ for this measure. Note that the map

$$u\mapsto\int_{[u]} f(y)\,d\underline{\alpha}^{[u]}(y)$$

is Borel.

Fix a quasi-invariant measure μ on $G^{(0)}$ and a generalized Bruhat approximate cross section b on $G*\Sigma_0$. As usual, let $\nu=\mu\circ\lambda$ and form the Radon measure $b\cdot\nu$ on G given by

$$b\cdot\nu(f)=\int_{G^{(0)}}\int_G f(\gamma)b(\gamma,G(s(\gamma)))\,d\lambda^u(\gamma)\,d\mu(u).$$

The push-forward $\pi_*(b\cdot\nu)$ is a measure κ on R given by

$$\begin{aligned}\kappa(f)&=\int_{G^{(0)}}\int_G f(\pi(\gamma))b(\gamma,G(s(\gamma)))\,d\lambda^u(\gamma)\,d\mu(u)\\ &=\int_{G^{(0)}}\int_R\int_{G(s(\gamma))} f(\pi(\gamma))b(\gamma t,G(s(\gamma)))\,d\beta^{s(\gamma)}(t)\,d\alpha^u(\pi(\gamma))\,d\mu(u)\\ &=\int_{G^{(0)}}\int_R f(\pi(\gamma))\,d\alpha^u(\pi(\gamma))\,d\mu(u).\end{aligned}\tag{11.6}$$

Since α is a Borel Haar system, and hence proper, it follows that κ is σ-finite.

PROPOSITION 11.6. *Let μ be a quasi-invariant measure on $G^{(0)}$ and κ the σ-finite measure on R given by (11.6). Let κ^{-1} be the push-forward of κ under the flip $\pi(\gamma)\mapsto\pi(\gamma)^{-1}$ on R. Then κ and κ^{-1} are equivalent, and*

$$D(\pi(\gamma))=\omega(\gamma,G(s(\gamma)))^{-1}\Delta(\gamma) \tag{11.7}$$

is a choice of Radom-Nikodym derivative for $d\kappa/d\kappa^{-1}$ where Δ is the modular function for μ, and ω is defined in Lemma 3.25 on page 82. In particular, we can assume that both Δ and D are homomorphisms into the multiplicative group $\mathbf{R}^{+,\times}$ of positive reals and that

$$\Delta(t) = \Delta_{G(s(t))}(t) \quad \textit{for any } t \in G'.$$

PROOF. Let $\nu = \mu \circ \lambda$. Since ν and ν^{-1} are equivalent, and since ω never vanishes, $b \cdot \nu$ is equivalent to $b\omega \cdot \nu^{-1}$ where the later is given by

$$(b\omega \cdot \nu^{-1})(f) = \int_{G^{(0)}} \int_G f(\gamma)b(\gamma, G(s(\gamma)))\omega(\gamma, G(s(\gamma)))\, d\lambda_u(\gamma)\, d\mu(u).$$

Thus κ is equivalent to the push-forward $\pi_*(b\omega \cdot \nu^{-1})$. But

$$\begin{aligned}
&\pi_*(b\omega \cdot \nu^{-1})(f)\\
&\quad= \int_{G^{(0)}} \int_G f(\pi(\gamma))b(\gamma, G(s(\gamma)))\omega(\gamma, G(s(\gamma)))\, d\lambda_u(\gamma)\, d\mu(u)\\
&\quad= \int_{G^{(0)}} \int_G f(\pi(\gamma)^{-1})b(\gamma^{-1}, G(r(\gamma)))\omega(\gamma^{-1}, G(r(\gamma)))\, d\lambda^u(\gamma)\, d\mu(u)
\end{aligned}$$

which, in view of the construction of α^u, is

$$\begin{aligned}
= \int_{G^{(0)}} \int_R \int_{G(s(\gamma))} &f(\pi(\gamma)^{-1})b(t^{-1}\gamma^{-1}, G(r(\gamma)))\omega(t^{-1}\gamma^{-1}, G(r(\gamma)))\\
&d\beta^{s(\gamma)}(t)\, d\alpha^u(\pi(\gamma))\, d\mu(u)
\end{aligned}$$

which, since $\omega(t^{-1}\gamma^{-1}, G(r(\gamma))) = \omega(\gamma^{-1}, G(r(\gamma)))\Delta_{G(s(\gamma))}(t)^{-1}$, is

$$\begin{aligned}
= \int_{G^{(0)}} \int_R \int_{G(s(\gamma))} &f(\pi(\gamma)^{-1})b(t\gamma^{-1}, G(r(\gamma)))\omega(\gamma^{-1}, G(r(\gamma)))\\
&d\beta^{s(\gamma)}(t)\, d\alpha^u(\pi(\gamma))\, d\mu(u)
\end{aligned}$$

which, using the fundamental property of ω, is

$$\begin{aligned}
&= \int_{G^{(0)}} \int_R \int_{G(r(\gamma))} f(\pi(\gamma)^{-1})b(\gamma^{-1}t, G(r(\gamma)))\, d\beta^{r(\gamma)}(t)\, d\alpha^u(\pi(\gamma))\, d\mu(u)\\
&= \int_{G^{(0)}} \int_R f(\pi(\gamma)^{-1})\, d\alpha^u(\pi(\gamma))\, d\mu(u)\\
&= \kappa^{-1}(f).
\end{aligned}$$

It follows that κ and κ^{-1} are equivalent as claimed. Let $\underline{D} : R \to (0, \infty)$ be a choice of Radon-Nikodym derivative for $d\kappa/d\kappa^{-1}$. By considering the integral

$$\int_R f(\pi(\gamma))\underline{D}(\pi(\gamma)^{-1})\underline{D}(\pi(\gamma))\, d\kappa(\pi(\gamma),$$

we conclude that $\underline{D}(\pi(\gamma))\underline{D}(\pi(\gamma)^{-1}) = 1$ for κ-almost all $\pi(\gamma))$. Then

$$D(\pi(\gamma)) := \underline{D}(\pi(\gamma))^{\frac{1}{2}}\underline{D}(\pi(\gamma)^{-1})^{-\frac{1}{2}}$$

is also a representative of $d\kappa^{-1}/d\kappa$ and which satisfies $D(\pi(\gamma)^{-1}) = D(\pi(\gamma))^{-1}$.

Consider the integral

$$\begin{aligned}
&\int_{G^{(0)}}\int_G D(\pi(\gamma)^{-1})\omega(\gamma, G(s(\gamma)))^{-1}\Delta(\gamma)f(\gamma)\,d\lambda_u(\gamma)\,d\mu(u)\\
&\quad=\int_{G^{(0)}}\int_G D(\pi(\gamma))^{-1}\omega(\gamma, G(s(\gamma)))^{-1}f(\gamma)\,d\lambda^u(\gamma)\,d\mu(u)\\
&\quad=\int_{G^{(0)}}\int_R\int_{G(s(\gamma))} D(\pi(\gamma))^{-1}\omega(\gamma t, G(s(\gamma)))^{-1}f(\gamma t)\\
&\qquad\qquad\qquad\qquad\qquad\qquad d\beta^{s(\gamma)}(t)\,d\alpha^u(\pi(\gamma))\,d\mu(u)\\
&\quad=\int_R\Big(\int_{G(s(\gamma))}\omega(\gamma t, G(s(\gamma)))^{-1}f(\gamma t)\,d\beta^{s(\gamma)}(t)\Big)D(\pi(\gamma))^{-1}\,d\kappa(\pi(\gamma))
\end{aligned}$$

which, in view of the defintion of D, is

$$\begin{aligned}
&\quad=\int_R\Big(\int_{G(r(\gamma))}\omega(\gamma^{-1}t, G(r(\gamma)))^{-1}f(\gamma^{-1}t)\,d\beta^{r(\gamma)}(t)\Big)\,d\kappa(\pi(\gamma))\\
&\quad=\int_{G^{(0)}}\int_R\int_{G(r(\gamma))}\Delta_{G(r(\gamma))}(t)^{-1}\omega(\gamma^{-1}, G(r(\gamma)))^{-1}f(\gamma^{-1}t)\\
&\qquad\qquad\qquad\qquad\qquad\qquad d\beta^{r(\gamma)}(t)\,d\alpha^u(\pi(\gamma))\,d\mu(u)\\
&\quad=\int_{G^{(0)}}\int_R\int_{G(r(\gamma))}\omega(\gamma, G(s(\gamma)))f(\gamma^{-1}t^{-1})\,d\beta^{r(\gamma)}(t)\,d\alpha^u(\pi(\gamma))\,d\mu(u)\\
&\quad=\int_{G^{(0)}}\int_R\int_{G(s(\gamma))}f(t^{-1}\gamma^{-1})\,d\beta^{s(\gamma)}(t)\,d\alpha^u(\pi(\gamma))\,d\mu(u)\\
&\quad=\int_{G^{(0)}}\int_G f(\gamma^{-1})\,d\lambda^u(\gamma)\,d\mu(u)\\
&\quad=\int_{G^{(0)}}\int_G f(\gamma)\,d\lambda_u(\gamma)\,d\mu(u).
\end{aligned}$$

Since the above computation holds for all $f\in C_c(G)$, we conclude that

$$D(\pi(\gamma)^{-1})\omega(\gamma, G(s(\gamma)))^{-1}\Delta(\gamma)=1\quad\text{for } \nu^{-1}\text{-almost all } \gamma.$$

If we let

$$D'(\pi(\gamma))=\omega(\gamma, G(s(\gamma)))^{-1}\Delta(\gamma),$$

then $D'(\pi(\gamma))=D(\pi(\gamma))$ for ν-almost all γ (since ν and ν^{-1} are equivalent). Hence we can replace D by D' so that (11.7) holds. Since we can assume that Δ is a homomorphism (Proposition 7.6 on page 162), it follows that D is as well (in view of the Lemma 3.25 on page 82). In particular, if $t\in G'$, then $D(\pi(t))=1$ and

$$\Delta(t)=\omega(t, G(t))D(\pi(t))=\Delta_{G(t)}(t)D(\pi(t))=\Delta_{G(t)}(t)\quad\square$$

Exercises

Ex 11.1.1. If G is a locally compact group, show that $C^*(G)$ is simple if and only if G is the trivial group.

—Answer on page 383

Ex 11.1.2. Let Y and Z be Borel spaces and $f : Y \to X$ a map. Let X have the quotient Borel stucture. Show that $\varphi : X \to Z$ is Borel if and only if the lift $\bar{\varphi} : Y \to Z$ is Borel.

—Answer on page 383

Ex 11.1.3. If $f \in \mathcal{B}^{+,e}(G)$, define $\beta(f) \in \mathcal{B}^{+,e}(R)$ as in Lemma 11.4. Show that for $f, g \in \mathcal{B}^{+}(G)$, we have

$$\lambda(f\beta(g)) = \lambda(\beta(f)g)$$

where we let $\mathcal{B}^{+,e}(R)$ act on $\mathcal{B}^{+,e}(G)$ in the obvious way. (Mimic the proof of Lemma 6.16.)

—Answer on page 383

11.2. Restricting to the Isotropy

We fix a nondegenerate representation $\underline{R}$ of $C^*(G)$. The Disintegration Theorem on page 169 allows us to assume that $\underline{R}$ is the integrated form of a unitary representation

$$(\mu, G^{(0)} * \mathscr{H}, \hat{V})$$

where $\hat{V}(\gamma) : G \to \operatorname{Iso}(G^{(0)} * \mathscr{H})$ is a Borel homomorphism of the form

$$\hat{V}(\gamma) = \big(r(\gamma), V_\gamma, s(\gamma)\big).$$

Hence the map $(u, H, t) \mapsto (u, V_t, u)$ is Borel from Σ to $\operatorname{Iso}(G^{(0)} * \mathscr{H})$. Thus if $F \in C_c(\Sigma)$ and $h, k \in L^2(G^{(0)} * \mathscr{H}, \mu)$, then we can form the integral

$$(11.8)\quad \big(r(F)h \mid k\big) := \int_{G^{(0)}} \int_{G(u)} F(u, G(u), t)\big(V_t(h(u)) \mid k(u)\big)\Delta_{G(u)}(t)^{-\frac{1}{2}}\, d\beta^u(t)\, d\mu(u)$$

with $|\big(r(F)h \mid k\big)| \le \|F\|_I \|h\| \|k\|$ (see the last part of the proof of Proposition 7.12 on page 163).

Moreover, for each $u \in G^{(0)}$, we obtain a Borel representation $r_u : G(u) \to U(\mathcal{H}(u))$ given by $r_u(t) = V_t$. It is well-known (see [**Wil07**, Theorem D.11]) that r_u must be continuous with respect to the weak-operator topology on $U(\mathcal{H}(u))$ so that r_u is a unitary representation of the group $G(u)$ in the classical sense. Following the lines of Proposition 7.12 on page 163 we have the following.

LEMMA 11.7. *The formula in* (11.8) *defines a representation* $r : C_c(\Sigma) \to B(L^2(G^{(0)} * \mathscr{H}, \mu))$. *Moreover, if* $u \in G^{(0)}$ *and* $f_u(t) = F(u, G(u), t)$, *then*

$$r(F)(h)(u) = r_u(f)(h(u))$$

where r_u *is the representation of* $C^*(G(u))$ *defined above.*[4]

REMARK 11.8. For motivation only, it is helpful to note that we can think of the representation r as the direct integral

$$r = \int_{G^{(0)}}^{\oplus} r_u \, d\mu(u)$$

[4]Since we are treating the group $G(u)$ as a groupoid, the modular function appears in the integrated form of its representations.

where we view each r_u as a representation of $C^*(\Sigma)$ in the obvious way. We want to construct a sort of induced reprersentation, $\operatorname{Ind} r$, of $C^*(G)$. Rather than build a proper Rieffel machine, it will be easier to proceed in an *ad hoc* manner thinking of $\operatorname{Ind} r$ as the direct integral

$$\int_{G^{(0)}}^{\oplus} \operatorname{Ind}_{G(u)}^{G} r_u \, d\mu(u)$$

as described below.

Recall that the representation of G induced from r_u on $G(u)$ (see Definition 5.12 on page 115) acts on the completion $\tilde{\mathcal{H}}(u)$ of $C_c(G_u) \odot \mathcal{H}(u)$ with respect to the inner product

$$\big(\varphi \otimes h \mid \psi \otimes k\big)_u = \big(r_u(\psi^* * \varphi)h \mid k\big) = \int_{G(u)} \psi^* * \varphi(t)\big(V_t(h) \mid k\big)\Delta_{G(u)}^{-\frac{1}{2}}(t)\, d\beta^u(t)$$

Then $\operatorname{Ind}_{G(u)}^{G}(f)$ acts on (the class of) $\varphi \otimes h$ by $f * \varphi \otimes h$.

There is no harm in replacing $C_c(G_u)$ with $C_c(G)$ in the above.

If $\varphi, \psi \in C_c(G)$ and $h, k \in L^2(G^{(0)} * \mathscr{H}, \mu)$, then

$$u \mapsto \int_{G(u)} \psi^* * \varphi(t)\big(V_t(h(u)) \mid k(u)\big)\Delta_{G(u)}(t)^{-\frac{1}{2}}\, d\beta^u(t)$$

is Borel, and we can define a sesquilinear form on $C_c(G) \odot L^2(G^{(0)} * \mathscr{H}, \mu)$ which satisfies

$$\big(\varphi \otimes h \mid \psi \otimes k\big) = \int_{G^{(0)}} \big(\varphi \otimes h(u) \mid \psi \otimes k(u)\big)_u \, d\mu(u). \tag{11.9}$$

Since each $\big(\cdot \mid \cdot\big)_u$ is a pre-inner product, so is (11.9):

$$\Big(\sum_i \varphi_i \otimes h_i \mid \sum_j \varphi_j \otimes h_j\Big) = \int_{G^{(0)}} \Big(\sum_i \varphi \otimes h_i(u) \mid \sum_j \varphi_j \otimes h_j(u)\Big)_u \, d\mu(u) \geq 0.$$

We are going to abuse notation and denote the Hilbert space completion of $C_c(G) \odot L^2(G^{(0)} * \mathscr{H}, \mu)$ by $L^2(G^{(0)} * \widetilde{\mathscr{H}}, \mu)$ to suggest it is a direct integral of the spaces $\tilde{\mathcal{H}}(u)$. (See Remark 11.8 on the preceding page.)

If $f \in C_c(G)$, then

$$u \mapsto \big(f * \varphi \otimes h(u) \mid \psi \otimes k(u)\big)$$

is certainly Borel and

$$\begin{aligned}|\big(f * \varphi \otimes h(u) \mid \psi \otimes k(u)\big)_u| &\leq \|\operatorname{Ind}_{G(u)}^{G}(f)\|\|\varphi \otimes h(u)\|\|\psi \otimes k(u)\| \\ &\leq \|f\|\|\varphi \otimes h(u)\|\|\psi \otimes k(u)\|.\end{aligned}$$

Hence it is not hard to see that we get a representation $\operatorname{Ind} r : C_c(G) \to B(L^2(G^{(0)} * \widetilde{\mathscr{H}}, \mu)$ where $\operatorname{Ind} r(f)$ is given by

$$\varphi \otimes h \mapsto f * \varphi \otimes h.$$

REMARK 11.9. It is clear from its construction that if $f \in C_c(G)$, then $\operatorname{Ind} r(f)h(u) = \operatorname{Ind}_{G(u)}^{G} r_u(f)(h(u))$ for all u. Hence if $a \in \ker \operatorname{Ind}_{G(u)}^{G} r_u$ for (almost) all u, then $a \in \ker \operatorname{Ind} r$.

In order to see how the representations r, $\operatorname{Ind} r$, and $\underline{R}$ are related, we need to enlarge the domain of the representation $\underline{R}$ to include Borel functions. Let $\mathcal{L}^\infty(G^{(0)}, L^I(G,\lambda))$ be the collection of Borel functions f on G such that $f \in \mathcal{L}^1(\lambda^u)$ and $\mathcal{L}^1(\lambda_u)$ for all $u \in G^{(0)}$, and such that

$$\|f\|_{I,r} = \sup_{u\in G^{(0)}} \int_G |f(\gamma)|\,d\lambda^u(\gamma) \quad \text{and} \quad \|f\|_{I,s} = \sup_{u\in G^{(0)}} \int_G |f(\gamma)|\,d\lambda_u(\gamma)$$

are both finite. Then as before we define

$$\|f\|_I = \max\{\, \|f\|_{I,r}, \|f\|_{I,s} \,\}.$$

Of course, with respect to $\|\cdot\|_I$, $C_c(G)$ sits isometrically inside of $\mathcal{L}^\infty(G^{(0)}, L^I(G,\lambda))$.

We would like to extend the $*$-algebra operations on $C_c(G)$ to all of $\mathcal{L}^\infty(G^{(0)}, L^I(G,\lambda))$. This is routine for the involution:

$$f^*(\gamma) = \overline{f(\gamma^{-1})}$$

is clearly back in $\mathcal{L}^\infty(G^{(0)}, L^I(G,\lambda))$. Unfortunately, associativity of the convolution becomes problematical without agreeing to identify functions on some sort of null set. For example, if $f, g \in \mathcal{L}^\infty(G^{(0)}, L^I(G,\lambda))$, then

$$(\eta,\gamma) \mapsto |f(\eta)g(\eta^{-1}\gamma)|$$

is certainly Borel and we can define $c \in \mathcal{B}^{+,e}(G)$ by

$$c(\gamma) = \int_G |f(\eta)g(\eta^{-1}\gamma)|\,d\lambda^{r(\gamma)}(\eta) = \int_G |f(\gamma\eta)g(\eta^{-1})|\,d\lambda^{s(\gamma)}(\eta).$$

But just as in the group case we are only guaranteed that c is finite for λ^u-almost all γ for a given $u \in G^{(0)}$. However, it either f or g is bounded, it is clear that c is finite everywhere and we can define the convolution in the usual way

$$f * g(\gamma) = \int_G f(\eta)g(\eta^{-1}\gamma)\,d\lambda^{r(\gamma)}(\eta).$$

In fact, $f * g$ itself is bounded in this case. Then the usual computations show that $f * g \in \mathcal{L}^\infty(G^{(0)}, L^I(G,\lambda))$. In fact, repeating the proof in Ex 1.4.4, we have $\|f * g\|_I \le \|f\|_I \|g\|_I$. Also $(f * g)^* = g * f$. Therefore if g is bounded, it is not so hard to check that $(f_1 * g) * f_2 = f_1 * (g * f_2)$. Fortunately, this will be enough for our purposes.

PROPOSITION 11.10. *We can extend the representation $\underline{R}$ to a $*$-map $\underline{R}'$: $\mathcal{L}^\infty(G^{(0)}, L^I(G,\lambda)) \to B(L^2(G^{(0)} * \mathscr{H}, \mu)$ by the formula*

$$\underline{R}'(f)h(u) = \int_G f(\gamma)V_\gamma(h(s(\gamma)))\Delta(\gamma)^{-\frac{1}{2}}\,d\lambda^u(\gamma).$$

Furthermore, $\|\underline{R}'(f)\| \le \|f\|_I$ and if $f, g \in \mathcal{L}^\infty(G^{(0)}, L^I(G,\lambda))$ with one of them bounded, then $\underline{R}'(f * g) = \underline{R}'(g)\underline{R}'(f)$. In particular, if $g \in C_c(G)$, then we have $\underline{R}'(f * g) = \underline{R}'(f)\underline{R}(g)$ and $\underline{R}'(g * f) = \underline{R}(g)\underline{R}'(f)$.

PROOF. We can invoke the proof of Proposition 7.12 on page 163 *mutatis mutandis* to see that $\underline{R}'(f)$ is an operator with norm at most $\|f\|_I$. Similarly, $\underline{R}'(f^*) = \underline{R}'(f)^*$ and if either f or g is bounded then $\underline{R}'(f * g) = \underline{R}'(f)\underline{R}'(g)$. □

LEMMA 11.11. *Suppose that* $f \in C_c(G)$ *and* $\Phi \in C_c(\Sigma)$. *Then we can define* $\Phi \cdot f \in \mathcal{B}^b_c(G) \subset \mathcal{L}^\infty(G^{(0)}, L^I(G,\lambda))$ *by*

$$\Phi \cdot f(\gamma) = \int_{G(r(\gamma))} \Phi(r(\gamma), G(r(\gamma)), t) f(t^{-1}\gamma)\, d\beta^{r(\gamma)}(t).$$

Furthermore, $\|\Phi \cdot f\|_I \le \|\Phi\|_I \|f\|_I$, *and* $\underline{R}'(\Phi \cdot f) = r(\Phi)\underline{R}(f)$ *where*

$$\|\Phi\|_I = \max\Bigl\{\sup_{H\in\Sigma_0} \int_H |\Phi(p(H), H, t)|\, d\beta^H(t), \sup_{H\in\Sigma_0} \int_H |\Phi(p(H), H, t^{-1})|\, d\beta^H(t)\Bigr\}.$$

PROOF. Note that

$$\begin{aligned}
&\int_G |\Phi \cdot f(\gamma)|\, d\lambda_u(\gamma) \\
&\quad\le \int_G \int_{G(r(\gamma))} |\Phi(r(\gamma), G(r(\gamma)), t) f(t^{-1}\gamma)|\, d\beta^{r(\gamma)}(t)\, d\lambda_u(\gamma) \\
&\quad= \int_G \omega(\gamma^{-1}, G(r(\gamma))) \int_{G(u)} |\Phi(r(\gamma), G(r(\gamma)), \gamma t \gamma^{-1}) f(\gamma t^{-1})|\, d\beta^u(t)\, d\lambda_u(\gamma)
\end{aligned}$$

which, after switching the order of integration twice and sending $\gamma \mapsto \gamma t$, is

$$= \int_G \int_{G(u)} \omega(t^{-1}\gamma^{-1}, G(r(\gamma))) |\Phi(r(\gamma), G(r(\gamma)), \gamma t \gamma^{-1})|\, d\beta^u(t) |f(\gamma)|\, d\lambda_u(\gamma)$$

which, since $\omega(t^{-1}\gamma^{-1}, G(r(\gamma))) = \omega(\gamma^{-1}, G(r(\gamma)))\Delta_{G(u)}(t)^{-1}$, is

$$\begin{aligned}
&= \int_G \omega(\gamma^{-1}, G(r(\gamma))) \int_{G(u)} |\Phi(r(\gamma), G(r(\gamma)), \gamma t^{-1} \gamma^{-1})|\, d\beta^u(t) |f(\gamma)|\, d\lambda_u(\gamma) \\
&= \int_G |f(\gamma)| \int_{G(r(\gamma))} |\Phi(r(\gamma), G(r(\gamma)), t^{-1})|\, d\beta^{r(\gamma)}(t)\, \lambda^u(\gamma) \\
&\le \|\Phi\|_I \|f\|_I.
\end{aligned}$$

Hence $\|\Phi \cdot f\|_{I,s} \le \|\Phi\|_I \|f\|_I$. A much easier computation shows that $\|\Phi \cdot f\|_{I,r} \le \|\Phi\|_I \|f\|_I$. Thus $\Phi \cdot f \in \mathcal{L}^\infty(G^{(0)}, L^I(G,\lambda))$.

Using some basic properties of vector-valued integrals as in [**Wil07**, §1.5], we have

$$\begin{aligned}
&\underline{R}'(\Phi \cdot f)(h(u)) \\
&\quad= \int_G \Phi \cdot f(\gamma) V_\gamma(h(s(\gamma))) \Delta^{-\frac{1}{2}}(\gamma)\, d\lambda^u(\gamma) \\
&\quad= \int_G \int_{G(u)} \Phi(u, G(u), t) f(t^{-1}\gamma) V_\gamma(h(s(\gamma))) \Delta(\gamma)^{-\frac{1}{2}}\, d\beta^u(t)\, d\lambda^u(\gamma)
\end{aligned}$$

which, in view of left invariance and Proposition 11.6, is

$$\begin{aligned}
&= \int_{G(u)} \Phi(u, G(u), t) \Bigl(\int_G f(\gamma) V_{t\gamma}(h(s(\gamma))) \Delta_{G(u)}(t)^{-\frac{1}{2}} \Delta(\gamma)^{-\frac{1}{2}}\, d\lambda^u(\gamma) \Bigr) \beta^u(t) \\
&= \int_{G(u)} \Phi(u, G(u), t) V_t\bigl(\underline{R}'(f) h(u)\bigr) \Delta_{G(u)}(t)^{-\frac{1}{2}}\, d\beta^u(t) \\
&= r(\Phi) \underline{R}'(f)(h(u)).
\end{aligned}$$

□

Dealing with R and Discontinuity. To paraphrase Renault from [**Ren91**, §1], we are going to need techniques to deal with the discontinuity of the isotropy map $u \mapsto G(u)$ and the potential for a pathological orbit space $G^{(0)}/G$.

Since we're assuming G is second countable, both $G^{(0)}$ and Σ_0 are Polish spaces. Hence Lusin's Theorem (such as stated in [**Par78**, Corollary 3.3.22]) guarantees that for every $\epsilon > 0$ and any finite measure μ on $G^{(0)}$, there is a compact set $L \subset G^{(0)}$ such that $u \mapsto G(u)$ is continuous on L and $\mu(G^{(0)} \setminus L) < \epsilon$.[5] The following comes from [**Ren91**, Proposition 1.11] which references [**GR79**, Theorem 1.4].

LEMMA 11.12. *Fix a diagonally compact symmetric neighborhood K of $G^{(0)}$ in G, a neighborhood M of the diagonal $\Delta \subset G^{(0)} \times G^{(0)}$, and a compact set $L \subset G^{(0)}$ such that $u \mapsto G(u)$ is continuous on L. Then each point $u_0 \in L$ has an open neighborhood V in L such that "$u \sim_K v$ if there is a $\gamma \in K$ such that $\pi(\gamma) = (u, v)$" is an equivalence relation on V such that*

(a) *$u \sim_K v$ implies $(u, v) \in M$ and*
(b) *if $U \subset V$ is open then the saturation $[U] := \{ w \in V : w \sim_K u \text{ and } u \in U \}$ is open in V. (That is, $\sim_K$ is an open equivalence relation on V.)*

PROOF. Let $u_0 \in L$, and let $K' = G' \cdot K \cap r^{-1}(L)$. We'll proceed by establishing a sequence of claims.

CLAIM 1. K' is a neighborhood of $G(u_0)$ in $r^{-1}(L)$.

PROOF. Let $\{\gamma_i\}$ be a net in $r^{-1}(L)$ converging to $t \in G(u_0)$. It will suffice to see that there is a subnet eventually in K'. Since the isotropy map is continuous on L, we have $G(r(\gamma_i)) \to G(u_0)$. By Lemma 3.22 on page 81, we can pass to a subnet, relabel, and assume that there are $t_i \in G(r(\gamma_i))$ such that $t_i \to t$. But then $t_i^{-1}\gamma_i \to u_0 \in K$. Hence $t_i^{-1}\gamma_i$ is eventually in K. That means $\gamma_i = t_i(t_i^{-1}\gamma_i)$ is eventually in K'. □

Since $G^{(0)}$ is metrizable, it is straightforward to find a neighborhood M_0 of $\Delta \subset G^{(0)} \times G^{(0)}$ such that $M_0^2 \subset M$ (Ex 11.2.1). In addition, Lemma 4.12 on page 102 allows us to find a diagonally compact neighborhood K_0 of $G^{(0)}$ in G such that $K_0^2 \subset K$.

CLAIM 2. There is an open neighborhood N of $G(u_0)$ in $\pi^{-1}(L)$ such that $N^2 \subset \pi^{-1}(M) \cap K'$.

PROOF. Let $K_0' = G' \cdot K_0 \cap r^{-1}(L)$ be as in Claim 1. Let N be an open neighborhood of $G(u_0)$ in $\pi^{-1}(M_0) \cap K_0'$. Suppose that $\gamma', \eta' \in N$ are composible. Say $\gamma' = t\gamma$ and $\eta' = s\eta$ with $t \in G(r(\gamma))$, $s \in G(r(\eta)) = G(s(\gamma))$, and $\gamma, \eta \in K_0$. Then $\gamma'\eta' = t(\gamma s \gamma^{-1})\gamma\eta$. Since $\gamma s \gamma^{-1} \in G(r(\gamma))$ and $\gamma\eta \in K$, we have $\gamma'\eta' \in K'$. Our choice of M_0 then implies that $\gamma'\eta' \in \pi^{-1}(M) \cap K'$. □

CLAIM 3. There is a neighborhood V of u_0 in L such that $\gamma \in K|_V = \{ \gamma \in K : r(\gamma) \in V \text{ and } s(\gamma) \in V \}$ implies $\gamma \in N$.

PROOF. If the claim were false, then for every neighborhood V of u_0 in L, there is a $\gamma_V \in K|_V \subset K \cap r^{-1}(L)$ such that $\gamma_V \notin N$. But $K \cap r^{-1}(L)$ is compact, so we

[5] This does not say that $u \mapsto G(u)$ is continuous at points of L as a map on $G^{(0)}$. The characteristic function of the rationals is nowhere continuous, but is continuous when restricted to the irrationals.

can pass to a subnet, relabel, and assume that $\gamma_V \to \gamma \in K$ with $r(\gamma) = u_0 = s(\gamma)$. That is, $\gamma \in G(u_0)$. Since N is a neighborhood of $G(u_0)$ in $r^{-1}(L)$, γ_V must eventually be in N. This contradicts our choices and completes the proof. □

We want to see that V has the required properties. Since K is a symmetric neighborhood of $G^{(0)}$, it is clear that $\sim_K$ is reflexive and symmetric. Suppose $u, v, w \in V$ are such that $u \sim_K v \sim_K w$. Then there are $\gamma, \eta \in K$ such that $\pi(\gamma) = (u, v)$ and $\pi(\eta) = (v, w)$. By Claim 3, $\gamma, \eta \in N$. Hence $\gamma\eta \in \pi^{-1}(M) \cap K'$. Then $\gamma\eta = t\gamma'$ with $\gamma' \in K$ and $\pi(\gamma') = (v, w)$. Thus $v \sim_K w$. This shows that $\sim_K$ is an equivalence relation. We also have shown part (a), as $u \sim_K v$ implies $(u, v) = \pi(\gamma)$ with $\gamma \in N \subset \pi^{-1}(M)$.

For part (b), note that $[U] = U \cdot K \cap V = U \cdot N \cap V$, and the latter is open in V. □

We also need a technical result about the support of the measures $\underline{\alpha}^u$ restricted to compact subsets of $G^{(0)}$. (The measures $\underline{\alpha}^u$ are given by (11.5).) Specifically, we want the following.

LEMMA 11.13. *Let K be a diagonally compact neighborhood of $G^{(0)}$ in G and $L \subset G^{(0)}$ compact. Then there is a Borel set $L' \subset L$—called the support of $\alpha|_L$—such that $\mu(L \setminus L') = 0$ and such that if $\psi \in C^+(L)$, $u \in L'$, and $\psi(u) > 0$, then*

$$\int_{[u]} \psi(y)\, d\underline{\alpha}^u(y) = \int_{\pi(K)} \psi(u \cdot \gamma)\, d\alpha^u(\pi(\gamma)) > 0.$$

PROOF. Let $N(u)$ be the union of relatively open sets $U \subset [u] \cap L$ such that $\underline{\alpha}^u(U) = 0$. Let $\iota_u : G^u \to [u]$ be the continuous surjection $\gamma \mapsto u \cdot \gamma$. Then $\iota_u^{-1}(N(u))$ is relatively open in G^u_L and hence σ-compact. Hence $N(u)$ is the increasing union of compact sets K_n in $[u] \cap L$. The Monotone Convergence Theorem implies that $\underline{\alpha}^u(N(u)) = 0$. Let $L(u)$ be the complement of $N(u)$ in $[u] \cap L$. Note that $L(u)$ has the property that if U is relatively open in $[u] \cap L$ and $U \cap L(u) \neq \emptyset$, then $\underline{\alpha}^u(U) > 0$. On the other hand, if $U \cap L(u) = \emptyset$, then $\underline{\alpha}^u(U) = 0$.

Let

$$L' = \{\, u \in L : u \in L(u) \,\}.$$

To see that L' is a Borel set, let d be a metric on $G^{(0)}$ compatible with the topology and let $N_i = \{\, (u, v) \in G^{(0)} \times G^{(0)} : d(u, v) < \frac{1}{i} \,\}$. Let $M_i = N_i \cap (G^{(0)} \times L)$. Then M_i is Borel and

$$u \mapsto \int_R \mathbb{1}_{M_i}(u, u \cdot \gamma)\, d\alpha^u(\pi(\gamma)) = \int_R \mathbb{1}_{M_i}(u, v)\, d\underline{\alpha}^u(v)$$

is a Borel function. Let

$$B_i = \{\, u \in L : \int_R \mathbb{1}_{M_i}(u, u \cdot \gamma)\, d\alpha^u(\pi(\gamma)) > 0 \,\}.$$

Then B_i is Borel, and since M_i is a neighborhood of (u, u) in $G^{(0)} \times L$, we have $L' \subset B_i$. On the other hand, if $u \notin L(u)$, then since $L(u)$ is closed in $[u]$, there is an open neighborhood U of u in $[u] \cap L$ disjoint from $L(u)$. Then $(M_i)_u = \{\, v : (u, v) \in M_i \,\} = \{\, v \in L : d(u, v) < \frac{1}{i} \,\}$ is a neighborhood of u in $[u] \cap L$ and must eventually be contained in U. It follows that u is eventually not in B_i and that

$$L' = \bigcap B_i$$

is Borel.

Let $\kappa = \mu \circ \alpha$ be as in (11.6). Then on the one hand due to the definition of $L(u)$,

$$\kappa(L \times L \setminus L') = \int_L \int_R \mathbb{1}_{L\times L\setminus L'}(u, u \cdot \gamma) \, d\alpha^u(\pi(\gamma)) \, d\mu(u) = 0.$$

Since κ is equivalent to κ^{-1}, we also have

$$\begin{aligned} 0 &= \kappa(L \setminus L' \times L) \\ &= \int_{G^{(0)}} \int_R \mathbb{1}_{L\setminus L'\times L}(u, u \cdot \gamma) \, d\alpha^u(\pi(\gamma)) \, d\mu(u) = \int_{L\setminus L'} \alpha^u(R) \, d\mu(u). \end{aligned}$$

Thus we can conclude that $\mu(L \setminus L') = 0$.

Now if $\psi \in C^+(L)$, $u \in L'$ and $\psi(u) > 0$, then $U = \{ v \in L : \psi(v) > 0 \} \cap [u]$ is a neighborhood of u in $[u] \cap L$. Hence the result. □

LEMMA 11.14. *Fix K, M, and L as in Lemma 11.12 on page 290. We let L' be the conull Borel subset of L coming from Lemma* 11.13. *Let $\mathscr{V} = \{ V_1, \dots, V_n \}$ be a cover of L by open sets such that $\sim_K$ is an open equivalence relation on each V_i as in Lemma* 11.12. *Also, let $\mathcal{P} = \{ \psi_1, \dots, \psi_n \}$ be a partition of unity in $C^+(L)$ subordinate to $\mathscr{V}$. Then there are $e_i \in \mathcal{B}^+(G^{(0)})$ such that e_i vanishes off V_i and for all $u \in L$ we have*

$$\sum_{j=1}^{n} e_j(u) \int_{\pi(K)} e_j(u \cdot \gamma) \, d\alpha^u(\pi(\gamma)) \leq 1 \tag{11.10}$$

with equality whenever $u \in L'$.

PROOF. Define a Borel function on $G^{(0)}$ by

$$\varphi_i(u) = \int_{\pi(K)} \psi_i(u \cdot \gamma) \, d\alpha^u(\pi(\gamma)).$$

Suppose $u \cdot \gamma \in V_i$ with $\gamma \in K$. Then I claim

$$u \cdot K \cap V_i = u \cdot \gamma \cdot K \cap V_i. \tag{11.11}$$

Suppose v belongs to the right-hand side. Then there is an $\eta \in K$ such that $v = u \cdot \gamma\eta$. In particular, $u \sim_K s(\gamma) = r(\eta) \sim_K s(\eta) = v$. Then $u \sim_K v$ which means v belongs to the left-hand side. On the other hand, if v belongs to the left-hand side, then there is an $\eta \in K$ such that $u \cdot \eta = v$. Thus $u \sim_K v = s(\eta)$ and $u \sim_K s(\gamma)$. Hence $s(\gamma) \sim_K v$ and there is an $\eta' \in K$ such that $r(\eta') = s(\gamma)$ and $s(\eta') = s(\eta) = v$. Therefore $v = (u \cdot \gamma) \cdot \eta'$ and v belongs to the right-hand side.

Since $\{ u \in G^{(0)} : \varphi_i(u) > 0 \}$ is Borel, we can define a Borel function by

$$e_i(u) = \begin{cases} \psi_i(u)\varphi_i(u)^{-\frac{1}{2}} & \text{if } \varphi_i(u) > 0, \text{ and} \\ 0 & \text{otherwise.} \end{cases}$$

Since e_i vanishes when ψ_i does, e_i vanishes off V_i.

I claim that

$$e_i(u \cdot \gamma) = \varphi_i(u)^{-\frac{1}{2}} \psi_i(u \cdot \gamma) \quad \text{if } \pi(\gamma) \in \pi(K). \tag{11.12}$$

This is clear if $\psi_i(u\cdot\gamma)=0$. Otherwise $u\cdot\gamma\in V_i$ and we can use (11.11) to compute that

$$\begin{aligned}\varphi_i(u)&=\int_{\pi(K)}\psi_i(u\cdot\gamma)\,d\alpha^u(\pi(\gamma))\\&=\int_{u\cdot K\cap V_i}\psi_i(v)\,d\underline{\alpha}^u(v)\\&=\int_{u\cdot\gamma\cdot K\cap V_i}\psi_i(v)\,d\underline{\alpha}^u(v)\end{aligned}$$

which, since $\underline{\alpha}^u=\underline{\alpha}^{u\cdot\gamma}$, is

$$=\varphi_i(u\cdot\gamma).$$

Hence (11.12) is always satisfied.

Thus if $\varphi_i(u)>0$, then

$$e_i(u)\int_{\pi(K)}e_i(u\cdot\gamma)\,d\alpha^u(\pi(\gamma))=\psi_i(u)\varphi_i(u)^{-1}\int_{\pi(K)}\psi_i(u\cdot\gamma)\,d\alpha^u(\pi(\gamma))=\psi_i(u).$$

If $\varphi_i(u)=0$, then the left-hand side is zero. Since $\sum\psi_i(u)=1$ if $u\in L$, we get the inequality in (11.10).

Moreover if $u\in L'$ and $\psi_i(u)>0$ then $\varphi_i(u)>0$ by Lemma 11.13 on page 291. Hence we get equality in (11.10) for all $u\in L'$ as required. □

Let $f\in C_c(G)$. Then we get a continuous function on $G*\Sigma_0$ given by

$$\Phi(\gamma,H)=\int_H f(\gamma t)\,d\beta^H(t).$$

Note that $\Phi(\gamma,H)=\Phi(\gamma t,H)$ for all $t\in H$. It is not hard to see that Φ is bounded: if it weren't, then we could find (γ_i,H_i) in the support of f such that $|\Phi(\gamma_i,H_i)|\to\infty$. Since $\operatorname{supp} f$ is compact, the continuity of Φ leads to a contradiction. This allows us to define a real constant

$$C(f,M)=\sup\Bigl\{\int_{G(s(\gamma))}|f(\gamma t)|\,d\beta^{s(\gamma)}:(r(\gamma),s(\gamma))\in M\Bigr\}.$$

When M is all of $G^{(0)}\times G^{(0)}$, we write simply $C(f)$.

Lemma 11.15. *Fix a triple (K,M,L) together with the ancillary data L', $\mathscr{V}$, $\mathscr{P}$ and corresponding $\{e_1,\dots,e_n\}$ as in Lemma 11.14 on the preceding page. For each $f\in C_c(G)$, define*

$$Q(f)(\gamma)=Q(K,M,L)(f)(\gamma)=\sum_{i=1}^n e_i(r(\gamma))f(\gamma)e_i(s(\gamma)).\tag{11.13}$$

If $f\in C_c(G)$ with $\operatorname{supp} f\subset K$, then

$$\|Q(f)\|_I\le\max\{C(f,M),C(f^*,M)\}.$$

PROOF. Note that $Q(f)(\gamma)=0$ if $r(\gamma)\notin L$. If $u\in L$, then

$$\begin{aligned}\lambda(|Q(f)|)(u) &\le \sum_i e_i(u)\int_K |f(\gamma)|e_i(s(\gamma))\,d\lambda^u(\gamma)\\ &= \sum_i e_i(u)\int_{\pi(K)}\int_{G(s(\gamma))} |f(\gamma t)|\,d\beta^{s(\gamma)}(t)e_i(s(\gamma))\,d\alpha^u(\pi(\gamma))\\ &\le C(f,M)\sum_i e_i(u)\int_{\pi(K)} e_i(s(\gamma))\,d\alpha^u(\pi(\gamma))\\ &\le C(f,M).\end{aligned}$$

Hence $\|Q(f)\|_{I,r}=\sup_{u\in L}\lambda(Q(f))(u)\le C(f,M)$.

The result follows since $\|Q(f)\|_{I,s}=\|Q(f^*)\|_{I,r}$. □

LEMMA 11.16. *Let $f\in C_c(G)$ and define $P(f)\in C_c(\Sigma)$ by*

$$P(f)(u,H,t)=f(t).$$

Let $\mathscr{Q}$ be the collection of maps $Q=Q(K,M,L)$ defined in (11.13) as K varies over diagonally compact symmetric neigbhorhoods of $G^{(0)}$ in G containing $\operatorname{supp} f$*, M varies over neighborhoods of the diagonal $\Delta\subset G^{(0)}\times G^{(0)}$ and L varies over compact subsets of $G^{(0)}$ such that the isotropy map $u\mapsto G(u)$ is continuous on L. Then the operator $r(P(f))$ on $L^2(G^{(0)}*\mathscr{H},\mu)$ is in the weak operator closure of $\{\,\underline{R}'(Q(f)) : Q\in\mathscr{Q}\,\}$.*

PROOF. Given $\epsilon'>0$ and $h_1',\dots,h_m',k_1,\dots,k_m\in L^2(G^{(0)}*\mathscr{H},\mu)$, it will suffice to find $Q\in\mathscr{Q}$ such that

$$\big|\big(\underline{R}'(Q(f))h_i'\mid k_i\big)-\big(r(P(f))h_i'\mid k_i\big)\big|<\epsilon'\quad\text{for all } i.$$

Since $\underline{R}$ is nondegerate and $\|Q(f)\|_I$ is bounded by the maximum of $C(f)$ and $C(f^*)$ independent of the choice of $Q\in\mathscr{Q}$, we can assume $h_i'=\underline{R}(g_i)h_i$ for $g_i\in C_c(G)$ and $h_i\in L^2(G^{(0)}*\mathscr{H},\mu)$.

Using Proposition 11.10 on page 288 and Lemma 11.11 on page 289, we have

$$\underline{R}'(Q(f))h_i'=\underline{R}'(Q(f)*g_i)h_i\quad\text{and}\quad r(P(f))h_i'=\underline{R}'(P(f)\cdot g_i)h_i.$$

Fix an $\epsilon>0$ to be named later. Use Lusin's Theorem to find $L\subset G^{(0)}$ compact such that $u\mapsto G(u)$ is continuous on L and

$$\int_{G^{(0)}\setminus L}\|k_i(u)\|^2\,d\mu(u)<\epsilon^2\quad\text{for all } i.$$

Define $\underline{\xi}_i$ on $G*_r G$ by

$$\underline{\xi}_i(\eta,\gamma)=\int_{G(s(\eta))} f(\eta t)g_i(t^{-1}\eta^{-1}\gamma)\,d\beta^{s(\eta)}(t). \tag{11.14}$$

Clearly $\underline{\xi}_i(\eta,\gamma)$ depends only on $\pi(\eta)$ and we'll write $\xi_i(\pi(\eta),\gamma)$ for (11.14) in the sequel. Using Lemma 3.25 on page 82 and that $r(\eta)=r(\gamma)$, we see that

$$\xi_i(\pi(\eta),\gamma)=\omega(\eta^{-1},G(r(\gamma)))^{-1}\int_{G(r(\gamma))} f(t\eta)g_i(\eta^{-1}t^{-1}\gamma)\,d\beta^{r(\gamma)}(t).$$

It follows from Lemma 3.29 on page 83 that ξ_i is continuous on $R*_r r^{-1}(L)$. Now we assume that K is a diagonally compact symmetric neighborhood of $G^{(0)}$ in G such that $\operatorname{supp} f\operatorname{supp} g_i\subset K$ for all i. I claim there is a neighborhood M of

$\Delta \subset G^{(0)} \times G^{(0)}$ such that $(\pi(\eta), \gamma) \in R *_r r^{-1}(L)$, $(\eta, \gamma) \in K *_r K$, and $\pi(\eta) \in M$ implies that

$$|\xi_i(\pi(\eta), \gamma) - \underline{\xi}_i(r(\gamma), \gamma)| < \epsilon.$$

If not, we can find neighborhoods $\{M_j\}$ and $(\eta_j, \gamma_j) \in K \cap r^{-1}(L) \times K \cap r^{-1}(L)$ such that

$$|\xi_i(\pi(\eta_j), \gamma_j) - \underline{\xi}_i(r(\gamma_j), \gamma_j)| \geq \epsilon.$$

However, since K is digaonally compact, we can pass to a subnet and assume $(\eta_j, \gamma_j) \to (\eta, \gamma)$. Since $\pi(\eta_j) \in M_j$ for all j and $r(\eta_j) = r(\gamma_j)$ for all j, we must have $\pi(\eta) = (r(\gamma), r(\gamma))$. Now the continuity of ξ_i leads to a contradiction.

We now proceed with these choices of K, M and L and fix $Q = Q(K, M, L)$ with appropriate ancillary data.

Then

$$\begin{aligned}\big(\underline{R}'(Q(f))h_i' \mid k_i\big) &- \big(r(P(f))h_i' \mid k_i\big) \\ &= \big(\underline{R}'(Q(f) * g_i - P(f) \cdot g_i)h_i \mid k_i\big) \\ &= \int_{G^{(0)}} \big(\underline{R}'(Q(f) * g_i - P(f) \cdot g_i)h_i(u) \mid k_i(u)\big)\, d\mu(u).\end{aligned}$$

Then we can write the last integral as $A+B$ where A is the integral over $G^{(0)} \setminus L$ and B is the integral over the support L' of $\alpha|_L$ from Lemma 11.13 on page 291 (since $\mu(L \setminus L') = 0$).

As in the remarks preceding Proposition 11.10 on page 288, we have

$$\|Q(f) * g_i\|_I \leq \|Q(f)\|_I \|g_i\|_I \leq \max\{\, C(f), C(f^*)\,\} \|g_i\|_I,$$

and

$$\|P(f) \cdot g_i\|_I \leq \|P(f)\|_I \|g_i\|_I$$

by Lemma 11.11 on page 289. Hence

$$|A| \leq C\|h_i\|\epsilon$$

where C depends only on f and g_i.

On the other hand,

$$|B| \leq \|\mathbb{1}_{r^{-1}(L')} \cdot (Q(f) * g_i - P(f) \cdot g_i)\|_I \|h_i\| \|k_i\|.$$

Since $\mathbb{1}_{r^{-1}(L')} \cdot (Q(f) * g_i - P(f) \cdot g_i)$ vanishes off $K \cap r^{-1}(L')$ and since K is symmetric,

$$\begin{aligned}\|\mathbb{1}_{r^{-1}(L')} &\cdot (Q(f) * g_i - P(f) \cdot g_i)\|_I \\ &\leq M \sup\{\, |Q(f) * g_i(\gamma) - P(f) \cdot g_i(\gamma)| : \gamma \in K \cap r^{-1}(L')\,\},\end{aligned}$$

where

$$M = \sup_{u \in L'} \lambda^u(K \cap r^{-1}(L')) \leq \sup_{u \in L} \lambda^u(K \cap r^{-1}(L)) < \infty$$

(because K is diagonally compact).

It follows that we just have to be able to make the supremum

$$\sup\{\, |Q(f) * g_i(\gamma) - P(f) \cdot g_i(\gamma)| : \gamma \in K \cap r^{-1}(L')\,\}$$

small. Towards this end we check that on the one hand,

$$
\begin{aligned}
Q(f) * g_i(\gamma)& \\
&= \int_G Q(f)(\eta) g_i(\eta^{-1}\gamma)\, d\lambda^{r(\gamma)}(\eta) \\
&= \sum_i e_i(r(\gamma)) \int_K f(\eta) e_i(s(\eta)) g_i(\eta^{-1}\gamma)\, d\lambda^{r(\gamma)}(\eta) \\
&= \sum_i e_i(r(\gamma)) \int_{\pi(K)} \int_{G(s(\eta))} f(\eta t) g_i(t^{-1}\eta^{-1}\gamma)\, d\beta^{s(\eta)}(t) e_i(s(\eta))\, d\alpha^{r(\gamma)}(\pi(\gamma)) \\
&= \sum_i e_i(r(\gamma)) \int_{\pi(K)} \xi_i(\pi(\eta),\gamma) e_i(s(\eta))\, d\alpha^{r(\gamma)}(\pi(\eta))
\end{aligned}
$$

On the other hand,

$$
P(f) \cdot g_i(\gamma) = \int_{G(r(\gamma))} f(t) g_i(t^{-1}\gamma)\, d\beta^{r(\gamma)}(t) = \underline{\xi}_i(r(\gamma),\gamma)
$$

which by Lemma 11.14, provided $r(\eta) = r(\gamma) \in L'$, is

$$
= \sum e_j(r(\gamma)) \int_{\pi(K)} e_j(s(\eta))\, d\alpha^{r(\gamma)}(\eta) \xi_i(r(\gamma),\gamma).
$$

Therefore if $\gamma \in K \cap r^{-1}(L')$,

$$
\begin{aligned}
|Q(f) * g_i(\gamma) - P(f) \cdot g_i(\gamma)|& \\
&\le \sum e_j(r(\gamma)) \int_{\pi(K)} e_j(s(\eta)) |\xi_i(\pi(\eta),\gamma) - \underline{\xi}_i(r(\gamma),\gamma)|\, d\alpha^{r(\gamma)}(\pi(\eta)) \\
&< \epsilon,
\end{aligned}
$$

where the last inequality is due to the observation that $\pi(\eta) \in V_j \times V_j$ implies $\pi(\eta) \in M$ and hence $|\xi_i(\pi(\eta),\gamma) - \underline{\xi}_i(r(\gamma),\gamma)| < \epsilon$.

This completes the proof. □

THEOREM 11.17. *Let $\underline{R}$ be a nondegenerate representation of $C^*(G)$ and $\operatorname{Ind} r$ the representation induced from the restriction of $\underline{R}$ to the isotropy as defined in Section 11.2. Then*

$$
\ker \underline{R} \subset \ker(\operatorname{Ind} r).
$$

PROOF. Let $a \in C^*(G)$. Fix $g_1 \otimes h$ and $g_2 \otimes k$ in $L^2(G^{(0)} * \widetilde{\mathscr{H}}, \mu)$ and $\epsilon > 0$. It will suffice to find $h_1, \dots, h_n, k_1, \dots, k_n$ in $L^2(G^{(0)} * \mathscr{H}, \mu)$ such that

$$
\Big| \big(\operatorname{Ind} r(a)(g_1 \otimes h) \mid g_2 \otimes h\big) - \sum_i \big(\underline{R}(a) h_i \mid k_i\big) \Big| < \epsilon.
$$

Then if $a \in \ker \underline{R}$, we have $\operatorname{Ind} r(a) = 0$ as required.

Let $f \in C_c(G)$ be such that

$$
\big| \big(\operatorname{Ind} r(a)(g_1 \otimes h) \mid g_2 \otimes k\big) - \big(\operatorname{Ind} r(f)(g_1 \otimes h) \mid g_2 \otimes k\big) \big| < \frac{\epsilon}{3} \tag{11.15}
$$

and such that

$$
C(g_1^* * g_1)^{\frac{1}{2}} C(g_2^* * g_2)^{\frac{1}{2}} \|f - a\| \|h\| \|k\| < \frac{\epsilon}{3}. \tag{11.16}
$$

Note that

$$
\big(\operatorname{Ind} r(f)(g_1 \otimes h) \mid g_2 \otimes k\big) = \big(r(P(g_2^* * f * g_1) h \mid k\big).
$$

Hence by Lemma 11.16 on page 294, we can find $Q \in \mathscr{Q}$ such that

$$\bigl|\bigl(\operatorname{Ind} r(f)(g_1 \otimes h) \mid g_2 \otimes k\bigr) - \bigl(\underline{R}'(Q(g_2^* * f * g_1))h \mid k\bigr)\bigr| < \frac{\epsilon}{3}. \tag{11.17}$$

Next we need to introduce the functions

$$\tilde{g}_{i,k}(\gamma) = g_i(\gamma)e_k(s(\gamma)).$$

We need to see that $\tilde{g}_{i,k} \in \mathcal{L}^\infty(G^{(0)}, L^I(G,\lambda))$. But

$$\|\tilde{g}_{i,k}\|_I^2 \leq \|\tilde{g}_{i,k}^* * \tilde{g}_{i,k}\|_I$$

which, since $g_{i,j}^* * \tilde{g}_{i,j}$ is self-adjoint, is

$$\begin{aligned}
&= \|\tilde{g}_{i,k}^* * \tilde{g}_{i,k}\|_{I,r} \\
&\leq \sum_{j=1}^n \|\tilde{g}_{i,j}^* * \tilde{g}_{i,j}\|_{I,r} \\
&\leq \sup_u \int_G \int_G \sum_{j=1}^n e_j(r(\gamma))|g_i(\eta^{-1})g_i(\eta^{-1}\gamma)|e_j(s(\gamma)\, d\lambda^u(\eta)\, d\lambda^u(\gamma) \\
&= \sup_u \int_G Q(|g_i^* * g_i|)(\gamma)\, d\lambda^u(\gamma) \\
&\leq \|Q(|g_i^* * g_i|)\|_I \\
&\leq C(|g_i^* * g_i|).
\end{aligned}$$

Then we can compute that

$$\begin{aligned}
\bigl(\underline{R}'(Q(g_2^* * f * g_1))h \mid k\bigr) &= \sum_{j=1}^n \bigl(\underline{R}'(\tilde{g}_{2,j}^* * f * \tilde{g}_{1,j})h \mid k\bigr) \\
&= \sum_{j=1}^n \bigl(\underline{R}(f)h_j \mid k_j\bigr),
\end{aligned}$$

where

$$h_j = \underline{R}'(\tilde{g}_{1,j})h \quad \text{and} \quad k_j = \underline{R}'(\tilde{g}_{2,j})k.$$

Moreover,

$$\begin{aligned}
\sum_j \|h_j\|^2 &= \sum_j \bigl(\underline{R}'(\tilde{g}_{1,j}^* * \tilde{g}_{1,j})h \mid h\bigr) \\
&= \bigl(\underline{R}'(Q(g_1^* * g_1))h \mid h\bigr) \\
&\leq C(g_1^* * g_1)\|h\|^2.
\end{aligned}$$

Similarly,

$$\sum_j \|k_j\|^2 \leq C(g_2^* * g_2)\|k\|^2.$$

It follows that for any $b \in C^*(G)$,

$$\begin{aligned}\Bigl|\sum_j \bigl(\underline{R}(b)h_i \mid k_i\bigr)\Bigr| &\le \|b\| \sum_{i=1}^n \|h_i\|\|k_i\| \\ &\le \|b\| \Bigl(\sum \|h_j\|^2\Bigr)^{\frac12} \Bigl(\sum \|h_j\|^2\Bigr)^{\frac12} \\ &\le \|b\| C(g_1^* * g_1)^{\frac12} C(g_2^* * g_2)^{\frac12} \|h\|\|k\|.\end{aligned}$$

Thus, letting $b = a - f$, in view of (11.15) we have

$$\Bigl|\sum \bigl(\underline{R}(a)h_i \mid k_i\bigr) - \sum \bigl(\underline{R}(f)h_i \mid k_i\bigr)\Bigr| < \frac{\epsilon}{3}. \tag{11.18}$$

Combining (11.16), (11.17) and (11.18) gives

$$\Bigl|\bigl(\operatorname{Ind} r(a)(g_1 \otimes h) \mid g_2 \otimes k\bigr) - \sum \bigl(\underline{R}(a)h_i \mid k_i\bigr)\Bigr| < \epsilon$$

as required. □

Exercises

Ex 11.2.1. Let X be a metric space and G the trivial groupoid $X \times X$. Let M be a neighborhood of the diagonal $\Delta \subset X \times X$. Show that there is a neighborhood M_0 of Δ such that $M_0^2 \subset M$.

—Answer on page 383

11.3. Renault's Simplicity Result

In order to prove Theorem 11.24 on page 302—and hence Theorem 11.1 on page 279—we are going to have to require the isotropy be discrete in a strong sense to be specified below. Fortunately, this will be automatic if the groupod is étale. Recall that a locally compact group $\mathcal{H}$ is discrete exactly when the identity element is open in $\mathcal{H}$. Thus the following definition of "discrete at v" is a bit stronger than simply requiring the isotropy groups $G(u) = G_u^u$ to be discrete near $v \in G^{(0)}$.

Definition 11.18 (Renault). We say that the isotropy of G is *discrete at* $v \in G^{(0)}$ if there is a neighborhood U of v in G such that $u \in U \cap G^{(0)}$ implies $U \cap G(u) = \{u\}$. We say that the isotropy is *discretely trivial at* $v \in G^{(0)}$ if given any compact set $K \subset G$ there is a neighborhood V of v in $G^{(0)}$ such that $u \in V$ implies $K \cap G(u) \subset \{u\}$.

We leave the following as an exercise (Ex 11.3.1).

Lemma 11.19. *Let G be a locally compact groupoid.*

(a) *If the isotropy is discretely trivial at v, then it is discrete at v.*
(b) *If the isotropy is discrete at v and $G(v) = \{v\}$, then the isotropy is discretely trivial at v.*
(c) *If G is étale and $G(v) = \{v\}$, then this isotropy is discretely trivial at v.*

Therefore for étale groupoids, having discretely trivial isotropy is that same as having trivial isotropy. In general, as we'll see in Example 11.23 on page 301, having trivial isotropy at u does not imply that the isotropy is discretely trivial at u.

PROPOSITION 11.20. *Suppose that G has discretely trivial isotopy at $u_0 \in G^{(0)}$. Let $\underline{R}$ be a nondegenerate representation of $C^*(G)$ and let* $\operatorname{Ind} r$ *be the associated representation of $C^*(G)$ obtained in Section* 11.2 *by restricting $\underline{R}$ to the isotropy groupoid G' and inducting. Suppose that $u_0 \in \operatorname{supp} M_{\underline{R}}$ where $M_{\underline{R}}$ is the representation of $C_0(G^{(0)})$ associated to $\underline{R}$ as in* (5.1)*. Then*

$$\ker(\operatorname{Ind} r) \subset \ker\bigl(\operatorname{Ind} \delta_{u_0}\bigr).$$

REMARK 11.21. We may as well assume that $\underline{R}$ is the integrated form of a unitary representation on $L^2(G^{(0)} * \mathscr{H}, \mu)$. Then $\operatorname{Ind} r$ acts on $L^2(G^{(0)} * \widetilde{\mathscr{H}}, \mu)$ for the same quasi-invariant measure μ. In particular, the support of $M_{\operatorname{Ind} r}$ and the support of $M_{\underline{R}}$ are both equal to the support of μ.

PROOF. For notational convenience, let $L = \operatorname{Ind} r$. Thus we can realize L as the representation on the completion $L^2(G^{(0)} * \overline{\mathscr{H}}, \mu)$ of $C_c(G) \odot L^2(G^{(0)} * \mathscr{H}, \mu)$ with respect to the inner product

$$\begin{aligned}(\varphi \otimes h \mid \psi \otimes k) &= \int_{G^{(0)}} \int_{G(u)} \psi^* * \varphi(t) \bigl(V_t(h(u)) \mid k(u)\bigr)\, \Delta_{G(u)}(t)^{-\frac{1}{2}} \\ &\qquad\qquad d\beta^u(t)\, d\mu(u). \\ &= \int_{G^{(0)}} \int_{G(u)} \int_G \overline{\psi(\gamma^{-1})} \varphi(\gamma^{-1}t) \bigl(V_t(h(u)) \mid k(u)\bigr)\, \Delta_{G(u)}(\eta)^{-\frac{1}{2}} \\ &\qquad\qquad d\lambda^u(\gamma)\, d\beta^u(t)\, d\mu(u),\end{aligned}$$

and $L(f)(\varphi \otimes h) = f * \varphi \otimes h$. In particular, the vector state corresponding to $\xi = \varphi \otimes h$ is given by

$$\begin{aligned}\omega_\xi(f) &= \bigl(L(f)\xi \mid \xi\bigr) \\ &= \int_{G^{(0)}} \int_{G(u)} \int_G \int_G \overline{\varphi(\gamma^{-1})} f(\eta) \varphi(\eta^{-1}\gamma^{-1}t) \bigl(V_t(h(u)) \mid h(u)\bigr)\, \Delta_{G(u)}(\eta)^{-\frac{1}{2}} \\ &\qquad\qquad d\lambda^{s(\gamma)}(\eta)\, d\lambda^u(\gamma)\, d\beta^u(t)\, d\mu(u) \qquad (11.19) \\ &= \int_{G^{(0)}} \int_{G(u)} \int_G \int_G \overline{\varphi(\gamma^{-1})} f(\gamma^{-1}\eta) \varphi(\eta^{-1}t) \bigl(V_t(h(u)) \mid h(u)\bigr)\, \Delta_{G(u)}(\eta)^{-\frac{1}{2}} \\ &\qquad\qquad d\lambda^u(\eta)\, d\lambda^u(\gamma)\, d\beta^u(t)\, d\mu(u)\end{aligned}$$

when $f \in C_c(G)$.

On the other hand, $\operatorname{Ind} \delta_{u_0}$ acts on $L^2(G, \lambda_{u_0})$ via convolution: $(\operatorname{Ind} \delta_{u_0}(f)(h) = f * h$ provided $f \in C_c(G)$. To show that $a \in \ker \operatorname{Ind} \delta_{u_0}$ it suffices to see that the vector state

$$\zeta(a) := \bigl((\operatorname{Ind} \delta_{u_0}(a))\varphi \mid \varphi\bigr) = 0 \qquad (11.20)$$

for all $\varphi \in C_c(G) \subset L^2(G, \lambda_{u_0})$ with $\|\varphi\|_2 = 1$. Note that if $f \in C_c(G)$, then computing as above

$$\zeta(f) = \int_G \int_G \overline{\varphi(\gamma^{-1})} f(\gamma^{-1}\eta) \varphi(\eta^{-1})\, d\lambda^{u_0}(\eta)\, d\lambda^{u_0}(\xi).$$

The idea of the proof is to choose a *bounded sequence* ξ_n so that $\omega_n := \omega_{\xi_n}$ converges weakly to ζ in (11.20); that is so that $\omega_n(a) \to \zeta(a)$ for all $a \in C^*(G)$. The result follows immediately from this. Since the ω_n are bounded, it suffices to do this for $f \in C_c(G)$. For the remainder of the proof, we fix $\varphi \in C_c(G)$ and $f \in C_c(G)$

Note that if W is diagonally compact and K is any compact subset of G, then $W \cup K$ is diagonally compact. Hence G has an increasing family $\{K_n\}$ of symmetric diagonally compact neighborhoods of $G^{(0)}$ whose interiors cover G. Since the isotropy is discretely trivial at u_0, it is discrete at u_0 and we can find a relatively compact open set V in G so that $V \cap G(u) = \{u\}$ for all $u \in V$. In particular, $G(u)$ is discrete for all $u \in V \cap G^{(0)}$. In addition, we can find a neighborhood V_n of u_0 in $G^{(0)}$ such that $u \in V_n$ implies $K_n^3 \cap G(u) \subset K_n^3 \cap r^{-1}(\overline{V \cap G^{(0)}}) \cap G(u) = \{u\}$.

There is no harm in assuming $V_{n+1} \subset V_n \subset V$ and that $\{V_n\}$ is a neighborhood basis at u_0.

Note that if $u \in V$, then $G(u)$ is discrete and β^u must be a multiple $\tau(u)$ of counting measure. Also, $\Delta_{G(u)} \equiv 1$ if $u \in V$. Furthermore, letting $\psi \in C_c^+(G)$ be such that $\operatorname{supp}\psi \subset V$ and $\psi(u_0) = 1$, it follows from Lemma 6.10 on page 145 that $\tau(u)\psi(u) = \int_{G(u)} \psi(t)\,d\beta^u(t) \to \int_{G(u_0)} \psi(t)\,d\beta^{u_0}(t) = \tau(u_0)$ if $u_i \to u$. Hence τ is continuous at u_0.

We can find n_0 such that $n \geq n_0$ implies that $\operatorname{supp}\varphi$ and $\operatorname{supp} f$ are subsets of K_n, and we assume $n \geq n_0$ from now on.

In view of Remark 11.21 on the preceding page, $u_0 \in \operatorname{supp}\mu$ and we can find $h_n \in B(G^{(0)} * \mathscr{H}, \mu)$ such that h_n vanishes off V_n and $u \mapsto h_n(u)$ has norm 1 in $L^2(G^{(0)} * \mathscr{H}, \mu)$. We let $\xi_n = \varphi \otimes h_n$.

Now, using (11.19),

$$\omega_n(f) = \int_{G^{(0)}} \sum_{t \in G(u)} \tau(u) \int_G \int_G \overline{\varphi(\gamma^{-1})} f(\gamma^{-1}\eta)\varphi(\eta^{-1}t)\big(V_t(h_n(u)) \mid h_n(u)\big) \\ d\lambda^u(\eta)\,d\lambda^u(\gamma)\,d\mu(u).$$

In the integrand, we can assume that $u \in V$, and that γ^{-1}, $\gamma^{-1}\eta$ and $\eta^{-1}t$ are all in $K_n = K_n^{-1}$. Hence $t \in K_n^3$ and we are only concerned with $t = u$. Hence we get

$$\omega_n(f) = \int_{G^{(0)}} \int_G \int_G \overline{\varphi(\gamma^{-1})} f(\gamma^{-1}\eta)\varphi(\eta^{-1})\,d\lambda^u(\eta)\,d\lambda^u(\gamma)\tau(u)\|h_n(u)\|^2\,d\mu(u)$$

provided $n \geq n_0$.

Let $\epsilon > 0$. Since λ is a Haar system,

$$u \mapsto \int_G \int_G \overline{\varphi(\gamma^{-1})} f(\gamma^{-1}\eta)\varphi(\eta^{-1})\,d\lambda^u(\eta)\,d\lambda^u(\gamma) := A(u)$$

is continuous. Since τ is continuous at u_0, it follows that for large n, $u \in V_n$ implies that

$$|\tau(u)A(u) - \tau(u_0)A(u_0)| < \epsilon.$$

Therefore for large n.

$$|\omega_n(f) - \zeta(f)| \leq \int_{G^{(0)}} |\tau(u)A(u) - \tau(u_0)A(u_0)|\|h_n(u)\|^2\,d\mu(u) \leq \epsilon.$$

This will complete the proof once we verify that the $\{\xi_n\}$ are bounded in the space of L. But by Lemma 6.10 on page 145, we can assume that

$$u \mapsto \|\varphi^* * \varphi\|_{L^1(G(u),\beta^u)}$$

is bounded by M if $u \in V_n$. Hence for large n

$$\begin{aligned}\|\xi_n\|^2 &= \|\varphi \otimes h_n\|^2 \\ &= \Big| \int_{G^{(0)}} \sum_{t \in G(u)} \varphi^* * \varphi(t)\big(V_t(h_n(u)) \mid h_n(u)\big)\tau(u)\, d\mu(u)\Big| \\ &\leq \int_{V_n} \|\varphi^* * \varphi\|_{L^1(G(u),\beta^u)} \|h_n(u)\|^2\, \mu(u) \\ &\leq M. \end{aligned}$$

□

By combining Proposition 11.20 on page 299 with Theorem 11.17 on page 296, we obtain the following.

THEOREM 11.22 ([**Ren91**, Theorem 4.3]). *Suppose that G has discretely trivial isotopry at $u \in G^{(0)}$. Let $\underline{R}$ be a nondegenerate representation of $C^*(G)$ such that $u \in \operatorname{supp} M_{\underline{R}}$. Then*

$$\ker \underline{R} \subset \ker \operatorname{Ind} \delta_u.$$

Recall that if G is étale, then the isotropy is discretely trivial at u exactly when $G(u)$ is trivial. The following example due to Renault shows that, for general G, assuming that $G(u)$ is trivial is not sufficient to conclude that $\ker \underline{R} \subset \ker \operatorname{Ind} \delta_u$ when u belongs to the support of $M_{\underline{R}}$.

EXAMPLE 11.23 ([**Ren91**, Remark 4.5]). Let $G = \{\, z \in C : |z| \leq 1 \,\}$ be closed disk in the complex plane viewed as a group bundle over $[0,1]$: here $(ae^{i\theta})(ae^{i\varphi}) = ae^{i(\theta+\varphi)}$. The isotropy groups are

$$G(a) = \begin{cases} a\mathbf{T} & \text{if } a \in (0,1], \text{ and} \\ \{0\} & \text{if } a = 0. \end{cases}$$

Hence, the isotropy is trivial at 0, but not discrete at 0 let alone discretely trivial.

It seems clear that the isotropy maps is continuous, and in any event,

$$\lambda^a(f) = \frac{1}{2\pi} \int_0^{2\pi} f(ae^{i\theta})\, d\theta$$

is a Haar system on G (of course $\lambda^0 = \delta_0$). The regular representation $\operatorname{Ind} \delta_0$ is just evaluation at 0: $\operatorname{Ind}(\delta_0)(f) = f(0)$. Note that if $f \in C(G)$, then

$$\|f\|_I = \sup_{a \in [0,1]} \frac{1}{2\pi} \int_0^{2\pi} |f(ae^{i\theta})|\, d\theta.$$

If $a \in (0,1]$ and $f \in C(G)$, let

$$\hat{f}_a(1) = \frac{1}{2\pi} \int_0^{2\pi} f(ae^{i\theta}) e^{-i\theta}\, d\theta$$

be the Fourier coefficient of $e^{i\theta} \mapsto f(ae^{i\theta})$. Then $|\hat{f}_a(1)| \leq \|f\|_I$, and $(f * g)_a^\wedge(1) = \hat{f}_a(1)\hat{g}_a(1)$. Hence we get a representation of $C(G)$ on $L^2[0,1]$ given by

$$\underline{R}(f)\xi(a) = \begin{cases} \hat{f}_a(1)\xi(a) & \text{if } a \in (0,1] \text{ and} \\ 0 & \text{if } a = 0. \end{cases}$$

Moreover, $M_{\underline{R}}$ is just the multiplication representation $M_{\underline{R}}(\varphi)\xi(a) = \varphi(a)\xi(a)$ and has support all of $[0,1]$. In particular, 0 belongs to the support of $M_{\underline{R}}$ and $G(0)$ is trivial.

Nevertheless the function $f(z) = 1$ for all $z \in G$ is in $\ker \underline{R}$, but not in $\ker \operatorname{Ind} \delta_0$. That is,

$$\ker \underline{R} \not\subset \ker \operatorname{Ind} \delta_0.$$

Therefore the conclusion of Theorem 11.22 can fail without the hypotheses of discrete triviality.

Recall that we say G acts minimally on $G^{(0)}$ if there are no nontrivial closed invariant subsets of $G^{(0)}$. This is equivalent to every orbit $[u]$ being dense.

THEOREM 11.24 ([**Ren91**, Corollary 4.9]). *Suppose that the action of G on $G^{(0)}$ is minimal and there is a $u \in G^{(0)}$ where the isotropy is discretely trivial. Then $C^*_r(G)$ is simple.*

PROOF. Let $\underline{R}$ be any representation of $C^*(G)$. Since $\operatorname{supp} M_{\underline{R}}$ is a closed invariant subset of $G^{(0)}$ by Proposition 5.5 on page 111, $\operatorname{supp} M_{\underline{R}} = G^{(0)}$. By Theorem 11.22 on the previous page, $\ker \underline{R} \subset \ker \operatorname{Ind} \delta_u$. However, if $v \in [u]$, then $\operatorname{Ind} \delta_v$ is equivalent to $\operatorname{Ind} \delta_u$ by Ex 1.4.8. It follows that $\ker \underline{R} \subset \ker \operatorname{Ind} \delta_v$ for a dense set of $v \in G^{(0)}$. Hence $\ker \underline{R}$ is contained in the kernel of the natural map $\iota : C^*(G) \to C^*_r(G)$ by Ex 5.3.3. This suffices as $\underline{R}$ was arbitrary. □

EXERCISES

Ex 11.3.1. Prove Lemma 11.19.

—Answer on page 384

11.4. The Amenable Case

In Theorem 11.17 on page 296, we showed that for any representation $\underline{R}$ of $C^*(G)$, we have $\ker \underline{R} \subset \ker \operatorname{Ind} r$. If G is amenable, we can sharpen this to assert that the kernels are equal.

THEOREM 11.25. *Suppose that G is a second countable* amenable *groupoid with a Haar system. Let $\underline{R}$ be a nondegenerate representation of $C^*(G)$ on a separable Hilbert space and* $\operatorname{Ind} r$ *the corresponding representation of $C^*(G)$ induced from the isotropy groupoid as in Section 11.2 on page 286. Then*

$$\ker \underline{R} = \ker \operatorname{Ind} r.$$

REMARK 11.26. With a bit more effort, we can weaken our assumption on G to assume only measurewise amenability. We sketch the necessary modifications to the proof for measurewise amenable G in Section 11.4.2, and proceed here with the amenable case for the sake of clarity.

In view of Theorem 11.17 on page 296, we just have to show that $\ker \operatorname{Ind} r \subset \ker \underline{R}$. As before, we realize $\underline{R}$ as the integrated form on a unitary representation $(\hat{V}, G^{(0)} * \mathscr{H}, \mu)$. But it will be handy to have a concrete realization of the space of $\operatorname{Ind} r$—or more precisely, the space of a representation equivalent to $\operatorname{Ind} r$. This will take some work. Once we completed these preliminaries, we'll prove Theorem 11.25 in Section 11.4.1 on page 308.

We start by forming the pull-back $G * \mathscr{H}_s$ of $G^{(0)} * \mathscr{H}$ via the source map $s : G \to G^{(0)}$ (see [**Wil07**, Example F.18]). Recall that if $\{f_i\}$ is a fundamental sequence for $G^{(0)} * \mathscr{H}$, then $\{f_i \circ s\}$ is a fundamental sequence for $G * \mathscr{H}_s$. Thus $\xi \in B(G * \mathscr{H}_s)$ if and only if $\xi(\gamma) \in \mathcal{H}(s(\gamma))$ for all γ and

$$\gamma \mapsto \big(\xi(\gamma) \mid f_i(s(\gamma))\big)$$

is Borel for all i. We let $B_V(G * \mathscr{H}_s)$ denote those $\xi \in B(G * \mathscr{H}_s)$ such that

$$\xi(\gamma t) = V_t^{-1}(\xi(\gamma)) \quad \text{for all } \gamma \in G \text{ and } t \in G(s(\gamma)). \tag{11.21}$$

We let $\mathcal{L}^2_V(G * \mathscr{H}_s, \kappa^{-1})$ be those $\xi \in B_V(G * \mathscr{H}_s)$ such that

$$\|\xi\|^2 = \int_R \|\xi(\gamma)\|^2 \, d\kappa^{-1}(\pi(\gamma)) = \int_{G^{(0)}} \int_R \|\xi(\gamma)\|^2 \, d\alpha_u(\pi(\gamma)) \, d\mu(u) < \infty.$$

We will also need to consider $\mathcal{L}^2_V(G_u * \mathscr{H}_s, \alpha_u)$ and $\mathcal{L}^2_V(G^u * \mathscr{H}_s, D^{-1} \cdot \alpha^u)$ for the functions on G_u and G^u, respectively, with

$$\int_G \|\xi(\gamma)\|^2 \, d\alpha_u(\pi(\gamma)) < \infty \quad \text{and} \quad \int_G \|\xi(\gamma)\|^2 D(\pi(\gamma))^{-1} \, d\alpha^u(\pi(\gamma)) < \infty.$$

LEMMA 11.27. *If $\xi, \zeta \in \mathcal{L}^2_V(G * \mathscr{H}_s, \kappa^{-1})$, then the set F for which $\pi(\gamma) \mapsto \big(\xi(\gamma) \mid \zeta(\gamma)\big)$ is α_u integrable is μ-conull and*

$$u \mapsto \begin{cases} \int_R \big(\xi(\gamma) \mid \zeta(\gamma)\big) \, d\alpha_u(\pi(\gamma)) & \text{if } u \in F, \text{ and} \\ 0 & \text{otherwise} \end{cases}$$

is in $\mathcal{L}^1(\mu)$. In particular,

$$(\xi \mid \zeta) := \int_R \big(\xi(\gamma) \mid \zeta(\gamma)\big) \, d\kappa^{-1}(\pi(\gamma)) = \int_{G^{(0)}} \int_R \big(\xi(\gamma) \mid \zeta(\gamma)\big) \, d\alpha_u(\pi(\gamma)) \, d\mu(u)$$

is a pre-inner product on $\mathcal{L}^2_V(G * \mathscr{H}_s, \kappa^{-1})$.

PROOF. Let $\hat{\xi}(\gamma) = \|\xi(\gamma)\|$ and $\hat{\zeta}(\gamma) = \|\zeta(\gamma)\|$. Then both $\hat{\xi}$ and $\hat{\zeta}$ are in $\mathcal{L}^2(\kappa^{-1})$. It then follows from the Cauchy-Schwarz inequality that $\gamma \mapsto \big(\xi(\gamma) \mid \zeta(\gamma)\big)$ is in $\mathcal{L}^1(\kappa^{-1})$. The rest follows from the observation that

$$\int_R g(\pi(\gamma)) \, d\kappa(\pi(\gamma)) = \int_{G^{(0)}} \int_R g(\pi(\gamma)) \, d\alpha^u(\pi(\gamma)) \, d\mu(u)$$

for all g for which both sides make sense (see Proposition 3.6 on page 71). □

REMARK 11.28. We'll need the following observations and notations concerning the α^u's and κ. If $\xi \in \mathcal{L}^2_V(G * \mathscr{H}_s, \kappa^{-1})$, then $\xi|_{G_u} \in \mathcal{L}^2_V(G_u * \mathscr{H}_s, \alpha_u)$ for almost all u, and

$$\|\xi\|^2 = \int_{G^{(0)}} \|\xi\|^2_{\mathcal{L}^2_V(G_u * \mathscr{H}_s, \alpha_u)} \, d\mu(u).$$

But we also have

$$\begin{aligned} \|\xi\|^2 &= \int_R \|\xi(\gamma)\|^2 \, d\kappa^{-1}(\pi(\gamma)) \\ &= \int_R \|\xi(\gamma)\|^2 D(\pi(\gamma))^{-1} \, d\kappa(\pi(\gamma)) \\ &= \int_{G^{(0)}} \int_R \|\xi(\gamma)\|^2 D(\pi(\gamma))^{-1} \, d\alpha^u(\pi(\gamma)) \, d\mu(u). \end{aligned}$$

Hence $\xi|_{G^u}$ is in $\mathcal{L}^2_V(G^u * \mathscr{H}_s, D^{-1}\cdot\alpha^u)$ for almost all u and

$$\|\xi\|^2 = \int_{G^{(0)}} \|\xi\|^2_{\mathcal{L}^2_V(G^u*\mathscr{H}_s, D^{-1}\cdot\alpha^u)}\,d\mu(u).$$

Then we'll write $L^2_V(G * \mathscr{H}_s, \kappa^{-1})$ for the completion of $\mathcal{L}^2_V(G * \mathscr{H}_s, \kappa^{-1})$ with respect to $\|\cdot\|$ defined in (11.4).[6] Of course, at this point, $L^2_V(G * \mathscr{H}_s, \kappa^{-1})$ could be the zero space. However, if $\varphi\otimes h \in C_c(G)\odot L^2(G^{(0)} * \mathscr{H}, \mu)$, then we can define $W(\varphi\otimes h) \in B_V(G * \mathscr{H}_s)$ by

$$W(\varphi\otimes h)(\eta) = \int_{G(s(\eta))} \varphi(\eta t)\omega(\eta t, G(s(\eta)))^{-\frac12} V_t(h(s(\gamma)))\,d\beta^{s(\eta)}(t).$$

We want to see that $W(\varphi\otimes h) \in \mathcal{L}^2_V(G * \mathscr{H}_s, \kappa^{-1})$. To this end, recall that the representation induced from the isotropy, $\operatorname{Ind} r$, corresponding to $\underline{R}$ acts on the completion of $C_c(G)\otimes L^2(G^{(0)} * \mathscr{H}, \mu)$ with respect to the inner-product

$$\begin{aligned}
&\big(\varphi\otimes h \mid \psi\otimes k\big)\\
&= \int_{G^{(0)}} \psi^* * \varphi(t)\Delta(t)^{-\frac12}\big(V_t(h(u)) \mid k(u)\big)\,d\beta^u(t)\,d\mu(u).\\
&= \int_{G^{(0)}}\int_{G(u)}\int_G \overline{\psi(\eta^{-1})}\varphi(\eta^{-1}t)\Delta(t)^{-\frac12}\big(V_t(h(u)) \mid k(u)\big)\,d\lambda^u(\eta)\,d\beta^u(t)\,d\mu(u)\\
&= \int_{G^{(0)}}\int_{G(u)}\int_R\int_{G(s(\eta))} \overline{\psi(s^{-1}\eta^{-1})}\varphi(s^{-1}\eta^{-1}t)\Delta(t)^{-\frac12}\big(V_t(h(u)) \mid k(u)\big)\\
&\qquad\qquad d\beta^{s(\eta)}(s)\,d\alpha^u(\pi(\eta))\,d\beta^u(t)\,d\mu(u)\\
&= \int_{G^{(0)}}\int_{G(u)}\int_R \omega(\eta^{-1}, G(u))^{-1}\int_{G(u)} \overline{\psi(\eta^{-1}s^{-1})}\varphi(\eta^{-1}s^{-1}t)\Delta(t)^{-\frac12}\\
&\qquad\qquad \big(V_t(h(u)) \mid k(u)\big)\,d\beta^u(s)\,d\alpha^u(\pi(\eta))\,d\beta^u(t)\,d\mu(u)
\end{aligned}$$

which, after Fubini and a change of variables, is

$$\begin{aligned}
&= \int_{G^{(0)}}\int_R\int_{G(u)}\int_{G(u)} \omega(\eta^{-1}, G(u))^{-1}\overline{\psi(\eta^{-1}s^{-1})}\varphi(\eta^{-1}t)\Delta(st)^{-\frac12}\\
&\qquad\qquad \big(V_t(h(u)) \mid V_{s^{-1}}(k(u))\big)\,d\beta^u(t)\,d\beta^u(s)\,d\alpha^u(\pi(\eta))\,d\mu(u)
\end{aligned}$$

which, after sending $s\mapsto s^{-1}$, is

$$\begin{aligned}
&= \int_{G^{(0)}}\int_R\int_{G(u)}\int_{G(u)} \omega(\eta^{-1}, G(u))^{-1}\overline{\psi(\eta^{-1}s)}\varphi(\eta^{-1}t)\Delta(s^{-1}t)^{-\frac12}\Delta(s)^{-1}\\
&\qquad\qquad \big(V_t(h(u)) \mid V_s(k(u))\big)\,d\beta^u(t)\,d\beta^u(s)\,d\alpha^u(\pi(\eta))\,d\mu(u)\\
&= \int_{G^{(0)}}\int_R\int_{G(u)}\int_{G(u)} \omega(\eta, G(u))^{-1}\overline{\psi(\eta s)}\varphi(\eta t)\Delta(t)^{-\frac12}\Delta(s)^{-\frac12}\\
&\qquad\qquad \big(V_t(h(u)) \mid V_s(k(u))\big)\,d\beta^u(t)\,d\beta^u(s)\,d\alpha_u(\pi(\eta))\,d\mu(u)\\
&= \int_{G^{(0)}}\int_G \big(W(\varphi\otimes h)(\eta) \mid W(\psi\otimes k)(\eta)\big)\,d\alpha_u(\eta)\,d\mu(u).
\end{aligned}$$

---

[6]Note the $\|\cdot\|$ is only a seminorm on $\mathcal{L}^2_V(G * \mathscr{H}_s, \kappa^{-1})$. Hence the completion process first mods out by the vectors of length zero. Of course, we'd like to claim that $L^2_V(G * \mathscr{H}_s, \kappa^{-1})$ *is* the set of almost-everywhere equivalence classes in $\mathcal{L}^2_V(G * \mathscr{H}_s, \kappa^{-1})$, but we want to avoid the overhead of proving that.

Therefore W defines an isometry of the space $L^2(G^{(0)} * \widetilde{\mathscr{H}}, \mu)$ into $L^2_V(G * \mathscr{H}_s, \kappa^{-1})$.

We'll need the following technical observation. Recall that if μ is a Radon measure on X, then $f \in \mathcal{B}(X)$ is *locally in* $L^2(\mu)$ if $f|_K \in \mathcal{L}^2(\mu)$ for every compact set $K \subset X$.

LEMMA 11.29. *If $\xi \in \mathcal{L}^2_V(G * \mathscr{H}_s, \kappa^{-1})$, then ξ is locally in $L^2(\nu^{-1})$.*

PROOF. Suppose $g \in C_c(G)$. Then

$$(\gamma, H) \mapsto \int_H g(\gamma t)\, d\beta^H(t)$$

is a continuous compactly supported function on $G * \Sigma_0/\sim$ (see Lemma 3.29 on page 83). Hence it is bounded as is

$$F_g(\gamma) = \int_{G(s(\gamma))} g(\gamma t)\, d\beta^{s(\gamma)}(t).$$

Let $K \subset G$ be compact and $g \in C_c^+(G)$ such that $g(\gamma) = 1$ for all $\gamma \in K^{-1}$. Then

$$\begin{aligned}
\|\xi|_K\|_{\mathcal{L}^2(\nu^{-1})} &\leq \int_{G^{(0)}} \int_G g(\gamma^{-1}) \|\xi(\gamma)\|^2 \, d\lambda_u(\gamma)\, d\mu(u) \\
&= \int_{G^{(0)}} \int_G g(\gamma) \|\xi(\gamma^{-1})\| \, d\lambda^u(\gamma)\, d\mu(u) \\
&= \int_{G^{(0)}} \int_R \|\xi(\gamma)^{-1}\|^2 \int_{G(s(\gamma)} g(\gamma t)\, d\beta^{s(\gamma)}(t)\, d\alpha^u(\pi(\gamma))\, d\mu(u) \\
&\leq \|F_g\|_\infty \int_{G^{(0)}} \int_R \|\xi(\gamma)\|^2 \, d\alpha_u(\pi(\gamma))\, d\mu(u) \\
&= \|F_g\|_\infty \|\xi\|_{\mathcal{L}^2_V(G * \mathscr{H}_s, \kappa^{-1})}.
\end{aligned}$$

□

LEMMA 11.30. *The image $W(C_c(G) \otimes L^2(G^{(0)} * \mathscr{H}, \mu))$ is dense in $L^2_V(G * \mathscr{H}_s, \kappa^{-1})$, and hence W extends to a unitary also denoted by W from $L^2(G^{(0)} * \widetilde{\mathscr{H}}, \mu)$ onto $L^2_V(G * \mathscr{H}_s, \kappa^{-1})$.*

PROOF. Suppose that ξ is in the orthogonal complement of the image of W. We just need to see that this forces ξ to be zero in $L^2_V(G * \mathscr{H}_s, \kappa^{-1})$. However, for any φ and h, we have

$$\big(W(\varphi \otimes h) \mid \xi\big) = \int_{G^{(0)}} \int_R \int_{G(s(\eta))} \varphi(\eta t)\omega(\eta t, G(s(\eta)))^{-\frac{1}{2}} \big(V_t(h(s(\eta))) \mid \xi(\eta)\big)\, d\beta^{s(\eta)}(t)\, d\alpha_u(\pi(\eta))\, d\mu(u)$$

which, since $\xi(\eta) = V_t(\xi(\eta t))$, is

$$\begin{aligned}
&= \int_{G^{(0)}} \int_R \int_{G(s(\eta))} \varphi(\eta t)\omega(\eta t, G(s(\eta)))^{-\frac{1}{2}} \big(h(s(\eta)) \mid \xi(\eta t)\big) \\
&\qquad\qquad d\beta^{s(\eta)}(t)\, d\alpha_u(\pi(\eta))\, d\mu(u) \\
&= \int_{G^{(0)}} \int_R \int_{G(r(\eta)))} \varphi(\eta^{-1} t)\omega(\eta^{-1} t, G(r(\eta)))^{-\frac{1}{2}} \big(h(r(\eta)) \mid \xi(\eta^{-1} t)\big) \\
&\qquad\qquad d\beta^{r(\eta)}(t)\, d\alpha^u(\pi(\eta))\, d\mu(u)
\end{aligned}$$

which, after sending $\eta^{-1}t$ to $t\eta^{-1}$, is

$$= \int_{G^{(0)}} \int_R \int_{G(s(\eta)))} \varphi(t\eta^{-1})\omega(\eta^{-1}, G(r(\eta)))\omega(t\eta^{-1}, G(r(\eta)))^{-\frac{1}{2}} \big(h(r(\eta)) \mid \xi(t\eta^{-1})\big)\, d\beta^{r(\eta)}(t)\, d\alpha^u(\pi(\eta))\, d\mu(u)$$

which, since $\omega(t, G(s(\eta))) = \Delta_{G(s(\eta))}(t)$, is

$$= \int_{G^{(0)}} \int_R \int_{G(s(\eta)))} \varphi(t\eta^{-1})\omega(t\eta^{-1}, G(r(\eta)))^{\frac{1}{2}} \Delta_{G(s(\eta))}(t)^{-1} \big(h(r(\eta)) \mid \xi(t\eta^{-1})\big)\, d\beta^{r(\eta)}(t)\, d\alpha^u(\pi(\eta))\, d\mu(u)$$

$$= \int_{G^{(0)}} \int_R \int_{G(s(\eta)))} \varphi(t^{-1}\eta^{-1})\omega(t^{-1}\eta^{-1}, G(r(\eta)))^{\frac{1}{2}} \big(h(r(\eta)) \mid \xi(t^{-1}\eta^{-1})\big)\, d\beta^{r(\eta)}(t)\, d\alpha^u(\pi(\eta))\, d\mu(u)$$

$$= \int_{G^{(0)}} \int_G \varphi(\eta^{-1})\omega(\eta^{-1}, G(r(\eta)))^{\frac{1}{2}} \big(h(r(\eta)) \mid \xi(\eta^{-1})\big)\, d\lambda^u(\eta)\, d\mu(u)$$

$$= \int_{G^{(0)}} \int_G \varphi(\eta)\omega(\eta, G(r(\eta)))^{\frac{1}{2}} \big(h(s(\eta)) \mid \xi(\eta)\big)\, d\lambda_u(\eta)\, d\mu(u). \tag{11.22}$$

Since $(H, \eta) \mapsto \omega(\eta, H)$ is continuous, $\eta \mapsto \omega(\eta, G(r(\eta)))$ is bounded on compact sets. Using Lemma 11.29 on the preceding page, we see that

$$\eta \mapsto \omega(\eta, G(r(\eta)))^{\frac{1}{2}} \big(h(s(\eta)) \mid \xi(\eta)\big)$$

is locally in $\mathcal{L}^1(\nu^{-1})$. But for any h, (11.22) is zero for all $\varphi \in C_c(G)$. Since ω never vanishes, this means for all $h \in L^2(G^{(0)} * \mathscr{H}, \mu)$,

$$\big(h(s(\eta)) \mid \xi(\eta)\big)$$

is zero ν^{-1}-almost everywhere. But if $\{u_i\}$ is a special orthogonal fundamental sequence for $L^2(G^{(0)} * \mathscr{H}, \mu)$ (see [**Wil07**, Proposition F.6 and Remark F.7]), then

$$\|\xi(\gamma)\|^2 = \sum_i \big(\xi(\gamma) \mid u_i(s(\gamma))\big)\big(u_i(s(\gamma) \mid \xi(\gamma)\big).$$

Thus $\xi(\gamma) = 0$ for ν^{-1}-almost all γ, and hence for ν-almost all γ (since μ is quasi-invariant). But using (11.4) and a Bruhat approximate cross section as in (11.6), we have

$$\|\xi\|^2 = \int_{G^{(0)}} \int_G \|\xi(\gamma)\|^2\, d\alpha^u(\pi(\gamma))\, d\mu(u) = \int_G \|\xi(\gamma)\|^2 b(\gamma, G(s(\gamma)))\, d\nu(\gamma) = 0$$

as required. □

Since W is a unitary, it implements an equivalence of $\operatorname{Ind} r$ with a representation L on $L^2_V(G * \mathscr{H}_s, \kappa^{-1})$. Since $\operatorname{Ind} r(f)(\varphi \otimes h) = f * \varphi \otimes h$ and

$$W(f*\varphi \otimes h)(\gamma)$$

$$= \int_{G(s(\gamma))} f * \varphi(\gamma t)\omega(\gamma t, G(s(\gamma)))^{-\frac{1}{2}} V_t(h(s(\gamma)))\, d\beta^{s(\gamma)}(t)$$

$$= \int_{G(s(\gamma))} \int_G f(\eta)\varphi(\eta^{-1}\gamma t)\, d\lambda^{r(\gamma)}(\eta)\omega(\gamma t, G(s(\gamma)))^{-\frac{1}{2}} V_t(h(s(\gamma)))\, d\beta^{s(\gamma)}(t)$$

$$= \int_G f(\eta) \int_{G(s(\gamma))} \varphi(\eta^{-1}\gamma t)\omega(\gamma t, G(s(\gamma)))^{-\frac{1}{2}} V_t(h(s(\gamma)))\, d\beta^{s(\gamma)}(t)\, d\lambda^{r(\gamma)}(\eta)$$

which, after some ω-gymnastics using Lemma 3.25 (Ex 11.4.1), is

(11.23)
$$= \int_G f(\eta)\omega(\eta, G(s(\eta)))^{-\frac{1}{2}} \int_{G(s(\gamma))} \varphi(\eta^{-1}\gamma t)\omega(\eta^{-1}\gamma t, G(s(\gamma)))^{-\frac{1}{2}} V_t(h(s(\gamma)))\, d\beta^{s(\gamma)}(t)\, d\lambda^{r(\gamma)}(\eta)$$
$$= \int_G f(\eta)\omega(\eta, G(s(\eta)))^{-\frac{1}{2}} W(\varphi \otimes h)(\eta^{-1}\gamma)\, d\lambda^{r(\gamma)}(\eta).$$

All this suggests that if $f \in C_c(G)$, then $L(f)$ is given by the formula

$$L(f)\xi(\gamma) = \int_G f(\eta)\omega(\eta, G(s(\eta))^{-\frac{1}{2}}\xi(\eta^{-1}\gamma)\, d\lambda^{r(\gamma)}(\eta) \tag{11.24}$$

where, as is usual, we define the left-hand side to be zero if the integrand on the right-hand side is not integrable. To make sense of this, we need a technical lemma. It is interesting to notice that we make use of the equivalence of κ and κ^{-1} to make it possible to use Fubini.

LEMMA 11.31. *For $f \in C_c(G)$ and $\xi \in \mathcal{L}^2_V(G * \mathscr{H}_s, \kappa^{-1})$, the integral in (11.24) is defined for ν-almost all γ. Furthermore, we have*

$$\|L(f)\xi\| \le \|f\|_I \|\xi\|.$$

PROOF. For the first assertion, it suffices to see that

$$\int_G |f(\eta)|\omega(\eta, G(s(\eta)))^{-\frac{1}{2}} \|\xi(\gamma)\|\, d\lambda^{r(\gamma)}(\eta) \tag{11.25}$$

is finite almost everywhere. But if $g \in C_c(\Sigma_0 * G)$, then

$$(H, \gamma) \mapsto \int_H g(\gamma t, H)\, d\beta^H(t)$$

is continuous with compact support on $\Sigma_0 * G/|!\sim$. In particular,

$$F_f(\gamma) = \int_{G(s(\gamma)} f(\gamma t)\omega(\gamma t, s(\gamma))^{-\frac{1}{2}}\, d\beta^{s(\gamma)}(t)$$

is bounded for any $f \in C_c(G)$. But (11.25) equals

$$\int_R \|\xi(\eta^{-1}\gamma)\| \int_{G(s(\eta)} f(\eta t)\omega(\eta t, G(s(\eta)))^{-\frac{1}{2}}\, d\beta^{s(\eta)}(t)\, d\alpha^{r(\gamma)}(\pi(\eta))$$
$$\le \|F_f\|_\infty \int_R \|\xi(\eta^{-1}\gamma)\|\, d\alpha^{r(\gamma)}(\pi(\eta))$$
$$= \|F_f\|_\infty \int_R \|\xi(\eta^{-1})\|\, d\alpha^{s(\gamma)}(\pi(\eta))$$
$$= \|F_f\|_\infty \|\xi\|_{\mathcal{L}^2_V(G_{s(\gamma)} * \mathscr{H}_s, \alpha_{s(\gamma)})}$$

But $\|\xi\|_{\mathcal{L}^2_V(G_u * \mathscr{H}_s, \alpha_u)}$ is finite for μ-almost all u. This suffices.

Let $\zeta \in \mathcal{L}^2_V(G * \mathscr{H}_s, \kappa^{-1})$. Consider

$$\begin{aligned}
&\int_{G^{(0)}} \int_R \big|\big(L(f)\xi(\gamma) \mid \zeta(\gamma)\big)\big| \, d\alpha_u(\pi(\gamma))\, d\mu(u) \\
&\quad \le \int_{G^{(0)}} \int_R \int_G |f(\eta)| \omega(\eta, G(s(\eta)))^{-\frac{1}{2}} \big|\big(\xi(\eta^{-1}\gamma) \mid \zeta(\gamma)\big)\big| \\
&\qquad\qquad\qquad\qquad\qquad\qquad d\lambda^{r(\gamma)}(\eta)\, d\alpha_u(\pi(\eta))\, d\mu(u) \\
&\quad = \int_{G^{(0)}} \int_R \big|\big(L(f)\xi(\gamma) \mid \zeta(\gamma)\big)\big| D(\pi(\gamma))^{-1}\, d\alpha^u(\pi(\gamma))\, d\mu(u) \\
&\quad \le \int_{G^{(0)}} \int_R \int_G |f(\eta)| \omega(\eta, G(s(\eta)))^{-\frac{1}{2}} \big|\big(\xi(\eta^{-1}\gamma) \mid \zeta(\gamma)\big)\big| \\
&\qquad\qquad\qquad\qquad\qquad\qquad d\lambda^u(\eta) D(\pi(\gamma))^{-1}\, d\alpha^u(\pi(\gamma))\, d\mu(u) \\
&\quad = \int_{G^{(0)}} \int_G |f(\eta)| \omega(\eta, G(s(\eta)))^{-\frac{1}{2}} \\
&\qquad\qquad \int_G \big|\big(\xi(\eta^{-1}\gamma) \mid \zeta(\gamma)\big)\big| D(\pi(\gamma))^{-1}\, d\alpha^u(\pi(\gamma))\, d\lambda^u(\eta)\, d\mu(u)
\end{aligned}$$

which, after employing gymnastics with the α's as in Remark 11.28, is

$$\begin{aligned}
&\le \int_{G^{(0)}} \int_G |f(\eta)| \omega(\eta, G(s(\eta)))^{-\frac{1}{2}} \\
&\qquad \|\xi\|_{\mathcal{L}^2_V(G^{s(\eta)} * \mathscr{H}_s, D^{-1} \cdot \alpha^{s(\eta)})} \|\zeta\|_{\mathcal{L}^2_V(G^u * \mathscr{H}_s, D^{-1} \cdot \alpha^u)}\, d\lambda^u(\eta)\, d\mu(u)
\end{aligned}$$

which, in view of our choices of D and Δ in Proposition 11.6, is

$$= \int_{G^{(0)}} \int_G |f(\eta)| \Delta(\eta)^{-\frac{1}{2}} \|\xi\|_{D^{-1} \cdot \alpha^{s(\eta)}} \|\zeta\|_{D^{-1} \cdot \alpha^u}\, d\lambda^u(\eta)\, d\mu(u)$$

which, after applying Cauchy-Schwartz, is

$$\begin{aligned}
&\le \Big(\int_{G^{(0)}} \|\xi\|^2_{D^{-1} \cdot \alpha^{s(\eta)}} \int_G |f(\eta)| \Delta(\eta)^{-1}\, d\lambda^u(\eta)\, d\mu(u) \Big)^{\frac{1}{2}} \\
&\qquad\qquad\qquad \Big(\int_{G^{(0)}} \|\zeta\|^2_{D^{-1} \cdot \alpha^u} \int_G |f(\eta)|\, d\lambda^u(\eta)\, d\mu(u) \Big)^{\frac{1}{2}} \\
&= \|f\|^{\frac{1}{2}}_{I,s} \|\xi\| \|f\|_{I,r} \|\zeta\| \\
&\le \|f\|_I \|\xi\| \|\zeta\|.
\end{aligned}$$

This implies $L(f)\xi$ acts as a linear functional on $L^2_V(G * \mathscr{H}_s, \kappa^{-1})$ of norm at most $\|f\|_I \|\xi\|$. Hence $L(f)\xi$ is in $L^2_V(G * \mathscr{H}_s, \kappa^{-1})$ with norm at most $\|f\|_I \|\xi\|$. □

Putting the above technicalities together, we have proved the following.

PROPOSITION 11.32. *The representation* $\operatorname{Ind} r$ *is equivalent to the representation* L *on* $L^2_V(G * \mathscr{H}_s, \kappa^{-1})$ *given on* $\xi \in \mathcal{L}^2_V(G * \mathscr{H}_s, \kappa^{-1})$ *by*

$$L(f)\xi(\gamma) = \int_G f(\eta) \omega(\eta, G(\eta))^{-\frac{1}{2}} \xi(\eta^{-1}\gamma)\, d\lambda^{r(\gamma)}(\eta).$$

11.4.1. Proof of Theorem 11.25. As we observed at the beginning of the section, in view of Theorem 11.17, it will suffice to prove that $\ker \operatorname{Ind} r \subset \ker R$. Since the representation L defined in Proposition 11.32 is equivalent to $\operatorname{Ind} r$, it will suffice to show that for every unit vector $h \in L^2(G^{(0)} * \mathscr{H}, \mu)$ there is a net $\{\xi^{h,K}\}$ of unit vectors in $L^2_V(G * \mathscr{H}_s, \kappa^{-1})$, indexed by compact subsets of G, such that

$$\big(L(f)\xi^{h,K} \mid \xi^{h,K}\big) \to \big(R(f)h \mid h\big)$$

for all $f \in C_c(G)$. Hence given a compact set K and $\epsilon > 0$, it will suffice to find ξ^h such that

$$\big|\big(L(f)\xi^h \mid \xi^h\big) - \big(R(f)h \mid h\big)\big| < \epsilon \|f\|_I$$

for every f with $\operatorname{supp} f \subset K$.

We define ξ^h to be

$$\xi^h(\gamma) = D(\pi(\gamma))^{\frac{1}{2}} \Theta(\pi(\gamma)) V_\gamma^{-1}(h(r(\gamma))) \tag{11.26}$$

where $\Theta \in \mathcal{B}^+(R)$ is to be described later. Since V comes from a unitary representation of G, it is clear that any such ξ^h belongs to $B_V(G * \mathscr{H}_s)$. Since

$$\begin{aligned}\|\xi^h\|^2 &= \int_{G^{(0)}} \int_R \Theta(\pi(\gamma))^2 \|h(r(\gamma))\|^2 D(\pi(\gamma))\, d\alpha_u(\pi(\gamma))\, d\mu(u) \\ &= \int_{G^{(0)}} \|h(u)\|^2 \int_R \Theta(\pi(\gamma))^2\, d\alpha^u(\pi(\gamma))\, d\mu(u),\end{aligned}$$

we are going to require that

$$\int_R \Theta(\pi(\gamma))^2\, d\alpha^u(\pi(\gamma)) = 1 \quad \text{for all } u \in G^{(0)}.$$

This will insure that $\|\xi^h\| = 1$ as required.

If ξ^h is as in (11.26), then

$$\begin{aligned}&\big(L(f)\xi^h \mid \xi^h\big) \\ &= \int_{G^{(0)}} \int_R \int_G f(\eta)\omega(\eta, G(s(\eta)))^{-\frac{1}{2}} \big(\xi^h(\eta^{-1}\gamma) \mid \xi^h(\gamma)\big) \\ &\qquad\qquad d\lambda^{r(\gamma)}(\eta)\, d\alpha_u(\pi(\gamma))\, d\mu(u)\end{aligned}$$

which, since D is a homomorphism, is

$$\begin{aligned}&= \int_{G^{(0)}} \int_R \int_G f(\eta)\omega(\eta, G(s(\eta)))^{-\frac{1}{2}} D(\eta)^{-\frac{1}{2}} \Theta(\pi(\eta^{-1}\gamma))\Theta(\pi(\gamma)) \\ &\quad \big(V_{\eta^{-1}\gamma}^{-1}(h(s(\eta))) \mid V_\gamma^{-1}(h(r(\gamma)))\big)\, d\lambda^{r(\gamma)}(\eta) D(\pi(\gamma))\, d\alpha_u(\pi(\gamma))\, d\mu(u)\end{aligned}$$

which, using the definition of D and (11.7), is

$$\begin{aligned}&= \int_{G^{(0)}} \int_R \int_G f(\eta)\Delta(\eta)^{-\frac{1}{2}} \Theta(\pi(\eta^{-1}\gamma))\Theta(\pi(\gamma)) \big(V_\eta(h(s(\eta))) \mid h(u)\big) \\ &\qquad\qquad d\lambda^u(\eta)\, d\alpha^u(\pi(\gamma))\, d\mu(u) \\ &= \int_{G^{(0)}} \int_G \Big(\int_R \Theta(\pi(\eta^{-1}\gamma))\Theta(\pi(\gamma))\, d\alpha^u(\pi(\gamma))\Big) \\ &\qquad f(\eta)\big(V_\eta(h(s(\eta))) \mid h(u)\big)\Delta(\eta)^{-\frac{1}{2}}\, d\lambda^u(\eta)\, d\mu(u)\end{aligned} \tag{11.27}$$

On the other hand,

$$\big(R(f)h \mid h\big) = \int_{G^{(0)}} \int_G f(\eta)\big(V_\eta(h(s(\eta))) \mid h(u)\big)\Delta(\eta)^{-\frac{1}{2}}\, d\lambda^u(\eta)\, d\mu(u). \tag{11.28}$$

Comparing (11.27) and (11.28) shows that

$$\begin{aligned}
&\big|\big(L(f)\xi^h \mid \xi^h\big) - \big(R(f)h \mid h\big)\big| \\
&= \Big|\int_{G^{(0)}}\int_G\Big(\int_R \Theta(\pi(\eta^{-1}\gamma))\Theta(\pi(\gamma))\,d\alpha^u(\pi(\gamma) - 1\Big) \\
&\qquad\qquad f(\eta)\big(V_\eta(h(s(\eta))) \mid h(u)\big)\Delta(\eta)^{-\frac12}\,d\lambda^u(\eta)\,d\mu(u)\Big| \\
&\le \int_{G^{(0)}}\int_G\Big|\int_R \Theta(\pi(\eta^{-1}\gamma))\Theta(\pi(\gamma))\,d\alpha^u(\pi(\gamma) - 1\Big| \\
&\qquad\qquad |f(\eta)|\|h(s(\eta))\|\|h(u)\|\Delta(\eta)^{-\frac12}\,d\lambda^u(\eta)\,d\mu(u)
\end{aligned}$$

which, after using Cauchy-Schwarz as well as our assumption that $\operatorname{supp} f \subset K$ and $\|h\| = 1$, is

$$\le \sup_{\eta\in K}\Big|\int_R \Theta(\pi(\eta^{-1}\gamma))\Theta(\pi(\gamma))\,d\alpha^u(\pi(\gamma) - 1\Big|\|f\|_I.$$

Consequently, to prove Theorem 11.25 on page 302, it will suffice to prove the following lemma.

LEMMA 11.33. *Let G be as in Theorem 11.25 on page 302 with K compact in G and $\epsilon > 0$. Then there is a $\Theta \in \mathcal{B}^+(R)$ such that*

(a) $\displaystyle\int_R \Theta(\pi(\gamma))^2\,d\alpha^u(\pi(\gamma)) = 1$ *for all $u \in G^{(0)}$ and*

(b) $\displaystyle\Big|\int_R \Theta(\pi(\eta^{-1}\gamma))\Theta(\pi(\gamma))\,d\alpha^u(\pi(\gamma) - 1\Big| < \epsilon$ *for all $\eta \in K$.*

REMARK 11.34. Of course if (b) holds for all $\eta \in K$, it holds for all $\eta \in \pi^{-1}(\pi(K))$, but we don't need this.

PROOF. Since G is amenable, $r : G \to G^{(0)}$ is amenable (by definition) and Proposition 9.42 on page 209 implies there is a $h \in C^+_{c,r}(G)$ such that

$$\int_G h(\gamma)\,d\lambda^u(\gamma) = 1 \quad \text{for all } u \in G^{(0)}$$

and such that

$$\int_G |h(\eta^{-1}\gamma) - h(\gamma)|\,d\lambda^{r(\eta)}(\gamma) < \epsilon \quad \text{for all } \eta \in K.$$

Define

$$\Theta(\pi(\gamma)) = \Big(\int_{G(s(\gamma))} h(\gamma t)\,d\beta^{s(\gamma)}(t)\Big)^{\frac12}.$$

Then

$$\begin{aligned}
\int_R \Theta(\pi(\gamma))^2\,d\alpha^u(\pi(\gamma)) &= \int_R\int_{G(s(\gamma))} h(\gamma t)\,d\beta^{s(\gamma)}(t)\,d\alpha^u(\pi(\gamma)) \\
&= \int_G h(\gamma)\,d\lambda^u(\gamma) = 1,
\end{aligned}$$

and (a) is satisfied.

On the other hand, using (a), we have

$$2\Big|\int_R \Theta(\pi(\eta^{-1}\gamma))\Theta(\pi(\gamma))\,d\alpha^{r(\eta)}(\pi(\gamma)) - 1\Big|$$
$$= \int_R \big|\Theta(\pi(\eta^{-1}\gamma) - \Theta(\pi(\gamma))\big|^2\,d\alpha^{r(\eta)}(\gamma) \tag{11.29}$$

So we are reduced to showing that (11.29) is bounded by 2ϵ. Of course, ϵ will also work just fine.

For this, let

$$g(\gamma) := \sqrt{h(\gamma)}.$$

Then

$$\Theta(\pi(\gamma)) = \|t \mapsto g(\gamma t)\|_{L^2(\beta^{s(\gamma)})} \quad \text{and} \quad \Theta(\pi(\eta^{-1}\gamma)) = \|t \mapsto g(\eta^{-1}\gamma t)\|_{L^2(\beta^{s(\gamma)})}.$$

Using the L^2-norm identity $\big(\|a\|_2 - \|b\|_2\big)^2 \leq \|a-b\|_2^2$, we have

$$\Big|\Theta(\pi(\eta^{-1}\gamma)) - \Theta(\pi(\gamma))\Big|^2 \leq \int_{G(s(\gamma)} \big|g(\eta^{-1}\gamma t) - g(\gamma t)\big|^2\,d\beta^{s(\gamma)}(t).$$

Thus

$$\int_R \big|\Theta(\pi(\eta^{-1}\gamma) - \Theta(\pi(\gamma))\big|^2\,d\alpha^{r(\eta)}(\gamma)$$
$$\leq \int_R \int_{G(s(\gamma))} \big|g(\eta^{-1}\gamma t) - g(\gamma t)\big|^2\,d\beta^{s(\gamma)}(t)\,d\alpha^{r(\eta)}(\gamma)$$
$$= \int_G \big|g(\eta^{-1}\gamma) - g(\gamma)\big|^2\,d\lambda^{r(\eta)}(\gamma)$$

which, using the identity $|a-b|^2 \leq |a^2 - b^2|$ if $a, b \geq 0$, is

$$\leq \int_G |h(\eta^{-1}\gamma) - h(\gamma)|\,d\lambda^{r(\eta)}(\gamma).$$

Since the final expression is bounded by ϵ if $\eta \in K$, this completes the proof of the lemma and of Theorem 11.25. □

We say that G is *strongly essentially free* if the set of points where the isotropy is discretely trivial is dense in every closed G-invariant subset F of $G^{(0)}$.[7] As an application of Theorem 11.25 on page 302, we have the following result due to Renault [**Ren91**, Corollary 4.9].

PROPOSITION 11.35. *Let G be a strongly essentially free amenable*[8] *second countable locally compact groupoid with a Haar system. Then the lattice $\mathfrak{I}(C^*(G))$ of ideals in $C^*(G)$ is in one-to-one correspondence with the lattice of open G-invariant subsets of $G^{(0)}$.*

---

[7]In [**Ren91**], Renault calls this "essentially free". Unfortunately that term has not been used consistently in the literature for even for étale groupoids let alone general groupoids. I've added the adjective "strongly" to emphasize this property is *a priori* much stronger than other uses of the term in the literature outside of the category of étale groupoids.

[8]As proved in the next subsection, it suffice for G to be measurewise amenable in order to apply Theorem 11.25 on page 302. Hence we can prove this result if G is only measurewise amenable provided we also assume that the isotropy groups $G(u)$ are amenable for all $u \in G^{(0)}$. I an not aware of a proof that this follows from measurewise amenability.

REMARK 11.36 (Ideal Lattices). Recall that the set of ideals in $C_0(X)$ is in one-to-one correspondence with the closed subsets of X by sending $J \in \mathfrak{I}(C_0(X))$ to its zero set F_J. Of course, we also have a correspondence with open subsets of X by mapping J to the complement $O_J := G^{(0)} \setminus F_J$. This is sometimes preferable as it is a lattice isomorphism: it takes unions to unions and intersections to intersections.

PROOF. Let J be an ideal in $C^*(G)$. Then if L is any nondegenerate representation of $C^*(G)$ with $\ker L = J$, the support of M_L is a closed G-invariant set $F \subset G^{(0)}$. Furthermore, F depends only on J and not on L.

If the isotropy is discretely trivial at $u \in F$ then

$$\ker L \subset \ker \operatorname{Ind} \delta_u$$

by Theorem 11.22 on page 301. Since such points are dense in F and since induction preserves containment,

$$\ker L \subset \ker \operatorname{Ind}_{G^{(0)}}^G M_L.$$

On the other hand, each $G(u)$ is an amenable group by Proposition 9.77 on page 235, and as a result

$$\ker r_u \supset \ker \operatorname{Ind}^{G(u)} \delta_u$$

for all $u \in G^{(0)}$. By induction in stages,

$$\ker \operatorname{Ind}_{G(u)}^G r_u \supset \ker \operatorname{Ind} \delta_u$$

for all $u \in G^{(0)}$. As in Remark 11.9 on page 287, this easily implies

$$\ker \operatorname{Ind} r \supset \ker \bigcap_{u \in \operatorname{supp} \mu} \operatorname{Ind} \delta_u = \ker \operatorname{Ind}_{G^{(0)}}^G M_L.$$

Since amenability allows us to assert that $\ker L = \ker \operatorname{Ind} r$ (by Theorem 11.25 on page 302), we have

$$\ker L = \ker \operatorname{Ind}_{G^{(0)}}^G M_L.$$

This makes it clear that map from J to F_J is a bijection. In order to make nice with lattices, we consider the bijection $J \mapsto G^{(0)} \setminus F_J$. □

11.4.2. Measurewise Amenable Case. If G is merely measurewise amenable Theorem 11.25 on page 302 is still valid. Then the quasi-invariant measure μ associated to the representation G is amenable by assumption. Hence Corollary 10.46 on page 273 implies there is a completely invariant mean $m : L^\infty G) \to L^\infty(G^{(0)})$. Then Proposition 10.37(c) on page 262 implies that there is a sequence $\{h_n\} \subset \mathcal{B}^+(G)$ asuch that

$$\int_G h_n(\gamma)\, d\lambda^u(\gamma) = 1 \quad \text{for all } u \in G^{(0)}$$

and such that for all $f \in L^1(G, \nu)$,

$$\lim_n \int_{G^{(0)}} \int_G \int_G \bigl|h_n(\eta^{-1}\gamma) - h_n(\gamma)\bigr| f(\eta)\, d\lambda^u(\gamma)\, d\lambda^u(\eta)\, d\mu(u) = 0.$$

With the $\{h_n\}$ in hand, we can define $\xi^{h,n}$ using a Θ defined using h_n. The same computations as in Lemma 11.33 on page 310 shows that

$$(11.30)\qquad \begin{aligned} &\big|\big(L(f)\xi^{h,n}\,|\xi^{h,n}\big) - \big(R(f)h \mid h\big)\big| \\ &\quad\le \int_{G^{(0)}}\int_G\Big(\int_R \Theta(\pi(\eta^{-1}\gamma)\Theta(\pi(\gamma))\,d\alpha^{r(\eta)}(\pi(\gamma)) - 1\Big) \\ &\qquad\qquad |f(\eta)|\|h(s(\eta))\|\|h(u)\|\Delta(\eta)^{-\frac12}\,d\lambda^u(\eta)\,d\mu(u) \\ &\quad\le \frac12\int_{G^{(0)}}\int_G\int_G |h_n(\eta^{-1}\gamma) - h_n(\gamma)|b(\eta) \\ &\qquad\qquad d\lambda^u(\gamma)\,d\lambda^u(\eta)\,d\mu(u), \end{aligned}$$

where $b \in L^1(G,\nu)$ is given by

$$b(\eta) = |f(\eta)|\|h(s(\eta))\|\|h(r(\eta))\|\Delta(\eta)^{-\frac12}.$$

Hence (11.30) tends to zero as required for any $f \in C_c(G)$.

REMARK 11.37. As Renault points out in [**Ren91**, Remark 3.7] one can weaken the assumption that G is measurewise amenable to a similar statement for the measured groupoid (R,α,μ). Then one can produce the Θ's directly.

EXERCISES

Ex 11.4.1. Verify the ω-gymnastics to establish (11.23).

—Answer on page 384

APPENDIX A

Duals and Topological Vector Spaces

The treatment of amenability taken from [**ADR00**] requires some delicate work involving the dual and double dual of $C_0(X, L^1(Y, \alpha))$ as well as some gymnastics with topological vector spaces. I have used this appendix to collect some of the relevant definitions and technical results for reference so as not to distract from the main flow of ideas in the main body of the text.

A.1. The Strict Dual

Recall that if X is a (second countable) locally compact Hausdorff space, then we write $M(X)$ for the space of complex Borel measures on X. Thus if $\mu \in M(X)$ is a positive measure (usually we just say measure), then $\mu(X) < \infty$. We take it as "well-known" that we can identify $M(X)$ with the (norm) dual of $C_0(X)$ (see [**Fol99**, Theorem 7.17]). Then $M(X)$ becomes a Banach space with respect to the *total variation norm* $\|\cdot\|_1$ where

$$\|\nu\|_1 = \sup_{\substack{f \in C_0(X) \\ \|f\|_\infty = 1}} |\nu(f)|.$$

Alternatively, there is a unique finite positive measure $|\nu|$, called the *total variation of* ν, and a unimodular Borel function φ defined uniquely $|\nu|$-almost everywhere, such that

$$\nu(E) = \int_E \varphi(x)\, d|\nu|(x) \quad \text{for all Borel sets } E.$$

Then $|\nu(E)| \leq |\nu|(E)$ and $\|\nu\|_1 = |\nu|(X)$.

If X is not compact, then the (norm) dual of $C_b(X)$ is $M(\beta X)$ where $\beta(X)$ is the Stone-Cech compactification of X. Suffice it to say, this is unpleasant to contemplate. But $C_b(X)$, viewed as the multiplier algebra of $C_0(X)$, is also a topological vector space in the strict topology. If $\mu \in M(X)$, then we get a linear functional

$$\varphi_\mu(f) := \int_X f(x)\, d\mu(x) \tag{A.1}$$

defined on $C_b(X)$. While φ_u is fairly clearly norm bounded, it is not so clearly continuous for the strict topology on $C_b(X)$. In fact it is, and we have the following result. Recall that if V is a topological vector space, then we write V^* for the set of *continuous* linear functionals on V. We refer to V^* as the topological dual of V. Thus the *strict dual* of $C_b(X)$ the collection of linear functionals on $C_b(X)$ which are continuous with respect to the strict topology on $C_b(X)$.

THEOREM A.1 ([**Buc58**, Theorem 2]). *The map $\mu \mapsto \varphi_\mu$, with φ_μ defined by* (A.1), *identifies $M(X)$ with the strict dual of $C_b(X)$.*

Before turning to the proof of the theorem, we need a couple of observations. The first is very straightforward (and unnecessary in the separable case). We will repeatedly make use of the fact that if $\mu \in M(X)$, then $\mu = \mu_1 - \mu_2 + i(\mu_3 - \mu_4)$ for finite positive measures μ_k.

LEMMA A.2. *Every $\mu \in M(X)$ has σ-compact support.*

PROOF. We can assume that μ positive. Let $f_n \in C_c^+(X)$ be such that $\|f_n\|_\infty = 1$ and $\varphi_u(f_n) \to \|\varphi_\mu\| = \mu(X)$. Let $S = \bigcup_n \operatorname{supp} f_n$. It will suffice to see that if $g \in C_c^+(X)$ with $\|g\|_\infty \le 1$ and $\operatorname{supp} g \cap S = \emptyset$, then $\varphi_u(g) = 0$. But $\|f_n \pm g\|_\infty = 1$ for all n. Thus

$$\varphi_\mu(f_n \pm g) = \varphi_\mu(f_n) \pm \varphi_\mu(g) \le \mu(X) \quad \text{for all } n.$$

The conclusion follows. □

PROPOSITION A.3. *If $\mu \in M(X)$, then φ_μ is continuous on $C_b(X)$ when the later is equipped with the strict topology. (Hence φ_μ is in the strict dual of $C_b(X)$)*

PROOF. It suffices to consider μ positive. By Lemma A.2, we can assume that the support of μ is contained in the union $S_\mu := \bigcup K_n$ of compact sets K_n with K_n in the interior of K_{n+1}. Let $a_n := \mu(K_{n+1} \setminus K_n)$. Then $a_n \ge 0$ and $\sum a_n = \mu(X)$. Let $\{b_n\}$ be a sequence of positive numbers such that $b_n \searrow 0$ and

$$\sum \frac{a_n}{b_n} < 2\mu(X).$$

Let $c_n := b_n - b_{n+1}$. Then $b_n = \sum_{k \ge n} c_k$. Let $\psi_n \in C_c^+(X)$ be such that $0 \le \psi_n \le 1$, $\psi_n(x) = 1$ if $x \in K_n$ and ψ_n vanishes off K_{n+1}. Define

$$\psi(x) := \sum_{n=1}^{\infty} c_n \psi_n(x).$$

It is not hard to check that $\psi \in C_0(X)$, that ψ vanishes off $\overline{S_\mu}$ and that $\psi(x) \ge b_n$ on $K_n \setminus K_{n-1}$. Since ψ never vanishes on S_μ, $1/\psi$ is continuous on S_μ and μ-integrable:

$$\int_X \frac{1}{\psi}\, d\mu < 2\mu(X).$$

Now $U = \{\, f \in C_b(X) : \|f\psi\|_\infty < \epsilon/(2\mu(X))\,\}$ is a strict neighborhood of 0 in $C_b(X)$. If $f \in U$, then

$$\Big|\int_X f\, d\mu\Big| \le \int_{S_\mu} |f\psi| \frac{1}{\psi}\, d\mu \le \frac{\epsilon}{2\mu(X)} \int_{S_\mu} \frac{1}{\psi}\, d\mu < \epsilon.$$

This shows φ_μ is continuous. □

PROOF OF THEOREM A.1. Let $\varphi : C_b(X) \to \mathbf{C}$ be strictly continuous. Then the restriction of φ to $C_0(X)$ is a *norm continuous* linear functional. Hence it must be of the form φ_μ for some $\mu \in M(X)$ by general nonsense. Since $C_0(X)$ is strictly dense in $C_b(X)$, the result follows from Proposition A.3. □

Let V be a topological vector space. The *weak topology*[1] on V is the $\sigma(V, V^*)$ topology induced by pairing with the topological dual V^*: this is the smallest topology making each $\varphi \in V^*$ continuous. In particular, $v_i \to v$ weakly if and only if $\varphi(v_i) \to \varphi(v)$ for all $\varphi \in V^*$. Note that the weak topology is always coarser than

[1] See [**Ped89**, §2.4] for a good treatment of weak topologies.

the original topology: weakly open sets are open in V, but the converse can fail. Thus closed sets in V need not be weakly closed. But this is true for convex sets.

PROPOSITION A.4. *If V is a topological vector space then every closed convex set in V is weakly closed.*

PROOF. This is a straightforward consequence of the Hahn Banach Separation Theorem [**Ped89**, Theorem 2.4.7]. □

Recall that the direct sum $\bigoplus_i X_i$ of (topological) vector spaces is the subset of the direct product in which only *finitely* many terms are non-zero.

PROPOSITION A.5. *Let $X = \prod X_i$ be the product of topological vector spaces X_i equipped with the product topology. Then X is a topological vector space and $X^* = \bigoplus X_i^*$. In particular, the weak topology on X is the product of the weak topologies on the X_i. That is, if we let X_w denote X equipped with the weak topology (aka the $\sigma(X, X^*)$ topology), then*

$$X_w = \prod_i X_{i,w}.$$

PROOF. This is Proposition 8 of [**Bou87**, II §6.6].[2] □

COROLLARY A.6. *In the notation of Proposition A.5, $X = \prod X_i$ and $X_w = \prod X_{i,w}$ have the same closed convex sets.*

PROOF. This is an immediate consequence of Proposition A.4 and Proposition A.5. □

A.2. Projective Tensor Products

If A and B are Banach spaces, then their *projective tensor product* $A \mathbin{\widehat{\otimes}} B$ is the completion of their algebraic tensor product $A \odot B$ with respect to the norm

$$\|t\|_\pi = \inf\{\, \sum \|a_i\|\|b_i\| : t = \sum a_i \otimes b_i \,\}.$$

Notice that if $\varphi \in B^*$, then $(a, b) \mapsto \varphi(b)a$ is a bilinear map. Hence there is a map $S_\varphi : A \odot B \to A$ given by $S_\varphi(a \otimes b) = \varphi(b)a$. (Such maps are called *slice maps*.) Thus $S_\varphi(t) = \sum \varphi(b_i)a_i$ for any representation $t = \sum a_i \otimes b_i$. In particular,

$$\|S_\varphi(t)\| \leq \inf\{\, \big\|\sum \varphi(b_i)a_i\big\| : t = \sum a_i \otimes b_i \,\} \leq \|\varphi\|\|t\|_\pi.$$

Therefore we can extend the slice map S_φ to a map on all of $A \mathbin{\widehat{\otimes}} B$ with norm $\|\varphi\|$. In particular, if $a \otimes b$ is an elementary tensor and if we choose $\varphi \in B^*$ of norm one with $\varphi(b) \sim \|b\|$, then

$$\|a\|\|b\| \sim \|S_\varphi(a \otimes b)\| \leq \|a \otimes b\|_\pi.$$

Thus the project tensor norm is a *cross norm* in that

$$\|a \otimes b\|_\pi = \|a\|\|b\|.$$

The following is well-known.

[2]In the English translation cited, it is "page" II.50.

PROPOSITION A.7. *Suppose that A and B are Banach spaces. Then the norm dual $(A \mathbin{\widehat{\otimes}} B)^*$ is isomorphic to the Banach space $B(A, B^*)$ of bounded linear maps from A to B^*. The pairing is given by*

$$\langle a \otimes b \,,\, m\rangle = \langle m(a) \,,\, b\rangle. \tag{A.2}$$

PROOF. Let $m \in B(A, B^*)$ and let T_m be the linear functional on $A \odot B$ determined by (A.2). If $t = \sum a_i \otimes b_i$, then

$$|T_m(t)| = \big|\sum \langle m(a_i) \,,\, b_i\rangle\big| \leq \|m\| \sum \|a_i\|\|b_i\|$$

Since the representation of t was arbitrary, we have

$$|T_m(t)| \leq \|m\|\|t\|_\pi.$$

But if $a \in A$ has norm 1, then we can choose $b \in B$ of norm one such that $|\langle m(a) \,,\, b\rangle| \sim \|m(a)\|$. Then $\|a \otimes b\|_\pi = 1$ and $|T_m(a \otimes b)| \sim \|m\|$. Hence $\|T_m\| = \|m\|$ and $B(A, B^*)$ embeds into $(A \mathbin{\widehat{\otimes}} B)^*$ isometrically.

On the other hand, suppose that $\varphi \in (A \mathbin{\widehat{\otimes}} B)^*$. If $a \in A$, define $m(a) \in B^*$ by

$$\langle m(a) \,,\, b\rangle = \varphi(a \otimes b).$$

It is not hard to see that $m \in B(A, B^*)$. Hence $\varphi = T_m$. □

Now suppose that A and B are Banach algebras. Then $A \odot B$ is an algebra with respect to the product determined by

$$(a \otimes b)(c \otimes d) = ac \otimes bd.$$

(For a proof, see [**RW98**, Lemma B.1].) It is not hard to verify that $\|ts\|_\pi \leq \|t\|_\pi \|s\|_\pi$ for $t, s \in A \odot B$. Hence $A \mathbin{\widehat{\otimes}} B$ becomes a Banach algebra. Suppose that C is a Banach algebra such that A is a right Banach C-module and B a left Banach C-module. Then one can form the closed ideal I in $A \mathbin{\widehat{\otimes}} C$ which is generated by terms of the form $a \cdot c \otimes b - a \otimes c \cdot b$ for $a \in A$, $b \in B$ and $c \in C$. Then we call the quotient $A \mathbin{\widehat{\otimes}}_C B := A \mathbin{\widehat{\otimes}} B / I$ the *balanced tensor product*.

Now we make the additional assumption that A and B are central Banach C-modules in that $(a \cdot c)a' = aa' \cdot c$ and $b'(c \cdot b) = c \cdot b'b$ for $c \in C$, $a, a' \in A$ and $b, b' \in B$. This entails that the ideal I above is also the *subspace* generated by terms of the form $a \cdot c \otimes b - a \otimes c \cdot b$. Note that this is automatic if A, B and C are all commutative to begin with (which is the only application here).

Note that if B is a left C-module, then so is B^*: $\langle c \cdot f \,,\, b\rangle := \langle f \,,\, c \cdot b\rangle$.

PROPOSITION A.8. *Now suppose that A and B and Banach algebras such that A is a central right C-module and B is a central left C-module. Then the norm dual $(A \mathbin{\widehat{\otimes}}_C B)^*$ is identified with $B_C(A, B^*)$ via* (A.2) *where $B_C(A, B^*)$ denotes the elements of $B(A, B^*)$ which are C-linear.*

PROOF. The dual of $A \mathbin{\widehat{\otimes}}_C B$ is identified with the part of the dual of $A \mathbin{\widehat{\otimes}} B$ which annihilates the balancing *subspace* I. But these clearly correspond to the C-linear elements of $B(A, B^*)$. □

A.3. The Dual of $C_0(X, L^1(Y, \alpha))$

Recall from Section 3.3 that a complex Radon measure on Y is a linear functional ν on $C_c(Y)$ that is continuous in the inductive limit topology. If $\nu(f) \geq 0$ when $f \geq 0$, then a complex Radon measure is a bonafide Radon measure and there is an associated set-function denoted by the same letter. For (positive) Radon measures $\mu_1 \leq \mu_2$ simply means $\mu_1(E) \leq \mu_2(E)$ for every Borel set E. Equivalently, $\mu_1 \leq \mu_2$ if $\mu_1(f) \leq \mu_2(f)$ for every $f \in C_c^+(Y)$ (Ex 9.6.1). Let X and Y be locally compact G-spaces and $\pi : Y \to X$ and continuous equivariant surjection. Let $\alpha = \{\alpha^x\}_{x \in X}$ be a (proper) equivariant π-system. We say that a complex Radon measure ν on Y is *α-bounded* if there is a finite measure μ on X such that $|\nu| \leq \mu \circ \alpha$. The collection of all α-bounded measures is denoted by $M(Y, \alpha)$. Using Ex 9.6.1, we see that $M(Y, \alpha)$ is a normed vector space with respect to the norm

$$\|\nu\| = \inf\{\, \|\mu\|_1 : |\nu| \leq \mu \circ \alpha \,\}. \tag{A.3}$$

We let $C_0(X, L^1(Y, \alpha))$ be the completion of $C_c(Y)$ with respect to the norm

$$\|f\|_{\alpha,1} = \sup_{x \in X} \lambda(|f|)(x) = \sup_{x \in X} \int_Y |f(y)|\, d\alpha^x(y).$$

Since $\|f * g\|_{\alpha,1} \leq \|f\|_{\alpha,1}\|g\|_{\alpha,1}$, the product extends to $C_0(X, L^1(Y, \alpha))$ making it into a Banach algebra.

EXAMPLE A.9 ($\mathcal{E}$ and $M(G, \lambda)$). We can view a Haar system λ on G as a (full) equivariant $r : G \to G^{(0)}$ system. Then $C_0(X, L^1(Y, \alpha))$ is the Banach algebra $\mathcal{E} = C_0(G^{(0)}, L^1(G, \lambda))$ introduced in Section 9.6.

Note that if $\nu \in M(Y, \alpha)$, $f \in C_c(Y)$, and $|\nu| \leq \mu \circ \alpha$, then we have

$$|\nu(f)| \leq |\nu|(f) \leq \mu \circ \alpha(f) \leq \|\mu\|_1 \|f\|_{\alpha,1}. \tag{A.4}$$

Hence $|\nu(f)| \leq \|\nu\|\|f\|_{\alpha,1}$ and ν defines a bounded linear functional on $C_0(X, L^1(Y, \alpha))$ of norm at most $\|\nu\|$. This leads to the following result from [**ADR00**, Proposition 1.1.5].

PROPOSITION A.10. *Each functional $\nu \in M(Y, \alpha)$ extends to a linear functional on $C_0(X, L^1(Y, \alpha))$ with norm given by* (A.3) *identifying $M(Y, \alpha)$ with the dual of $C_0(X, L^1(Y, \alpha))$.*

PROOF. Suppose that $\varphi \in C_0(X, L^1(Y, \alpha))^*$. If $f_i \to f$ in the inductive limit topology on $C_c(Y)$, then $\|f_i - f\|_{\alpha,1} \to 0$. Hence φ defines (by restriction) a complex Radon measure ν on $C_c(Y)$ such that

$$|\nu(f)| \leq \|\varphi\|\|f\|_{\alpha,1} \quad \text{for } f \in C_c(Y).$$

It will suffice to see that ν is α-bounded—so that $\nu \in M(Y, \alpha)$—and that $\|\nu\| = \|\varphi\|$. But if $\nu = \rho \cdot |\nu|$ for a unimodular function ρ, we have

$$\big||\nu|(f)\big| = |\nu(\bar{\rho} f)| \leq \|\varphi\|\|\bar{\rho} f\|_{\alpha,1} = \|\varphi\|\|f\|_{\alpha,1}.$$

Since we also have $\||\nu|\| = \|\nu\|$, we can replace ν by $|\nu|$ as assume that ν is a (positive) Radon measure.

If $h \in C_c^+(X)$, let

$$\mu_0(h) = \sup\{\, \nu(f) : \alpha(f) = h \text{ with } f \in C_c^+(Y) \,\}. \tag{A.5}$$

Let $h_1, h_2 \in C_c^+(X)$ and set $h = h_1 + h_2$. Assume that $f_i \in C_c^+(Y)$ are such that $\alpha(f_i) = h_i$ for each i. Then $\mu_0(h) \geq \nu(f_1) + \nu(f_2)$ and hence

$$\mu_0(h_1 + h_2) \geq \mu_0(h_1) + \mu_0(h_2). \tag{A.6}$$

Now let $f \in C_c^+(Y)$ be such that $\alpha(f) = h$. Let $g \in C_{\pi,c}^+(Y)$ be such that $\alpha(g)(x) = 1$ for all $x \in X$ (see Proposition 3.18 on page 77). If $\epsilon > 0$, let

$$f_i(y) = \frac{(f(y) + \epsilon g(y))(h_i(\pi(y)))}{h(\pi(y)) + \epsilon}.$$

Check that $f_i \in C_c^+(Y)$ and that $\alpha(f_i) = h_i$. Also, observe that

$$f(y) = f_1(y) + f_2(y) + \epsilon k_1(y) - \epsilon k_2(y), \tag{A.7}$$

where

$$k_1(y) = \frac{f(y)}{h(\pi(y)) + \epsilon} \quad \text{and} \quad k_2(y) = \frac{h(\pi(y))g(y)}{h(\pi(y)) + \epsilon}.$$

Since it's not hard to check that $\|k_i\|_{\alpha,1} \leq 1$ for each i, (A.7) gives

$$\nu(f) \leq \nu(f_1) + \nu(f_2) + 2\epsilon\|\varphi\|.$$

Since f and ϵ were arbitrary,

$$\mu_0(h_1 + h_2) \leq \mu_0(h_1) + \mu_0(h_2).$$

Combining with (A.6), we see that μ_0 is additive on $C_c^+(X)$.

However, if $\alpha(f) = h$, then $\|f\|_{\alpha,1} = \|h\|_\infty$. Hence

$$\mu_0(h) \leq \|\varphi\|\|h\|_\infty \quad \text{for all } h \in C_c^+(X). \tag{A.8}$$

Therefore μ_0 extends to a finite measure on X (viewed as a $\|\cdot\|_\infty$-bounded linear functional on $C_c(X)$). If $f \in C_c^+(Y)$, then $\nu(f) \leq \mu_0(\alpha(f))$ and it follows that $\nu \leq \mu_0 \circ \alpha$ (see Ex 9.6.1). That is, $\nu \in M(Y, \alpha)$ as required. Furthermore, (A.8) implies

$$\|\nu\| \leq \|\mu_0\|_1 \leq \|\varphi\|.$$

Since we always have $\|\varphi\| \leq \|\nu\|$ by (A.4), $\|\nu\| = \|\varphi\|$ and we're done. □

A.4. The Dual of $L^\infty(Y) \odot_{L^\infty(X)} L^1(X)$

In this section, we work in what is called the "standard set-up" in Chapter 10. Thus X and Y will be second countable locally compact G-spaces and $\pi : Y \to X$ and continuous equivariant surjection. We let α be a (proper) equivariant π-system, and let $M(Y, \alpha)$ be the space of α-bounded complex Radon measures on Y as in Section A.3. If μ is finite measure on X, we let $M(Y, [\mu] \circ \alpha)$ be the subspace of $M(Y, \alpha)$ consisting of $\nu \in M(Y, \alpha)$ with the property that there is a finite measure μ_1 such that $\mu_1 \ll \mu$ and $|\nu| \leq \mu_1 \circ \alpha$. (Alternatively, $\mu_1 = h\mu$ with $h \in L^1(X, \mu)$.)

We'll write just $L^\infty(X)$ for $L^\infty(X, \mu)$, $L^\infty(Y)$ for $L^\infty(Y, \mu \circ \alpha)$, and $L^1(X)$ for $L^1(X, \mu)$. Note that $L^\infty(Y)$ is naturally a $L^\infty(X)$-module: $h \cdot \varphi(y) = \varphi \cdot h(y) = h(\pi(y))\varphi(y)$. (If N is μ-null, then $\pi^{-1}(N)$ is $(\mu \circ \alpha)$-null so we truly get a $L^\infty(X)$-action.)

If $\varphi \in L^\infty(Y)$ and $h \in L^1(X)$, the we get a Complex Radon measure $\nu = \varphi(h\mu \circ \alpha)$ on Y given by

$$\nu(f) = \int_X \int_Y f(y)\varphi(y)\, d\alpha^x(y) h(x)\, d\mu(x) \quad \text{for } f \in C_c(Y).$$

Since $|\nu| = |\varphi|(|h|\mu\circ\alpha)$ and $|\nu|\le \mu_1\circ\alpha$ where $\mu_1 = \|\varphi\|_\infty|h|\mu$, we get a map

$$\Psi : L^\infty(Y)\odot L^1(X)\to M(Y,[\mu]\circ\alpha).$$

This map Ψ clearly vanishes on the subspace V generated by $\varphi\cdot f\otimes h-\varphi\otimes f\cdot h$ with $f\in L^\infty(X)$. (In fact, V is also the ideal generated by these differences.) Hence Ψ factors through the balanced tensor product $L^\infty(Y)\odot_{L^\infty(X)} L^1(X)$. Since this algebraic balanced tensor product plays an important role in the sequel, we need to examine it in some detail.

Note that every elementary tensor in $L^\infty(Y)\odot_{L^\infty(X)} L^1(X)$ has a representation $\varphi\otimes h$ with $h\in L^1(X)^+$. But then the sum of two such elements can also be written in that form:

$$\varphi\otimes h+\psi\otimes k = \big(\varphi\cdot\frac{h}{h+k}+\psi\cdot\frac{k}{h+k}\big)\otimes(h+k).$$

Thus every element of $L^\infty(Y)\odot_{L^\infty(X)} L^1(X)$ is represented as an elementary tensor $\varphi\otimes h$ with $h\ge 0$. If p is the projection onto $\{\,x\in X: h(x)=0\,\}$, then modulo V, $\varphi\otimes h = (\varphi(1-p))\otimes(h+p)$. Thus we can assume h never vanishes. But then $\Psi(\varphi\otimes h)$ is the zero functional exactly when $\varphi=0$ $\mu\circ\alpha$-almost everywhere. It follows that the kernel of Ψ is exactly V. Hence we can identify $L^\infty(Y)\odot_{L^\infty(X)} L^1(X)$ with its image in $M(Y,[\mu]\circ\alpha)\subset M(Y,\alpha) = C_0(X,L^1(Y,\alpha))^*$.

Since the dual of $L^1(X)$ can be identified with $L^\infty(X)$, Proposition A.8 on page 318 implies that we can identify the norm dual of the balanced projective tensor product $L^\infty(Y)\ \widehat{\otimes}_{L^\infty(X)}\ L^1(X)$ with the bounded $L^\infty(X)$-linear functions, $B_{L^\infty(X)}\big(L^\infty(Y),L^\infty(X)\big)$, from $L^\infty(Y)$ to $L^\infty(X)$: given $m\in B_{L^\infty(X)}\big(L^\infty(Y),L^\infty(X)\big)$ and $\nu=\varphi\otimes h$, then

$$\langle m\,,\nu\rangle = \langle m\,,\varphi\otimes h\rangle = \langle m(\varphi)\,,h\rangle,$$

where the later pairing is the usual one between $L^\infty(X)$ and $L^1(X)$.

We write $B_{L^\infty(X)}\big(L^\infty(Y),L^\infty(X)\big)_1$ for the maps of norm at most one, and we say that $m\in B_{L^\infty(X)}\big(L^\infty(Y),L^\infty(X)\big)^+$ if $m\big(L^\infty(Y)^+\big)\subset L^\infty(X)^+$.

To get a better understanding of the projective norm on $L^\infty(Y)\odot_{L^\infty(X)} L^1(X)$, it will be convenient to observe that we can extend functionals in $L^\infty(Y)\odot_{L^\infty(X)} L^1(X)\subset C_0(X,L^1(Y,\alpha))^*$ to the space $\mathcal{L}^\infty\big(X,L^1(Y,\alpha)\big)$ of Borel functions f on Y such that $x\mapsto\alpha(|f|)(x)$ is in $L^\infty(X)$. With $\|f\|_{\infty,\alpha}=\|\alpha(|f|)\|_\infty$, we get an isometric embedding of $\mathcal{L}^\infty\big(X,L^1(Y,\alpha)\big)$ into $B_{L^\infty(X)}\big(L^\infty(Y),L^\infty(X)\big)$ via $f\mapsto m_f$ with $m_f(\varphi)=\alpha(f\varphi)$. In particular, any $\nu=\varphi\otimes h$ defines a linear functional on $\mathcal{L}^\infty\big(X,L^1(Y,\alpha)\big)$ given by

$$\nu(f) = m_f(\nu) = \big\langle\alpha(f\varphi)\,,h\big\rangle = \int_X\int_Y f(y)\varphi(y)\,d\alpha^x(y)h(x)\,d\mu(x). \tag{A.9}$$

Conversely, we have the following.

LEMMA A.11 ([**ADR00**, Lemma 1.2.3]). *Suppose that L is a bounded linear functional on $\mathcal{L}^\infty\big(X,L^1(Y,\alpha)\big)$ such that there is a $h\in L^1(X)^+$ such that*

$$|L(f)|\le\int_X\int_Y|f(y)|\,d\alpha^x(y)h(x)\,d\mu(x) = \big\langle\alpha(|f|)\,,h\big\rangle. \tag{A.10}$$

Then there exists $\varphi\in L^\infty(Y)$ such that $\|\varphi\|_\infty\le 1$ and L is the functional defined by $\nu=\varphi\otimes h$ as in (A.9).

PROOF. Let Y be the increasing union $\bigcup Y_n$ of compact sets Y_n. Hence $x \mapsto \alpha^x(Y_n)$ is bounded for each n and $h\mu(\pi(Y_n))$ is finite. Let σ_n be the finite measure on Y_n given by the restriction of $\sigma = h\mu \circ \alpha$.[3] Let γ_n be the finitely additive finite set function on the Borel sets of Y_n given by

$$\gamma_n(E) = L(\mathbb{1}_E).$$

By hypothesis, $\gamma_n(E) \leq \sigma_n(E)$. Since σ_n is finite and countably additive, it follows that γ_n is actually countably additive and absolutely continuous with respect to σ_n. Let φ_n be the Radon-Nykodym derivative. Note that $0 \leq \varphi_n \leq 1$. Since $\varphi_n = \varphi_{n+1}$ almost everywhere on Y_n, we can assume there is a $\varphi \in L^\infty(Y)$ with $\|\varphi\|_\infty \leq 1$ such that $\varphi_n = \varphi|_{Y_n}$. If f is a simple function on Y, then

$$L(f\mathbb{1}_{Y_n}) = \int_X \int_Y f(y)\mathbb{1}_{Y_n}(y)\varphi(y)\, d\alpha^x(y)\, h(x)\, d\mu(x). \tag{A.11}$$

If f is a bounded Borel function on Y, then there are simple functions f_k converging uniformly to f on Y. It follows from (A.10) that $L(f_k\mathbb{1}_{Y_n}) \to L(f\mathbb{1}_{Y_n})$. Hence (A.11) holds for all bounded Borel functions f on Y. If $f \in \mathcal{L}^\infty\big(X, L^1(Y,\alpha)\big)$, let $f_n = \mathbb{1}_{D_n} f$ where $D_n = \{\, y \in Y : |f(y)| \geq n \,\}$. Note the $f_n \in L^\infty(Y)$. By the Dominated Convergence Theorem

$$\int_X \int_Y |f(y) - f_n(y)|\, d\alpha^x(y) h(x)\, d\mu(x) \to 0.$$

It follows that

$$L(f) = \int_X \int_Y f(y)\varphi(y)\, d\alpha^x(y) h(x)\, d\mu(x)$$

as required. □

COROLLARY A.12. *The map* $\Psi : L^\infty(Y) \odot_{L^\infty(X)} L^1(X) \to M(Y, [\mu] \circ \alpha)$ *is a bijection.*

PROOF. We have already observed the Ψ is injective. If $\nu \in M(Y, [\mu] \circ \alpha)$, then just as in (A.4), ν is $\|\cdot\|_{\infty,\alpha}$-bounded and extends to a linear functional on $\mathcal{L}^\infty\big(X, L^1(Y,\alpha)\big)$ which clearly satisfies the hypotheses of Lemma A.11 on the preceding page for an appropriate h. Hence ν is in the image of Ψ. □

In the sequel, we will use Corollary A.12 to tacitly identify $L^\infty(Y) \odot_{L^\infty(X)} L^1(X)$ with $M(Y, [\mu] \circ \alpha)$ as sets. To see that we can identify them as normed spaces takes some additional work (see [**ADR00**, Lemma 1.2.5]).

LEMMA A.13. *Every element* $\nu \in L^\infty(Y) \odot_{L^\infty(X)} L^1(X)$ *has a representative*

$$\nu = \varphi_0 \otimes h_0$$

with $\|\varphi_0\|_\infty = 1$, $h_0 \in L^1(X)^+$ *and* $\|\nu\| = \|h_0\|_1$. *Furthermore,*

$$\|\nu\| = \sup\{\, |\nu(f)| : f \in \mathcal{L}^\infty\big(X, L^1(Y,\alpha)\big)_1 \,\}$$

where $\mathcal{L}^\infty\big(X, L^1(Y,\alpha)\big)_1 = \{\, f \in \mathcal{L}^\infty\big(X, L^1(Y,\alpha)\big) : \|f\|_{\infty,\alpha} \leq 1 \,\}$.

---

[3]Since Y and X are second countable, $h\mu \circ \alpha$ is a σ-finite measure. In particular, we can apply Proposition 3.6 on page 71 as well as the corresponding comments in Section 3.1.

PROOF. Let $\nu = \varphi \otimes h$ with $h \geq 0$. Then $|\nu| = |\varphi| \otimes h$. Since we can write $\varphi = g|\varphi|$ for a unimodular function $g \in L^{\infty}(Y)$ and since $f \in \mathcal{L}^{\infty}\big(X, L^1(Y,\alpha)\big)_1$ implies $gf \in \mathcal{L}^{\infty}\big(X, L^1(Y,\alpha)\big)_1$, it is clear that

$$\sup\{\, |\nu(f)| : f \in \mathcal{L}^{\infty}\big(X, L^1(Y,\alpha)\big)_1 \,\} = \sup\{\, \big||\nu|(f)\big| : f \in \mathcal{L}^{\infty}\big(X, L^1(Y,\alpha)\big)_1 \,\}.$$

On the other hand, we clearly have $\|\nu\| = \big\||\nu|\big\|$ by (A.3) Hence we may as well assume $\varphi \geq 0$.

For each $g \in L^{\infty}(X)^+$, define

$$\mu_0(g) = \sup\{\, \nu(f) : f \in \mathcal{L}^{\infty}\big(X, L^1(Y,\alpha)\big)^+ \text{ and } \alpha(f) = g \,\}.$$

As in the proof of Proposition A.10 (see (A.5) on page 319), μ_0 extends to a positive linear functional on all of $L^{\infty}(X)$. Notice that if $\nu \leq \mu_1 \circ \alpha$ and $\alpha(f) = 1$, then $\nu(f) \leq \mu_1(1) = \|\mu_1\|_1$. Hence $\|\mu_0\| = \mu_0(1) \leq \|\nu\|$ as the later is the infimum of such $\|\mu_1\|_1$. I claim that $\mu_0 \in L^{\infty}(X)_* \cong L^1(X)$. Such functionals are called *normal*—see [**Bla06**, III.2]. But it is not hard to see that μ_0 is dominated by the normal functional $\mu_{00}(g) = \|\varphi\|_{\infty} \int_X gh\, d\mu$. Hence μ_0 itself is normal (see [**Bla06**, Definition III.2.1.2] and [**Bla06**, III.2.1.4]). That means there is a $h_0 \in L^1(X)^+$ such that $\|\mu_0\| = \|h_0\|_1$ and

$$\mu_0(g) = \int_X h_0(x) g(x)\, d\mu(x).$$

Hence if $f \in \mathcal{L}^{\infty}\big(X, L^1(Y,\alpha)\big)$,

$$|\nu(f)| \leq \nu(|f|) \leq \mu_0(\alpha(|f|) \leq \int_X \int_Y |f(y)|\, \alpha^x(y) h_0(x)\, d\mu(x).$$

Hence Lemma A.11 on page 321 implies that there is a $\varphi_0 \in L^{\infty}(Y)_1$ such that $\nu = \varphi_0 \otimes h_0$. But

$$\|h_0\|_1 = \|\mu_0\| \leq \|\nu\| \leq \|\varphi_0\|_{\infty} \|h_0\|_1 \leq \|h_0\|_1.$$

Thus $\|h_0\|_1 = \|\nu\|$.

To check the last assertion, note that if $f \in \mathcal{L}^{\infty}\big(X, L^1(Y,\alpha)\big)_1$, then on the one hand,

$$|\nu(f)| \leq \mu_0(\alpha(|f|)) \leq \mu_0(1) = \|\nu\|.$$

On the other hand,

$$\|\nu\| = \mu_0(1) = \sup\{\, \nu(f) : f \in \mathcal{L}^{\infty}\big(X, L^1(Y,\alpha)\big)^+ \text{ with } \alpha(f) = 1. \}$$

This proves the assertion.

To sum up, we've shown that we can write ν as $\varphi \otimes h_0$ with $\|\varphi\|_{\infty} \leq 1$ and $\|h_0\|_1 = \|\nu\|$. To show that $\|\varphi\|_{\infty} = 1$, we will need the next corollary. □

Corollary A.12 on the preceding page allows us to identify $L^{\infty}(Y) \odot_{L^{\infty}(X)} L^1(X)$ with $M(Y, [\mu] \circ \alpha) \subset M(Y, \alpha)$. But we can also map $L^{\infty}(Y) \odot_{L^{\infty}(X)} L^1(X)$ into the projective tensor product $L^{\infty}(Y) \mathbin{\widehat{\otimes}}_{L^{\infty}(X)} L^1(X)$. For clarity, let $\psi : L^{\infty}(Y) \odot_{L^{\infty}(X)} L^1(X) \to L^{\infty}(Y) \mathbin{\widehat{\otimes}}_{L^{\infty}(X)} L^1(X)$ denote the natural map. Since the projective tensor norm is a cross norm and since we proved in Lemma A.13 on the facing page that we can write any $\nu \in L^{\infty}(Y) \odot_{L^{\infty}(X)} L^1(X)$ as an elementary tensor $\varphi \otimes h_0$ with $\|\varphi\|_{\infty} \leq 1$ and $\|h_0\|_1 = \|\nu\|$, it follows that ψ is norm reducing: $\|\psi(\nu)\| \leq \|\nu\|$. The next corollary asserts that ψ is isometric. This forces $\|\varphi\|_{\infty} = 1$ as claimed in the Lemma A.13 on the preceding page and completes the proof of that result.

COROLLARY A.14. *The natural map*

$$\psi : L^\infty(Y) \odot_{L^\infty(X)} L^1(X) \equiv M(Y, [\mu] \circ \alpha) \to L^\infty(Y) \widehat{\otimes}_{L^\infty(X)} L^1(X)$$

is isometric. In particular, the closure $X([\mu])$ *of* $M(Y, [\mu] \circ \alpha)$ *in* $M(Y, \alpha)$ *has dual isomorphic to* $B_{L^\infty(X)}\big(L^\infty(Y), L^\infty(X)\big)$.

PROOF. We observed above that ψ is norm reducing. Proposition A.8 on page 318 allows to identify the dual of $L^\infty(Y) \ \widehat{\otimes}_{L^\infty(X)} \ L^1(X)$ with $B_{L^\infty(X)}\big(L^\infty(Y), L^\infty(X)\big)$, and we have

$$\begin{aligned}\|\psi(\nu)\| &= \sup\{\, |\langle \psi(\nu)\ ,\ m\rangle| : m \in B_{L^\infty(X)}\big(L^\infty(Y), L^\infty(X)\big)_1 \,\} \\ &\geq \sup\{\, |\langle \psi(\nu)\ ,\ m_f\rangle| : f \in \mathcal{L}^\infty\big(X, L^1(Y,\alpha)\big)_1 \,\} \\ &= \sup\{\, |\nu(f)| : f \in \mathcal{L}^\infty\big(X, L^1(Y,\alpha)\big)_1 \,\} \\ &= \|\nu\|\end{aligned}$$

where the last equality comes from Lemma A.13 on page 322. Thus ψ is isometric and clearly has dense range. It follows that $X([\mu])$ is isomorphic as a Banach space to $L^\infty(Y) \ \widehat{\otimes}_{L^\infty(X)} \ L^1(X)$ which has $B_{L^\infty(X)}\big(L^\infty(Y), L^\infty(X)\big)$ as dual by Proposition A.8. □

The following summary will be useful.

PROPOSITION A.15 ([**Ren15**, Proposition 2.9]). *If* $\nu \in C_0(X, L^1(Y,\alpha))^*$, *then there is a* $\mu \in M(X)^+$ *and a* $\varphi \in L^\infty(Y, \mu \circ \alpha)$ *such that* ν *is given by* $\varphi(\mu \circ \alpha)$ *so that*

$$\nu(f) = \int_X \int_Y f(y)\varphi(y)\, d\alpha^x(y)\, d\mu(x) \quad \textit{for all } f \in C_c(Y).$$

Furthermore,

$$\|\nu\| = \inf \|\mu\|_1 \|\varphi\|_\infty$$

where the infimum is taken over all representations $\nu = \varphi(\mu \circ \alpha)$.

PROOF. If $\nu = \varphi(\mu \circ \alpha)$, then $\|\nu\| \leq \|\nu\|_1 \|\varphi\|_\infty$. However, every ν lies in some $M(Y, [\mu] \circ \alpha)$, and Lemma A.13 on page 322 implies that $\nu = \varphi_0(\mu_0 \circ \alpha)$ with $\|\varphi_0\|_\infty = 1$ and $\|\mu_0\|_1 = \|h_0 \cdot \mu\|_1 = \|h_0\|_1 = \|\nu\|$. The result follows. □

By Lemma A.13 on page 322 we can (isometrically) view elements of $L^\infty(Y) \odot_{L^\infty(X)} L^1(X)$ as elements of the dual of $\mathcal{L}^\infty\big(X, L^1(Y,\alpha)\big)$. More precisely, if $\nu = \varphi \otimes h$, then ν determines a function ψ on Y given by $y \mapsto \varphi(y)h(\pi(y))$. The pairing of ν acting on $g \in \mathcal{L}^\infty\big(X, L^1(Y,\alpha)\big)$ is given by

$$\langle \nu\ ,\ g\rangle = \int_Y \psi(y)g(y)\, d(\mu \circ \alpha)(y) = \int_X \int_Y g(y)\varphi(y)\, d\alpha^x(y) h(x)\, d\mu(x).$$

Since the map of $L^\infty(Y) \odot_{L^\infty(X)} L^1(X)$ onto $M(Y, [\mu] \circ \alpha)$ is bijective, it follows that the map sending $\varphi \otimes h$ to the function $\psi(y) = \varphi(y)h(\pi(y))$ is injective provided we identify functions that agree $\mu \circ \alpha$-almost everywhere. It will be convenient to be able to identify functions in the range of this map.

LEMMA A.16 ([**ADR00**, Lemma 1.2.4]). *A Borel function* ψ *on* Y *is in (the image of)* $L^\infty(Y) \odot_{L^\infty(X)} L^1(X)$ *if and only if* $\alpha(|g\psi|) \in L^1(X)$ *for all* $g \in \mathcal{L}^\infty\big(X, L^1(Y,\alpha)\big)$ *and there is a* $h \in L^1(X)^+$ *such that*

$$\Big| \int_X \alpha(g\psi)(x)\, d\mu(x) \Big| \leq \int_Y |g(y)| h(\pi(y))\, d(\mu \circ \alpha)(y).$$

PROOF. Suppose ψ is as in the statement of the lemma. Define a linear functional on $\mathcal{L}^\infty\big(X, L^1(Y,\alpha)\big)$ by $L(g) = \int_X \alpha(g\psi)(x)\,d\mu(x)$. Lemma A.11 on page 321 implies there is a $\varphi \in L^\infty(Y)_1$ such that

$$\int_X \int_Y g(y)\psi(y)\,d\alpha^x(y)\,d\mu(x) = \int_X \int_Y g(y)\varphi(y)h(\pi(y))\,d\alpha^x(y)\,d\mu(x)$$

for all $g \in \mathcal{L}^\infty\big(X, L^1(Y,\alpha)\big)$. It follows that $\psi(y) = \varphi(y)h(\pi(y))$ for $\mu \circ \alpha$-almost all y.

The other direction is left as Ex A.1. □

Suppose that $f \in C_c(G)$ and that $g \in \mathcal{L}^\infty\big(X, L^1(Y,\alpha)\big)$, then since α is equivariant, it is not hard to check that

$$f * g(y) = \int_G f(\gamma)g(\gamma^{-1}\cdot y)\,d\lambda^{r(y)}(\gamma)$$

yields another element of $\mathcal{L}^\infty\big(X, L^1(Y,\alpha)\big)$: in fact, since $r(y) = r(x)$ if $y \in \operatorname{supp}\alpha^x$,

$$\alpha(f * g)(x) = \int_Y \int_G f(\gamma)g(\gamma^{-1}\cdot y)\,d\lambda^{r(x)}(\gamma)\,d\alpha^x(y)$$

which, by Fubini and the equivariance of α, is

$$\begin{aligned} &= \int_G f(\gamma)\alpha(g)(\gamma^{-1}\cdot x)\,d\lambda^{r(y)}(\gamma) \\ &\le \|\alpha(g)\|_\infty \|f\|_I. \end{aligned}$$

Let the map of $\mathcal{L}^\infty\big(X, L^1(Y,\alpha)\big)$ to itself given by convolution by $f \in C_c(G)$ be $M(f)$. (Of course $M(f)$ is just the extension of $L(f)$ on $C_0(X, L^1(Y,\alpha))$ defined in Section 9.4 to all of $\mathcal{L}^\infty\big(X, L^1(Y,\alpha)\big)$, but having a separate notation makes it easier to sort out the details in the sequel.) Let ${}^tM(f)$ be the transpose map on the dual of $\mathcal{L}^\infty\big(X, L^1(Y,\alpha)\big)$.

LEMMA A.17. *If we view $L^\infty(Y) \odot_{L^\infty(X)} L^1(X)$ as a subspace of the dual of $\mathcal{L}^\infty\big(X, L^1(Y,\alpha)\big)$, then ${}^tM(f)$ maps the subspace $L^\infty(Y) \odot_{L^\infty(X)} L^1(X)$ to itself. If $\nu = \varphi \otimes h$ and $\psi(y) = \varphi(y)h(\pi(y))$, then as a function on Y,*

$$ {}^tM(f)\psi(y) = \int_G f(\gamma^{-1})\psi(\gamma^{-1}\cdot y)\delta(\gamma^{-1}, \gamma^{-1}\cdot\pi(y))\,d\lambda^{r(y)}(\gamma), \tag{A.12}$$

where δ is the Raydon-Nikodym derivative $\frac{d(\mu\circ\lambda)}{d(\mu\circ\lambda)^{-1}}$.

PROOF. Let $\psi = \varphi \cdot h$. We want to use Lemma A.16 on the preceding page to verify that the function $y \mapsto \zeta(y)$ given by the right-hand side of (A.12) is in the image of $L^\infty(Y) \odot_{L^\infty(X)} L^1(X)$. As in Section A.4, we can assume $h \ge 0$.

Let $g \in \mathcal{L}^\infty\big(X, L^1(Y,\alpha)\big)$. Then we compute that

$$\begin{aligned} &\Big|\int_X \alpha(g\zeta)(x)\,d\mu(x)\Big| \\ &\quad\le \int_X \int_Y \int_G |g(y)f(\gamma^{-1})|\|\varphi\|_\infty h(\gamma^{-1}\cdot x)\delta(\gamma^{-1},\gamma^{-1}\cdot x)\,d\lambda^{r(x)}(\gamma)\,d\alpha^x(y)\,d\mu(x) \\ &\quad= \int_X \int_Y |g(y)|\|\varphi\|_\infty H(x)\,d\alpha^x(y)\,d\mu(x), \end{aligned}$$

where

$$H(x) := \int_G |f(\gamma^{-1})| h(\gamma^{-1}\cdot x)\delta(\gamma^{-1},\gamma^{-1}\cdot x)\,d\lambda^{r(x)}(\gamma).$$

Therefore we just need to see that H is integrable. But

$$\int_X H(x)\,d\mu(x) = \int_X\int_G |f(\gamma^{-1})| h(\gamma^{-1}\cdot x)\delta(\gamma^{-1},\gamma^{-1}\cdot x)\,d\lambda^{r(x)}(\gamma)\,d\mu(x)$$

which, using the definition of δ, is

$$\begin{aligned}&= \int_X\int_G |f(\gamma)| h(x)\,d\lambda^{r(x)}(\gamma)\,d\mu(x)\\ &= \int_X \lambda(|f|)(r(x))h(x)\,d\mu(x) < \infty.\end{aligned}$$

Thus ζ is in the image of $L^\infty(Y)\odot_{L^\infty(X)} L^1(X)$ as claimed.

We compute

$$\begin{aligned}\langle \nu\,,\,M(f)g\rangle &= \int_X\int_Y\int_G \psi(y)f(\gamma)g(\gamma^{-1}\cdot y)\,d\lambda^{r(x)}(\gamma)\,d\alpha^x(y)\,d\mu(x)\\ &= \int_X\int_G F(\gamma,x)\,d\lambda^{r(x)}(\gamma)\,d\mu(x),\end{aligned}$$

where

$$F(\gamma,x) = \int_Y f(\gamma)\psi(y)g(\gamma^{-1}\cdot y)\,d\alpha^x(y).$$

Using the definition of δ,

$$\begin{aligned}&\langle \nu\,,\,M(f)g\rangle\\ &\quad= \int_X\int_G F(\gamma^{-1},\gamma^{-1}\cdot x)\delta(\gamma^{-1},\gamma^{-1}\cdot x)\,d\lambda^{r(x)}(\gamma)\,d\mu(x)\\ &\quad= \int_X\int_G\int_Y f(\gamma^{-1})\psi(y)g(\gamma\cdot y)\,d\alpha^{\gamma^{-1}\cdot x}(y)\delta(\gamma^{-1},\gamma^{-1}\cdot x)\,d\lambda^{r(x)}(\gamma)\,d\mu(x)\end{aligned}$$

which, by equivariance of α, is

$$\begin{aligned}&= \int_X\int_Y\int_G f(\gamma^{-1})\psi(\gamma^{-1}\cdot y)g(y)\,d\alpha^x(y)\delta(\gamma^{-1},\gamma^{-1}\cdot x)\,d\lambda^{r(x)}(\gamma)\,d\mu(x)\\ &= \int_X\int_Y \zeta(y)g(y)\,\alpha^x(y)\,d\mu(x)\\ &= \int_Y \zeta(y)g(y)\,d(\mu\circ\alpha)(y).\end{aligned}$$

Hence ${}^tM(f)\psi$ is given by (A.12) as claimed. □

If X is any normed vector space viewed as a subset of X^{**}, it follows from the Hahn Banach Separation Theorem that the unit ball of X is weak-$*$ dense in the unit ball of X^{**}. A slight variation allows us to take positivity into account. Let $C_c^+(Y)_{\alpha,1} = \{\, f \in C_c^+(Y) : \alpha(f)(x) \le 1 \text{ for all } x\in X \,\} = C_c^+(Y)\cap \mathcal{L}^\infty\big(X, L^1(Y,\alpha)\big)_1$.

LEMMA A.18. *Let $\pi : Y\to X$, μ, α and G be as in our standard set-up and view $C_c(Y)$ as a subset of $B_{L^\infty(X)}\big(L^\infty(Y), L^\infty(X)\big)$ by identifying $f\in C_c(Y)$ with the map $m_f(\varphi)=\alpha(f\cdot\varphi)$. Then $C_c^+(Y)_{\alpha,1}$ is weak-$*$ dense in $B_{L^\infty(X)}\big(L^\infty(Y), L^\infty(X)\big)_1^+$.*

PROOF. We follow the outline of the proof of [**ADR00**, Lemma 1.2.6]. Let C be the weak-$*$ closure of $C_c^+(Y)_{\alpha,1}$ in $B_1^+ := B_{L^\infty(X)}\bigl(L^\infty(Y),L^\infty(X)\bigr)_1^+$. If $C\neq B_1^+$, then there is a $m\in B_1^+\setminus C$. The Hahn-Banach Separation Theorem [**Ped89**, Theorem 2.4.7] implies there is $\chi\in X([\mu])=\overline{M(Y,[\mu]\circ\alpha)}$ and an element $s\in\mathbf{R}$ such that

$$\operatorname{Re}\chi(f)\le s<\operatorname{Re}\chi(m)\quad\text{for all } f\in C_c^+(Y)_{\alpha,1}.$$

Since $0\in C_c^+(Y)_{\alpha,1}$, $s\ge 0$. Let $d=\operatorname{Re}\chi(m)-s$ and $r=s+d/3$. If $\nu\in L^\infty(Y)\odot_{L^\infty(X)}L^1(X)$ is such that $\|\chi-\nu\|<d/3$, then since every $f\in C_c^+(Y)_{\alpha,1}$ has norm at most one, we have

$$\operatorname{Re}\nu(f)\le r<\operatorname{Re}\nu(m)\quad\text{for all } f\in C_c^+(Y)_{\alpha,1}.$$

Let $\nu=\varphi\otimes h$ with $h\ge 0$. Since $m\ge 0$, $\operatorname{Re}m(\varphi)=m(\operatorname{Re}\varphi)$. Since $\nu(m)=\langle m(\varphi)\,,\,h\rangle$, we can replace φ by its real part and assert that

$$\nu(f)\le r<\nu(m)\quad\text{for all } f\in C_c^+(Y)_{\alpha,1}.$$

Let $\nu=(\varphi^+-\varphi^-)\otimes h$ where $\varphi^+\cdot\varphi^-=0$ and $\varphi^\pm\ge 0$. Let $\nu^+:=\varphi^+\otimes h$. I claim it will suffice to prove that

$$\nu^+(f)\le r\quad\text{for all } f\in C_c^+(Y)_{\alpha,1},\text{ and}\tag{A.13}$$

$$\|\nu^+\|=\sup_{f\in C_c^+(Y)_{\alpha,1}}\nu^+(f).\tag{A.14}$$

To check the claim, note that (A.13) and (A.14) imply that $\|\nu^+\|\le r$ while $\nu^+(m)>r$ in view of (A.13). This forces $\|m\|>1$ which contradicts our choice of m.

To verify (A.13), recall that $\nu(f)=\langle m_f\,,\,\nu\rangle$ and that

$$m_f(\varphi)=\alpha(f\varphi)=\alpha(f\varphi^+-f\varphi^-).$$

Let $E:=\{\,y\in Y:\varphi^+(y)>0\,\}$. Then by construction, $\varphi^-(E)=\{0\}$. Fix $\epsilon>0$ and find a closed set C and an open set V such that $C\subset E\subset V$ and such that $V\setminus C$ has $h\cdot\mu\circ\alpha$-measure less than ϵ. Let $g\in C_c^+(Y)$ be such that $0\le g\le 1$, $g(E)=\{1\}$ and $g(y)=0$ off V. Then if $f\in C_c^+(Y)_{\alpha,1}$,

$$\begin{aligned}\nu^+(f)&=\langle\alpha(f\varphi^+)\,,\,h\rangle\\&=\langle\alpha(f\mathbb{1}_E\varphi)\,,\,h\rangle\\&=\bigl\langle\alpha(f\mathbb{1}_C\varphi)+\alpha(f\mathbb{1}_{E\setminus C}\varphi)\,,\,h\bigr\rangle\\&=\bigl\langle\alpha(fg\mathbb{1}_C\varphi)+\alpha(f\mathbb{1}_{V\setminus C}\varphi)-\alpha(f\mathbb{1}_{V\setminus E}\varphi)\,,\,h\bigr\rangle\\&=\bigl\langle\alpha(fg\varphi)+\alpha((f-g)\mathbb{1}_{V\setminus C}\varphi)-\alpha(f\mathbb{1}_{V\setminus E}\varphi)\,,\,h\bigr\rangle\\&\le\nu(fg)-2\epsilon\|\varphi\|_\infty\end{aligned}$$

which, since $fg\in C_c^+(Y)_{\alpha,1}$, is

$$\le r-2\epsilon\|\varphi\|_\infty.$$

Since $\epsilon>0$ was arbitrary, (A.13) follows.

It only remains to verify (A.14) and this follows almost exactly as in the proof of Lemma A.13 on page 322. We'll show that if $\nu=\varphi\otimes h$ with both φ and h nonnegative, then $\|\nu\|=\sup_{f\in C_c^+(Y)_{\alpha,1}}\nu(f)$. To start, for each $g\in C_c^+(X)$, let

$$\mu_0(g)=\sup\{\,\nu(f):f\in C_c^+(Y)\text{ and }\alpha(f)=g\,\}.$$

As in the proof of Proposition A.10 (see (A.5) on page 319), μ_0 is additive and extends to a linear functional on $C_c^+(X)$ with $\mu_0(g) \leq \|\nu\|\|g\|_\infty$ for all $g \in C_c^+(X)$. Since

$$\mu_0(g) \leq \|\varphi\|_\infty \int_X g(x)h(x)\,d\mu(x),$$

$\mu_0 \ll \mu$. Thus $\mu_0 = h \cdot \mu$ with $h \in L^1(X)$. If $f \in C_c(Y)$, then

$$|\nu(f)| \leq \nu(|f|) \leq \mu_0(\alpha(|f|)).$$

Thus ν is a Radon measure on $C_c(Y)$ with $\nu \ll \mu_0 \circ \alpha$, and there is a Radon-Nykodym derivative φ_0 with $\|\varphi_0\| \leq 1$ and $\nu = \varphi_0 \cdot (\mu_0 \circ \alpha) = \varphi_0 \cdot (h \cdot \mu \circ \alpha)$. Thus $\nu = \varphi_0 \otimes h_0$. Since $\|h_0\|_1 = \|\mu_0\|_1 \leq \|\nu\|$, we have

$$\|h_0\|_1 \leq \|\nu\| \leq \|\varphi_0\|\|h_0\|_1 \leq \|h_0\|_1.$$

Hence

$$\|\nu\| = \|h_0\|_1 \quad \text{and} \quad \|\varphi_0\|_\infty = 1.$$

Let $\epsilon > 0$. Then there is a g in the unit ball of $C_c^+(X)$ such that $\mu_0(g) > \mu_0(1) - \epsilon$. But there is a $f \in C_c^+(Y)_{\alpha,1}$ such that $\alpha(f) = g$. Then

$$\nu(f) > \mu_0(1) - \epsilon = \|h_0\|_1 - \epsilon.$$

Since ϵ was arbitrary, $\|h_0\|_1 = \|\nu\| \geq \sup_{f \in C_c^+(Y)_{\alpha,1}} \nu(f)$. Since the other direction is easy, we have established (A.14) and completed the proof. □

Exercises

Ex A.1. Provide the proof of the other half of Lemma A.16.

—Answer on page 384

A.5. A Dense Subspace of $\mathcal{E}^{**}$

As usual, we suppose that $\pi : Y \to X$ is a continuous G-equivariant surjection and that α is a (proper) equivariant π-system. Let $C_0(X, L^1(Y, \alpha))$ be the Banach algebra introduced in Section A.4. Proposition A.10 on page 319 allows us to identify the norm dual $C_0(X, L^1(Y, \alpha))^*$ with the space of α-bounded complex Radon measures $M(Y, \alpha)$. We let $C_0(X, L^1(Y, \alpha))^{**}_+$ be the collection of $m \in C_0(X, L^1(Y, \alpha))^{**}$ such that $m(\nu) \geq 0$ if $\nu \geq 0$. If $f \in C_c^+(Y)_{\alpha,1} \subset C_0(X, L^1(Y, \alpha))$, then let $m_f \in C_0(X, L^1(Y, \alpha))^{**}_+$ be the natural injection: $m_f(\nu) = \nu(f)$. Notice that if $\nu = \varphi \otimes h \in L^\infty(Y) \odot_{L^\infty(X)} L^1(X)$, then $m_f(\varphi \otimes h) = \langle \alpha(\varphi f)\, , h \rangle$ just as in Lemma A.18 on page 326.

Proposition A.19. *With respect to the natural inclusion $f \mapsto m_f$ of $C_c^+(Y)_{\alpha,1}$ into $C_0(X, L^1(Y, \alpha))^{**}$, $C_c^+(Y)_{\alpha,1}$ is weak-$*$ dense in the unit ball of $C_0(X, L^1(Y, \alpha))^{**}_+$.*

Proof. By Corollary A.14 on page 324, $\overline{M(G, [\mu] \circ \lambda)} = X([\mu])$ is a Banach subspace of $C_0(X, L^1(Y, \alpha))^*$, and

$$C_0(X, L^1(Y, \alpha))^* = \bigcup_{[\mu]} X([\mu])$$

where the union is taken over all measures classes of finite measures on X. The union is directed in the sense that given μ_1 and μ_2, $X([\mu_k]) \subset X([\mu_1 + \mu_2])$.

Let $R_{[\mu]}$ be the map from $C_0(X, L^1(Y, \alpha))^{**}$ to $X([\mu])^*$ given by restriction. Let $m \in C_0(X, L^1(Y, \alpha))^{**}_+$. Then a basic weak-$*$ open neighborhood of m is given by

$$U = \{\, n \in C_0(X, L^1(Y, \alpha))^{**} : |n(\nu_k) - m(\nu_k)| < \epsilon \text{ for } k = 1, \dots n \,\}$$

where $\{\, \nu_1, \dots \nu_n \,\} \subset C_0(X, L^1(Y, \alpha))^*$. We can assume that there is a μ such that $\{\, \nu_1, \dots, \nu_n \,\} \subset X([\mu])$. Then Lemma A.18 on page 326 implies that there is a $f \in C_c^+(Y)_{\alpha,1}$ such that

$$|R_{[\mu]}(m_f)(\nu_k) - R_{[\mu]}(m)(\nu_k)| < \epsilon \quad \text{for all } k.$$

Thus $m_f \in U$. This completes the proof. □

We are most interested in the case where π is the map $r : G \to G^{(0)}$ and α is a Haar system λ for G. Recall that in Section 9.6 we wrote $\mathcal{E}$ for $C_0(G^{(0)}, L^1(G, \lambda))$ and identified $\mathcal{E}^*$ with λ-bounded complex Radon measures $M(G, \lambda)$ on G. As above, we'll write $\mathcal{E}^{**}_+$ for the those $m \in \mathcal{E}^{**}$ such that $m(\nu) \geq 0$ if $\nu \geq 0$. An important example of such a m being a topological invariant mean. As an immediate corollary of Proposition A.19 on the preceding page we have the following.

PROPOSITION A.20. *We respect to the natural inclusion $f \mapsto m_f$ of $C_c^+(G)_{\lambda,1}$ into $\mathcal{E}^{**}$, $C_c^+(G)_{\lambda,1}$ is weak-$*$ dense in the unit ball of $\mathcal{E}^{**}_+$.*

A.6. An Alternative Topology

Recall that each $\nu \in L^\infty(Y) \odot_{L^\infty(X)} L^1(X)$ acts as a linear functional on $\mathcal{L}^\infty(X, L^1(Y, \alpha))$ via the pairing defined in (A.9). Hence we can equip $\mathcal{L}^\infty(X, L^1(Y, \alpha))$ with the weak topology τ_w induced by this pairing (see [**Ped89**, §2.3.8] for more details). Notice that if $\nu = \varphi \otimes h \in L^\infty(Y) \odot_{L^\infty(X)} L^1(X)$ with $h \in L^1(X)^+$ and if $f \in \mathcal{L}^\infty(X, L^1(Y, \alpha))$, then

$$\begin{aligned} |\langle f \,, \nu \rangle| &= \Big| \int_X \int_Y f(y)\varphi(y) \, d\alpha^x(y) h(x) \, d\mu(x) \Big| \\ &\leq \|\varphi\|_\infty \int_X \int_Y |f(y)| \, d\alpha^x(y) h(x) \, d\mu(x) \\ &= \|\varphi\|_\infty \|f\|_h, \end{aligned}$$

where we've defined

$$\|f\|_h := \int_X \int_Y |f(y)| \, d\alpha^x(y) h(x) \, d\mu(x).$$

Thus we can just as well equip $\mathcal{L}^\infty(X, L^1(Y, \alpha))$ with the topology τ generated by the seminorms $\| \cdot \|_h$ for all $h \in L^1(X)^+$. Since every ν in $L^\infty(Y) \odot_{L^\infty(X)} L^1(X)$ has a representative with $h \geq 0$, the τ-topology is finer than the τ_w-topology.

LEMMA A.21. *Let L be a linear functional on $\mathcal{L}^\infty(X, L^1(Y, \alpha))$ that is continuous for the τ-topology. Then L is continuous for the τ_w-topology and is given by paring with some $\nu \in L^\infty(Y) \odot_{L^\infty(X)} L^1(X)$.*

PROOF. There must be $h_1, \dots, h_n \in L^1(X)^+$ and $\epsilon > 0$ such that $\|f\|_{h_k} < \epsilon$ for all k implies $|L(f)| \leq 1$. If we let $h = (\frac{1}{\epsilon})(h_1 + \cdots + h_n)$, then $\|f\|_h \leq 1$ implies $|L(f)| \leq 1$. Thus $|L(f)| \leq \|f\|_h$ for all $f \in \mathcal{L}^\infty(X, L^1(Y, \alpha))$. The result now follows from Lemma A.11 on page 321. □

COROLLARY A.22. *Let $C \subset \mathcal{L}^\infty\big(X, L^1(Y,\alpha)\big)$ be a convex set with 0 in its τ_w-closure. Then there is a net $\{f_i\} \subset C$ such that for all $h \in L^1(X)$ we have*

$$\lim_i \int_X \int_Y |f_i(y)|\, d\alpha^x(y) h(x)\, d\mu(x) = 0. \tag{A.15}$$

PROOF. Since it suffices to prove (A.15) for $h \geq 0$. For this, it suffices to see that 0 is in the τ-closure of C. If not, we can separate C and 0 with a τ-continuous linear functional (by the Hahn-Banach Separation Theorem as in [**Ped89**, Theorem 2.4.7]). But since such functionals are τ_w-continuous by Lemma A.21 on the previous page, this is impossible. □

APPENDIX B

Remarks on Blanchard's Theorem

To complete the proof of Lemma 3.20 on page 77, we need to see that if $\pi : Y \to X$ is continuous and open, then there is a full π-system $\alpha = \{\alpha^x\}$ consisting of probability measures. This is a consequence of (1)$\Longrightarrow$(3) of [**Bla96**, Theorem 3.3]. The purpose of this section is to see how the result can be extracted from [**Bla96**, Theorem 3.3] and at the same time to repair a slight glitch in the proof.

Recall that A is called a $C_0(X)$-algebra if there is a homomorphism

$$\Phi_A : C_0(X) \to ZM(A)$$

into the center of the multiplier algebra which is nondegenerate in the sense that the linear span of $\Phi_A(C_0(X))(A)$ is dense in A. (More precisely, we should talk about the pair (A, Φ_A).) We discussed $C_0(X)$-algebras briefly at the beginning of Section 5.6 and more details can be found in [**Wil07**, §C.1]. We will retain the terminology and notations from Section 5.6 and [**Wil07**, §C.1].

EXAMPLE B.1. Let $\pi : Y \to X$ be a continuous map. Then $C_0(Y)$ is a $C_0(X)$ algebra with $\Phi_{C_0(Y)}(f) = f \circ \pi$. If in addition, π is open, then $C_0(Y)$ is a continuous field of C^*-algebras with fibres $C_0(Y)(x) = C_0(\pi^{-1}(\{x\}))$.

A positive linear functional of norm one on a C^*-algebra A is called a *state*, and we write $\mathcal{S}(A)$ for the collection of all states on A. If A is a unital $C_0(X)$-algebra, then we can restrict a state $\varphi \in \mathcal{S}(A)$ to a state $\varphi' \in \mathcal{S}(C_0(X))$ via

$$\varphi'(f) = \varphi(\Phi_A(f)).$$

In general, we need a bit more care. If $\varphi \in \mathcal{S}(A)$, then the GNS-construction (see [**RW98**, Proposition A.6]) provides a representation $\pi : A \to B(\mathcal{H}_\varphi)$ and a unit vector h_φ so that

$$\varphi(a) = \big(\pi_\varphi(a)h_\varphi \mid h_\varphi\big).$$

If we also have

$$\varphi(a) = \big(\pi(a)h \mid h\big)$$

for a representation π and unit vector h, then there is a unitary $U : H_\varphi \to \mathcal{H}_\pi$ intertwining π_φ and π and taking h_φ to h.

The GNS-representation π_φ extends to a representation $\bar{\pi}_\varphi$ of $M(A)$. Then we can define a state φ' on $C_0(X)$ by

$$\varphi'(f) = \big(\bar{\pi}_\varphi(\Phi_A(f))h_\varphi \mid h_\varphi\big).$$

The state φ' does not depend on our choice of GNS-representation and we follow Blanchard and call it the *restriction of* φ *to* $C_0(X)$.

DEFINITION B.2. If A is a $C_0(X)$-algebra, then $\mathcal{S}_X(A)$ is the set of states φ on A such that restriction φ' of φ to $C_0(X)$ is a point evaluation. If $\varphi \in \mathcal{S}_X(A)$, we let $p(\varphi) \in X$ be the unique point in X such that $\varphi'(f) = f(p(\varphi))$. We equip $\mathcal{S}_X(A)$ with the relative weak-$*$ topology.

REMARK B.3. If φ' is the restriction of φ and $\{e_i\}$ is an approximate identity for A, then

$$\varphi'(f) = \lim_i \varphi(f \cdot e_i).$$

Moreover, if $f \geq 0$, then $\{\varphi(f \cdot e_i)\}$ is increasing in i (Ex B.1).

LEMMA B.4. *We have $\varphi \in \mathcal{S}_X(A)$ if and only if there is a $x \in X$ such that $\varphi(f \cdot a) = f(x)\varphi(a)$ for all $f \in C_0(X)$ and $a \in A$ in which case $x = p(\varphi)$. Hence there is a one-to-one correspondence between $\varphi \in \mathcal{S}_X(A)$ with $p(\varphi) = x$ and states in $\mathcal{S}(A(x))$.*

PROOF. Let π_φ be the GNS-representation for φ and φ' its restriction to $C_0(X)$ as above. If $f \in C_0(X)$ and $a \in A$, then by Cauchy-Schwarz

$$\begin{aligned} |\varphi(f \cdot a)|^2 &= \big|\big(\bar{\pi}_\varphi(\Phi_A(f))h_\varphi \mid \pi_\varphi(a^*)h_\varphi\big)\big|^2 \\ &\leq \|\bar{\pi}_\varphi(\Phi_A(f))h_\varphi\|^2 \|\pi_\varphi(a^*)h_\varphi\|^2 = \varphi'(|f|^2)\varphi(aa^*). \end{aligned}$$

Thus if $\varphi \in \mathcal{S}_X(A)$, then φ factors through $A(p(\varphi))$, and $\varphi(f \cdot a) = f(p(\varphi))\varphi(a)$.

If φ factors through $A(x)$, then clearly $\varphi(f \cdot a) = f(x)\varphi(a)$. If $\{e_i\}$ is an approximate identity for A, then $\varphi'(f) = \lim_i \varphi(f \cdot e_i)$. It follows that $\varphi \in \mathcal{S}_X(A)$. □

Using Lemma B.4, we'll often identfy $\mathcal{S}(A(x))$ with the appropriate subset of $\mathcal{S}_X(A)$.

REMARK B.5. If $A = C_0(Y)$ as in Example B.1, then states $\varphi \in \mathcal{S}_X(A)$ with $p(\varphi) = x$ correspond to probability measures α^x with $\operatorname{supp} \alpha^x \subset \pi^{-1}(\{x\})$.

LEMMA B.6. *Suppose that A is a $C_0(X)$ algebra. Then $p : \mathcal{S}_X(A) \to X$ is continuous.*

PROOF. Suppose that $\varphi_j \to \varphi$ in $\mathcal{S}_X(A)$. We need to show that $p(\varphi_j) \to p(\varphi)$ in X.

But if $f \in C_0^+(X)$, then I claim $\liminf_j \varphi_j'(f) \geq \varphi'(f)$. To see this, let $\{e_i\}$ be an approximate identity for A and $\epsilon > 0$. Using Remark B.3, there is an i such that $\varphi(e_i f) > \varphi'(f) - \epsilon$. By continuity, we eventually have $\varphi_j(e_i f) > \varphi'(f) - \epsilon$. Since $\varphi_j'(f) \geq \varphi_j(e_i f)$, we eventually have $\varphi_j'(f) > \varphi'(f) - \epsilon$. Since ϵ is arbitrary, the claim follows.

Now if $p(\varphi_i) \not\to p(\varphi)$, then there is a neighborhood U of $p(\varphi)$ such that $p(\varphi_i)$ is not eventually in U. But if we choose $f \in C_c^+(X)$ with $f(p(\varphi)) = 1$ and $\operatorname{supp} f \subset U$, then $\varphi_j'(f) = f(p(\varphi_j))$ takes the value zero for arbitrarily large j. But $\varphi'(f) = 1$. This leads to a contradiction. □

The following version of the Hahn Banach Theorem will be useful.

LEMMA B.7. *Let $1 \in S \subset A_{s.a.}$ be a real subspace of the self-adjoint part of a unital C^*-algebra A. Let φ be a real linear functional on S such that $\varphi(1) = 1$ and $\varphi(S^+) \geq 0$ where $S^+ = S \cap A^+$. Then there is a state $\overline{\varphi} \in \mathcal{S}(A)$ extending φ.*

PROOF. Suppose that $a \in S$. Then $\|a\| \cdot 1 - a \in S \cap A^+$. Hence $\|a\| \geq \varphi(a)$. Replacing a by $-a$ shows that φ has norm 1. Hence the Hahn-Banach extension theorem gives us a real linear functional $\psi : A_{\text{s.a.}} \to \mathbf{R}$ of norm 1 extending φ. But it is not hard to see that

$$\overline{\varphi}(a + ib) = \psi(a) + i\psi(b) \quad \text{when } a, b \in A_{\text{s.a.}},$$

yields a (complex) linear functional on all of A; for example,

$$i\overline{\varphi}(a + ib) = i\psi(a) - \psi(b) = \overline{\varphi}(-b + ia) = \overline{\varphi}(i(a + ib)).$$

Also $\overline{\varphi}$ has the same norm as ψ. To see this, fix $z \in A$ and let $\tau \in \mathbf{T}$ be such that $\tau\overline{\varphi}(z) = |\overline{\varphi}(z)|$. Then, since $\tau\overline{\varphi}(z)$ is real, we have

$$|\overline{\varphi}(z)| = \tau\overline{\varphi}(z) = \overline{\varphi}(\tau z) = \psi\Big(\frac{1}{2}(\tau z + \bar{\tau} z^*)\Big) \leq \|\psi\|\|z\|.$$

Thus $\overline{\varphi}$ is linear functional of norm one on A with $\overline{\varphi}(1) = 1$. Hence $\overline{\varphi}$ is a state (cf., [**RW98**, Lemma A.9]). □

In [**Bla96**, §3], Blanchard makes use of the topological vector space A^*_c which is the usual norm dual of A but equipped with the topology of uniform convergence on compact sets. It is key to his argument that A^*_c is a Fréchet space.[1] (See the proof of [**Bla96**, Lemme 3.8].) Unfortunately, as shown in the next lemma, A^*_c is not a Fréchet space. I'm taking the opportunity to fix this minor gap here. Lemma B.8 is not used elsewhere and can be skipped without loss of context.

LEMMA B.8. *Let A be a separable C^*-algebra. Let A^*_c be the usual norm dual of A equipped with the topology of uniform convergence on compact subsets of A. Then A^*_c is a complete topological vector space. However, unless A is finite dimensional, A^*_c is not metrizable. Hence it is not a Fréchet space unless A is finite dimensional.*

PROOF. Although it would be easier just to show that A^*_c is sequentially complete, we go the extra foot and show its complete. Let $\{\varphi_\alpha\}_{\alpha \in A}$ be a net in A^*_c which is Cauchy in the sense that given any neighborhood V of 0 in A^*_c, there is a $\alpha(V)$ such that $\alpha, \beta \geq \alpha(V)$ implies that $\varphi_\alpha - \varphi_\beta \in V$. Thus given $a \in A$, we can let

$$V = \{\, \psi \in A^*_c : |\psi(a)| < \epsilon \,\}.$$

Then $\alpha, \beta \geq \alpha(V)$ implies $|\varphi_\alpha(a) - \varphi_\beta(a)| < \epsilon$. Since $a \in A$ was arbitrary and $\mathbf{C}$ is complete, there is a linear map $\varphi : A \to \mathbf{C}$ such that $\varphi_\alpha(a) \to \varphi(a)$ for all $a \in A$.[2] General nonsense implies that $\varphi_\alpha \to \varphi$ uniformly on compact sets.[3] To see that $\varphi \in A^*$, it suffices to see that φ is continuous at 0. Suppose that $a_n \to 0$. Let $\epsilon > 0$. Since $\{a_n\} \cup \{0\}$ is compact in A, there is an α_0 such that $|\varphi_{\alpha_0}(a_n) - \varphi(a_n)| < \epsilon/2$ for all n. Let N be such that $n \geq N$ implies $|\varphi_{\alpha_0}(a_n)| < \epsilon/2$. Then $n \geq N$ implies $|\varphi(a_n)| \leq |\varphi(a_n) - \varphi_{\alpha_0}(a_n)| + |\varphi_{\alpha_0}(a_n)| < \epsilon$. This proves completeness.

Notice that if ρ were any metric on the set A^* compatible with the topology on A^*_c, then ρ would be complete and A^*_c a Fréchet space. But it is easy to see that $B_n = \{\, \psi : \|\psi\| \leq n \,\}$ is a closed subset of A^*_c and that the relative topology

[1] A Fréchet space is a completely metrizable locally convex topological vector space. Alternatively, it is a topological vector space admitting a translation invariant complete metric.

[2] Note that if $\{\varphi_\alpha\}$ was sequence, we'd have $\varphi \in A^*$ by uniform boundedness. Then we'd essentially be done.

[3] Let K in A be compact and $\epsilon > 0$. Let $V = \{\, \psi : |\psi(a)| < \epsilon/2 \text{ for all } a \in K \,\}$. Let $\alpha \geq \alpha(V)$ and $a \in K$. Let β be such that $\beta \geq \alpha(V)$ and $|\varphi_\beta(a) - \varphi(a)| < \epsilon/2$. Then $|\varphi_\alpha(a) - \varphi(a)| < \epsilon$.

is the weak-$*$ topology. (See [**Bou87**, IV.5.24 Theorem 1] or [**Sch71**, Theorem IV.6.3] for more on this.) Thus B_n is not only closed but compact in A^*_c. But $A^*_c = \bigcup B_n$. Then the Baire Category Theorem would imply that some, and hence all, B_n had interior. But this is impossible since no nonempty open set in A^*_c can have bounded norm. To see this, let V be nonempty and open in A^*_c with $k = \sup\{\, \|\psi\| : \psi \in V \,\} < \infty$. We can normalize and assume that $k = 1$. Then V is open in B_1 and there are $a_i \in A$ such that

$$\{\, \psi \in B_1 : |\psi(a_i) - \varphi(a_i)| < \epsilon \text{ for } i = 1, 2, \ldots, n \,\} \subset V.$$

Let S be the subspace spanned by the a_i. Since we've assumed $S \neq A$, there are $\psi \in A^*$ of arbitrary norm satisfying $\psi(a_i) = 0$ for all i. Hence V contains a functional φ of norm one. Since V is open in A^*_c, there is a compact set $K \subset A$ such that $\{\, \psi : |\psi(a) - \varphi(a)| < \epsilon \text{ for all } a \in K \,\} \subset V$. But then $(1 + \delta)\varphi \in V$ for small δ. This contracts $V \subset B_1$. □

Since A^*_c is not Fréchet, we need a different version of Blanchard's [**Bla96**, Lemme 3.6]. Suppose that E is a locally convex topological vector space whose topology if given by a countable, separating family of seminorms $\{\, \|\cdot\|_k \,\}$. This topology is metrizable for the translation invariant metric

$$\rho(a, b) = \sum_{k=1}^{\infty} 2^{-k} \frac{\|a - b\|_k}{1 + \|a - b\|_k}. \tag{B.1}$$

A subspace B of E is complete if ρ restricts to a complete metric on B.[4] Note that a sequence $\{e_n\}$ is Cauchy in E if and only if it is Cauchy for each $\|\cdot\|_k$. Moreover, if $\{V_n\}$ is a fundamental system of convex neighborhoods of 0 in E, then $\{e_n\}$ is Cauchy if and only if given k there is a N such that $n, m \geq N$ implies $a_n - a_m \in V_k$.

LEMMA B.9. *Let E be a locally convex topological vector space whose topology is given by a countable separating family of seminorms as above. Let B be a complete convex subset. We suppose that T is a subset of B and that $p : T \to X$ is a continuous open surjection onto a paracompact space X such that $p^{-1}(x) \subset B$ is convex for each $x \in X$.*

(a) *For every convex neighborhood V of 0 in E, there is a continuous map $\varphi : X \to B$ such that*
$$\varphi(x) \in p^{-1}(x) + V \quad \textit{for all } x \in X.$$
(b) *There is a continuous map $\varphi : X \to B$ such that $\varphi(x) \in \overline{p^{-1}(x)}$ for all $x \in X$.*

REMARK B.10. The proof is a straightforward modification of the argument in [**Bla96**] which in turn is based on the proof of (a)$\Longrightarrow$(b) from [**Mic56**, Theorem 3.2″].

PROOF. Since the interior of V is convex (see [**Ped89**, E 2.4.3]), we can replace V by its interior and assume that V is open. Then $V' = V \cap B$ is open in B and for each $t \in T$, $U_t = (t - V') \cap T$ is open in T. Then $\mathscr{U} = \{\, p(U_t) \,\}_{t \in T}$ is an open cover of X. Since X is paracompact, there is a locally finite partition of unity

[4] If E is complete, then E is a Fréchet space.

$f = \{f_i\}_{i\in I}$ for X subordinate to $\mathscr{U}$ (see [**Wil70**, 20C]). Thus given $i \in I$, there is a $t(i) \in T$ such that $\operatorname{supp} f_i \subset p(U_{t(i)})$. Now define

$$\varphi(x) = \sum_{i\in I} f_i(x)t(i).$$

By assumption on f, each $x \in X$ has a neighborhood W so that f_i vanishes on W for all but finitely many i. Thus φ is continuous from X into E. But if $f_i(x) \neq 0$, then $t(i) \in p^{-1}(x) + V'$. Since $\sum_i f_i(x) = 1$ and $p^{-1}(x) + V'$ is convex, $\varphi(x) \in p^{-1}(x) + V' \subset B$. Therefore φ is B-valued; this establishes part (a).

(b) Let $\{V_n\}$ be a fundamental system of symmetric convex open neighborhoods of 0 in E such that $V_n + V_n \subset V_{n-1}$. There is no harm in replacing V_0 by E.

I claim there are continuous functions $\varphi_i : X \to B$ such that

(i) $\varphi_i(x) \in \varphi_{i-1}(x) + V_{i-2}$ for $i \geq 2$ and
(ii) $\varphi_i(x) \in p^{-1}(x) + V_i$ for $i \geq 1$.

We can define φ_1 using part (a) and proceed inductively. Assume we have defined $\varphi_1, \ldots, \varphi_k$ satisfying (i) and (ii). (We can let φ_0 be the zero map.) Let

$$T_{k+1} = \{\, t \in T : \varphi_k(p(t)) \in t - V_k \,\}.$$

Then T_{k+1} is open in T. Thus $p_{k+1} = p|_{T_{k+1}}$ is open. If $x \in X$, then $\varphi_k(x) \in p^{-1}(x) - V_k$, and there is a $t \in p^{-1}(x)$ such that $\varphi_k(x) \in t - V_k$. That is, $t \in T_{k+1}$. Hence p_{k+1} is surjective. Furthermore, I claim that

$$p_{k+1}^{-1}(x) = \big(\varphi_k(x) + V_k\big) \cap p^{-1}(x). \tag{B.2}$$

Thus p_{k+1} satisfies the hypotheses of part (a) provided we establish the claim. Suppose that t belongs to the right-hand side of (B.2). Then $t \in \varphi_k(p(t)) + V_k$ and $\varphi_k(p(t)) \in t - V_k$. Hence $t \in T_{k+1}$ and $t \in p_{k+1}^{-1}(x)$. On the other hand, if $t \in p_{k+1}^{-1}(x)$, then $p(t) = x$ and $\varphi_k(x) \in t - V_k$. Thus t belongs to the right-hand side of (B.2).

Hence there is a continuous function $\varphi_{k+1} : X \to B$ such that

$$\begin{aligned}\varphi_{k+1}(x) &\in p_{k+1}^{-1}(x) + V_{k+1} \\ &\in \varphi_k(x) + V_k + V_{k+1} \\ &\in \varphi_k(x) + V_{k-1}.\end{aligned}$$

This completes the induction argument.

Another induction argument implies that

$$\varphi_{n+k}(x) - \varphi_n(x) \in V_{n-2}.$$

If ρ is the associated metric as in (B.1), then $\{\varphi_k(x)\}$ is uniformly ρ-Cauchy in B. Since (B, ρ) is complete, it follows that $\varphi_n \to \varphi$ uniformly for some continuous function $\varphi : X \to B$. Moreover,

$$\varphi(x) = \varphi(x) - \varphi_n(x) + \varphi_n(x) \in \varphi(x) - \varphi_n(x) + p^{-1}(x) + V_{n-1} \quad \text{for all } n.$$

Hence $\varphi(x) \in \overline{p^{-1}(x)}$. This completes the proof. □

In order to able to invoke Lemma B.9 on the facing page, we are going to have to restrict to *separable* C^*-algebras for the rest of this section.

Recall that a state φ on a C^*-algebra A is called faithful if $\varphi(a^*a) = 0$ implies that $a = 0$.

LEMMA B.11 ([**Bla96**, Lemme 3.8]). *Suppose that X is a second countable compact Hausdorff space and that A is a* unital *separable $C(X)$-algebra with $A(x) \neq \{0\}$ for all x. Suppose that $p : \mathcal{S}_X(A) \to X$ is open. Then A admits a* continuous family of faithful states *in the sense that there is a continuous cross section $\Psi : X \to \mathcal{S}_X(A)$ for p with each $\Psi(x)$ faithful state on $A(x)$. In particular, for each $a \in A$, the map $x \mapsto \Psi(x)(a)$ is continuous.*

Since each fibre $A(x)$ is nonzero and A is unital, it follows that $C(X)$ embeds in A. Hence we can view it as a subalgebra of A. For the proof we need the following.

LEMMA B.12. *Suppose that as in the statement of Lemma* B.11, *A is a separable unital $C(X)$-algebra with $p : \mathcal{S}_X(A) \to X$ open. Then A is a continuous field over X. (That is, $x \mapsto \|a(x)\|$ is continuous for all $a \in A$.)*

PROOF. We need to show that $x \mapsto \|a(x)\|$ is continuous. Since A is a $C(X)$-algebra, it suffices to see that $\{ x : \|a(x)\| > \epsilon \}$ is open. If not, there is a sequence $x_n \to x$ such that $\|a(x_n)\| \leq \epsilon$ while $\|a(x)\| > \epsilon$. Since $a(x)$ is positive, there is a $\varphi \in \mathcal{S}_X(A)$ with $p(\varphi) = x$ and $\varphi(a) = \|a(x)\|$. But after passing to a subsequence and relabeling we can assume there are $\varphi_n \to \varphi$ in $\mathcal{S}_X(A)$ with $p(\varphi_n) = x_n$. Then we eventually have $\varphi_n(a) > \epsilon$. This contradicts the fact that $\varphi_n(a) \leq \|a(x_n)\|$. □

PROOF OF LEMMA B.11. If $\{b_n\}$ is a countable dense subset of the unit sphere $\{ b \in A : \|b\| = 1 \}$ of A, then

$$\rho(\varphi, \psi) := \sum_n \frac{1}{2^n} |\varphi(b_n) - \psi(b_n)|$$

is a metric on $E := A^*$ which is compatible with the locally convex topology induced by the seminorms $\|\varphi\|_n = |\varphi(b_n)|$. If $B = \{ \psi \in A^* : \|\psi\| \leq 1 \}$, then the relative topology on B is the weak-$*$ topology. In particular, (B, ρ) is complete. If we let T be an open subset of $\mathcal{S}_X(A)$, then we can apply Lemma B.9 to the restriction of $p : \mathcal{S}_X(A) \to X$ to T. Using this, Blanchard's proof can be followed *mutatis mutandis.*

Let $\{a_n\}$ be a countable dense set in the positive part of the unit sphere $\{ a \geq 0 : \|a\| = 1 \}$ in A. We can suppose that $a_1 = 1$. By Lemma B.12, for each $a \in A$ we can let $P(a) \in C(X)$ be the function $x \mapsto \|a(x)\|$.

For each n, then set

$$\Omega_n := \{ \varphi \in \mathcal{S}_X(A) : \varphi\big(a_n - \tfrac{1}{2}P(a_n)\big) > 0 \}$$

is open in $\mathcal{S}_X(A)$. Hence $p|_{\Omega_n} : \Omega_n \to X$ is a continuous open surjection with convex fibres. Hence we can apply Lemma B.9(b) to get a section $\Psi_n : X \to \mathcal{S}_X(A)$ such that $\Psi_n(x)(a_n) \geq \frac{1}{2}P(a_n)(x) = \frac{1}{2}\|a_n(x)\|$.

Let Ψ be the field of states given by $\sum 2^{-n}\Psi_n$. Let $x \in X$ and a' a nonzero positive element of $A(x)$. Multiplying a' by a nonzero constant if necessary, we may as well assume there is a norm one positive element $a \in A$ with $a(x) = a'$. Choose n such that $\|a(x) - a_n(x)\| < \frac{1}{3}\|a(x)\|$. But this forces $\frac{2}{3}\|a(x)\| < \|a_n(x)\|$. Hence $\|(a - a_n)(x)\| < \frac{1}{2}\|a_n(x)\|$. Therefore

$$\big|\Psi_n(x)(a_n) - \Psi_n(x)(a)\big| = \big|\Psi_n(x)(a_n - a)\big| < \frac{1}{2}\|a_n(x)\|.$$

Since we also have $\Psi_n(x)(a_n) \geq \frac{1}{2}\|a_n(x)\|$. Hence $\Psi_n(x)(a) > 0$. Of course, this means $\Psi(x)(a) > 0$ and each $\Psi(x)$ is faithful as required. □

To get a version of Lemma B.11 for a not necessarily unital $C_0(X)$-algebra A, we follow Blanchard. Let X^+ be the one-point compactification of X and view $C(X^+)$ as a subalgebra of $C^b(X) \cong M(C_0(X))$ so that we can extend the central $C_0(X)$-action on A to a central $C(X^+)$-action. We let $\mathcal{A}(X^+)$ be the vector space $A + C(X^+)$ with a $*$-algebra structure given by

$$(a+f)(b+g) = (ab + f\cdot b + g\cdot a) + fg \quad \text{and} \quad (a+f)^* = a^* + \bar{f}.$$

We need to equip $\mathcal{A}(X^+)$ with a suitable C^*-norm. To do this, consider the $*$-isomorphism u of $A + C(X^+)$ into $M(A \oplus C(X^+))$ where

$$u(a)(b,h) = (ab, 0) \qquad u(f)(b,h) = (f\cdot b, fh).$$

Thus $a + f$ acts on $A \oplus C(X^+)$ via the matrix

$$a + f \mapsto \begin{pmatrix} a+f & 0 \\ 0 & f \end{pmatrix}.$$

I claim that the image of $A + C(X^+)$ is closed in $M(A \oplus C(X^+))$. To see this, note that we can identify $u(a)$ with A in $M(A \oplus C(X^+))$. Suppose that $u(a_n + f_n) \to T$. Let $q : M(A \oplus C(X^+)) \to M(A \oplus C(X^+))/(A \oplus C(X^+))$ be the quotient map. Then $q(u(a_n)) = 0$ and $q(u(f_n))$ must converge to $q(u(g))$ for some $g \in C(X^+)$. (The range of $C(X^+)$ must be closed.) But then $u(a_n + f_n - g) \to T - u(g)$, and $T - u(g)$ is in the kernel of q. Say $T - u(g) = \left(\begin{smallmatrix} a & 0 \\ 0 & h \end{smallmatrix}\right)$. Thus $f_n - g \to h$, and as multipliers on A, $a_n \to a + h$. Since the image of A is closed, $h = 0$ and $u(a_n + f_n) \to u(a + g)$. Hence we can pull back the C^*-norm to get a norm on $\mathcal{A}(X^+)$. Now it is straightforward to check that $\mathcal{A}(X^+)$ is a $C(X^+)$-algebra with respect to $f \cdot (a,h) = (f\cdot a, fh)$.

LEMMA B.13 ([**Bla96**, Proposition 3.2]). *If A is a $C_0(X)$-algebra which is a continuous field over X, then $\mathcal{A}(X^+)$ is a continuous field over X^+.*

PROOF. Let $\alpha \in \mathcal{A}(X^+)$. We need to see that $x \mapsto \|\alpha(x)\|$ is continuous on X^+. Since $\|\alpha(x)\|^2 = \|\alpha^*(x)\alpha(x)\|$, we can assume that $\alpha \geq 0$. Let $\alpha = a + f$. Since f is the image of α in $C(X^+)$ under the quotient map, we have $f \geq 0$. Since α is self-adjoint, so is a. Let $a = a_+ - a_-$ with $a_\pm \geq 0$ and $a_+ a_- = 0$.

Let $B(x)$ be the unital subalgebra of $\mathcal{A}(X^+)(x)$ generated by $\alpha(x)$ (see Ex B.3). Let Ψ be the Gelfand Transform mapping $B(x)$ onto some $C(\Delta)$. Then it follows that

$$\|\alpha(x)\| = \|\Psi(\alpha(x))\|_\infty = \|\Psi(a(x)) + f(x)\|_\infty = \|\Psi(a_+(x))\|_\infty + f(x).$$

The result follows from the assumed continuity of $x \mapsto \|a_+(x)\| = \|\Phi(a_+(x))\|_\infty$. □

LEMMA B.14. *Suppose that A is a $C_0(X)$-algebra and $\varphi \in \mathcal{S}_X(A)$. Then φ extends to a state φ' in $\mathcal{S}_{X^+}(\mathcal{A}(X^+))$ given by $\varphi'(a+f) = \varphi(a) + f(p(\varphi))$. Furthermore, if $\varphi_n \to \varphi$ in $\mathcal{S}_X(A)$, then $\varphi'_n \to \varphi'$ in $\mathcal{S}_{X^+}(\mathcal{A}(X^+))$.*

PROOF. Let $p(\varphi) = x$. Note that $\varphi(f\cdot a) = f(x)\varphi(a)$. Define $\varphi'(a+f) = \varphi(a) + f(x)$. To see that φ' is a state, it will suffice to see that φ' is positive on $\mathcal{A}(X^+)$—then it will have norm $\varphi'(1) = 1$. But we just compute as in [**RW98**, Lemma A.7]:

$$\begin{aligned}\varphi'\big((a+f)^*(a+f)\big) &= \varphi'(a^*a + \bar{f}\cdot a + f\cdot a^* + |f|^2) \\ &= \varphi(a^*a) + 2\operatorname{Re}(\overline{f(x)}\varphi(a)) + |f(x)|^2\end{aligned}$$

which, since $\varphi(a^*a) \geq |\varphi(a)|^2$ and $\varphi(a^*) = \overline{\varphi(a)}$, is

$$\geq |f(x)|^2 - 2|f(x)||\varphi(a)| + |\varphi(a)|^2$$
$$\geq (|f(x)| - \varphi(a))^2 \geq 0$$

It is clear by construction that φ' is multiplicative on the image of $C(X^+)$ in $\mathcal{A}(X^+)$.

The last statement is clear since $p : \mathcal{S}_X(A) \to X$ is continuous and $a_n + f_n \to a + f$ implies that $a_n \to a$ and $f_n \to f$ as in the discussion preceding Lemma B.13 on the previous page. □

We need to see that the openness of $p : \mathcal{S}_X(A) \to X$ follows whenever A is a continuous field. The proof comes from [**Bla96**, Lemme 3.4].

LEMMA B.15. *Let A be a separable* unital *$C(X)$-algebra with $A(x) \neq \{0\}$ for all x. If A is a continuous field over X, then $p : \mathcal{S}_X(A) \to X$ is open.*

PROOF. Let $O \subset \mathcal{S}_X(A)$ be open with $x \in p(O)$. By Ex B.4, there is a faithful state $\varphi \in O \cap \mathcal{S}(A(x))$. We'll use the notation

$$U(c_1, \dots, c_n; \delta)$$

for the basic open neighorhood of φ in $\mathcal{S}_X(A)$ given by

$$\{\, \psi \in \mathcal{S}_X(A) : |\psi(c_k) - \varphi(c_k)| < \delta \text{ for all } 1 \leq k \leq n \,\}. \tag{B.3}$$

Using this notation and Ex B.5, we can find self-adjoint elements $a_1, \dots, a_m$ such that

$$V := U(a_1, \dots, a_m; 2\epsilon) \subset O.$$

Since we're working with states, there is no harm in assuming $a_1 = 1_A$.

After reordering if need be, we can assume that for some $n \leq m$, $\{\, a_1(x), \dots, a_n(x) \,\}$ is a basis for the *real* subspace spanned by $\{\, a_1(x), \dots, a_m(x) \,\}$. Thus if $n < m$ and $n < j \leq m$, there are real constants λ_k^j such that

$$a_j(x) = \sum_{k=1}^{n} \lambda_k^j a_k(x).$$

We let

$$U = \{\, y \in X : \|a_j(y) - \sum_k \lambda_k^j a_k(y)\| < \epsilon \text{ for } n < j \leq m \,\}.$$

(If $n = m$, let $U = X$.) Let

$$V' = U(a_1, \dots, a_n; \frac{\epsilon}{nR})$$

where $R = \max\{\, 1; |\lambda_k^j| \,\}$. Since $\psi \in \mathcal{S}_X$ implies that $|\psi(a)| \leq \|a(p(\psi))\|$, we have

$$V' \cap p^{-1}(U) \subset V \subset O.$$

For each $y \in X$, define $\alpha_y : \mathbf{R}^n \to A(y)$ by

$$\alpha_y(\lambda) = \sum_{k=1}^{n} \lambda_k a_k(y).$$

Thus the range of α_y is a real-linear subspace $S(y)$ of $A(y)$.

Let

$$\mathbf{S}^n = \{\, \lambda \in \mathbf{R}^n : \max_{1 \leq k \leq n} |\lambda_k| = 1 \,\}.$$

Since the set $E = \{ (y, \lambda) \in U \times \mathbf{S}^n : \alpha_y(\lambda) = 0 \}$ is closed, Ex B.6 implies that the projection $\pi(E)$ is closed in U. But $x \notin p(E)$. Hence we can shrink U if necessary so that U is still a neighborhood of x, and $y \in U$ implies that

$$\{ a_1(y), \dots, a_n(y) \}$$

is linearly independent (over $\mathbf{R}$). This means that for all $y \in U$, the map $\alpha_y(\lambda) \mapsto \alpha_x(\lambda)$ is a well-defined real-linear map of $S(y)$ onto $S(x)$.

Since $S(x)$ is the real span of self-adjoint elements, $S(x) \subset A(x)_{\text{s.a.}}$ and φ is real-valued on $S(x)$. Hence we can define

$$F = \{ (y, \lambda) \in U \times \mathbf{S}^n : \alpha_y(\lambda) \geq 0 \text{ and } \varphi(\alpha_x(\lambda)) \leq 0 \}.$$

To see that F is closed, notice that $\alpha_y(\lambda) \in A(y)_{\text{s.a.}}$. Hence $\alpha_y(\lambda) \geq 0$ in $A(y)$ if and only if

$$\big\| \alpha_y(\lambda) - \|\alpha_y(\lambda)\| 1_y \big\| \leq \|\alpha_y(\lambda)\|. \tag{B.4}$$

Since A is continuous field, $(a, y) \mapsto \|a(y)\|$ is continuous. Using (B.4) it is straightforward to see that F is closed.

Note that if $\lambda \in \mathbf{S}^n$, then $\alpha_x(\lambda) \neq 0$. Since φ is faithful, it follows that $x \notin \pi(F)$ where π is the projection onto U. By Ex B.6, there is a neighborhood $U' \subset U$ of x such that $y \in U'$ implies $y \notin \pi(F)$.

It follows that if $y \in U'$ and if $\lambda \in \mathbf{R}^n$ is such that $\alpha_y(\lambda) \geq 0$, then $\varphi\big(\alpha_x(\lambda)\big) \geq 0$. Thus we can apply Lemma B.7 on page 332 to the functional ψ' given by $\alpha_y(\lambda) \mapsto \varphi\big(\alpha_x(\lambda)\big)$ to a state $\psi \in \mathcal{S}(A(y))$. By construction, $\psi(a_k) = \varphi(a_k)$ and we trivially have $\psi \in V'$. Hence $U' \subset p(V) \subset p(O)$. This completes the proof. □

PROPOSITION B.16 (Fields of Faithful States). *Let A be a separable $C_0(X)$-algebra such that $A(x) \neq \{0\}$ for all x and such that the continuous surjection $p : \mathcal{S}_X(A) \to X$ is open. Then A admits a continuous family $\{\varphi_x\}$ of faithful states.*

PROOF. The idea is to apply Lemma B.11 to $\mathcal{S}_{X^+}(\mathcal{A}(X^+))$. We verify the hypotheses of Lemma B.11. Suppose $x \in X$. Then there is a state $\varphi \in \mathcal{S}_X(A)$ such that $p(\varphi) = x$. Then φ' is a state on $\mathcal{A}(X^+)$ factoring through $\mathcal{A}(X^+)(x)$. Hence $\mathcal{A}(X^+)(x) \neq \{0\}$ and the fibre over the point at infinity is clearly nontrivial. We still need to see that $p^+ : \mathcal{S}_{X^+}(\mathcal{A}(X^+)) \to X^+$ is open. By Lemma B.12 on page 336, A is a continuous field over X. Hence $\mathcal{A}(X^+)$ is a continuous field over X^+ by Lemma B.13 on page 337. Hence p^+ is open by Lemma B.15 on the preceding page.

Now we can apply Lemma B.11 to produce a continuous family $\{\varphi_x\}_{x \in X^+}$ of faithful states on $\mathcal{A}(X^+)$. Unfortunately, the collection of restrictions to A inside $\mathcal{A}(X^+)$ need not have norm one, nor is it clear that there norms vary continuously. Instead, we do what Blanchard does. Let $\{a_n\}$ be a countable dense set in the unit ball of A. Define ψ on A by

$$\psi_x(a) := \sum 2^{-n} \varphi_x(a_n a a_n^*).$$

Since the uniform sum of continuous functions is continuous, $\{\psi_x\}$ is a continuous family of nonzero positive linear functionals on A with continuously varying norms $\psi_x(1) = \sum 2^{-n} \varphi_x(a_n a_n^*)$. Thus we get a continuous family of states via $\Psi_x = \psi_x(1)^{-1}\psi_x$. Note that if $a \geq 0$ and $\Psi_x(a) = 0$, then $a_n(x)a(x)a_n(x)^* = 0$ for all n. Hence $a(x) = 0$. □

PROPOSITION B.17. *Suppose that A is a separable $C_0(X)$-algebra which is a continuous field over X and such that $A(x) \neq \{0\}$ for all x. Then $p : \mathcal{S}_X(A) \to X$ is open.*

PROOF. If A is unital, this is just Lemma B.15 on page 338. Otherwise, let $\mathcal{A}(X^+)$ the associated field over X^+ which is a continuous field by Lemma B.13 on page 337. Hence $p^+ : \mathcal{S}_{X^+}(\mathcal{A}(X^+)) \to X^+$ is open by Lemma B.15. Define $\chi : \mathcal{S}_{X^+}(\mathcal{A}(X^+)) \to [0,1]$ by letting $\chi(\psi)$ equal the norm of the restriction of ψ to A. Note that if $\chi(\psi) > d$, then there is a $a \in A$ of norm one such that $|\psi(a)| > d$. Hence if $\psi_i \to \psi$, then eventually $|\psi_i(a)| > d$. Hence χ is lower-semicontinuous.

Let $\varphi \in \mathcal{S}_X(A)$ and let $U(a_1, \dots, a_n; \epsilon)$ be a basic open neighborhood of φ. We can assume that $\|a_k\| \leq 1$ for all k. Let φ' be the extension of φ to $\mathcal{A}(X^+)$ given by Lemma B.14 on page 337. Then

$$W := \{\, \psi' \in \mathcal{S}_{X^+}(\mathcal{A}(X^+)) : \\ |\psi'(a_k) - \varphi'(a_k)| < \frac{\epsilon}{2} \text{ for } 1 \leq k \leq n \,\} \cap \chi^{-1}\big((\frac{2}{2+\epsilon}, 1]\big)$$

is an open neighborhood of φ' in $\mathcal{S}_{X^+}(\mathcal{A}(X^+))$.

If $\psi' \in W$, then $\chi(\psi')^{-1}\psi'$ restricts to state $\psi \in \mathcal{S}_X(A)$ belonging to $U(a_1, \dots, a_n; \epsilon)$. Clearly $p(\psi) = p^+(\psi')$. Therefore $p\big(U(a_1, \dots, a_n; \epsilon)\big)$ contains $p^+(W)$ which is a neighborhood of $p(\varphi)$ in X. □

Now we can complete the proof of Lemma 3.20 on page 77 as a corollary.

COROLLARY B.18. *Suppose that Y is second countable and that $\pi : Y \to X$ be a continuous, open surjection. Then there is a full π-system $\alpha = \{\alpha^x\}$ consisting of probability measures.*

PROOF. As we have observed earlier, $A = C_0(X)$ is a separable $C_0(X)$-algebra which is a continuous field over X. Since π is surjective, $A(x) \neq \{0\}$ for all x. Moreover $p : \mathcal{S}_X\big(C_0(Y)\big) \to X$ is open by Proposition B.17. Therefore Proposition B.16 on the previous page implies that there is a continuous family of faithful states $\{\varphi_x\}$. These states correspond to measures α^x, and since these states are faithful, these measures must have full support: $\operatorname{supp} \alpha^x = \pi^{-1}(\{x\})$. Moreover

$$x \mapsto \int_Y f(x)\, d\alpha^x(y) = \varphi_x(f)$$

is continuous. This completes the proof. □

EXERCISES

Ex B.1. Suppose that A is a $C_0(X)$-algebra and that φ' is the restriction of $\varphi \in \mathcal{S}(A)$ to $C_0(X)$. Show that for any approximate identity $\{e_i\}$ in A, we have $\varphi'(f) = \lim_i \varphi(f \cdot e_i)$. Furthermore, if $f \in C_0^+(X)$, it follows that $\{\varphi(f \cdot e_i)\}$ is increasing with limit $\varphi'(f)$.

—Answer on page 384

Ex B.2. Let A^*_c be A^* equipped with the topology of uniform convergence on compact sets. Show that the relative topology on $B_n = \{\, \varphi \in A^*_c : \|\varphi\| \leq n \,\}$ is the weak-$*$ topology.

—Answer on page 385

Ex B.3. Let A be a $C_0(X)$-algebra and $\mathcal{A}(X^+)$ the corresponding $C(X^+)$-algebra as in Lemma B.13 on page 337. Let $\mathcal{A}(x)$ be the $*$-subalgebra of $M(A(x) \oplus \mathbf{C})$ generated by the operators $\begin{pmatrix} a+\lambda & 0 \\ 0 & \lambda \end{pmatrix}$ in $A(x) \oplus \mathbf{C}$. (Here $A(\infty) := \{0\}$.) Show that the map φ given by $a + \varphi \mapsto \begin{pmatrix} a(x)+f(x) & 0 \\ 0 & f(x) \end{pmatrix}$ factors through a $*$-isomorphism of $\mathcal{A}(X^+)(x)$ onto $\mathcal{A}(x)$.

—Answer on page 385

Ex B.4. Let A be a separable C^*-algebra. Then the collection of faithful states is dense in $\mathcal{S}(A)$.

—Answer on page 385

Ex B.5. Each $\varphi \in \mathcal{S}(A)$ has a neighborhood basis consisting of open sets $U(a_1, \ldots, a_n; \delta)$ (as defined in (B.3)) with each a_k self-adjoint.

—Answer on page 385

Ex B.6. Suppose that Y is compact. Then the projection $\pi : X \times Y \to X$ is a closed map.

—Answer on page 385

APPENDIX C

The Inductive Limit Topology

The inductive limit topology makes $C_c(X)$ into a locally convex topological vector space and is described in detail in [**RW98**, §D.2].[1] The term comes from viewing $C_c(X)$ as the union of $C_c(X;K) = \{ f \in C_c(X) : \operatorname{supp} f \subset K \}$ over compact subsets K of X. This topology is the unique largest locally convex topology on $C_c(X)$ such that the inclusions $\iota_K : C_c(X;K) \hookrightarrow C_c(X)$ are continuous (see [**Wil07**, Propositions D.1 & D.7]). Moreover, any linear map $F : C_c(X) \to M$ into a locally convex space M is continuous if and only if $F \circ \iota_K : C_c(X;K) \to M$ is continuous for all compact K. A local base for the topology (that is, a neighborhood basis at 0) is given by the collection β of balanced convex sets $U \subset C_c(X)$ such that $\iota_K^{-1}(U) = U \cap C_c(X;K)$ is open in $C_c(X;K)$ for all compact sets K. (Here, $C_c(X;K)$ is given the usual norm topology from the supremum norm.)

In particular, $W = \{ f \in C_c(X) : \|f\|_\infty < \epsilon \}$ is in β. It follows that if $\{f_n\}$ is a net converging to f in the inductive limit topology, then $f_n \to f$ uniformly. Conversely, if $f_n \to f$ uniformly and there is a compact set K such that eventually $\operatorname{supp} f_n \subset K$, then the f_n are eventually in $C_c(X;K)$ and it follows that $f_n \to f$ in the inductive limit topology. We say that such nets are *eventually compactly supported*. Even though it is widely used here and elsewhere in the literature, it is just a lousy topology. We want to think of it as the topology were a net $\{ f_n \}$ converges to f if and only if $f_n \to f$ uniformly and is eventually compactly supported. As we just noted, the "if" direction is true. Sadly, as examples such as [**RW98**, Example D.9] show, *the converse is false*! Worse, in [**RW98**, Example D.9], the space X is just **Z**, so we can't blame the pathology on X not being second countable. But we do have the following.

LEMMA C.1. *Suppose that X is a second countable locally compact Hausdorff space. Then a* sequence *$\{f_n\}$ converges to f in the inductive limit topology on $C_c(X)$ if and only if $f_n \to f$ uniformly and is eventually compactly supported.*

PROOF. We only need to demonstrate the "only if" part. We suppose $f_n \to 0$ in the inductive limit topology. Since X is second countable, $X = \bigcup K_n$ with each K_n compact and contained in the interior of K_{n+1}. Hence every compact set in X is contained in some K_n. If the assertion is false, we can pass to a subsequence, relabel, and assume $\operatorname{supp} f_n \not\subset K_n$. Hence there is a $x_n \notin K_n$ such that $f_n(x_n) \neq 0$. Then

$$U = \{ g \in C_c(X) : |g(x_j)| < |f_j(x_j)| \text{ for all } j \}$$

[1] A topological vector space V is *locally convex* if there is a neighborhood basis of 0—called a *local basis* consisting of convex open sets. It follows from [**Rud73**, Theorem 1.14] that V has a local basis of *balanced* open convex sets.

is a balanced convex set. If K is compact, then $K \subset K_n$ for some n. Then

$$\iota_K^{-1}(U) = \{ g \in C_c(X;K) : |g(x_j)| < |f_j(x_j)| \text{ for } 1 \le j < n \}$$

is open in $C_c(X;K)$. Hence U is a neighborhood of 0 in the inductive limit topology. But no f_n is in U. This is a contradiction.

In general, if $f_n \to f$, we can replace f_n with $f_n - f$ and the result follows. □

Since any subsequence of the net in [**RW98**, Example D.9] would not be eventually compactly supported, it means that no subsequence of that net can converge. Hence the inductive limit topology on $C_c(X)$ is not necessarily first countable even when X is second countable. This means sequences can not be used to describe the topology. Nevertheless, [**RW98**, Lemma D.10] says that to check whether a linear map $F : C_c(X) \to M$ is continuous from $C_c(X)$ with the inductive limit topology to any locally convex space M, it just suffices to check that $F \circ \iota_K$ is continuous on $C_c(X;K)$. Since sequences do suffice to determine the topology on $C_c(X;K)$ we can show that F is continuous by showing that $F(f_n) \to F(f)$ for any *sequence* $\{ f_n \}$ converging uniformly to f which is eventually compactly supported.[2]

If X is second countable, we can also see that $C_c(X)$ is separable. (Recall that a topological space is separable if it has a countable dense subset.)

LEMMA C.2 ([**FD88**, §II.14.10]). *Suppose that X is a second countable locally compact Hausdorff space. Then $C_c(X)$ is separable in the inductive limit topology.*

Lemma C.2 is a consequence of the next result which is slightly stronger.

PROPOSITION C.3. *Suppose that X is a second countable locally compact Hausdorff space. Then there is a countable dense subset $D \subset C_c(X)$ in the inductive limit topology. Furthermore, if $f \in C_c(X)$, then there there is a* sequence $\{f_n\} \subset D$ *such that $f_n \to f$ uniformly with supports eventually contained in a compact set K.*

PROOF. Let $X = \bigcup K_n$ with each K_n compact and contained in the interior of K_{n+1}. Note that every $f \in C_c(X)$ is contained in some $C_c(X;K_n)$. Let D_n' be a countable dense subset of $C(K_{n+1})$ and fix $h_n \in C_c^+(X)$ such that $0 \le h \le 1$, $h(x) = 1$ for all $x \in K_n$ and such that h_n vanishes off K_{n+1}. If $g \in D_n'$, then $hg \in C_c(X;K_{n+1})$. Moreover if $f \in C_c(X;K_n)$ there is a sequence $\{g_k\} \subset C(K_{n+1})$ such that $g_k \to f$ uniformly on K_{n+1}. Then $hg_k \to f$ in the inductive limit topology. Let $D_n = \{ hg : g \in D_n' \}$, and let $D = \bigcup D_n$. □

EXERCISES

Ex C.1. Suppose that X is second countable. Show that $C_0(X)$ is a separable Banach algebra.

—Answer on page 385

[2]This is just [**RW98**, Lemma D.10] together with the above observation that sequences suffice.

APPENDIX D

Ramsay Almost Everywhere

It is well known that if Σ is a conull subsemigroup of a locally compact group $\mathcal{G}$, then $\Sigma = \mathcal{G}$ (see [**Wil07**, Lemma D.36]). We also know that if $\pi : \mathcal{G} \to \mathcal{H}$ is a Borel map between locally compact groups such that $\{ (g, g') \in \mathcal{G} \times \mathcal{G} : \pi(gg') = \pi(g)\pi(g') \}$ is conull, then there is a Borel homomorphism $\pi_0 : \mathcal{G} \to \mathcal{H}$ such that $\pi_0 = \pi$ almost everywhere (see [**Zim84**, Theorem B.2]).[1] In this section, we want to give Arlan Ramsay's generalization of these results to groupoids—namely Proposition D.2 on the following page and Theorem D.8 on page 350. The material here is based on [**Ram71**], [**Ram82**] and [**Muh99**]. Ramsay worked in the more general setting of Borel groupoids, so we can take a few shortcuts in the locally compact case.

To maintain decorum, we assume that all our measures are all σ-finite and that Borel spaces are analytic. Throughout, G will be a second countable locally compact Hausdorff groupoid with a Haar system $\lambda = \{\lambda^u\}_{u \in G^{(0)}}$. We'll fix a quasi-invariant measure μ on $G^{(0)}$. As usual, we write ν for the Radon measure

$$\nu(f) = \int_{G^{(0)}} \int_G f(\gamma)\, d\lambda^u(\gamma)\, d\mu(u). \tag{D.1}$$

We will use the fact the decompositions such as in (D.1) pass to non-negative Borel functions (Proposition 3.6 on page 71).

We fix a probability measure ρ which is equivalent to ν. (We will write $\rho \sim \nu$ to indicate equivalent measures.) Note that since μ is quasi-invariant, we have $\nu \sim \nu^{-1}$ and $\rho \sim \rho^{-1}$. Replacing ρ by $\frac{1}{2}(\rho + \rho^{-1})$, we can assume that $\rho^{-1} = \rho$. The push forward, $r_*\rho$, by the range map $r : G \to G^{(0)}$ is also a probability measure. Since $\rho = \rho^{-1}$, we have $r_*\rho = s_*\rho$. Hence $r_*\rho$ is what we called the saturation of μ as defined in Lemma 7.3 on page 158. Since μ is quasi-invariant, $r_*\varphi \sim \mu$ by Lemma 7.3.

We can use [**Wil07**, Theorem I.5] and decompose ρ as

$$\rho(f) = \int_{G^{(0)}} \int_G f(\gamma)\, \nu^u(\gamma)\, dr_*\rho(u) \tag{D.2}$$

where the $\{\nu^u\}$ are a Radon family (see Definition 3.1 on page 70) of measures with ν^u a probability measure for μ-almost all u and $\operatorname{supp} \nu^u \subset G^u$. Furthermore, the ν^u are uniquely determined μ-almost everywhere (since $r_*\rho \sim \mu$).

[1] Note that as a Borel homormophism, π_0 is actually continuous by [**Wil07**, Theorem D.3].

Note that

$$\begin{aligned}\rho(f) &= \int_G f(\gamma)\frac{d\rho}{d\nu}(\gamma)\,d\nu(\gamma)\\ &= \int_{G^{(0)}}\int_G f(\gamma)\frac{d\rho}{d\nu}(\gamma)\,d\lambda^u(\gamma)\,d\mu(u)\\ &= \int_{G^{(0)}}\int_G f(\gamma)\frac{d\rho}{d\nu}(\gamma)\,d\lambda^u(\gamma)\frac{d\mu}{d(r_*\rho)}(u)\,dr_*\rho(u)\\ &= \int_{G^{(0)}}\int_G f(\gamma)\,d\lambda_1^u(\gamma)\,dr_*\rho(u),\end{aligned}$$

where

$$\lambda_1^u = \frac{d\rho}{d\nu}\cdot\frac{d\mu}{d(r_*\rho)}\cdot\lambda^u.$$

Since we can assume both $\frac{d\rho}{d\nu}$ and $\frac{d\mu}{d(r_*\rho)}$ take values in $(0,\infty)$, each $\lambda_1^u \sim \lambda^u$. The uniqueness in (D.2) implies that $\lambda_1^u = \nu^u$ for a μ-conull set $U \subset G^{(0)}$. Hence

$$\gamma\cdot\nu^{s(\gamma)} \sim \gamma\cdot\lambda^{s(\gamma)} = \lambda^{r(\gamma)} \sim \nu^{r(\gamma)} \quad \text{provided } \gamma\in G|_U. \tag{D.3}$$

LEMMA D.1. *Let G be a second countable locally compact Hausdorff groupoid with a Haar system $\{\lambda^u\}_{u\in G^{(0)}}$ and quasi-invariant measure μ. Let ν, ρ and ν^u be as above. Then the Radon measure*

$$\rho^{(2)}(F) := \int_{G^{(0)}}\int_G\int_G F(\gamma,\eta)\,d\nu_u(\gamma)\,d\nu^u(\eta)\,dr_*\rho(u)$$

is equivalent to the Radon measure

$$\nu^{(2)}(F) := \int_{G^{(0)}}\int_G\int_G F(\gamma,\eta)\,d\lambda_u(\gamma)\,d\lambda^u(\eta)\,d\mu(u)$$

PROOF. Use the Radon Nikodym derivatives for $\nu\sim\rho$ and $r_*\rho\sim\mu$. □

Note that if $U\subset G^{(0)}$ is μ-conull, then $r^{-1}(U) = G^U$ is ν-conull and $s^{-1}(U) = G_U$ is ν^{-1}-conull. Since μ is quasi-invariant, G_U is also ν-conull. Hence the reduction $G|_U = G_U\cap G^U$ is ν-conull.[2] We can now state and prove the first of Ramsay's results.

PROPOSITION D.2. *Let G be a second countable locally compact Hausdorff groupoid with a Haar system $\{\lambda^u\}_{u\in G^{(0)}}$ and quasi-invariant measure μ. Suppose that Σ is a subset of G containing a ν-conull set and such that $\Sigma^2\subset\Sigma$. Then there is a μ-conull set $U\subset G^{(0)}$ such that $G|_U\subset\Sigma$.*

PROOF. Let ν, ρ and ν^u be as above. Replacing Σ with $\Sigma\cap\Sigma^{-1}$, we can assume that Σ is a subgroupoid of G (with possibly smaller unit space). Let B be a ν-conull Borel subset of Σ. Hence

$$1 = \rho(B) = \int_{G^{(0)}}\nu^u(B)\,dr_*\rho(u).$$

Since $r_*\rho$ is a probability measure, it follows that $\nu^u(B) = \nu^u(G^u\cap B) = 1$ for μ-almost all u. In view of (D.3), there is a μ-conull set U such that for all $\gamma\in G|_U$ we have $\nu^{s(\gamma)}(B) = \nu^{r(\gamma)}(B) = 1$ and $\gamma\cdot\nu^{s(\gamma)}\sim\nu^{r(\gamma)}$.

[2] We called such reductions inessential earlier.

I claim that $G|_U \subset \Sigma$. Let $\gamma \in G|_U$. Then $\nu^{r(\gamma)}(G^{r(\gamma)} \cap B) = 1$ and $\gamma \cdot \nu^{s(\gamma)} \sim \nu^{r(\gamma)}$. Thus $\nu^{s(\gamma)}(\gamma^{-1} \cdot (G^{r(\gamma)} \cap B)) = 1$. But we also have $\nu^{s(\gamma)}(G^{s(\gamma)} \cap B) = 1$. Hence

$$\gamma^{-1} \cdot (G^{r(\gamma)} \cap B) \cap (G^{s(\gamma)} \cap B) \neq \emptyset.$$

Thus there is an $\eta \in G^{r(\gamma)} \cap B$ and a $\zeta \in G^{s(\gamma)} \cap B$ such that $\gamma^{-1}\eta = \zeta$. But then $\gamma = \eta\zeta^{-1} \in BB \subset \Sigma$. This completes the proof. □

Now we state and prove a stepping stone to the second of Ramsay's results.

LEMMA D.3. *Let G be a second countable locally compact Hausdorff groupoid with a Haar system $\{\lambda^u\}_{u\in G^{(0)}}$ and quasi-invariant measure μ. Let $\varphi : G \to H$ be a Borel map into an analytic Borel groupoid which is an* almost everywhere homomorphism *in that*

$$B = \{ (\gamma, \eta) \in G^{(2)} : \varphi(\gamma)\varphi(\eta) = \varphi(\gamma\eta) \}$$

is $\nu^{(2)}$-conull. Then there is a Borel map $\psi : G \to H$ and a μ-conull set $U \subset G^{(0)}$ such that $\psi = \varphi$ ν-almost everywhere and the restriction of ψ to $G|_U$ is a homomorphism.

REMARK D.4. Since $G|_U$ is ν-conull, this result can be interpreted as almost saying that "an almost everywhere homomorphism is equal almost everywhere to a homomorphism".

PROOF. Let ν, ρ and ν^u be as above. Lemma D.1 implies $\nu^{(2)} \sim \rho^{(2)}$. Hence B is $\rho^{(2)}$-conull and

$$1 = \rho^{(2)}(B) = \int_{G^{(0)}} \nu_u \times \nu^u(B)\, r_*\rho(u).$$

Hence there is a μ-conull set $U_1 \subset G^{(0)}$ such that $\nu_u \times \nu^u(B) = 1$ if $u \in U_1$. Hence if $u \in U_1$, then

$$\varphi(\gamma\eta) = \varphi(\gamma)\varphi(\eta) \quad \nu_u \times \nu^u\text{-almost everywhere on } G_u \times G^u.$$

Since

$$\nu_u \times \nu^u(B) \le \nu_u \times \nu^u(G_u \times G^u) = \nu_u(G_u)\nu^u(G^u) \le 1,$$

we see that $u \in U_1$ implies $\nu_u(G_u) = 1 = \nu^u(G^u)$. We can shrink U_1 if necessary and also assume that $\gamma \cdot \nu^{s(\gamma)} \sim \nu^{r(\gamma)}$, or equivalently $\nu_{r(\gamma)} \cdot \gamma \sim \nu_{s(\gamma)}$, for any $\gamma \in G|_{U_1}$.

Since H is analytic, we can use Kuratowski's Theorem (see [**Wil07**, Theorem I.40]) to assume that H is an analytic subset of $[0, 1]$, and in particular, that φ is a Borel map of G into $[0, 1]$.

Fix once and for all an element $h_0 \in H$.

Define $f : G^{(2)} \to H$ by

$$f(\gamma, \eta) = \begin{cases} \varphi(\gamma)^{-1}\varphi(\gamma\eta) & \text{if } \varphi(\gamma)^{-1} \text{ and } \varphi(\gamma\eta) \text{ are composible, and} \\ h_0 & \text{otherwise.} \end{cases}$$

To see that f is a Borel map, note that $f_1(\gamma, \eta) = (\varphi(\gamma)^{-1}, \varphi(\gamma\eta))$ is certainly Borel from $G^{(2)}$ to $H \times H$. But $A := f_1^{-1}(H^{(2)})$ is Borel. Since multiplication is Borel on H^2,

$$f(\gamma, \eta) = \begin{cases} \varphi(\gamma)^{-1}\varphi(\gamma\eta) & \text{if } (\gamma, \eta) \in A \text{ and} \\ h_0 & \text{otherwise} \end{cases}$$

is Borel.

Now define $\varphi_1 : G \to [0,1]$ by

$$\varphi_1(\eta) = \int_G f(\gamma, \eta)\, d\nu_{r(\eta)}(\gamma).$$

We can use Proposition 3.3 on page 70 to see that φ_1 is Borel.

Since

$$1 = \int_{G^{(0)}} \nu_u \times \nu^u(B)\, dr_*\rho(u) = \int_G \int_{G_{r(\eta)}} \mathbb{1}_B(\gamma, \eta)\, d\nu_{r(\eta)}(\gamma)\, d\rho(\eta),$$

we see that for ν-almost all η,

$$D(\eta) = \{\, \gamma \in G_{r(\eta)} : (\gamma, \eta) \in G^{(2)} \text{ and } \varphi(\gamma\eta) = \varphi(\gamma)\varphi(\eta) \,\}$$

is $\nu_{r(\eta)}$-conull. One consequence of this is that

$$\varphi = \varphi_1 \quad \nu\text{-almost everywhere.}$$

Let G_2 be the set of $\eta \in G$ such that there is a constant $\varphi_2'(\eta)$ such that

$$\{\, \gamma \in G_{r(\eta)} : \varphi(\gamma)^{-1}\varphi(\gamma\eta) \text{ is defined and equal to } \varphi_2'(\eta) \,\}$$

is $\nu_{r(\eta)}$-conull. Note that G_2 contains the set

$$B_2 = \{\, \eta \in G : D(\eta) \text{ is conull} \,\}.$$

It follows that G_2 is ν-conull.

Define $\varphi_2 : G \to H$ by

$$\varphi_2(\eta) = \begin{cases} \varphi_2'(\eta) & \text{if } \gamma \in G_2 \text{ and} \\ h_0 & \text{otherwise.} \end{cases}$$

Let G_1 be the ν-conull reduction $G|_{U_1}$, and set $\Sigma = G_1 \cap G_2$. Since $B_2 \cap G_1 \subset \Sigma$, it follows that Σ contains a Borel ν-conull set. We want to see that $\Sigma^2 \subset \Sigma$.

Suppose that $(\eta, \zeta) \in G^{(2)} \cap (\Sigma \times \Sigma)$. Since $\Sigma \subset G_1 = G|_{U_1}$, we clearly have $\eta\zeta \in G_1$. Showing that $\eta\zeta \in G_2$ is trickier.

Since $\eta \in G_2$,

$$\{\, \gamma \in G_{r(\eta)} : \varphi(\gamma)\varphi_2(\eta) = \varphi(\gamma\eta) \,\} \quad \text{is } \nu_{r(\eta)}\text{-conull.} \tag{D.4}$$

Similarly, $\zeta \in G_2$ implies

$$\{\, \beta \in G_{r(\zeta)} : \varphi(\beta)\varphi_2(\zeta) = \varphi(\beta\zeta) \,\} \quad \text{is } \nu_{r(\zeta)}\text{-conull.}$$

Since $G_{r(\eta)} \cdot \eta = G_{r(\zeta)}$,

$$\begin{aligned} \{\, \gamma \in G_{r(\eta)} : \varphi(\gamma\eta)\varphi_2(\zeta) = \varphi(\gamma\eta\zeta) \,\} \\ = \{\, \beta \in G_{r(\zeta)} : \varphi(\beta)\varphi_2(\zeta) = \varphi(\beta\zeta) \,\} \cdot \eta^{-1}. \end{aligned} \tag{D.5}$$

Since $\nu_{r(\eta)} \cdot \eta \sim \nu_{s(\eta)} = \nu_{r(\zeta)}$ (as $\eta \in G_1$), it follows that (D.5) is $\nu_{r(\eta)}$-conull.

Hence the function $\gamma \mapsto \varphi(\gamma)^{-1}\varphi(\gamma\eta\zeta)$ is equal to $\gamma \mapsto \varphi(\gamma)^{-1}\varphi(\gamma\eta)\varphi_2(\zeta)$ $\nu_{r(\eta)}$-almost everywhere by (D.5). But this is equal to $\gamma \mapsto \varphi_2(\eta)\varphi_2(\zeta)$ $\nu_{r(\eta)}$-almost everwhere by (D.4). Since $r(\eta) = r(\eta\zeta)$, we have shown that $\eta\zeta \in G_2$. Thus $\Sigma^2 \subset \Sigma$ as claimed. We have also shown that $\varphi_2(\eta\zeta) = \varphi_2(\eta)\varphi_2(\zeta)$ for $(\gamma, \eta) \in \Sigma^{(2)}$.

Now we can evoke Proposition D.2 on page 346 to find a μ-conull set $U_0 \subset G^{(0)}$ such that $G_0 := G|_{U_0} \subset \Sigma$. Then we define $\varphi_0 : G \to H$ by

$$\varphi_0(\eta) = \begin{cases} \varphi_2(\eta) & \text{if } \eta \in G_0 \text{ and} \\ h_0 & \text{otherise.} \end{cases}$$

To see that φ_0 is Borel, it suffices to see that $\varphi_0|_{G_0}$ is Borel. But if $\eta \in G_0$, then $\eta \in G_2$ and $\varphi_1(\eta) = \varphi_2(\eta)$ since $\nu_{r(\eta)}$ is a probability measure. But φ_1 is Borel on all of G. This suffices.

We saw that φ_2 is a homomorphism on Σ, so it follows that $\varphi_0|_{G_0}$ is a homomorphism. But $\varphi_0 = \varphi_1$ on the conull set G_0 and $\varphi = \varphi$ almost everywhere. Hence $\varphi = \varphi_0$ almost everywhere, and we're done. □

Producing a homomorphism φ_0 on a ν-conull reduction seems tantalizingly close to producing a homomorphism on all of G. Notice that if U were *invariant*—that is, if $U = G \cdot U$ for the natural G-action on $G^{(0)}$—then G is the disjoint union of $G|_U$ and $G|_F$ where $F = G^{(0)} \setminus U$. Then we could extend φ_0 anyway we please on $G|_F$. The remainder of this section is devoted to some hocus-pocus that lets us get an invariant U.

DEFINITION D.5. Let $U \subset G^{(0)}$ be Borel and $\theta : G^{(0)} \to G^U$ a Borel cross section for $s : G \to G^{(0)}$ (so that $s(\theta(u)) = u$ for all $u \in G^{(0)}$) such that $\theta(u) = u$ for all $u \in U$. Then

$$\psi_\theta(\gamma) = \theta(r(\gamma))\gamma\theta(s(\gamma))^{-1}$$

is called the *reduction homomorphism* corresponding to θ.

Notice that the reduction homomorphism ψ_θ is a bona fide groupoid homomorphism of G into G^U. Furthermore, ψ_θ is the identity on $G|_U$.

LEMMA D.6. *Suppose that $A \subset G$ is σ-compact. Then $s(A)$ is σ-compact and there is a Borel section $\alpha : s(A) \to A$ for $s|_A : A \to G^{(0)}$.*

PROOF. Let $A = \bigcup K_n$ with each K_n compact and $K_n \subset K_{n+1}$. Then $s(A) = \bigcup s(K_n)$ is certainly σ-compact and therefore Borel. The map

$$s|_{K_n} : K_n \to s(K_n)$$

is a continuous map (hence Borel) between Polish spaces. Since every open set in K_n is σ-compact, its image under s is Borel. Furthermore, $s^{-1}(u) \cap K_n$ is closed for all $u \in s(K_n)$. It follows from [**Arv76**, Theorem 3.4.1] that $s|_{K_n}$ has a Borel cross section $\alpha_n : s(K_n) \to K_n$. Define $\alpha : s(A) \to A$ by $\alpha(u) = \alpha_n(u)$ if $u \in s(K_n) \setminus s(K_{n-1})$. This suffices. □

LEMMA D.7. *Let G be a second countable locally compact Hausdorff groupoid with a Haar system $\{\lambda^u\}$ and quasi-invariant measure μ. Suppose that $U_0 \subset G^{(0)}$ is μ-conull. Then there is a σ-compact, μ-conull, invariant set $U \subset G^{(0)}$ and a reduction homomorphism $\psi : G|_U \to G|_{U \cap U_0}$.*

PROOF. Since μ is a Radon measure on the second countable space $G^{(0)}$, it is σ-finite. Hence there is a conull F_σ-set $U_1 \subset U_0$ (see Ex D.2). Hence $A = r^{-1}(U_1)$ is the countable union of closed sets and hence σ-compact. Then $s(A) = s(r^{-1}(U_1))$ is σ-compact and saturated (Ex 5.1.1). Since it contains U_1, it is also conull. By Lemma D.6, there is a Borel section $\alpha : U \to r^{-1}(U_1)$. Define

$$\theta(u) = \begin{cases} u & \text{if } u \in U \cap U_0 \text{ and} \\ \alpha(u) & \text{if } u \in U \setminus U_0. \end{cases}$$

Then θ is a section yielding the required reduction homomorphism $\psi_\theta : G|_U \to G|_{U \cap U_0}$. □

Now we can give the second of Ramsay's results which make precise the assertion that "an almost everywhere homomorphism is equal almost everywhere to a homomorphism" for locally compact groupoids. (The term "almost everywhere homomorphism was formalized in Lemma D.3 on page 347.")

THEOREM D.8. *Let G be a second countable locally compact Hausdorff groupoid with a Haar system λ and quasi-invariant measure μ. Let H be an analytic Borel groupoid and $\varphi : G \to H$ be a Borel $\nu^{(2)}$-almost everywhere homomorphism. Then there is a Borel homomorphism $\varphi_1 : G \to H$ such that $\varphi = \varphi_1$ ν-almost everywhere.*

PROOF. By Lemma D.3 on page 347, there is a conull Borel set U_0 and a Borel map $\varphi_0 : G \to H$ whose restriction to $G|_{U_0}$ is a homomorphism and $\varphi = \varphi_0$ almost everywhere.

By Lemma D.7 on the preceding page, there is a Borel conull invariant set $U \subset G^{(0)}$ and a reducing homomorphism $\psi : G|_U \to G|_{U\cap U_0}$. Let $v \in H^{(0)}$. Define

$$\varphi_1(\gamma) = \begin{cases} \varphi_0\big(\psi(\gamma)\big) & \text{if } \gamma \in G|_U \text{ and} \\ v & \text{if } \gamma \notin G|_U. \end{cases}$$

Clearly, φ_1 is Borel. Since U is invariant, φ_1 will be a homomorphism provided its restriction to $G|_U$ is a homomorphism. However, φ_0 is a homomorphism on $G|_{U_0}$ and ψ maps $G|_U$ into $G|_{U_0}$. Hence φ_1 is a homomorphism. Since ψ is the identity on $G|_{U\cap U_0}$, $\varphi_1 = \varphi_0$ on $G|_{U\cap U_0}$ which is ν-conull. Hence $\varphi_1 = \varphi$ almost everywhere are required. □

An application of Ramsay's Theorem D.8 is the following.

COROLLARY D.9. *Let $L = (\mu, G^{(0)} * \mathscr{H}, U, \hat{L})$ be an almost everywhere representation of a locally compact groupoid G with a Haar system $\{\lambda^u\}_{u\in go}$. Let $\nu = \mu \circ \lambda$. Then there is an essential representation $L' = (\mu, G^{(0)} * \mathscr{H}, V, \hat{L}')$ such that $L = L'$ ν-almost everywhere.*

PROOF. By Ramsay's Theorem D.8, there is a Borel homomorphism $L' : G \to \operatorname{Iso}(G^{(0)} * \mathscr{H})$ which coincides with L ν-almost everywhere. The rub is that L' might not be a representation in that we might have $r(L'(\gamma)) \neq r(\gamma)$ for some γ. But on $G|_U$, $r(L(\gamma)) = r(\gamma) = r(L'(\gamma)) = r(L'(r(\gamma)))$ for ν-almost all, and hence ρ-almost all, γ. Hence there must be a μ-conull set $V \subset U$ such that $r(L'(\gamma)) = r(\gamma)$ for all $\gamma \in G|_V$. This implies $s(L'(\gamma)) = s(\gamma)$ for $\gamma \in G|_V$. Hence $(\mu, G^{(0)} * \mathscr{H}, V, \hat{L}')$ is an essential representation (and still equal to L almost everywhere). □

EXERCISES

Ex D.1. Let μ be a Radon probability measure on X and suppose that $f \in \mathcal{B}^+(X)$ takes values in $[0, 1]$ and is such that

$$\int_X f(x)\, d\mu(x) = 1.$$

Show that $f(x) = 1$ for μ-almost all x.

—Answer on page 386

Ex D.2. Let μ be a σ-finite Radon measure on X. So in particular, for every measurable set $E \subset X$, we have

$$\mu(E) = \inf\{\, \mu(V) : V \text{ is open and } E \subset V \,\}.$$

Show that for every measurable set E and $\epsilon > 0$ there is an open set V and a closed set F such that $F \subset E \subset V$ and $\mu(V \setminus F) < \epsilon$. In particular, there is a F_σ-set U and a G_δ set G such that $U \subset E \subset G$ and such that $\mu(G \setminus U) = 0$.

—Answer on page 386

Answers to Some of the Exercises

Ex 1.2.1. (a) Suppose $(\gamma,\eta)\in G^{(2)}$. Then

$$r(\eta)=\eta\eta^{-1}=((\gamma^{-1}\gamma)\eta)\eta^{-1}=\gamma^{-1}((\gamma\eta)\eta^{-1})=\gamma^{-1}\gamma=s(\gamma).$$

On the other hand, suppose that $s(\gamma)=r(\eta)$. Then $(\gamma,\gamma^{-1}\gamma)=(\gamma,\eta\eta^{-1})\in G^{(2)}$. But we also have $(\eta\eta^{-1},\eta)\in G^{(2)}$. Thus by associativity, $(\gamma,(\eta\eta^{-1})\eta)=(\gamma,\eta)\in G^{(2)}$.

(b) We have $(\eta^{-1},\gamma^{-1})\in G^{(2)}$ by part (a). Similarly, $(\eta^{-1}\gamma^{-1},\gamma\eta)\in G^{(2)}$. But then $\eta^{-1}\gamma^{-1}=\eta^{-1}\gamma^{-1}(\gamma\eta)(\gamma\eta)^{-1}=(\gamma\eta)^{-1}$.

Ex 1.2.3. To show that $\varphi(\gamma)^{-1}=\varphi(\gamma^{-1})$ it suffices to see that $\varphi(\gamma^{-1})\varphi(\gamma)=\varphi(\gamma)^{-1}\varphi(\gamma)$. But $\varphi(\gamma)\varphi(\gamma^{-1})\varphi(\gamma)=\varphi(\gamma)$. Now let $u\in G^{(0)}$. Then $u=u^2=u^{-1}$. Then $\varphi(u)=\varphi(u)^2=\varphi(u)^{-1}$, and $\varphi(u)\in H^{(0)}$.

Ex 1.2.4. Suppose that $(a,b)\in H^{(2)}$. Say $(a,b)=(\varphi(\gamma),\varphi(\eta))$. Then $\varphi(\gamma^{-1}\gamma)=\varphi(\gamma)^{-1}\varphi(\gamma)=a^{-1}a=bb^{-1}=\varphi(\eta)\varphi(\eta)^{-1}=\varphi(\eta\eta^{-1})$. Then $(\varphi^{-1}(a),\varphi^{-1}(b))=(\gamma,\eta)\in G^{(2)}$. Furthermore $\varphi^{-1}(a)\varphi^{-1}(b)=\gamma\eta=\varphi^{-1}(ab)$. Thus φ^{-1} is a homomorphism as claimed.

Ex 1.2.5. Let $\{\alpha_i\beta_i\}$ be a net in AB. It will suffice to see it has a convergent subnet. Since A is compact, we can pass to a subnet, relabel, and assume that $\alpha_i\to\alpha$. Since B is compact, we can pass to another subnet, relabel and assume $\beta_i\to\beta$. But since multiplication is continuous $\alpha_i\beta_i\to\alpha\beta$.

Ex 1.2.6. Suppose that $\gamma\beta\in UV$. Let $\{\gamma_i\}$ be a net in G converging to $\gamma\beta$. It suffices to prove a subnet is is eventually in UV.

We must have $r(\gamma_i)\to r(\gamma)$. Since r is an open map, we can pass to a subnet, relabel, and assume that there are $\zeta_i\to\gamma$ with $r(\zeta_i)=r(\gamma_i)$ (using Fell's Criterion). Since U is open, we eventually have ζ_i in U. But $\zeta_i^{-1}\gamma_i\to\beta$. Since V is open, we can eventually have $\zeta_i^{-1}\gamma_i$ in V. But then $\gamma_i=\zeta_i(\zeta_i^{-1}\gamma_i)$ is eventually in UV.

Ex 1.2.7. Let G be the space X. (That is, G is the diagonal in $X\times X$ viewed as an equivalence relations—we called this the co-trivial equivalence relations in the text.) Then $UV=U\cap V$ and it is easy to produced an appropriate example.

Ex 1.2.8. The product of closed sets AB need not be closed even in a group. Let $G=\mathbf{R}^2$ with respect to addition. Let $A=\{\,(0,x):x\leq -1\,\}$ and $B=\{\,(x,\frac{1}{x}):x\in(0,1]\,\}$. Then $(0,-n)+(\frac{1}{n},n)=(\frac{1}{n},0)$ is in $A+B$. But $(0,0)\notin A+B$.

But if, say A is compact, and $a_ib_i\to c$. Then we can pass to a subnet, relabel, and assume $a_i\to a\in A$. Then $b_i\to a^{-1}c$ and $a^{-1}c\in B$. Then $c=a(a^{-1}c)\in AB$.

Ex 1.2.10. $H^{(0)}=\{\,\gamma\in H:\gamma=\gamma^{-1}=\gamma^2\,\}\subset G^{(0)}$.

Ex 1.2.11. Suppose that $\operatorname{supp}F\subset K\times K$ with K compact in G^u. Let N be a compact neighborhood of u in G. Then KN is compact by Ex 1.2.5. If the assertion were false, then for every open neighborhood V of u contained in N we can find $\gamma_V, \gamma'_V, \eta_V, \eta'_V \in G^u$ such that $\gamma_V^{-1}\gamma'_V, \eta_V^{-1}\eta'_V \in V$ and

$$|F(\gamma_V, \eta_V) - F(\gamma'_V, \eta'_V)| \geq \epsilon. \tag{D.6}$$

We must have $\gamma_V, \gamma'_V, \eta_V, \eta'_V \in KN$. Hence we can pass to a subnet, relabel, and assume that there are $\gamma, \eta \in KN$ such that $\gamma_V \to \gamma$, $\gamma'_V \to \gamma$, $\eta_V \to \eta$ and $\eta'_V \to \eta$. This eventually contradicts (D.6).

Ex 1.3.1. This is an immediate consequence of the observation that inversion is a homeomorphism and $r(\gamma) = s(\gamma^{-1})$.

Ex 1.3.2. The given map is a homeomophism of G^u onto G^v and $\{u\}$ is clearly open in G^u. Hence $\{\eta\}$ is open in G^v. (This is [**Ren80**, Lemma I.2.7(i)].)

Ex 1.3.3. Let $u \in G^{(0)}$. We need to see that $\lambda(f)$ is continuous at u. Let K be a compact neighborhood of u in $G^{(0)}$. Let $\varphi \in C_c^+(G)$ be such that $\varphi \equiv 1$ on the compact set $r^{-1}(K) \cap \operatorname{supp} f$. Then $g = \varphi f$ is in $C_c(G)$ and $\lambda(g)$ is necessarily continuous at u. But $\lambda(f)(v) = \lambda(g)(v)$ if $v \in K$, so this suffices.

Ex 1.3.4. (a) Let φ be the characteristic function of $G^{(0)}$ in $C(G)$. Then φ has r-compact support and $\alpha := \lambda(\varphi)$ is a continuous function on $G^{(0)}$ such that $\alpha(u) = \lambda^u(\{u\})$. Since $\{u\}$ is open in G^u, $\alpha(u) > 0$ since λ^u has full support. By left invariance of λ,

$$\int_G \varphi(\gamma^{-1}\beta)\, d\lambda^{r(\gamma)}(\beta) = \int_G \varphi(\beta)\, d\lambda^{s(\gamma)}(\beta),$$

and it follows that $\lambda^u(\{\gamma\}) = \alpha(s(\gamma))$ for all $\gamma \in G^u$.

(b) Let $\sigma = \{\sigma^u\}$ be the appropriate family of counting measures on G. Define $\Phi : C_c(G, \lambda) \to C_c(G, \sigma)$ by $\Phi(f)(\gamma) = \alpha(s(\gamma))f(\gamma)$. Observe that $\lambda(f) = \sigma(\Phi(f))$. The assertion follows.

Ex 1.3.5. Since $G^x = \{x\} \times \mathcal{G} \times X$, it is clear that $\operatorname{supp} \lambda^x = G^x$. Let $\xi(x) \in C_c(\mathcal{G} \times X)$ be given by $\xi(x)(g, y) = f(x, g, y)$. Suppose that $\operatorname{supp} f \subset K_1 \times C \times K_2$ with each K_i compact in X and C compact in $\mathcal{G}$. If $x_i \to x$ in X, then $\xi(x_i) \to \xi(x)$ pointwise on $\mathcal{G} \times X$. Furthermore $|\xi(x_i)| \leq \|f\|_\infty \mathbb{1}_{C \times K_2}$. Hence $\lambda(f)(x_i) \to \lambda(f)(x)$ by the Dominated Convergence Theorem. For left-invariance, let $\eta = (x, g, y)$. Then

$$\begin{aligned} \int_G f(\eta\gamma)\, d\lambda^y(\gamma) &= \int_{\mathcal{G}} \int_X f(x, g, y)(y, h, z)\, d\mu(h)\, d\mu(z) \\ &= \int_{\mathcal{G}} \int_X f(x, gh, z)\, d\mu(h)\, d\mu(z) \\ &= \int_{\mathcal{G}} \int_X f(x, h, z)\, d\mu(h)\, d\mu(z) \\ \int_G f(\gamma)\, d\lambda^x(\gamma). \end{aligned}$$

Ex 1.3.6. It is clear that the support of $\lambda^x \subset G^x = \{\, (x, g, g^{-1} \cdot x) : g \in G \,\}$. On the other hand if $f \in C_c^+(G)$ is nonzero on G^x, then $g \mapsto f(x, g, g^{-1} \cdot x)$ is positive, continuous and nonzero on G. Hence $\lambda^x(f) > 0$. Thus $\operatorname{supp} \lambda^x = G^x$. Continuity and left-invariance follow as in the previous problem.

Ex 1.3.7. Let $f \in C_c^+(G)$ be such that $f(\gamma) = 1$ for all $\gamma \in K$. Then we can let $M = \|\lambda(f)\|_\infty$.

Ex 1.3.8. Since $\operatorname{supp} \lambda^x = \{x\} \times X$, we can write $\lambda^x = \delta_x \times \mu^x$ for a Radon measure μ^x on X with full support. Let $x, y \in X$ and $f \in C_c(X \times X)$. Then by left-invariance,

$$\int_{\{y\}\times X} f\big((x,y)(y,z)\big)\, d\lambda^y(y,z) = \int_X f(x,z)\, d\mu^y(z) = \int_X f(x,z)\, d\mu^x(z).$$

Letting $f(x,y) = g(x)h(y)$ we see that $\mu^x = \mu^y$. Since x and y are arbitrary, we're done.

Ex 1.4.1. We have

$$(f^*)^*(\gamma) = (\overline{f^*(\gamma^{-1})}) = (\overline{\overline{f((\gamma^{-1})^{-1})}}) = f(\gamma).$$

Hence $f^{**} = f$.

Then we compute

$$\begin{aligned} g^* * f^*(\gamma) &= \int_G g^*(\eta) f^*(\eta^{-1}\gamma) d\lambda^{r(\gamma)}(\eta) = \int_G \overline{f(\gamma^{-1}\eta) g(\eta^{-1})} d\lambda^{r(\gamma)}(\eta) \\ &= \overline{\int_G f(\gamma^{-1}\eta) g(\eta^{-1}) d\lambda^{r(\gamma)}(\eta)} = \overline{\int_G f(\gamma^{-1}(\gamma\eta)) g((\gamma\eta)^{-1}) d\lambda^{s(\gamma)}(\eta)} \\ &= \overline{\int_G f(\eta) g(\eta^{-1}\gamma^{-1}) d\lambda^{s(\gamma)}(\eta)} = \overline{f * g(\gamma^{-1})} = (f*g)^*(\gamma) \end{aligned}$$

Thus, $g^* * f^* = (f * g)^*$.

Ex 1.4.2. Let K be a compact set such that $\operatorname{supp} f \subset K$ and $K = K^{-1}$. Then we have already seem there is a M such that $\lambda^u(K) \leq M$ for all u. Since K is symmetric, $\lambda_u(K) \leq M$ for all u. But then

$$\|f\|_{I,r} = \sup_{u \in G^{(0)}} \int_G |f(\gamma)|\, d\lambda^u(\gamma) \leq \|f\|_\infty M.$$

Similarly, $\|f\|_{I,s} \leq \|f\|_\infty M$. Thus $\|f\|_I \leq \|f\|_\infty M < \infty$.

If μ is a Radon measure on $G^{(0)}$, then since $\lambda(f) \in C_c(G^{(0)})$ (with $\operatorname{supp} \lambda(f) \subset r(\operatorname{supp} f)$,

$$|\nu(f)| \leq \int_{G^{(0)}} \lambda(|f|)(u)\, d\mu(u) \leq \|\lambda(|f|)\|_\infty \mu\big(r(\operatorname{supp} f)\big)$$

which is bounded since Radon measures are finite on compact subsets.

Ex 1.4.3. If $f_i \to f$ in the inductive limit topology, then $\|f_i - f\|_I \to 0$. On the other hand,

$$\big|\big(L(f_i)h \mid k\big) - \big(L(f)h \mid k\big)\big| \leq \|L(f_i - f)\|\|h\|\|k\| \leq \|f_i - f\|_I \|h\|\|k\|.$$

Hence the result.

Ex 1.4.4. On the one hand,

$$\|f * g\|_{I,r} = \sup_u \int_G |f * g(\gamma)|\, d\lambda^u(\gamma)$$

which, after interchanging the order of integration, is

$$\le \sup_u \int_G \int_G |f(\eta)||g(\eta^{-1}\gamma)|\,d\lambda^u(\gamma)\,d\lambda^u(\eta)$$

which, in view of left-invariance, is

$$\begin{aligned}&\le \|g\|_I \sup_u \int_G |f(\eta)|\,d\lambda^u(\eta)\\&\le \|g\|_I\|f\|_I.\end{aligned}$$

Similarly,

$$\|f*g\|_{I,s} = \|(f*g)^*\|_{I,r} = \|g^* * f^*\|_{I,r} \le \|g^*\|_I\|f^*\|_I = \|f\|_I\|g\|_I.$$

The result follows.

Ex 1.4.5. Let K be a compact set (eventually) containing the supports of the f_i and g_i. Note that if $h,k\in C_c(G)$ have supports in K and M is such that $\lambda^u(K)\le M$ for all u (Ex 1.3.7), then

$$\|h*k\|_\infty \le M\|h\|_\infty\|k\|_\infty.$$

Let i_0 be such that $i\ge i_0$ implies that the supports of f_i and g_i are in K and that f_i and g_i are uniformly bounded. Then

$$\|f_i*g_i - f*g\|_\infty \le \|(f_i-f)*g_i\|_\infty + \|f*(g_i-g)\|_\infty.$$

The assertion follows from this.

Ex 1.4.6. It is clear that $L(f)$ maps $\{\,L(g)h : g\in C_c(G) \text{ and } h\in\mathcal{H}_L\,\}$ to itself. Hence $L(f)(\mathcal{H}_{\mathrm{ess}})\subset\mathcal{H}_{\mathrm{ess}}$ for any $f\in C_c(G)$. That is, $\mathcal{H}_{\mathrm{ess}}$ is invariant and we can form $\operatorname{ess}(L) := L|_{\mathcal{H}_{\mathrm{ess}}}$. If $h_0\in\mathcal{H}_{\mathrm{ess}}^\perp$ and $h\in\mathcal{H}$, then $L(f^*)h\in\mathcal{H}_{\mathrm{ess}})$ and

$$\big(L(f)h_0 \mid k\big) = \big(h_0 \mid L(f^*)k\big) = 0.$$

Hence $L(f)$ annihilates $\mathcal{H}_{\mathrm{ess}}^\perp$. It follows that the natural map $U:\mathcal{H}\to\mathcal{H}_{\mathrm{ess}}\oplus\mathcal{H}_{\mathrm{ess}}^\perp$ intertwines L and $\operatorname{ess}(L)\oplus 0$ as required.

Ex 1.4.7. Let $\|f\|_0$ be the supremum over all nondegenerate representations. Let L be any representation. Observe that $\|L(f)\| = \|\operatorname{ess}(L)(f)\| \le \|f\|_0$. Hence $\|f\|\le\|f\|_0$. This suffices.

Ex 1.4.8. Define $U : C_c(G_{s(\gamma)},\lambda_{s(\gamma)}) \to C_c(G_{r(\gamma)},\lambda_{r(\gamma)})$ by $U(h)(\eta) = h(\eta\gamma)$. Then check that U is isometric for the L^2-norms:

$$\begin{aligned}\big(U(h)\mid U(k)\big)_{L^2(G_{r(\gamma)},\lambda_{r(\gamma)})} &= \int_G h(\eta\gamma)\overline{k(\eta\gamma)}\,\lambda_{r(\gamma)}(\eta)\\&= \int_G h(\eta^{-1}\gamma)\overline{k(\eta^{-1}\gamma)}\,d\lambda^{r(\gamma)}(\eta)\\&= \int_G h(\eta^{-1})\overline{k(\eta^{-1})}\,d\lambda^{r(\gamma)}(\eta)\\&= (h\mid k)_{L^2(G_{s(\gamma)},\lambda_{s(\gamma)})}.\end{aligned}$$

Since U maps onto a dense subspace, U is actually a unitary. It is then easy to check that

$$\operatorname{Ind}\delta_{r(\gamma)}(f)\circ U = U\circ\operatorname{Ind}\delta_{s(\gamma)}(f).$$

We're done.

Ex 1.4.9. Let $L : C(G) \to A$ be a $*$-isomorphism. Then $C(G)$ is a C^*-algebra with respect to $\|f\|_A = \|L(f)\|$. Since L is a representation, $\|f\| \geq \|f\|_A$. But if L' is any representation of $C(G)$, then L' is a representation of $(C(G), \|\cdot\|_A)$ and $\|L'(f)\| \leq \|f\|_A$. Hence $\|f\| \leq \|f\|_A$. This suffices.

Ex 1.4.10. The map $\Phi : C_c(G, \lambda) \to C_c(G, \sigma)$ defined in the solution to Ex 1.3.4 on page 354 is easily seen to be a $*$-isomorphism which is isometric for the respective I-norms. This suffices.

Ex 1.4.11. Note that $G_x = \{ (s, s \cdot x) : s \in \mathcal{G} \}$. Hence the regular representation $L^x = \operatorname{Ind} \delta_x$ acts on $L^2(G_x, \lambda_x)$ by

$$L^x(f)h(r, r \cdot x) = \int_{\mathcal{G}} f(s, r \cdot x)h(s^{-1}r, s^{-1}r \cdot x)\Delta(s)^{-1}\, d\mu_{\mathcal{G}}(s).$$

We get a unitary $U : L^2(G_x, \lambda_x) \to L^2(\mathcal{G}, \mu_{\mathcal{G}})$ given by $U(h)(r) = \Delta(s)^{\frac{1}{2}}h(r, r \cdot x)$. It is not hard to see that $R^x(f) = UL^x(f)U^*$. Now the result follows.

Ex 1.4.12. Choose i_0 such that $i \geq i_0$ implies $\|f * e_i - f\|_\infty < \frac{\epsilon}{2}$ and $\|e_i * f - f\|_\infty < \frac{\epsilon}{4}$. Since we can assume $\|e_j\|_I \leq 2$, it follows that

$$\begin{aligned}\|e_i * f * e_j - f\|_\infty &\leq \|(e_i * f - f) * e_j\|_\infty + \|f * e_j - f\|_\infty \\ &< \|e_i * f - f\|_\infty \|e_j\|_I + \frac{\epsilon}{2} < \epsilon\end{aligned}$$

Ex 1.4.13. If $\varphi \in \mathcal{B}^b(X)$, then there is a bounded sequence $\{\varphi_i\} \in C_0(X)$ converging μ-almost everywhere to φ. If $h \in \mathcal{L}^2(X * \mathscr{H}, \mu)$, then $\|\varphi_i(x)h(x) - \varphi(x)h(x)\|^2 \to 0$ pointwise almost everywhere. Then $\|M(\varphi_i)h - M(\varphi)h\| \to 0$ by the Dominated Convergence Theorem.

Ex 1.4.14. We have

$$\begin{aligned}&\int_G f(s, x)\Delta(s)d(s, x)^{-1}\, d\nu^{-1}(s, x) \\ &\qquad = \int_{\mathcal{G}} \int_X f(s^{-1}, s^{-1} \cdot x)\Delta(s^{-1})d(s^{-1}, s^{-1} \cdot x)^{-1}\, d\mu(x)\, d\mu_{\mathcal{G}}(s) \\ &\qquad = \int_{\mathcal{G}} \int_X f(s, s \cdot x)d(s, s \cdot x)^{-1}\, d\mu(x)\, d\mu_{\mathcal{G}}(s)\end{aligned}$$

which, using Lemma 1.58, is

$$\begin{aligned}&= \int_{\mathcal{G}} \int_X f(s, x)\, d\mu(x)\, d\mu_{\mathcal{G}}(s) \\ &= \int_G f(s, x)\, d\nu(s, x).\end{aligned}$$

Ex 2.1.1. (a) π is continuous as r and s are. If $u \in G^{(0)}$, then $\pi(u) = (u, u)$ and $\Delta \subset \pi(G)$. Since $\pi(\gamma)^{-1} = \pi(\gamma^{-1})$, $\pi(G)^{-1} = \pi(G)$. To check transitivity, suppose (u, v) and $(v, w) \in \pi(G)$. Say $\pi(\gamma) = (u, v)$ and $\pi(\eta) = (v, w)$. Then $(\gamma, \eta) \in G^{(2)}$ and $\pi(\gamma\eta) = (v, w)$.

Similarly, $\pi(\gamma\eta) = \pi(\gamma)\pi(\eta)$ since $s(\gamma) = r(\eta)$.

(b) Suppose G is principal and $\pi(\gamma) = \pi(\eta) = (u, v)$. Then $\gamma \cdot v = \eta \cdot v$ and $\gamma = \eta$. On the other hand, if π is injective and $\gamma \cdot v = \eta \cdot v$, the $\pi(\gamma) = \pi(\eta)$ and $\gamma = \eta$.

(c) Suppose that G is transitive and $u, v \in G^{(0)}$. Then there is a $\gamma \in G$ such that $\gamma \cdot v = u$. Then the $\pi(\gamma) = (u, v)$ and π is surjective. On the other hand, if π is surjective and $u, v \in G^{(0)}$, then there is a γ such that $\pi(\gamma) = (u, v)$. Then $u = \gamma \cdot v$. Hence G is transitive.

Ex 2.1.2. Let $\gamma \in G$ be such that $s(\gamma) = u$ and $r(\gamma) = v$. Then it is not hard to check that $\gamma G(u)\gamma^{-1} = G(v)$ so that $\eta \mapsto \gamma\eta\gamma^{-1}$ is the required group isomorphism.

Ex 2.1.3. By definition the topology on $G\backslash X$ is the largest topology such that q is continuous. In particular, q is continuous so $q^{-1}(U)$ is open if U is open in $G\backslash X$. But the set $\tau = \{ U \subset G\backslash X : q^{-1}(U) \text{ is open in } X \}$ is a topology on $G\backslash X$ and any topology for which q is continuous must be contained in τ. Hence the result.

Ex 2.1.4. Since q is always continuous, $\tilde{f}$ is continuous when f is. Now suppose that $\tilde{f}$ is continuous and U is open in Y. Since $q^{-1}(f^{-1}(U)) = (\tilde{f})^{-1}(U)$ is open, $f^{-1}(U)$ is open by Ex 2.1.3. Hence f is continuous.

Ex 2.1.5. Clearly, we get a well defined map $\tilde{s} : G\backslash G \to G^{(0)}$ given by $\tilde{s}(G \cdot \gamma) = s(\gamma)$. This map is surjective because s is and continuous by general nonsense (Ex 2.1.4). Since s is open, it follows easily that $\tilde{s}$ is as well. (In fact, it is not hard to see that $\tilde{s}$ is open if and only if s is.) It only remains to see that $\tilde{s}$ is injective. But if $s(\gamma) = s(\eta)$, then $\eta = \eta\gamma^{-1} \cdot \gamma$. Hence $G \cdot \gamma = G \cdot \eta$.

Ex 2.1.6. Suppose G is principal and $\gamma, \eta \in G_v^u$. Then $\gamma \cdot v = u = \eta \cdot v$. Thus $\gamma = \eta$. Conversely, if $\gamma \cdot v = u = \eta \cdot v$, then $\gamma, \eta \in G_v^u$ and $\gamma = \eta$.

Suppose G is principal and $\gamma \in G(u)$. Then $\gamma \cdot u = u$ and we must have $\gamma = u$. If $G(u) = \{u\}$, and $\gamma \cdot u = \eta \cdot u$, then $\gamma^{-1}\eta \in G(u)$ and $\gamma = \eta$.

Ex 2.1.7. Since $(x, \gamma, \gamma^{-1} \cdot x) \mapsto (\gamma, x)$ is a topological groupoid isomorphism of the action groupoid $A = \{ (\gamma \cdot x, \gamma, x) : r(x) = s(\gamma) \}$ onto $G \rtimes X$, it will suffice to see that

$$\beta_1^x(f) = \int_G f(\gamma, x)\, d\lambda^{r(x)}(\gamma)$$

is a Haar system on $G \rtimes X$. Note that $\operatorname{supp} \beta_1^x = G^{r(x)} \times \{x\} = (G \rtimes X)^x$. Furthermore, if $f(\gamma, x) = \varphi(\gamma)\psi(x)$ with $\gamma \in C_c(G)$ and $\psi \in C_c(X)$, then $x \mapsto \beta_1^x(f)$ is continuous since λ is a Haar system on G. Since sums of functions of this form are dense in $C_c(G \rtimes X)$ in the inductive limit topology, it follows that $x \mapsto \beta_1^x(f)$ is continuous for all $f \in C_c(G \rtimes X)$. It only remains to check left-invariance.

Note that in $G \rtimes X$, $(\eta, \eta \cdot x)(\gamma, x) = (\eta\gamma, \eta \cdot x)$. Then since $s(\eta, \eta \cdot x) = x = s(\eta)$,

$$\begin{aligned}
\int_{G \rtimes X} f((\eta, \eta \cdot x)(\gamma, x))\, d\beta_1^{s(\eta, \eta \cdot x)}(\gamma, x) &= \int_G f(\eta\gamma, \eta \cdot x)\, d\lambda^{s(\eta)}(\gamma) \\
&= \int_G f(\gamma, \eta \cdot x)\, d\lambda^{r(\eta)}(\gamma) \\
&= \int_{G \rtimes X} f(\gamma, \eta \cdot x)\, d\beta_1^{r(\eta, \eta \cdot x)}(\gamma, \eta \cdot x)
\end{aligned}$$

as required.

Ex 2.1.8. Let $A = \{ (x, \gamma, x \cdot \gamma) \in X \times G \times X : r(x) = s(\gamma) \}$, etc.

Ex 2.1.9. If $\gamma \cdot x = \eta \cdot x$, then $\eta^{-1}\gamma \in H$. Therefore we get a well defined continuous bijection $\varphi : G_{r(x)}/H \to G \cdot x$ defined by $\varphi(\gamma \cdot H) = \gamma \cdot x$. To see that

φ is a homeomorphism, assume $\gamma_i \cdot x \to \gamma \cdot x$. It suffices to see that every subnet of $\{\gamma_i \cdot H\}$ has a subnet converging to $\gamma \cdot H$. Since the G-action is proper, (P3) of Proposition 2.17 on page 32 implies that we can pass to a subnet relabel and assume that $\gamma_i \to \eta$ in G. But then $\gamma \cdot x = \eta \cdot x$ and $\gamma \cdot H = \eta \cdot H$. But then $\gamma_i \cdot H \to \gamma \cdot H = \eta \cdot H$, and we are done.

Ex 2.1.10. Suppose $G\backslash G^{(0)}$ is Hausdorff and $(r(\gamma_i), s(\gamma_i)) \to (u, v)$. Then $G \cdot s(\gamma_i) = G \cdot r(\gamma_i)$ and $\{G \cdot s(\gamma_i)\}$ converges to both $G \cdot u$ and $G \cdot v$. Since the orbit space is Hausdorff, $G \cdot u = G \cdot v$ and (u, v) is in the range of π.

Now suppose that $\pi(G)$ is closed. Suppose that $G \cdot u_i$ converges to $G \cdot u$ and $G \cdot v$. Thus we can pass to a subnet, relabel, and assume that there are $\gamma_i, \eta_i \in G$ such that $\gamma_i \cdot u_i \to u$ and $\eta_i \cdot u_i \to v$. But then $\pi(\gamma_i \eta_i^{-1}) \to (u, v)$. Hence $(u, v) \in \pi(G)$ and it follows that $G \cdot u = G \cdot v$. This shows the orbit space is Hausdorff.

Ex 2.1.11. Let $\gamma \cdot x \in V \cdot U$ with $\gamma \in V$ and $x \in U$. It will suffice to see that any net $x_i \to \gamma \cdot x$ has a subnet which is eventually in $V \cdot U$. But $r(x_i) \to r(\gamma \cdot x) = r(\gamma)$. Since $r : G \to G^{(0)}$ is open, we can pass to a subnet, relabel, and assume that there are $\gamma_i \in V$ with $\gamma_i \to \gamma$ with $r(\gamma_i) = r(x_i)$. But then

$$\gamma_i^{-1} \cdot x_i \to x,$$

and $\gamma_i^{-1} \cdot x_i$ is eventually in U. Hence

$$x_i = \gamma_i \cdot (\gamma_i^{-1} \cdot x_i)$$

is eventually in $V \cdot U$.

Ex 2.1.12. We use Fell's net lifting criterion to show that the source map $s_A : A \to X$ is open. Suppose that $x_i \to x = s(\gamma \cdot x, \gamma, x)$. Then $r_X(x_i) \to r_X(x) = s(\gamma)$. Since $s : G \to G^{(0)}$ is open, we can, after passing to a subnet and relabeling, find $\gamma_i \to \gamma$ such that $s(\gamma_i) = r_X(x_i)$. But the $(\gamma_i \cdot x_i, \gamma_i, x_i) \to (\gamma \cdot x, \gamma, x)$ as required.

Ex 2.1.13. Suppose G is proper. Let $K \subset G^{(0)} \times G^{(0)}$ be compact and let $\{\gamma_i\} \subset \pi^{-1}(K)$. It will suffice to see that $\{\gamma_i\}$ has a convergent subnet. But after passing to a subnet and relabeling, we can assume $s(\gamma_i) \to u$ and $r(\gamma_i) \to v$ with $(v, u) \in K$. Then $\gamma_i \dot{s}(\gamma_i) \to v$, and the assertion follows from (PA3) of Proposition 2.17 on page 32.

Now suppose that π is proper. Suppose that $u_i \to u$ while $\gamma_i \cdot u_i \to v$. Let K be a compact neighborhood of (v, u) in $G^{(0)} \times G^{(0)}$. Then we eventually have $\{\gamma_i\} \subset \pi^{-1}(K)$. Then as above, $\{\gamma_i\}$ must have a convergent subnet.

Ex 2.1.14. Suppose $z_i \to z$ and $\gamma_i \cdot z_i \to w$. Note that $\gamma_i \cdot z_i = (\gamma \cdot z_i, \gamma_i, z_i) \cdot z_i$. Now apply (PS3) of Proposition 2.17 on page 32.

Ex 2.1.15. Let $y_i \to y$ and $\gamma_i \cdot y_i \to z$ be convergent nets in Y. Then in X, $\pi(y_i) \to \pi(y)$ and $\gamma_i \cdot \pi(y_i) \to \pi(z)$. Then Proposition 2.17 on page 32 implies that $\{\gamma_i\}$ has a convergent subnet. Thus the action on Y is proper by the same proposition.

Ex 2.1.16. Since g is strictly positive on K and each λ^u has support equal to G^u, it follows that $\lambda(g)$ is strictly positive on the compact set $q(K)$. Hence there exists $\delta > 0$ such that $\lambda(g)(G \cdot z) \geq \lambda$ if $G \cdot z \in q(K)$. Then we can define $f \in C_c(Z)$ by

$$f(z) = \begin{cases} g(z)\lambda(g)(G \cdot z)^{-1} d(G \cdot z) & \text{if } \lambda(g)(G \cdot z) > \delta/2, \text{ and} \\ 0 & \text{if } \lambda(g)(G \cdot z) < \delta. \end{cases}$$

Clearly, $\lambda(f) = d$.

Ex 2.1.17. Suppose that X is a Cartan G-space. If $x \in X$, let K be a compact neighborhood of x such that $\Theta^{-1}(K \times K)$ is compact. Then $P(K, K)$ is the projection of $\Theta^{-1}(K \times K)$ onto its first factor. Hence $P(K, K)$ is compact.

On the other hand, let K be a compact neighborhood of $x \in X$ such that $P(K, K)$ is compact. Let $\{(\gamma_i, x_i)\}$ be a net in $\Theta^{-1}(K \times K)$. Then $\{\gamma_i\} \subset P(K, K)$ and $\{x_i\} \subset K$. Hence we can pass to a subnet, relabel, and assume $\gamma_i \to \gamma$ and $x_i \to x$. Since $\Theta^{-1}(K \times K)$ is closed, we've shown that every net in $\Theta^{-1}(K \times K)$ has a subnet converging in $\Theta^{-1}(K \times K)$. This suffices.

Ex 2.1.18. (a)$\Longrightarrow$(b): This is Proposition 2.18 on page 33.

(b)$\Longrightarrow$(c): Suppose $(\gamma_i \cdot x_i, x_i) \to (y, x)$. Let $q : X \to G\backslash X$ be the orbit map. Then $q(x_i)$ converges to both $q(y)$ and $q(x)$. Since $G\backslash X$ is Hausdorff, $q(y) = q(x)$.

(c)$\Longrightarrow$(a): We'll use (PA3) of Proposition 2.17 on page 32 Suppose that $x_i \to x$ and $\gamma_i \cdot x_i \to y$. Since $R(G, X)$ is closed, $(y, x) \in R(G, X)$ and $(\gamma_i \cdot x_i, x_i) \to (y, x)$ in $R(G, X)$. Since the translation map is continuous, we must have $\gamma_i \to \tau(y, x)$.

Ex 2.1.19. We use Lemma 2.23 on page 34 to find K as in the statement of the problem. Then we can find a net $\{\gamma_i \cdot x\} \subset K$ converging to y. Fix i_0 and note that $\gamma_i \cdot x = (\gamma_i \gamma_{i_0}^{-1})\gamma_{i_0} \cdot x$. Then $\{\gamma_i \gamma_{i_0}^{-1}\}$ is in the compact set $P(K, K)$. Hence we can pass to a subnet, relabel, and assume $\{\gamma_i \gamma_{i_0}^{-1}\}$ converges to some $\eta \in G$. But then $\gamma_i \to \gamma := \eta\gamma_{i_0}$. Since X is Hausdorff, we have $y = \gamma \cdot x \in G \cdot x$.

Ex 2.2.1. It will suffice to see that if V is open in $G_{r_X(x)}$, then so is $V \cdot H$. It suffices to show that every net $\gamma_n \to \gamma \cdot h$ (for $\gamma \in V$ and $h \in H$) is eventually in $V \cdot H$. But $\gamma_n h^{-1} \to \gamma$ must eventually be in V. Hence $\gamma_n = (\gamma_n h)h^{-1}$ is eventually in $V \cdot H$

Ex 2.2.2. Let d be a compatible metric on X. If F is closed, let $U_n = \{\, x \in X : d(x, F) < \frac{1}{n} \,\}$. Then U_n is open and $F = \bigcup_n U_n$.

If X is separable and U is open in X, then U is the union of countably many balls $B_n = B_{\epsilon_n}(x)$ such that $\overline{B_{\epsilon_n}(x)} \subset U$. Then $U = \bigcup_n \overline{B_n}$.

Ex 2.2.3. Let p and q be distinct points in X. If X is T_0, we may as well assume that there is an open set U containing p and not q. Hence $\overline{\{q\}} \subset X \setminus U$. But $p \in \overline{\{p\}}$ means that $\overline{\{p\}} \neq \overline{\{q\}}$.

Now suppose that p and q have distinct closures. Observe that if $p \notin \overline{\{q\}}$, then $U = X \setminus \overline{\{q\}}$ is an open set containing p but not q. But if $p \in \overline{\{q\}}$, then $\overline{\{p\}} \subset \overline{\{q\}}$.

If also $q \in \overline{\{p\}}$, then the closures coincide contradicting our assumptions. Hence $q \notin \overline{\{p\}}$ and we can let $U = X \setminus \overline{\{p\}}$.

Ex 2.2.4. Suppose that $U \subset \bigcap F_i$ with each F_i closed in X. Then $U = \bigcap U \cap F_i$, and for some i, $U \cap F_i$ has nonempty interior in U. Since U is open, F_i must have interior in X.

Ex 2.2.5. Suppose that U is open in X and that $\emptyset \neq U \cap D \subset A \subset D$. Then $\overline{U \cap D} \subset \overline{A}$. But $U \setminus \overline{U \cap D}$ is open. It it were nonempty, it would intersect D nontrivially. Hence $U \subset \overline{U \cap D} \subset \overline{A}$.

Ex 2.2.6. If A is of first category in $\overline{X}$, then $A \subset \bigcup F_i$ with each F_i closed and nowhere dense in $\overline{X}$. But then

$$A = \bigcup F_i \cap A.$$

By assumption, some $F_i \cap A$ must have interior in X. But the previous exersice, $\overline{F_i \cap A} \subset F_i \cap \overline{A} \subset F_i$ must have interior in $\overline{X}$. This is a contradiction.

Ex 2.2.7. Since q is continuous, $\overline{G \cdot x} \subset q^{-1}\big(\overline{q(x)}\big)$. But $X \setminus \overline{G \cdot x}$ is an invariant open set. Hence $q(X \setminus \overline{G \cdot x})$ is an open set in $G \backslash X$ disjoint from $q(x)$. Hence $\overline{q(x)} \subset q(\overline{G \cdot x})$. This suffices.

Ex 2.3.1. The argument in Example 2.32 hinges on the openness of the restriction of the range map to G_x for any/some $x \in \mathbf{R}$ (identified with $G^{(0)}$). But $G_x = \{\, (y, y + x)\mathbf{R}_d \times \mathbf{R} : y \in \mathbf{R}_d \,\} \cong \mathbf{R}_d$ and $(y, y + x) \mapsto y + x$ is continuous but certainly not open as a map into $\mathbf{R}$

Ex 2.3.2. The only details remaining are the openness of the moment maps $r_W : W \to Z$ and $s_W : W \to G^{(0)}$, together with the observations that the moment maps induce bijections on the respective orbit spaces.

To see that r_W is open, we use Fell's lifting criterion. To that end, let $(z, \gamma) \in W$ and suppose that $z_n \to z$. Then $f(z_n) \to f(z) = r(\gamma)$. Since r is open, we can pass to a subnet, relabel, and assume that there are $\gamma_n \in G$ such that $r(\gamma_n) = f(z_n)$ and $\gamma_n \to \gamma$. Then $(z_n, \gamma_n) \in W$ and $(z_n, \gamma_n) \to (z, \gamma)$ as required.

To see that s_W is open, we proceed similarly. Let $(z, \gamma) \in W$ and suppose that $u_n \to s(\gamma)$ in $G^{(0)}$. Since s is open, we can pass to a subnet, relabel, and assume there are $\gamma_n \in G$ such that $\gamma_n \to \gamma$. Then $r(\gamma_n) \to r(\gamma) = f(z)$. Since f is open, after passing to a subnet and relabeling, we can assume that there are $z_n \to z$ in Z with $f(z_n) = r(\gamma_n)$. Then $(z_n, \gamma_n) \in W$ and $(z_n, \gamma_n) \to (z, \gamma)$. This shows s_W is open.

If (z, γ) and (z, η) are both in W, then $r(\gamma) = r(\eta)$ and

$$(z, \gamma) \cdot \gamma^{-1}\eta = (z, \eta).$$

Hence r_W induces a bijection of W/G onto Z are required.

If (z, γ) and (w, η) are in W and such that $s(\gamma) = s(\eta)$, then $(z, \gamma\eta^{-1}, w) \in G[Z]$ and

$$(z, \gamma\eta^{-1}, w) \cdot (w, \eta) = (z, \gamma).$$

Thus s_W induces a bijection of $G[Z] \backslash W$ onto $G^{(0)}$.

Ex 2.3.3. Note that we can identify $G^{(0)}$ with $\mathcal{G} \backslash P$ and $H^{(0)}$ with $P/\mathcal{H}$. The moment maps on P are the orbit maps which are always open for group actions.

Note that

$$(g, \mathcal{G}p) \cdot p = g \cdot p \quad \text{and} \quad p \cdot (p\mathcal{H}, h) = p \cdot h.$$

Clearly, the moment maps define bijections of $G\backslash P$ onto $\mathcal{G}\backslash P$ and P/H onto $P/\mathcal{H}$, respectively. This suffices.

Ex 2.3.4. We identify $R^{(0)}$ with X. We define moment maps on X by $r_X = \text{id}$ and letting $s_X : X \to R\backslash X$ be the quotient map. Then r_X is obviously open and s_X is open since its the orbit map for a groupoid with open range and source maps. The actions are given by

$$(x, y) \cdot y = x \quad \text{and} \quad y \cdot Ry = y.$$

The rest is straightforward.

Ex 2.3.5. The moment maps are $r_Z = r|_Z$ and $s_Z = s|_Z$. Since $Z = G_B^A$ is open in G and G has open range and source mpas, r_Z and s_Z are clearly open. Furthermore, the left- and right-actions are clearly free and proper, and commute since multiplication is associative in G.

To see that r_Z maps onto A, let $a \in A$. Let $\gamma \in G$ be such that $r(\gamma) = a$. Since B is full, there is an $\eta \in G_B$ with $r(\beta) = s(\gamma)$. Then $\gamma\beta \in G_B^A$ and $r_Z(\gamma\beta) = r(\gamma) = a$. On the other hand, if $r_Z(\gamma) = r_Z(\gamma')$, then $\gamma^{-1}\gamma' \in G(B)$ and $\gamma' = \gamma(\gamma^{-1}\gamma')$. Hence r_Z induces a homeomorphism of $Z/G(B)$ onto $G^{(0)}$. The proof for s_Z is similar.

Ex 2.3.6. We'll write $[z, y]$ for the class of (z, y) in $(Z * Y)/H$. We can equip $(Z * Y)/H$ with the moment maps r'_Z and s'_Y where $r'_Z([z, y]) = r_Z(z)$ and $s'_Y([z, y]) = s_Y(y)$. This clearly gives us free actions of G and P, respectively. For example, if $g \cdot [z, y] = [z, y]$, then for some h, $g \cdot z \cdot h = z$ and $h^{-1} \cdot y = y$. But the second equation forces $h = r_Y(y)$. Then $g = r_Z(z)$. A similar argument applies to the P-action.

To see that the actions are proper, we use (PA3) of Proposition 2.17 on page 32. Suppose that $[z_i, y_i] \to [z, y]$ while $g_i \cdot [z_i, y_i] \to [z', y']$. Since orbit maps are open, we may as well assume, after passing to subnets and relabeling, that $(z_i, y_i) \to (z, y)$ and that there are h_i such that $(g_i \cdot z_i \cdot h_i, h_i^{-1} \cdot y_i) \to (z', y')$. Since the left H-action on Y is proper, this means that after passing to a subnet and relabeling, we can assume $h_i \to h$ in H. In particular, $g_i \cdot z_i \to z' \cdot h^{-1}$. But then the properness of the left G-action on Z implies $\{g_i\}$ has a convergent subnet. Thus the G-action is proper and the properness of the right P-action is proved similarly.

Now suppose that $[z, y]$ and $[z', y']$ are such that $r_Z(z) = r_Z(z')$. To see that r'_Z induces a bijection we need to see that there is a p such that $[z', y'] = [z, y \cdot p]$. But $r_Z(z) = r_Z(z')$ implies there is a $h \in H$ such that $z' = z \cdot h$. But then $r_Y(h^{-1} \cdot y) = s(h) = s_Z(z \cdot h) = s_Z(z') = r_Y(y')$. Hence there is a $p \in P$ such that $y' = h^{-1} \cdot y \cdot p$. But then

$$[z', y'] = [z \cdot h, h^{-1} \cdot y \cdot p] = [z, y \cdot p]$$

as required.

To see that r'_Z is open, we again employ Fell's lifting criterion. So, let $[z, y] \in (Z * Y)/H$ and suppose $u_i \to r_Z(z)$ in $G^{(0)}$. Since r_Z is open, we can pass to subnet, relabel, and assume that there are $z_i \to z$ in Z such that $r_Z(z_i) = u_i$. Then $s_Z(z_i) \to s_Z(z) = r_Y(y)$. Since r_Y is open, we can pass to a subnet, relabel, and assume that there are $y_i \to y$ such that $r_Y(y_i) = s_Z(z_i)$. Then $[z_i, y_i] \to [z, y]$ as required. Hence r'_Z defines the required homeomorphism.

The proof of the corresponding assertions for s'_Y is similar.

Ex 2.3.7. For convenience, let $u = r(z)$ and $v = s(z)$. Define $\varphi : G(u) \to H(v)$ by $\varphi(g) = \tau_H(z, g \cdot z)$ where τ_H is the continuous map coming from Lemma 2.42 on page 46. Because the G- and H-actions are free, φ is one-to-one. Since $\varphi(g)$ is the unique element of $H(v)$ such that $z \cdot \varphi(g) = g \cdot z$, it is not hard to see that φ is a homomorphism. If $h \in H(v)$, then $g = \tau_G(z \cdot h, z)$ satisfies $\varphi(g) = h$. Hence φ is an isomorphism.

Ex 2.3.8. Let Z be a (G, H)-equivalence. By Lemma 2.44 on page 47, we can assume $G = H_Z$. Then the range map is given by $[z, w] \mapsto z \cdot H$. To see that this map is open, we use Fell's Criterion. Suppose $z_i \cdot H \to z \cdot H = r([z, w])$. Since the H has an open range map, the orbit map $z \mapsto z \cdot H$ is open. Hence we can pass to a subnet, relabel, and assume there are η_i such that $z_i \cdot \eta_i \to z$. We may as well replace $z_i \cdot \eta_i$ with z_i. Since the moment maps on Z are open, we can pass to a subnet, relabel, and assume that there are $w_i \in Z$ converging to $s(z)$ such that $s(w_i) = s(z_i)$. But then $[z_i, w_i] \to [z, w]$ and $r([z_i, w_i]) = z_i \cdot H$. Hence the range map on H_Z is open.

Ex 2.4.1. To see that $r : Z \to G^{(0)}$ is open, we use Fell's criterion. Suppose that $u_i \to u = r(u, \eta)$. Then $\varphi(u_i) \to \varphi(u)$ in $H^{(0)}$. Since the source map on H is open, we can pass to a subnet, relabel, and assume that there are $\eta_i \in H$ such that $\eta_i \to \eta$ and $s(\eta_i) = \varphi(u_i)$. Then $(u_i, \eta_i) \in \mathrm{Gr}(\varphi)$ and $(u_i, \eta_i) \to (u, \eta)$.

If (u, η) and (u, η') are both in $\mathrm{Gr}(\varphi)$, then $(u, \eta') = (u, \eta) \cdot \eta^{-1}\eta'$. Thus r factors through Z/H. Since the orbit map $q : Z \to Z/H$ is open (Proposition 2.12 on page 31), we have established (M4). The remaining conditions are straightforward to check.

Ex 2.8.1. (a) This follows from [**Wil07**, Lemma D.4]. Alternatively, we can take advantage of the fact that $C_c(X)$ is separable in the inductive limit topology (Lemma C.2 on page 344) and that $C_c(X)$ is dense in $L^2(X, \mu)$.

(b) Suppose $L^2(X, \mu)$ has dimension $n < \infty$. If X is infinite, we can find $n+1$ distinct points. Since X is Hausdorff, we can find pairwise disjoint nonempty open sets $V_0, \dots, V_n$. Let $f_k = \mathbb{1}_{V_k}$. The set $\{f_k\}$ is linearly dependent in $L^2(X, \mu)$. We can renumber so as to assume f_0 is equal in $L^2(X, \mu)$ to a nontrivial linear combination $g := \sum_{k=1}^{n} a_k f_k$. Then $f_0 - g$ would have to be zero μ-almost everywhere. Since $f_0 - g$ is nonzero on V_0, we must have $\mu(V_0) = 0$ which contradicts the assumption that μ has full support.

(c) If X is finite, then points are open and an appropriate scaling of the characteristic functions of points is an orthonormal basis.

Ex 2.8.2. (a) We have

$$\langle W(x) \,, y \rangle_\star = \langle W(x) \,, WW^{-1}(y) \rangle_\star = \langle x \,, W^{-1}(y) \rangle_\star,$$

which shows that W is adjointable with adjoint W^{-1}. Hence $W \in \mathcal{L}(\mathsf{X}, \mathsf{Y})$ as claimed.

(b) By part (a), $WTW^{-1} \in \mathcal{L}(Y)$ and $S \mapsto W^{-1}SW$ is an inverse. Hence the first assertion. For the second, just compute that

$$W\Theta_{x,y}W^{-1} = \Theta_{W(x),W(y)}.$$

Ex 2.8.3. Since r is continuous and open, $r(K_n)$ is compact, and hence closed, and contained in the interior of $r(K_{n+1})$. Hence every compact set in $G^{(0)}$ is contained in some $r(K_n)$ as claimed. If $u \in r(K_n) \setminus r(K_{n-1})$, define $\rho(u) = r_{K_n}(u)$. Then $\rho : G^{(0)} \to G_u$ is a Borel section for r. If $C \subset G^{(0)}$ is compact, then $C \subset r(K_n)$ for some n and

$$\rho(C) \subset \bigcup_{m \leq n} r_{K_m}(C \cap r(K_m))$$

which has compact closure since each $r_{K_m}(C \cap r(K_m)) \subset K_m \subset K_n$.

Ex 2.8.4. Note that if $f \in C_c^+(G_u)$ and if $f(\rho(v)) > 0$, then $\beta(f)(v)$ is finite and nonzero. In particular, if V is a pre-compact open subset of $G^{(0)}$, then $\rho(V)$ is pre-compact and there is a $f \in C_c^+(G_u)$ such that $\beta(f)(v) > 0$ for all $v \in V$. Thus if $\{V_i\}$ is a locally finite cover of $G^{(0)}$ and $f_i \in C_c^+(G_u)$ is such that $\beta(f_i)(v) > 0$ for all $v \in V_i$, we can define

$$f(\gamma) = \sum_i \mathbb{1}_{V_i}(r(\gamma)) f_i(\gamma). \tag{D.7}$$

Since $\{V_i\}$ is locally finite, (D.7) is a finite sum for all γ and f is in $\mathcal{B}^+(G_u)$. Furthermore, $0 < \beta(f)(v) < \infty$ for all v. Now let

$$d(\gamma) = \beta(f)(r(\gamma))^{-1} f(\gamma).$$

Ex 2.8.5. Parts (a) and (b) are routine.

(c) We compute that

$$\begin{aligned} \upsilon_*(\lambda^u)(\xi) &= \int_G \xi\big(r(\gamma), \rho(r(\gamma))^{-1}\gamma\rho(s(\gamma)), s(\gamma)\big)\, d\lambda^u(\gamma) \\ &= \int_G \xi\big(s(\gamma), \rho(s(\gamma))^{-1}\gamma^{-1}\rho(r(\gamma)), r(\gamma)\big)\, d\lambda_u(\gamma) \end{aligned}$$

which, in view of Proposition 2.82, is

$$= \int_{G^{(0)}} \int_{G(u)} \xi\big(u, \rho(u)^{-1}s, v\big)\, d\beta^u(s)\, d\mu(u)$$

which, by left invariance of β^u, is

$$= \int_{G^{(0)}} \int_{G(u)} \xi\big(u, s, v\big)\, d\beta^u(s)\, d\mu(u)$$

Ex 2.8.6. Since μ is additive, (2.16) holds for simple functions almost by definition. Then if $b_0 \in \mathcal{B}^+(G^{(0)})$, there is a sequence of simple functions $b_i \nearrow b_0$ pointwise. We can assume that $b_i = \beta(f_i)$ for some $f_i \in \mathcal{B}^+(G_u)$ for all $i \geq 0$. Then using (2.15),

$$\mu(b_0) = \mu(\beta(f_0)) = \lambda_u(f_0) = \lambda_u(d \cdot \beta(f_0) \circ r).$$

But the Monotone Convergence Theorem implies

$$\begin{aligned} \lambda_u(d \cdot \beta(f_0) \circ r) &= \lim_i \lambda_u(d \cdot \beta(f_i) \circ r) \\ &= \lim_i \mu(b_i) = \lim_i \int_{G^{(0)}} b_i(u)\, d\mu(u) = \int_{G^{(0)}} b_0(u)\, \mu(u). \end{aligned}$$

Ex 3.2.1. Let $f \in C_c^+(Y)$ be such that $f(y) = 1$ for all $y \in K$. Then $\beta(f)$ is in $C_c(X)$ and $M = \|\beta(f)\|_\infty$ satisfies $\beta^x(K) \leq \beta(f)(x) \leq M$.

Ex 3.2.2. Suppose $y, z \in \pi^{-1}(x)$ and $G \cdot y = G \cdot z$. Then $z = \gamma \cdot y$ and hence $x = \gamma \cdot x$. This forces $\gamma = r(x) = r(y)$. Thus $z = y$ and the map is injective. Suppose $\dot{\pi}(G \cdot y) = G \cdot x$. Then $\pi(y) = \gamma \cdot x$. Then $\gamma^{-1} \cdot y \in \pi^{-1}(x)$ and $G \cdot (\gamma^{-1} \cdot y) = G \cdot y$. Therefore, the restriction of q_Y is surjective and defines a continuous bijection.

To see that it is a homeomorphism, suppose that $G \cdot y_i \to G \cdot y$ with each y_i and y in $\pi^{-1}(x)$. It will suffice to see that every subnet of $\{y_i\}$ has a subnet converging to y. But after passing to a subnet, we can pass to another subnet, relabel, and assume that there are γ_i such that $\gamma_i \cdot y_i \to y$ (since q_Y is open). But then $\gamma_i \cdot x \to x$. By properness, we may as well assume that $\gamma_i \to \gamma$ in G. But then $x = \gamma \cdot x$ and $\gamma = r(x) = r(y)$. It follows that $y_i \to y$. This suffices.

Ex 3.2.3. Let $\Phi(y) := (G \cdot y, \pi(y))$. Clearly, Φ is a continuous map into $G\backslash Y \times_{G\backslash X} X$. If $(G \cdot y, x) \in G\backslash Y \times_{G\backslash X} X$, then $\pi(y) = \gamma \cdot x$ and $\Phi(\gamma^{-1} \cdot y) = (G \cdot y, x)$. Hence Φ is surjective.

Suppose that $(G \cdot y, \pi(y)) = (G \cdot z, \pi(z))$. Then $z = \gamma \cdot y$ and $\pi(y) = \gamma \cdot \pi(y)$. Thus $\gamma = r(y)$ and $y = z$. Thus, Φ is a continuous bijection.

Suppose that $(G \cdot y_i, \pi(y_i)) \to (G \cdot y, \pi(y))$. It will suffice to see that every subnet of $\{y_i\}$ has a subnet converging to y. But the orbit map is open, so after passing to a couple of subnets, we can assume $\gamma_i \cdot y_i \to y$. Thus $\gamma_i \cdot \pi(y_i) \to \pi(y)$. By properness and freeness of the action of G on X, we may as well assume that $\gamma_i \to r(y) = r(x)$. But then $y_i \to y$.

Ex 3.2.4. Fix $x_0 \in X$ and let K be a compact neighborhood of x_0. By assumption, $\pi^{-1}(K) \cap \operatorname{supp} f$ is compact. Let $g \in C_c^+(Y)$ be such that $g(y) = 1$ for all $y \in \pi^{-1}(K) \cap \operatorname{supp} f$. Then $h = gf$ is in $C_c(Y)$ and $\beta(h) = \beta(f)$ near x_0. Since $\beta(h)$ is continuous at x_0 by assumption, so is $\beta(f)$.

Ex 3.3.1. It suffices to show that if $\{f_i\}$ is net of functions converging to f with supports all contained in K, then $\mu(f_i) \to \mu(f)$ (see [**RW98**, Lemma D.10]). Since μ is a Radon measure, $\mu(K) < \infty$. Hence $|\mu(f_i - f)| \leq \mu(K)\|f_i - f\|_\infty$. The result follows.

Ex 3.4.1. Suppose that $H_i \to H$ in Σ_0. To show continuity, we need to see that $p_0(H_i) \to p_0(H)$. It suffices to show that every subnet has a subnet converging to $p_0(H)$. Let $t \in H$. Assume we have already passed to a subnet and relabeled. After passing to another subnet we can assume there $t_i \in H_i$ such that $t_i \to t$. But then $p(H_i) = r(t_i) \to r(t) = p(H)$. Thus p_0 is continuous.

Ex 3.4.2. Suppose that $H_i \in \Sigma_0$ and $H_i \to C \neq \emptyset$ in $\mathscr{C}(G)$. It suffices to see that $C \in \Sigma_0$. Since p_0 is continuous, $C \subset G_{p_0(C)}^{p_0(C)} := G(p_0(C)) \in \Sigma_0$. It will suffice to see that C is a subgroup. If $a, b \in C$, we can pass to subnets, relabel, and find $a_i, b_i \in H_i$ such that $a_i \to a$ and $b_i \to b$. Since $a_i b_i \in H_i \to ab$, we have $ab \in C$. Similarly, if $a \in C$, then $a^{-1} \in C$. Since C is nonempty by assumption, it must be a subgroup of $C(p_0(C))$. Thus $C \in \Sigma_0$ as required.

Ex 3.4.3. Let $\{H_i\} \subset p_0^{-1}(K) \cap \Sigma_0$. Since $\mathscr{C}(G)$ is compact, we can pass to a subnet, relabel, and assume $H_i \to C$ in $\mathscr{C}(G)$. Since $p_0(H_i) \in K$, we can pass to another subnet, relabel, and assume $p_0(H_i) \to u \in K$. But $p_0(H_i) \in H_i$. Hence $u \in C$ and $C \neq \emptyset$. It follows from Ex 3.4.2 that $C \in \Sigma_0$. This suffices.

Ex 3.4.4. Since Σ has the relative product topology, p is clearly continuous.

To see that p is open, we will use Fell's criterion. Suppose that $H_i \to H$ and that $(p_0(H), H, t) \in \Sigma$. Passing to a subnet and relabeling, we can assume we have $t_i \in H_i$ such that $t_i \to t$. But then $(p_0(H_i), H_i, t_i) \to (p_0(H), H, t)$.

Ex 3.4.5. Suppose that $(\gamma_i, H_i) \to (\gamma, H)$ in $G * \Sigma_0$. We want to show that $\gamma_i \cdot H_i \to \gamma \cdot H$ in Σ_0. We use the criteria in Lemma 3.22 on page 81. For part (a), assume that $\gamma_i h_i \gamma_i^{-1} \to \eta$. Then $h_i \to \gamma^{-1}\eta\gamma$ and we must have $\gamma^{-1}\eta\gamma \in H$. Thus $\eta \in \gamma \cdot H$.

Let $h \in H$ and consider $\gamma h \gamma^{-1}$. After passing to a subnet and relabeling, we can assume there are $h_i \in H$ such that $h_i \to h$. But then $\gamma_i h_i \gamma_i^{-1} \to \gamma h \gamma^{-1}$.

Ex 3.6.1. Since $\mathcal{H}$ is a separable metric space, it is second countable. Hence it suffices to consider V in a countable basis for the topology. Hence it suffices to see that $f^{-1}(B_\epsilon(k))$ is Borel if $B_\epsilon(k) = \{\, h \in \mathcal{H} : \|h-k\| < \epsilon \,\}$. But there are countably many unit vectors k_n such that $\|h - k\| < \epsilon$ if and only if $|(h - k \mid k_n)| < \epsilon$ for all n. Thus

$$f^{-1}(B_\epsilon(k)) = \bigcap_n \{\, x \in X : |\big(f(x) - k \mid k_n\big)| < \epsilon \,\}.$$

This suffices.

Ex 3.6.2. (a) This follows from Ex 3.6.1 since the inner product is continuous on $\mathcal{H} \times \mathcal{H}$. Alternatively, notice that

$$\big(h(x) \mid k(x)\big) = \sum_n \big(h(x) \mid e_n\big)(e_n|k(x))$$

for any (necessarily) countable orthonormal basis $\{e_n\}$.

(b) As in [**Wil07**, Lemma I.19], the Cauchy-Schwarz inequality gives $|\big(h(x) \mid k(x)\big)| \le \|h(x)\|\|k(x)\|$. Hence the integrand in (3.19) in integrable by Hölder. Just as in the scalar case, (3.19) defines a positive sesquilinear form on $\mathcal{L}^2(X, \mu, \mathcal{H})$. Cauchy-Schwarz implies that (3.19) induces a well-defined inner product on $L^2(X, \mu, \mathcal{H})$.

(c) We follow [**Wil07**, Lemma I.20]: given a Cauchy sequence $\{h_n\}$ in $\mathcal{L}^2(X, \mu, \mathcal{H})$, it suffices to find a subsequence $\{h_{n_k}\}$ and a $h \in \mathcal{L}^2(X, \mu, \mathcal{H})$ such that $h_{n_k}(x) \to h(x)$ off a Borel null set N and such that $\|h_{n_k} - h\|_2 \to 0$. So we pass to a subsequence, relabel, and assume $\|h_{n+1} - h_n\|_2 \le 2^{-n}$. We let

$$z_n(x) = \sum_{k=1}^{n} \|h_{k+1}(x) - h(x)\| \quad \text{and} \quad z(x) := \sum_{n=1}^{\infty} \|h_{n+1}(x) - h_n(x)\|$$

with $z \in \mathcal{B}^{+,e}(X)$. By the triangle inequality, $\|z_n\|_2 \le 1$. The usual estimates using the Monotone Convergence Theorem implies that $\|z\|_2^2 = 1 < \infty$. Hence z is finite off a Borel null set N and $\sum_n h_{n+1}(x) - h_n(x)$ is absolutely convergent off N. Since $\mathcal{H}$ is complete, we can let

$$h(x) = \begin{cases} h_1(x) + \lim_n \sum_{k=1}^{n} h_{n+1}(x) - h_n(x) = \lim_n h_{n+1}(x) & \text{if } x \notin N, \text{ and} \\ 0 & \text{otherwise.} \end{cases}$$

Since $h_n'(x) = \mathbb{1}_{X \setminus N}(x) h_n(x)$ is Borel and $h_n'(x) \to h(x)$ for all x, h is Borel.

We still need to see that $h \in \mathcal{L}^2(X, \mu, \mathcal{H})$ and that $h_n \to h$ in $L^2(X, \mu, \mathcal{H})$. Let $\epsilon > 0$. Let N be such that $m, n \geq N$ implies $\|h_m - h_n\| < \epsilon$. Off N we have $\|h(x) - h_m(x)\| = \lim_n \|h_n(x) - h_m(x)\|$. By Fatou,

$$\|h - h_m\|_2^2 \leq \liminf_n \|h_n - h_m\|_2^2 \leq \epsilon^2.$$

It follows that $h \in \mathcal{L}^2(X, \mu, \mathcal{H})$ and that $h_m \to h$.

Ex 3.6.3. Suppose that $T = 0$. (The other implication is easy.) Then if $\{e_n\}$ is an orthonormal basis for $\mathcal{H}$, we have

$$0 = \int_X \varphi(x)\psi(x)\big(T(x)e_n \mid e_m\big)\,d\mu(x)$$

for all m and n and all $\varphi, \psi \in C_c(X)$. Hence for each m and n, $\big(T(x)e_n \mid e_m\big) = 0$ for almost all x. Thus for each n, $T(x)e_n = 0$ for almost all x. Thus $T(x) = 0$ for almost all x.

Ex 3.6.4. Let $\{v_n\}$ be a countable dense subset of the unit ball of $\mathcal{H}$. Let $\varphi_{n,m}(x) = \big(T(x)v_n \mid v_m\big)$. Then $\|\varphi_{n,m}\|_\infty$ is the supremum of the values

$$\Big|\int_X f(x)\varphi_{n,m}(x)\,d\mu(x)\Big| \tag{D.8}$$

where f varies over the unit ball of $L^1(X)$. But any such f is the product of two functions f_1 and f_2 in the unit ball of $L^2(X)$. Let $h_{k,n}(x) = f_k(x)v_n$. Then $h_{k,n}$ is in the unit ball of $L^2(X, \mu, \mathcal{H})$ and $\big|\big(Th_{1,n} \mid h_{2,m}\big)\big|$ is given by (D.8). It follows that $\|T\|$ dominates $\|\varphi_{n,m}\|_\infty$ for all n and m. Hence there is a Borel null set N such that

$$\|T\| \geq |\varphi_{n,m}(x)| \quad \text{for all } n, m \text{ and all } x \notin N.$$

Hence $\|T\| \geq \operatorname{ess\,sup} \|T(x)\|$.

Ex 3.7.1. Let $\{\xi_n\}$ be a Cauchy sequence in $\mathcal{L}^2_\pi(G_u, \sigma_u, \mathcal{H}_\pi)$. We can pass to a subsequence and assume that

$$\|\xi_{n+1} - \xi_n\|_2 < \frac{1}{2^n}.$$

Let

$$z_n(\dot\gamma) = \sum_{k=1}^n \|\xi_{k+1}(\gamma) - \xi_k(\gamma)\| \quad \text{and} \quad z(\dot\gamma) = \sum_{k=1}^\infty \|\gamma_{k+1}(\gamma) - \xi_k(\gamma)\|,$$

where $\dot\gamma = q(\gamma) \in G_u/G(u)$. For convenience, let $\|\cdot\|_\sigma = \|\cdot\|_{L^2(\sigma_u)}$ and $\hat\xi(\dot\gamma) = \|\xi(\gamma)\|$.

The triangle inequality implies that $\|z_n\|_\sigma \leq 1$. Therefore the Monotone Convergence Theorem implies that $\|z\|_\sigma \leq 1$. Hence there is a σ_u-null set N such that

$$\sum_{k=1}^\infty \xi_{k+1}(\gamma) - \xi(\gamma)$$

is absolutely convergent if $\dot{\gamma} \notin N$. Since $\mathcal{H}$ is complete, for all $\gamma \notin N$, there is a $\xi'(\gamma) \in \mathcal{H}$ such that

$$\xi'(\gamma) = \lim_{n\to\infty} \sum_{k=1}^{n} \xi_{k+1}(\gamma) - \xi_k(\gamma) = \lim_{n\to\infty} \xi_{n+1}(\gamma) - \xi_1(\gamma).$$

Thus if we let $\xi(\gamma) := \xi'(\gamma) + \xi_1(\gamma)$, then $\xi(\gamma) = \lim_{n\to\infty} \xi_n(\gamma)$ provided $\dot{\gamma} \notin N$. We can let $\xi(\gamma) = 0$ if $\dot{\gamma} \in N$. Then

$$\xi(\gamma h) = \pi(h)^{-1}\big(\xi(\gamma)\big) \quad \text{for all } \gamma \in G_u \text{ and all } h \in G(u).$$

Note that $q^{-1}(N)$ is λ_u-null by Corollary 3.44 on page 91. Therefore $\xi_n \to \xi$ λ_u-almost everywhere.

Give $\epsilon > 0$, there is a N such that $n, m \geq N$ implies that

$$\|\xi_n - \xi_m\|_2 < \epsilon.$$

If $\gamma \notin N$, then

$$\|\xi(\gamma) - \xi_m(\gamma)\| = \lim_{n\to\infty} \|\xi_n(\gamma) - \xi_m(\gamma)\|.$$

Thus if $m \geq N$, Fatou's Lemma implies that

$$\|\xi - \xi_m\|_2^2 \leq \liminf_n \|\xi_n - \xi_m\|_2^2 \leq \epsilon^2.$$

On the other hand,

$$\|\xi(\gamma)\|^2 \leq \big(\|\xi(\gamma) - \xi_N(\gamma)\| + \|\xi_N(\gamma)\|\big)^2 \leq 4\|\xi(\gamma) - \xi_N(\gamma)\|^2 + 4\|\xi_N(\gamma)\|^2.$$

It follows that $\xi \in \mathcal{L}^2_\pi(G_u, \sigma_u, \mathcal{H}_\pi)$ and that $\xi_n \to \xi$ in $L^2_\pi(G_u, \sigma_u, \mathcal{H}_\pi)$.

Ex 4.2.1. (a) Of course,

$$F * K(l') = \int_L F(l)K(l^{-1}l')\, d\kappa^{r(l')}(l).$$

Suppose that $l' = \gamma \in G$. Then using the definition of κ from Lemma 2.68 on page 58, we have

$$F * K(\gamma) = \int_G F_{11}(\eta)K_{11}(\eta^{-1}\gamma)\, d\lambda^{r(\gamma)}(\eta) + \int_Z F_{12}(w)K_{21}(\overline{w}\gamma)\, d\sigma_Z^{r(\gamma)}(w)$$

which, if $r(\gamma) = r(z)$, is

$$\begin{aligned}
&= F_{11} * K_{11}(\gamma) + \int_H F_{12}(z\cdot h)K_{21}(\overline{z\cdot h}\cdot\gamma)\, d\beta^{s(z)}(h) \\
&= F_{11} * K_{11}(\gamma) + \int_H F_{12}(z\cdot h)\overline{K_{21}^*(\gamma^{-1}\cdot z\cdot h)}\, d\beta^{s(z)}(h) \\
&= F_{11} * K_{22}(\gamma) + {}_\star\langle F_{12}\,,\, K_{21}^*\rangle(\gamma).
\end{aligned}$$

This gives us formula claimed for the upper left-hand corner.

To examine the upper right-hand entry, let $l' = z \in Z$. Then

$$\begin{aligned} F * K(z) &= \int_G F_{11}(\eta)K_{12}(\eta^{-1} \cdot z)\,d\lambda^{r(z)}(\gamma) + \int_Z F_{12}(w)K_{22}(\overline{w}z)\,d\sigma_Z^{r(z)}(w) \\ &= F_{11} \cdot F_{12}(z) + \int_H F_{12}(z \cdot h)K_{22}(\tau_H(z \cdot \eta, z))\,d\beta^{s(z)}(h) \\ &= F_{11} \cdot K_{12}(z) + \int_H F_{12}(z \cdot h)K_{22}(h^{-1})\,d\beta^{s(z)}(h) \\ &= F_{11} \cdot K_{12}(z) + F_{12} \cdot K_{22}(z) \end{aligned}$$

as required.

The lower left-hand entry requires some gyrations with adjoints. Let $l' = \overline{z} \in Z^{\mathrm{op}}$. Then

$$\begin{aligned} F * K(\overline{z}) &= \int_{Z^{\mathrm{op}}} F_{21}(\overline{w})K_{11}(w\overline{z})\,d\sigma_{Z^{\mathrm{op}}}^{s(z)}(\overline{w}) + \int_H F_{22}(h)K_{21}(h^{-1}\overline{z})\,d\beta^{s(z)}(h) \\ &= \int_G F_{21}(\overline{z} \cdot \gamma)K_{11}(\tau_H(\gamma^{-1} \cdot z, x))\,d\lambda^{r(z)}(\gamma) + \int_H F_{22}(h)K_{21}(\overline{z \cdot h})\,d\beta^{s(z)}(h) \\ &= \overline{\int_G \overline{K_{11}(\gamma^{-1})F_{21}(\overline{\gamma^{-1} \cdot z})}\,d\lambda^{r(z)}(\gamma) + \int_H \overline{K_{21}(\overline{z \cdot h})F_{22}(h)}\,d\beta^{s(z)}(h)} \\ &= \overline{\int_G K_{11}^*(\gamma)F_{21}^*(\gamma^{-1} \cdot z)\,d\lambda^{r(z)}(\gamma) + \int_H K_{21}^*(z \cdot h)F_{22}^*(h^{-1})\,d\beta^{s(z)}(h)} \\ &= \overline{K_{11}^* \cdot F_{21}^*(z)} + \overline{K_{21}^* \cdot F_{22}^*(z)} \\ &= (K_{11}^* \cdot F_{21}^*)^*(\overline{z}) + (K_{21}^* \cdot F_{22}^*)^*(\overline{z}) \end{aligned}$$

as required.

For the final lower right-hand entry, let $l' = h \in H$. Then

$$\begin{aligned} F * K(h) &= \int_{Z^{\mathrm{op}}} F_{21}(\overline{w})F_{12}(wh)\,d\sigma_{Z^{\mathrm{op}}}^{r(h)}(\overline{w}) + \int_H F_{22}(k)K_{22}(k^{-1}h)\,d\beta^{r(h)}(k) \\ &= \int_G F_{21}(\overline{z} \cdot \gamma)K_{12}(\gamma^{-1} \cdot z \cdot h)\,d\lambda^{r(z)}(\gamma) + F_{22} * K_{22}(h) \\ &= \int_G \overline{F_{21}^*(\gamma^{-1} \cdot z)}K_{12}(\gamma^{-1} \cdot z \cdot h)\,d\lambda^{r(z)}(\gamma) + F_{22} * K_{22}(h) \\ &= \langle F_{21}^*\,,\, K_{12}\rangle_{\star}(h) + F_{22} * K_{22}(h). \end{aligned}$$

This establishes the formula in part (a).

(b) These follows from part (a).

Ex 4.2.2. Let $\varphi \in C_c(Z)$. We can assume $U \subset M$ with M diagonally compact and symmetric. Then $\mathrm{supp}(e \cdot \varphi)$ is contained in the compact set $M \cdot \mathrm{supp}\,\varphi = (M \cap r_G^{-1}(r_Z(\mathrm{supp}\,\varphi)) \cdot \mathrm{supp}\,\varphi$, and it suffices to prove uniform convergence. Fix $\epsilon > 0$. I claim that there is a conditionally compact neighborhood U_0 of $G^{(0)}$ such that $\gamma \in U_0$ implies

$$|\varphi(\gamma^{-1} \cdot z) - \varphi(z)| < \epsilon \quad \text{for all } z \in Z.$$

If not, then for all $U \subset M$, we can find $\gamma_U \in U$ and $z_U \in M \cdot \mathrm{supp}\,\varphi$ such that

$$|\varphi(\gamma_U^{-1} \cdot z_U) - \varphi(z_U)| \geq \epsilon. \tag{D.9}$$

Since $M \cdot \mathrm{supp}\,\varphi$ is compact, we can pass to a subnet, relabel and assume $z_U \to z$ in Z. Similarly, the properness of the G-action on Z implies we can also assume $\gamma_U \to r(z)$. This eventually contradicts (D.9).

Now if $U \subset U_0$ and $r(M \cdot \operatorname{supp} \varphi) \subset K$, and if $e = e_{K,U,\epsilon}$, then $e \cdot \varphi(z) - \varphi(z) = 0$ if $z \notin M \cdot \operatorname{supp} \varphi$. Otherwise $r(z) \in K$ and

$$\begin{aligned}
&|e \cdot \varphi(z) - \varphi(z)| \\
&= \Big| \int_G e(\gamma) z(\gamma^{-1}) \, d\lambda^{r(z)}(\gamma) - \varphi(z) \Big| \\
&\le \Big| \int_G e(\gamma)(\varphi(\gamma^{-1} \cdot z) - \varphi(z)) \, d\lambda^{r(z)}(\gamma) \Big| + \Big| \int_G e(\gamma) \, d\lambda^{r(z)}(\gamma) - 1 \Big| |\varphi(z)| \\
&\le \epsilon(1 + \epsilon) + \epsilon \|\varphi\|_\infty .
\end{aligned}$$

This suffices.

Ex 4.2.3. We have

$$M = \sup_{u \in G^{(0)}} \sum_{i=1}^n \underline{\lambda}^u(\overline{V_i}).$$

This is bounded in view of Ex 3.2.1.

Ex 4.2.4. This is mostly a matter of untangling identifications. Let $\Phi_{11} : C^*(G) \to \mathfrak{p}_G C^*(L) \mathfrak{p}_G$ be the isomorphism guaranteed by Theorem 4.9 on page 99. Similarly, let $\Phi_{22} : C^*(H) \to \mathfrak{p}_H C^*(L) \mathfrak{p}_H$ be the isomorphism onto the lower right-hand corner. Let $\mathsf{X} = \overline{C_c(Z)}$ be the imprimitivity bimodule from Theorem 4.9 and $\Phi_{12} : \mathsf{X} \to \mathfrak{p}_G C^*(L) \mathfrak{p}_H$ the imprimitivity bimodule isomorphism provided in that result. Then

$$F \mapsto \begin{pmatrix} \Phi_{11}^{-1}(\mathfrak{p}_G F \mathfrak{p}_G) & \Phi_{12}^{-1}(\mathfrak{p}_G F \mathfrak{p}_H) \\ (\Phi_{12}^{-1}(\mathfrak{p}_G F^* \mathfrak{p}_H))^\flat & \Phi_{22}^{-1}(\mathfrak{p}_H F \mathfrak{p}_H) \end{pmatrix}$$

is a homomorphism of $C^*(L)$ onto $L(\mathsf{X})$. The map

$$\begin{pmatrix} a & x \\ y^\flat & b \end{pmatrix} \mapsto \Phi_{11}(a) + \Phi_{12}(x) + \Phi_{12}(y)^* + \Phi_{22}(b)$$

is an inverse. Since we're working at the level of C^*-algebras, this suffices.

Ex 4.2.5. Since the action is proper, the orbit space is Hausdorff. This certainly implies that the orbits are closed. Now either apply Theorem 2.27 on page 38 or note that the map $\gamma \mapsto \gamma \cdot z$ is clearly a continuous bijection. We just have to see that if $\gamma_i \cdot z \to \gamma \cdot z$, then $\gamma_i \to \gamma$. If not, we can pass to a subnet, relabel, and assume γ has a neighborhood V such that $\gamma_i \notin V$ for all i. But since the action is proper, we can pass to a subnet, relabel, and assume $\gamma_i \to \eta$. Then $\{\gamma_i \cdot z\}$ converges to both $\eta \cdot z$ and $\gamma \cdot z$. Hence $\gamma = \eta$, and we have a contradiction.

Ex 5.1.1. Suppose that $u \in A$ and $\gamma \in G_u$. Then there is a $\eta \in r^{-1}(B)$ such that $s(\eta) = u$. Then $\eta\gamma^{-1} \in r^{-1}(B)$. Hence $r(\gamma) = s(\eta\gamma^{-1}) \in A$ as required.

Ex 5.1.2. Let M be the multiplication representation on $L^2(G^{(0)} * \mathscr{H})$ given in (5.2). Let $J(C)$ be the ideal of functions in $C_0(G^{(0)})$ vanishing on the closed set C and suppose that $J(C) = \ker M$ so that C is the support of M. The support of μ is the complement D of the open set formed by the union of open sets $V \subset G^{(0)}$ such that $\mu(V) = 0$. Since $\mu(G^{(0)} \setminus D) = 0$, we clearly have $J(D) \subset \ker M = J(C)$ and $C \subset D$.

If $D \setminus C \neq \emptyset$, then let $u \in D \setminus C$. Then there is a $\varphi \in C_c^+(G^{(0)})$ such that $\varphi \equiv 1$ on a neighborhood V of u and $\varphi \in J(C)$. Note that $\mu(V) > 0$. Let $\{e_k\} \subset B(G^{(0)} * \mathscr{H})$ be a special orthogonal fundamental sequence as in [**Wil07**, Proposition F.6]. Then $e_1(v)$ is unit vector for all v (since $\dim \mathcal{H}(y) \geq 1$ for all v). Let $h \in \mathcal{L}^2(G^{(0)} * \mathscr{H}, \mu)$ be such that $h(v) = e_1(v)$ for all $v \in V$.

$$\|M(\varphi)h\|^2 \geq \mu(V) > 0.$$

But $\varphi \in J(C) = \ker M$. This is a contradiction and $C = D$ as required.

Ex 5.1.3. Let $Z = \{ x \in X : f(x) = 0 \text{ for all } f \in J \}$. It will suffice to show that $J = J(Z)$. Let $f \in J(Z)$. Fix $\epsilon > 0$. Since $J \subset J(Z)$ and J is closed, it will suffice to produce $g \in J$ such that $\|f - g\|_\infty < \epsilon$. Let $K = \{ x : |f(x)| \geq \epsilon/2 \}$. For each $y \in K$, let $g_y \in J$ be such that $g_y(y) = f(y)$. Let V_y be an pre-compact open neighborhood of y such that $|f(x) - g_y(x)| < \epsilon/2$ if $x \in V_y$. Let $\{ V_{y_1}, \dots, V_{y_n} \}$ be a cover of K and $\{ \varphi_k \} \subset C_c(X)$ such that $0 \leq \varphi_k(x) \leq 1$ for all x, $\operatorname{supp} \varphi_k \subset V_k$, $\sum_k \varphi_k(x) \leq 1$ for all x and equal to 1 if $x \in K$ (see [**Wil07**, Lemma 1.43]). Let $g(x) = \sum_k \varphi_k(x) g_{y_k}(x)$. Then $g \in J$. Notice that

$$|f(x) - g(x)| \leq \big|f(x)\big(1 - \sum_k \varphi_k(x)\big)\big| + \big|\sum_k \varphi_k(x)\big(f(x) - g_{y_k}(x)\big)\big| < \epsilon.$$

Ex 5.1.4. If A is saturated, then $(G|_A)^u = G^u$ and $(G|_A)_u = G_u$ for all $u \in A$. Then $\lambda(|f|)(u)$ is the same as $(\lambda|_A)(|f|)(u)$ for any u and any $f \in C_c(G|_A)$. Thus, parts (a) and (b) are immediate.

(c) Both representations act on $L^2(G_u, \lambda_u)$ by convolution.

(d) As in part (c), the two representations act on $L^2(G_u, \lambda_u)$ and if $h \in C_c(G_u)$, then $j(f) * g(\gamma) = f * g(\gamma)$ for any $\gamma \in G_u$.

Ex 5.1.5. Suppose $U : \mathcal{H}_L \to \mathcal{H}_R$ implements the equivalence. Then

$$\begin{aligned} UM_L(\varphi)L(f)h &= UL(V(\varphi)f)h \\ &= R(V(\varphi)f)Uh = M_R(\varphi)R(f)Uh = M_R(\varphi)UL(f)h. \end{aligned}$$

Since vectors of the form $L(f)h$ span $\mathcal{H}_L$, this suffices.

Ex 5.1.6. It suffices to see that $G \cdot \overline{A} \subset \overline{A}$. But if $u \in \overline{A}$, then there is a net $u_i \to u$ with $u_i \in A$. Let $\gamma \in G_u$. Since the source map is open, we can pass to a subnet, relabel, and assume that there are $\gamma_i \in G_{u_i}$ such that $\gamma_i \to \gamma$. Then

$r(\gamma_i) = \gamma_i \cdot u_i \to r(\gamma) = \gamma \cdot u$. By assumption, $r(\gamma_i) \in A$. Hence $r(\gamma) \in \overline{A}$. This suffices.

Ex 5.2.1. Suppose that $f \in C_c(G)$. Let $K = \operatorname{supp}\varphi \subset U$. Then $[K] = G \cdot K \subset G \cdot U = [U]$. If $\gamma \notin G|_{[K]}$, then $r(\gamma) \notin K$. Hence $V(\varphi)f(\gamma) = 0$. Thus $\operatorname{supp} V(\varphi)f \subset G|_{[K]} \subset G|_{[U]}$.

Ex 5.2.2. If $U \cap F \neq \emptyset$, then $U \setminus F$ is a nonempty open set and there is a nonzero $\varphi \in C_c^+(G^{(0)})$ with $\operatorname{supp}\varphi \subset U \setminus F$. Then $M_L(\varphi) \neq 0$, while $V(\varphi)f \in \operatorname{Ex}(U)$ for all $f \in C_c(G)$. Thus $L(\operatorname{Ex}(U)) \neq \{0\}$.

Ex 5.2.3. Recall that if $f, g \in C_c(G)$, then $\operatorname{supp}(f * g) \subset (\operatorname{supp} f)(\operatorname{supp} g)$. Thus if $f \in C_c(G|_{U_1})$ and $g \in C_c(G|_{U_2})$, we have $\operatorname{supp}(f * g) \subset G_{U_2}^{U_1}$. But the later is equal to $G|_{U_1 \cap U_2}$ since the U_i are saturated.

Ex 5.3.1. Here $G * \mathscr{H} = G \times \mathbf{C}$, and $L^2(G * \mathscr{H}_s, \nu^{-1})$ can be identified with $L^2(G, \nu^{-1})$. Then

$$\big((\operatorname{Ind} L)(f)k \mid k'\big) = \int_{G^{(0)}} \int_G \int_G f(\eta\gamma)k(\gamma^{-1})\overline{k'(\eta)}\, d\lambda^u(\gamma)\, d\lambda_u(\eta)\, d\mu(u). \tag{D.10}$$

On the other hand,

$$\begin{aligned}\big((\operatorname{Ind}\mu)(f)k \mid k'\big) &= \int_{G^{(0)}} \int_G (\operatorname{Ind}\mu)(f)k(\eta)\overline{k'(\eta)}\, d\lambda_u(\eta)\, d\mu(u)\\ &= \int_{G^{(0)}} \int_G \int_G f(\gamma)k(\gamma^{-1}\eta)\overline{k'(\eta)}\, d\lambda^{r(\eta)}(\gamma)\, d\lambda_u(\eta)\, d\mu(u),\end{aligned}$$

which after sending γ to $\eta\gamma$ equals (D.10).

Ex 5.3.2. This is a special case of Ex 5.3.1.

Ex 5.3.3. $L = \bigoplus_{u \in D} \operatorname{ev}_u$ is a faithful representation of $C_0(G^{(0)})$.

Ex 5.4.1. Let $\gamma \in G_A$. Then $r(\gamma) \in [A] \subset B$. Hence $\gamma \in G_A^B$.

Ex 5.5.1. Suppose points are closed in X and $x, y \in X$. Then $U := X \setminus \{y\}$ is an open set containing x and not y. Hence X is T_1.

Suppose X is T_1. Let $x \in X$, and $y \in X \setminus \{x\}$. By assumption, there is an open set U such that $x \notin U$ and $y \in U$. Hence $X \setminus \{x\}$ is open. Thus $\{x\}$ is closed

Ex 5.5.2. If $q(u)$ is closed, then $[u] = q^{-1}(q(u))$ is closed (since q is continuous).

Suppose that $[u]$ is closed. Then $G^{(0)} \setminus [u] = q^{-1}((G\backslash G^{(0)}) \setminus \{q(u)\})$ is open, and by definition of the quotient topology, $(G\backslash G^{(0)}) \setminus \{q(u)\}$ is open. Hence $\{q(u)\}$ is closed.

Ex 5.6.1. Suppose that $f \in C_c(G|_{U(t)})$. Since $K = p(r(\operatorname{supp} f))$ is compact in T and does not contain t, there is a $\varphi \in J_t$ such that $\varphi(s) = 1$ for all $s \in K$. Then $\varphi \cdot f = f$ and $f \in I_t^0$.

Suppose that $\varphi \in J_t$ and $f \in C_c(G)$. Then $\varphi \cdot f \in C_c(G)$ and vanishes off $G|_{p^{-1}(t)}$. Hence $\varphi \cdot f \in C_c(G|_{U(t)})$. This suffices.

Ex 5.6.2. Suppose that $f \in \bigcap J_{t_n}$. Then $f(u) = 0$ whenever $p(u) \in \bigcup_n \{t_n\}$. Now suppose that $p(u) = t$. Since p is open, we can find a subnets $\{t_{n_k}\}$ and $\{u_{k_k}\}$ such that $p(u_{n_k}) = t_{n_k}$ and $u_{n_k} \to u$. Since $f(u_{n_k}) = 0$ for all k, we must have $f(u) = 0$. Thus $f \in J_t$.

Ex 5.8.1. Suppose that $I_{F_i} \to I_F$ and $t \in F$. Suppose that $t \in F$ and let $\mathcal{N}(t)$ be the set of open neighborhoods of t directed by containment. If $U \in \mathcal{N}(t)$, then $\mathcal{O}_{F_{X\setminus U}}$ is a neighborhood of I_F: $I_F \not\supset I_{F_{X\setminus U}}$. Hence $\{I_{F_i}\}_{i\in I}$ is eventually in $\mathcal{O}_{F_{X\setminus U}}$. Therefore

$$\Lambda = \{\, (U,i) \in \mathcal{N}(t) \times I : I_{F_i} \in \mathcal{O}_{F_{X\setminus U}} \,\}$$

is a directed set. If $\lambda = (U,i) \in \Lambda$ and we let $I_{F_\lambda} = I_{F_i}$ and find $t_\lambda \in U \cap F_i$. Then $\{t_\lambda\}_{\lambda\in\Lambda}$ is our subnet. (This argument applies to any subnet of $\{I_{F_i}\}$.)

Now suppose that $\{I_{F_i}\}$ is a net with the given property for all its subnets. Suppose that $I_{F_i} \not\to I_F$. Then after passing to a subnet and relabeling, we can assume there is a $J = I_E$ such that $I_F \in \mathcal{O}_J$ and $I_{F_i} \notin \mathcal{O}_J$ for all i. In particular, $F_i \subset E$ for all i. Given $t \in F$, we can pass to another subnet and find $t_i \in F_i \subset E$ such that $t_i \to t$. Hence $t \in E$ and $F \subset E$. This contradicts $I_F \in \mathcal{O}_J$.

Ex 6.2.1. The set of $x \in \ell^\infty$ such that $\lim_k x_{n_k}$ exists is a subspace M of ℓ^∞ containing $\mathfrak{c}$. The corresponding limit functional on M has a norm one extension to ℓ^∞.

Ex 6.2.2. Since Λ is positive, we have $\mathrm{Re}\big(\Lambda(x)\big) = \Lambda\big(\mathrm{Re}(x)\big)$ where $\mathrm{Re}(x) = \big(\mathrm{Re}(x_n)\big)$. Let $\tau \in \mathbf{C}$ be such that $|\Lambda(x)| = \tau\Lambda(x) = \Lambda(\tau x)$. Since the latter is real, $\Lambda(\tau x) = \Lambda(\mathrm{Re}(\tau x))$. Since Λ is positive, $\Lambda(\mathrm{Re}(\tau x)) \le \Lambda(|\mathrm{Re}(\tau x)|) \le \Lambda(|x|)$.

Ex 6.2.3. Note that if we eventually have $x_n = y_n$, then $x - y$ is eventually zero and $\Lambda(x) = \Lambda(y)$. Hence if x_n is eventually nonnegative, then $\Lambda(x) \ge 0$. Thus if $d = \limsup_n |x_n|$, then $\Lambda(|x|) \le d + \epsilon$ for all $\epsilon > 0$. Hence the result follows from Ex 6.2.2.

Ex 6.2.4. If the limit does not exist or is not 0, then there is a $\epsilon > 0$ and a subsequence $u_{n_k} \to u$ such that

$$\sup_{t\in G(u_{n_k})} |f(t) - g(t)| > \epsilon.$$

Thus there is a $t_{n_k} \in G(u_{n_k})$ such that

$$|f(t_{n_k}) - g(t_{n_k})| > \epsilon.$$

Since we can assume each $t_{n_k} \in K$, we can assume a subsequence of $\{t_{n_k}\}$ converges to $t \in G(u)$. This leads to a contradiction as f and g agree on $G(u)$.

Ex 6.3.1. By Proposition 2.17 on page 32, it suffice to show that if $\gamma_i \to \gamma$ in $G_{H^{(0)}}$ and $\gamma_i h_i \to \eta$, then $\{h_i\}$ has a convergent subnet. But $h_i = \gamma_i^{-1}(\gamma_i h_i) \to \gamma^{-1}\eta$.

Ex 6.3.2. Part (a) follows immediately from (6.2) on page 147.

For (b), we know that q is open if r' is by Theorem 6.12 on page 146 and Proposition 2.12 on page 31. For the converse, assume that $q : G \to G/G'$ is open. Let $h \in G(u) \subset G'$. Using Fell's criterion, it will suffice to see that if $u_n \to u$, then after passing to a subnet, and relabeling, there are $h_n \in G'$ such that $h_n \to h$ and $r'(h_n) = u_n$. Since $s : G \to G^{(0)}$ is open, we may as well assume that there are $\gamma_n \in G$ such that $\gamma_n \to h^{-1}$ and $s(\gamma_n) = u_n$. Then $q(\gamma_n) \to q(h^{-1})$. Since q is open and $q(h^{-1}) = q(u)$, we can, after passing to a subnet and relabeling, find $h_n \in G'$ such that $\gamma_n h_n \to u$. But then $h_n = \gamma_n^{-1}(\gamma_n h_n) \to h$ as required.

Ex 6.4.1. Observe that on the one hand,

$$\begin{aligned}\Phi(f * g)(\gamma) &= \alpha(r(\gamma))^{\frac{1}{2}} f * g(\gamma)\alpha(s(\gamma))^{\frac{1}{2}} \\ &= \alpha(r(\gamma))^{\frac{1}{2}}\Big(\sum_{r(\eta)=r(\gamma)} f(\eta)g(\eta^{-1}\gamma)\alpha(s(\gamma))\Big)\alpha(s(\gamma))^{\frac{1}{2}}.\end{aligned}$$

On the other,

$$\begin{aligned}\Phi(f) * \Phi(g)(\gamma) &= \sum_{r(\eta)=r(\gamma)} \Phi(f)(\eta)\Phi(g)(\eta^{-1}\gamma) \\ &= \alpha(r(\gamma))^{\frac{1}{2}} \sum_{r(\eta)=r(\gamma)} f(\eta)\alpha(s(\eta))^{\frac{1}{2}}\alpha(s(\eta))^{\frac{1}{2}} g(\eta^{-1}\gamma)\alpha(s(\gamma))^{\frac{1}{2}}.\end{aligned}$$

It follows that Φ is multiplicative. Showing that it preserves the involutions is straightforward.

Ex 6.6.1. As is Ex 2.1.7, we realize the action groupoid as pairs $G \rtimes X$. Let $\alpha = \{\alpha^x\}_{x\in X}$ be a full r_X-system. Then we define measures β^x on $G \rtimes X$ by

$$\beta^x(F) = \int_G F(\gamma, x)\, d\alpha^x(\gamma).$$

Then β can be shown to be a Haar system almost exactly as in Ex 2.1.7.

Conversely, if $\beta = \{\beta^x\}$ is a Haar system, then β^x must be of the form $\alpha^x \times \delta_x$ where α^x is a Radon measure on G with support $G^{r(x)}$. Letting F be of the form $F(\gamma, x) = f(x)\psi(x)$, we see easily that $x \mapsto \alpha(f)(x)$ is continuous. Invariance follows from $\gamma \cdot \beta^{s(\gamma)} = \beta^{r(\gamma)}$.

Ex 6.6.2. For $z \in Z$, the map $(g, w) \mapsto f(z, g, w)$ is in $C_c(G^{f(z)} * Z)$. Hence

$$g \mapsto \int f(z, g, w)\, d\beta^{s(g)}(w)$$

is in $C_c(G^{f(z)})$. Hence κ^z is a Radon measure with

$$\operatorname{supp}\kappa^z = \{\, (z, g, w) : f(z) = r(g) \text{ and } s(g) = f(w) \,\} = r_{G[Z]}^{-1}(z).$$

If f is separable, then clearly $z \mapsto \kappa^z(f)$ is continuous. Then the usual approximation arguments in the inductive limit topology suffice for continuity.

Now observe that

$$\begin{aligned}&\int_{G[Z]} f\big((w', g', z)(z, g, w)\big)\, d\kappa^z(z, g, w) \\ &\qquad\qquad = \int_G \int_Z f(w', g'g, w)\, d\beta^{s(g)}(w)\, d\lambda^{f(z)}(g)\end{aligned}$$

which, since $f(w) = r(g) = s(g')$, is

$$= \int_G \int_Z f(w', g'g, w)\, d\beta^{s(g'g)}(w)\, d\lambda^{s(g')}(g)$$

which, after sending $g'g$ to g and using the left-invariance of λ, is

$$= \int_G \int_Z f(w', g, w)\, d\beta^{s(g)}(w)\, d\lambda^{r(g')}(g)$$
$$= \int_{G[Z]} f(w', g, w)\, d\kappa^{w'}(w', g, w).$$

This completes the proof.

Ex 7.1.1. By (7.3), we have

$$\int_{G^{(0)}} \int_G f(\gamma)\, d\lambda_u(\gamma)\, d\mu(u) = \int_{G^{(0)}} \int_G f(\gamma)\Delta(\gamma)\, d\lambda^u(\gamma)\, d\mu(u). \tag{D.11}$$

But we can also compute that

$$\begin{aligned}\int_{G^{(0)}} \int_G f(\gamma)\, d\lambda_u(\gamma)\, d\mu(u) &= \int_{G^{(0)}} \int_G f(\gamma^{-1})\, d\lambda^u(\gamma)\, d\mu(u)\\ &= \int_{G^{(0)}} \int_G f(\gamma^{-1})\Delta(\gamma)^{-1}\, d\lambda_u(\gamma)\, d\mu(u)\\ &= \int_{G^{(0)}} \int_G f(\gamma)\Delta(\gamma^{-1})^{-1}\, d\lambda^u(\gamma)\, d\mu(u) \qquad \text{(D.12)}\end{aligned}$$

Comparing (D.11) and (D.12) gives the result.

Ex 7.3.1. Let $k = \overline{\varphi}$ and let h in $C_c(G)$ be such that $h \equiv 1$ on $\eta^{-1} \operatorname{supp} \varphi$. Let $\{h_i\}$ and $\{k_i\}$ be sequences in D such that $h_i \to h$ and $k_i \to k$ uniformly with supports in a fixed compact set K (see Proposition C.3 on page 344).

Ex 8.2.1. For fixed g and η, the right-hand side of (8.3) is bilinear in f and h. Therefore, by the universal properties of the algebraic tensor product, (8.3) defines linear map $m(g, \eta) : C_c(G) \odot \mathcal{H}_0 \to \mathbf{C}$. Then $(g, \eta) \mapsto \overline{m(g, \eta)}$ is a bilinear map into the space $\mathrm{CL}(C_c(G) \odot \mathcal{H}_0)$ of conjugate linear functionals on $C_c(G) \odot \mathcal{H}_0$. Then we get a linear map $N : C_c(G) \odot \mathcal{H}_0 \to \mathrm{CL}(C_c(G) \odot \mathcal{H}_0)$. We can then define $\langle \alpha\,,\, \beta \rangle := \overline{N(\beta)(\alpha)}$. Clearly $\alpha \mapsto \langle \alpha\,,\, \beta \rangle$ is linear and it is not hard to check that $\overline{\langle \alpha\,,\, \beta \rangle} = \langle \beta\,,\, \alpha \rangle$.

Ex 8.2.2. We need to find (a necessarily countable) orthonormal set $\{e_i\} \subset V$ with dense span. Since $\mathcal{H}$ is second countable in its metric topology, V is second countable and has a countable dense subset D. By eliminating vectors in the span of those vectors that precede it, we get a linearly independent set $D' = \{v_i\}$ such that $\operatorname{span} D = \operatorname{span} D'$. Given n, we use Gram-Schmidt to find an orthonormal set $\{e_i\}$ such that

$$\operatorname{span}\{ v_1, \dots, v_n \} = \operatorname{span}\{ e_1, \dots, e_n \} \tag{D.13}$$

Then we can use Gram-Schmidt to find $e_{n+1} \in \{ e_1, \dots, e_n \}^\perp$ such that

$$\operatorname{span}\{ v_1, \dots, v_{n+1} \} = \operatorname{span}\{ e_1, \dots, e_n, v_{n+1} \} = \operatorname{span}\{ e_1, \dots, e_{n+1} \}.$$

Continuing, we find a sequence $\{e_i\}$ of orthonormal vectors such that (D.13) holds for all n. Since

$$D = \operatorname{span}\{v_i\} = \operatorname{span}\{e_i\},$$

we are done.

Ex 9.2.1. We certainly have $h \in C_c(G)$. Let $\gamma_1, \dots, \gamma_n \in G^u$ and $c_1, \dots, c_n \in \mathbf{C}$. Then

$$\begin{aligned}\sum_{ij} f * f^*(\gamma_i^{-1}\gamma_j)\overline{c_i}c_j &= \sum_{ij} \int_G f(\eta) f^*(\eta^{-1}\gamma_i^{-1}\gamma_j)\overline{c_i}c_j \, d\lambda^{s(\gamma_i)}(\eta) \\ &= \sum_{ij} f(\eta)\overline{f(\gamma_j^{-1}\gamma_i\eta)}\overline{c_i}c_j \, d\lambda^{s(\gamma_i)}(\eta) \\ &= \int_G \Bigl(\sum_{ij} f(\gamma_i^{-1}\eta)\overline{f(\gamma_j^{-1}\eta)}\overline{c_i}c_j\Bigr)\, d\lambda^u(\eta) \\ &= \int_G \Bigl|\sum_i \overline{c_i} f(\gamma_i^{-1}\eta)\Bigr|^2 \, d\lambda^u(\eta) \\ &\geq 0.\end{aligned}$$

Ex 9.2.2. We have

$$\sum_{ij} \overline{c_i}c_j\tilde{h}(\gamma_i^{-1}\gamma_j) = \sum_{ij} \overline{c_i\varphi(s(\gamma_i))}c_j\varphi(s(\gamma_j))h(\gamma_i^{-1}\gamma_j) \geq 0.$$

Ex 9.2.3. We proceed exactly as in Remark 9.4 on page 190. Since $u \in G^{(0)}$ implies $u = u^{-1}u$, we have $h_i(u) \geq 0$ for all $u \in G^{(0)}$. Let

$$\varphi_i(u) = \begin{cases} 1 & \text{if } h_i(u) \leq 1, \text{ and} \\ h_i(u)^{-1} & \text{otherwise.}\end{cases}$$

Note that $\{\varphi_i\}$ must converge to 1 uniformly on compact subsets of $G^{(0)}$. Then let $h_i'(\gamma) = \varphi_i(r(\gamma))^{\frac{1}{2}} h_i(\gamma)\varphi_i(s(\gamma))^{\frac{1}{2}}$ and use Ex 9.2.2.

Ex 9.2.4. To see that $k = k^*$, note that the matrix

$$\begin{pmatrix} k(u) & k(\gamma) \\ k(\gamma^{-1}) & u \end{pmatrix}$$

must be positive.

To establish (9.6), fix $u \in G^{(0)}$ and $\epsilon > 0$. Let $F(\gamma, \eta) = k(\gamma^{-1}\eta)\overline{f(\gamma)}f(\eta)$, and let $K = \operatorname{supp} f$. By Ex 1.2.11 there is an open neighborhood V of u in G such that

$$|F(\gamma,\eta) - F(\gamma',\eta')| < \epsilon \quad \text{if } (\gamma,\eta), (\gamma',\eta') \in \gamma_0 V \times \eta_0 V \text{ for any } \gamma_0, \eta_0 \in G^u.$$

We can find a Borel partition of K by finitely many nonempty Borel sets E_i each contained in a translate $\eta_i V$. Let $\gamma_i \in E_i$. Then (9.6) equals

$$\begin{aligned}&\sum_{ij} \int_{E_i}\int_{E_j} k(\gamma^{-1}\eta)\overline{f(\gamma)}f(\eta)\, d\lambda^u(\eta)\, d\lambda^u(\gamma) \\ &\qquad\qquad = \sum_{ij} \overline{f(\gamma_i)\lambda^u(E_i)}f(\gamma_j)\lambda^u(E_j)k(\gamma_i^{-1}\gamma_j) + R\end{aligned}$$

where

$$R = \sum_{ij} \int_{E_i} \int_{E_j} \big(F(\gamma,\eta) - F(\gamma_i,\gamma_j)\big)\, d\lambda^u(\eta)\, d\lambda^u(\gamma).$$

But $|R| \le \epsilon\lambda^u(K)^2$.

It follows that the real part of (9.6) is bounded below by $-\epsilon\lambda^u(K)^2$ and the modulus of the complex part is bounded by $\epsilon\lambda^u(K)^2$. Since ϵ is arbitrary, (9.6) must be non-negative.

Ex 9.2.5. Let α_i be the right-hand side of (9.1) on page 191. (Since $e_i f$ is not necessarily in $C_c(G)$, we can no longer write $\omega_h(e_i f)$.) Then just as in the proof of Proposition 9.6, and retaining the notations in that proposition,we have

$$\alpha_i = \big((\operatorname{Ind} M)(f)k_i \mid k_i\big).$$

But assumption (b)″ implies that $\alpha_i \to \omega_h(f)$. On the other hand, assumption (a)″ implies that

$$\|k_i\|_2^2 \to \|h\|^2 = 1.$$

Therefore,

$$\begin{aligned} |\omega_h(f)| &= \lim_i |\alpha_i| \\ &\le \limsup_i \big|\big((\operatorname{Ind} M)(f)k_i \mid k_i\big)\big| \\ &\le \limsup_i \|f\|_r \|k_i\|_2^2 \\ &\le \|f\|_r. \end{aligned}$$

Just as in the proof of Proposition 9.6, this suffices.

Ex 9.2.6. If $f \in C_c(G)$, then $f * f^*(x) = |f(x)|^2$. By Lemma 9.5 on page 190 we only need to let f tend to 1 uniformly on compacta.

Ex 9.2.7. Let $\{\tau_n\}$ be a sequence in $C_c^+(X)$ converging uniformly to 1 on compacta. Let $\sigma \in C_c^+(X)$ be such that $\|\sigma\|_2 = 1$. Then let $f_n(x,y) = \tau_n(x)\sigma(y)$. Then

$$\begin{aligned} f_n * f_n^*(x,y) &= \int_X f(x,z)\overline{f(y,z)}\, d\mu(z) = \tau_n(x)\tau_n(y) \int_X |\sigma(z)|^2\, d\mu(z) \\ &= \tau_n(x)\tau_n(y) \end{aligned}$$

converges to 1 uniformly on compacta.

Ex 9.2.8. Proposition 3.3 on page 70 implies that $f * \varphi$ is a bounded Borel function on G whose supremum norm is bounded by $\|f\|_I \cdot \sup_{\gamma\in G} |\varphi(\gamma)|$. Since $f*(\varphi+\psi) = f*\varphi + f*\psi$, it will suffice to see that $f*\varphi$ is zero ν-almost everywhere whenever φ is zero ν-almost everywhere.

But we have

$$\int_G f * \varphi(\eta)\, d\nu(\eta) = \int_{G^{(0)}} \int_G \int_G f(\gamma)\varphi(\gamma^{-1}\eta)\, d\lambda^u(\gamma)\, d\lambda^u(\eta)\, d\mu(u)$$

which, using Fubini, is

$$= \int_{G^{(0)}} \int_G \int_G f(\gamma)\varphi(\gamma^{-1}\eta)\, d\lambda^{r(\gamma)}(\eta)\, d\lambda^u(\gamma)\, d\mu(u)$$

which, since μ is quasi-invariant, is

$$= \int_{G^{(0)}} \int_G \int_G f(\gamma^{-1})\varphi(\gamma\eta)\, d\lambda^{s(\gamma)}(\eta)\Delta(\gamma^{-1})\, d\lambda^u(\gamma)\, d\mu(u)$$
$$= \int_{G^{(0)}} \int_G \int_G f(\gamma^{-1})\varphi(\eta)\, d\lambda^{u}(\eta)\Delta(\gamma^{-1})\, d\lambda^u(\gamma)\, d\mu(u)$$

which, using Fubini, is

$$= \int_G \int_G f(\gamma^{-1})\Delta(\gamma^{-1})\, d\lambda^u(\gamma)\varphi(\eta)\, d\nu(\eta)$$

which is zero if φ is zero ν-almost everywhere.

The argument for $f * h$ is similar.

Ex 9.3.1. Let $x_0 \in A$. Then there is a compact subset K of $\pi^{-1}(x_0)$ such that $\alpha^{x_0}(K) > \frac{1}{2}$. Let $f_{x_0} \in C_c^+(Y)$ be such that $0 \leq f_{x_0}(y) \leq 1$ and $f_{x_0}(y) = 1$ for all $y \in K$. Then $\alpha(f_{x_0})(x) > \frac{1}{2}$ for all x in a neighborhood V_{x_0} of x_0. Suppose that $V_{x_1}, \dots, V_{x_n}$ cover A. Let $\{\varphi_i\}_{i=1}^n \subset C_c^+(X)$ be such that $\operatorname{supp}\varphi_i \subset V_{x_i}$ and such that $\sum_{i=1}^n \varphi_i(x) \leq 1$ for all $x \in X$ with equality if $x \in A$. Then $f(y) = \sum_{i=1}^n \varphi_i(\pi(x))f_{x_i}(y)$ will do.

Ex 9.3.2. (This is almost identical to Ex 3.2.3.) Because the G-action on X is free, the map is clearly a bijection. Suppose that $(\pi(y_i), G\cdot y_i) \to (\pi(y), G\cdot y)$. After passing to a subnet, and relabeling, it suffices to see that a subnet of $\{y_i\}$ converges to y. Since the orbit map is open, we can pass to a subnet and assume that there are γ_i such that $\gamma_i \cdot y_i \to y$. But then $\gamma_i \cdot \pi(y_i) \to \pi(y)$. Since the G-action on X is proper, we may as well assume $\gamma_i \to \gamma$ in G. But then $y_i \to \gamma^{-1}\cdot y$ and $\gamma^{-1}\cdot\pi(y) = \pi(y)$. Hence $\gamma^{-1}\cdot y = y$ and we're done.

Ex 9.4.1. Let T be a compact set in $G \rtimes X$. Fix $\epsilon > 0$. Then we can find i_0 such that $i \geq i_0$ implies that

$$\|\gamma^{-1}\cdot p_i^x - p_i^{\gamma^{-1}\cdot x}\|_1 < \epsilon \tag{D.14}$$

provided $(\gamma, x) \in T$. Let $g \in C_c(Y)$ with $\|g\|_\infty \leq 1$. Now if $(\gamma, x) \in T$ and $i \geq i_0$, and using that ρ is G-equivariant,

$$\begin{aligned}
&\bigl|\gamma^{-1}\cdot m_i^x(g) - m_i^{\gamma^{-1}\cdot x}(g)\bigr| \\
&\qquad = \Bigl|\int_Z g(\rho(\gamma^{-1}\cdot z))\, dp_i^x(z) - \int_Z g(\rho(z))\, dp_i^{\gamma^{-1}\cdot x}(z)\Bigr| \\
&\qquad = \Bigl|\int_Z g(\rho(z))\, d(\gamma^{-1}\cdot p_i^x)(z) - \int_Z g(\rho(z))\, dp_i^{\gamma^{-1}\cdot x}(z)\Bigr|
\end{aligned}$$

which, since $\|g\circ\rho\|_\infty \leq 1$, is, by (D.14)

$$< \epsilon.$$

This suffices.

Ex 9.4.2. Note that

$$p^x(f) = \int_Y f(y)\varphi(y)\, dn^x(y).$$

If $L \subset X$ is compact, then $K = \pi^{-1}(L) \cap \operatorname{supp} \varphi$ is compact in Y and $\operatorname{supp} p^x \subset K$.

Ex 9.4.3. Fix $y_0 \in Y$. Let L be a compact neighborhood of $\pi(y_0)$. As shown in the text, if $y \in \pi^{-1}(L)$ there is compact set L' such that $g_i(y) = 0$ if $y \notin L'$. If $\varphi \in C_c^+(Y)$ is such that $\varphi \equiv 1$ on L', we have

$$g_i(y) = \int_Y F(y,z)\, dm_i^{\pi(y)}(z) \quad \text{for all } y \in \pi^{-1}(L).$$

where $F \in C_c(Y \times Z)$ is given by $F(y,z) = \varphi(y)f(y,z)$. Now g_i is clearly continuous if F is of the form $F(y,z) = \psi(y)\xi(z)$ with $\psi \in C_c(Y)$ and $\xi \in C_c(Z)$. But we can approximate F with sums of such functions.

Ex 9.4.4. Let $h = (\xi_i, \xi_i)$ as in the proof. If $u \in G^{(0)}$, then $h_i(u) = \lambda(|\xi_i|^2)(u)$. Thus $\|\xi_i\|_{r,2} \leq 1$. Since $C_c(G)$ is a dense subspace, for any $\epsilon > 0$ we can find $\xi_{i,\epsilon} \in C_c(G)$ be such that $\|\xi_i - \xi_{i,\epsilon}\|_{r,2} < \epsilon$ and $\|\xi_{i,\epsilon}\|_{r,2} \leq 1$. Then using (9.17),

$$\begin{aligned}
|(\xi_{i,\epsilon}, \xi_{i,\epsilon})(\gamma) - (\xi_i, \xi_i)(\gamma)| &\leq |(\xi_{i,\epsilon} - \xi_i, \xi_{i,\epsilon})(\gamma)| + |(\xi_i, \xi_{i,\epsilon} - \xi_i)(\gamma)| \\
&\leq \|\xi_{i,\epsilon} - \xi_i\|_{r,2}\|\xi_{i,\epsilon}\|_{r,2} + \|\xi_i\|_{r,2}\|\xi_{i,\epsilon} - \xi_i\|_{r,2} \\
&\leq 2\epsilon.
\end{aligned}$$

Hence $h_{i,\epsilon} = (\xi_{i,\epsilon}, \xi_{i,\epsilon})$ suffices. In the case of a sequence, we can let $h_i = (\xi_{i,\frac{1}{i}}, \xi_{i,\frac{1}{i}})$.

Ex 9.5.1. Suppose that $f \in C_c(Z)$. Then

$$\left|p_i^{x_1}(f) - p_i^x(f)\right| = \left|\int_H f(x_1 \cdot \eta')\, dm_i^{s(x_1)}(\eta') - \int_H f(x \cdot \eta')\, dm_i^{s(x)}(\eta')\right|$$

which, since $x_1 = x \cdot \eta$, is

$$\begin{aligned}
&= \left|\int_H f(x \cdot \eta\eta')\, dm_i^{s(\eta)}(\eta') - \int_H f(x \cdot \eta')\, dm_i^{r(\eta)}(\eta')\right| \\
&= \left|\int_H f(x \cdot \eta')d\,(\eta \cdot m_i^{s(\eta)} - m_i^{r(\eta)})(\eta')\right|.
\end{aligned}$$

The result now follows from (9.22).

Ex 9.5.2. If $f \in C_c(Z)$ with $\|f\|_\infty \leq 1$, then

$$\begin{aligned}
\left|w_i^{[x_1]}(f) - p_i^{x_1}(f)\right| &= \left|\int_Z \left(p_i^y(f) - p^{x_1}(f)\right) dn^{[x_1]}(y)\right| \\
&\leq \left|\int_{Z\setminus L} \left(p_i^y(f) - p^{x_1}(f)\right) dn^{[x_1]}(y)\right| \\
&\qquad + \left|\int_L \left(p_i^y(f) - p^{x_1}(f)\right) dn^{[x_1]}(y)\right| \\
&\leq 2(\epsilon/8) + \epsilon/4 = \epsilon/2.
\end{aligned}$$

Ex 9.6.1. (a) If $\mu_1 \leq \mu_2$, then $\nu = \mu_2 - \mu_1$ is a Radon measure on Y and $\nu(f) \geq 0$ for all $f \in C_c^+(Y)$.

Conversely, suppose that $\mu_1(f) \leq \mu_2(f)$ for all $f \in C_c^+(Y)$. Suppose that $\mu_1(E) = \infty$. If $\mu_2(E) = \infty$, then there is nothing prove. If $\mu_2(E) < \infty$, then given $N > 0$ we can find a compact set $K \subset E$ and an open set $V \supset E$ such that $\mu_1(K) > \mu_2(E) + 2$ and $\mu_2(V) \leq \mu_2(E) + 1$. Let $f \in C_c^+(Y)$ be such that $0 \leq f \leq 1$, $f(y) = 1$ for all $y \in K$ and $\operatorname{supp} f \subset V$. Then $\mu_2(f) \leq \mu_2(E) + 1$. But $\mu_1(f) \geq \mu_2(E) + 2$. This is a contradiction. Hence $\mu_2(E) = \infty$ if $\mu_1(E) = \infty$. If

$\mu_1(E) < \infty$, we may as well assume that $\mu_2(E) < \infty$. Then given any $\epsilon > 0$, we can pick K, V, and f as above such that $|\mu_k(f) - \mu_k(E)| < \epsilon$. The result follows easily from this.

(b) Let $\mu = |\nu_1|+|\nu_2|$. Then $|\nu_k| \ll \mu$ and there is a Radon-Nykodym derivative $\alpha_k : Y \to (0,\infty)$ such that

$$|\nu_k|(f) = \int_Y f(y)\alpha_k(y)\,d\mu(y).$$

Hence there are essentially unique unimodular functions ρ_k such that

$$\nu_k(f) = \int_Y f(y)\rho_k(y)\,d|\nu_k|(y) = \int_Y f(y)\rho_k(y)\alpha_k(y)\,d\mu(y).$$

Therefore $|\nu_1 + \nu_2| = |\rho_1\alpha_1 + \rho_2\alpha_2| \cdot \mu$. Thus if $f \in C_c^+(Y)$, we have

$$|\nu_1 + \nu_2|(f) \le \int_Y f(y)(\alpha_1(y) + \alpha_2(y))\,d\mu(y) = |\nu_1|(f) + |\nu_2|(f).$$

This suffices by part (a).

Ex 9.6.2. Since Theorem 9.58 on page 222 implies G is amenable if and only if there is a Borel approximate invariant density, we just have to show that the we have Borel approximate invariant density on G if and only if there is a sequence $\{\xi_i\}$ as in the problem.

But given $\{\xi_i\}$ we can proceed exactly as in the proof of (c)$\Longrightarrow$(b) in Theorem 9.43 on page 211. On the other hand, given a Borel approximate invariant density $\{g_i\}$, we can let $\xi_i(\gamma) = \sqrt{g_i(\gamma)}$. Since the ξ_i are everywhere positive, we have

$$\int_G |\xi_i(\gamma^{-1}\eta) - \xi_i(\eta)|^2\,d\lambda^{r(\gamma)}(\eta) \le \int_G |g_i(\gamma^{-1}\eta) - g_i(\eta)|\,d\lambda^{r(\gamma)}(\eta).$$

This suffices.

Ex 9.7.1. By assumption, $G = \bigcup A_n$ such that $u \mapsto \lambda^u(A_n) < \infty$ is bounded for each n. Then there are positive constants $\alpha_n \le 2^{-n}$ such that

$$\alpha_n\lambda^u(A_n) \le 2^{-n} \quad \text{for all } u \in G^{(0)}.$$

Let

$$g(\gamma) = \sum_{n=1}^{\infty} \alpha_n \mathbb{1}_{A_n}(\gamma).$$

Then g is Borel with values in $(0,\infty)$. Let $p_0^u = g \cdot \lambda^u$. Then $p_0 = \{p_0^u\}$ is a Borel r-system of finite nonzero measures on G. Let $p^u = \|p_0^u\|_1^{-1} p_0^u$. Then $p = \{p_u\}$ suffices.

Ex 9.7.2. The only issue is to observe that a free and proper G-space Z is also a free and proper Borel G-space. But the quotient map $q : Z \to G\backslash Z$ has a Borel cross section by [**Arv76**, Theorem 3.4.1]. Hence the result follows from Lemma 9.65 on page 229.

Ex 9.8.1. Recall that a subnet of x is determined by a function $M : E \to D$ on a directed set E such that for all $d \in D$ there is a $e' \in E$ such that $e \ge e'$ implies $M_e \ge d$. Then $y = \{y_e\}$ is subnet with $y_e = x_{M_e}$.

Let $D = \{d_n\}_{n\in\mathbf{N}}$. Inductively, choose d'_n in D such that $d'_n \geq d'_{n-1}$ and $d'_n \geq d_n$. Then $M : \mathbf{N} \to D$ is given by $M_n = d'_n$.

Ex 9.8.2. Let $u_n \to u = r(\gamma)$ with $\{u_n\} \subset F$. It will suffice to find a subnet $\{\gamma_{n_k}\}$ in $G(F)$ converging to γ with $r(\gamma_{n_k}) = u_{n_k}$. But since r is open, there is such a subnet in G. Since F is saturated, that subnet must be in $G(F)$.

Ex 10.1.1. We consider $f * \varphi$ as the case for $f * \psi$ is similar. We can assume f and φ are non-negative and then observe that since $r(y) = r(x)$ for $y \in \operatorname{supp} \alpha^x$

$$\begin{aligned}\int_Y f * \varphi(y)\, d(\mu\circ\alpha)(y) \\ &= \int_X\int_Y\int_G f(\gamma)\varphi(\gamma^{-1}\cdot y)\, d\lambda^{r(x)}(\gamma)\, d\alpha^x(y)\, d\mu(x) \\ &= \int_X\int_G\int_Y f(\gamma)\varphi(\gamma^{-1}\cdot y)\, d\alpha^x(y)\, d\lambda^{r(x)}(\gamma)\, d\mu(x)\end{aligned}$$

which, by quasi-invariance, is

$$= \int_G\int_G\int_Y f(\gamma^{-1})\varphi(\gamma\cdot y)\, d\alpha^{\gamma^{-1}\cdot x}(y)\delta(\gamma^{-1},\gamma^{-1}\cdot x)\, d\lambda^{r(x)}(\gamma)\, d\mu(x)$$

which, by the equivariance of α, is

$$\begin{aligned}&= \int_X\int_G\int_Y f(\gamma^{-1})\varphi(y)\, d\alpha^x(y)\delta(\gamma^{-1},\gamma^{-1}\cdot x)\, d\lambda^{r(x)}(\gamma)\, d\mu(x) \\ &= \int_X\int_Y\int_G f(\gamma^{-1})\delta(\gamma^{-1},\gamma^{-1}\cdot x)\, d\lambda^{r(x)}(\gamma)\varphi(y)\, d\alpha^x(y)\, d\mu(x).\end{aligned}$$

This is clearly zero if $\varphi(y) = 0$ for $\mu\circ\alpha$-almost all y.

Ex 10.2.1. Since m is invariant, $f\star m = (\lambda(f)\circ r)\cdot m$. Thus for all $\nu \in M(G, [\mu]\circ\lambda)$, we have

$$\langle m\,,\, {}^tM(f)\nu\rangle = \langle f\star m\,,\,\nu\rangle = \langle(\lambda(f)\circ r)\cdot m\,,\,\nu\rangle = \langle m\,,\,(\lambda(f)\circ r)\cdot\nu\rangle. \tag{D.15}$$

By Corollary 10.17 on page 252, there is a normalized approximate weakly invariant mean $\{\,g_i\,\}$ for $r : G \to G^{(0)}$. Then $m_{g_i} \to m$ in the weak-$*$ topology. The assertion now follows readily from (D.15).

Ex 10.3.1. Let $\varphi \in \mathcal{B}^b(Y*T)$ and let h be an integrable Borel function on $X*T$. If $\Theta(|\varphi|,|h|)$ is given by the the right-hand side of (10.9), then $\Theta(|\varphi|,|h|) \in \mathcal{B}^{+,e}(Y*T)$ by Proposition 3.3 on page 70, and is bounded off a $(\mu\circ\alpha)\circ\mu$-null set by

$$\|\varphi\|_\infty \int |h(\pi(y),t)|\, d\beta^{r(y)}(t). \tag{D.16}$$

Since the right-hand side of (D.16) is finite off a μ-null set N and since $\pi^{-1}(N)$ is $\mu\circ\alpha$-null, $\Theta(|\varphi|,|h|)(y)$ is finite $\mu\circ\alpha$-almost everywhere. Hence we can define $\Theta(\varphi,h)(y)$ using the right-hand side of (10.9) if $\Theta(|\varphi|,|h|)(y)$ is finite and set it to 0 otherwise. The estimates above show that $\Theta(\varphi,h)$ is 0 $\mu\circ\alpha$-everywhere if either φ or h is zero almost everywhere. The assertion follows since the right-hand side of (10.9) is bilinear in φ and h.

Ex 10.3.2. (a) If $\nu \in L^\infty(Y*G)\odot L^1(X*G)$, then

$$\langle m_{g_i}\otimes 1_G\,,\,\nu\rangle = \langle m_{g_i}\,,\,c_T(\nu)\rangle \to \langle m\,,\,c_T(\nu)\rangle = \langle m\otimes 1_G\,,\,\nu\rangle.$$

(b) By part (a), it suffices to show that if $m = \lim_i m_i$, then $\mathfrak{I}_X \circ m = \lim_i \mathfrak{I}_X \circ m_i$. Let $\nu = \varphi \otimes f$ with $\varphi \in L^\infty(Y * G)$ and $f \in L^1(X * G)$. Then

$$\langle \mathfrak{I}_X \circ m_i \,,\, \varphi \otimes f \rangle = \int_X \int_G m_i(\varphi)(\gamma^{-1} \cdot x, \gamma^{-1}) f(x,\gamma)\, d\lambda^{r(x)}(\gamma)\, d\mu(x)$$

which, since μ is quasi-invariant, is

$$= \int_X \int_G m_i(\varphi)(x,\gamma) f(\gamma^{-1} \cdot x, \gamma^{-1}) \delta(\gamma^{-1}, \gamma^{-1} \cdot x)$$
$$d\lambda^{r(x)}(x)\, d\mu(x)$$
$$= \langle m_i \,,\, \varphi \otimes \tilde{f} \rangle,$$

where $\tilde{f}(x,\gamma) = f(\gamma^{-1} \cdot x, \gamma^{-1}) \delta(\gamma^{-1}, \gamma^{-1} \cdot x)$. Hence

$$\lim_i \langle \mathfrak{I}_X \circ m_i \,,\, \varphi \otimes f \rangle = \langle m \,,\, \varphi \otimes \tilde{f} \rangle = \langle \mathfrak{I}_X \circ m \,,\, \varphi \otimes f \rangle.$$

This proves (b), and the proof of (c) is similar but more straightforward.

Ex 10.3.3. The first assertion amounts to

$$\lim_i \int_X h_i(x) \lambda(f)(x)\, d\mu(x) = \int_X \lambda(f)(x)\, d\mu(x),$$

which follows by assumption (since $\lambda(f) \in L^1(X)$). The second assertion follows similarly since

$$\int_X \int_G h_i(\gamma^{-1} \cdot x) f(x,\gamma)\, d\lambda^x(\gamma)\, d\mu(x)$$
$$= \int_X \int_Y h_i(x) f(\gamma^{-1} \cdot x, \gamma^{-1}) \delta(\gamma^{-1} \cdot x, \gamma^{-1})\, d\lambda^x(\gamma)\, d\mu(x)$$
$$= \int_X h_i(x) \lambda(\tilde{f}\delta)\, d\mu(x).$$

Ex 10.3.4. Suppose that $h \in L^1(Z)$. Then

$$\lim_i \int_Z (\operatorname{Re} e_i(z)) h(z)\, d\mu(z) = \frac{1}{2} \lim_i \Big(\int_Z e_i(z) h(z)\, d\mu(z) + \overline{\int_Z e_i(z) \overline{h(z)}\, d\mu(z)} \Big).$$

Ex 10.4.1. Recall from (10.10), that if $\varphi \in L^\infty(Z * Z * Z)$ and $g \in C_c(Z * Z)$ then we write

$$[\![\varphi, g]\!](z) = \int_Z \int_Z \varphi(z, w, w')\, d\beta^{r(z)}(w)\, d\beta^{r(z)}(w').$$

Then if $h \in L^1(G^{(0)})$

$$\langle p \otimes 1_{Z*Z}(\varphi) \,,\, h \otimes g \rangle = \langle p([\![\varphi, g]\!]) \,,\, h \rangle.$$

If $\psi \in L^\infty(G^{(0)} * Z * Z)$ and $f \in L^1(G^{(0)} * Z * Z)$, then we have

$$\langle I_{G^{(0)}}(\psi) \,,\, f \rangle = \langle \psi \,,\, F(f) \rangle$$

where F is the flip: $F(f)(z,w) = f(w,z)$.

Hence on the one hand,

$$\langle p \otimes 1_{Z*Z}(I_Z(\varphi)) \,,\, h \otimes g \rangle = \langle p([\![I_Z(\varphi), g]\!]) \,,\, h \rangle,$$

while

$$\langle I_{G^{(0)}}(p \otimes 1_{Z*Z}(\varphi)) \,,\, h \otimes g \rangle = \langle p([\![\varphi, F(g)]\!]) \,,\, h \rangle.$$

But we easily check

$$[\![I_Z(\varphi), g]\!] = [\![\varphi, F(g)]\!].$$

This completes the proof.

Ex 10.4.2. (a) We have

$$\|\Phi_{\mathfrak{h}}(f)\|_{L^1(X)} = \int_X |f(\mathfrak{h}(x))|\,d\mu(x) = \int_Y |f(y)|\,d\mathfrak{h}_*\mu(y) = \|f\|_{L^1(Y,\mathfrak{h}_*\mu)}.$$

(b) By definition of the transpose

$$\begin{aligned}\big\langle {}^t\Phi_{\mathfrak{h}}(\varphi)\,,\, f\big\rangle = \big\langle \varphi\,,\, \Phi_{\mathfrak{h}}(f)\big\rangle &= \int_X \varphi(x) f(\mathfrak{h}(x))\,d\mu(x)\\ &= \int_Y \varphi(\mathfrak{h}^{-1}(y)) f(y)\,d\mathfrak{h}_*\mu(y).\end{aligned}$$

Ex 11.1.1. If G is not the trivial group, then there is a $f \in C_c(G)$ such with integral zero. Hence it is a nontrivial element in the kernel of the trivial representation. Hence $C^*(G)$ is not simple.

Ex 11.1.2. Note that $\bar{\varphi}^{-1}(B) = f^{-1}(\varphi^{-1}(B))$ for any Borel set in Z. If φ is Borel, then $\varphi^{-1}(B)$ is Borel in the quotient Borel structure. This means that $f^{-1}(\varphi^{-1}(B))$ is Borel in Y. On the other hand, if $\bar{\varphi}$ is Borel, then $\bar{\varphi}^{-1}(B)$ is Borel in Y. Then $\varphi^{-1}(B)$ is Borel in the quotient Borel structure on X.

Ex 11.1.3. We have

$$\begin{aligned}\lambda(f\beta(g))(u) &= \int_G \int_{G(s(\gamma))} f(\gamma)g(\gamma t)\,d\beta^{s(\gamma)}(t)\,d\lambda^u(\gamma)\\ &= \int_G \int_{G(r(\gamma))} \omega(\gamma, G(s(\gamma))) f(\gamma) g(t\gamma)\,d\beta^u(t)\,d\lambda^u(\gamma)\end{aligned}$$

which, after two applications of Fubini and translation by t^{-1}, is

$$\begin{aligned}&= \int_G \int_{G(r(\gamma))} \omega(t^{-1}\gamma, G(s(\gamma))) f(t^{-1}\gamma) g(\gamma)\,d\beta^u(t)\,d\lambda^u(\gamma)\\ &= \int_G \int_{G(r(\gamma))} \omega(\gamma, G(s(\gamma)))\Delta_{G(r(\gamma))}(t)^{-1} f(t^{-1}\gamma) g(\gamma)\,d\beta^u(t)\,d\lambda^u(\gamma)\\ &= \int_G \int_{G(r(\gamma))} \omega(\gamma, G(s(\gamma))) f(t\gamma) g(\gamma)\,d\beta^u(t)\,d\lambda^u(\gamma)\\ &= \int_G \int_{G(s(\gamma))} f(\gamma t) g(\gamma)\,d\beta^{s(\gamma)}(t)\,d\lambda^u(\gamma)\\ &= \lambda(\beta(f)g)(u)\end{aligned}$$

Ex 11.2.1. Let d be a metric on $G^{(0)}$ compatible with the topology. Let

$$D_\epsilon(u) = \{\, (v,w) \in G^{(0)} \times G^{(0)} : d(u,v) < \epsilon \text{ and } d(u,w) < \epsilon \,\}.$$

For each $u \in G^{(0)}$, let ϵ_u be such that $D_{3\epsilon_u}(u) \subset M$ and define

$$M_0 = \bigcup_{u \in G^{(0)}} D_{\epsilon_u}(u).$$

Now suppose that (u,v) and (v,w) are in M_0. Let $u_i \in G^{(0)}$ be such that $(u,v) \in D_{\epsilon_{u_1}}(u_1)$ and $(v,w) \in D_{\epsilon_{u_2}}(u_2)$. We can assume $\epsilon_{u_2} \leq \epsilon_{u_1}$. Then

$$\begin{aligned} d(u_1,w) &\leq d(u_1,v) + d(v,w) \leq d(u_1,v) + d(v,u_2) + d(u_2,w) \\ &< \epsilon_{u_1} + \epsilon_{u_2} + \epsilon_{u_3} \leq 3\epsilon_{u_1}. \end{aligned}$$

Hence $(u,w) \in D_{3\epsilon_{u_1}} \subset M$ as required.

Ex 11.3.1. (a) Let U be a relatively compact neighborhood of v. By assumption, there is a neighborhood V of v such that $\overline{U} \cap G(u) = \{u\}$ for all $u \in V$.

(b) Let $K \subset G$ be compact. Let U be such that $u \in U \cap G^{(0)}$ implies $U \cap G(u) = \{u\}$. It will suffice to find an open set $V \subset G^{(0)}$ such that $u \in V$ implies $K \cap G(u) \subset U$. If the result is false, then for every open set $V \subset U \cap G^{(0)}$ there is a $u_V \in V$ and $k_V \in \big(K \cap G(u_V)\big) \setminus U$. Passing to a subnet and relabeling, we have $k_V \to k \notin U$. At the same time, $r(k_V) = u_v = s(k_V) \to v$. Hence $k \in G(v) \setminus \{v\}$ which is a contradiction.

(c) It suffices to see that the isotropy is discrete at v. Since G is étale, $G^{(0)}$ is open and v has an open neighborhood U in G such that $U \subset G^{(0)}$. That clearly suffices.

Ex 11.4.1. We need to see that

$$\omega(\eta, G(s(\eta)))^{-\frac{1}{2}} = \omega(\gamma t, G(s(\gamma)))^{-\frac{1}{2}} \omega(\eta^{-1}\gamma t, G(s(\gamma)))^{\frac{1}{2}}.$$

But after factoring out $\Delta_{G(s(\gamma))}(t)$, the right-hand side is

$$\begin{aligned} &\omega(\gamma, G(s(\gamma)))^{-\frac{1}{2}} \omega(\eta^{-1}\gamma, G(s(\gamma)))^{\frac{1}{2}} \\ &\qquad = \omega(\gamma, G(s(\gamma)))^{-\frac{1}{2}} \omega(\gamma, G(s(\gamma)))^{\frac{1}{2}} \omega(\eta^{-1}, G(r(\gamma)))^{\frac{1}{2}} \end{aligned}$$

which, since $r(\eta) = r(\gamma)$, is

$$\begin{aligned} &= \omega(\eta^{-1}, G(r(\eta)))^{\frac{1}{2}} \\ &= \omega(\eta, G(s(\eta)))^{-\frac{1}{2}} \end{aligned}$$

As required.

Ex A.1. Let $\varphi \otimes h' \in L^\infty(Y) \odot_{L^\infty(X)} L^1(X)$ and let $\psi(y) = \varphi(y)h(\pi(y))$. Suppose that $g \in \mathcal{L}^\infty\big(X, L^1(Y,\alpha)\big)$ and set $h = \|\varphi\|_\infty |h'|$. Then

$$\begin{aligned} \int_X |\alpha(g\psi)(x)|\,d\mu(x) &\leq \int_X \int_Y |g(y)\varphi(y)|\,d\alpha^x(y)|h'(x)|\,d\mu(x) \\ &\leq \|\varphi\|_\infty \int_Y |g(y)||h'(\pi(x))|\,d(\mu\circ\alpha)(y) \\ &= \int_Y |g(y)|h(\pi(x))\,d(\mu\circ\alpha)(y). \end{aligned}$$

Ex B.1. Let π be a GNS representation of φ. Then

$$\begin{aligned} \lim_i \varphi(f\cdot e_i) &= \lim_i \big(\pi(f\cdot e_i)h \mid h\big) = \lim_i \big(\bar{\pi}(\Phi_A(f))\pi(e_i)h \mid h\big) = \big(\bar{\pi}(\Phi_A(f))h \mid h\big) \\ &= \varphi'(f). \end{aligned}$$

The set $\{\varphi(f \cdot e_i)\}$ is increasing because $\{e_i\}$ is, f acts centrally, and φ is positive.

Ex B.2. We can assume $n = 1$. Clearly if $\varphi_i \to \varphi$ in A_c^*, then $\varphi_i(a) \to \varphi(a)$ for all $a \in A$. Hence $\varphi_i \to \varphi$ in the weak-$*$ topology. Now suppose that $\varphi_i \to \varphi$ in the weak-$*$ topology and $\|\varphi_i\| \leq 1$ for all i.

Suppose that C is a compact subset for which $\varphi_i \not\to \varphi$ uniformly on C. Then there is a $\epsilon > 0$ for which the statement

$$|\varphi_i(a) - \varphi(a)| < \epsilon \quad \text{for all } a \in C \text{ and all } i \geq i_0$$

is false for all i_0. Hence the set

$$D = \{\, i : \text{there exists } a \in C \text{ such that } |\varphi_i(a) - \varphi(a)| \geq \epsilon \,\}$$

is directed. For each $d \in D$, let $a_d \in C$ be such that $|\varphi_d(a_d) - \varphi(a_d)| \geq \epsilon$. We can pass to a subnet, relabel, and assume that $a_d \to c$ in C. But

$$|\varphi_d(c) - \varphi(c)| \geq \epsilon - 2\|c - a_d\|.$$

This eventually leads to a contradiction.

Hence the topologies are the same.

Ex B.3. Let J_x be the ideal of functions in $C(X^+)$ vanishing at x and I_x the ideal generated by $f \cdot m$ with $m \in \mathcal{A}(X^+)$ so that $\mathcal{A}(X^+)(x) = \mathcal{A}(X^+)/I_x$. As with $\mathcal{A}(X^+)$, it is not hard to see that $\mathcal{A}(x)$ is closed in $M(A(x) \oplus \mathbf{C})$ and hence a C^*-algebra. Since we clearly have $I_x \subset \ker\varphi$, φ factors through $\mathcal{A}(X^+)(x)$ as claimed. On the other hand, if $\varphi(a + f) = 0$, then $f(x) = 0$ which forces $a(x) = 0$. Thus we can approximate $a + f$ with elements of I_x, and $a + f \in I_x$. Thus $\ker\varphi = I_x$.

Ex B.4. Since A is separable, it has a faithful state τ. If $\delta > 0$ and $\varphi \in \mathcal{S}(A)$, let $\varphi_\delta = \frac{1}{\delta+1}(\varphi + \delta\tau)$. Then φ_δ is a faithful state. Also $\bigl|\varphi_\delta(a) - \varphi(a)\bigr| < 2\delta\|a\|$. Thus for sufficiently small δ, φ_δ is in the basic open neighborhood

$$U = \{\, \psi \in \mathcal{S}(A) : \bigl|\psi(a_k) - \varphi(a_k)\bigr| < \epsilon, \text{ for } 1 \leq k \leq n \,\}.$$

Ex B.5. Notice that if $c = a + ib$ with a and b self-adjoint, then for any state ψ, $\psi(a)$ and $\psi(b)$ are real. In particular,

$$|\psi(c) - \varphi(c)| = \sqrt{(\psi(a) - \varphi(a))^2 + (\psi(b) - \varphi(b))^2}.$$

Thus if $c_i = a_i + ib_i$, then

$$U(a_1, \dots, a_n, b_1, \dots, b_n; \frac{\delta}{\sqrt{2}}) \subset U(c_1, \dots, c_n; \delta).$$

The result follows.

Ex B.6. Let $F \subset X \times Y$ be closed. Suppose that $\{(x_i, y_i)\} \subset F$ and $x_i \to x$. After passing to a subnet and relabeling, we can assume that $y_i \to y$ in Y. But then $(x, y) \in F$. Hence $x \in \pi(F)$.

Ex C.1. We can view $C_0(X)$ as a subspace of $C(X^+)$, where X^+ is the one-point compactification of X. If $C(X^+)$ is separable, then since it is a metric space, $C(X^+)$ is second countable. Hence $C_0(X)$ is second countable and hence separable. Since X is σ-compact, X^+ is second countable if X is, so it suffices to consider X compact.

Since X is second countable, there is a countable family of functions $D \subset C(X)$ that separate points of X. We can certainly assume that $1 \in D$ and that D

is closed under complex conjugation. Since $\mathbf{N} \times D$ is countable, $D' := \{ f^n : f \in D \text{ and } n \in \mathbf{N} \}$ is countable.

Let $\mathbf{F}$ be a subfield of $\mathbf{C}$. The then $\mathbf{F}$-subalgebra of $C_0(X)$ generated by D' is

$$\mathbf{F}[D'] = \{ \sum_{i=1}^{n} \lambda_i f_i g_i : n \in \mathbf{N},\ \lambda_i \in \mathbf{F}, \text{ and } f_i, g_i \in D' \}.$$

The Stone-Weierstrass Theorem [**Ped89**, Theorem 4.3.4] implies that $\mathbf{C}[D']$ is dense in $C(X)$. If $\mathbf{F} := \mathbf{Q} + i\mathbf{Q}$ then $\mathbf{F}[D']$ is countable and dense in $\mathbf{C}[D']$. This suffices.

Ex D.1. Let $A_n = \{ x : f(x) \leq 1 - \frac{1}{n} \}$. Then

$$1 \leq (1 - \frac{1}{n})\mu(A_n) + \mu(X \setminus A_n) = 1 - \frac{1}{n}\mu(A_n).$$

Hence $\mu(A_n) = 0$. But $\{ x : f(x) < 1 \} = \bigcup A_n$.

Ex D.2. Suppose $\mu(E) < \infty$. Then there is an open set $V \supset E$ such that $\mu(V) - \mu(E) < \epsilon/2$. Since $\mu(E) < \infty$, $\mu(V \setminus E) < \epsilon/2$. Now in general, $X = \bigcup_n X_n$ with $\mu(X_n) < \infty$ for each n. Let $E_n = E \cap X_n$. Then there are open sets $V_n \subset E_n$ such that $\mu(V_n \setminus E_n) < \frac{\epsilon}{2^{n-1}}$. Let $V = \bigcup V_n$. Then V is open and contains E. Furthermore,

$$\mu(V \setminus E) = \mu\Big(\bigcup V_n \setminus \bigcup E_n\Big) \leq \mu\Big(\bigcup_n (V_n \setminus E_n)\Big) \leq \epsilon/2.$$

But the above reasoning shows that there is an open set W containing E^c such that $\mu(W \setminus E^c) < \epsilon/2$. Then $F = W^c$ is a closed subset of E, and $\mu(E \setminus F) < \epsilon/2$. Then, since $V \setminus F = V \setminus E \cup E \setminus F$, we have $\mu(V \setminus F) < \epsilon$ as required.

Let V_n and F_n be as above such that $\mu(V_n \setminus F_n) < \frac{1}{n}$. Let $G = \bigcap V_n$ and $A = \bigcup F_n$. But for any n, $\mu(G \setminus A) \leq \mu(V_n \setminus F_n) < \frac{1}{n}$. Hence $\mu(G \setminus A) = 0$.

Bibliography

[ADR00] Claire Anantharaman-Delaroche and Jean N. Renault, *Amenable groupoids*, Monographies de L'Enseignement Mathématique [Monographs of L'Enseignement Mathématique], vol. 36, L'Enseignement Mathématique, Geneva, 2000. With a foreword by Georges Skandalis and Appendix B by E. Germain. MR1799683

[Arv76] William Arveson, *An invitation to C^*-algebras*, Springer-Verlag, New York-Heidelberg, 1976. Graduate Texts in Mathematics, No. 39. MR0512360

[BCFS14] Jonathan H. Brown, Lisa Orloff Clark, Cynthia Farthing, and Aidan Sims, *Simplicity of algebras associated to étale groupoids*, Semigroup Forum **88** (2014), 433–452, DOI 10.1007/s00233-013-9546-z. MR3189105

[BCS15] Jonathan H. Brown, Lisa Orloff Clark, and Adam Sierakowski, *Purely infinite C^*-algebras associated to étale groupoids*, Ergodic Theory Dynam. Systems **35** (2015), 2397–2411. MR3456600

[BCSS16] Jonathan H. Brown, Lisa Orloff Clark, Adam Sierakowski, and Aidan Sims, *Purely infinite simple C^*-algebras that are principal groupoid C^*-algebras*, J. Math. Anal. Appl. **439** (2016), 213–234, DOI 10.1016/j.jmaa.2016.02.055. MR3474359

[BdlHV08] Bachir Bekka, Pierre de la Harpe, and Alain Valette, *Kazhdan's property (T)*, New Mathematical Monographs, vol. 11, Cambridge University Press, Cambridge, 2008. MR2415834

[BDNH76] R. Brown, G. Danesh-Naruie, and J. P. L. Hardy, *Topological groupoids. II. Covering morphisms and G-spaces*, Math. Nachr. **74** (1976), 143–156, DOI 10.1002/mana.3210740110. MR0442937

[BH76] Ronald Brown and J. P. L. Hardy, *Topological groupoids. I. Universal constructions*, Math. Nachr. **71** (1976), 273–286. MR0412333

[Bla06] Bruce Blackadar, *Operator algebras*, Encyclopaedia of Mathematical Sciences, vol. 122, Springer-Verlag, Berlin, 2006. Theory of C^*-algebras and von Neumann algebras; Operator Algebras and Non-commutative Geometry, III. MR2188261

[Bla96] Étienne Blanchard, *Déformations de C^*-algèbres de Hopf*, Bull. Soc. Math. France **124** (1996), 141–215 (French, with English and French summaries). MR1395009

[Bou87] Nicolas Bourbaki, *Topological vector spaces. Chapters 1–5*, Elements of Mathematics (Berlin), Springer-Verlag, Berlin, 1987. Translated from the French by H. G. Eggleston and S. Madan. MR910295

[Bra27] Heinrich Brandt, *Über eine Verallgemeinerung des Gruppenbegriffes*, Math. Ann. **96** (1927), 360–366, DOI 10.1007/BF01209171 (German). MR1512323

[Bro87] Ronald Brown, *From groups to groupoids: a brief survey*, Bull. London Math. Soc. **19** (1987), 113–134, DOI 10.1112/blms/19.2.113. MR872125

[Buc58] R. Creighton Buck, *Bounded continuous functions on a locally compact space*, Michigan Math. J. **5** (1958), 95–104. MR0105611

[Bun01] Mădălina Roxana Buneci, *C^*-algebras associated to the transitive groupoids*, An. Univ. Craiova Ser. Mat. Inform. **28** (2001), 79–92. MR1902038

[Bun06] Mădălina Roxana Buneci, *Groupoid C^*-algebras*, Surv. Math. Appl. **1** (2006), 71–98. MR2274294

[Bun19] Mădălina Roxana Buneci, *A second countable locally compact transitive groupoid without open range map*, 2019. (preprint arXiv:1811.02692).

[Cla07] Lisa Orloff Clark, *CCR and GCR groupoid C^*-algebras*, Indiana Univ. Math. J. **56** (2007), 2087–2110, DOI 10.1512/iumj.2007.56.2955. MR2359724

[Con79] Alain Connes, *Sur la théorie non commutative de l'intégration*, Algèbres d'opérateurs (Sém., Les Plans-sur-Bex, 1978), Lecture Notes in Math., vol. 725, Springer, Berlin, 1979, pp. 19–143 (French). MR548112

[Con82] Alain Connes, *A survey of foliations and operator algebras*, Operator algebras and applications, Part I (Kingston, Ont., 1980), Proc. Sympos. Pure Math., vol. 38, Amer. Math. Soc., Providence, R.I., 1982, pp. 521–628. MR679730

[Con85] John B. Conway, *A course in functional analysis*, Graduate Texts in Mathematics, vol. 96, Springer-Verlag, New York, 1985. MR768926

[Dea95] Valentin Deaconu, *Groupoids associated with endomorphisms*, Trans. Amer. Math. Soc. **347** (1995), 1779–1786, DOI 10.2307/2154972. MR1233967

[Dei18] Anton Deitmar, *On Haar systems for groupoids*, Z. Anal. Anwend. **37** (2018), 269–275, DOI 10.4171/ZAA/1613. MR3826616

[Dix77] Jacques Dixmier, *C^*-algebras*, North-Holland Publishing Co., Amsterdam-New York-Oxford, 1977. Translated from the French by Francis Jellett; North-Holland Mathematical Library, Vol. 15. MR0458185

[Dix81] Jacques Dixmier, *von Neumann algebras*, North-Holland Mathematical Library, vol. 27, North-Holland Publishing Co., Amsterdam-New York, 1981. With a preface by E. C. Lance; Translated from the second French edition by F. Jellett. MR641217

[dlH07] Pierre de la Harpe, *On simplicity of reduced C^*-algebras of groups*, Bull. Lond. Math. Soc. **39** (2007), 1–26. MR2303514

[Edw65] Robert E. Edwards, *Functional analysis. Theory and applications*, Holt, Rinehart and Winston, New York-Toronto-London, 1965. MR0221256

[EH67] Edward G. Effros and Frank Hahn, *Locally compact transformation groups and C^*-algebras*, Memoirs of the American Mathematical Society, No. 75, American Mathematical Society, Providence, R.I., 1967. MR0227310

[FD88] J. M. G. Fell and R. S. Doran, *Representations of *-algebras, locally compact groups, and Banach *-algebraic bundles. Vol. 1*, Pure and Applied Mathematics, vol. 125, Academic Press, Inc., Boston, MA, 1988. Basic representation theory of groups and algebras. MR936628

[FM75] Jacob Feldman and Calvin C. Moore, *Ergodic equivalence relations, cohomology, and von Neumann algebras*, Bull. Amer. Math. Soc. **81** (1975), 921–924, DOI 10.1090/S0002-9904-1975-13888-2. MR0425075

[FM77a] Jacob Feldman and Calvin C. Moore, *Ergodic equivalence relations, cohomology, and von Neumann algebras. I*, Trans. Amer. Math. Soc. **234** (1977), 289–324, DOI 10.2307/1997924. MR0578656

[FM77b] Jacob Feldman and Calvin C. Moore, *Ergodic equivalence relations, cohomology, and von Neumann algebras. II*, Trans. Amer. Math. Soc. **234** (1977), 325–359, DOI 10.2307/1997925. MR0578730

[FMY05] Cynthia Farthing, Paul S. Muhly, and Trent Yeend, *Higher-rank graph C^*-algebras: an inverse semigroup and groupoid approach*, Semigroup Forum **71** (2005), 159–187, DOI 10.1007/s00233-005-0512-2. MR2184052

[Fol95] Gerald B. Folland, *A course in abstract harmonic analysis*, Studies in Advanced Mathematics, CRC Press, Boca Raton, FL, 1995. MR1397028

[Fol99] Gerald B. Folland, *Real analysis*, 2nd ed., Pure and Applied Mathematics (New York), John Wiley & Sons, Inc., New York, 1999. Modern techniques and their applications; A Wiley-Interscience Publication. MR1681462

[Gli62] James Glimm, *Families of induced representations*, Pacific J. Math. **12** (1962), 885–911. MR0146297

[Goe09] Geoff Goehle, *Groupoid crossed products*, ProQuest LLC, Ann Arbor, MI, 2009. Thesis (Ph.D.)–Dartmouth College. MR2941279

[Goo73] Elliot C. Gootman, *The type of some C^* and W^*-algebras associated with transformation groups*, Pacific J. Math. **48** (1973), 93–106. MR0335681

[GR79] Elliot C. Gootman and Jonathan Rosenberg, *The structure of crossed product C^*-algebras: a proof of the generalized Effros-Hahn conjecture*, Invent. Math. **52** (1979), 283–298, DOI 10.1007/BF01389885. MR537063

[Gre69] Frederick P. Greenleaf, *Invariant means on topological groups and their applications*, Van Nostrand Mathematical Studies, No. 16, Van Nostrand Reinhold Co., New York-Toronto, Ont.-London, 1969. MR0251549

[Gre78] Philip Green, *The local structure of twisted covariance algebras*, Acta Math. **140** (1978), 191–250, DOI 10.1007/BF02392308. MR0493349

[Gre80] Philip Green, *The structure of imprimitivity algebras*, J. Funct. Anal. **36** (1980), 88–104, DOI 10.1016/0022-1236(80)90108-1. MR568977

[Hah78] Peter Hahn, *Haar measure for measure groupoids*, Trans. Amer. Math. Soc. **242** (1978), 1–33, DOI 10.2307/1997726. MR496796

[HR63] Edwin Hewitt and Kenneth A. Ross, *Abstract harmonic analysis. Vol. I: Structure of topological groups. Integration theory, group representations*, Die Grundlehren der mathematischen Wissenschaften, Bd. 115, Academic Press, Inc., Publishers, New York; Springer-Verlag, Berlin-Göttingen-Heidelberg, 1963. MR0156915

[IW09a] Marius Ionescu and Dana P. Williams, *The generalized Effros-Hahn conjecture for groupoids*, Indiana Univ. Math. J. **58** (2009), 2489–2508, DOI 10.1512/iumj.2009.58.3746. MR2603756

[IW09b] Marius Ionescu and Dana P. Williams, *Irreducible representations of groupoid C^*-algebras*, Proc. Amer. Math. Soc. **137** (2009), 1323–1332, DOI 10.1090/S0002-9939-08-09782-7. MR2465655

[Jec78] Thomas Jech, *Set theory*, Academic Press [Harcourt Brace Jovanovich, Publishers], New York-London, 1978. Pure and Applied Mathematics. MR506523

[KL17] Alexander Kumjian and Hui Li, *Twisted topological graph algebras are twisted groupoid C^*-algebras*, J. Operator Theory **78** (2017), 201–225, DOI 10.7900/jot.2016jun23.2136. MR3679613

[KMRW98] Alexander Kumjian, Paul S. Muhly, Jean N. Renault, and Dana P. Williams, *The Brauer group of a locally compact groupoid*, Amer. J. Math. **120** (1998), 901–954. MR1646047

[KPRR97] Alexander Kumjian, David Pask, Iain Raeburn, and Jean N. Renault, *Graphs, groupoids, and Cuntz-Krieger algebras*, J. Funct. Anal. **144** (1997), 505–541, DOI 10.1006/jfan.1996.3001. MR1432596

[Kum85] Alexander Kumjian, *Diagonals in algebras of continuous trace*, Operator algebras and their connections with topology and ergodic theory (Buşteni, 1983), Lecture Notes in Math., vol. 1132, Springer, Berlin, 1985, pp. 297–311, DOI 10.1007/BFb0074890. With an appendix by Toshikazu Natsume. MR799574

[Kum86] Alexander Kumjian, *On C^*-diagonals*, Canad. J. Math. **38** (1986), 969–1008, DOI 10.4153/CJM-1986-048-0. MR854149

[KW99] Eberhard Kirchberg and Simon Wassermann, *Permanence properties of C^*-exact groups*, Doc. Math. **4** (1999), 513–558. MR1725812

[LaL18] Scott M. LaLonde, *On some permananence properties of exact groupoids*, preprint, 2018. (arXiv: math.OA.1703.05190v2).

[Lan02] Serge Lang, *Algebra*, 3rd ed., Graduate Texts in Mathematics, vol. 211, Springer-Verlag, New York, 2002. MR1878556

[Lan95] E. Christopher Lance, *Hilbert C^*-modules*, London Mathematical Society Lecture Note Series, vol. 210, Cambridge University Press, Cambridge, 1995. A toolkit for operator algebraists. MR1325694

[LGTX07] Camille Laurent-Gengoux, Jean-Louis Tu, and Ping Xu, *Chern-Weil map for principal bundles over groupoids*, Math. Z. **255** (2007), 451–491, DOI 10.1007/s00209-006-0004-4. MR2270285

[Mac05] Kirill C. H. Mackenzie, *General theory of Lie groupoids and Lie algebroids*, London Mathematical Society Lecture Note Series, vol. 213, Cambridge University Press, Cambridge, 2005. MR2157566

[Mac66] George W. Mackey, *Ergodic theory and virtual groups*, Math. Ann. **166** (1966), 187–207, DOI 10.1007/BF01361167. MR0201562

[Mac68] George W. Mackey, *Virtual groups*, Topological Dynamics (Symposium, Colorado State Univ., Ft. Collins, Colo., 1967), Benjamin, New York, 1968, pp. 335–364. MR0241567

[Mas91] William S. Massey, *A basic course in algebraic topology*, Graduate Texts in Mathematics, vol. 127, Springer-Verlag, New York, 1991. MR1095046

[Mic56] Ernest Michael, *Continuous selections. I*, Ann. of Math. (2) **63** (1956), 361–382, DOI 10.2307/1969615. MR0077107

[MM03] Ieke Moerdijk and Janez Mrčun, *Introduction to foliations and Lie groupoids*, Cambridge Studies in Advanced Mathematics, vol. 91, Cambridge University Press, Cambridge, 2003. MR2012261

[MRW87] Paul S. Muhly, Jean N. Renault, and Dana P. Williams, *Equivalence and isomorphism for groupoid C^*-algebras*, J. Operator Theory **17** (1987), 3–22. MR873460

[Muh99] Paul S. Muhly, *Coordinates in operator algebra*, CMBS Conference Lecture Notes (Texas Christian University 1990), Unpublished manuscript, 1999.

[Mur90] Gerard J. Murphy, *C^*-algebras and operator theory*, Academic Press, Inc., Boston, MA, 1990. MR1074574

[MW08a] Paul S. Muhly and Dana P. Williams, *Equivalence and disintegration theorems for Fell bundles and their C^*-algebras*, Dissertationes Math. (Rozprawy Mat.) **456** (2008), 1–57. MR2446021

[MW08b] Paul S. Muhly and Dana P. Williams, *Renault's equivalence theorem for groupoid crossed products*, New York Journal of Mathematics. NYJM Monographs, vol. 3, State University of New York, University at Albany, Albany, NY, 2008. MR2547343

[MW95] Paul S. Muhly and Dana P. Williams, *Groupoid cohomology and the Dixmier-Douady class*, Proc. London Math. Soc. (3) **71** (1995), 109–134, DOI 10.1112/plms/s3-71.1.109. MR1327935

[Pal61] Richard S. Palais, *On the existence of slices for actions of non-compact Lie groups*, Ann. of Math. (2) **73** (1961), 295–323, DOI 10.2307/1970335. MR0126506

[Par78] Kalyanapuram R. Parthasarathy, *Introduction to probability and measure*, Springer-Verlag New York Inc., New York, 1978. MR0651013

[Pat02] Alan L. T. Paterson, *Graph inverse semigroups, groupoids and their C^*-algebras*, J. Operator Theory **48** (2002), 645–662. MR1962477

[Pat88] Alan L. T. Paterson, *Amenability*, Mathematical Surveys and Monographs, vol. 29, American Mathematical Society, Providence, RI, 1988. MR961261

[Pat99] Alan L. T. Paterson, *Groupoids, inverse semigroups, and their operator algebras*, Progress in Mathematics, vol. 170, Birkhäuser Boston, Inc., Boston, MA, 1999. MR1724106

[Ped79] Gert K. Pedersen, *C^*-algebras and their automorphism groups*, London Mathematical Society Monographs, vol. 14, Academic Press, Inc. [Harcourt Brace Jovanovich, Publishers], London-New York, 1979. MR548006

[Ped89] Gert K. Pedersen, *Analysis now*, Graduate Texts in Mathematics, vol. 118, Springer-Verlag, New York, 1989. MR971256

[PR89] Judith A. Packer and Iain Raeburn, *Twisted crossed products of C^*-algebras*, Math. Proc. Cambridge Philos. Soc. **106** (1989), 293–311, DOI 10.1017/S0305004100078129. MR1002543

[Ram71] Arlan Ramsay, *Virtual groups and group actions*, Advances in Math. **6** (1971), 253–322 (1971), DOI 10.1016/0001-8708(71)90018-1. MR0281876

[Ram76] Arlan Ramsay, *Nontransitive quasi-orbits in Mackey's analysis of group extensions*, Acta Math. **137** (1976), 17–48, DOI 10.1007/BF02392411. MR0460531

[Ram82] Arlan Ramsay, *Topologies on measured groupoids*, J. Funct. Anal. **47** (1982), 314–343, DOI 10.1016/0022-1236(82)90110-0. MR665021

[Ram90] Arlan Ramsay, *The Mackey-Glimm dichotomy for foliations and other Polish groupoids*, J. Funct. Anal. **94** (1990), 358–374, DOI 10.1016/0022-1236(90)90018-G. MR1081649

[Ren15] Jean N. Renault, *Topological amenability is a Borel property*, Math. Scand. **117** (2015), 5–30, DOI 10.7146/math.scand.a-22235. MR3403785

[Ren80] Jean N. Renault, *A groupoid approach to C^*-algebras*, Lecture Notes in Mathematics, vol. 793, Springer, Berlin, 1980. MR584266

[Ren82] Jean N. Renault, *C^*-algebras of groupoids and foliations*, Operator algebras and applications, Part I (Kingston, Ont., 1980), Proc. Sympos. Pure Math., vol. 38, Amer. Math. Soc., Providence, R.I., 1982, pp. 339–350. MR679714

[Ren87] Jean N. Renault, *Représentation des produits croisés d'algèbres de groupoïdes*, J. Operator Theory **18** (1987), 67–97 (French). MR912813

[Ren91] Jean N. Renault, *The ideal structure of groupoid crossed product C^*-algebras*, J. Operator Theory **25** (1991), 3–36. With an appendix by Georges Skandalis. MR1191252

[Rie72] Marc A. Rieffel, *On the uniqueness of the Heisenberg commutation relations*, Duke Math. J. **39** (1972), 745–752. MR0412340

[Rie74] Marc A. Rieffel, *Induced representations of C^*-algebras*, Advances in Math. **13** (1974), 176–257, DOI 10.1016/0001-8708(74)90068-1. MR0353003

[Rie82] Marc A. Rieffel, *Applications of strong Morita equivalence to transformation group C^*-algebras*, Operator algebras and applications, Part I (Kingston, Ont., 1980), Proc. Sympos. Pure Math., vol. 38, Amer. Math. Soc., Providence, R.I., 1982, pp. 299–310. MR84k:46046

[Rud73] Walter Rudin, *Functional analysis*, McGraw-Hill Book Co., New York-Düsseldorf-Johannesburg, 1973. McGraw-Hill Series in Higher Mathematics. MR0365062

[Rud87] Walter Rudin, *Real and complex analysis*, 3rd ed., McGraw-Hill Book Co., New York, 1987. MR924157

[RW17] Jean N. Renault and Dana P. Williams, *Amenability of groupoids arising from partial semigroup actions and topological higher rank graphs*, Trans. Amer. Math. Soc. **369** (2017), 2255–2283, DOI 10.1090/tran/6736. MR3592511

[RW98] Iain Raeburn and Dana P. Williams, *Morita equivalence and continuous-trace C^*-algebras*, Mathematical Surveys and Monographs, vol. 60, American Mathematical Society, Providence, RI, 1998. MR1634408

[Sau77] Jean-Luc Sauvageot, *Idéaux primitifs de certains produits croisés*, Math. Ann. **231** (1977/78), 61–76, DOI 10.1007/BF01360030 (French). MR473355

[Sau79] Jean-Luc Sauvageot, *Idéaux primitifs induits dans les produits croisés*, J. Funct. Anal. **32** (1979), 381–392, DOI 10.1016/0022-1236(79)90047-8 (French). MR538862

[Sch71] Helmut H. Schaefer, *Topological vector spaces*, Springer-Verlag, New York-Berlin, 1971. Third printing corrected; Graduate Texts in Mathematics, Vol. 3. MR0342978

[Sed76] Anthony Karel Seda, *Haar measures for groupoids*, Proc. Roy. Irish Acad. Sect. A **76** (1976), 25–36. MR0427598

[Sed86] Anthony Karel Seda, *On the continuity of Haar measure on topological groupoids*, Proc. Amer. Math. Soc. **96** (1986), 115–120, DOI 10.2307/2045664. MR813822

[STY02] Georges Skandalis, Jean-Louis Tu, and Guoliang Yu, *The coarse Baum-Connes conjecture and groupoids*, Topology **41** (2002), 807–834, DOI 10.1016/S0040-9383(01)00004-0. MR1905840

[SW12] Aidan Sims and Dana P. Williams, *Renault's equivalence theorem for reduced groupoid C^*-algebras*, J. Operator Theory **68** (2012), 223–239. MR2966043

[SW13] Aidan Sims and Dana P. Williams, *Amenability for Fell bundles over groupoids*, Illinois J. Math. **57** (2013), 429–444. MR3263040

[Tu00] Jean-Louis Tu, *The Baum-Connes conjecture for groupoids*, C^*-algebras (Münster, 1999), Springer, Berlin, 2000, pp. 227–242. MR1798599

[Tu12] Jean-Louis Tu, *The coarse Baum-Connes conjecture and groupoids. II*, New York J. Math. **18** (2012), 1–27. MR2891024

[TXLG04] Jean-Louis Tu, Ping Xu, and Camille Laurent-Gengoux, *Twisted K-theory of differentiable stacks*, Ann. Sci. École Norm. Sup. (4) **37** (2004), 841–910, DOI 10.1016/j.ansens.2004.10.002 (English, with English and French summaries). MR2119241

[vEY17] Erik van Erp and Robert Yuncken, *On the tangent groupoid of a filtered manifold*, Bull. Lond. Math. Soc. **49** (2017), 1000–1012, DOI 10.1112/blms.12096. MR3743484

[vEY19] Erik van Erp and Robert Yuncken, *A groupoid approach to pseudodifferential operators*, J. reine angew. Math. (2019), in press. (arXiv: math.DG.1511.01041v5).

[vW18a] Daniel W. van Wyk, *The orbit spaces of CCR groupoid C^*-algebras*, pre-print, 2018.

[vW18b] Daniel W. van Wyk, *The orbit spaces of groupoids whose C^*-algebras are GCR*, J. Operator Theory **80** (2018), 167–185, DOI 10.7900/jot.2017sep11.2185. MR3835454

[Wes67] Joel J. Westman, *Locally trivial C^r groupoids and their representations*, Pacific J. Math. **20** (1967), 339–349. MR0211351

[Wes68] Joel J. Westman, *Harmonic analysis on groupoids*, Pacific J. Math. **27** (1968), 621–632. MR0244443

[Wil07] Dana P. Williams, *Crossed products of C^*-algebras*, Mathematical Surveys and Monographs, vol. 134, American Mathematical Society, Providence, RI, 2007. MR2288954

[Wil15] Rufus Willett, *A non-amenable groupoid whose maximal and reduced C^*-algebras are the same*, Münster J. Math. **8** (2015), 241–252. MR3549528

[Wil16] Dana P. Williams, *Haar systems on equivalent groupoids*, Proc. Amer. Math. Soc. Ser. B **3** (2016), 1–8, DOI 10.1090/bproc/22. MR3478528

[Wil70] Stephen Willard, *General topology*, Addison-Wesley Publishing Co., Reading, Mass.-London-Don Mills, Ont., 1970. MR0264581

[Wil81] Dana P. Williams, *The topology on the primitive ideal space of transformation group C^*-algebras and C.C.R. transformation group C^*-algebras*, Trans. Amer. Math. Soc. **266** (1981), 335–359, DOI 10.2307/1998427. MR617538

[Yee07] Trent Yeend, *Groupoid models for the C^*-algebras of topological higher-rank graphs*, J. Operator Theory **57** (2007), 95–120. MR2301938

[Zim84] Robert J. Zimmer, *Ergodic theory and semisimple groups*, Monographs in Mathematics, vol. 81, Birkhäuser Verlag, Basel, 1984. MR776417

Notation and Symbol Index

Index